BASIC, CLINICAL, AND THERAPEUTIC ASPECTS OF ALZHEIMER'S AND PARKINSON'S DISEASES
Volume 2

ADVANCES IN BEHAVIORAL BIOLOGY

Recent Volumes in this Series

Volume 27 THE BASAL GANGLIA: Structure and Function
Edited by John S. McKenzie, Robert E. Kemm, and Lynette N. Wilcock

Volume 28 BRAIN PLASTICITY, LEARNING, AND MEMORY
Edited by B. E. Will, P. Schmitt, and J. C. Dalrymple-Alford

Volume 29 ALZHEIMER'S AND PARKINSON'S DISEASES: Strategies for Research and Development
Edited by Abraham Fisher, Israel Hanin, and Chaim Lachman

Volume 30 DYNAMICS OF CHOLINERGIC FUNCTION
Edited by Israel Hanin

Volume 31 TOBACCO SMOKING AND NICOTINE: A Neurobiological Approach
Edited by William R. Martin, Glen R. Van Loon, Edgar T. Iwamoto, and Layten Davis

Volume 32 THE BASAL GANGLIA II: Structure and Function—Current Concepts
Edited by Malcolm B. Carpenter and A. Jayaraman

Volume 33 LECITHIN: Technological, Biological, and Therapeutic Aspects
Edited by Israel Hanin and G. Brian Ansell

Volume 34 ALTERATIONS IN THE NEURONAL CYTOSKELETON IN ALZHEIMER' DISEASE
Edited by George P. Perry

Volume 35 MECHANISMS OF CEREBRAL HYPOXIA AND STROKE
Edited by George Somjen

Volume 36 NOVEL APPROACHES TO THE TREATMENT OF ALZHEIMER'S DISEASE
Edited by Edwin M. Meyer, James W. Simpkins, and Jyunji Yamamoto

Volume 37 KINDLING 4
Edited by Juhn A. Wada

Volume 38A BASIC, CLINICAL, AND THERAPEUTIC ASPECTS OF ALZHEIMER'S AND PARKINSON'S DISEASES
Volume 1
Edited by Toshiharu Nagatsu, Abraham Fisher, and Mitsuo Yoshida

Volume 38B BASIC, CLINICAL, AND THERAPEUTIC ASPECTS OF ALZHEIMER'S AND PARKINSON'S DISEASES
Volume 2
Edited by Toshiharu Nagatsu, Abraham Fisher, and Mitsuo Yoshida

BASIC, CLINICAL, AND THERAPEUTIC ASPECTS OF ALZHEIMER'S AND PARKINSON'S DISEASES
Volume 2

Edited by

Toshiharu Nagatsu

Nagoya University School of Medicine
Nagoya, Japan

Abraham Fisher

Israel Institute for Biological Research
Ness-Ziona, Israel

and

Mitsuo Yoshida

Jichi Medical School
Tochigi, Japan

PLENUM PRESS • NEW YORK AND LONDON

Library of Congress Cataloging-in-Publication Data

International Conference on Alzheimer's and Parkinson's Diseases: Basic and Therapeutic Strategies (2nd : 1989 : Kyoto, Japan)
Basic, clinical, and therapeutic aspects of Alzheimer's and Parkinson's diseases / edited by Toshiharu Nagatsu, Abraham Fisher, and Mitsuo Yoshida.
p. cm. -- (Advances in behavioral biology ; v. 38)
"Proceedings of the Second International Conference on Alzheimer's and Parkinson's Diseases: Basic and Therapeutic Strategies, held November 6-10, 1989, in Kyoto, Japan"--T.p. verso.
Includes bibliographical references.
Includes index.
ISBN 978-1-4684-5849-7 ISBN 978-1-4684-5847-3 (eBook)
DOI 10.1007/978-1-4684-5847-3
1. Alzheimer's disease--Congresses. 2. Parkinsonism--Congresses. I. Nagatsu, Toshiharu. II. Fisher, Abraham. III. Yoshida, Mitsuo. IV. Title. V. Series.
[DNLM: 1. Alzheimer's Disease--therapy--congresses. 2. Parkinson Disease--therapy--congresses. W3 AD215 v. 38 / WM 220 I593]
RC523.I577 1989
616.8'31--dc20
DNLM/DLC
for Library of Congress 90-14291
CIP

Proceedings of the Second International Conference on Alzheimer's and Parkinson's Diseases: Basic and Therapeutic Strategies, held November 6–10, 1989, in Kyoto, Japan

ISBN 978-1-4684-5849-7

Softcover reprint of the hardcover 1st edition 1990

A Division of Plenum Publishing Corporation
233 Spring Street, New York, N.Y. 10013

PREFACE

This two-volume work contains the full text of the oral and poster presentations and the general discussion at the round table discussion of the Second International Conference on Alzheimer's and Parkinson's Diseases: Basic and Therapeutic Strategies, held at the Kyoto Park Hotel in Kyoto, Japan, on November 6-10, 1989.

The First Conference was held at the Aviya Sonesta Hotel in Eilat, Israel, on March 24-27, 1985. The record of this First Conference was published by Plenum Press in 1986 as Volume 29 in Advances in Behavioral Biology, under the title "Alzheimer's and Parkinson's Diseases: Strategies for Research and Development." We are happy that the comprehensive texts of the oral and poster presentations of the Second Conference could again be published within the framework of this series.

Since the First Conference in 1985, rapid progress has been made in both basic and therapeutic aspects of these diseases. About 700 scientists from all over the world participated in the Second Conference, and 300 papers were presented in oral and poster sessions.

Many people and organizations have helped to organize this multidisciplinary international conference and hence have contributed to the scientific quality of these two volumes. We thank the members of the organizing committee, the organizations that provided financial support, and the contributing scientists for their enthusiastic participation.

These two volumes follow the same publishing philosophy as the volume derived from the First Conference. They span a broad spectrum of topics and bridge preclinical and clinical concepts related to these diseases. They review the old literature and emphasize new findings. They pose important questions, which have yet to be answered, as summarized in the round table discussion in Volume 2.

Alzheimer's and Parkinson's diseases are the most frequent age-related neurodegenerative diseases. The idea of bridging basic and therapeutic research and of comparing Alzheimer's and Parkinson's diseases is fascinating. The two diseases have similarities in etiology, in clinical and pathological features, and in drug development strategies like the neurotransmitter-supplementation therapy.

Research on these two most common age-related diseases will present important clues for the elucidation of the mechanism and therapy of brain aging in general.

We hope that the two books will mark another milestone on the road to effective treatment of Alzheimer's and Parkinson's diseases, and that they will be a valuable addition to the libraries of many students and scientists interested in these two devastating neuropsychiatric disorders, in neuroscience in general, and in brain aging in particular.

Toshiharu Nagatsu
Abraham Fisher
Mitsuo Yoshida

ACKNOWLEDGEMENTS

The 2nd International Conference on Alzheimer's and Parkinson's Diseases acknowledges with cordial thanks the generous support of the following (in alphabetical order):

Abbott Laboratories, Abbott Park, IL, U.S.A.
American Cyanamid Company, Pearl River, NY, U.S.A.
Banyu Pharmaceutical Co. Ltd., Tokyo, JAPAN
Daiichi Pharmaceutical Co. Ltd., Tokyo, JAPAN
Dainippon Pharmaceutical Co. Ltd., Osaka, JAPAN
Eisai Co. Ltd., Tokyo, JAPAN
E.I. du Pont de Nemours & Co. Inc., Wilmington, DE, U.S.A.
E.R. Squibb & Sons Inc., Princeton, NJ, U.S.A.
Eli Lilly and Company, Indianapolis, IN, U.S.A.
Eli Lilly Japan K.K., Kobe, JAPAN
F. Hoffmann-La Roche Ltd., Nutley, NJ, U.S.A.
Fujirebio Inc., Tokyo, JAPAN
Fujisawa Pharmaceutical Co. Ltd., Osaka, JAPAN
Glaxo Group Research Limited, London, U.K.
Hoechst Japan Limited, Tokyo, JAPAN
Kissei Pharmaceutical Co. Ltd., Matsumoto, JAPAN
Kowa Shinyaku Co. Ltd., Tokyo, JAPAN
Kyorin Pharmaceutical Co Ltd., Tokyo, JAPAN
Kyowa Hakko Kogyo Co. Ltd., Tokyo, JAPAN
Merck Sharp & Dohme Research Laboratories, Rahway, NJ, U.S.A.
Nihon Iyakuhin Kogyo Co. Ltd., Toyama, JAPAN
Nihon Schering K.K., Osaka, JAPAN
Nippon Boehringer Ingelheim Co. Ltd., Hyogo, JAPAN
Nippon Chemiphar Co. Ltd., Tokyo, JAPAN
Nippon Roche K.K., Tokyo, JAPAN
Otsuka Pharmaceutical Co. Ltd., Tokyo, JAPAN
Sandoz Yakuhin K.K., Tokyo, JAPAN
Sankyo Co. Ltd., Tokyo, JAPAN
Smith Kline & Fujisawa KK, Tokyo, JAPAN
Shionogi & Co. Ltd., Osaka, JAPAN
Snow Brand Milk Products Co. Ltd., Tokyo, JAPAN
Sumitomo Pharmaceuticals Cooperation, Osaka, JAPAN
Suntory Limited, Tokyo, JAPAN
Taiho Pharmaceutical Co. Ltd., Tokushima, JAPAN
Taisho Pharmaceutical Co. Ltd., Tokyo, JAPAN
Takeda Chemical Industries Ltd., Osaka, JAPAN
Tanabe Seiyaku Co. Ltd., Osaka, JAPAN
The Upjohn Company, Kalamazoo, MI, U.S.A.

Tokyo Tanabe Co. Ltd., Tokyo, JAPAN
Toyama Chemical Co. Ltd., Tokyo, JAPAN
Tsumura & Co., Tokyo, JAPAN
Warner-Lambart Company, Ann Arbor, MI, U.S.A.
Wyeth Ayerst Laboratories Inc., Philadelphia, PA, U.S.A.
Yamanouchi Pharmaceutical Co. Ltd., Tokyo, JAPAN

CONTENTS

MARKER

IMAGING

CLINICAL ASPECT-1 (ALZHEIMER)

CLINICAL ASPECT-2 (PARKINSON)

THERAPY-1(ALZHEIMER)

THERAPY-2(PARKINSON)

MODIFICATIONS OF NEUROENDOCRINE FUNCTION DURING ALZHEIMER TYPE DEMENTIA

*H. Parvez, *Y. Kiuchi, **S. Parvez, *V. Casimiri, *I. Nevo, ***H. Ollat, ****K. Oguchi and ****H. Yasuhara

*Neuropharmacology Unit, University of Paris XI, Bat 440, 91405 Orsay, France
**Department of Physiology, University of Reims, UFR - Sciences, 51 Reims, France
***French Association for Neuro Psycho Pharmacology, Paris, France, ****Showa University School of Medicine, Department of Pharmacology, Tokyo, Japan

INTRODUCTION

Alzheimer type dementia (ATD) is characterized by the presence of numerous neuritic plaques and neurofibrillary tangles particularly in the cerebral cortex. Neurochemical studies on post-mortem brain tissue from ATD patients have demonstrated cholinergic deficits. Choline acetyltransferase (ChAT) activity was found to be markedly reduced in the cerebral cortex of the patients. However, the loss of ChAT activity was shown to be very similar throughout the cortical areas, without a relationship to areas of functional deficit, i.e., the temporal cortex or parietal cortex as suggested by PET following fluorodeoxyglucose injection. Besides the cholinergic deficits, loss of locus coeruleus cells and reduced noradrenaline concentrations and dopamine-ß-hydroxylase activity have been reported in ATD. Recently, many neuropeptides including pituitary hormones, pituitary hormone releasing or inhibiting factors, brain-gut hormones (e.g. CCK, VIP), and other nervous system specific peptides (e. g. neurotensin, neuropeptide Y) have been found in the mammalian central nervous system and there is an evidence that they might act as neurotransmitters or neuromodulators. Several peptides in the hypothalamus, as well as a variety of classical neurotransmitters, are known to modulate secretions of peptide hormones from the pituitary gland. Therefore, it would be expected to find neuroendocrinological abnormalities in ATD patients, since pathological changes have also been described in important hypothalamic nuclei during progression of this disease. Some of the neuropeptides most commonly used as markers of neuroendocrine function would be looked upon for any correlation that may exist with ATD.

<u>Somatostatin</u>: The reduction of somatostatin-like immunoreactivity (SLI) in the cerebral cortical areas was reported by a number of groups.[1-3] This was the first observation for the loss of a neuronal peptide in ATD. The temporal and parietal cortex, which often show pronounced pathological degenerations, have constantly shown a reduction in the concentration of SLI. In less affected cortical areas (the frontal and cingulate cortex), the contents were often not reduced. On the other hand, findings on the contents of SLI in subcortical regions, such as the hippocampus, septum, and amygdala were controversial. Increase in SLI of the hypothalamus or substantia innominata was also reported in ATD.[2] A significant correlation between ChAT and SLI levels was found in the frontal and temporal cortex, but not in other areas of ATD patients.[1,2] However, the reduction of SLI observed in the brain is unlikely to be directly associated with the deficit of cholinergic system, because the distribution and magnitude of the reduction of SLI was much less than that of ChAT[1,2] and no correlation between distribution of

Basic, Clinical, and Therapeutic Aspects of Alzheimer's and Parkinson's Diseases, Volume 2
Edited by T. Nagatsu *et al.*, Plenum Press, New York, 1990

atrophic somatostatin immunoreactive neurons and cholinergic staining pattern was found in the amygdala of ATD.[4] These findings suggest that alterations in somatostatin neurons are not characteristics of the primary pathogenic process that underlie cholinergic cell loss in ATD. SLI was also reduced in lumbar CSF of ATD (Table 1). However, correlation between SLI levels in CSF and the severity of dementia or cholinesterase activity in CSF was also controversial. Reduction of SLI in CSF is unlikely to help in the differential diagnosis of ATD from depression with cognitive impairment or other degenerative brain diseases, because lower SLI concentrations in CSF of patients with depression, multi-infarct dementia, multiple sclerosis, Parkinson's disease and Huntington's disease were also reported. Somatostatin receptor binding in several areas of the cerebral cortex was reported to be normal or reduced according to the choice of the ligand in ATD as compared to controls. In the latter study,[5] the reduced binding was due to a decrease in Bmax rather than a change in Kd and SLI contents were linearly related to somatostatin-receptor binding in the cortices of ATD patients. Initial attempt to treat ATD patients with a peripherally administered somatostatin analogue (L363,586) has failed to achieve sufficient CSF levels of drug and to produce significant memory improvements.[6]

Growth Hormone: The secretion of growth hormone (GH) is known to be affected by cholinergic input and somatostatin, i. e., cholinergic drugs stimulate GH secretion whereas it is inhibited by somatostatin. On the other hand, noradrenaline increases GH secretion by stimulating alpha adrenoreceptors. In ATD, all of above mentioned systems were reported to be markedly affected in the brain. The increase in morning plasma GH concentrations, reduced sleep-related GH secretion or no change of GH levels in ATD patients compared with controls were reported (Table 1). A significant positive correlation between duration of disease and basal GH levels[7] and a significant negative correlation between evening GH levels and psychological test scores[8] were found. Administration of peripheral cholinesterase inhibitors, such as edrophonium, increases plasma GH. Thienhaus et al.[9] demonstrated reduced GH response to edrophonium challenge in ATD patients as compared to controls and a correlation between magnitude of GH blunting and congnitive deficits, but these findings were not confirmed by other studies. In ATD patients, dopaminergic function was shown to be declined at least in some studies. Therefore, abnormal response to dopaminergic stimulation would be expected. However, plasma levels of GH after stimulation with oral administration of l-dopa[7] or apomorphine[10] were reported to be similar in ATD and controls. The increase of blood GH levels by TRH injection, which is remarkable in certain states, such as acromegaly, depression, anorexia nervosa, was reported to be normal[11] in ATD.

CRF: There are some evidences that CRF correlates significantly with ChAT activity in the cortex and can modulate cholinergic behavioral effects as well as cortical somatostatin release in rats. CRF contents in the hypothalamus, amygdala, substantia innominata and hippocampus were similar in ATD and controls but significant reduction in cerebral cortical areas and caudate nucleus in ATD was found (>50 % reduction in the frontal, temporal and occipital cortex)(reviewed by Fine[12]). The contents of CRF in CSF were reported to be normal or reduced in ATD (Table 1). A reciprocal increase in CRF receptor levels (Bmax) in areas with reduced levels of CRF was also observed.[12]

ACTH: Facchinetti et al.[13] observed a reduced level of ACTH in CSF of demented patients, but no correlation was found between ACTH levels and dementia score or information score. ATD patients showed normal levels of plasma ACTH but exhibited significantly higher levels of ACTH after dexamethasone administration,[14] suggesting abnormality of the hypothalamus-pituitary-adrenal axis in this disease. N-terminal fragments of ACTH have positive effects on learning and memory tasks in animals and might improve mood and motivation in older human patients. $ACTH_{4-9}$ analog, Org 2766, or $ACTH_{1-17}$ improved motivation and ward behavior but had no effect on memory function and on CSF levels of cholinesterase, dopamine-ß-hydroxylase, SLI, HVA and cAMP in ATD patients.[15,16]

Cortisol: Plasma cortisol levels of ATD patients were reported to be unchanged or increased in the evening, night or morning (Table 1). A correlation between plasma cortisol levels and severity of dementia was also found in some studies.[8] Resistance of cortisol to suppression by

Table 1 Reported alterations of contents of neuropeptides and hormones in the hypothalamus, CSF and blood in Alzheimer-type dementia

	Hypothalamus	CSF	Blood
Somatostatin	→ (3) ↑ (1)	↓ (16)	
GH			→ (6) ↑ (2)[a] ↓ (1)[b]
Somatomedin		↑ (1)	↑ (1)
CRF	→ (2)	→ (1) ↓ (1)	
ACTH		↓ (1)	→ (1)
Cortisol			→ (5) ↑ (5)[c]
ß-Lipotropin		→ (2)	→ (1)
ß-Endorphin		→ (2) ↓ (3)	↑ (1)
α-MSH	→ (1)	↓ (1)	
TRH	→ (1)	↓ (1)	
TSH			→ (8) ↑ (1)[d]
T_3, T_4			→ (5)
Vasopressin	→ (1) ↑ (1)	↑ (1) ↓ (7)	→ (2) ↓ (3)
Oxytocin		→ (1)	
Prolactin			→ (6) ↑ (2) ↓ (2)
LH-RH	→ (1)	↓ (1)	
LH			→ (2)
FSH			→ (1)
CCK	→ (1)		
VIP	→ (2)	→ (2)	
Substance P	→ (2)		
Neuropeptide Y	→ (2)	→ (1)	
Neurotensin	→ (1) ↑ (1)		

The contents of neuropeptides and hormones of patients with Alzheimer-type dementia were reported to be : ↑ significantly higher, ↓ significantly lower, → normal, as compared with those of control subjects. Numbers of reports we surveyed are in parentheses. Arrows with remarks mean reports of : [a] significantly higher GH level in the morning, [b] significantly lower GH level in the night, [c] higher cortisol level in the evening, in the night or in the morning, [d] higher TSH level in female but not in male patients, in comparison with the corresponding values of control subjects.

dexamethasone was also documented by several authors, although the incidence of nonsuppression was controversial (Table 2). However, ATD patients can not be differentiated from depression or other types of dementia by plasma cortisol levels or dexamethasone suppression test, because increased cortisol levels and high incidence of nonsuppression in both of latter patients were also observed.

TSH: Somatostatin is reported to serve an inhibitory role in modulating TSH secretion. Therefore, the decrease in SLI observed in the brain of ATD might affect the regulation of TRH-TSH axis. However, no significant difference of basal blood TSH levels was found between ATD patients and controls except one report[8] which showed the increase in plasma TSH in only female patients (Table 1). TSH response to TRH in ATD patients was controversial, i.e., lower than controls[17] or normal.[7,11] Heyman et al.[18] reported higher frequencies of prior tyroid diseases in women with ATD than in control women, but this finding was not replicated in other studies.

Vasopressin: Extrahypothalamic projections of vasopressin are thought to play a role in learning and memory processes and exogenously administered vasopressin appears likely to act directly on the central nervous system to facilitate the consolidation and retrieval of memory. Reduction of vasopressin contents in some subcortical areas was reported in ATD patients,[19] whereas the contents in the hypothalamus were documented to be normal or increased (Table 1). Decrease in vasopressin contents in CSF and blood was also found (Table 1). Weingartner et al.[20] observed marginal cognitive enhancement by 1-desamino-8-D-arginine vasopressin in demented patients. However, other clinical trials revealed no significant effect of vasopressin on memory impairment.

Prolactin: Cholinergic drugs and somatostatin are known to modify the secretion of prolactin, suggesting the posibility of abnormal secretion of prolactin in ATD patients. Findings on blood levels of prolactin in ATD patients have been controversial (Table1). Prolactin response to TRH was reported to be augmented,[11] reduced[21] or unchanged,[7] whereas prolactin response to dopamine receptor antagonist (metoclopramide)[7] was found to be unchanged in ATD patients as compared with controls. Pouplard et al.[22] detected high incidence (97%) of prolactin cell autoantibodies in ATD patients, suggesting that autoimmune factors might be included in ATD, but this finding could not be replicated by other groups.

Neuropeptide Y: There is controversy about the change of neuropeptide Y (NPY) in the brain of ATD patients. Beal et al.[23] reported significant decrease of the contents of NPY in cerebral cortical areas, hippocampus and locus coeruleus of ATD patients, while an increase in NPY in the substantia innominata was also observed. In either case, in spite of well-documented co-localization of this peptide with somatostatin in cortical areas, the changes of NPY in brains of ATD do not coincide with somatostatin. NPY levels in the hypothalamus and CSF were similar in ATD and controls (Table 1).

CCK and Vasoactive Intestinal Polypeptide: Both of CCK and vasoactive intestinal polypeptide (VIP), as well as somatostatin, are known to be located in the intrinsic neurons in the cerebral cortex. Contrary to the well-documented reduction of somatostatin contents in the brain, however, most studies could not find any significant changes of CCK- or VIP-like immunoreactivity in cortical and subcortical regions of ATD patients as compared with controls. Normal levels of VIP in CSF have also been documented in ATD patients (Table 1).

Other Peptides and Related Hormones: We have tried to analyse some correlation with ATD progression with some of the hormones mostly affected in this disorder. However, a complete review of all hormones showing some involvement either directly or indirectly with ATD, is beyond the space permitted for this contribution. Neurotensin, substance P, LH, FSH, LH-RH, and TRH have all been investigated in most of clinical and experimental conditions to find some sort of relevance with ATD. Table 1 summarizes the entire spectrum of peptidergic modifications that might exist in the hypothalamus, CSF and blood. Tissue levels are very often different from CSF or blood concentrations and do not permit to make an index with histophysiological changes that have taken place in ATD patients.

Table 2 Dexamethasone Suppression Test in Patients with Alzheimer-type Dementia

Authors		Incidence of Nonsuppressors [a] in ATD	Controls	Others	Correlation of Response with Dementia Score	Depression Score
Spar and Garner	1982	53 % [c]			(-)	(-)
Raskind et al.	1982	47 % [b]	0 % [b]			(-)
Balldin et al.	1983	57 %	7 %	MID 73 %	(-)	(-)
Coppen et al.	1983	47 %	11 %	Dep 70 %, Sch 22%	(-)	
Castro et al.	1983	13 %				
Carnes et al.	1983	18 %		ATD with Dep 40 %	(-)	(-)
McKeith et al.	1984	58 %		MID100 %, Dep 54 %	(-)	(-)
Jenike and Albert	1984	28 %			(+)	
Katona and Aldridge	1985	50 %			(-)	(+)
Greenwald et al.	1986	50 %			(-)	(±)
Georgotas et al.	1986	17 %	4 %	MID 17 % Dep 38 %	(-)	(+)
Serby et al.	1986	80 % [b]				
Charles et al.	1986	10 %		MID 63 %		
Davous et al.	1988	26 %	36 %		(±)	
Serbey et al.	1988	56 %			(±)	
Ferrier et al.	1988	53 %			(-)	

[a] defined as postdexamethasone cortisol level > 5 μg/dl ([b] > 6 μg/dl). [c] including other types of dementia than Alzheimer-type dementia (ATD). MID, muti-infarct dementia;Dep, major depression; Sch, schizophrenia. (+) significant correlation, (-) no correlation, (±) partial correlation.

DISCUSSION AND GENERAL CONCLUSIONS

Apart from the well-known cholinergic deficit in ATD, the reduction in brain and CSF somatostatin-like immunoreactivity and high incidence of dexamethasone nonsuppression are probably the best documented biochemical and endocrinological abnormalities in this disease. However, these abnormalities have also been found in patients with depression and Parkinson's disease, and dementia of other types. On the other hand, abnormalities of other neuropeptides or other neuroendocrine systems in ATD patients were considerably conflicting, and seem to be less affected than cholinergic and somatostatinergic systems, indicating that the loss of neurons in ATD may not affect all categories of cells equally. In brain areas with marked Alzheimer-type pathology (e.g. the temporal cortex, hippocampus, and amygdala), in CSF, or in blood, most peptides seem to be unchanged, or less modified in both distribution and magnitude than parameters of cholinergic systems. Furthermore, most alterations in neuropeptides except somatostatin observed in ATD may occur only at later stages of the disease. Therefore, the difference in the population of patients from which tissues were obtained might be the main cause of discrepancies in data from different groups. In reference to above mentioned variations in several peptides and other neuroendocrine parameters, it can be concluded that somatostatin seems to be the main indicator of ATD patients. However, certain caution must be exercised in drawing conclusions regarding the implication of some neuropeptides in memory processes from findings of low concentrations in the brain tissue or CSF of patients with ATD. The reduction of neuropeptides may reflect the nonspecific diffuse loss of cells within CNS or the increase in CSF volume accompanying cerebral atrophy.

REFERENCES

1. P. Davies and R. D. Terry, Cortical somatostatin-like immunoreactivity in cases of Alzheimer's disease and senile dementia of the Alzheimer type, Neurobiol. Aging 2:9 (1981).
2. I. N. Ferrier, A. J. Cross, J. A. Johnson, G. W. Roberts, T. J. Crow, J. A. N. Corsellis, Y. C. Lee, D. O'Shaughnessy, T. E. Adrian, G. P. McGregor, A. J. Baracese-Hamilton and S. R. Bloom, Neuropeptides in Alzheimer type dementia, J. Neurol. Sci. 62:159 (1983).
3. M. F. Beal, M. F. Mazurek, C. N. Svendsen, E.D. Bird and J. B. Martin, Widespread reduction of somatostatin-like immunoreactivity in the cerebral cortex in Alzheimer's disease, Ann. Neurol. 20:489 (1986).
4. J. W. Unger, T. H. McNeill, L. L. Lapham and R. W. Hamill, Neuropeptides and neuropathology in the amygdala in Alzheimer's disease: relationship between somatostatin, neuropeptide Y and subregional distribution of neuritic plaques, Brain Res. 452:293 (1988).
5. M. F. Beal, M.F. Mazurek, V. T. Tran, G. Chattha, E. D. Bird and J. B. Martin, Reduced numbers of somatostatin receptors in the cerebral cortex in Alzheimer's disease, Science 229:289 (1985).
6. N. R. Cutler, J. V. Haxby, P. K. Narang, C. May, C. Burg and S. A. Reins, Evaluation of an analogue of somatostatin (L363,586) in Alzheimer's disease, New Engl. J. Med. 312:725 (1985).
7. M. Franceschi, L. Perego, L. Ferini-Strambi, S. Smirne and N. Canal, Neuroendocrinological function in Alzheimer's disease, Neuroendocrinology 48:367 (1988).
8. J. E. Christie, L. J. Whalley, J. Bennie, H. Dick, I.M. Blackburn, D.H.R. Blackwood and G. Fink, Characteristic plasma hormone changes in Alzheimer's disease, Br. J. Psychiat. 150:674 (1987).
9. O. J. Thienhaus, F. P. Zemlan, D. Bienenfeld, J. T. Hartford and H.B. Bosmann, Growth hormone response to edrophonium in Alzheimer's disease, Am. J. Psychiat. 144:1049 (1987).
10. S. Lal, N. P. V. Nair, J. X. Thavundayil, V. Tawar, Y. Tesfaye, D. Dastoor, S. Gauthier and H. Guyda, Growth hormone response to apomorphine, a dopamine receptor agonist, in normal aging and in dementia of the Alzheimer type, Neurobiol. Aging 10:227 (1989).
11. C. A. Peabody, J. R. Minkoff, H. D. Davies, C. H. Winograd, J. Yesavage and J. R. Tinklenberg, Thyrotropin-releasing hormone stimulation test and Alzheimer's disease, Biol. Psychiat. 21:553 (1986).
12. A. Fine, Peptides and Alzheimer's disease, Nature 319:537 (1986).
13. F. Facchinetti, G. Nappi, F. Petraglia, E. Martignoni, E. Sinforiani and A. R. Genazzani, Central ACTH deficit in degenerative and vascular dementia, Life Sci. 35:1691 (1984).
14. I. N. Ferrier, J. Pascual, B.G. Charlton, C. Wright, A. Leake, H.W. Griffiths, A. F. Fairbairn and J. A. Edwardson, Cortisol, ACTH, and dexamethasone concentrations in a psychogeriatric population, Biol. Psychiat. 23:252 (1988).
15. J. T. Jolkkonen, H. S. Soininen and P. J. Riekkinen, The effect of ACTH4-9 analog (Org2766) on some cerebrospinal fluid parameters in patients with Alzheimer's disease, Life Sci. 37:585 (1985).
16. G. Nappi, F. Facchinetti, E. Martignoni, F. Petraglia, E. Sinforiani, G. Bono and A. R. Genazzani, N-terminal ACTH fragments increase the CSF ß-EP content in Alzheimer type dementia, Acta Neurol. Scand. 78:146 (1988).
17. T. Sunderland, P. N. Tariot, E. A. Mueller, P. A. Newhouse, D. L. Murphy and R. M. Cohen, TRH stimulation test in dementia of the Alzheimer type and elderly controls, Psychiatry Res. 16:269 (1985).
18. A. Heyman, W. E. Wilkinson, J. A. Stafford, M. J. Helms, A. H. Sigmon and T. Weinberg, Alzheimer's disease: a study of epidemiological aspects, Ann. Neurol. 15:335 (1984).
19. M.N. Rossor, L. L. Iversen, C. Q. Mountjoy, M. Roth, J. Hawthorn, V. Y. Ang and J. S. Jenkins, Arginine vasopressin and choline acetyltransferase in brains of patients with Alzheimer type senile dementia, Lancet ii:1367 (1980).
20. H. Weingartner, W. Kaye, P. Gold, S. Smallberg, R. Peterson, J. C. Gillin and M. Ebert, Vasopressin treatment of cognitive dysfunction in progressive dementia, Life Sci. 29:2721 (1981).
21. P. A. Newhouse, T. Sunderland, P. Tariot, E.A. Mueller, D.L. Murphy and R.M. Cohen, Prolactin response to TRH in Alzheimer's disease and elderly controls, Biol. Psychiat. 21:963 (1986).
22. A. Pouplard, J. Emile and F. Vincent-Pineau, Autoanticorps circulants anti cellules à prolactine de l'hypophyse humaine et maladie d'Alzheimer, Rev. Neurol. (Paris) 139:187 (1983).
23. M.F. Beal, M. F. Mazurek, G. K. Chattha, C. N. Svendsen, E. D. Bird and J. B. Martin, Neuropeptide Y immunoreactivity is reduced in cerebral cortex in Alzheimer's disease, Ann. Neurol. 20:282 (1986).

PREPARATION OF MONOCLONAL ANTIBODY AGAINST A BETA PROTEIN SUBPEPTIDE AND SCREENING FOR IMMUNOREACTIVITY IN THE SERA OF PATIENTS WITH ALZHEIMER'S DISEASE

Tomotaka Shinoda,[1] Yoshie Kametani,[1] Kazuo Miyanaga,[2] and Kazuso Iinuma[3]

[1]Department of Chemistry, Tokyo Metropolitan University Setagaya-ku, Tokyo; [2]Department of Neuropsychiatry, Gunma University School of Medicine, Maebashi; [3]National Children's Hospital, Setagaya-ku, Tokyo, Japan

INTRODUCTION

β protein, a 42-residue peptide, has been reported to be a principal constituent of the brain amyloid of patients with Alzheimer's disease and also of adult Dow's syndrome brain amyloid.[1-3] Recentry, the base sequence of a cDNA for a much larger precursor amyloid protein, consisting of 695 amino acid residues, has been determined; and the gene has subsequently been mapped within band 21q21 on chromosome 21. Two additional putative amyloid β proteins, consisting of 751 and 770 amino acid residues, have also been reported more recently. These are the result of alternative mRNA splicing products coding for much larger precursor amyloid proteins with protease inhibitor activity.

Several questions may be raised regarding the nature and the role of these precursor amyloid proteins in relation to the pathogenesis of Alzheimer's disease and related neurological disorders. First, from where do the precursors arise? Ifprimarily from brain, then it may be of value to test cerebrospinal fluid for its presence; alternatively if it arises from blood, the plasma may be of primary importance for examination. Second, if the precursor(s) and its metabolites exist in the plasma, at what levels? Third, is there any difference in the level of the precursor(s) in the plasmas between patients with Alzheimer's disease and normal individuals? These are all basic problems to be solved.

In order to investigate further the role of the precursor protein(s) in Alzheimer's disease and in some other groups with neurological disorders, we measured the levels of the immunoreactivity in the sera obtained from them using a sandwich ELISA method employing a monoclonal antibody against a β protein subpeptide.

MATERIALS AND METHODS

Preparation of Antigen

A 10-residue peptide correesponding to the first 10 amino acid residues of the amyloid β protein(sequence: DAEFRHDSGY; referred to as SP01), which was coupled with an additional cysteine residue at the carboxy terminus was synthesized and purified by reverse-phase HPLC under the reported conditions. The purified peptide was covalently coupled to either KLH (SP01-KLH) or rabbit serum albumin (SP01-RSA). Numbers of SP01 introduced into these carrier proteins were estimated spectrophotometrically according to a conventional procedure.

Polyclonal Antiserum

Two ten-week-old female rabbits(New Zealand White) were injected subcutaneously in tghe back with 500 g od SP01-RSA in complete Freund's adjuvant. The animals were boosted in every 10 days with the same amount of the antigen until titers of the antisera become constant. The antusera were incubated with the free carrier protein to remove any minute amount of crossreactivity against the carrier protein.

Monoclonal Antibody

Four five-week-old female Balb/c mice were injected subsutaneously into several parts of the body with a total of 100 g of SP01-KLH in complete Freund's adjuvant. The animals were boosted biweekly by injection of the same amount of the antigen until antibody titer in the serum reached a constant value. Hybridomas were prepared conventionally by mixing spleen cells(3 x 10^6) and mouse myeloma cells(P3U1,3 x 10^6) with polyethylene glycol. Of these, the SP01-4 clone produced the antibody with the highest titler; thus it was used for subsequent study.

Assay for Immunoreactivity

Specificities of polyclonal antisera and the monoclonal antibody were characterized by an ELISA and also by an immunoblot technique. Titers of these antisera were determined by conventional procedures. The immunoreactivity assay of sera was performed by a sandwich ELISA technique using 40 1 of test sera. The plates were incubated with a peroxidase-conjugated rabbit anti mouse IgG (final dilution: 1 ; 2000); and after a final washing with 0.01M Tris-0.2% Tween 20, the presence of the antibody was visualized by incubation of the sample with 0.2M diaminobenzidine and 0.44mM hydrogen peroxide in 10mM Tris, pH 7.4. The capacity of the immunoreactivity was expressed as ng equivalent to SP01-RSA/ml in the test sera.

Immunoblot

For immunoblot analysis, a modified two-dimensional electrophoresis was employed. In the first dimension, isoelectric focusing in the absence of denaturant was done. For the second dimension, a gradient gel(4-17% of polyacrylamide) containing 1% SDS was used. Following the electrotransfer to a nitrocellulose membrane, the immunoreactivity in the sera or other test samples was detected essentially as described.[8]

RESULTS

Antibody Specificity

Average numbers of the SP01 introduced into the carrier proteins were found to be 7.5mol eq/mol of RSA and 38mol eq/mol of KLH. On ELISA assay, titers for polyclonal antisera and for the monoclonal antibody (aSP01) were x 10000 and x 20000, respectively. Specificity of these antibodies were examined by ELISA and also by Western blot technique using SP01-RSA as the antigen. We found that both of them recognized the antigen SP01 but not the

Table 1. Property of the monoclonal antibody aSP01

Titer (ELISA)	20000
Heavy chain type	IgG1
Light chain type	kappa
pI range	6-6.5

free carrier protein. The specificity was also examined after the absorption with the antigen, which resulted in parallel displacement of the immunoreactivity from the antibody.

Sensitivity of the Assay System

Crossreactivity of the aSP01 against SP01-RSA was examined at various antigen concentrations. A linear increase in the OD 490nm was attained over the range of antigen concentrations of 5-100ng eq/ml in the test system employed.

Distribution of Immunoreactivity in Patients' Sera

Using aSP01 in the sandwich ELISA technique, we determined crossreactivities in the sera for four different groups, three neurolosically affected groups (Aizheimer's disease and senile dementia of the Alzheimer's disease[AD/SDAT], multi infarct dementia[MID], senile dementia not otherwise specified[NOS]) and a non-affected group. Number of serum specimens available for each group was as follows: AD/SDAT, 24; MID, 51; NOS, 25 and the non-affected, 94. Practically no significant difference in the mean value of the level of the crossreactivity was seen between male and female groups(male: 16.04 ±3.51 ng eq/ml, female: 14.99 ±2.47 ng eq/ml). No correlation was seen between the age and the mean value of the immunoreactivity in the sera from either sex. The mean values of the immunoreactivities are summarized in Table 2 for the four sample groups.

DISCUSSION

As summarized in Table 1, the monoclonal antibody aSP01 had atiter of 20000 in a ELISA procedure, which was sufficiently effective in sandwich

Table 2. Mean Concentration of the Immunoreactivity in the Sera of Four Different Groups

Group	Number of samples	Concentration (mean ± s.d.)
AD/SDAT	24	35.11 ± 7.91
MID	51	16.76 ± 2.58
NOS	25	13.28 ± 2.20
Control	94	14.41 ± 1.91

ELISA assay system for conventional screening. We showed by Western blot analysis that the antibody reacted preferentially with the SP01 moiety of the antigen-carrier conjugate. The antibody was also shown todecorate the senile plaques of different stages in the brains of patients with Alzheimer's disease and somewhat less intensely the ones in the brain of an adult Down's syndrome patient (findings to be published elesewhere).

As indicated in Table 2, when the mean values of the immunoreactivity were compared, the highest value was attained with Alzheimer's disease cases. The difference was significant between AD/SDAT and MID(p 0.01), NOS(p 0.025), or the control(p 0.001). The increase in the level of the immunoreactivity in the sera of patients with Alzheimer's disease does not necessarily mean that the brain is the principal source of this immunoreactivity, since mRNA for amyloid precursor(s) has recently been found in many other organs. If these mRNAs are indeed transcribed, the products would more likely be transported into the blood circulation. Even if its transcription is within the normal level, if there are certain metabolic disturbances in the Alzheimer's disease patients,the level of such immunoreactivity in the serum would be expected to increase with time.

Although our data presented here suggest that there is a significant difference in the level of immunoreactivity between the group with Alzheimer's disease and the control without dementia, futher studies are obviously needed in many respects before we can arrive at an overall conclusion.

ACKNOWLEDGMENT

This work was supported in part by grants-in-aid from the Primary Amyloidosis Research Committee of the Ministry of Health and Welfare, and by the Human science Foundation.

REFERENCES

1. G. G. Glenner and C. W. Wong, Alzheimer'sdisease and Down's syndrome; sharing of a unique cerebrovascular amyloid protein, Biochem. Biophys. Res. Comm. 122: 1131 (1984).
2. G. G. Glenner and C. W. Wong, Alzheimer's disease: initial report of the purification and characterization of a novel cerebrovascular amyloid protein, Biochem. Biophys. Res. Comm. 120: 885 (1984).
3. C. L. Masters, G. Simms, N. A. Weinman, G. Multihaup, B. L. MacDonald and K. Beyreuther, Amyloid plaque core protein in Alzheimer's disease and Down's syndrome, Proc. Natl.Acad. Sci. USA 82: 4245 (1985).
4. J. Kang, H. G. Lemaire, A. Unterbeck, et al., The precursor of Alzheimer's disease amyloid A4 protein resembles a cell-surface receptor, Nature 325; 733 (1987).
5. P. H. St.G-Hyslop, R. E. Tanzi, R. J. Polinski, et al., The genetic defect causing familial Alzheimer's disease maps on chromosome 21, Science 235: 885 (1987).
6. R. E. Tanzi, A. I. MaClatchey, E. D. Lamperti, L. V. Komaroff, J. F. Gusella, and R. L. Neve, Protease inhibitor domain encoded by an amyloid protein precursor mRNA associated with Alzheimer's disease, Nature 331: 528 (1988).
7. T. Shinoda, F. Kametani, H. Tonoike, and S. Kito, Sailent structural features of low molecular weight amyloid fibril protein in FAP of Japanease origin, in :"Amyloidosis" G. G. Glenner et al. eds., Plenum, New york (1986).
8. H. Towbin, T. Staeherin and J. Goedon,Electrophoretic transefer of protein from gels to nitrocellulose sheet, Proc. Natl. Acad. Sci. USA 76: 4350 (1979).

STUDY ON VITAMIN B12 LEVELS IN SERUM AND CEREBROSPINAL FLUID OF PATIENTS WITH ALZHEIMER'S DISEASE

Toshimi Ikeda, Yuichiro Furukawa, Shigeki Mashimoto,
Yasushi Mizuki and Michio Yamada

Department of Neuropshychiatry
Yamaguchi University School of Medicine
Ube, Yamaguchi, Japan

INTRODUCTION

Vitamin B12 (V-B12) is present *in vivo* mainly in 2 forms, *i.e.*, coenzyme type V-B12 or dimethylbenzimidazole cobamide coenzyme (DBCC) and methylcobalamin (CH_3-B12). Their physiological roles have been elucidated step by step. Recent studies reported that DBCC functions as a storage form of V-B12, while CH_3-B12 is involved in transmethylation as an existing form in the serum and cerebrospinal fluid (CSF). Transmethylation is a reaction in which the methyl group of mecobalamin is transferred to methylate another substrate, and it plays an important role in protein synthesis by promoting methylation of mRNA, an important factor in the protein synthesis process. Transmethylation is also important in the synthesis of phospholipids such as phosphatidylcholine, choline and acetylcholine. Phosphatidylcholine (or lecithin) is an essential phospholipid and consists of approximately 40% in the gray matter of the brain. It is also a structurally important membrane lipid: The membrane of nerve cells, nucleus, organella such as mitochondria and lysosomes, and the medullary sheath are composed of a bimolecular layer of lipid (mainly lecithin), and various proteins are associated with the lipid. Therefore, transmethylation needs to be activated in order to create the nervous system and to repair demyelination, and the methyl group is transferred from CH_3-B12 to phosphatidylcholine, choline and acetylcholine.

Recently, various studies on dementia have been conducted. One study reported inactivation of cholinergic neurons in patients with Alzheimer's disease (AD) and senile dementia of Alzheimer's type (SDAT). This report indicates that V-B12 plays in certain role in the pathology of these diseases: Because of the close relationship between V-B12 and the cholinergic system *in vivo*, the inactivation of the cholinergic system in these patients may be attributable to changes in the V-B12 level. Inada *et al.*[1] reported that lowering of the coupling activity of V-B12 with protein causes a functional V-B12 deficiency in aged persons. Van Tiggelen[2] also reported that the V-B12 level in the CSF decreased in SDAT patients.

In this study, we monitored the V-B12 level in the serum and CSF in AD and SDAT patients, and compared the findings in the control group (mainly consisting of multi-infarct-dementia (MID) patients).

Basic, Clinical, and Therapeutic Aspects of Alzheimer's and Parkinson's Diseases, Volume 2
Edited by T. Nagatsu *et al.*, Plenum Press, New York, 1990

Table 1. Summary of the Study Results

Mean Value Group	Age (Year Old)	Disease Period (Year)	Reisberg Dementia Scale	Vitamin B12 Level (pg/ml) Serum	CSF
AD (n=13)	65.69	4.77	4.85	761.00	28.77
Control (n=13)	64.46	--	--	1643.85	83.62
(MID sub-group; n=10)	(66.70)	(2.09)	(5.50)	(1257.30)	(43.60)

TEST PATIENTS AND METHOD

The test subjects were 13 patients clinically diagnosed as having AD (7 persons) or SDAT (6 persons). For diagnosis, all necessary examinations such as blood analysis, CT scan, MRI, EEG, and psychological tests were conducted. The test group (AD group) consisted of 3 males and 10 females with an average age of 65.7 years. Their average score on the Reisberg Dementia Scale (RDS) was 4.85. The control group consisted of 10 MID patients (MID sub-group: 6 males and 4 females; average age: 66.7 years; average score on RDS: 5.5) and 3 patients with other diseases (1 male and 2 females; average age: 57 years; 1 case each of Parkinson's disease, Pick's disease and hypertensive encephalopathy). No mental disorders were observed in the 3 patients with other diseases. The control group and the AD group had a similar age distribution.

We measured the V-B12 level in 1 ml of CSF (collected by lumbar puncture) and 1 ml of serum, which were collected simultaneously. The examination of the cerebrospinal fluid was conducted only after the patient or his/her family had agreed to the objectives of study. All subjects had received medication for each disorder, but they had not been administered V-B12 for at least 3 months before the study.

Competitive protein binding assay was performed to quantify the V-B12 level. For statistical analysis, Wilcoxon's rank sum test and the non-paired *t*-test were used.

RESULTS

The test results are summarized in Table 1. Because the smallest measurable value by the test method is 25 pg/ml, figures below this level for the cerebrospinal fluid were assumed as value of 25 pg/ml for the sake of convenience. The mean V-B12 level in the serum was 761.0 pg/ml for the AD group, 1643.85 pg/ml for the control group and 1257.3 pg/ml for the MID sub-group. The level in the AD group was significantly lower than the control group ($p<0.01$) and the MID sub-group ($p<0.05$).

The mean V-B12 level in the cerebrospinal fluid was 28.77 pg/ml for the AD group, 83.62 pg/ml for the control group and 43.60 pg/ml for the MID sub-group. The AD group value was significantly lower than in the control group and MID sub-group ($p<0.01$).

We further classified the AD group into 2 sub-groups, i.e.,AD (less than 65 year old) and SDAT (older than 65 year old) sub-groups, and compared their V-B12 levels in the serum and CSF. However, no difference was observed.

Patients in the AD group showed a positive correlation between the disease period and the RDS score ($r = 0.85$). However, no correlation was detected between the V-B12 levels in the serum and CSF.

DISCUSSION

The major forms of V-B12 *in vivo* are known to be DBCC (dimethylbenzimidazole cobamide coenzyme) and mecobalamin (CH_3-B12). It has gradually become apparent that other forms of V-B12, cyanocobalamin (CN-B12) and hydroxycobalamin (OH-B12) which had long been thought of as natural forms, rarely exist *in vovo* and are quickly converted to either DBCC or CH_3-B12 (within a few days) after administration. Among them, mecobalamin is thought to have special importance. It plays an important role in the peripheral and central nervous systems through transmethylation, in which CH_3-B12 donates its methyl group to other substrates. Most of the V-B12 existing in the human brain and blood is thought to be in the form of CH_3-B12.

CH_3-B12 is thought to act directly on impaired nervous tissue and improve neuropathy through (a) enhancement of nucleic acid and protein syntheses in nerve cells, (b) acceleration of axoplasmic transport, (c) stimulation of axonal regeneration, (d) promotion of myelinization and (e) acceleration of the recovery of synaptic transmission. Many authors have reported similar effects of CH_3-B12 on the central nervous system. Deana *et al.* demonstrated that V-B12-deficient rats showed a significant decrease in noradrenaline in the brain and of plasma acetylcholinesterase. Tashiro *et al.*[3] hypothesized the changes of mecobalamin → methionine → lecithin → choline → acetylcholine through observations of $^{14}CH_3$-B12 absorption in the tissue of rat brain. Hakim *et al.*[4] demonstrated this change experimentally: In their experimental animal model using nitrous oxide-induced V-B12 deficient rats, the speed of methionine synthesis decreased, transmethylation in the liver was disturbed, the volume of synthesized choline (precursor of acetylcholine) in the brain became insufficient, and the noradrenergic system and locus ceruleus were finally affected. The close relationships between V-B12 and the cholinergic neurotransmitter system are being steadily clarified.

The cholinergic system now attracts great attention because of its possible relation to the memory system. In Alzheimer's disease patients, the cholinergic system is reported to be inactivated. It is possible to presume that inactivation of the cholinergic neurotransmitter system has some relationship with the memory disturbance of AD patients, even though it may not be a single cause of Alzheimer's disease. Therefore, inactivation of the cholinergic system may relate to inactivation of V-B12, and, furthermore, AD patients might show inactivation of V-B12.

Based on these assumptions, we determined the V-B12 levels in the serum and cerebrospinal fluid of AD and SDAT patients (AD group) and compared them with the figures of the control group. As a result, the AD group showed a significantly lower V-B12 level in both the serum and CSF than the control group and the MID sub-group of the control, even though the serum level of each group was within the normal range. The CSF V-B12 level cannot be discussed in detail because normal range has not been established. However, the results also indicate the difference of the AD group from the others. These findings indicate that the V-B12 levels in the serum and CSF can serve as a biological marker for Alzheimer's disease.

In daily practice, we are required to diagnose dementia patients mainly as cerebrovascular dementia and dementia of Alzheimer's type. This is because the prognosis of the patients will differ greatly as a function of the treatment and medication. Differential diagnosis of dementia is made by various examinations such as MRI, CT-scan, EEG, blood analysis and psychological tests. However, we are often unable to make a differential diagnosis of patients. In such cases, Hachinski's ischemic score is applied as a standard procedure.

Our study indicates that the V-B12 level can be a useful guideline for diagnosis. The serum V-B12 level presents some difficulty to use as a biological marker because the figures for all groups were within the normal range. On the other hand, the CSF V-B12 levels of 10 patients in the AD group were below 25 pg/ml. Such a low level was far more frequent in the AD group than in the control, which had only 4 cases out of 13. Therefore, a CSF V-B12 level of 25 pg/ml or lower can indicate AD or SDAT.

If we assume the CSF V-B12 level of AD patients to be lower than 25 pg/ml, false-negative cases in the AD group and the false-positive cases in the MID sub-group become a controversial problem. The false-negative cases in the AD group were the 2 youngest patients and the oldest patient. This may indicate the variety of Alzheimer's disease. In the AD group, there was no difference in the serum V-B12 level of patients above and below 65 years old. Therefore, it might be recommended to classify AD patients into (a) a young group, (b) a middle-aged group and (c) an advanced age group. For a false-positive case, we should consider (a) mixed type dementia and (b) impairment or hypofunction of the cholinergic nervous system due to cerebrovascular disorders. In spite of these points, the CSF V-B12 level (lower than 25 pg/ml) can be a useful biological marker in differential diagnosis of AD from the other types of dementia. However, the severity of the dementia did not correlate with the V-B12 levels in the serum or CSF.

In this paper, we have emphasized the usefulness of the serum and CSF V-B12 levels as a biological marker. However, it also implies a new medical treatment for AD patients. If we assume that a V-B12 deficiency in the brain of AD patients related to the intellectual disturbance, activation of the neurotransmitter system, mainly the cholinergic nervous system, by V-B12 administration could alleviate the severity of the disturbance. Furthermore, the CSF V-B12 level could be a key to establishing a therapy for mixed-type dementia, for which researchers have not established a common definition. We should continue studies on Alzheimer's disease from the viewpoint of the neurotransmitter system and vitamin B12.

CONCLUSION

We measured the vitamin B12 levels in the serum and cerebrospinal fluid of patients with Alzheimer's disease and senile dementia of Alzheimer's type (AD group, n=13), and compared the data with the levels of 13 control patients with multi-infarct dementia (n=10) and other diseases (n=3). The findings were as follows.

1. The levels of V-B12 in the serum and cerebrospinal fluid in the AD group were significantly lower than those in the control group and MID sub-group of the control.
2. The serum V-B12 levels in all the subjects were within the standard range.
3. The cerebrospinal V-B12 level in 10 of 13 patients from the AD group was lower than 25 pg/ml. In the control group, only 4 of 13 patients recorded such a low level. Therefore, a low CSF V-B12 level(below 25 pg/ml) might be a characteristic feature of Alzheimer's disease.

REFERENCES

1. Inada, M and Kameyama M. "Study on nervous system and vitamin B12," *Brain and Nerve*, **31**:963-972, 1979.
2. Cees, J.M. and Van Tiggelen. "Alzheimer's disease/alcohol dementia: Association with zinc deficiency and cerebral vitamin B12 deficiency," *J of Orthomolecular Psychiatry*, **13**:97-104, 1984.
3. Tashiro S. Metabolism and effect of mecobalamin in the brain and nervous system, *in* "Nervous system and methyl B12," G. Kuroiwa, ed., Kyowa Kikaku Tsushin, Tokyo (1981).
4. Hakim, A.M.; Cooper B.A.; Rosenblatt, D.S. and Pappius, H.M. "Local cerebral glucose utilization in two models of B12 deficiency," *J. Neurochem*, **40**:1155-1160, 1983.

SERUM AND CEREBROSPINAL FLUID VITAMIN B12 LEVELS IN DEMENTED PATIENTS TREATED WITH CH3-B12

Yoshio Mitsuyama

Department of Psychiatry, Miyazaki Medical College
Miyazaki, Japan

INTRODUCTION

Reduced activity of methionin synthetase causes defective methyllation in the liver, and leads to an insufficient formation of choline, the neccessary precursor of acetylcholine in the brain.[1] The results of studies of serum and cerebrospinal fluid(CSF) level of VB12 are conflicting. Uncorrected VB12 deficiency may result in dementia. The connection between VB12 and dementia has been accepted.[2,3,4] It is demonstrated that patients with senile dementia(SD) and Alzheimer's disease(AD) show high incidence of pathologically low levels of VB12 in CSF, despite the normal serum VB12 levels.[5] This indicates an abnormal functioning of the choroid plexus, the main transport route for VB12 through the blood-brain barrier,[6] in AD and SD. Little is known about the relationship between the serum VB12 and CSF-VB12. Moreover, the VB12 status of demented patients during the treatment with CH3-B12 has not been reported.

The purpose of the present study was to examine the effect of CH3-B12 on the central nervous system and its clinical effectiveness for organic mental symptoms. The author also set out to determine the correlation between the serum VB12 and CSF-VB12.

SUBJECTS AND METHODS

Subjects for this study were 28 patients suffering from dementia and were selected from the Department of Psychiatry, Miyazaki Medical College Hospital. These patients fulfilled the clinical criteria of organic dementia(DSM-III), confirmed in all the patients by the finding of cerebral atrophy and widening of the ventricles on the CT-scan. Criteria for the study also included the absence of any disorder known to affect the VB12 metabolism, such as acute physical disease, malnutrition, severe anemia, and myeloproliferative disorders. All were normal on the liver and kidney function tests. These patients were judged to be mildly to severely demented based on evaluation by GBS neuropsychological scale, which was developed by Gottfries et al.[7] Venous blood specimens for serum VB12 assays were taken at 7:00 a.m. before and after the administration of CH3-B12 which was given orally after each meal(7:00 a.m., 12:00 p.m., 18:30 p.m.) and intramuscular administration was given at 10:30 a.m.. On the same day, at 10:00 a.m. CSF was obtained by lumbar puncture. The CSF

Basic, Clinical, and Therapeutic Aspects of Alzheimer's and Parkinson's Diseases, Volume 2
Edited by T. Nagatsu *et al.*, Plenum Press, New York, 1990

and serum samples were kept frozen(-20°C) until the day of assay. The radioassay used is a very sensitive for both serum and CSF-VB12 ranging between 25 and 1,600 pg/ml. Nineteen patients(Table 1) were given intramuscular administration(500 µg. per day) of VB12 for 30-70 days and nine patients(Table 2) were given only the oral administration of CH3-B12(2.0 mg. per day) for 40-60 days.

Table 1. VB12 IN CSF AND SERUM(500 µG/DAY, I.M.)

CASES	SEX	AGE	CDR	PRE(PG/ML) CSF	PRE(PG/ML) SERUM	POST(PG/ML) CSF	POST(PG/ML) SERUM	ADMINISTRATION	CLINICAL EFFECTIVE
1. AD	F	35	2	25	1,259.1	418	14,354.0	40 DAYS	2
2. AD	M	49	1	25	626.7	431	9,150.1	60 DAYS	2
3. AD	M	56	2	34	626.0	239	9,876.2	40 DAYS	0
4. AD	M	61	3	25	942.5	110	11,491.0	44 DAYS	0
5. AD	F	62	2	38	635.0	556	24,541.0	34 DAYS	2
6. SD	F	68	2	25	892.1	387	6,101.0	40 DAYS	1
7. SD	F	74	2	37	509.0	354	6,203.2	50 DAYS	2
8. SD	M	77	2	82	1,673.0	692	33,405.0	50 DAYS	1
9. SD	F	77	2	43	4,700.0	531	26,356.0	70 DAYS	1
10. VD	M	59	1	47	293.8	526	19,000.0	40 DAYS	2
11. VD	M	62	1	56	382.5	394	12,039.0	46 DAYS	2
12. VD	F	72	2	74	467.5	626	15,898.0	63 DAYS	1
13. PD	M	46	3	25	664.0	214	15,316.0	60 DAYS	0
14. PD	M	53	2	25	881.9	432	9,098.0	60 DAYS	1
15. PD	F	64	2	38	1,003.1	481	17,214.0	60 DAYS	1
16. WERNICKE KORSAKOW	F	34	1	75	1,243.2	428	15,758.0	60 DAYS	2
17. M.N.D	F	56	1	83	869.5	266	12,711.0	37 DAYS	1
18. TRAUMATIC DEMENTIA	F	57	2	94	449.7	185	8,725.4	30 DAYS	2
19. TRAUMATIC DEMENTIA	M	72	1	77	1,158.4	749	8,066.0	32 DAYS	2
AVERAGE				48.84	1,014.58	422.05	18,790.49		

Abbreviations; AD:Alzheimer's disease, SD:senile dementia, M.N.D:Motor Neuron Disease, PD:Pick's disease

Table 2. VB12 IN CSF AND SERUM (2 MG/DAY, ORAL)

CASES	SEX	AGE	CDR	PRE(PG/ML) CSF	PRE(PG/ML) SERUM	POST(PG/ML) CSF	POST(PG/ML) SERUM	ADMINISTRATION	CLINICAL EFFECT
1. AD	M	53	1	25	636.4	25	638.4	60 DAYS	0
2. AD	F	66	2	36	626.6	41.1	1,408.6	40 DAYS	0
3. SD	F	79	2	25	610.9	45	988.2	60 DAYS	0
4. VD	M	64	2	25	654.7	25	856.3	40 DAYS	1
5. VD	M	70	2	38.1	632.2	55	1,638.1	40 DAYS	1
6. PD	F	46	3	40	1,023.0	42	1,226.3	60 DAYS	0
7. SD	M	78	1	25	316.3	25	353.8	60 DAYS	0
8. PARKINSON DIS.	M	55	1	25	680.4	25	686.8	60 DAYS	0
9. ALCOHOLIC DEMENTIA	M	55	1	25	1,210.0	25	1,320.0	60 DAYS	0
AVERAGE				29.34	710.06	34.23	1,036.03		

RESULTS

The results of the study indicated a tendency towards low levels of CSF-VB12, despite normal levels of VB12 in the serum(500-1,300 pg/ml). The concentration of serum and CSF-VB12 did not correlate with the sevrity of dementia without the CH3-B12 administration. The concentration of serum and CSF-VB12 was markedly elevated in all cases after the treatment with CH3-B12 by intramuscular administration(Table 1). However, no difference were seen in the CSF-B12 level after the oral administration of CH3-B12, 2.0 mg per day, for 40-60 days. A positive relationship between the serum VB12 level and CSF-B12 level was seen with the i.m. route(Fig. 1). CH3-B12 administration did not improve the memory or any intellectual dysfunction observed in the demented patients. However, patients with high elevation of CSF-VB12 level after the CH3-B12 administration showed a reduction in anxiety, improvement of mood, increased communication, increased interaction with other people and the environment, and enhanced self-ratings of competencies. They became more sociable as well.

DISCUSSION

Tiggelen et al.[8] recorded reduced levels of CSF-VB12 in patients with organic mental disorder. Frenkel[9] and Tiggelen et al.[8] reported a discrepancy between the normal levels of serum VB12 and pathologically low levels of VB12 in CSF in their patients with organic mental symptoms. Worm-Peterson[10] found a geometric mean of 389.5 pg/ml in the serum and 20.5 pg/ml n the CSF of the patients without dementia. The results in this

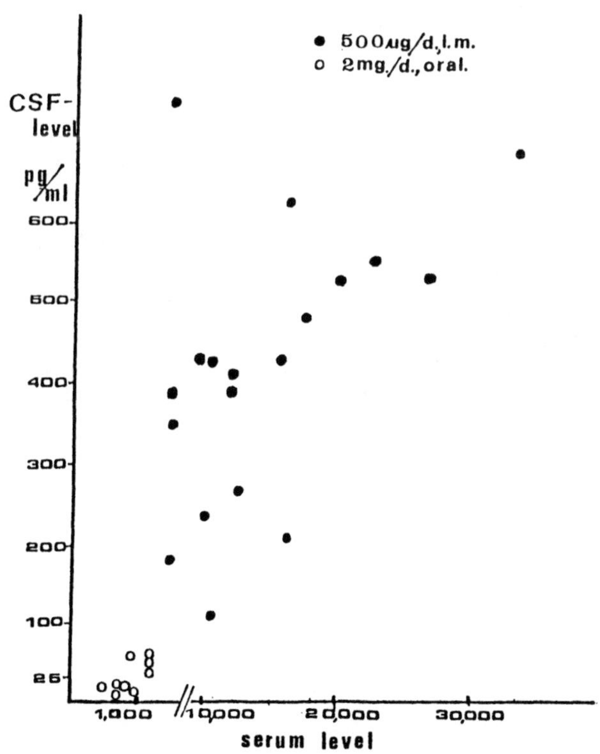

Fig. 1. VB12 in serum vs. CSF after administration of CH3-B12

study are generally in agreement with the findings presented by others. The cause of low levels(less than 200 pg/ml for CSF-VB12 after the CH3-B12 intramuscular administration(500 ug per day) in the severly demented patients is unclear. One possible cause could be selectively in the blood-brain-barrier function, permitting a preferential transfer of CH3-B12 into the CSF of the patients. Throughout this study, it can be assumed that a serum VB12 level of more than 9,000 pg/ml is neccessary for obtaining high CSF-VB12 levels, which might be effective in improving the organic mental symptoms. However, the CSF level of VB12 did not increase with the oral administration of 2.0 mg. per day(CH3-B12). Intramuscular administrations of CH3-B12(500 ug per day) suggest a hopeful improvement in the treatment of organic affective syndromes. The cerebral VB12 deficiency may result in the development of an organic affective syndrome, and early treatment with CH3-B12 given i.m. is recommended to prevent the development of irreversible damage.

REFERENCES

1. A.M. Hakim, B.A. Cooper, D.S. Rosenblatt and H.H. Pappius, Local cerebral glucose utilization in two models of B12 deficiency, J. Neurochem. 40:1155(1983).
2. P.M. Dreyfus and S.E. Geel, Vitamin and nutritional deficiencies, in: "Basic Neurochemistry", G.J. Siegel, R.W. Alberts, B.W. Agranoff and R. Katzman, Little and Brown, Boston(1981).
3. D. Roos and R. Willanger,Various degrees of dementia in a selected group of gastromised patients with low serum B12, Acta. Neurol. Scand. 55:363(1977).
4. C.J.M. Tiggelen, Alzheimer's disease/alcohol dementia:Association with zinc deficiency and cerebral vitamin B12 deficiency, J. Orthomlecular Psychiat. 13:97(1984).
5. M.G. Cole and J.F. Prchal, Low serum vitamin B12 in Alzheimer-type dementia, Age Aging 13:;01(1984).
6. L.A. Ordonez, Control of the availability in the brain of folic acid, vitamin B12 and choline, in:"Nutrition and the Brain", R.J. Wurtman and J.J. Wurtman,eds, Raven Press, New York(1977).
7. C.G. Gottfries, G. Brane, B. Gullberg and G. Steen, A new rating scale for dementia syndrome, Arch. Gerontol. Geriatr.1:311(1982).
8. C.J.M. Tiggelen, J.P.C. Peperkamp and J.F.W. Tertoolen, Vitamin B12 levels in cerebrospinal fluid from patients with organic mental disorder, J. Orthomolecular Psychiat. 12:305(1983).
9. E.P. Frenkel, M.S. MacCall and R.G. Sheehan, Cerebrospinal fluid folate and vitamin B12 in anticonvulsant-induced megaloblastosis, J. Lab. Clin. Med. 81:105(1973).
10. J. Worm-Petersen, Vitamin B12 in the serum and cerebrospinal fluid in neurological disease, Acta. Neurol. Scand. 38:241(1962).

NEUROCHEMICAL MARKERS IN THE CEREBROSPINAL FLUID OF PATIENTS WITH ALZHEIMER-TYPE DEMENTIA

Shinobu Kawakatsu,[1] Shigeru Morinobu,[1] Masao Shinohara,[1] Katsuo Sagawa,[1] Arata Oiji,[1] Shiro Totsuka,[1] and Kyoichi Kobashi[2]

[1]Department of Neuro-Psychiatry, Yamagata University School of Medicine Yamagata; and [2]Department of Pharmaceutical Science, Toyama Medical and Pharmaceutical University, Toyama, Japan

INTRODUCTION

As possible ante-mortem markers of the neurochemical deficits in Alzheimer-type dementia (ATD), neurotransmitter metabolites and some related enzymes in cerebrospinal fluid (CSF) have been investigated, but the results reported are inconsistent. If these metabolites or enzymes actually reflect neurochemical status in the brain, they might be beneficial not only for diagnosis but also for evaluation of therapeutic consequences. We measured acetylcholinesterase (AChE), butyrylcholinesterase (BuChE) and monoamine metabolites 3-methoxy-4-hydroxy phenylglycol (MHPG), 5-hydroxy-indole acetic acid (5HIAA), and homovanillic acid (HVA). Since neurochemical and histopathologic heterogeneity(Bondareff et al., 1987) in ATD has been argued, we focused on early-onset ATD (before age 65 years), i.e., Alzheimer's disease (AD) and also studied late-onset ATD, i.e., senile dementia (SD) and multi-infarct dementia (MID).

SUBJECTS

Our study involved 25 patients with AD (16 males and 9 females; mean age, 56.2 years) who were stricken with the disease before age 65, 6 patients with SD (3 males and 3 females; mean age, 72.1 years) who developed the disease after 65, and 4 patients with MID (3 males and a female; mean age, 68.3 years). Diagnosis was made on the basis of detailed histories, electroencephalograms, computer-assisted tomographic scans, and magnetic resonance imaging. ATD patients included in this study met the DSM-III criteria for primary degenerative dementia, and the NINCDS-ADRDA criteria for probable AD. We categorized AD patients into three groups: Mild AD patients were living at home and capable of self care. Moderate AD patients were living at home with constant help from guardians. Severe AD patients were chair-bound or bed-ridden. Wechesler Adult Intelligence Scale (WAIS), Hasegawa's dementia scale (HDS), and GBS rating scale were used to evaluate the degree of dementia. All subjects or their guardians gave informed consent for participation after the nature of the procedure had been fully explained. Twenty-four subjects (11 male and 13 female; mean age, 51.8 years) who underwent lumbar puncture in the Department of Orthopediatrics and had no neuropsychiatric diseases served as controls. Controls and AD patients did not receive any neuroleptics for at least one week before CSF sampling. CSF samples were collected between 10 and

Table 1. CSF Neurochemical Markers in Demented Patients

	AChE (nmol/min/ml)	BuChE (nmol/min/ml)	MHPG (ng/ml)	5-HIAA (ng/ml)	HVA (ng/ml)
Control	12.6±2.1 (21)	4.3±1.6 (21)	6.3±1.0 (15)	10.2±2.5 (17)	21.6±5.6 (17)
AD	10.0±2.3** (25)	4.3±2.0 (25)	7.3±2.5 (22)	6.7±3.5** (22)	15.8±11.2** (22)
AD 1	12.1±1.7 (6)	3.6±1.9 (6)	7.1±3.2 (6)	4.9±3.2** (6)	16.7±12.4 (6)
AD 2	9.9±1.8** (12)	3.9±1.5 (12)	7.6±1.9 (10)	7.5±3.2** (10)	19.6±11.4 (10)
AD 3	8.3±2.1** (7)	5.5±2.4 (7)	7.2±3.0 (6)	7.4±4.3* (6)	8.8±6.8* (6)
SD	8.4±3.4** (6)	4.6±2.8 (6)	6.7±1.0 (6)	11.3±3.7 (6)	29.7±18.9 (6)
MID	1.4±0.5 (4)	3.6±1.2 (4)	7.5±1.3 (4)	7.8±2.8 (4)	14.3±5.9** (4)

Values are means±S.D. *p<0.05, **P<0.01 vs. respective values in controls
AChE: acetylcholinesterase, BuChE: butyrylcholinesterase
( ): number of patients
AD 1: mildly demented, AD 2: moderately demented, AD 3: severely demented

Table 2. CSF Neurochemical Markers in AD/SD Patients with and without Parkinsonism.

	AChE (nmol/min/ml)	BuChE (nmol/min/ml)	MHPG (ng/ml)	5-HIAA (ng/ml)	HVA (ng/ml)
Without Parkinsonism	10.0±2.5 (23)	4.1±2.0 (23)	7.5±2.5 (20)	8.6±4.3 (20)	23.1±14.1 (20)
With Parkinsonism	8.4±2.3 (8)	5.4±2.1 (8)	6.5±1.1 (8)	5.3±1.8* (8)	8.2±5.8* (8)

Values are means±S.D. *p<0.01 vs. respective values in patients without parkinsonism. ( ):number of patients

12 AM to avoid diurnal fluctuations. After an hour of bed rest, lumbar puncture was performed in the recumbent position. CSF samples were immediately centrifuged and stored at -70° C until assayed.

ASSAY PROCEDURES

Cholinesterase activity was measured by the method developed by Ellman et al. (1961). We used benzoylphosphoric triamide (BPA) as a selective inhibitor of BuChE (developed by Kobashi et al., 1985). The CSF levels of MHPG, 5-HIAA and HVA were measured by high-performance liquid chromatography (HPLC) in an apparatus equipped with an electrochemical detector.

RESULTS

1. CSF neurochemical markers in demented patients (Table 1)

AChE activities were significantly lower (p<0.01) in AD patients than in controls. Within the AD group, the severe AD patients had the lowest AChE activities. AChE activities were also lower in SD patients, but there were no changes in MID patients. There was a tendency for BuChE activity to be higher in severe AD patients, but statistical analysis did not show this difference to be significant.

MHPG levels in AD patients were slightly higher than those in controls, but did not reach a significant level. 5-HIAA levels in AD patients were significantly lower ($p<0.01$) than those in controls. Within the AD group, the mild AD patients had the lowest 5-HIAA levels. There were no significant changes in SD and MID patients. HVA levels in AD patients were significantly lower than those in controls. The severe AD patients had the lowest HVA levels. HVA levels were also lower in MID patients than in controls.

2. Correlations between CSF neurochemical markers and clinical features in AD patients

In the AD patient group, CSF AChE activity correlated with the values on the Hasegawa dementia scale ($r=0.57$, $p<0.05$), WAIS ($r=0.55$, $p<0.05$), and the GBS rating scale ($r=-0.52$, $p<0.05$). CSF monoamine metabolite levels did not correlate with the scores on the dementia tests. In AD/SD patients, patients with parkinsonian symptoms, such as rigidity and akinesia showed lower 5-HIAA and HVA levels compared with those without these symptoms(Table 2). These motor impairments were seen in the later period of the disease.

DISCUSSION

Before interpretation of the alteration of CSF cholinergic enzyme activity, we have to consider whether or not they reflect central cholinergic activity. AChE is less specific than choline acetyltransferase for cholinergic neurons, and AChE activity in brains of patients with ATD is reported to be decreased. AChE is considered to be transported by axonal flow and secreted from the brain into the CSF (Chubb et al. 1976). Furthermore, the main component of CSF AChE was shown to be the 10S form (Nakano et al., 1986), of which selective loss has been demonstrated in the brain of ATD patients (Atack et al., 1983). These lines of evidence may increase the significance of CSF AChE as a cholinergic marker.

While some studies showed reduced CSF AChE activity in AD patients, others did not. Huff et al. (1986) reported that AChE activity was significantly correlated with visual contrast sensitivity, but not with dementia severity. Wester et al. (1988) reported the degree of dementia, as assessed by global deterioration score and activity of daily life status, as well as dyspraxia and dysphasia, were associated with low AChE activities in CSF. It is likely that patients selected influenced the results, since patients in the present study were mainly early onset (before 65 years) ATD in whom more severe involvement of cholinergic denervation is suspected. Mesulam et al (1987) showed that many cortical plaques and tangles in patients with AD were BuChE rich and suggested that assays of BuChE in the CSF may provide diagnostic information for identification of patients with AD. Our results do not support their suggestion.

It is interesting to note that CSF 5-HIAA levels were lowest in mild AD patients. Serotonergic disturbance is reported to occur at a relatively early stage of the disease (Palmer et al., 1987). It is possible that lower 5-HIAA level may be associated with depressive mood or loss of interest often seen in the early stage of AD. CSF HVA levels were lower in the severe AD patients who showed muscle rigidity and/or primitive reflexes, and were also lower in MID patients, implying that the parkinsonism seen in MID and in the later stage of AD may be related to the decreased activity of dopaminergic neurons. Recently,

Kaye et al. (1988) reported CSF HVA was significantly decreased in ATD with extrapyramidal signs compared with the level in those without them.

Deficits in several neurotransmitter systems have been observed in the brains of ATD patients. The present study showed that cholinergic, serotonergic and dopaminergic deficits may be reflected in the CSF. And our data also suggest that in AD patients cholinergic and monoaminergic deficits may coexist, while in SD patients neurochemical deficits may be restricted to the cholinergic system. These findings suggest that the combined therapy of cholinergic and monoaminergic drugs is worth trying and that CSF measurements would be helpful not only for identification of a population more likely to be responsive to neurotransmitter enhancers but also for evaluation of therapeutic consequences.

ACKNOWLEDGMENTS

This research was supported in part by grants 63570500 and 01770810 from the Ministry of Education, Japan.

REFERENCES

Atack, J. R., Perry, E. K., Bonham, J. R., Perry, R. H., Tomlinson, B. E., Blessed, G., and Fairbairn, A., 1983, Molecular forms of acetylcholinesterase in senile dementia of Alzheimer type: selective loss of the intermediate (10S) form, Neurosci. Lett., 40:199.

Bondareff, W., Mountjoy, C. Q., Roth, M., Rossor, M. N., Iversen, L. L., and Reynolds, G. P., 1987, Age and histopathologic heterogeneity in Alzheimer's disease, Evidence for subtypes, Arch. Gen. Psychiatry, 44:412.

Chubb, I. W., Goodman, S., Smith, A. D., 1976, Is acetylcholinesterase secreted from central neurons into the cerebrospinal fluid? Neuroscience, 1:57.

Ellman, G. L., Courtney, K. D., Andres, V., and Featherstone, R. M., 1961, A new and rapid colorimetric determination of acetylcholinesterase activity, Biochem. Pharmacol., 7:88.

Huff, F. J., Maire J-C, Growdon, J. H., Corkin, S., and Wurtman, R. J., 1986, Cholinesterases in cerebrospinal fluid. Correlations with clinical measures in Alzheimer's disease. J. Neurol. Sci., 72:121.

Kaye, J. A., May, C., Daly, E., Atack, J. R., Sweeney, D. J., Luxenberg, J. S., Kay, A. D., Kaufman, S., Milstien, S., Friedland, R.P., and Rapoport, S. I., 1988, Cerebrospinal fluid monoamine markers are decreased in dementia of the Alzheimer type with extrapyramidal features, Neurol., 38:554.

Kobashi, K., Takebe, S., and Numata, A., 1985, Specific inhibition of urease by N-acylphosphoric triamides, J. Biochem., 98:1681.

Mesulam, M.M., Geula, C., and Moran, A., 1987, Anatomy of cholinesterase inhibition in Alzheimer's disease: Effect of physostigmine and tetrahydroaminoacridine on plaques and tangles, Ann. Neurol., 22:683

Nakano, S., Kato, T., Nakamura, S., and Kameyama, M., 1986, Acetylcholinesterase activities in cerebrospinal fluid of patients with Alzheimer's disease and senile dementia, J. Neurol. Sci., 75:213.

Palmer, A. M. Francis, P.T., Bowen, D. M., Benton, J.S., Neary, D., Mann, D.M.A., and Snowden, J.S., 1987, Catecholaminergic neurons assessed ante-mortem in Alzheimer's disease, Brain Res., 414:365.

Wester, P., Eriksson, S., Forsell, A, Puu, G., and Adolfsson, R., 1988, Monoamine metabolite concentrations and cholinesterase activities in cerebrospinal fluid of progressive dementia patients: relation to clinical parameters, Acta. Neurol. Scand., 77:12.

THE VALUE OF CHOLINERGIC RECEPTORS IN BLOOD LYMPHOCYTES MEASURED BY (^{3}H)QUINUCLIDINYL BENZILATE(3H)QNB FOR THE DIAGNOSIS OF CHOLINERGIC RELATED DEMENTIA

J.M. Rabey, E. Grynberg, and E. Graff

Depts. of Neurology and Chemical Pathology, Tel-Aviv Medical Center, Sackler School of Medicine, Tel-Aviv University, Tel-Aviv, Israel

INTRODUCTION

A derangement in the basal forebrain cholinergic system has been reported in Alzheimer's disease (AD) and Parkinson's Disease (PD) (1-2). In addition, deficits in presynaptic cholinergic markers were well documented in the cortices of both diseases (3-5). On the other hand, studies on cholinergic muscarinic receptors were inconsistent (2,5-7). We have demonstrated that peripheral lymphocytes derived from patients with "probable" AD showed a marked reduction in binding capacity as measured by (^{3}H) Quinuclinidyl benzilate, a muscarinic receptor antagonist (8). In the present report, we examined the binding characteristics of lymphocytes in PD with and without dementia compared to age-matched controls.

Patients and Methods

Fourty-four PD patients were included in the study. Twenty of them suffered also from dementia according to DSM IIIR (9) criteria, and were scored according to the Short Mental Test(S.M.T.)(10). All of them were on levodopa therapy alone or combined with bromocriptine, except six non-demented subjects who were also on anticholinergics. These 44 patients were compared with 23 age-matched healthy volunteers.

Blood was collected by venipuncture using heparin as an anti-coagulant. Human lymphocytes were isolated on a Ficoll Hypaque gradient.

QNB binding assay: One million lymphocytes were incubated at room temperature for 45 min in 0.5ml of RPMI nutrient medium containing different concentrations of (3H) quinuclidinyl benzilate (QNB:5-300nM). Specific binding was determined by the preincubation of tubes with 100(uM) atropine for 15 min before adding QNB. The reaction was stopped by adding 3ml of ice-cold saline, followed by rapid filtration on pre-wetted Whathman GF/B filters. After being washed with 6ml saline, the radioactivity on the filters was measured by liquid scintillation spectrometry.

Specific binding constitutes 50 to 70% of total binding measured in the absence of atropine. One hundred femtomoles bound corresponds to a difference of about 2.000 cpm. Experiments were performed according to

Basic, Clinical, and Therapeutic Aspects of Alzheimer's and Parkinson's Diseases, Volume 2
Edited by T. Nagatsu *et al.*, Plenum Press, New York, 1990

Scatchard (11). The protein content in the lymphocyte samples was determined by the method of Lowry et al. (12).

Results

We found an age-dependent increase in binding capacity (Bmax) indicating an age-dependent increase in the number of binding sites per lymphocyte (Fig. 1). All 20 patients with Parkinson and dementia showed low binding compared to age-matched controls (Fig. 1). Among 24 patients without dementia, six showed low binding, 12 showed normal binding and six (on anticholinergic therapy) showed very high binding (Fig. 2). The Kd did not differ significantly among the different groups. The low binding observed in PD with dementia did not correlate with the scoring of the S.M.T. Because of the changes observed with age in binding capacity, statistics were applied calculating the prediction interval of the curve (Figs. 1, 2).

DISCUSSION

There is currently no definitive peripheral marker for the clinical diagnosis of AD and PD (13). Over recent years, it has been demonstrated that circulating blood lymphocytes respond to cholinergic agonists with increased cyclic-GMP levels, enhance RNA and protein synthesis, and alter immune function (14,15). These induced alterations were diminished or blocked by atropine. Indeed, cholinergic muscarinic receptors were demonstrated and characterized in peripheral lymphocytes (16).

These findings suggest that lymphocytes may prove useful as a peripheral marker for the cholinergic muscarinic system. Recently, we demonstrated functional analogy between muscarinic binding in the brain and in circulating lymphocytes of rats (17). Moreover, we have also shown that lymphocytes from patients with "probable" AD have lower binding than age-matched controls (8), also confirmed by other groups (18,19). The present report demonstrates that lymphocytes from PD patients with dementia also show low binding suggesting a similar change in the muscarinic receptors, as described in AD. It is also of interest to note that some PD patients without dementia show low binding, and it will be useful to follow these patients in order to determine if they later develop dementia. In this case, the binding properties of plasma lymphocytes acquire an important role as an early predictive marker for dementia.

The apparent discrepancy of different studies on brain cholinergic receptors (2,5-7) may be explained by the recent finding that muscarinic receptors in the brain can be classified into at least two major subtypes: M1 labeled by (^{3}H) pirenzepine (20) and M2 labeled by (^{3}H)N-methylscopolamine (21), and (^{3}H) oxotremorine-M (22). According to recent reports (7,23), M2 receptors are markedly reduced in the cortex and hippocampus of AD, suggesting a presynaptic location while M1 may not be much affected, perhaps implying a post synaptic ubiquity (7). Our present study performed with a non-specific muscarinic marker (^{3}H) QNB does not allow a final conclusion, although recent data (19) using (^{3}H)-N-methylscopolamine suggests that peripheral lymphocytes may have mainly M2 receptors.

The reduction in cholinergic binding in lymphocytes of PD dementia indicates that the dysfunction of the cholinergic system described is not confined to the brain but is also expressed in the periphery; and it also implies that the immune system may be affected,primarily of secondary. However, the changes observed, could in principle, originate from different populations of B and T lymphocytes (24). In fact, a recent report suggests that (^{3}H)QNB binding is greater in B-cell than in T-cell murine lymphocytes (25). In conclusion, our results suggest that cholinergic binding by circulating blood lymphocytes may serve as a useful peripheral marker reflecting drug-induced changes and alterations associated with aging and cholinergic hypofunction as observed in AD and PD.

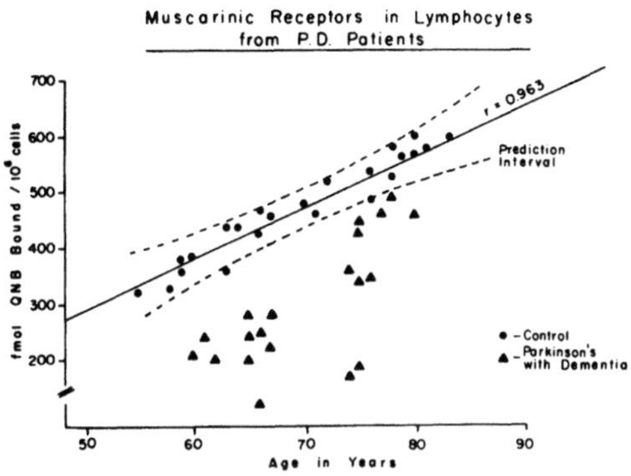

Figure 1

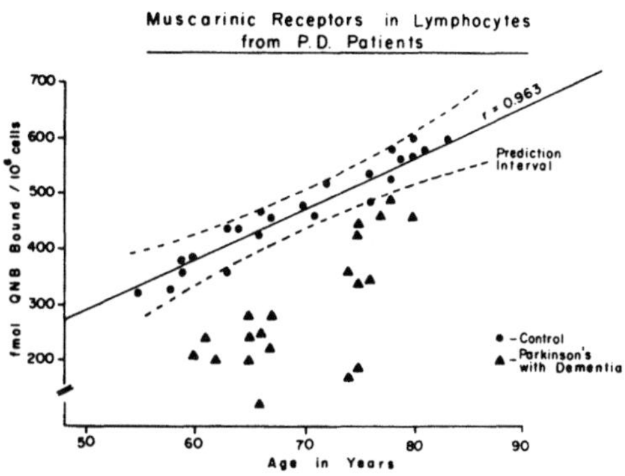

Figure 2

Scatter diagram of maximum binding capacity (Bmax) determined by Scatchard analysis of (^{3}H)QNB binding to lymphocytes from normal volunteers of various ages (black circles) and from patients with PD and dementia (triangles) (Fig. 1) and from patients with PD without dementia (black squares) (Fig. 2). The solid line is a regression line of the data from normal subjects. Bmax values demonstrated a high positive correlation with age (r:0.963).

References

1. Candy, J.M., Perry, R.H., and Perry, E.K. Pathological changes in the nucleus of Meynert in Alzheimer's disease and Parkinson's disease. J. Neurol. Sci. 59:277-289 (1983).
2. Whitehouse, P.J., Price, D.L., Struble, R.G., Clark, A.W., Coyle, J.T., De Long, M.R. Alzheimer's disease and senile dementia: loss of neurons in the basal forebrain. Science 215:123-129 (1984).
3. Davis, P., Maloney, A.J.F. Selective loss of central cholinergic neurons in Alzheimer's disease and senile dementia. Nature 288:29-280 (1977).
4. Perry, E.K., Perry, R.H., Gibson, P.H., Blessed, G., Tomlinson, B.E. A cholinergic connection between normal aging and senile dementia in the human hippocampus. Neurosci. Lett. 6:85-89 (1977).
5. Ruberg, M., Ploska, A., Javoy-Agid, F., Agid, Y. Muscarinic binding and choline acetyl-transferase activity in Parkinsonian subjects with reference to dementia. Brain Res. 232:129-139 (1982).
6. Reisine, T.D., Yamamura, H.I., Bird, E.D., Spokes, E. Pre- and post-synaptic neurochemical alterations in Alzheimer's disease. Brain Res. 159:477-485 (1978).

7. Mash, D.C., Flynn, D.D., Potter, L.T. Loss of M_2 muscarinic receptors in the cerebral cortex in Alzheimer's disease and experimental cholinergic denervation. Science 228:1115-1117 (1985).
8. Rabey, J.M., Shenkman, L., Gilad, G.M. Cholinergic muscarinic binding by human lymphocytes: changes with aging, antagonist treatment and senile dementia of the Alzheimer type. Ann. Neurol. 20:628-623 (1986).
9. Diagnostic and Statistical Manual of Mental Disorders, 1987, third ed. revised, American Psychiatric Association, Washington DC.
10. Korczyn, A.D., Inzelberg, R., Treves, T., Neufeld, M., 1986, Dementia of Parkinson's disease. in:"Advances in Neurology," vol. 45, Yahr MD, Bergmann KJ, eds. Raven Press, New York.
11. Scatchard, G. The attractions of proteins for small molecules and ions. Ann. N.Y. Acad. Sci. 51:660-672 (1949).
12. Lowry, O.H., Rosebrough, N.J., Farr, A.L., Randall, R.J. Protein measurement with the Folin phenol reagent. J. Biol. Chem. 193:265-275 (1951).
13. Atack, J.A., May, C., Kayre, J.A., Kay, A.D. Cerebrospinal fluid cholinesterases in aging and in dementia of the Alzheimer type. Ann. Neurol. 23(2):161-167 (1988).
14. Illiano, G., Tell, G.P.E., Segel, M.I., Cuatrecasas, P. Guanosine 3`5`-cyclic monophosphate and the action of insulin and acetylcholine. Proc. Natl. Acad. Sci. USA 70:2443-2447 (1973).
15. Schreiner, G.R., Ununue, E.R. The modulation of spontaneous and anti-Ig stimulated motility of lymphocytes by cyclic nucleotides and adrenergic and cholinergic agents. J. Immunol. 114:802-808 (1975).
16. Atweh, S.F., Grayhack, J.J., Richman, D.P. A cholinergic receptor on murine lymphocytes with novel binding characteristics. Life Sci. 35:2459-2465 (1984).
17. Shenkman, L., Rabey, J.M., Gilad, G.M. Cholinergic muscarinic binding by rat lymphocytes: effects of antagonist treatment, strain and aging. Brain Res. 380:303-308 (1986).
18. Adam, A., Nordverg, A., Bucht, G., Winblad, B., 1986, Nicotinic and muscarinic binding sites on lymphocytes from Alzheimer's patients. in:"Alzheimer's and Parkinson's disease. Strategies for research and development." Fisher A, Hanin I, Lachman Ch. eds. Plenum Press, N.Y.
19. Ferrero, P., Rocca, P., Eva, C., Genazzani, E. Decrease of [^{3}H]-N-methylscopolamine binding to human lymphocytes muscarinic receptors in patients with Alzheimer's disease. 1988 (Abst) Psychopharmacology Suppl, 96:140.
20. Watson, M., Vickroy, T.W., Roeske, W.R., Yamamura, H.I. Subclassification of muscarinic receptors based upon the selective antagonist pirenzepine. Trends Pharmacol. Sci. 5 (Suppl): 9-11, (1984).
21. Palacios, J.M., Cortes, R., Probst, A., Karobath, M. Mapping of subtypes of muscarinic receptors in the human brain with receptor autoradiographic techniques. Trends Pharmacol. Sci. 7 (Suppl) 56-65, (1986).
22. Bevan, P. (^{3}H) Oxtremorine-M binding to membranes prepared from rat brain and heart. Evidence for subtypes of muscarinic receptors. Eur. J. Pharmacol, 101;101-110 (1984).
23. Araujo, D.M., Lapchak, P.A., Robitaille, Y., Gauthier, C., Quirion, R. Differential alteration of various cholinergic markers in cortical and subcortical regimens of human brain in Alzheimer's Disease. J. Neurochem. 50(6):1914-1923 (1988).
24. Ben Zvi, A., Galili, U., Russell, A., Schlesinger, M. Age associated changes in subpopulations of human lymphocytes. Clin. Immunol. Immunopathol. 2:139-145 (1977).
25. Miller, A.E., Neighbour, P.A., Katzman, Immunological studies in senile dementia of the Alzheimer type: evidence for enhanced suppressor cell activity. Ann. Neurol., 10: 506-510 (1981).

REDUCED NATURAL KILLER CELL ACTIVITY IN DEMENTIA OF THE ALZHEIMER TYPE

Shigeru Araga, Hiroshi Kagimoto, Kouji Funamoto
Kazuhiko Inoue and Kazuro Takahashi

Division of Neurology, Institute of Neurological Sciences
Tottori University School of Medicine
86 Nishimachi, Yonago 683, Japan

INTRODUCTION

Recent investigation have shown that Alzheimer's disease is involved, not only in the central nervous systems, but also in systemic organs including the abnormalities of the red blood cells[1)], lymphocytes[2)], platelets[3)] and fibroblasts[4)]. Recently, the gene for amyloid precursor protein (APP) has been reported to be located in chromosome 21[5,6)]. Patients with trisomy of chromosome 21 (Down's syndrome), which exhibited characteristics similar to Alzheimer's changes in the brain, have been reported abnormal natural killer cell (NK) activity[7)]. We examined the NK activity and its subsets in dementia of the Alzheimer type (DAT).

MATERIALS AND METHODS

Subjects

Forty five patients with DAT (8 men and 37 women), ranging in age from 66-93 years (median 79.4 years) and 34 normal healthy controls (4 men and 30 women), ranging in age from 65-88 years (median 77.4 years), were examined. DAT was diagnosed on the basis of a history of progressive dementia with an insidious onset according to the diagnostic criteria of the Diagnostic and Statistical Manual of Mental Disorders III-R[8)]. To asses the patients, we used mini-mental state examinations (MMSE)[9)] and Hachinski ischemic score[10)]. Computed tomographic scans were performed on all patients with DAT. Cases with mixed type and multi-infarct dementia were not accepted for this project. Mean MMSE score for patients with DAT was 6.44 (SD, 6.56; range, 0 to 23); Mean ischemic score, 1.53 (SD, 1.26; range, 0 to 4).

Cell preparation

Heparinized peripheral blood was collected from the subjects mentioned above. Lymphocytes were separated by Ficoll-Hypaque density centrifugation at room temperature for 20 minutes with 800g[11)]. Adherent cells were removed by plastic plate adherence.

Basic, Clinical, and Therapeutic Aspects of Alzheimer's and Parkinson's Diseases, Volume 2
Edited by T. Nagatsu *et al.*, Plenum Press, New York, 1990

NK cell assay

K-562 cells susceptible to human NK cells, were used for target cells. Assay for NK cell activity was performed by ^{51}Cr releasing method[5]. Effector : target (ET) ratios were 10:1, 20:1 and 50:1. NK cell activity (%NK) in each ET ratio was calculated by the formula:

$$\%NK = \frac{\text{test cpm - spontaneous cpm}}{\text{maximum cpm - spontaneous cpm}} \times 100$$

Data were reported in terms of lytic unit (LU), defined as 10 units per 10^6 effector cells needs to cause 50% cytotoxicity of target cells. For assessment of effects of NK activity by interferon-alpha (IFN-alpha), lymphocytes were incubated with 1000 U/ml of IFN-alpha for 18 hours and then were assayed for NK cell activity. NK activity was reported in terms of %NK at ET ratio of 10:1.

NK cell subsets

To assay the subsets of lymphocytes used for NK cell, two-color analysis by using monoclonal antibodies of phycoerythrin labeled Leu11 and fluorescein isothiocyanate labeled Leu7 were performed with an automatched fluorescence-activated cell sorter (FACS)[12].

Statistical analysis

Statistical analysis was performed using Wilcoxon's rank sum test.

RESULTS

LUs of patients in DAT (4.00±3.29%) were significantly lower than in the normal controls (8.66±10.33%)[p<0.01]. The %NKs induced by IFN-alpha in DAT (55.00±19.82%) were significantly lower than in the normal controls (71.96±15.69%)[p<0.01].

The two-color analysis of NK cell subsets showed no significanty differences in the percentage of $Leu11^+Leu7^-$ cells, $Leu11^+Leu7^+$ cells and $Leu11^-Leu7^+$ cells between DAT (3.40±3.10%, 4.81±5.55% and 35.41±16.4%, respectively) and normal controls (3.58±2.56%, 4.91±4.66% and 34.75±12.37%, respectively).

COMMENT

It is interesting that DAT showd the same reduced NK activity, as Down's syndrome did. Taken together with the fact that the gene of the receptor of IFN-alpha has been mapped to the long arm of chromosome 21[13], reduced NK activity induced by IFN-alpha may be linked to the gene abnormality. The molecular basis of abnormal NK activity is unknown. However, our results suggest that abnormal NK cell activity may also be one of the systemic abnormalities of the non-neuronal tissues in DAT.

REFERENCES

1. J. P. Blass, I. Hanin, L. Barclay, U. Kopp and M. Reding, Elevated red cell to plasma cholin ratios in Alzheimer's disease, in:"Dynamics of cholinergic function," I. Hanin, ed., Plenum Press, New York, (1984).
2. A. E. Miller, P. A. Neighbour, K. Ratzman, M. Aronson and R. Lipkowits, Immunological studies in senile dementia of the Alzheimer type - Evidence for enhanced suppressor cell activity, Ann Neruol, 10: 506 (1981).

3. H. Ksiezak-Reding, C. Murphy and J. P. Blass, Enzyme activities in platelets from Patients with Alzheimer disease, Age, 6: 11 (1983).
4. A. M. Andria-Waltenbaugh and T. T. Puck, Alzheimer's disease: Further evidence of a microtubular defect, J cell Biol, 75(supp.): 279a (1977).
5. R. E. Tanzi, J. F. Gusella, P. C. Watkins, G. A. Bruns, Amyloid (β) protein gene: cDNA, mRNA distribution, and genetic linkage near the Alzheimer locus, Science 1987; 253:880-884.
6. D. Goldgaber, M. I. Lerman, O. W. McBride, V. Saffiotti and D.C. encoding brain amyloid of Alzheimer's disease. Science 235: 887 (1987).
7. M. P. N. Nair and S. A. Schwarts, Association of decreased T-cell-mediated natural cytotoxicity and interferon production in Down's syndrome, Clin Immunol Immunopathol 33: 412 (1984).
8. American Psychaiatric Association: Quick reference to the Diagnostic criteria, in Diagnostic and Statistical Manual of Mental Disorders, ed. 3 revised, Washington D. C., American Psychiatric Association, (1987).
9. M. F. Folstein and S. E. Folstein, 'Mini-mental State': A proctical method for grading the cognitive state of patients for the clinician, J Psychiatr Res 12: 189 (1976).
10. V. C. Hachinski, L. D. Illif, E. Dilhaka et al., Cerebral blood flow in dementia. Arch Neurol 1975; 32:632-637.
11. A. Boyum, Isolation of mononuclear cells and granulocytes from human blood, Scand J Clin Lab Invest 21: 77 (1977).
12. L. L. Lanier, A. M. Le, J. H. Philps, N. L. Warner and J. E. Babcock, Subpopulations of human natural killer cells defined by expression of the Leu¯7 (HNK-1) and Leu¯11 (NKp-15) antigens. J Immunol 131: 1789 (1983).
13. L. B. Epstein and C. J. Epstein, Localization of the gene for the antiviral expression of immune and classical interferon to the distal portion of the long arm of chromosome 21, J Infect Dis 133 (supp.): a56 (1976).

CSF OF PATIENTS WITH ALZHEIMER'S DISEASE CONTAIN ANTIBODIES RECOGNIZING CHOLINERGIC CELLS IN THE RAT CNS, AND CAN PROTECT CHOLINERGIC NEURONAL CULTURES

Amanda McRae[1,3], Annelie Wigander[1], Kerstin Lundmark[1], Kaj Blennow[2], Carl-Gustav Gottfries[2], Ronald J. Polinski[4], and Annica Dahlström[1]

Inst. of Neurobiology[1] and Depts. of Psychiatry and Neurochemistry at St. Jörgen's Hospital[2], NRCG, University of Göteborg, Göteborg, Sweden, INSERM, Unité 259[3], Bordeaux France, and NIH[4], Bethesda Md, USA

INTRODUCTION

A number of recent reports have indicated the presence of antibodies in the cerebrospinal fluid (CSF) and serum of AD patients which recognize cholinergic neuronal elements (e.g. Fillit et al., 1985; Chapman et al., 1986; Singh and Fudenberg, 1986; McRae-Degueurce et al., 1987, 1988). Furthermore, the CSF of a majority of AD patients contain antibodies recognizing thyroglobuline (TG) in an ELISA assay (McRae-Degueurce et al., 1988). In view of the fact that IgG species in the CSF from AD patients recognize cholinergic neurons both in culture (Dahlström et al., 1989) as well as in the rat CNS, it appeared of interest to examine the influence of the IgG in the CSF on cholinergic neuronal populations in culture. This seemed to be an ideal system to investigate the presence of neurotoxic elements or possible neurotrophic factors in the CSF from AD patients, as compared to other types of dementias.

The basis of the present experiments were the following observations: Nerve growth factor-receptor (NGF-r) antibody recognizes cholinergic neurons (e.g. Hartikka and Hefti, 1988), and so do the CSF antibodies (see above). Also, our recent data indicate that incubation of cholinergic neurons with a-NGF-r antibody destroys the cells (Wigander et al., 1990). The question which we sought to answer was: Will CSF samples from AD patients also destroy cholinergic neuronal populations, or is there, as previously proposed (cf. Hyman et al., 1987), an abundance of neurotrophic activity in the CSF from AD patients that will overide the effects of a-NGF-r antibody towards cholinergic cells?

The goal of this article will be first to describe the influence of CSF samples <u>or</u> a-NGF-r alone on cultured cholinergic neurons, and then to present evidence that CSF in the presence of a-NGF-r antibodies has a protective influence on the cultures.

Basic, Clinical, and Therapeutic Aspects of Alzheimer's and Parkinson's Diseases, Volume 2
Edited by T. Nagatsu *et al.*, Plenum Press, New York, 1990

MATERIAL AND METHODS

Cholinergic neurons were harvested from the medial septum, a region rich in cholinergic neurons, of E17 fetal rats and dispersed in a culture medium of nutrient mixture F 12 and MEM (1:1) with 8% inactivated calf serum, 5% glucose, insulin 5µg/ml, L-glutamine (5mM), penicillin (100 IU/ml) and streptomycin (100µg/ml).

The cultures were incubated in the presence of a-NGF-r at different dilutions varying from 1:1 ,1:10, and 1:200. The dilution of 1:1 destroyed the cells over-night whereas the dilutions of 1:10 and 1:200 destroyed cells at 3 and 7 days, respectively. In view of these results we selected to perform the remainder of the study with the dilution 1:200.

CSF samples were selected from 15 AD patients, 6 patients with other types of dementias, and 4 control (healthy) patients. The CSF samples were sterilized by Millipore-filtration and diluted 1:1 with the culture medium. The cells were incubated for 1-3 weeks with the CSF samples. The cultures were regularly observed and photographed in a Nikon Optiphot inverted microscope. Monoclonal a-NGF-r (Johnson et al., 1986) was added to the culture medium in a dilution of 1:200 and the cells were incubated for an additional one week.

Immunocytochemical investigations were performed on cultures fixed with 4% paraformaldehyde. The cells were then preincubated with 2% normal goat serum (NGS) in PBS for 30 min and then processed for immunocytochemistry using a-neuron specific enolase (NSE)/a-NF, or a-glial fibrillary acidic protein (GFA) antiserum, to study neurons or glial cells, respectively. These primary antisera were diluted in PBS containing 1% normal goat or horse serum and 0.1 % Triton X-100. The overnight incubation was carried out at room temperature. The cultures were then washed and the corresponding biotinylated secondary antisera were applied and were visualized with streptavidin-FITC (Vector Lab, Burlingame, CA).

RESULTS

In the initial experiments CSF samples from the various patients were added to the cell culture medium for an incubation period of 3 weeks. The presence of CSF from these patients did not modify the cultures. There were no differences between control cultures incubated with standard neuronal medium compared to cultures incubated with the CSF samples (Fig. 1a).

A number of reports indicate that a-NGF-r antibody recognizes cholinergic neurons in culture as well as in the rats CNS, and we therefore investigated if this antibody would destroy cholinergic neurons in culture. Cultures were incubated with a-NGF-r at different dilutions. Dilution 1:1 destroyed the cells over night, whereas dilutions of 1:10 and 1:200 had destroyed cells at 3 and 7 days, respectively (Fig. 1b). In view of the fact that there are antibodies in the CSF of AD patients which recognizes both cholinergic neurons in culture as well as in the rat CNS, and since we have noticed a cross reactivity between NGF-r and some CSF samples, the next question was: Would incubation of CSF on cholinergic cultures prevent a-NGF-r from destroying the cholinergic neurons? CSF samples (1:1) added to cholinergic cultures for 3 weeks alone did not influence the survival of the cells. Anti-NGF-r was then added at a dilution of 1:200 and the cells were observed daily. On day 7 the cells of the cholinergic cultures that were preincubated with CSF from some AD patients,

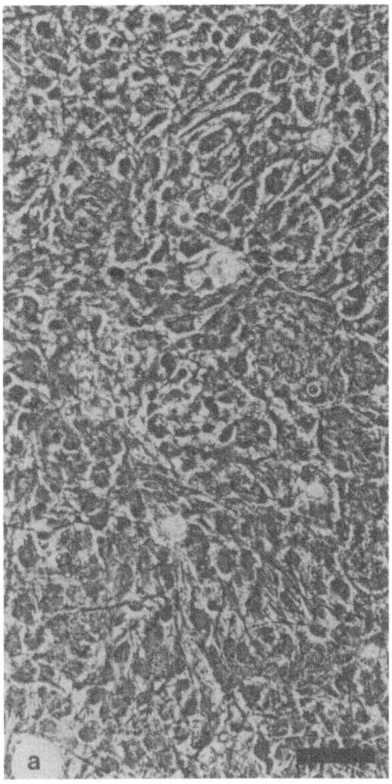

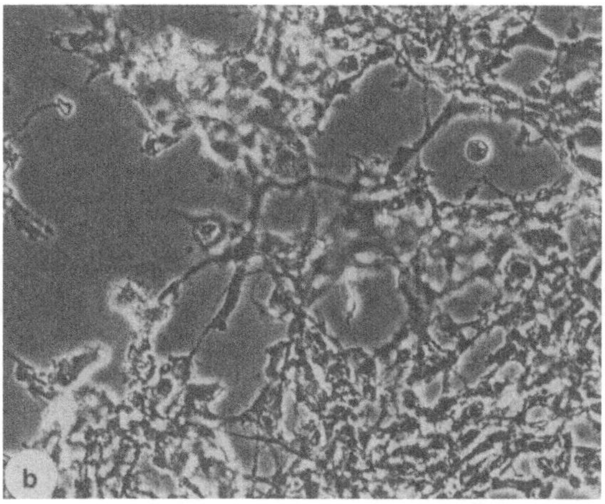

Fig. 1. a) Phase contrast micrograph from cholinergic neuronal culture incubated for 2 weeks with CSF from an AD patient. The same apparance aacontrol cultures or cultures incubated with CSF from healthy control patient. Over a carpet of glial cells (compare fig 2 c) neurons are extending their processes (compare fig 2 b). b) Cholinergic neuronal culture incubated with anti-NGF-r antibodies (1:200) at day 6. Both glia and neurons are degenerating. Bar is 50 μm.

were not destroyed by the NGF-r antibody (Fig. 2a-c). However, cultures preincubated with CSF samples from controls (Fig. 2d-f), or from patients with other forms of dementias (Fig. 2 g-i), were destroyed or in the process of degenerating (Table 1). Both glia and neurons were effected, in most cases the glia showed signs of degeneration somewhat earlier than did neurons.

TABLE 1

CSF SAMPLES	**PROTECTION** against a-NGF-r	
AD	**n=15**	**11**
Other dementias	**n= 6**	**0**
Control patients	**n= 4**	**0**

DISCUSSION

As demonstrated in this study the CSF from some AD patients had a clear capacity to protect the cholinergic cultures from the destruction induced by incubating in a-NGF-r antibodies. It is also evident that neither CSF from control patients nor CSF from other types of dementias, like vascular dementia or multi-infarct dementia, had this preventive capacity.

The results may indicate 1) that CSF from the AD patients contain a protecting trophic factor which can counteract or balance the effect of the a-NGF-r

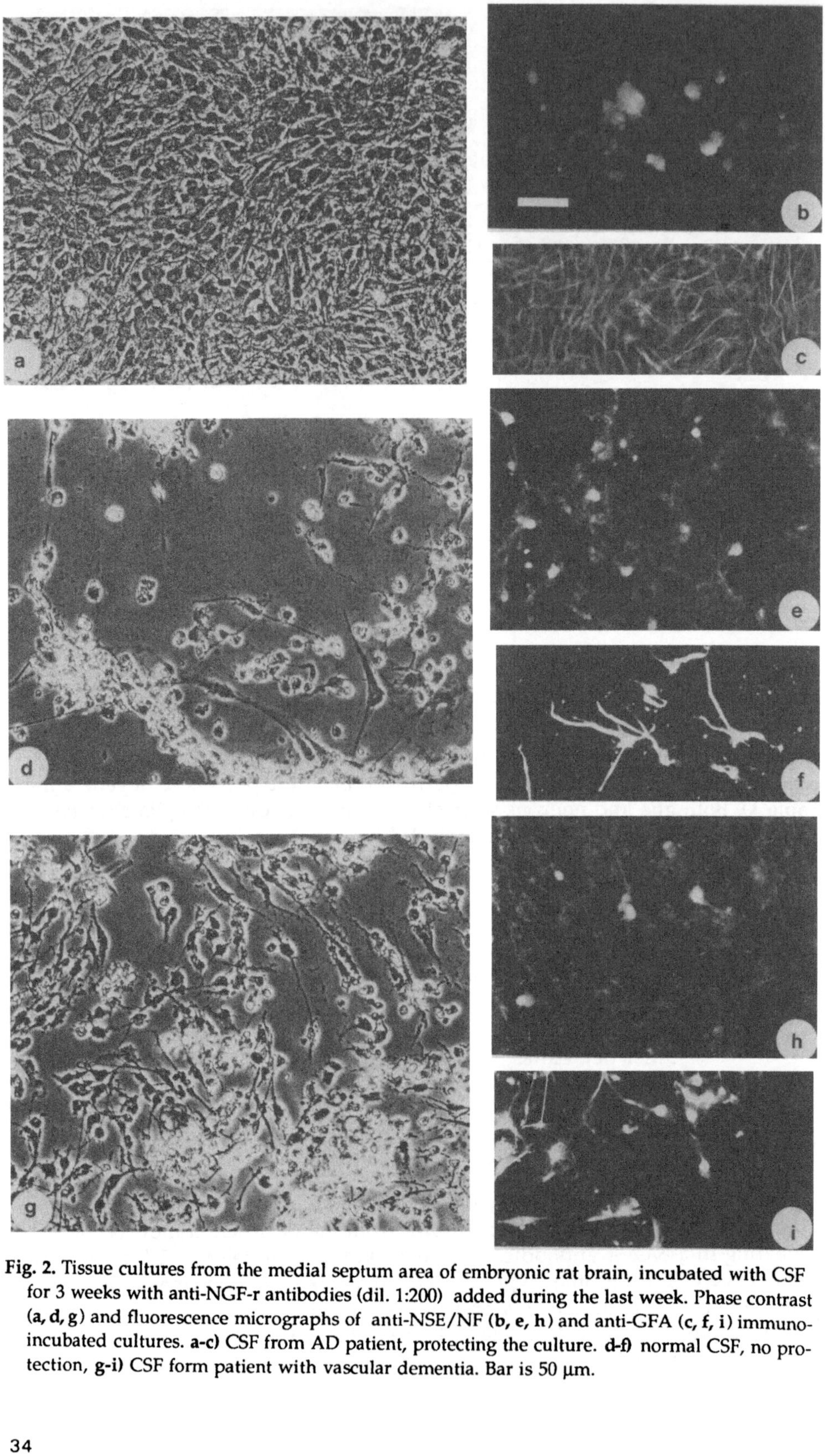

Fig. 2. Tissue cultures from the medial septum area of embryonic rat brain, incubated with CSF for 3 weeks with anti-NGF-r antibodies (dil. 1:200) added during the last week. Phase contrast (**a, d, g**) and fluorescence micrographs of anti-NSE/NF (**b, e, h**) and anti-GFA (**c, f, i**) immuno-incubated cultures. **a-c)** CSF from AD patient, protecting the culture. **d-f)** normal CSF, no protection, **g-i)** CSF form patient with vascular dementia. Bar is 50 μm.

antibodies. Alternatively, 2) the CSF may contain an antibody (an anti-idiotypic antibody?) which neutralizes the a-NGF-r antibodies in the culture medium.

Ad. 1: In the AD brain degenerative processes are ongoing, but it may well be that the brain, in an effort to repair the damage, turns on the synthesis of neurotrophic factors which normally are not expressed in the adult CNS and thus not present in the control CNS samples tested in this study. Geddes et al., (1986) have suggested that neurons grow as a way of compensating for the degeneration that occurs in AD and that in fact the neuritic plaques represent abberant sprouting in the AD brain. Furthermore, Whitson et al. (1989) recently reported that ß-amyloid protein 1-28 stimulates the growth of hippocampal neurons in culture. An observation indicating that the AD brain expresses macromolecules which are present normally only during development is, that CSF from AD patients contain IgG species that strongly recognize axonal components of the embryonic E18 rat brain, but not of the P30 rat brain (McRae et al., 1989a). Also, developmentally regulated microtubule associated protein (MAP) species are re-expressed in the AD brain (McKee et al., 1989).

Ad 2: In the degenerative processes in the AD brain the dying neurons release their contents which then are exposed and may be presented to cells of the immune system which invade the diseased brain. Thus, soluble molecules, organelles and membranes, including membrane bound receptors, are exposed to antigen presenting cells in the parenchyma. It is known that under pathological conditions the astrocytes turn on the expression of MHC class II antigens (Fontana et al., 1984; Lampson, 1987) and may thus act as antigen presenting cells to invading T-lymphocytes. Thus, it would not be surprising if AD patients produce a spectrum of antibodies against a number of neuronal components. One such antibody may be the a-TG, present in high titer in CSF specimens of many AD patients, because TG-immunoreactive material is present in cholinergic neurons of the rat forebrain (McRae-Degueurce et al., 1988). It is interesting that the level of a-TG in the CSF is related to the blood brain barrier function of the patients (McRae et al., 1989b), indicating that there is extratecal production of antibodies which are then passing into the brain. Neuronal antigens released during degeneration of neurons may also include receptor molecules, and antibodies directed against receptor populations may therefore occur. Anti-receptor-antibodies may then trigger the production of anti-idiotypic antibodies (c.f. Kennedy et al., 1986).

Thus, both alternatives are possible, and future experiments, using tissue cultures as well as embryonic and adult rat brain to "fish out" and characterize antibodies, will be employed to explore this question.

ACKNOWLEDGEMENTS

Supported by the Swedish MRC (2207, 5002), Riksbankens Jubileumsfond, Stiftelsen Gamla Tjänarinnor, Handl. Hjalmar Svenssons Forskningfond, A. & M. Ax:son Johnssons Stiftelse för Allmännyttiga Ändamål, Greta och Johan Kocks Stiftelser, and Torsten och Ragnar Söderbergs Stiftelser. The technical assistance of Ms Anne Kling-Petersen and Mr. George McFarlane is acknowledged.

For generous donation of antibodies we thank Dr. Eugene Johnson, St. Louis (anti NGF-r antibodies) and Dr. Kenneth Haglid (anti-NSE and Nti GFA antisera).

REFERENCES

Chapman, J., Korczyn, A.D., Hareuveni, M., and Michaelson, D.M. ,1986, Antibodies to cholinergic cell bodies in Alzheimer's disease, in: "Alzheimer's and Parkinson's Diseases, Strategies for Research and Development", A. Fisher, I. Hanin and C. Lachman, eds., pp. 329-336, Plenum Press, New York.

Dahlström, A., Wigander, A., Lundmark, K., Gottfries, C.G., and McRae, A., 1989, Investigations on autoantibodies in Alzheimer's and Parkinson's diseases, using defined neuronal cultures, J. Neural Transm. (in press).

Fillit, H.M., Luine, V.N., Reisberg, B., Amador, R., McEwen, B., and Zabriskie, J.B., 1985, Studies of the specificity of antibrain antibodies in Alzheimer's disease. in: "Senile Dementia", J.T. Hutton, and A.D. Kenny eds., pp. 307-318, Alan R Liss, New York.

Fontana, A., Fierz, W., and Wekerle, H., 1984, Astrocytes present myelin basic protein to encephalogenic T cell lines. Nature, 307:273-276.

Geddes, J. W., Andersen, K. J, and Cotman, C. W., 1986, Senile plaques as aberant sprout-stimulating structures, Exp. Neurol., 94: 767-776.

Hyman, B.T., Kromer, L.J., and van Hoesen, 1987, Reinnervation of the hippocampal perforant pathway zone in Alzheimer's disease, Ann. Neurol., 21:259.

Kennedy, R. C., Melnick, J.L., and Dreesman, G. R., 1986, Anti-idiotypes and immunity, Sci.Am., 255:48-56.

McKee, A., Binder, L., Kosik, K., Kowall, N.A. ,1989, Developmentally regulated microtubule-associated protein is reexpressed in Alzheimer's disease, Ann. Neurol., 26, 122.

McRae-Degueurce, A., Bööj, S., Haglid, K., Rosengren, L., Karlsson, J.E., Karlsson, I., Wallin, A., Svennerholm, L., Gottfries, C.G., and Dahlström, A., 1987, Antibodies in the cerebrospinal fluid of some Alzheimer's diseases patients recognize cholinergic neurons in the rat central nervous system, PNAS 84:9214-9218.

McRae-Degueurce, A., Haglid, K., Rosengren, L., Wallin, A., Blennow, K., Gottfries, C.G., and Dahlström, A. ,1988, Antibodies recognizing cholinergic neurons and thyroglobuline are found in the cerebrospinal fluid of a subgroup of patients with Alzheimer's disease, Drug Dev. Res. 15:153-163.

McRae, A., Blennow, K., Gottfries, C.-G., Wallin, A., and Dahlström, A., Brain specific antibodies in the CSF of patients with Alzheimer's disease and other types of dementias, in: "Biological Markers in Dementia", L.A. Carlson, C. Fowler, C.-G. Gottfries, and B. Winblad, eds., Smith-Gordon Co Ltd (in press).

McRae, A., Blennow, K., Wallin, A., Fredman, P., Gottfries, C.-G., and Dahlström, A., 1989b, The presence of antibrain antibodies in the CSF of some Alzheimer disease patients: correlation with CSF parameters, J. Neural Transm., (in press).

Singh,V., and Fudenberg, H.H., 1986, Detection of brain autoantibodies in the serum of patients with Alzheimer's disease but not with Down's syndrome, Immunol. Lett., 12: 277-280.

Whitson, J.S., Selcoe, D.J., and Cotman, C. W., 1989, Amyloid ß protein enhances the survival of hippocampal neurons in vitro, Science, 243:1488-1490.

Wigander, A., Lundmark, K., McRae, A., Mölne, J., Dahlström, A., and Ahlman, H., 1990, Production of a transferrable neuronotrophic factor by human midgut carcinoid tumour cells: studies using rat fetal cholinergic neurons. Acta Physiol. Scand., submitted.

ALTERATIONS OF TRANSFERRIN AND PREALBUMIN IN CEREBROSPINAL FLUID IN PARKINSON'S DISEASE

Masanobu Arai[1,2] Satoshi Orimo[1], Takashi Kurosawa[1], Eizo Hiyamuta[1], Tomoko Nakayama[2], Masami Nakatsu[2], and Kiyoko Sano Shiba[3]

[1]Department of Neurology, and [2]2nd Department of Research Divisions, Kanto Teishin Hospital, Tokyo, and [3]Department of Clinical Biochemistry, Faculty of Medicine, Tokyo Medical and Dental University, Tokyo, Japan

INTRODUCTION

The pathogenetic mechanism of degeneration of neurons underlying Parkinson's desease (PD) and Alzheimer's disease (AD) is not as yet clarified and remains a challenging issue to be settled only by multidisciplinary research efforts. Prealbumin (PA) and transferrin (TF), among cerebrospinal fluid (CSF) proteins, are unique, because a portion of each of these CSF proteins is of brain tissue origin (Weisner and Bernhardt, 1978). However, few studies have so far been undertaken on these CSF proteins (Stibler and Kjellin, 1976) despite interesting reports made on the relationship of PA to neuritic plaques and neurofibrillary tangles seen in AD and occasionally in PD (Shirahama et al., 1982) as well as on the potential action of TF to promote nerve growth (Toran-Allerand, 1980) and to function in the transport of iron to the brain (Swaiman and Machen, 1984). Therefore we determined CSF levels of PA and TF and also studied by electrophoretic techniques the isoelectric point (pI) of PA and the microheterogeneity (MH) of TF in untreated PD patients and compared the results with those of the control group.

MATERIALS AND METHODS

The subjects used in this study were 14 untreated PD patients (6 male, 8 female) aged between 56 and 84 years.(70.1±9.4, mean±SD). Fourteen subjects (male 7, female 7) who were found to have no organic brain disease served as controls. These control subjects were between 35 and 81 years of age (56.1±15.2), hence significantly younger than the PD group ($p<0.05$) and not age-matched. Informed consent was obtained from all patients. The CSF was taken at fasting early in the morning after overnight bed rest and frozen until used. The following determinations were performed on each sample within 2 weeks of the tapping:

PA in CSF was estimated by an electroimmunodiffusion method (Laurell, 1966). In order to investigate the influence of the blood-CSF barrier, we quantified total protein (TP) and albumin (Alb) in CSF by the sulphosalicylic acid method and radial immunodiffusion method utilizing LC-Partigen-Albumin (Hoechst), respectively, to calculate PA/TP and PA/Alb therefrom.

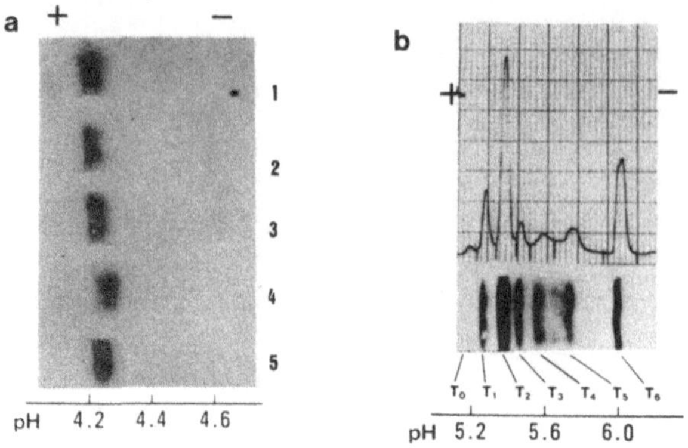

Figure 1. Prealbumin(a) and transferrin(b) by isoelectric focusing and immunofixation. See text.

TF in CSF was measured by radical immunodiffusion using LC-Partigen-Transferrin (Hoechst).

Identification of PA in CSF was accomplished by isoelectric focusing according to the method of one of us (Sano et al., 1984), immunofixation by a modification of the method of Whicher et al. (1980), and thereafter silver-staining according to Kerényi and Gallyas (1972). Finally, the pI of PA in CSF was determined by the method of Sano et al. (1984).

Identification and quantification of TF in CSF were accomplished by the same assay procedures as used for PA except for the type of antiserum. As 7 different types (MH) of TF were detected in the CSF, densitometry of electrophoretic bands was performed. Percent contents of MH of TF were automatically calculated and their concentrations in CSF were determined as the total TF concentration multiplied by the percentages. Statistical analyses of data were carried out by the F test and t-test.

RESULTS

The concentration of TP in CSF was 32.4±7.4 mg/dl for the control group and 40.7±18.9 mg/dl for the PD group, with the difference failing to achieve statistical significance. Alb concentrations in CSF were 18.9±4.9

Table 1. Transferrin Microheterogeneity in CSF (mean±SD)

	Controls(%)	PD	Controls(mg/dl)	PD
T_0	5.2±2.8	5.4±3.2	0.13±0.07	0.14±0.07
T_1	11.2±1.7	10.8±1.7	0.30±0.10	0.28±0.08
T_2	30.8±5.2	32.9±4.0	0.79±0.18	0.84±0.26
T_3	10.5±2.2	10.0±2.1	0.29±0.11	0.26±0.10
T_4	10.6±2.2	14.5±4.1**	0.28±0.10	0.37±0.13
T_5	16.9±5.9	11.8±6.4*	0.46±0.20	0.30±0.18*
T_6	15.0±3.4	14.7±5.8	0.39±0.11	0.36±0.11

*: P<0.05 **: P<0.01

mg/dl for the control group and 24.9±8.8 mg/dl for the PD group, which were significantly different ($p<0.05$). The concentration of PA in CSF, on the other hand, was 1.57±0.17 mg/dl in the control group as against 1.69±0.34 mg/dl in the PD group, thus being higher, though not significantly, in the latter group ($p<0.1$). PA/TP x 100 was 4.9±1.4% for the control group and 5.0±2.2% for the PD group; PA/Alb x 100 was 8.6±2.6% for the control group and 7.9 4.0% for the PD group, hence there was no significant difference between the two groups for neither of the parameters. The concentration of TF in CSF was 2.64±0.66 mg/dl and 2.59±0.69 mg/dl, respectively, for the control and PD groups, the difference again being not statistically significant.

Immunofixation electrophoretic patterns of PA in CSF obtained with anti-human PA rabbit serum are indicated in the figure (a), wherein lanes 1 and 2 represent the control group and lanes 3 through 5, the PD group, with the anodal side being at the left and the cathodal side, at the right. PA in CSF was identified as a thick immunoprecipitation band located at around pH 4.2-4.3 and its pI values were 4.20±0.02 and 4.19±0.04, respectively, for the control and PD groups at the anodal end and 4.32±0.03 and 4.31±0.03, respectively, for the control and PD groups at the cathodal end, the differences between the two groups being not statistically significant.

Immunofixation electrophoresis of CSF using anti-human TF goat serum yielded 7 distinct TF bands (MH) of varying thickness in the pH range of 5.2-6.0 as shown in the figure (b), where a densitogram of the bands from one PD patient is represented (left: anodal side, right: cathodal side). These 7 MH of TF separated were arbitrarily and serially designated from the anodal side toward the cathodal as T_0, T_1, T_2, T_3, T_4, T_5 and T_6 and their pI values were 5.2, 5.3, 5.4, 5.5, 5.6, 5.8, and 6.0, respectively. Comparisons of the percent content of these MH bands of TF between the two groups showed that T_4 was significantly higher in the PD group than in the control group ($p<0.01$), while T_5 was significantly lower in the former than in the latter group ($p<0.05$) (Table). When compared in terms of absolute concentration in CSF, T_5 was significantly lower in the PD than in the control group ($p<0.05$). There were no significant differences between the two groups in the other MH bands of TF with respect to either their percent content or absolute concentration in CSF.

In the control group, the comparative studies above showed no significant differences between 8 subjects below 60 years of age (younger subgroup) and 6 subjects ages 60 or above (older subgroup) with respect to TP, Alb, PA, and TF (data not shown).

DISCUSSION

Although age-matched controls were not available in the present study, there were no significant differences between the younger and older subgroups of control subjects in either of its CSF proteins studied. It seems, therefore, that the age differences in the present series were negligible and by no means invalidate intergroup comparisons.

In their recent study Shirahama et al. (1982) reported PA as a component of amyloid that constitutes neuritic plaques and neurofibrillary tangles, although some researchers dispute this. It is known that part of the PA present in the CSF is derived from brain tissue (Weisner and Bernhardt, 1978). Under such circumstances we felt it worthwhile to investigate into this issue. The results of our present study indicate that PD patients, though tending to show a higher concentration of PA in their CSF as compared with control subjects, had increased CSF levels of Alb and, consequently, gave values of PA/Alb that were not significantly different from those of the latter group. It seemes inferable, therefore, that the increased PA levels in the CSF of PD patients are due to non-specific disorder of the blood-CSF barrier seen so often in PD. The pI of PA in CSF showed no significant difference between the two groups. The study also yielded no findings suggestive of abnormal changes in PA in the brain of PD patients.

It is known that the globus pallidus, putamen, and substantia nigra, among other brain structures, have noticeably high tissue levels of iron (Hallgren and Sourander, 1958); and excessive deposition of iron in brain tissues occurs in PD and AD. Accumulation of iron in tissues facilitates the formation of peroxides from oxygen and leads to the formation of lipofuscin. On the other hand, TF, which is already known to promote nerve growth (Toran-Allerand, 1980) has recently been demonstrated to mediate iron transport to nerve tissue as well as to other tissues (Swaiman and Machen, 1984). Stibler and Kjellin (1976) analyzed the CSF of PD patients by isoelectric focusing, and reported that there was no noticeable change except for a protein of pI 5.8 which, though not identified as TF, revealed thicker precipitation band in 17% of the PD patients studied. Our closer study revealed that in PD patients as compared with control subjects, T_4 (pI 5.6) was significantly higher and T_5 (pI 5.8) was significantly lower. The true cause of these PD-related changes in CSF TF composition is obscure, but it might be likely at least that alterations in neuraminic acid residues and or peptides in the TF molecule account for the observed changes. Whether or not such fine changes in TF affect the mechanism of iron transport to and the formation of peroxides in the brain, and thereby play a role in the degeneration of nerve cells occurring in the basal ganglia and substantia nigra, remains to be clarified by further studies.

REFERENCES

Hallgren, B. and Sourander, P., 1958, The effect of age on the non-haemin iron in the human brain, J. Neurochem., 3:47.

Kerényi, L. and Gallyas, F., 1972, A highly sensitive method for demonstrating proteins in electrophoretic, immunoelectrophoretic and immunodiffusion preparations, Clin. Chim. Acta, 38:465.

Laurell, C. B., 1966, Quantitative estimation of proteins by electrophoresis in agarose gel containing antibodies, Anal. Biochem., 15:45.

Sano, K., Nakao, M., Shiba, A., and Kobayashi, K., 1984, An ultra-microassay for proteins in biological fluids other than blood using a combination of agarose gel isoelectric focusing and silver staining, Clin. Chim. Acta, 137:115.

Shirahama, T., Skinner, M., Westermark, P., Rubinow, A., Cohen, A. S., Brun, A., and Kemper, T. L., 1982, Senile cerebral amyloid; Prealbumin as a common constituent in the neuritic plaque, in the neurofibrillary tangle, and in the microangiopathic lesion, Am. J. Pathol., 107:41.

Stibler, H. and Kjellin, K. G., 1976, Isoelectric focusing and electrophoresis of the CSF proteins in tremor of different origins, J. Neurol. Sci., 30:269.

Swaiman, K. F. and Machen, V. L., 1984, Iron uptake by mammalian cortical neurons, Ann. Neurol., 16:66.

Toran-Allerand, C. D., 1980, Coexistence of α-fetoprotein, albumin and transferrin immunoreactivity in neurones of developing mouse brain, Nature, 286:733.

Weisner, B. and Bernhardt, W., 1978, Protein fractions of lumbar, cisternal and ventricular cerebrospinal fluid: Separate areas of reference, J. Neurol. Sci., 37:205.

Whicher, J. T., Higginson, J., Riches, P. G. and Radford. S., 1980, Clinical applications of immunofixation: Detection and quantitation of complement activation, J. Clin. Pathol., 33:781.

CLINICAL INVESTIGATION OF NEUROTRANSMITTERS IN EXTRAPYRAMIDAL SYSTEM DISEASES

Kazuko Hasegawa and Hisayuki Kowa

Kitasato University School of Medicine
Asamizodai 2-1-1, Sagamihara, Kanagawa 228, Japan

INTRODUCTION

In the degenerative disorders of the extrapyramidal system(EPS) has progressed owing to various the clinicopathological studies of these diseasess. From pathologic observation, such disorders are known to involve the degeneration of neurons, accompanied by gliosis, as well as alternations of the neurotransmitter(NTM) systems. However, as many of the reports on quantitative determinations of NTM used post-mortem brains, it is difficult to make exact comparison with the antemortem clinical state. In the present study, we determined in cerebrospinal fluid(CSF) NTM's such as 5-HIAA and HVA, which are metabolites of serotonin and dopamine respectively, and compared their values with the clinical state in various EPS disorders. Such kind of study has been done extensively in Parkinson's disease(PD), but our new aspect of study gave rise to interesting results. Therefore,we proceeded the study further to evaluate the clinical applicability of the detection of biological amines in CSF.

SUBJECTS ABD METHODS

a) Patients with PD

A total of 37 patients diagnosed clinically as PD were examined (symptomatic or secondary parkinsonism cases were excluded as much as possible). The patients were 20 males and 17 of females, and their age were ranged from 40 to 78 years, with the average being 60. The number of cases in terms of Yahr score were as follows: II,7; III, 10; IV, 13; V, 7. No significant sex difference was seen among the scores. As a rule,the clinical conditions under which CSF were taken were assigned as untreated PD, PD after the withdrawal of antiparkinson drugs, or PD after drug adjustment,if possible.

b) Other EPS Disorders

The disease groups subjected to the study showing EPS symptoms other than PD are listed below. These 4 groups were diagnosed by clinical, pathological and roentgenographical findings. 1) Shy-Drager syndrome(SDS): 3 cases; 2 males and 1 female; ages of 37-55 years; the average, 46. The 2 males were bedridden cases and

Basic, Clinical, and Therapeutic Aspects of Alzheimer's and Parkinson's Diseases, Volume 2
Edited by T. Nagatsu *et al.*, Plenum Press, New York, 1990

were considered the SND type.2) Olivo-Ponto-Cerebellar atrophy (OPCA): 12 cases; 7 males and 5 females; ages of 37-67 years old; the average, 56. ADL were various from ambulatory to bedridden. 3)Progressive Spranuclear Palsy(PSP): 4 cases; 2 males and 2 females, ranging from 56 to 63 years in age; the average, 61. As to ADL, 1 was bedridden and the other 3 required half-time care was; 4) Dentate-RubroPallido-Luysian atrophy(DRPLA): 2 cases(brother and sister),a brother of 62 and the sister of 61 years. The brother was bedridden and the sister required half-time care. Both cases were considered disease of the choreoathetoide form.

c) CSF Sampling

CSF was withdrawn by a routine method. For the measurement of NTM as in CSF in this study,the first 5ml was used. After taking the next 5ml for routine examination, we took another 5ml as "ventricular fluid(VF). After deproteinization the CSF was stored frozen at -80°C until the determination of NTM's. The deproteinized frozen stored CSF samples were analyzed for NTM levels by means of HPLC. 5-HIAA was determined colorimetrically according to the method of Goldenberg; and HVA was measured fluorometrically according to the method of Sato.

RESULTS

a) Comparison among disease groups

First of all, the concentrations of 5-HIAA and HVA in CSF from untreated PD and PD before treatment were compared with other EPS disorders(Table 1). As for the HVA value by disease, the PSP group showed a lower trend of the average value, 18.5ng/ml, while the OPCA group showed a higher(61.4ng/ml). On the other hand, when the PD group was compared with the others, no significant difference was noted between the PD and the SDS groups, while there was a significant difference($p<0.05$) between the PD and the OPCA in that of the latter was higher. As for the comparison between the PD and PSP groups, the latter showed a lower tendency, but no statistical significance. With regard to 5-HIAA value, the similar tendency to that of the HVA value was seen: was the OPCA was higher, and the PCP was lower, than the PD; but there was no statistical significance in either case. Since the number in the DRPLA group comprised only 2 cases, no statistical analysis was made; but the values are shown just for reference. The choreoathetoid form of DRPLA was identified to be similar to OPCA pathologically. However, according to this study, the DRPLA cases seemed closer to PD rather than to OPCA.

b) Comparison of the values of NTM metabolites in "CSF" and "VF" in PD

As a matter of convenience,"CSF" and "VF" were taken separately and NTM metabolites in 16 cases of PD were compared (Table 2). As to HVA values (ng/ml),it was 32.0 25.4 in CSF and 36.3 21.3 in VF; but this difference was not statistically significant. On the other hand, 5-HIAA value of CSF was 15.9 10.0 while that of the VF was 18.0 12.0($p<0.1$). This result is the reverse of the normal pattern of 5-HIAA levels in CSF and VF.

c) Examination by the severity grade of PD

The next study was to examine how the concentrations of NTM's in CSF changed with the progression of PD. The value were those of the untreated PD or after the drug withdrawal(Table 3). In the examination of the correlation between the Yahr score of PD and the values of NTM metabolites in CSF, the changes with the severity were not so clear as expected. However, between Yahr scores of III and IV,a significant difference in HVA values ($p<0.02$) and a decreasing trend in 5-HIAA values($p<0.10$)were observed, suggesting a large difference between the score III or less and the score IV or more. Further, for the HVA value, a statistically significant difference between score II and III was also noted. However,it was difficult to find a certain trend because of the great fluctuation of values in the score III patients.

c) Values of NTM metabolites in CSF before and after treatment of PD

Among the PD patients, quantitative analyses of NTM metabolites could be performed in 11 cases. The Yahr scores of them before the treatment are shown in Table 4, and the assayed values are shown in Table 5. Although the PD patients were given the various anti-parkinsonian drugs, the reference among the kind of drugs were not clear in this study, even or the anti-cholinergic drugs. Therefore the comparison was made simply between before and after the treatment simply. As a result, it became apparent that HVA and 5-HIAA increased after treatment. In proportion to the increase in the metabolites, clinical improvement was also observed. Thus it appears that the values of the metabolites are a reflection of the clinical state. Next,examinations by clinical varieties and by severity grade were performed, and an interesting result was obtained, as shown in Table 6. In the juvenile PD(JPD), a marked increase in HVA value was observed. For the explanation of this result, we assume that the JPD patients respond well to treatment and that the long-term levodopa syndrome is apt to develop. In addition, the mild cases of PD showed higher rates of increase in NTM metabolites while the advanced PD showed only a slight increase. So, these results suggest the difference in spare capacity of NTM in PD and ease of development of long-term levodopa syndrome.

DISCUSSION

a) NTM and various EPS disorders

It has been often difficult to diagnose clinically in the very early stage of these EPS disorders,especially in PSP and SND. It thus seems of value to investigate whether the determinations of these NTM metabolites are useful or not. From the described aspects, the present study was conducted and as shown in Table 1, differences between PD and each of the PSP and OPCA groups were observed. Further, by taking all of these accumulated data into consideration, we may be able to determine the extent of the lesion and make a correlation between and the severity grade. Because all of the SDS cases were the SND type, no significant difference from PD was seen. Therefore,it will be important to compare PD with the OPCA-type SDS. On the other hand, the EP signs of the OPCA group were all parkinsonian. So, it was interesting that, in spite of the expected HVA values being low, those of this group were high. Although this is only a hypothesis, the fact of the high value of HVA may be interpreted as the result of hyperactivity of dopamine neurons against the insufficiency of that receptor. In this respect, it may be necessary to accumulate many more cases of ? in the future. DRPLA has been classified into 3 forms by Hirayama, and our 2 cases were considered to be the choreoathetoid form. As the number of cases was only 2, the interpretation of result is difficult, but the contrast with OPCA group seems interesting in differentiating the lesions. These EPS disorders are sometimes not always identical between those of the early and the terminal stages during long-term observation. In view of these facts, the periodical determinations of HVA and 5-HIAA contents in CSF of the same patient would seem useful for the elucidation of the pathogenesis of degenerative disorders, and we are going to continue the study hereafter.

b) Change patterns of Change in NTM's in PD

Firstly, we tried to find any difference in NTM's between CSF and VF, but the statistical analyses by clinical stage could not be done because the number of cases was too small. Although it has been well known that substrates present in the CSF show a negative concentration gradient between ventricle and spinal subarachnoid space, it was noted that in PD patients with low Yahr scores the NTM concentration in VF was higher than that in CSF. In this point, we must investigate further, collecting data or a greater number of patients. It is not clear why along with the increase of severity the concentration gradient between ventiricle and aubarathnoid space of the spinal column is broken down. It is possible that some impairment is present in CSF circulation or absorption. A result in this study that deserves much attention is the change in 5-HIAA. Concentrations of 5-HIAA were decreased, statistically significantly so, in parallel with the worsening of PD. However, it was difficult to find a definite trend in HVA because of the much fluctuation in value,especially Yahr III PD. It may be necessary to investigate what symptom correlates with the NTMs metabolites with regard to the severity of PD. It has so far been said that in PD impairment of the dopamine system is primarily involved pathogenetically, but recently, attention has been given to the fact that along with the worsening of PD, impairments in other NTM's are also occurred. Our results seem to suggest this fact. If the deterioration of other metabolites become possible we will proceed with our study in this direction. Finally,the possibility of evaluating the therapeutic efficacy not only by the clinical findings but also by NTM values was examined. As shown in Tables 5 and 6, the results obtained indicate that the value is useful for the evaluation of prognosis. Not only HVA but also 5-HIAA increased in value significantly after the treatment. It is not certain whether antiparkinson drug activates the serotonin system or not; but since the dopamine and serotonin systems have the same enzymes, it may be natural to assume that both systems are activated together. The therapeutic direction of PD hereafter should utilize drugs that activate NTM's in multiple systems. In addition,a useful result was obtained to evaluate the prognosis by the type of disease and by severity. According to the examination of the rate of increase of HVA value after the treatment, the rate was markedly high in JPD; and in PD patients in Yahr II and III the rate was double, so that the therapeutic effect was expectable. While in PD in Yahr IV and V, drug efficacy was not expected. The 2 cases of JPD examined were those in which the "wearing-off" was marked. These results suggest that the determination of HVA value is extremely useful for the judgment of therapeutic effect in PD or for preventing development of the long-term levodopa syndrome. Such an examination may possibly to give rise to a reasonable selection of the drug to be used. We must proceed to examine this aspect in the future. (We thank Bristol-Meyers for the collabolation in the determinations of HVA and 5-HIAA in CSF.)

Table 1. Values of 5-HIAA and HVA in CSF by Diseases

Disease Group	HVA(ng/ml)			5-HIAA(ng/ml)		
PD	39.2	27.0*	(n=29)	24.9	30.7	(n=28)
SDS	30.9	17.4	(n= 3)	20.3	13.4	(n= 3)
PSP	18.5	7.3	(n= 4)	17.0	11.3	(n= 4)
OPCA	61.4	33.8*	(n= 8)	74.7	101.8	(n= 8)
DRPLA	31.4	16.6	(n-2)	19.9	5.2	(n= 2)

*: $p<0.05$

Table 2. Changes of NTM in CSF and VF in PD(ng/ml)

	HVA (n=16)		5-HIAA (n=16)	
PD CSF	32.9	25.4	15.9	10.0*
PD VF	36.3	21.3	18.9	12.0*

*: $p<0.01$

Table 3. Values Analyzed by Severity Grade

Yahr score		HVA(ng/ml)		5-HIAA(ng/ml)	
Score II	(n=7)	29.7	15.2*	18.9	10.2
Score III	(n=8)	55.4	29.3*#	43.5	45.1"
Score IV	(n=7)	20.9	13.5#	7.4	4.3"
Score V	(n=4)	14.7	9.2	11.8	5.2

*: $p<0.05$, #: $p<0.02$, ": $p<0.10$

Table 4. Yahr Score of the Analyzed Patients before the Treatment

Yahr score	No. of patients
Score II	3
Score III	2
Score IV	5
Score V	1

total: 11

Table 5. Values Analyzed before and after Treatment (n=11)

Yahr	HVA(ng/ml)		5-HIAA(ng/ml)	
before treat.	18.3	10.7*	19.6	13.8#
after treat.	36.9	19.4*	35.5	17.7#

*: $p<0.01$, #: $p<0.05$

Table 6. Rates of Increasing HVA Values by type of disease and by severity

$$\text{Rate of Increase (\%)} = \frac{\text{pre-treat. value-post-treat. value}}{\text{pre-treat. value}} \times 100$$

	Juvenile PD(n=2)	Yahr II III(n=5)	Yahr IV V(n=3)
	444.4	105.0	-4.0
	660.0	140.0	-34.9
		37.5	77.0
		77.5	
		124.1	
mean(%)	552.2	96.8	12.7

TOTAL BIOPTERIN AND NEOPTERIN LEVELS IN THE VENTRICULAR CSF OF PATIENTS WITH PARKINSON'S DISEASE

Y. Furukawa[1], T. Kondo[1], K. Nishi[1], F. Yokochi[1], K. Tanabe[1], Y. Mizuno[1] and H. Narabayashi[2]

[1]Dept. of Neurology, Juntendo Univ. School of Med.
[2]Neurological Clinic; Tokyo, Japan

INTRODUCTION

(6R)-L-erythro-5,6,7,8-tetrahydrobiopterin (BH_4) is the natural cofactor for tyrosine hydroxylase which catalyzes the rate-limiting reaction in the synthesis of catecholamines. D-erythro-neopterin is derived from D-erythro-7,8-dihydroneopterin triphosphate, an intermediate of BH_4 biosynthesis from guanosine triphosphate. Total L-erythro-biopterin (T-BP) in tissue and the cerebrospinal fluid (CSF) exists mostly as BH_4. D-erythro-7,8-dihydroneopterin triphosphate can be measured after its oxidation as total D-erythro-neopterin (T-NP).

Previous studies have demonstrated that T-BP levels are low in the lumbar CSF[1] and the autopsied brains[2] of patients with Parkinson's disease (PD). It has also been reported that T-BP levels in the lumbar CSF decrease with age[3], while T-NP levels increase[4]. To our knowledge, however, no investigation has specifically compared T-BP and T-NP levels with the clinical features of PD. In this study, therefore, T-BP and T-NP levels in the ventricular CSF were measured and correlated with the features which PD manifests.

PATIENTS and METHODS

Patients and Controls: This study consisted of 25 patients with PD (59.4±8.4 years) and 7 age-matched neurological controls (58.7±10.3 years). At the time of operation, none of the patients or controls had infectious or malignant diseases. The neurological control group consisted of 6 patients with essential tremor (ET) and 1 with torticollis. Clinical stages (I-V) described by Hoehn and Yahr[5] were used to classify all the PD patients according to their parkinsonian disabilities (Table).

CSF Collection: Samples of ventricular CSF were obtained during ventriculography before stereotaxic surgery. All medications were withdrawn one day prior to surgery, which was performed under local anesthesia. The samples were immediately frozen and stored at -70°C until assayed.

TABLE 1

Total biopterin and neopterin levels in the ventricular CSF

	No. of cases	T-BP[a] (pmol/ml)	T-NP[b] (pmol/ml)
(1) Parkinson's disease	25	14.71 (3.55)[c]]*	14.88 (8.16)
Controls	7	23.11 (5.15)	14.25 (4.34)
(2) Parkinson's disease			
1] Male	10	15.21 (3.22)	15.19 (9.65)
Female	15	14.37 (3.83)	14.68 (7.36)
2] Right tap	13	14.99 (3.27)	16.11 (9.39)
Left tap	12	14.40 (3.96)	13.56 (6.73)
3] Stage I	4	15.97 (2.46)	13.01 (2.87)
II	4	14.53 (6.04)	23.39 (11.30)
III[d]	14	14.80 (3.41)	12.51 (7.34)
IV	3	12.82 (1.86)	17.14 (6.82)

a: Total L-erythro-biopterin
b: Total D-erythro-neopterin
c: Data are expressed as mean (SD)
d: Classified according to the Hoehn and Yahr disability scale
* $P<0.001$, difference between PD patients and controls

<u>Assay for T-BP and T-NP</u>: Total L-erythro-biopterin and T-NP levels were determined by HPLC (Yanaco L-2000) with fluorometric detection (Shimadzu RF 530: Ex 360 nm; Em 445 nm) after treatment of the CSF samples according to the method of Fukushima and Nixon[6] with slight modifications.

<u>Statistical Analysis</u>: The results were analyzed statistically using the unpaired T test and one way ANOVA. The criterio of statistical significance was $P<0.05$ and the results expressed as the mean value ±SD.

RESULTS

1] In the ventricular CSF, the mean value of T-BP in the PD patients was 14.71±3.55 (pmol/ml), significantly lower than that in the neurological controls (23.11±5.15) ($P<0.001$). However, no significant difference was noted in the mean value of T-NP between the PD cases and controls (Table).

2] No significant correlations could be demonstrated between concentrations of T-BP and T-NP in the PD cases and either sex; right and left ventricular taps; or the disability stage of each patient (T-BP concentrations of the patients in Stage IV were relatively low) (Table).

3] In the PD cases, no significant correlation was found between T-BP levels and either age at operation or age at onset. Significant positive correlations, however, were observed betweer T-NP levels and age at operation (Figure 1) and age at onset ($r=0.54$; $P<0.005$).

4] Remarkably, in the PD cases, a significant negative correlation could be demonstrated between duration of illness

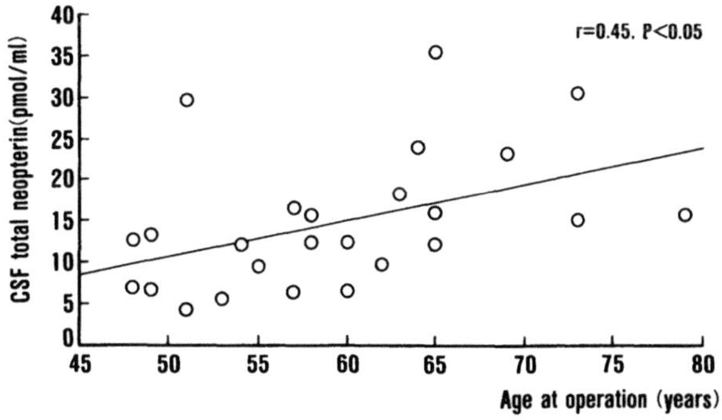

Figure 1. Ventricular CSF total neopterin levels as a function of age in patients with Parkinson's disease.

and T-BP levels (Figure 2), while no such correlation was found for the T-NP levels.

DISCUSSION

This investigation has demonstrated that the T-BP level in the ventricular CSF of the patient with PD is markedly diminished, while the T-NP level remains normal. It has been reported that the T-BP values in the lumbar CSF of patients with ET are lower than those for normal controls[7]. Thus, since 6 out of the 7 neurological controls in this study had ET, the inference can be made that the difference in the T-BP values between PD cases and normal subjects is even more remarkable.

It is known that, in the human brain, T-BP is concentrated in the striatum[8]. But the distribution of T-NP shows no correlation with that of T-BP[8]. In the striatum of the parkinsonian brain, TH, T-BP[2], and guanosine triphosphate cyclohydrolase I, which catalyzes the initial step of BH_4 synthesis, are all decreased. A correlation between decreases in T-BP concentrations and reduced TH activity has been demonstrated[2]. Moreover, it has been shown that, in the CSF, a correlation

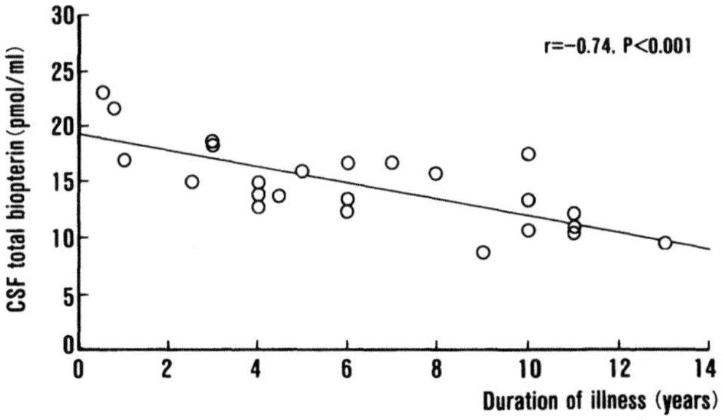

Figure 2. Ventricular CSF total biopterin levels as a function of duration of illness in patients with Parkinson's disease.

exists between T-BP levels and homovanillic acid values[1]. This evidence suggests, therefore, that the observed decrease in the T-BP content in the ventricular CSF represents the degeneration of the nigrostriatal dopaminergic system in patients with PD. The origin of T-NP in the CSF is still unknown. Its increase with age may indicate some age-related alteration in the pterin metabolism. Although T-NP is an intermediate metabolite in the BH_4 biosynthesis, its function appears to be different from that of T-BP.

In this study, T-BP levels in the ventricular CSF of the patients with PD could be correlated only with duration of illness. By extrapolating the data presented in Figure 2, it is conceivable that, at onset, the T-BP levels of PD patients have already decreased. Sufficient progress in the degeneration of the nigrostriatal dopaminergic system occurs during a period before onset, causing the appearance of the initial symptoms at onset.

We consider that it is important to investigate further the correlations between T-BP levels and the clinical features of PD. Such research may provide a basis for developing a more effective regimen using BH_4 administration combined with L-DOPA therapy in the future.

References

1. W. Lovenberg, R. A. Levine, D. S. Robinson, M. Ebert, A. C. Williams and D. B. Calne. Hydroxylase cofactor activity in cerebrospinal fluid of normal subjects and patients with Parkinson's disease. Science 204: 624 (1979).
2. T. Nagatsu, T. Yamaguchi, T. Kato, T. Sugimoto, S. Matsuura, M. Akino, I. Nagatsu, R. Iizuka and H. Narabayashi. Biopterin in human brain and urine from controls and parkinsonian patients: application of a new radioimmunoassay. Clin. Chim. Acta. 109: 305 (1981).
3. A. Williams, J. Ballenger, R. Levine, W. Lovenberg, and D. Calne. Aging and CSF hydroxylase cofactor. Neurology 30: 1244 (1980).
4. P. LeWitt, R. Levine, W. Lovenberg, N. Pomara, M. Stanley, D. Gurevich, P. Schlick, and R. Roberts. Monoamine neurotransmitter metabolites and hydroxylase cofactor in Alzheimer-type dementia and normals, in:"Alzheimer's and Parkinson's diseases: Strategies for research and development." A. Fisher, I. Hanin and C. Lachman, eds., Plenum Press, New York (1986).
5. M. M. Hoehn and M. D. Yahr. Parkinsonism: onset, progression, and mortality. Neurology 17: 427 (1967).
6. T. Fukushima and J. C. Nixon. Analysis of reduced forms of biopterin in biological tissues and fluids. Anal. Biochem. 102: 176 (1980).
7. A. C. Williams, R. A. Levine, T. N. Chase, W. Lovenberg, and D. B. Calne. CSF hydroxylase cofactor levels in some neurological diseases. J. Neurol. Neurosurg. Psychiatry 43: 735 (1980).
8. M. Sawada, Y Hirata, H. Arai, R. Iizuka, and T. Nagatsu. Tyrosine hydroxylase, tryptophan hydroxylase, biopterin, and neopterin in the brains of normal controls and patients with senile dementia of Alzheimer type. J. Neurochem. 48: 760 (1987).

MANGANESE CONTENT IN HAIR OF PATIENTS WITH PARKINSON'S DISEASE: MANGANESE DEFICIENCY AND SUSCEPTIBILITY TOWARDS PARKINSON'S DISEASE

Biao Chen, Zhou-Lin Liu, Xiu-Ling Liang, Caroline M. Tanner, and Xi-Bang Pan

Department of Neurology, Sun Yat-Sen University of Medical Sciences, Guangzhou, China (Drs. Chen, Liu, Liang, and Pan) the Department of Neurological Sciences, Rush-Presbyterian-St. Luke's Medical Center, Chicago, IL, USA (Dr. Tanner)

INTRODUCTION

Until now, the cause of Parkinson's Disease (PD) is unknown, but it is considered to be attributed to the interaction of a genetic factor, ageing and environmental factors. There is a loss of pigmented neurons distributed identically in neuromelanin-containing tissues such as substantia nigra, locus ceruleus, and dorsal vagal nucleus, and a selective depletion of dopamine occurring in these structures and even in adrenal medulla. Though the findings of numerous newly discovered Parkinsonism-causing neurotoxins may indicate their effects on the development of PD, it is still difficult to explain why just a few people in the population have PD. It has been found that melanin has an affinity to manganese (Mn) (1) and other heavy metals and some evidences suggested that the development of PD may be related to heavy metals.

METHOD

In a recent study which was one of our serial studies on environmental and genetic factors for PD conducted in China (3,4), we measured the content of Mn and other heavy metals in hair of patients with PD and matched controls. The PD cases and controls have lived in Guangzhou city and its outskirts for at least the past ten years and were all selected from the neurological clinic or ward of the First Affiliated Hospital of our University. Those with dyed or permed hair were eliminated from the study. PD cases were selected if they had onset of symptoms after the age of 25 and had at least two of the following signs: resting tremor, bradykinesia, cogwheel rigidity and postural reflex impairment. Controls were matched for sex, age (± 3 years) hair color and race and were among patients diagnosed as: cervical spondylosis, tension headache, migraine headache, Bell's palsy, low back pain, sciatica, spinal stenosis and herniated nucleolus pulposus. The hair was cut as close to the scalp as possible from the occipital area and only the proximal 2 cm section were retained for analysis with the method of Flame Atomic Absorption Spectrophotometer (Jarrell-Ash model 810). The samples of PD cases and controls were collected in the same season, and prepared with the same method by the same individuals. In the procedures of preparation and analysis,

Basic, Clinical, and Therapeutic Aspects of Alzheimer's and Parkinson's Diseases, Volume 2
Edited by T. Nagatsu *et al.*, Plenum Press, New York, 1990

the samples of PD case and matched control counterpart were processed successively at the same time in order to get a high comparability.

RESULT

The content of Mn in hair of 57 pairs of each PD cases and controls including 43 pairs of men and 14 pairs of women were measured. The mean age was 58.28 years for PD case (range 34-74) and 58.16 years for control (range 37-75). For PD cases, the mean onset age was 52.28 years (range 30-68); the mean disease duration was 6.12 years (range 0.08-20.50); and the mean Hoehn & Yahr Stage was 3.19 (range 1-5). In quantitative Student's paired samples t-test, it was shown that the concentration of Mn in the hair of PD patients (2.34±1.16ppm) was significantly lower than that of matched controls (2.95 ±1.61ppm)(t=2.56, p=0.013). Furthermore, in the comparisons of groups divided according to onset age, disease duration, and Hoehn & Yahr Stage, no significant correlations were achieved. Conditional Logistic Regression showed that the relative risk for the decreased Mn content in hair of PD cases was 0.6528 (confidence limit 0.4513-0.9442). We did not find that the Mn content was affected by other factors such as the contents of copper, lead, cadmium and mercury in hair, onset age, disease duration and disease stage in a multiple regression analysis. We also did not find any significant difference of Mn content between L-dopa administration group (42 cases, 2.35±1.24ppm) and non-L-dopa administration group (15 cases, 2.33±0.92ppm, t=0.06, p=0.956) and any correlation between decreased Mn content and the dosages of L-dopa administration analyzed as gram-year (r=0.014, F=0.464). At the same time, no significant differences in the contents of copper, lead, cadmium and mercury were found between PD cases and controls.

DISCUSSION

We have presented here that PD patients had a significantly lower Mn content in hair than that of matched controls with a relative risk less than one. Recently, another laboratory in the same city studied 47 PD cases and 50 controls and found that the Mn content in hair of PD (1.91±1.90ppm) was significantly lower than that of controls (3.94±4.09ppm). However, Pall (2) did not find a significant difference of Mn content in CSF between 9 PD cases and 29 controls. It might be related to the fact that the Mn concentration of CSF can not reflect the small difference of Mn in the human body which originally has a trace amount of Mn. On the other hand, hair is a melanin-containing tissue which has a high affinity to Mn and may be susceptible to the change of Mn metabolism defect in human body. A genetically pigmentary mutant pallid mouse, which suffers from an irreversible phenotypic "ataxia" associated with a defect of otolith development due to a local Mn metabolism defect in melanin-containing labyrinth, has a normal Mn level in its liver and kidney (5). Cotzias (6) has already reported that the turnover of radioactive Mn was significantly slower in the Parkinsonian miners than in the "healthy" Mn-exposed miners. Therefore, we suggest that the decreased Mn content in hair of patients with PD may indicate that the PD might have a defect of Mn metabolism, at least in those pigmented tissues.

In our previous one to two matched case control study on environmental factors for PD conducted in China, we found that the occupational and residen-tial exposure to industrial chemicals (OR=2.39), printing plants (OR=2.4) or quarries (OR=4.5) was associated with an increased risk of developing PD. In contrast, living in villages (OR=0.57), wheat growing (OR=0.40), and pig raising (OR=0.17), were associated with a decreased risk (3). From a further analysis of 80 PD cases and 160 controls conducted in Guangzhou, a decreased risk for PD was associated with having been born in Chinese villages (OR= 0.3846, confidence limit 0.2445-0.6051). In our genetic study on PD, each of the 88 PD cases and controls were investigated. We found that a positive family history of PD in first degree relatives was 3.14% for PD patients and 1.41% for controls and the prevalences of PD in their families were 0.83% for PD patients and 0.13% for controls separately. The estimated heritability was 38±8.34% which is very similar to those reported by Kondo of 22% and Martin of 40±11%. These heritabilities were larger than the value of congenital heart diseases, but less than those for hypertension, asthma, and diabetes. We proposed that PD is likely to have a multifactorial etiology (4).

Therefore, those negative correlations may reflect the fact that those people who were born in Chinese villages 50 to 70 years ago might be insusceptible to developing PD and there may exist a factor which occured during the gestation and infancy to protect them against PD. It is probable that the development of PD may begin in utero and is in favor of a combined effect of inheritary factor and extragenous agent acting in utero. This may be contributed to the susceptibility towards PD.

In a twin study of PD, it was shown that during early life, there was a personality difference between the affected and unaffected twins. This difference was similar to the behavioral difference between different strains of mice due to their variant initial number of dopaminergic neurons. Moreover, Myrianthopoulos (7) showed that the prevalence of PD in the relatives of patients who developed Parkinsonism after taking antipsychotropics was significantly greater than that of patients who did not develop Parkinsonism. Hence, the pathological basis for PD may exist in early life making them more susceptible to developing PD. Eldridge and Ince (8) suggested that those affected twins might have a low initial number of dopaminergic neurons and then have a predisposition to PD. They raised a hypothesis that in utero a factor, which, present in a critical amount, might influence the total number or ultimate survival of dopaminergic neurons of the substantia nigra. This factor is crucial for the eventual development of PD. We deduce that the protective factor might be Mn.

In the investigations on the pallid mouse which is particularly sensitive to a normal or low Mn level diet, it was found that the otolith development is Mn dependent, but the supply of Mn depends upon the existence and completed development of localized pigmented cells in the membranous labyrinth. The pleiotropic effects of the pallid gene on the Mn-dependent otolith are partially penetrant and highly variable, even showing evidence for environmental modification, as for example the otolith is well developed in the offspring when the pregnant pallid mice were supplied with large amount Mn (5).

Cotzias (1) found that Mn was abundant in black pigmented tissues and suggested a close relationship among Mn, biogenic amines, neuromelanin, and extrapyramidal disorders. It was shown that Mn can cause either increased or decreased dopamine in the rat brain according to variant Mn dosages used and affected in different life stages (9). Cotzias and his colleques found that pallid mice had a low Mn content in the brain and less storage and slower transportation of Mn, L-dopa and L-tryptophan which are the precursors of melanins and dopamine (10). Pallid mice seem insensitive to the cerebral effect of L-dopa. In neonatal mice, a high correlation between cerebral dopamine and Mn content was found and they suggested that the dopaminergic appratus might be Mn dependent for its maturation and function (11). Many studies have shown that a relatively large Mn exposure increases the dopamine content in neonates, in contrast to adult in whom such exposure results in a decrease in brain dopamine content (12). The fact of the inability of fetal liver to assimilate Mn together with low Mn concentrations in milk suggest that the prenatal fetus and newborns may be predisposed to Mn deficiency, particularly when they are influenced by other factors such as genetic factor which can cause the disturbance of Mn transportation and storage. Therefore, we suggest that the development and maturation of dopaminergic neurons, just like the otolith, are affected by the pleiotropic effects of pigmentary genes and trace element Mn. When they are defective in early life, a low initial number of dopaminergic neurons will result. It has been found that in rodents, the dopaminergic systems begin developing around gestation days 12-13 and continue nearly one month postnatally. Thus, these systems may be susceptible to insult during either gestation or neonatal development.

The relationship among Mn, pigmentation, biogenic amines, dopaminergic neurons, and genetic factor mentioned above very likely corresponds to that which happens in PD. It has been revealed that PD patients had a defect of neuromelanin in the substantia nigra and that patients with familial juvenile Parkinsonism have a congenital absence of neuromelanin. The coincidence of decreased dopamine in the neuromelanin-containing tissues, such as substantia nigra, locus ceruleus, and adrenal medulla in PD patients may suggest that PD is a disorders of neuromelanin system.

We are also intrigued by the rather identical differences of PD distribution and Mn met bolism. First, it was suggested there is a racial difference in the prevalence of PD generally (13): highest in the white races of caucasians; second in the Japanese and Chinese; and lowest in the negroid races of Africa. This is consistent with the finding that, in animals, the strain difference in Mn requirements was associated with genetic difference in pigmentation, while an increased prevalence of PD was associated with the dilution of the skin color. In addition, the differences on the cultural nutrition, in which caucasians are likely to eat more meat, bread, seafoods, dairy products and coffee, but Chinese and Japanese obtain the calories mainly from grains, fresh vegetables and like drinking tea, seem to indicate that the former have a lower Mn intake than that of the later in the daily diet. It is known that the Mn intake is decreased tens times in those who take meat, seafoods, dairy products, and bread as the main source of calories than in those who obtain their calories principally from grains, fruits, nuts, tea and fresh vegetables. A recent study on early life dietary factors for PD have shown that eating more nuts, salad oil or dressing, and plums which contain a high Mn level was associated with decreased risk for PD (14).

CONCLUSION

From the fact mentioned above, we suggest that the susceptibility towards PD is determined by the interaction of a genetic factor which affects the pigmentation and Mn metabolism, and the relatively Mn deficiency in gestation and infancy. They would influence the development and maturation of dopaminergic neurons within neuromelanin tissues and produce a lower initial number of these neurons. They could also be modified by a intake of appropriate amount of Mn in early life. After ageing and encounting the effect of environmental factors which can cause a preferential damage to the dopaminergic neurons in the neuromelanin tissues, together with the defect of Mn related antioxidative systems, the dopaminergic neurons and dopamine will be further lost and eventually the clinical symptoms appear.

ACKNOWLEDGEMENT: This work was supported partially by a grant from the SANDOZ FOUNDATION FOR GERONTOLOGICAL RESEARCH. We thank Mr. Liu Qing for the statistic analysis.

REFERENCES

1. G. C. Cotzias, P. S. Papavasiliou, and S. T. Miller, Manganese in melanin, Nature. 201:1228 (1964)

2. H. S. Pall, A. C. William, D. R. Blake, et al, Raised cerebrospinal fluid copper concentration in Parkinson's disease, Lancet. ii(8553):238 (1987)
3. C. M. Tanner, B. Chen, W. C. Wang, et al, Environmental factors and Parkinson's disease: a case control study in China, Neurol. 39:660 (1989)
4. B. Chen, Z. L. Liu, X.L. Liang, and X. B. Pan, Genetic study on Parkinson's disease, Heredity and Diseases. 6:87 (1989)
5. L. C. Erway, Genetic and developmental implications for trace metal metabolism from mutant and inbred strains of animals, in: "Metabolism of Trace Metals in Man," O. M. Rennert, and W. Y. Chan, eds., CRC Press Inc., Boca Raton, Florida (1984)
6. G. C. Cotzias, M. H. Van Woert, and L. M. Schiffer, Aromatic amino acid and modification of Parkinson's disease, N.Engl.J.Med. 276:374 (1967)
7. N. C. Myrianthopoulos, A. A, Kurland, and L. T. Kurland, Hereditary predisposition in drug-induced Parkinsonism, Arch.Neurol. 6:5 (1962)
8. R. Eldridge, and S. E. Ince, The low concordance rate for Parkinson's disease in twins: a possible explanation, Neurol. 34:1354 (1984)
9. P. K. Seth, and S. V. Chandra, Neurotransmitters and neurotransmitter receptors in developing and adult rats during manganese poisoning, Neurotoxicol. 5:67 (1984)
10. G. C. Cotzias, L. C. Tang, S. T. Miller, and D. Seadic-Simic, A mutation influencing the transportation of manganese, L-dopa, and L-tryptophan, Science. 176:410 (1972)
11. G. C.Cotzias, S. T. Miller, P. S. Papavasiliou, and L. C. Tang, Interaction between manganese and brain dopamine, Med.Clin.North.Am. 60:729 (1976)
12. S. F. Ali, and R. R. Holson, Agents that are not dopamine agonists or antogonists can affect the developing dopamine system (abstract), Teratology. 36:510 (1989)
13. M. R. Lerner, and R. S. Go, Skin colour, MPTP, and Parkinson's disease, Lancet. i:212 (1987)
14. L. I. Golbe, T. M. Farrell, and P. H. Davis, Case control study of early life dietary factors in Parkinson's disease, Arch.Neurol. 45:1350 (1988)

SINGLE PHOTON EMISSION COMPUTED TOMOGRAPHY IN ALZHEIMER-TYPE DEMENTIA

M. Hamamoto, M. Yamazaki, H. Igarashi, T. Miyazaki, K. Isse[1], M. Uchiyama[1], K. Tanaka[1], K. Chiba[2], and A. Terashi[3]

Department of Neurology, Psychiatry[1], Radiology[2], Tokyo Metropolitan Tama Geriatric Hospital and Second Department of Internal Medicine[3], Nippon Medical School, Tokyo, Japan

INTRODUCTION

To diagnose early changes of Alzheimer's-type dementia (ATD) is important for treatment and investigation of the disease. Recently, radionuclide techniques, such as positron emission tomography (PET) and single photon emission computed tomography (SPECT), have been regarded as the most hopeful techniques for diagnosis.

Using I 123 iofetamine (IMP) SPECT studies, Johnson et al.[1] studied the relationship between abnormal regional cerebral I 123 IMP uptake and the severity of ATD. However, they observed few obvious changes along with dementia progression, except for the parietal region.

In this study we attempted to make a new classification of severity in ATD patients and, using this classification, we evaluated the relationship between I 123 IMP uptake changes and the severity of ATD.

STAGING OF ALZHEIMER-TYPE DEMENTIA

Exact evaluations about the focal symptoms and character deterioration in dementia patients are difficult to quantify clinically, so there is some uncertainty in classifying the stages. We have made detailed observations on the clinical course of ATD for a long time, and tried to classify its stages according to the following criteria: Mild case (stage 1), in which only memory disturbance is the clinical symptom; moderate case (stage 2), with signs of higher cortical dysfunction, such as aphasia, apraxia, agnosia; severe case (stage 3), in which no communication can be made with face patterns, words, and actions; severest case (stage 4), in which the above signs proceed and patients become bedridden (Table 1).

PATIENTS AND METHODS

Sixty-two patients who met DSM-III-R criteria[2,3] for the diagnosis of probable ATD were studied with I 123 IMP SPECT. There were 18 men and 44 women; their mean (±SD) age was 75.0 ± 8.2 years (range, 55-90 years). According to above stage classifications, patients were divided into stages 1-3; among them 20 cases were in stage-1, 26 in stage-2, and 16 in stage-3. Stage-1 patients were all diagnosed retrospectively. Controls were eleven healthy aged persons (stage-0), of ages 60 to 80, with a mean age of 71.6 ± 6.0 years.

SPECT were performed 15 minutes after 6 mCi I 123 IMP injection intravenously, using a rolling-type gamma camera GE 400AC, 12x12 mm box regions of interest (ROI) were set in 8 mm-thick horizontal slices parallel to the orbitomeatal line, to count ROI in each cortex and basal nucleus (fig.1). According to Johnson et al.[4] method, relative I 123 IMP uptake was calculated as the ratio of each cortical ROI activity to the mean (right and left) cerebellar ROI counts (cortico-cerebellar ratio; CCR). Correlation between staging of severity and CCR changes in each respective cortical and basal nucleus region were analyzed with one-way analysis of variance (ANOVA).

RESULTS

CCRs in the bilateral lower temporal and left middle frontal cortex regions were significantly lower in stage-1 ATD patients than in controls. In stage-2, reduction in CCR was also seen in the frontal, parieto-temporal, and parietal cortex areas. In stage-3, reduction in CCR spread widely to bilateral hemispheris, except for the primary sensorimotor cortex (fig. 2). Along with dementia progression, changes in SPECT image spread progressively from the lower temporal to the parieto-temporal and frontal areas.

In a comparison of stage-1 and stage-2, reduction in CCR was observed in the temporal and parietal regions in the right hemisphere, but not in the left hemisphere. In a comparison of stage-2 and stage-3, a distinct decrease of CCR was seen in the left temporo-parietal area and the frontal area, but in the right thalamus CCR increased (fig. 2).

DISCUSSION

In this study, we found that, along with dementia progression in patients with ATD, changes in I 123 IMP SPECT image spread progressively from the lower temporal to the parieto-temporal and frontal cortex regions. There was a significant difference between each stage, which had a relationship with the clinical classification. Johnson et al.[1] reported that relative I 123 IMP uptake in patients with ATD differed significantly from uptake in control subjects in all cortical regions, and a significant association between reduced I 123 IMP uptake and dementia severity was found in the parietal region, but not in the temporal, frontal, or occipital regions. However, in this study we found that there were significant associations between I 123 IMP uptake and dementia severity in the temporal, parietal, and upper frontal regions. And the reductions in CCR that were found in the temporal cortex regions correlated with memory impairment, which is the main symptom of ATD. The reason for the difference between their results and our data depend on the staging of severity.

By observing the clinical symptoms of ATD patients precisely, we invented a level classification: at first memory disturbance occurred, then higher cerebral dysfunction was observed, and finally the personality deteriorated. Such a classifications, based on the clinical course, has been widely accepted[5], but no discussion of ATD using this classification has yet been reported. We used SPECT as a trial case to evaluate the cerebral disfuncitons of ATD patients using this classification.

In particular, the changes observed in mild dementia (stage-1), which were obtained from retrospective diagnoses, showed the possibility that I 123 IMP SPECT are useful for the early diagnosis of ATD.

Table 1. Stage of Severity in Alzheimer-Type Dementia.

Severity	(Stage)	Symptoms		
Normal	Stage 0	None		
Mild	Stage 1	Memory disturbance		
Moderate	Stage 2	Memory disturbance	+ Higher Cortical Dysfunctions	
Severe	Stage 3	Memory disturbance	+ Higher Cortical Dysfunctions	+ Personality deterioration

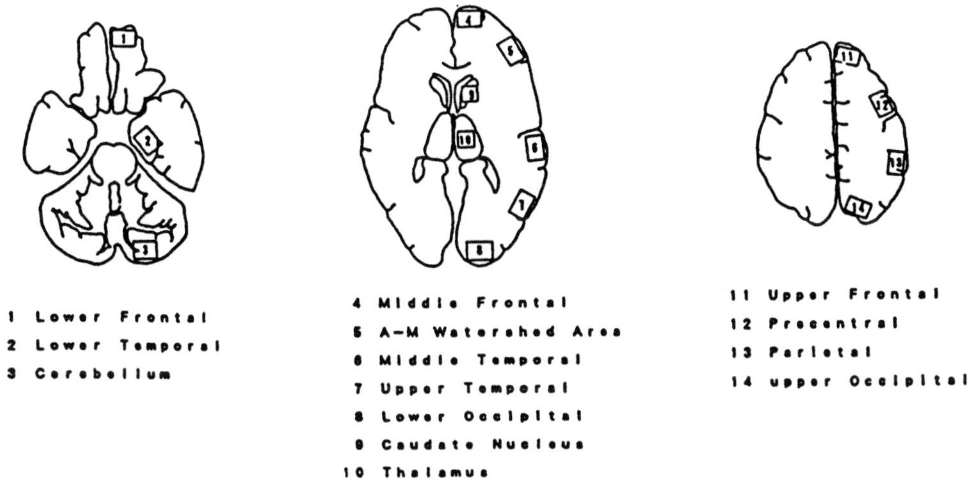

Figure 1. This figure shows the placement of regions of interest on transaxial images using single photon emission computed tomography and I 123 IMP in three different planes.

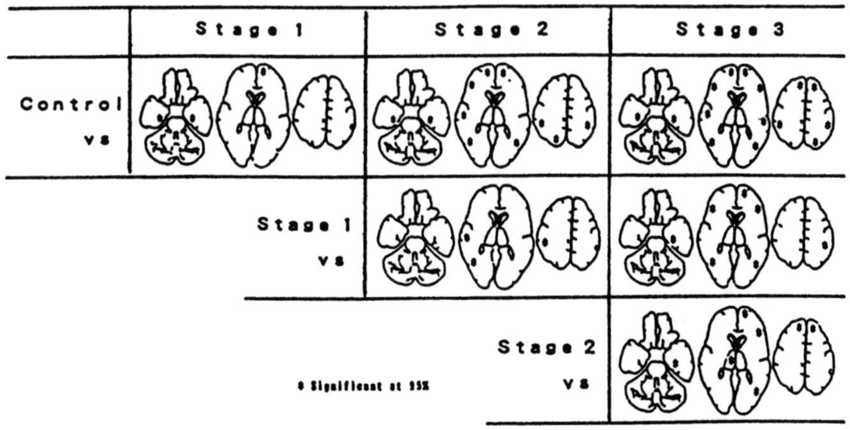

Figure 2. This figure shows the results of ANOVA, in comparison between the control and each stage of Alzheimer-Type dementia patients. Asterisks indicate a significant reduction in cortico-cerebellar ratio in I 123 IMP uptake.

REFERENCES

1. K. A. Johnson, B. L. Holman, et al., Single Photon Emission Computed Tomography in Alzheimer's Disease. Arch. Neurol., 45:392 (1988).
2. American Psychiatric Association Task Force on Nomenclature and Statics; Diagnostic and statistical manual of mental disorders 3rd. ed. American Psychiatric Association. Washington D.C.:124 (1980).
3. A. F. Jorm, A. S. Henderson, Pissible Improvements to the Diagnostic Criteria for Dementia in DSM-III. Brit. J. Psychiat., 147:394 (1985).
4. K. A. Johnson, S. T. Mueller, T. M. Walshe, et al., Cerebral perfusion imaging in Alzheimer's disease; Use of single photon emission computed tomography. Arch. Neurol., 44:258 (1987).
5. Psychiatric Diagnosis & Therapy ('88/89) (A LANGE clinical manual) edited by J. A. Flaherty, R. A. Chamsan, J. A. Davis.

APPLICATION OF MAGNETIC RESONANCE IMAGING TO A PRIMATE MODEL OF THE CHOLINERGIC DEFICIT OF ALZHEIMER'S DISEASE

Thomas Aigner, Richard Saunders, Chantal Chavoix, and Joseph Frank

Laboratory of Neuropsychology, National Institute of Mental Health and Department of Radiology National Institutes of Health, Bethesda, MD

INTRODUCTION

The neurotransmitter acetylcholine is now considered to play an important role in processes of memory. In primates, the major source of acetylcholine for the cortex and amygdala is the nucleus basalis of Meynert (nbM), and that for the hippocampus is the medial septum (ms) and the diagonal band of Broca (dbB) (Amaral and Cowan, 1980; Mesulam et al., 1983). These nuclei are the subject of intense investigation since it is now believed that their degeneration is responsible, at least in part, for the mnemonic and other cognitive impairments observed in dementia of the Alzheimer type (Perry et al., 1978; Whitehouse et al., 1982). Although an animal model of the cholinergic deficit of this type would be a valuable research tool, there are several major problems that have hindered efforts to produce such a model in nonhuman primates. Indeed, these cholinergic nuclei are located deep in the brain, their anterior-posterior position varies greatly between animals, and they are formed in an extremely complex shape. These factors make it difficult to locate and selectively damage the basal forebrain cholinergic system with conventional stereotactic methods. These cells do, however, lie in proximity to other neural structures that are easier to locate, such as the anterior commissure and the ventral portion of the globus pallidus, which were used as landmarks in earlier studies employing electrophysiological recording as a method to map the region of the basal forebrain in monkeys for injection of the excitatory neurotoxin ibotenic acid (Aigner et al., 1984, 1987). Although this method did produce impairments in recognition memory and reductions in cortical cholinergic markers, there were 3 major limitations to the continued use of this procedure: (1) the lesions were not complete and were highly variable, possibly due to the difficulty in aligning the injection needle so that it traveled the same path as that of the recording electrode; (2) to attain the best results, mapping the basal forebrain and injecting the neurotoxin had to be done during the same surgery, which greatly increased an animal's risk to complications resulting from an extended period of anesthesia; (3) the accuracy of the lesions could not be determined before the animal was sacrificed, which was usually several months after surgery due to the behavioral testing. To avoid these problems, we used magnetic resonance to produce images of the brain that could be used to calculate target coordinates for ibotenic acid injections and also to evaluate the accuracy of these placements premorbidly.

METHODS

Imaging was done with a General Electric Signa 1.5 T scanner. Four rhesus monkeys weighing 4-6 kg were scanned before and after surgery. During scanning, the animal was anesthetized with a combination of Ketamine (7 mg/kg) and Xylazine (.6 mg/kg) and placed in a specially designed nonferromagnetic-stereotactic headholder (Saunders et al., in press). The headholder was then carefully positioned so that its coronal, sagittal, and axial planes coincided with those of the scanner. A 12-cm general purpose receive-only surface coil was centered directly above the animal's head. For all scans, a 10-cm field of view, 2 excitations, and 128 phase-encoded steps were used.

Prior to surgery, a series of scans was obtained through the coronal and sagittal planes using an inversion recovery (IR) technique with an inversion time = 600 msec, repetition time (TR) = 2500 msec, and echo time (TE) = 25 msec. With an in plane resolution of 0.78 x 0.39 mm resolution, these parameters provided excellent visualization of the cerebral cortex, the anterior commissure, and the sagittal sinus, which were then used as reference points for determining the coordinates for the injection sites. Films of the images were printed and measurements (corrected for image enlargement) were made from structures that could be visualized later during surgery. The depth of the injection was the distance from the top of cortex to the target sites. The lateral distance was measured from a vertical line drawn through the sagittal sinus to the target sites. The anterior distance was measured from the stereotactic earbars, which contained a contrast agent, to the region ventral to the anterior commissure.

Surgery was performed within 2 weeks after initial scanning. The animal was immobilized with Ketamine (10 mg/kg) and deeply anesthetized with sodium pentobarbital. The animal was carefully re-positioned in the stereotactic frame so that it was in the same orientation as it had been for the scan, and then draped for aseptic surgery. An opening approximately 10 mm in diameter was made in the skull and 1 ul of ibotenic acid (10-12 ug/ul) was injected into each of 18 sites in each hemisphere to damage the ms, the dbB, and the nbM . Bilateral injections were made in 2 stages (left side first), separated by a two-week intraoperative period.

Within 1 week after each surgery, the animals were rescanned with 2 different procedures. To visualize the tracks made by the injection needle, a 3-dimensional (3-D) volume scan was obtained throughout the basal forebrain region using TR = 100 msec, TE = 20 msec, flip angle = 75^0, and a slice thickness of 1 mm. To visualize the injection sites, a T2-weighted series of images was obtained with TR = 2000 msec, TE = 80 msec, and flip angle = 20^0.

RESULTS AND DISCUSSION

The use of magnetic resonance imaging to visualize and locate target areas of the basal forebrain in monkeys eliminated the problems of earlier procedures. First, the MRI procedure is a much faster method of determining the location of these areas. The scanning procedure requires 2-3 hrs to obtain all of the images necessary to calculate the injection coordinates for both hemispheres and can be done several days before surgery. The previous method required 6-10 hrs of electrophysiological recording per hemisphere and had to be done during the same surgical procedure as the neurotoxin injections to achieve the best results. By using the MRI procedure, the duration of surgery has been reduced by more than half, from 10-15 hrs required previously to 4-5 hrs with the present method, thus greatly reducing the surgical stress on the animals. A second advantage provided by the MRI procedure is that an estimate of the location of the injection sites and the tracks made by the injection needle can now be visualized in the brain after surgery. Fig. 1 shows an example of a T2-weighted image obtained 5 days after ibotenic acid was injected into the left hemisphere of a monkey. The area of high signal intensity ventral to the anterior commissure in the substantia innominata is an example of an accurate placement. The high signal intensity may represent localized edema following the injections or it may be a direct effect of the neurotoxin itself. A 3-D volume image (Fig. 2) through the basal forebrain region obtained 5 days after the second surgery shows the vertical tracks left by the needle penetrating the brain to the caudolateral portions of the nbM bilaterally. Since the needle has a small diameter and must pass through 20-30 mm of brain tissue to the target sites, there is the possibility that it can be deflected, which could be determined with a postoperative scan. Now, using MRI, the target regions can be accurately located and the success of the surgery can be estimated noninvasively soon after surgery. For these reasons, MRI promises to be an especially valuable tool for basal forebrain lesions, with applications to many other regions of the brain and areas of study.

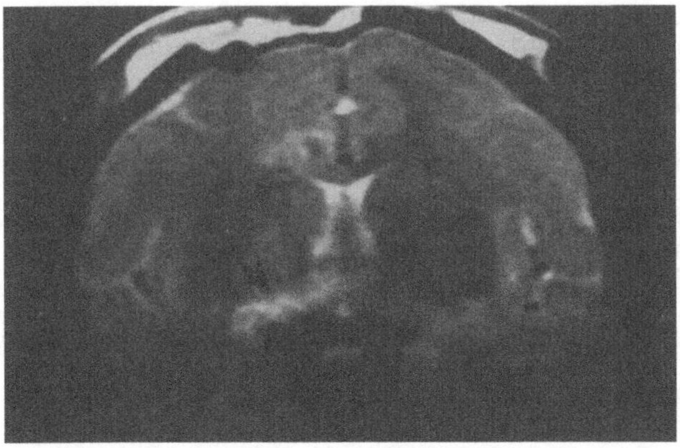

Fig. 1. Coronal T-2 weighted magnetic resonance image through the basal forebrain of a rhesus monkey 5 days after left hemisphere injections of ibotenic acid. Arrow points to injection sites in the nucleus basalis of Meynert.

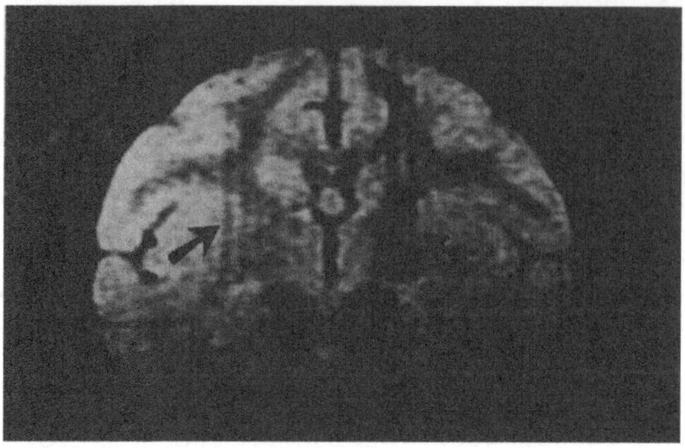

Fig. 2. Coronal 3-D volume image through the basal forebrain of a monkey 5 days after bilateral injections of ibotenic acid into the nucleus basalis of Meynert. Arrows indicate the penetration tracks left by the injection needle.

REFERENCES

Aigner, T., Mitchell, S., Aggleton, J., DeLong, M., Struble, R., Wenk, G., Price, D., and Mishkin, M., 1984, Recognition deficit in monkeys following neurotoxic lesions of the basal forebrain, Soc. Neurosci. Abstr. 10:386.

Aigner, T., Mitchell, S., Aggleton, J., DeLong, M., Struble, R., Wenk, G., Price, D., and Mishkin, M., 1987, Effects of scopolamine and physostigmine on recognition memory in monkeys with ibotenic-acid lesions of the nucleus basalis of Meynert, Psychopharmacology, 92:292-300.

Amaral, D. and Cowan, W., 1980, Subcortical afferents to the hippocampal formation in the monkey, J. Comp. Neurol., 189: 573-591.

Mesulam, M.-M., Mufson, E., Levey, A., and Wainer, B., 1983, Cholinergic innervation of cortex by the basal forebrain: Cytochemistry and cortical connections of the septal area, diagonal band nuclei, nucleus basalis (substantial innominata), and hypothalamus in the rhesus monkey, J. Comp. Neurol., 214: 170-197.

Perry, E., Tomlinson, B., Blessed, G., Bergman, K., Gibson, P., and Perry, R., 1978, Correlation of cholinergic abnormalities with senile plaques and mental test scores in senile dementia, Br. Med. J., 2: 1457-1459.

Saunders, R., Aigner, T., and Frank, J., (In press), Magnetic resonance imaging of the rhesus monkey brain: Use for stereotactic neurosurgery, Exp. Brain Res.

Whitehouse, P., Price, D., Struble, R., Clark, A., Coyle, J., and DeLong, M., 1982, Alzheimer's disease and senile dementia: Loss of neurons in the basal forebrain. Science 215: 1237-1239.

IN VIVO PHOSPHOROUS-31 NMR SPECTROSCOPY IN ALZHEIMER'S DISEASE

Toshiro Fujimoto,[1] Shigeji Noguchi,[1] Toshihiko Nakano,[1] Tsunehisa Uchida,[1] Kaneo Arimitsu,[1] and Tetsuhiko Asakura[2]

[1]Department of Neuropsychiatry,South Japan Health Science Centre Fujimoto Hospital,17-4,Hayasuzucho,Miyakonojoshi,Miyazaki 885 Japan.,and [2]Department of Neurosurgery,Kagoshima University Kagoshima 890,Japan

INTRODUCTION

Thanks to the advances in magnetic resonance imaging (MRI), phosphorus(^{31}P) nuclear magnetic resonance chemical shift imaging in the brain can be obtained non-invasively, thereby facilitating the in vivo observation of ^{31}P metabolism in brain tissue. Some reports have already shown in vivo abnormalities in the ^{31}P metabolism by ^{31}P NMR spectroscopy in Alzheimer's disease.[1,2,3]

However,the acquisition of localized ^{31}P-NMR spectra has involved a technically difficult problem. That is, the NMR signal of ^{31}P is small because the overall NMR sensitivity of ^{31}P is actually 2×10^{-4} that of proton.[4] In spite of this, it is necessary to obtain a signal from an area as small as possible in the brain, thereby necessitating special innovations. In order to solve this problem, ^{31}P chemical shift imaging using four-dimensional Fourier transformation (4-DFT) was proposed by Maudsley et al.,[5] and 4-DFT ^{31}P multivoxel spectroscopy was developed by Miyazaki et al.[6] in Japan.

Using this method, we examined a number of primary degenerative dementias of the Alzheimer type, presenile onset (AD) patients, and obtained their ^{31}P-NMR spectra. Phosphomonoester (PME), inorganic orthophosphate (Pi), phosphodiester (PDE), phosphocreatine(PCr), γ-ATP (γ-phosphate of nucleotide triphosphates), α-ATP (α-phosphate of nucleotide triphosphates) and β-ATP(β-phosphate of nucleotide triphosphates) were compared between the AD patients and healthy volunteers.

SUBJECTS AND METHODS

Primary degenerative dementia of the Alzheimer type with presenile onset was diagnosed according to DSM-Ⅲ-R,and the absence of any abnormal density area showing a cerebrovascular disorder was confirmed by means of a CT examination. The study was conducted on five female patients (mean age:67 $\pm$ 4 years) suffering from a presenile onset of AD and five healthy women (mean age:71 $\pm$ 6 years) who were designated as the control group.

The MRIS (ASAHI SUPER 200) used has a static magnetic field of 2.0 tesla, enabling ^{31}P chemical shift imaging by means of 4-DFT. The scan matrix was 8 $\times$

Edited by T. Nagatsu *et al.*, Plenum Press, New York, 1990

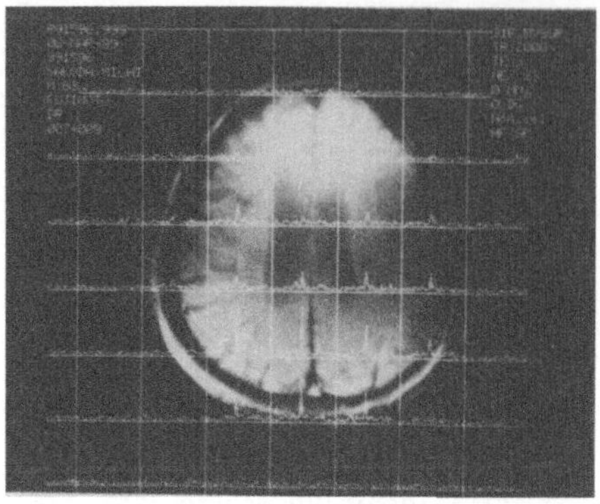

Figure 1A. A superimposed grid on a ^{1}H imaging, which gives the boundaries of the voxels.

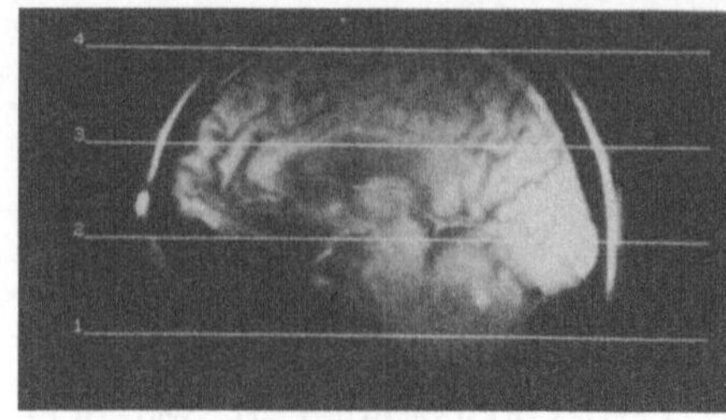

Figure 1B. Volume localized ^{31}P signal was obtained from an area with a 7cm thickness between 2 and 4.

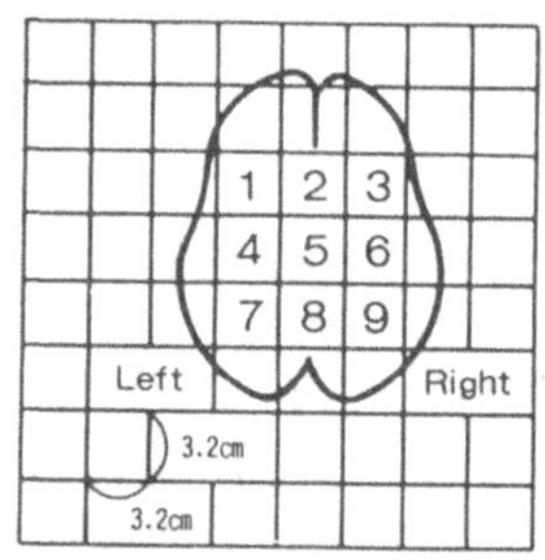

Figure 2. ^{31}P spectra were studied from the voxels numbered 1 to 9.

8 × 1, the repetition time was 2,000 msec, spectral resolution was 0.45ppm, and the total acquisition time was about 1 hour. The ^{31}P-NMR spectra of 8×8 voxels are obtained with this system. The size of each voxel was 3.2cm × 3.2cm ×7cm. Since 3 × 3 voxels are present in the area corresponding to the position of the brain, as shown in Figures 1 and 2, these 9 voxels were studied. The ^{31}P-NMR spectrum of each voxel was investigated for the areas at six peaks, (PME+Pi, PDE, PCr, γ-ATP, α-ATP, and β-ATP, determined with a digitizer), and the ratio of each peak to that of α-ATP were calculated.

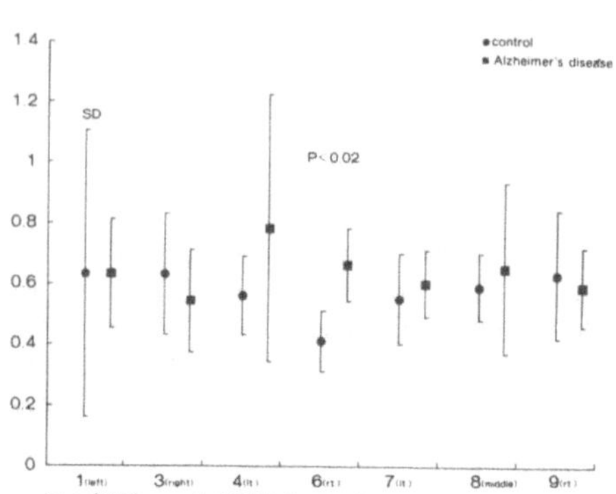

A. PME+Pi/α-ATP ratio.

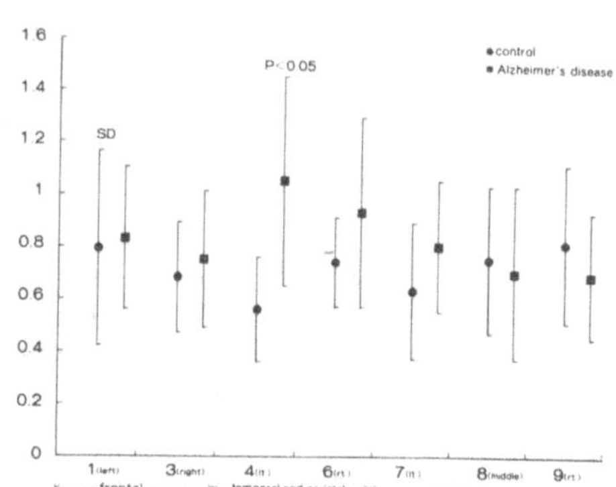

B. PDE/α-ATP ratio.

Figure 3.

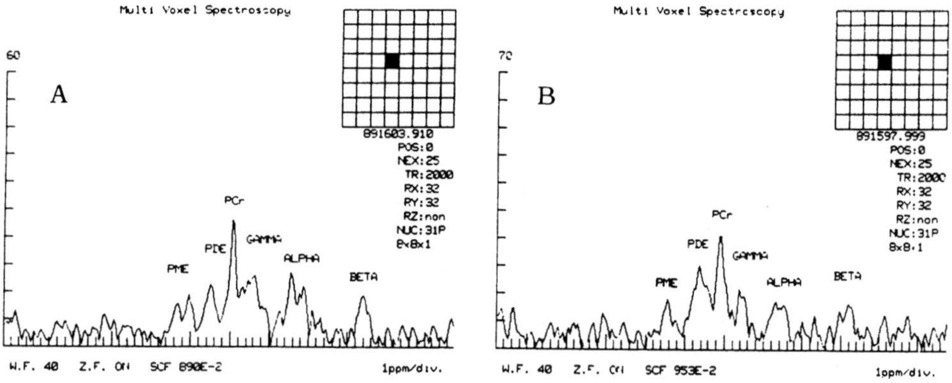

Figure 4. ^{31}P NMR spectra of the left temporal and parietal area(4).
The PDE peak of AD patient is larger than that of the normal control.
A:normal control(female-age 72) B:AD patient(female-age 69)

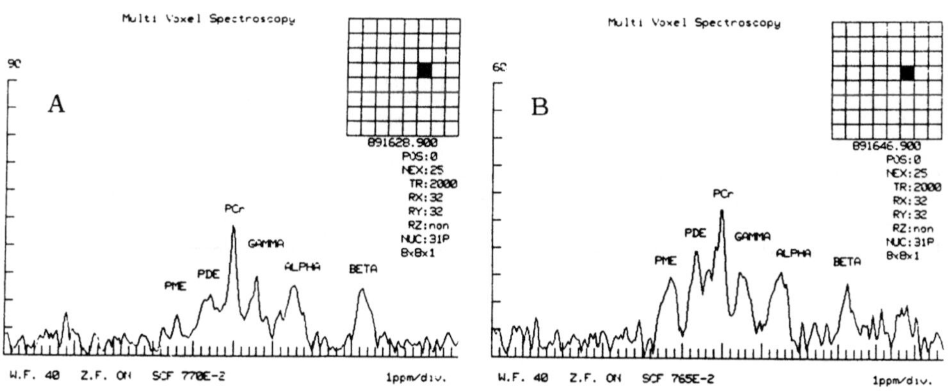

Figure 5. ^{31}P NMR spectra of the right temporal and parietal area(6).
The PME+Pi peak of AD patient is larger than that of the normal control.
A:normal control(female-age 78) B:AD patient(female-age 62)

Table 1 ratios of peaks of ^{31}P NMR spectra
Alzheimer's disease, ( ):normal control

			PME+Pi /α−ATP		PDE /α−ATP		PCr /α−ATP		γ−ATP /α−ATP		β−ATP /α−ATP	
			MEAN	SD	MEAN	SD	MEAN	SD	MEAN	SD	MEAN	SD
frontal	left	1	0.63	0.18	0.83	0.27	1.84	0.43	0.79	0.22	0.85	0.13
			(0.63)	(0.47)	(0.79)	(0.37)	(1.70)	(1.08)	(0.90)	(0.57)	(0.80)	(0.28)
area	right	3	0.54	0.17	0.75	0.26	1.09	0.19	0.85	0.42	0.63	0.11
			(0.63)	(0.20)	(0.68)	(0.21)	(1.22)	(0.41)	(0.73)	(0.27)	(0.83)	(0.39)
temporal	left	4	0.78	0.44	1.05 *	0.40	1.26	0.49	0.92	0.44	0.90	0.25
			(0.56)	(0.13)	(0.56)	(0.20)	(1.11)	(0.28)	(0.70)	(0.12)	(0.51)	(0.31)
and parietal area	right	6	0.66**	0.12	0.93	0.36	1.09	0.27	1.00	0.23	0.74	0.22
			(0.41)	(0.10)	(0.74)	(0.17)	(1.16)	(0.43)	(0.74)	(0.21)	(0.63)	(0.10)
occipital	left	7	0.60	0.11	0.80	0.25	1.28	0.39	0.87	0.17	0.85	0.21
			(0.55)	(0.15)	(0.63)	(0.26)	(1.15)	(0.42)	(0.71)	(0.24)	(0.67)	(0.23)
	middle	8	0.65	0.28	0.70	0.33	1.43	0.50	0.56	0.20	0.85	0.19
			(0.59)	(0.11)	(0.75)	(0.28)	(1.24)	(0.53)	(0.75)	(0.24)	(0.79)	(0.17)
area	right	9	0.59	0.13	0.69	0.24	1.25	0.25	0.81	0.22	0.69	0.14
			(0.63)	(0.21)	(0.81)	(0.30)	(1.28)	(0.48)	(0.77)	(0.13)	(0.80)	(0.17)

* $P<0.05$ ** $P<0.02$ N=5

RESULTS

Table 1 and Figure 3 show the results. Figures 4 and 5 show typical ^{31}P-NMR spectra obtained. In the AD patients, the PME + Pi/α-ATP ratio in the right temporal and parietal area (area 6), 0.66(SD, 0.12), was significantly higher than the 0.41 (SD, 0.10) ratio in the control group; and the PDE/α-ATP ratio in the left temporal and parietal area (area 4), 1.05 (SD, 0.40), was significantly higher than the 0.56 (SD, 0.20) ratio in the control.

DISCUSSION

The in vivo ^{31}P-NMR spectra of AD, which we obtained using ^{31}P chemical shift imaging by means of 4-DFT, revealed significant elevation of ratios of PME + Pi/α-ATP and PDE/α-ATP as compared with the control values. Although these are in vivo findings, there have already been reports on in vitro studies showing abnormalities in the ^{31}P-NMR spectra in the brain of patients with Alzheimer's disease. Pettegrew et al.[7], using high-resolution ^{31}P-NMR, reported that PME (phosphocholine, phosphoethanolamine) and PDE (glycerol 3-phosphocholine, glycerol 3-phosphoethanolamine) regions were significantly elevated in the postmortem brain of patients with Alzheimer's disease. These facts are consistent with the increase in phospholipase activity in degenerating neurons, and it is believed that such a change can be monitored by ^{31}P-NMR.

Barany et al.,[8] who studied by means of ^{31}P-NMR the postmortem cerebral cortex in the patients with neuropathologically confirmed Alzheimer's disease, reported that glycerol 3-phosphorylcholine (GPC) was three times higher in the Alzheimer's brain than in the matched normal control. According to Barany, the increased GPC suggests that degradation of GPC to glycerol phosphate and choline is decreased. Miatto et al.[9] found that relative levels of PDE were increased in frontal and parietal regions of patients with Alzheimer's disease. Smith et al.[10] also reported that PME and PDE levels were greater than control values for both white and grey matter in the brain of Alzheimer's disease. It is of great interest that in vitro results on PME and PDE elevations such as those just mentioned are consistent with the in vivo results. The in vivo data also suggest that abnormalities in phospholipid metabolic pathways contribute to membrane dysfunction in AD. These findings show that in vivo ^{31}P 4DFT multivoxel spectroscopy can be a valuable aid in the study of AD.

However, this method involved a worse signal to noise (S/N) ratio in the case of interference caused by the movement of the subject's head, possibly leading to a change in the absolute value of the signal intensity. There were also other problems such as the differences in the NMR signal intensity between the voxels, due to the lack of homogeneity in the magnetic field. It was confirmed with ATP phantom, however, that if the ratio is determined between the peak values, it will be constant for each voxel. The accuracy of ^{31}P-NMR spectroscopy is expected to increase with the advances made in MRIS; and in the future it will be possible to determine from each peak the absolute levels of phospholipid and high-energy phosphate.

REFERENCES

1. T.Fujimoto, T.Nakano, S.Noguchi, T.Uchida, and Y.Yokoyama, Study of dementia of Alzheimer type using ^{31}P chemical shift imaging and ^{31}P-NMR spectroscopy, Jap J Psychi Treat 4 : 1249 (1989).
2. L.S.Smith, P.A.Bottomley, B.P.Drayer, and R.J.Herfkens, Localized clinical ^{31}P NMR spectroscopy in Huntington's, Parkinson's, Alzheimer's and Binswanger's diseases, in: "Society of Magnetic Resonance in Medicine 5th Annual Meeting," Montreal (1986).

3. G.Withers, A.J.Martinez, G.R.Rao, J.Moossy, and J.Pettegrew, Applications of ^{31}P NMR studies in Alzheimer's disease, Neurology 37 (Suppl.1): 331 (1987).
4. D.M.Kramer, Imaging of elements other than hydrogen, in: "Nuclear magnetic resonance imaging in medicine," L.Kaufman, L.E.Crooks, A.R.Margulis, eds., Igaku-shoin, Tokyo (1981).
5. A.A.Maudsley, S.K.Hilal, W.H.Perman, and H.E.Simon, Spatially resolved high resolution spectroscopy by "four-dimensional" NMR, J Mag Reson 51:147 (1983).
6. T.Miyazaki, Multinuclear imaging (including chemical shift imaging) and volume selective spectroscopy, Innervision 4:28 (1989).
7. J.W.Pettegrew, S.J.Kopp, N.J.Minshew, T.Glonek, J.M.Feliksik, J.P.Tow, and M.M.Cohen, ^{31}P nuclear magnetic resonance studies of phosphoglyceride metabolism in developing and degenerating brain: preliminary observations, J Neuro Exp Neul 46:419 (1987).
8. M.Barany, Y-C.Chang, C.Arus, T.Rustan, and W.H.Frey Ⅱ, Increased glycerol-3-phosphorylcholine in post-mortem Alzheimer's brain, Lancet i:517 (1985).
9. O.Miatto, R.G.Gonzalez, F.Buonanno, and J.H.Growdon, In vitro ^{31}P NMR spectroscopy detects altered phospholipid metabolism in Alzheimer's disease, Can J Neurol Sci 13:535 (1986).
10. C.D.Smith, L.G.Gallenstein Ⅲ, and W.R.Markesbery, ^{31}P spectroscopy in Alzheimer's and Pick's disease using post-mortem perchloric acid brain extracts, in: "Society of Magnetic Resonance in Medicine 7th Annual Meeting," San Francisco (1988).

WHITE MATTER ABNORMALITIES ON MRI

IN PROBABLE ALZHEIMER'S DISEASE

Masashi Takita[1], Akito Nohtomi[1], Masahiro Okayama[1], Yasuyuki Uike[1], and Katsuya Goto[2]

[1]Imazu Red Cross Hospital and [2]Department of Radiology
Fukuoka University School of Medicine
Fukuoka, Japan

ABSTRACT

Magnetic resonance imaging (MRI) of the brain was carried out to evaluate white matter changes in patients with dementia of the Alzheimer type, fufilled DSM-III-R criteria for primary progressive dementia and NINCDS-ADRDA criteria for "probable Alzheimer's disease". MRI showed moderate to marked white matter changes in 14 out of 30 patients (46.7%) with dementia of Alzheimer type: 3 out of 15 patients with Alzheimer's disease (presenile onset) and 11 out of 15 patients with senile dementia of the Alzheimer type (senile onset). The clinical significance of the white matter changes is discussed.

INTRODUCTION

In recent years, it has been suggested that white matter lesions are frequent incidental findings on magnetic resonance imaging (MRI) in elderly demented patients, and may represent an index of cerebrovascular diseases in such patients. Our preliminary experience with MRI showed white matter changes not only in patients with vascular dementia but also in patients with dementia of the Alzheimer type. In the present study, 30 patients with dementia of the Alzheimer type were studied by MRI in order to evaluate white matter changes in this type of dementia.

PATIENTS AND METHODS

Thirty patients with dementia of the Alzheimer type, 10 males and 20 females, 59-81 years old (mean age, 68.6), diagnosed on the basis of DSM-III-R and NINCDS-ADRDA criteria, were studied. Fifteen patients had senile dementia of the Alzheimer type (SD, senile onset); and 15 patients, Alzheimer's disease (AD, presenile onset).

Patients were clinically examined by two psychiatrists. Diagnostic assessment included mediacal history, physical and neurological investigation, screening labolatory tests, neuropsychological test battery, and brain CT. Dementia was assessed by Hasegawa's Dementia Rating Scale (HDRS) and the Mini-Mental State examination (MMS). Hachinski's Ischemic

Basic, Clinical, and Therapeutic Aspects of Alzheimer's and Parkinson's Diseases, Volume 2
Edited by T. Nagatsu *et al.*, Plenum Press, New York, 1990

Table 1. Clinical Features

	AD n=15	SD n=15	Total n=30
Age(y)	63.7 ± 3.3	73.7 ± 4.2*	68.7 ± 6.3
Duration(y)	3.7 ± 2.7	3.3 ± 3.2	3.6 ± 2.9
HDRS	15.7 ± 7.5	16.8 ± 8.7	16.2 ± 8.0
MMS	15.5 ± 6.0	16.0 ± 6.8	15.7 ± 6.3
HIS	1.9 ± 1.2	3.1 ± 1.5**	2.5 ± 1.5

Values represent mean ± S.D.
*: $p<0.01$, **: $p<0.05$ vs. AD value

Score (HIS) was also given to the patients. The clinical features of the patients are summarized in Table 1.

All patients were imaged with the Hitachi G-50 MRI unit with a superconducting magnet to generate a field strength of 0.5-T. T2-weighted images were obtained with the spin-echo sequence (repetition time of 2,000 msec and time to echo of 60 msec).

MRI hyperintensities were classified as periventricular hyperintensity (PVH) and deep white matter hyperintensity (DWMH) according to Fazekas' method (Fazekas et al., 1987). PVH were graded as 0 (absent), 1 (caps or pencil-thin lining), 2 (smooth halo), and 3 (irregular PVH extending into the deep white matter). DWMH were graded as 0 (absent), 1 (punctuate foci), 2 (beginning confluence of foci), and 3 (large confluent areas).

Statistical significance was evaluated by means of Wilcoxon's test.

RESULTS

MRI revealed a variety of signal abnormalities in patients with dementia of the Alzheimer type (Table 2).

PVH was seen in all patients (Table 3). In patients with AD, grade 1 was seen most frequently (12/15); while in patients with SD, grade 2 or 3 was the most frequent (11/15). DWMH was seen in 22 of 30 patients (Table 4). In patients with AD, grade 0 was seen most frequently (8/15), whereas in patients with SD, grade 2 or 3 was seen frequently (8/15). Thus the degrees of PVH and DWMH in patients with SD were more severe than those in patients with AD.

Moderate to marked periventricular lucency (PVL) on CT was seen in 6 of 30 patients (20%), while grade 2 or 3 white matter changes (PVH and/ or DWMH) was seen in 14 patients (46.7%).

There was a significant difference of the mean age between patients with grade 2 or 3 white matter changes and patients without them ($p<0.05$, 71.1 ± 5.4 vs. 66.6 ± 6.4). There were no other differences of clinical features, including severity of dementia, Hachinski's Ischemic Score, and duration of the illness, between patients with and without white matter changes.

Table 2. Profile of Patients

NO	DX	Sex	Age	Duration	HDRS	MMS	HIS	PVL(CT)	PVH(MR)	DWMH(MR)
1	AD	F	65	2	9.0	11	3	1	1	0
2	AD	F	59	2	13.0	7	1	0	1	0
3	AD	F	66	2	20.5	24	2	1	1	1
4	AD	M	67	9	20.5	21	2	0	1	0
5	AD	F	60	3	13.5	11	0	0	1	0
6	AD	M	69	9	12.5	14	2	0	1	0
7	AD	M	65	6	12.0	14	1	0	1	0
8	AD	M	61	5	28.5	23	3	0	1	0
9	AD	F	60	3	10.5	13	2	1	1	0
10	AD	F	68	3	16.0	16	2	1	1	1
11	AD	F	60	2	6.5	8	2	1	1	1
12	AD	F	65	2	2.0	7	1	0	3	2
13	AD	M	61	1	26.5	21	5	3	3	2
14	AD	M	66	6	21.5	20	1	3	3	3
15	AD	F	63	2	23.0	22	2	1	1	1
16	SD	F	72	1	23.5	19	3	1	2	1
17	SD	F	67	2	25.5	14	3	0	2	1
18	SD	F	81	1	20.5	16	5	0	1	1
19	SD	M	76	3	10.5	8	4	0	1	1
20	SD	F	69	1	21.0	25	1	0	3	2
21	SD	M	71	2	21.0	19	4	0	1	1
22	SD	F	79	5	0.0	2	6	3	3	3
23	SD	M	75	1	6.0	13	2	2	3	3
24	SD	F	68	2	25.5	27	1	0	2	2
25	SD	F	72	7	5.0	12	2	3	3	2
26	SD	M	74	1	22.5	25	2	0	1	1
27	SD	F	71	11	18.5	16	4	3	2	2
28	SD	M	75	3	9.0	12	2	1	3	3
29	SD	M	76	2	28.5	20	5	1	2	2
30	SD	F	79	9	14.5	12	2	1	2	1

Table 3. Degree of PVH on MRI

grade	AD n=15	SD n=15	Total n=30
0	0	0	0
1	12	4	16
2	0	6	6
3	3	5	8

Table 4. Degree of DWMH on MRI

grade	AD n=15	SD n=15	Total n=30
0	8	0	8
1	4	7	11
2	2	5	7
3	1	3	4

DISCUSSION

Moderate to marked periventricular lucency was seen in 6 patients (20%) on CT, while grade 2 or 3 white matter change was seen in 14 patients (46.7%) on MRI, suggesting that MRI is more sensitive than CT to demonstrate white matter lesions.

White matter changes are known to occur in a variety of circumstances, for example, in disease states such as multiple sclerosis, normal pressure hydrocephalus, irradiation, and leukodystrophy. In the present study, patients with other causes of dementia were clinically excluded.

It has been reported that MRI showed white matter changes in all 29 patients with vascular dementia and in eight out of 22 patients with Alzheimer's disease, suggesting that white matter changes are a useful diagnostic aid in the differential diagnosis between vascular dementia and Alzheimer's desease (Erkinjuntti et al., 1987). Moreover, it has been shown that the incidence and severity of white matter changes are correlated with a previous history of ischemic cerebrovascular disease and with hypertension (Awad et al., 1986). Therefore, our patients with white matter changes might have ischemic or hypertensive changes in the brain. However, in the present study, there were no correlations between white matter changes and Hachinski's Ischemic Score.

On the other hand, it has been reported that MRI frequently shows white matter changes both in Alzheimer's disease patients and in normal control subjects, suggesting that some white matter changes are age related (Brant-Zawadzki et al., 1985). The present finding that the degree of white matter changes were more severe in patients with SD than in those with AD might indicate an age-related change. Another possibility might be that older Alzheimer patients with white matter changes on MRI constitute its own diagnostic entity.

In conclusion, although white matter changes on MRI were seen frequently in patients with dementia of the Alzheimer type, the clinical significance of the white matter changes remains conjectural until pathological examination.

REFERENCES

1. Awad, I. A., Spetzler, R. F., Hodak, J. A., Awad, C. A. and Carey, R., 1986, Incidental subcortical lesions identified on magnetic resonance imaging in the elderly, Stroke, 17: 1084.
2. Brant-Zawadzki, M., Fein, G., Van Dyke, C., Keirnan, R., Davenport, L. and De Groot, J., 1985, MR imaging of the aging brain: Patchy white matter lesions and dementia, A.J.N.R., 6: 675.
3. Erkinjuntti, T., Kentonen, L., Sulkava, R., Sipponen, J., Vuorialho, M. and Iivanainen, M., 1987, Do white matter changes on MRI and CT differentiate vascular dementia from Alzheimer's disease?, J. Neurol. Neurosurg. Psychiat., 50: 37.
4. Fazekas, F., Chawluk, J. B., Alavi, A., Hurtig, H. I. and Zimmerman, R. A., 1987, MR signal abnormalities at 1.5T in Alzheimer's dementia and normal aging, A.J.N.R., 8: 421.

PROGRESSIVE APHASIA WITHOUT GLOBAL DEMENTIA EXAMINED BY DYNAMIC PET SCAN

H. Fukuyama, K. Harada, H. Yamauchi, S. Yamaguchi
T. Miyoshi, J. Kimura, Y. Yonekura*, and Y. Iwasaki*

Department of Neurology, Nuclear Medicine and Radiology*
Kyoto University School of Medicine, Kyoto, JAPAN

INTRODUCTION

Progressive aphasia without global dementia (PA) was proposed by Dr. Mesulum in 1982, but it has not yet been established as a disease entity. We think this disorder is very interesting from three points of view. It is very similar to the aphasic type of Alzheimer's disease, but lacks dementia or any other cortical disturbances after it's fully developed stage. Closer comparison of progressive aphasia with Alzheimer's disease will provide clues to solve the pathophysiological mechanisms involving memory disturbances in Alzheimer's disease. Another point of view is that this disorder shows various types of aphasia, and that in contrast to aphasia combining cerebrovascular disorders, the responsible lesions are not clearly restricted and the brain suffers from incomplete tissue damage. Therefore, mechanisms producing aphasia depend on the degree of damages of the brain related to speech function, and there might be a relationship between the aphasic type and the degree of hypometabolism in progressive aphasia. The last point is that it has been unsettled whether this disorder is distinctive from any other known disorders or it is belonged to a atypical subtype of known disorders such as Alzheimer's disease or Pick's disease. Remarkable metabolic changes in the cortex of Alzheimer's patients has been fully elucidated in the previous works using Positron Emission Tomography (PET) with 18F-fluorodeoxyglucose (FDG) (Friedland et al 1982). On the other hand, little information has been reported about the metabolic derrangement of Pick's disease. The metabolic changes in PA should be examined precisely to reveal the similarity to those of Alzheimer's disease.

We performed dynamic PET scan on 5 cases with this disorder, and found that there is some difference from Alzheimer's disease as well as normal controls.

PATIENTS AND METHODS

Five cases of progressive aphasia without generalized dementia were examined and compared with 6 cases of Alzheimer's disease and 5 cases of normal control. These cases showed various types of aphasia, such as Wernicke's aphasia, anomic aphasia, conduction aphasia, semantic aphasia and transcortical sensory aphasia. The duration of aphasia was 2 to 7 years. They did not show apparent dementia at the time of PET examination.

Our PET scanner Positologica III was constructed with the idea that irregularly arranged detectors around the ring will be able to obtain uniform sampling densities in emission data with continuous rotation. The machine's characteristics are as follows: 5 rings with 192 BGO detectors providing 7 slices, FWHM is 7.6 mm in the center of the cross-plane and slice thickness is about 10 mm, slice pitch 16mm, and spacing between the rings is 8 mm (Senda et al 1985).

Dynamic PET scan using FDG is able to provide the glucose metabolic state of the brain, and also kinetic rate constants. Therefore, we can assess a closer metabolic state of the brain in pathological states than with the usual calculating method for glucose metabolic rate deriving from Sokoloff's method (1977).

We hypothesized a three compartment 4 parameter model on the glucose metabolism in our study. Each metabolic rate, such as glucose transfer between blood pool and nervous tissue (K1 & k2), and phosphorylation, dephosphorylation of glucose (k3 & k4) was calculated by the non-linear least square technique (Newton-Raphson method). Details of the calculation methods will be published elsewhere (Fukuyama et al 1989). Arterial blood sampling and 60 minute dynamic scan data were used to calculate rate constants pixel by pixel.

The CMRGlu was calculated by the equation 1. This is the direct calculation from obtained rate constants. Half life of FDG and FDG distribution volume were calculated from obtained rate constants using equations 2 and 3 respectively (Phelps et al 1979).

$$CMRGlu = \frac{Cp*K1*k3}{k2+k3} \qquad (1) \qquad (Cp: plasma\ glucose)$$

$$Half\ life = \frac{\ln(2)}{k2+k3} \qquad (2)$$

$$Distribution\ volume = \frac{K1}{k2+k3} \qquad (3)$$

We did statistical analysis by SAS software, using multivariate analysis. We considered a probability of less than 5 % as significant.

Regions of interest (ROI) were place as shown in Figure 1. Every ROI was consisted of 33 pixels.

RESULTS (Table)

Cerebral Metabolic Rate of Glucose

Three cases out of 5 PA cases showed localized hypometabolic state in the left temporal lobe, while the other 2 revealed the parietal and frontal lobe hypometabolism of glucose in the left hemisphere as well as in the left temporal lobe.

There was no difference in CMRglu between progressive aphasias and normal controls except for the left temporal lobe. But there were more widespread abnormalities in Alzheimer's disease, in the bilateral parietal and left temporal regions.

There was a tendency that the hypometabolic regions restricted around the perisylvian region coincided with paraphasia or amnestic aphasia, while larger hypometabolic regions including parietal and frontal regions coincided with transcortical aphasia.

Table1. Cerebral Metabolic Rate of Glucose and Rate Constants in Progressive aphasia, Alzheimer's disease and Normal Controls

LOCATION		FR	FL	PR	PL	TR	TL	OR	OL
CMRGlu	Aph	41.21	38.54	39.01	36.73	37.12	30.73*	38.67	40.14
	Alz	34.13	34.74	28.12#	28.20#	30.53	28.38#	33.57	35.48
	Nor	41.23	41.48	40.03	38.72	38.60	38.27	40.62	40.94
K1	Aph	0.072	0.068	0.073	0.072	0.074	0.066	0.079	0.082
	Alz	0.072	0.070	0.064#	0.066	0.069	0.067	0.077	0.082
	Nor	0.081	0.078	0.080	0.078	0.080	0.078	0.084	0.085
k2	Aph	0.089	0.071	0.095	0.103	0.088	0.113	0.102	0.115
	Alz	0.092	0.088	0.094	0.095	0.092	0.092	0.092	0.094
	Nor	0.107	0.104	0.109	0.109	0.111	0.109	0.101	0.105
k3	Aph	0.093	0.093	0.080	0.078	0.078	0.072	0.071	0.081
	Alz	0.073	0.073	0.060$	0.060#	0.061	0.059	0.062	0.061
	Nor	0.081	0.084	0.078	0.080	0.076	0.077	0.075	0.079
H.L.	Aph	3.875	4.360	3.972	3.880	4.259	3.951	4.042	3.660
	Alz	4.251	4.351	4.562	4.550	4.600	4.658	4.600	4.577
	Nor	3.700	3.677	3.687	3.680	3.694	3.722	3.975	3.882
D.V.	Aph	0.405	0.420	0.422	0.398	0.452	0.365	0.456	0.430
	Alz	0.439	0.440	0.418	0.431	0.449	0.444	0.501	0.537
	Nor	0.432	0.418	0.430	0.415	0.430	0.423	0.482	0.479

*, # and $ represent 5% difference between normal controls and progressive aphasia, normal controls and Alzheimer's disease, and progressive aphasia and Alzheimer's disease respectively

Abbreviations

FR:frontal right, FL:frontal left, PR:parietal right, PL:parietal left
TR:temporal right, TL:temporal left, OR:occipital right, OL:occipital left
Alz:Alzheimer's disease, Aph:Progressive aphasia, Nor:normal controls
CMRGlu:Cerebral Metabolic Rate of Glucose (micro mol/100g/min)
K1 :ml/100g/min, k2 :1/min, k3 :1/min
H.L.:Half Life of FDG (1/min), D.V.:Distribution Volume of FDG

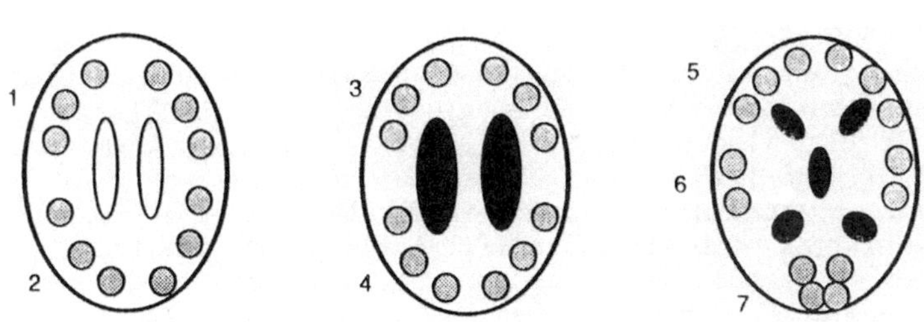

1 ... Frontal 1 2 ... Parietal 1 3 ... Frontal 2
4 ... Parietal 2 5 ... Frontal 3 6 ... Temporal
7 ... Occipital

Figure 1

Rate constants

k1 showed no statistically significant differences except for in the right parietal regions between normal control and Alzheimer's disease. K2 revealed no differences between PA, Alzheimer's disease and normal controls. There is the reduced value of k3 in Alzheimer's disease compared with normal control or progressive aphasia at the parietal regions. This means the reduced activity of hexokinase in Alzheimer's disease.

There are no remarkable changes in distribution volume of FDG directly calculated from obtained rate constants. The membrane permeability to glucose, or glucose transport mechanism from the blood vessels to the brain is considered to be intact in progressive aphasia and Alzheimer's disease. There is no remarkable prolongation of half life in the cerebral cortex. In Alzheimer's group, however, there is a tendency of half life to be long in the parietal and temporal regions. This indicates the metabolic speed slowed by the disorder.

CONCLUSION

The main difference of glucose metabolism between progressive aphasia and Alzheimer's disease compared with normal control is that the hypometabolic region is localized to the left temporal region in PA, while there is a widespread hypometabolism covering parietal and temporal regions in Alzheimer's disease. The hypometabolism confined to the left hemisphere in PA might be causative for maintaining other cortical functions than a verbal skill. The types of aphasia may be related to the hypometabolic regions of glucose tin the brain. Rate constant analyses on glucose metabolism show no specific alterations in PA.

So far, we can conclude from this study that progressive aphasia is different from Alzheimer's disease in the glucose metabolism and may be a distinct disorder from Alzheimer's disease. Therefore PA might be a subtype of Pick's disease or any other unknown disorder presenting a progressive aphasia.

REFERENCES

Friedland RP, Brun A, Budinger TF: Pathological and positron emission tomographic correlations in Alzheimer's disease. Lancet 1:228, 1985.
Fukuyama H, Kameyama M, Harada K, et al:Glucose metabolism and rate constants in Alzheimer's disease examined with dynamic PET scan. Acta Neurol Scand in press.
Mesulam M-M: Slowly progressive aphasia without generalized dementia. Ann Neurol 11:592-598, 1982.
Phelps ME, Huang SC, Hoffman EJ, Selin C, Sokoloff L, Kuhl DE: Tomographic measurement of local cerebral glucose metabolic rate in humans with (F-18)2-fluoro-2-deoxy-D-glucose: Validation of method. Ann Neurol 6:371-388, 1979.
Senda M, Tamaki N, Yonekura Y, Tanada S, Murata K et al : Performance characteristics of positrogica III: A whole-body positron emission tomograph. J Comput Assist Tomogr 9:940-946, 1985.
Sokoloff L, Reivich M, Kennedy C, Des Rosiers MH et al: The [14C]deoxyglucose method for the measurement of local cerebral glucose utilization: Theory, procedure, and normal values in the conscious and anesthetized albino rat. J Neurochem 28:897-916, 1977.

POSITRON EMISSION TOMOGRAPHY STUDIES USING ^{11}C-LABELLED DOPA AND EFFECT OF TETRAHYDROBIOPTERIN ON DOPAMINE RELEASE

Yasuyoshi Watanabe,[1] Peter Bjurling,[2] Per Hartvig,[2] Joakim Tedroff,[2] Nobuko Mataga,[1] Kazuyuki Imamura,[1] Shogo Oka,[1] Kunio Koshimura,[3] Soichi Miwa,[3] and Bengt Långström[2]

[1]Osaka Bioscience Institute, 6-2-4 Furuedai, Suita-shi, Osaka 565 Japan; [2]University of Uppsala, Uppsala, Sweden; and [3]Kyoto University Faculty of Medicine, Sakyo-ku, Kyoto 606, Japan

INTRODUCTION

To develop novel drugs for neurological disorders such as Alzheimer's and Parkinson's diseases, the assessment of neurotransmitters and their receptors in patients is of particular importance. From this viewpoint, positron emission tomography (PET) studies using short-lived radionuclides have been performed. For the dopaminergic system, N-[^{11}C]methyl-spiperone/[^{11}C]raclopride, [^{11}C]SCH-23390, and [^{11}C]nomifensine have been developed and used for dopamine D2-, D1-receptor, and presynaptic uptake site, respectively.[1] 6-Fluoro-L-DOPA labelled with ^{18}F has been extensively used as a tracer for study of dopamine formation and distribution under both normal and pathophysiological conditions.[2] We recently developed a position-specific labelling method for L-[^{11}C]DOPA for similar applications and further elucidation of partial metabolic pathway.[3,4] Another challenge using L-[^{11}C]DOPA might be assessment by PET of the capacity of dopamine release from the nerve terminals.

6R-L-Erythro-5,6,7,8-tetrahydrobiopterin (6R-BH_4) is the common cofactor for three aromatic amino acid hydroxylases, namely, tyrosine (Tyr) hydroxylase, tryptophan (Trp) hydroxylase, and phenylalanine (Phe) hydroxylase. The limited concentration of 6R-BH_4 in the brain as compared with the Km value of Tyr hydroxylase and Trp hydroxylase for 6R-BH_4 is hypothesized to regulate the <u>in vivo</u> activities of these enzymes. In fact, intracerebroventricular injection of the cofactor resulted in increased levels of dopamine and serotonin metabolites.[5,6] 6R-BH_4 is biosynthesized from guanosine triphosphate (GTP) via the enzymatic route shown in Fig. 1. Defect in the biosynthetic pathway or recycling enzyme of 6R-BH_4 causes a severe neurological disorder, which was first reported as atypical phenylketonuria by several groups in 1975. Since then, clinical studies on tetrahydrobiopterin contents in the brain have been done. Concerning Parkinson's and Alzheimer's diseases, there are several reports of a decrease in tetrahydrobiopterin content in various specimens. A summary of the reports is shown in Fig. 2. In 1981, Nagatsu et al.[7] showed a reduction in the total biopterin content in the caudate nucleus of postmortem Parkinson's patient brain, its level being one fourth of the normal control value. Then, Narabayashi and co-workers[8] used 6R-BH_4 for therapeutics after L-DOPA and saw an improvement in their symptoms. Similar results of a decrease in tetrahydrobiopterin content were

Basic, Clinical, and Therapeutic Aspects of Alzheimer's and Parkinson's Diseases, Volume 2
Edited by T. Nagatsu *et al.*, Plenum Press, New York, 1990

obtained in the postmortem brain of Alzheimer's patients.[9] And more recently, the therapeutic effect of tetrahydrobiopterin was also observed in depression[10] and infantile autism.[11] Naruse et al.[11] obtained improvement of autistic behavior by oral administration of 6R-BH_4.

In this way, 6R-BH_4 can be developed as a pharmaceutical for several neuropsychiatric disorders with some defect of the monoaminergic pathway. However, through the clinical study, we faced the puzzle that DOPA and tetrahydrobiopterin showed an additive effect on the improvement of many symptoms. To solve this problem, we investigated the effect of 6R-BH_4 on dopamine release from nerve terminals by using brain microdialysis and found a novel function of 6R-BH4, the promotion of dopamine release.[12]

In this paper, the data obtained by microdialysis experiments are summarized and our PET study using position-specific labelled L-[^{11}C]DOPA and 6R-BH_4 is described.

MATERIALS AND METHODS

Microdialysis experiments

Microdialysis experiments were performed using male Wister rats (200-250 g, body weight) and dialysis probes (CMA/10, 3 mm-length membrane, 20,000-dalton cutoff, Carnegie Medicine), in the free-moving state (for dopamine and DAHP experiment) and under urethane anesthesia (for dopamine, serotonin, and glutamate). Ringer solution was perfused (2.0 µl/min), using Teflon tubing connected to a microinfusion pump (CMA/100). After a 3-h equilibration period, the perfusates were collected every 30 min. After treatment for analysis of each compound, the fractions collected were applied to HPLC equipped with electrochemical or fluorescence detection. 2,4-Diamino-6-hydroxypyrimidine (DAHP), a potent inhibitor of GTP cyclohydrolase (rate-limiting enzyme of 6R-BH4 biosyn-thesis in rats), was injected intraperitoneally 4 times at 4-h intervals during microdialysis experiments. As described by us,[13] the content of 6R-BH4 4h after injections of DAHP in the brain was less than 50% of the control value.

PET study

L-[3-^{11}C]DOPA and L-[1-^{11}C]DOPA were synthesized as described.[3,4] After preparative LC and sterilization by filtration, the desired tracer (120-180 MBq) was

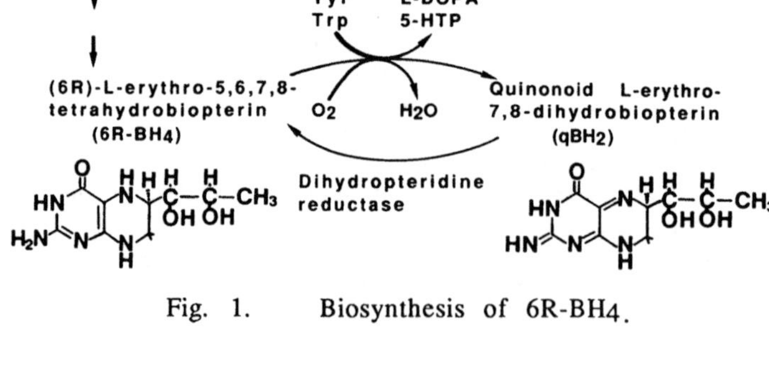

Fig. 1. Biosynthesis of 6R-BH_4.

injected intravenously into a rhesus monkey (6-8 kg body weight). PET scan was performed immediately after tracer injection by use of a PC384-3B positron emission tomograph (Scanditronix, Uppsala, Sweden). Experimental protocols were as follows: Three successive experiments using [3-^{11}C]DOPA on the same position of the monkey were conducted at 90-min intervals. Vehicle, 2 mg/kg of 6R-BH_4, and 20 mg/kg of 6R-BH_4 were injected intravenously in this order 10 min after (Protocol 1) or 20 min before (Protocol 2) each [^{11}C]DOPA injection. The studies were approved by the Ethics Protocol Committee for Clinical Research and for Animal Research, and the Isotope Committee at the University Hospital, Uppsala.

RESULTS AND DISCUSSION

Microdialysis

Enhancement of extracellular dopamine content by 6R-BH_4 applied by perfusion in the dialysis tubing was observed in the striatum of free-moving rats[22] and also under urethane anesthesia. The enhancement occurred in a dose-dependent manner and in the range of the physiological concentration of 6R-BH_4. The promotion of dopamine release by 6R-BH_4 occurred also after complete inhibition of tyrosine hydroxylase by α-methyltyrosine. The effect of 6R-BH_4 was not competitive with the presynaptic dopamine uptake site inhibited by nomifensine. Further, tetrodotoxin treatment ceased the enhancement of dopamine release by 6R-BH_4. These results forced us to the experiment in the opposite direction: to see if the lowering of endogenous 6R-BH_4 results in reduction of dopamine output from the striatal neurons. As expected, the reduction of 6R-BH_4 (to ~50% of the control level) by multiple administration of 2,4-diamino-6-hydroxypyrimidine, a potent inhibitor of the rate-limiting enzyme of BH_4 biosynthesis, resulted in marked reduction (to 30~40% of the control) of extracellular dopamine. Thus, we could demonstrate the possibility of the regulatory role of 6R-BH_4 in dopamine release in the physiological state (Fig. 3). Then, we investigated the effect of 6R-BH_4 on the release of other neurotransmitters. Positive results were obtained on serotonin release in the striatum in a similar manner. Concerning the glutamate release, two dialysis probes were implanted into both sides of the striatum, and one side was lesioned with 6-hydroxy-dopamine. An increase in glutamate release by 6R-BH_4 was observed exclusively in the intact side and little effect was observed in the lesioned side. The extracellular content of glutamine and other amino acids than glutamate and aspartate was not affected by 6R-BH_4 treatment. Obviously, glutamate release occurred via stimulation of dopamine release by 6R-BH_4.

	Brain (post mortem)	Plasma	Urine	CSF
Parkinson	↓	↓	↘	↓
Alzheimer	↓	→/↓	-	↓
Depression	↓	↑	↓	→
Autism	-	↓ (Blood cell)	-	-

Fig. 2. Total biopterin levels in patients.

PET study

To assess the content and capacity of the neurotransmitter amine in vivo, we developed the labelling method of L-DOPA, L-Trp, L-5-HTP, and L-Trp shown in Fig. 4. Since the physical half-life of ^{11}C is short (20.4 min), the effective use of a multi-enzyme system is desirable. ^{11}C-Carbon dioxide generated in the target was conveyed to the synthesis laboratory and then introduced to DL-^{11}C-alanine by chemical synthesis. Then, DL-^{11}C-alanine was converted to ^{11}C-pyruvate and immediately ^{11}C-pyruvate was converted to ^{11}C-DOPA or ^{11}C-Tyr by use of ß-tyrosinase, by a one-pot reaction. Analogous synthesis of ^{11}C-labelled tryptophan and 5-hydroxytryptophan was also achieved by use of tryptophanase and indole derivatives. The yield of radioactivity from DL-^{11}C-alanine was 40 to 50%, decay corrected, and the radiochemical purity after preparative LC was better than 97%.

[3-^{11}C]DOPA was applied to a rhesus monkey and then to a healthy volunteer. Reconstituted PET image at the late phase (20-50 min) after [^{11}C]DOPA injection showed the highest uptake in the striatum and frontal cortex. To determine the molecular species carrying radioactivity in the striatum, we did two experiments. In one we used position-specific labelled [^{11}C]DOPA, and in the other, the label in combination with a peripheral- or central-type of aromatic amino acid decarboxylase (AADC) inhibitor. The metabolic fate of the different positions of carbon and nitrogen atoms is distinct. This means that in well-designed experiments combining the compounds labelled in different positions, we can extract each metabolic or enzymatic pathway from the dynamics of metabolism and also that we can determine the molecular species carrying radioactivity. We labelled L-DOPA in the side-chain 1 position or the side-chain 3 position using the same method as shown in Fig. 4. Then, we injected [1-^{11}C] and [3-^{11}C]L-DOPA into the same subject at 2-h interval. As a result, the striatal structure was clear only in the case of [3-^{11}C]DOPA. This indicates that the radioactivity in the striatum is derived from not DOPA itself but from dopamine or further metabolites. Supporting data were obtained with [3-^{11}C]DOPA and decarboxylase inhibitor. By pretreatment with a peripheral type of AADC

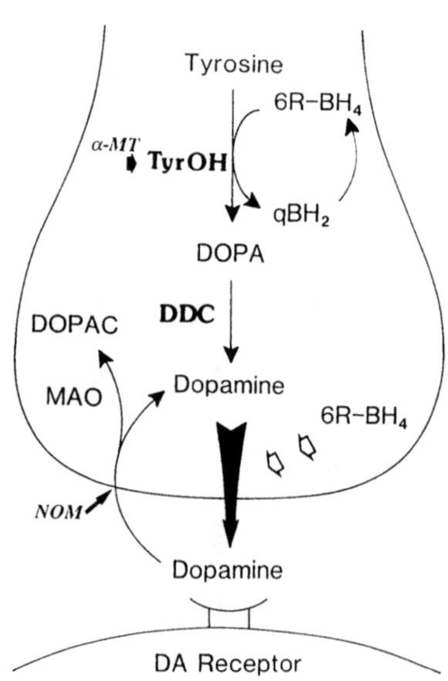

Fig. 3. Dual function of 6R-BH4.

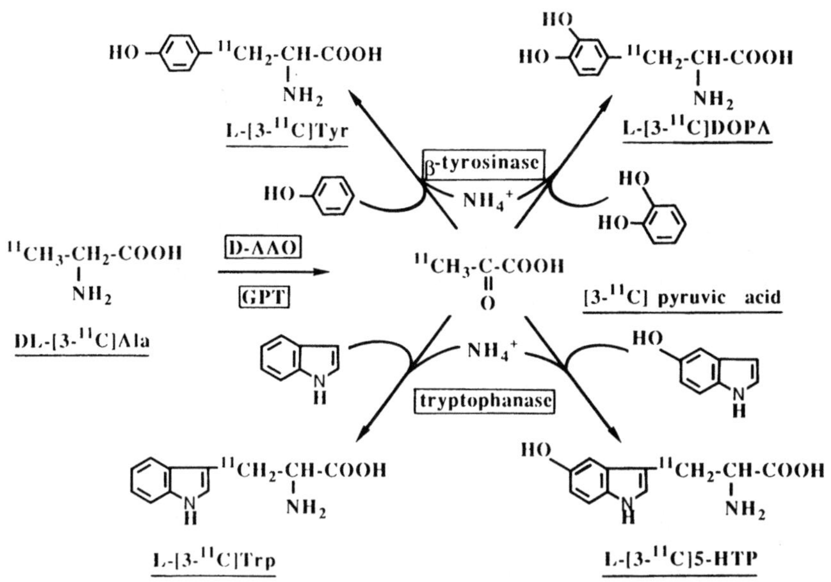

Fig. 4. Synthesis of ^{11}C-labeled amine precursors.

inhibitor (Benzeraside), a relatively higher uptake into the striatum than seen in the control was observed, while pretreatment with a central type of AADC inhibitor (NSD1015) resulted in much less accumulation of radioactivity in the striatum. These results and faster clearance of the acidic metabolites from the brain than that of dopamine itself indicate that the radioactivity accumulated in the striatum is mainly derived from dopamine.

To investigate if the effect of 6R-BH4 on dopamine biosynthesis and/or release could be evaluated in Parkinson's or Alzheimer's patients, we gave 6R-BH4 intravenously to a rhesus monkey with [3-^{11}C]DOPA according to the two injection protocols in "Materials and Methods". Essentially similar results were obtained with both protocols. Namely, as shown in Fig. 5, the uptake ratio of ^{11}C-DOPA and also the value itself were low in the case of 20 mg/kg 6R-BH4 injection compared with the ratio or value in the case of vehicle or 2 mg/kg 6R-BH4 injection. These results indicate some effects of 6R-BH4 on the turnover of dopamine in the

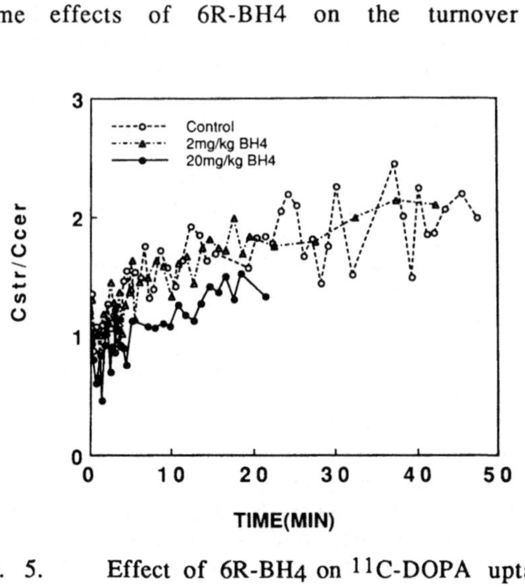

Fig. 5. Effect of 6R-BH4 on ^{11}C-DOPA uptake.

monkey striatum. This type of examination could be applied to the patients in the near future for understanding the functional status of the dopaminergic system and further for elucidating the effectiveness of therapeutics.

ACKNOWLEDGMENTS

We are indebted to Dr. O. Hayaishi, Director of our institute, Prof. M. Fujiwara of Kyoto University, Faculty of Medicine, and Dr. T. Noguchi of Suntory Institute for Biomedical Research for useful discussions and encouragement. We also thank Dr. P. Malmborg for radionuclide production. This work was supported in part by a Grant-in-Aid for New Drug Development from the Ministry of Health and Welfare of Japan, and by grants from the Japanese Foundation of Metabolism and Diseases, Scandinavia-Japan Sasakawa Foundation, and Japan Foundation for Applied Enzymology.

REFERENCES

1. B. Långström, G. Antoni, P. Bjurling, A. Gee, K. J. Fasth, J. Ulin, J. Fowler, R. MacGregor, and A. P. Wolf, Proceedings of the Third International Symposium on Synthesis and Applications of Isotopically Labelled Compounds, Baillie T. A. and Jones, J. R. eds., pp.465-469, Elsevier, Amsterdam (1989)
2. G. Firnau, E. S. Garnett, R. Chirakal, S. Sood, C. Nahmias, and G. Schrobilgen, [^{18}F]Fluoro-L-dopa for the in vivo study of intracerebral dopamine, Appl. Radiat. Isot. 37:669 (1986)
3. P. Bjurling, Y. Watanabe, S. Oka, T. Nagasawa, H. Yamada, and B. Långström, Multi-enzymatic synthesis of β-^{11}C-labelled L-tyrosine and L-DOPA, Acta Chem. Scand., in press (1989)
4. P. Bjurling, G. Antoni, Y. Watanabe, and B. Långström, Enzymatic synthesis of carboxy-^{11}C-labelled L-tyrosine, L-DOPA, L-tryptophan, and 5-hydroxy-L-tryptophan, Acta Chem. Scand., in press (1989)
5. R. Kettler, G. Bartholini, and A. Pletscher, In vivo enhancement of tyrosine hydroxylation in rat striatum by tetrahydrobiopterin, Nature (Lond.), 249:476 (1974)
6. S. Miwa, Y. Watanabe, and O. Hayaishi, 6R-L-Erythro-5,6,7,8-tetrahydrobiopterin as a regulator of dopamine and serotonin biosynthesis in the rat brain, Arch. Biochem. Biophys., 239:234 (1985)
7. T. Nagatsu, T. Yamaguchi, T. Kato, T. Sugimoto, S. Matsuura, M. Akino, I. Nagatsu, R. Iizuka, and H. Narabayashi, Biopterin in human brain and urine from controls and Parkinsonian patients: Application of a new radioimmuno-assay, Clin. Chim. Acta, 109:305 (1981)
8. H. Narabayashi, T. Kondo, T. Nagatsu, T. Sugimoto, and S. Matsuura, Tetrahydrobiopterin administration for Parkinsonian symptoms, Proc. Japan Acad, 58B: 283 (1982)
9. C. Morar, S. B. Whitburn, J. A. Blair, R. J. Leeming, and G. K. Willcock, Tetrahydrobiopterin metabolism in senile dementia of Alzheimer type, J. Neurol. Neurosurg. Psychiatry, 46:582 (1983)
10. H.-Ch. Curtius, A. Niederwieser, R. A. Levine, W. Lovenberg, B. Woggon, and J. Angst, Successful treatment of depression with tetrahydrobiopterin, Lancet, i:657 (1983)
11. H. Naruse, T. Hayashi, M. Takesada, A. Nakane, K. Yamazaki, T. Noguchi, Y. Watanabe, and O. Hayaishi, Therapeutic effect of tetrahydrobiopterin in infantile autism, Proc. Japan Acad., 63B:231 (1987)
12. K. Koshimura, S. Miwa, K. Lee, M. Fujiwara, and Y. Watanabe, Enhancement of in vivo dopamine release from the rat striatum by dialysis perfusion of 6R-L-erythro-5,6,7,8-tetrahydrobiopterin, J. Neurochem., in press (1989)
13. S. Suzuki, Y. Watanabe, S. Tsubokura, H. Kagamiyama, and O. Hayaishi, Decrease in tetrahydrobiopterin content and neurotransmitter amine biosynthesis in rat brain by an inhibition of guanosine triphosphate cyclohydrolase, Brain Res., 446:1 (1988)

STRIATAL PRESYNAPTIC DOPAMINERGIC FUNCTION AND CEREBRAL BLOOD FLOW IN THE MPTP PRIMATE MODEL OF PARKINSON'S DISEASE BY PET

Doris J. Doudet, Hitoshi Miyake, Richard Suddath, Thomas G. Aigner, Richard Wyatt and Robert M. Cohen

Clinical Brain Imaging Section/LCM, Neuropsychiatry Branch
Laboratory of Neuropsychology, NIMH, Bethesda MD, USA

INTRODUCTION

Parkinson's disease is characterized by a loss of dopaminergic (DA) neurons in the zona compacta of the substantia nigra and a reduction in striatal DA concentrations. A selective lesion of the DA nigro-striatal system, mimicking most of the motor features of the idiopathic disease, i.e. brady/akinesia and rigidity, can be induced in primates by administration of the neurotoxic compound 1-methyl-4-phenyl-1,2,3,6-tetrahydropyridine (MPTP)[1]. Depending on the dose and route of administration, it is possible to induce, in rhesus monkeys, different degrees of degeneration which can be correlated to striatal DA deficiency and behavioral impairment.

The study of intracerebral dopamine function in vivo has recently been made possible by the advent of positron emission tomography (PET). Using this newly developed technology, it is possible to determine in a live subject, the three dimensional distribution of a positron-labelled molecule. The regional distribution of 6-(^{18}F)-L-3,4-dihydroxyphenylalanine (6-^{18}F-DOPA), a ^{18}F labelled precursor of DA, and its metabolites was examined in normal monkeys and monkeys with lesions of the DA nigro-striatal system. Animals in two clinical stage were examined: monkeys in a sub-clinical, neurologically normal stage and monkeys in a midly affected, unilateral condition. In addition, the effects of striatal DA depletion on regional cerebral blood flow (CBF) in the basal ganglia was investigated using (^{15}O) labelled water.

METHODS

Eighteen rhesus monkeys, matched for age and sex, were studied. Five animals had received i.v. injections of MPTP (total cumulative dose: 2 to 4.8 mg/kg) 18 to 20 months before the PET study, at the time of which they were evaluated as normal by investigators unaware of their prior treatment. Four animals had received MPTP injections in the right internal carotid artery (total cumulative dose: .6 to 1.5 mg/kg), one year before the study. At this time, these 4 monkeys presented mild increases in muscle tone and a clear under-utilization of their contralateral side. For comparative purposes, a group of 9 normal control monkeys was studied. Cerebrospinal fluid (CSF) was obtained from each animal by cisternal tap in the two weeks preceding or following the PET study, and analyzed for their concentration of homovanillic acid (HVA), the main metabolite of DA in primates.

Basic, Clinical, and Therapeutic Aspects of Alzheimer's and Parkinson's Diseases, Volume 2
Edited by T. Nagatsu *et al.*, Plenum Press, New York, 1990

Scanning procedures and data analysis have been described in an earlier paper[2]. Briefly, a CBF study (10-20 mCi of (^{15}O) labelled water) with a slice at the level of the striatum was performed. CBF data were normalized to whole brain average. Between 90 and 180 min after administration of the 6-^{18}F-DOPA, anatomical data were acquired at 4 different Z axis translations of the scanner (interslice interval: 3.5 mm). A striatal region of interest (ROI) was drawn on the slice containing both the largest extent of striatal tissue <u>and</u> the greatest ^{18}F accumulation. The 2 main constituents of the striatum were then identified separately in the slice directly below the striatal slice (putamen) and directly above the striatal slice (caudate nucleus). In each slice considered, a DA-poor area in the occipito-parietal cortex was chosen as reference. Ratios between striatal areas and cortical areas were calculated, interpolated at 120 min and compared using Student't-tests. For the hemi-lesioned animals, the left and right striatum/cortex ratios were averaged for statistical correlation with CSF data.

RESULTS

The 6-^{18}F-DOPA data are summarized in Fig. 1 and 2. The striatum of the asymptomatic animals and the ipsilateral side of the hemi-treated animals presented a dramatic loss of ^{18}F accumulation compared to the normal controls and the contralateral side of the hemi-treated monkeys. No significant difference in ^{18}F accumulation and putamen or caudate ratios was found between the left and right side of the normal controls and asymptomatic animals. The ratios striatum/cortex of the 18 animals were strongly correlated to the CSF HVA concentrations (Fig. 3).

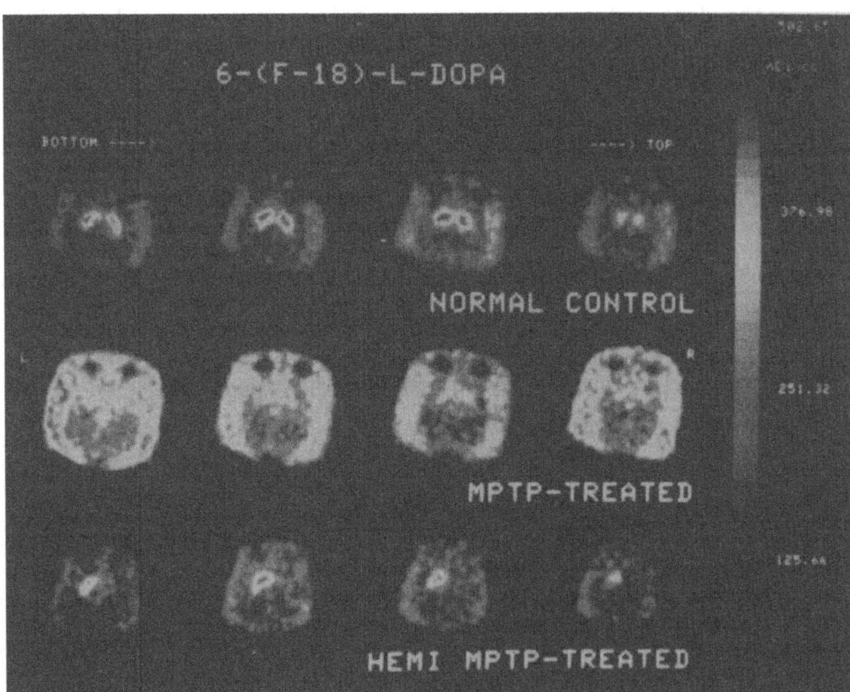

Fig. 1. Portions of the 6-^{18}F-DOPA anatomical set containing the striatum in normal (top), asymptomatic MPTP-treated (middle) and right MPTP-treated monkeys.

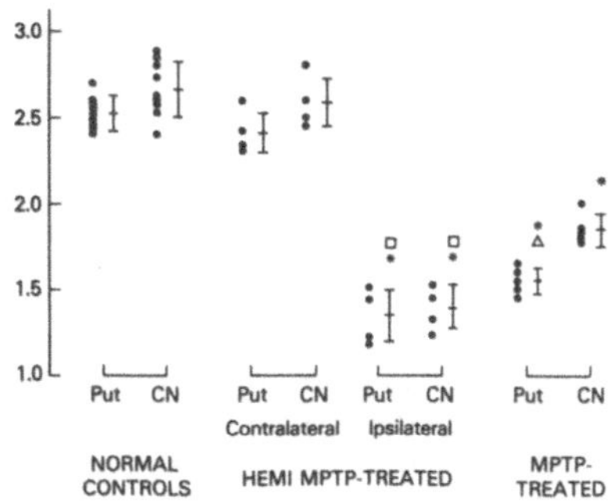

Fig. 2. Ratio putamen/cortex (Put) and caudate/cortex (CN) at 120 min after tracer injection. Bars indicates mean +-SD. There was a significant difference (p<.01) between MPTP-treated and normal controls (*), between CN and Put in the asymptomatic group (Δ) and between the ipsilateral side of the hemi-lesioned and the MPTP-treated animals (□).

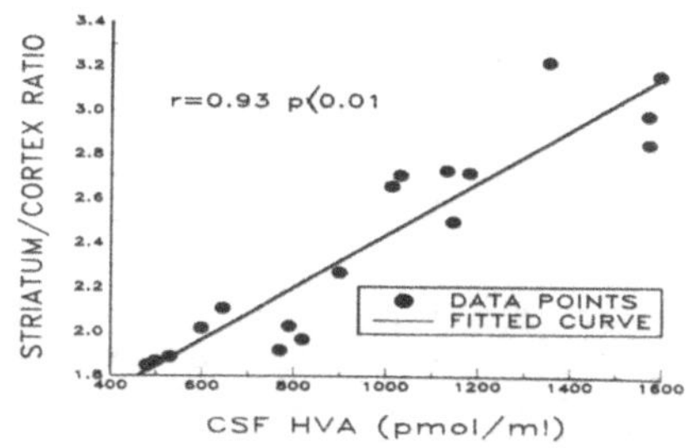

Fig. 3. Scatter plot of the averaged left/right ratios striatum/cortex and the CSF concentration of HVA, the main metabolite of dopamine in primates. The fitted curve was determined using the least square analysis method.

Table 1. CBF data in normal controls and MPTP-treated monkeys.

	Striatum		Occipital cortex	
	Left	Right	Left	Right
Normal controls	1.20+-0.10	1.17+-0.06	1.37+-0.11	1.33+-0.10
MPTP-treated	1.33+-0.10*	1.33+-0.11*	1.34+-0.06	1.38+-0.10
Hemi MPTP-treated	1.29+-0.08*	1.06+-0.07*	1.35+-0.12	1.35+-0.09

Averaged normalized values +- standard deviation. *: significant difference (p<0.05) between MPTP-treated animals and normal controls. *: significant difference left/right (p<0.05).

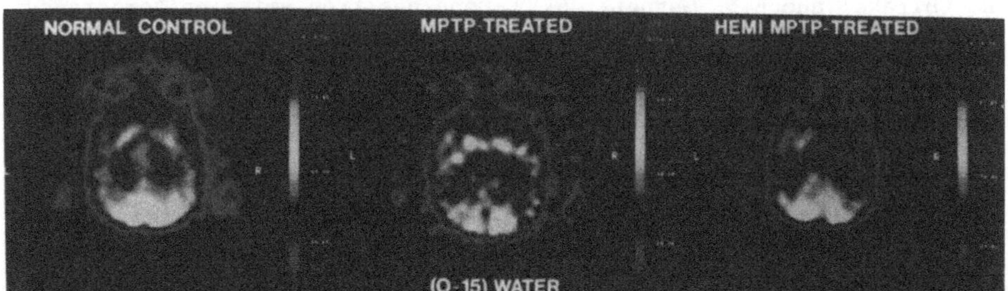

Fig. 4. Striatal CBF in normal, asymptomatic MPTP-treated and hemi MPTP-treated monkeys.

The striatal CBF data are summarized in Fig. 4 and Table 1. Striatal CBF in the asymptomatic monkeys was significantly increased compared to normal controls monkeys. The contralateral striatal CBF in the hemi-lesioned animals was increased while the ipsilateral striatal CBF was decreased. Occipital CBF was similar in the three groups of animals.

CONCLUSION

Our findings using 6-^{18}F-DOPA are consistent with reported data in human subjects[3] and non-human primates[4]. ^{18}F accumulation preferentially occurs in the striatal areas of both normal and MPTP-treated animals. The decrease in striatal ^{18}F accumulation and in the ratios of striatal areas/cortex are consistent with the observed behavior and HVA concentration in CSF. The greater caudate nucleus/cortex ratio in the i.v. MPTP-treated animal is consistent with anatomical findings showing a primary degeneration of the nigro-striatal neurons projecting to the more motor areas of the striatum, the putamen. On the other hand, caudate-projecting neurons, initially more resistant to MPTP, become increasingly affected with exposure to MPTP, as demonstrated by the decrease in caudate/cortex ratios, similar to the putamen/cortex ratios in the ipsilateral side of the hemi-lesioned, symptomatic animals.

The CBF data are also consistent with data reported in early and late stages of Parkinson's disease[5]. The increased blood flow in the asymptomatic MPTP-treated monkeys and the contralateral side of the hemi-lesioned animals may represent an adaptation to limited nigro-striatal degeneration. The decreased ipsilateral striatal CBF may represent the inability to compensate for further loss of striatal DA innervation of blood vessels.

REFERENCES

1. J.W. Langston and I. Irwin, MPTP: current concepts and controversies, Clin.Neuropharm., 9:485 (1986).

2. D.J. Doudet, H. Miyake, R.T. Finn, C.A. McLellan, T.G. Aigner, R.Q. Wan, H.R Adams and R.M. Cohen, 6-18F-L-DOPA imaging of the dopamine neo striatal system in normal and neurologically normal MPTP-treated rhesus monkeys, Exp.Brain Res., in press:(1989).

3. K.L. Leenders, A.J. Palmer, N. Quinn, J.C. Clark, G. Firnau, E.S. Garnett, C. Nahmias, T. Jones and C.D. Marsden, Brain dopamine metabolism in patients with Parkinson's disease measured with positron emission tomography, J.Neurol.Neurosurg. Psychiatry, 49:853 (1986).

4. C.C. Chiueh, R.S. Burns, I.J. Kopin, K.L. Kirk, G. Firnau, C. Nahmias, R. Chirakal and E.S. Garnett, 6-^{18}F-Dopa/positron emission tomography visualized degree of damage to brain dopamine in basal ganglia of monkeys with MPTP-induced parkinsonism, in: MPTP: a neurotoxin producing a parkinsonian syndrome, S.P. Markey, N. Castagnoli,Jr., A.J. Trevor and I.J. Kopin, eds., Academic Press, London (1986).

5. L.I. Wolfson, K.L. Leenders, L.L. Brown and T. Jones, Alterations of regional cerebral blood flow and oxygen metabolism in Parkinson's disease, Neurology, 35:1399 (1985).

CEREBRAL BLOOD FLOW (CBF) PATTERN IN PROGRESSIVE SUPRANUCLEAR PALSY AND CLINICO-CBF-MRI CORRELATIONS

Kazunori Okada, Shotai Kobayashi, Nobuo Suyama,
Hitoshi Fukuda, Kazuya Yamashita, and Tokugoro Tsunematsu

The 3rd Division of Internal Medicine
Shimane Medical University
Izumo, 693, Japan

OBJECTS

It is known that progressive supranuclear palsy(PSP) shows subcortical dementia.[1,2] The diagnosis of PSP is sometimes difficult because this disorder mimicks the clinical features of parkinsonian syndrome.[3] As a PET study[4,5] demonstrated frontal hypo-metabolism in PSP patients, we studied the cerebral blood flow (CBF) pattern by single-photon-emission-computed tomography (SPECT) using ^{123}I-isopropyl-iodoamphetamine (IMP) and clinico-CBF-MRI correlations in 3 PSP patients who were at different clinical stages.

SUBJECTS

Three PSP patients were Case I(60 y.o. male; Steele's stage I), Case II(60 y.o. male; stage II), and Case III(69 y.o. male; stage III). Their brief clinical features are summarized in Table 1. Control groups were normal control (Normal), Parkinson's disease(PD), olivo-ponto-cerebellar atrophy (OPCA), and dementia of Alzheimer's type (DAT). Their number and mean age of each study are presented in Table 2.

METHODS

Cerebral blood flow(CBF) study: CBF was measured by ^{123}I-IMP SPECT using 400AC/T Starcam(GE), and evaluated by semi-quantitative methods. Regions of Interest(ROI's) were located at cerebellar, basal ganglion, and lateral ventricle levels. Each count of ROI (Cerebellum:CE, Basal ganglia :BG, Frontal:F, Frontotemporal:FT, Temporal:T, Temporo-Occipital:TO,) was divided by the count of Occipital(O) ROI.

Magnetic Resonance Image (MRI) study: 0.15 Tesla MRI (TOSHIBA;short SE image;TR=500msec, TE=40msec) was used for the assessment of brain-stem atrophy. The degree of atrophy of the pretectum and tegmentum of the pons was determined in the sagittal plane by MRI.

Edited by T. Nagatsu *et al.*, Plenum Press, New York, 1990

Table 1. Clinical feature of PSP patients

	Age	Sex	Duration (yr)	Stage	Gaze disturbance	Dementia	Parkin-sonism	Pseudobulbar palsy
Case I	60	M	4	I	+	+	+	+
Case II	60	M	4	II	++	+	++	+
Case III	69	M	7	III	+++	+++	+++	+++

+:mild, ++:moderate, +++:severe

Table 2. Control groups

	CBF study	MRI study
Normal Control:	3 cases (59.3)	19 cases (72.9)
Parkinson's disease(PD):	7 cases (54.3)	11 cases (67.1)
Olivo-ponto-cerebellar atrophy(OPCA):	3 cases (60.0)	5 cases (48.4)
Dementia of Alzheimer's disease(DAT):	4 cases (66.5)	-

( );mean age

RESULTS

1)CBF study

The severely demented stage-III patient showed a marked and extensive decrease in CBF in the frontal lobe(F/O ratio, 34% vs. control, 99%) and basal ganglia(BG/O ratio, 51% vs. normal, 101%). This case showed typical frontal and prefrontal CBF reduction. The stage-II patient, who showed moderate dementia, showed mild frontal reduction(81%); and the stage-I patient, who showed mild dementia, showed an almost normal CBF pattern. In PSP patients, the relative blood flow of the frontal and fronto-temporal regions were most markedly affected, showing reduction rates of 67.2% and 67.1% of normal subjects. Other neurological diseases did not produce a similar pattern.[Fig.1] The results indicate that the CBF study with SPECT is a useful method to confirm the diagnosis of PSP. The sagittal image of Case III patient is presented in Fig.2.

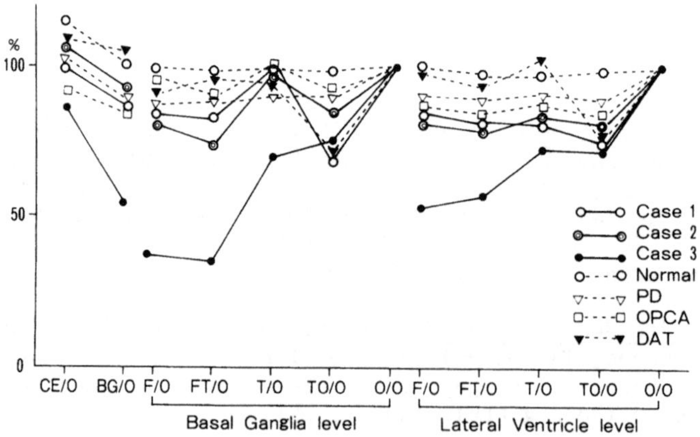

Fig.1. CBF pattern in PSP compared with that in other neurological disease

2)<u>MRI study</u>

The diameter of the midbrain and dorsal pons on MRI significantly decreased in order of advancing stage. On MRI study, characteristic findings of the PSP consisted of atrophy of the pretectum of the midbrain and tegmentum of the pons. These findings were made even in the early stage of the PSP, and the atrophy of the midbrain pretectum was related to the degree of neurological deficit.[Fig.3]

3)<u>Pathological study</u>

Case 1 patient was pathologically proved to be a PSP case. Brain weight was 1,375 g. Macroscopic examination showed brainstem atrophy and microscopic examination showed symmetrical neuronal loss with gliosis affecting the striatum, substantia nigra, tectum, and peri-aqueductal grey matter. In the striatum and periaqueductal area, nerve cells showed the globose type of Alzheimer's neurofibrillary change.

DISCUSSION

The diagnosis of PSP is sometimes difficult because the disease mimicks clinical features of Parkinson's disease.[3] Earlier CBF studies using PET demonstrated frontal hypometabolism in PSP patients.[4,5] This frontal hypometabolism is due to disconnection between frontal lobe and subcortex due to subcortical lesions and is related to a neuro-psychological feature called subcortical dementia in PSP patients.[2] Goffinet reported that there was no significant difference in the posterior cortical metabolic rate between PSP patients and normal volunteers in his PET study. Therefore, posterior cortical regions could be used as reference regions.[5] Our results in this study correspond to these of former PET studies, and the patient with advanced stage showed more severe hypometabolism in the frontal cortex. The results indicate that CBF study with SPECT is useful to confirm the diagnosis of PSP.

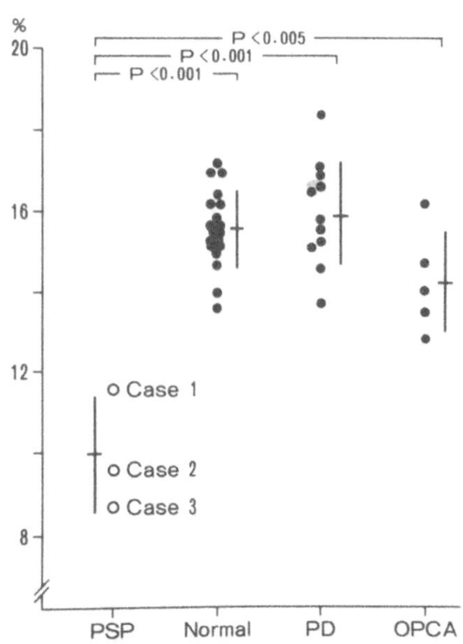

Fig.2. 123I-IMP SPECT image of case I (sagittal plane)

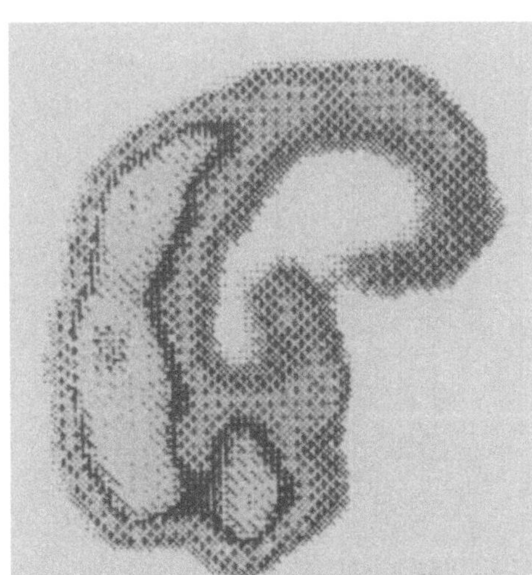

Fig.3. Pretectum index on MRI

Ambrosetto studied CT of PSP patients and reported that midbrain and brain atrophy is specific changes for PSP.[6] Our results in this study also demonstrated that atrophy of the pretectum of the midbrain and the tegmentum of the pons were characteristic findings in PSP. Such atrophy could be found even in the early stage of the PSP, and that of the midbrain pretectum was related to the degree of neurological deficit.

Frontal hypo-perfusion and midbrain atrophy develop with advancing stage in PSP patients. These findings are very helpful for the diagnosis of PSP.

REFERENCES

1. J. C. Steele, J. C. Richardson, and J. Olszewski, Progressive Supranuclear Palsy. Brain 10:333-359(1964).
2. J. L. Cummings, D. F. Benson, Subcortical dementia. Arch Neurol 41:874-879(1984).
3. R. C. Duvisin, L. I. Golbe, F. E. Lepore, Progressive Supranuclear Palsy. Can J Neurol Sci 14:547-554(1987).
4. R. D'Antona, J. C. Baron, Y. Samson, M. Serdaru, F. Viader, Y. Agid, J. Cambier, Subcortical dementia: Frontal cortex hypometabolism detected by positron tomography in patients with progressive supranuclear palsy. Brain 108:785-799(1985).
5. A. M. Goffinet, A. G. De Volder, C. Gillain, D. Rectem, A. Bol, C. Michel, M. Cogneau, D. Labar, C. Laterre, Positron Tomography demonstrates frontal lobe hypometabolism in Progressive Supranuclear Palsy. Ann Neurol 25:131-139(1989).
6. P. Ambrosetto, R. Michelucci, A. Forti, C. A. Tassinari, CT finding in progressive supranuclear palsy. J Comput Assist Tomogr 8:406-409(1984).

NUCLEAR RESONANCE IMAGING IN PARKINSONISM

ANALYTICAL STUDY OF "PARS COMPACTA"

J. Cruz Maurício, Helena Coelho, João de Sá,
and António Martins

Ressonância Magnética Nuclear, Lisboa
Servico de Neurologia, Hospital de Santa Maria, Lisboa

INTRODUCTION

Magnetic Resonance (MR), permitting the detailed visualization of affected brain structures, emerged recently as a helpful instrument in the diagnosis of extrapyramidal disorders.

The purpose of this study is to evaluate the MR capacity in providing more accuracy to the clinical diagnosis of idiopathic Parkinson disease, and to ascertain some anatomoclinical correlations.

MATERIAL AND METHODS

We studied by MR 11 patients affected by idiopathic Parkinson disease (8 males, 3 females) with a mean age of 55 ± 10,24 years (mean ± SD), and 19 normal controls (14 males, 5 females), so that there was no statistically significant difference between the ages of patients and controls.

Parkinson patients had 6 ± 8,84 years of evolution of the disease and were distributed as follows by Yahr stages: stage I - 2 patients, stage II - 6 patients, stage III - 3 patients.

MR was performed on a high-field GE Imaging Unit (Signa 1,5 T, GE, Milwaukee). Image acquisition was performed with a spin-echo technique (TR = 2600 m.sec and TE = 80 m.sec). The slices thickness was 5 mm with a gap of 1,5 mm, matrix 256^2, FOV 24 cm (pixel size 0.9 mm) and flow compensation program to mimimize CSF cisternal artifacts.

Pars Compacta breadths were measured on axial planes of midbrain, with scan orientation between + 0º - 15º to Virchov anatomic plane, using optical density profiles of the scans photographed on X-ray films obtained with an X-Rite densitometer (Model 301 X, Mich. USA), we applied. Duguid's method (4), in which optical densities were obtained in all the millimeters of a line traced between red nucleus and pars compacts centers and evaluating the mean of 3 parallel profiles for each hemimidbrain, through pars compacta centers and 1 mm to each side. All midbrain images were magnified 3 x with the same density window (width and level), (Fig.1).

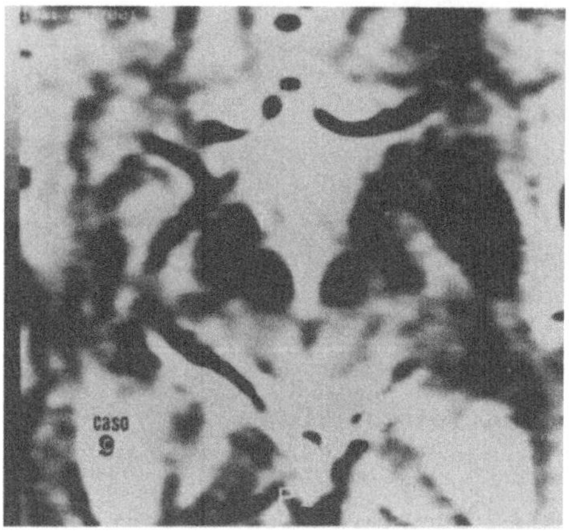

Fig. 1 - MR axial plan of the midbrain (TR 2600, TE 80 msec.) of one Parkinson's patient. Severe reduction of thickness of pars compacta mainly one the right side.

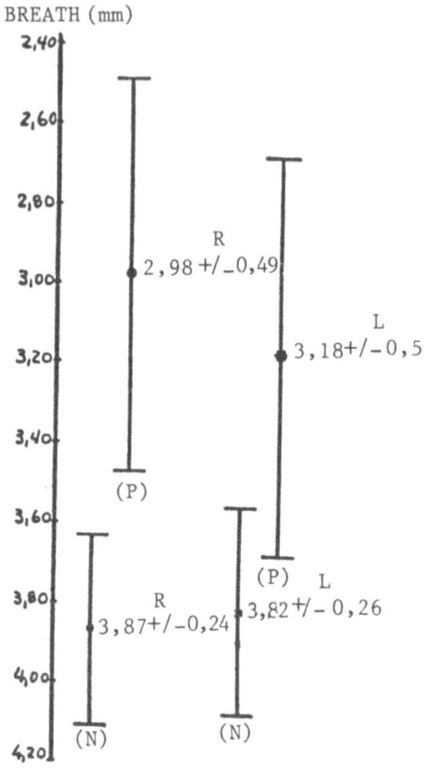

Fig. 2 - Results of MR breadth of pars compactas. N - Controls (N=19)
P - Parkinsonian patients (N=11)

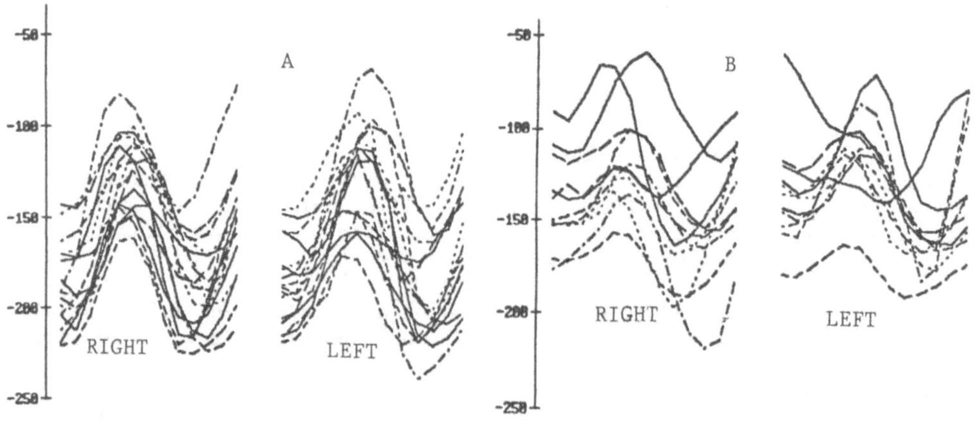

Fig. 3 - A - Optical densities of Controls (N=19). B - Optical density profiles of 11 Parkinsonian patients.

RESULTS AND DISCUSSION

The pars compacta breadth of controls was 3.87 ± 0,24 mm on the right and 3.82 ± 0,26 on the left side. In the Parkinsonian patients the values we obtained were 2.96 ± 0,4 mm on the right and 3.18 ± 0,5 mm on the left side ($p = 0{,}01$), (Fig. 2,3 A-B).

There was no correlationship between any clinical variables we studied (i.e., age of beginning, and duration of the extrapyramidal symptoms or the severity or the clinical form of the disease), and the uni- or bilateral reduction of the pars compacta width, although we found a prevalence of rigid akinetic forms when this breadth was extremely reduced (3 of 8 patients).

MR studies on extrapyramidal disorders and nucleus have developed since a probable relationship between the ferritine distribution and the hyposignal in T2 weighted sequences was established (2).

The signal differences between the two nigra pars should then be attributed both to a distinct histology and to a great iron content of the reticularis which is relatively hyperintense in T2 sequences (6,9). It is admissible that the pars compacta breadth reduction evaluated by this method would reflect not only the neuronal loss, atrophy and gliosis, but also the iron deposition in these structures (5,7). This hypothesis, however, requires further evaluation by correlating postmortem findings with the midbrain MR images (4).

Corroborating other published data (1,4), the narrowing of the pars compacta signal of Parkinson patients relative to controls was the only statistical significant finding in our study. It was not possible to demonstrate any relation with either the clinical form, age of onset, disease duration or the Yahr stage of severity. Although not significant, we found a prevalence of lower signal breadth among rigid akinetic forms.

In conclusion, MR is expected to become a useful instrument in the management of Parkinson diagnosis. If the reduction of the pars compacta breadth is, at the moment, the only relevant finding disclosed by this technique in Parkinson disease, it is admissible that the signal variation of the basal and midbrain nucleus in T2 weighted sequences will also become an important clue in further evaluations, particulary in what concerns the diagnosis between idiopathic Parkinson disease and the other Parkinsonian syndromes (3,8).

REFERENCES

1. Braffmam BH, Groosman RI, Goldberg HI, MR imaging of Parkinson disease with spin-echo and gradient-echo sequences, AJNR 9: 1093-1099 (1988)
2. Drayer BP, Burger P, Darwin R, Magnetic Resonance imaging of brain iron, AJNR 7: 373-380 (1986)
3. Drayer BP, Olanow, Burger P, Parkinson plus syndrome: diagnosis using high field MR imaging of brain iron, Radiology: 493-498 (1986)
4. Duguid JR, De La Paz R, Degroot J, Magnetic Resonance imaging of the midbrain in Parkinson's disease, Ann Neurol 20: 744-747 (1986)
5. Earle KM, Studies on Parkinson's disease including X-ray fluorescent spectroscopy of formalin - fixed brain tissue, J Neuropathol Exp Neurol 27: 1-14 (1968)
6. Emson PC, Chemical Neuroanatomy, p. 164, Raven Press (1983)
7. Oppenheimer DR, Diseases of Basal Ganglia, Cerebellum and Motor Neuro, <u>in</u>"Adams HJ", Greenfield's Neuropathology, ed 4, New York (1984)

8. Rutledge JN, Hillal SH, Silver AJ, Study of movement disorders and brain iron by MR, AJNR 8: 397-411 (1987)
9. Spatz H, Uber den Eisennachweis im Gehirn, besonders in Zentren des Extra-pyramidal-Motorischen Systems, Z Gesamte Neurol Psychiatr 77: 261-390 (1922)

MAGNETIC RESONANCE IMAGING OF PARKINSON'S DISEASE

Kenya Murata, Yukio Mano, Masaaki Konagaya, and Tetuya Takayanagi

Department of Neurology, Nara Medical University
Kasihara, Nara, Japan

INTRODUCTION

Recently the progress in magnetic resonance imaging (MRI) has been remarkable,affording non-invasive diagnosis of the central nervous system. Using 1.5T high field MRI, we studied patients with Parkinson's disease and age-matched controls and examined the distribution of lacunae, the signal intensity of the putamen, and the width of the pars compacta of the substantia nigra.

SUBJECTS AND METHODS

Patients with Parkinson's disease were recruited from the Department of Neurology, Nara Medical University. They were 47-75 years old and comprised 7 men and women. All had cogwheel rigidity, bradykinesia, and resting tremor, and were diagnosed as idiopathic Parkinson's disease. Controls were 53-70 years old:12 men and 2 women. They were patients with tension headache and peripheral neuropathy in the absence of a history of stroke or brain tumor. They had no complaints related to any disorder of the basal ganglia. All patients and controls were scanned with a Picker MRI system operating at 1.5T. The condition for the T1-weighted image was 2500 msec/500-600 msec(TR/TI), and that for the T2 weighted image, 2000 msec/100 msec(TR/TI). All scans represented 7-mm-thick sampling volumes.

Parkinson's disease patients were compared with controls for the existence of lacunae in the putamen, hemisphere, nucleus caudatus, pallidum, thalamus, and substantia nigra. Sites that showed low intensity on the T1-weighted image and high intensity on the T2-weighted one were defined to reveal infarctions in this study.We measured the signal intensity of the putamen on the T2-weighted image using Braffman's grading system (1):Grade 0 was no hypointensity related to cortical gray matter, Grade 1, for hypointensity limited to the lateral margin of the putamen, Grade 2, for hypointensity extending through part of the body of the putamen, and Grade 3, for diffuse hypointensity of the entire putamen. We used the T2-weighted MRI at the level of the red nucleus and pars reticularis of the substantia nigra. On the T2-weighted image the pars compacta was the width between the hypointensity of the pars reticularis and that of the red nucleus. We measured the width between the red nucleus and pars reticularis and defined it as the width of the pars compacta (2).

RESULTS

The distribution of lacunae is given in Table 1. Fourteen cases of Parkinson's disease (87.5%) and 12 cases of controls (85.7%) had lacunae in the putamen. Seven cases of Parkinson's disease and 2 cases of controls had lacunae in the nucleus caudatus. There were no lacunae in the substantia nigra in either group. Thus most of lacunae were found in the putamen, but without any significant difference between the two groups. The intensity grade of putamen is listed in Table 2. Fifty percent of Parkinson's disease patients and 78.6% of controls were Grade 1. The cases of Grade 1 were the most frequent in both groups. The signal intensity of the putamen on the T2-weighted image tended to demonstrate slight hypointensity in Parkinson's disease, but there was no statistical difference between the two groups. The mean width of the pars compacta on either side is listed in Figure 3. It was 1.59 mm (SD, 0.73 mm) on right side and 1.67 mm (SD 0.64 mm) on left side in Parkinson's disease,but was 2.21 mm (SD 0.61 mm) on the right and 2.39 mm(SD 0.73 mm) on the left in controls. The width of the pars compacta of Parkin-

Table 1. Distribution of Lacunae

	Parkinson's disease (n=16)	Controls (N=14)
Putamen	14 cases (87.5%)	12 cases (85.7%)
Hemisphere	12 cases (75.0%)	10 cases (71.4%)
N.caudatus	7 cases (43.8%)	2 cases (14.3%)
Pallidum	6 cases (37.5%)	6 cases (42.9%)
Thalamus	1 case (6.3%)	3 cases (21.4%)
Sub.nigra	0 case (0.0%)	0 case (0.0%)

Table 2. Intensity Grade of Putamen

	Parkinson's disease (n=14)	Controls (n=14)
Grade 0	1 case (7.1%)	1 case (7.1%)
Grade 1	7 cases (50.0%)	11 cases (78.6%)
Grade 2	5 cases (35.7%)	2 cases (14.2%)
Grade 3	1 case (7.1%)	0 case (0.0%)

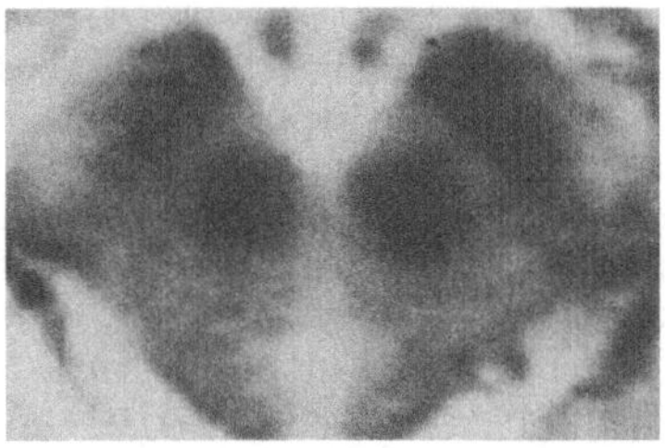

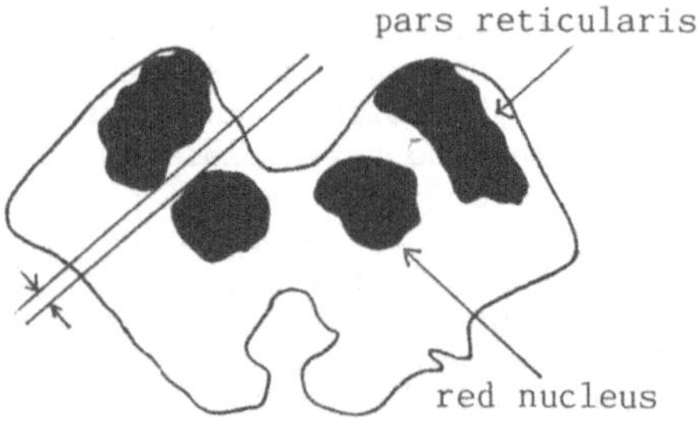

Fig.1. The width between the red nucleus and pars reticularis was defined as the width of the pars compacta.

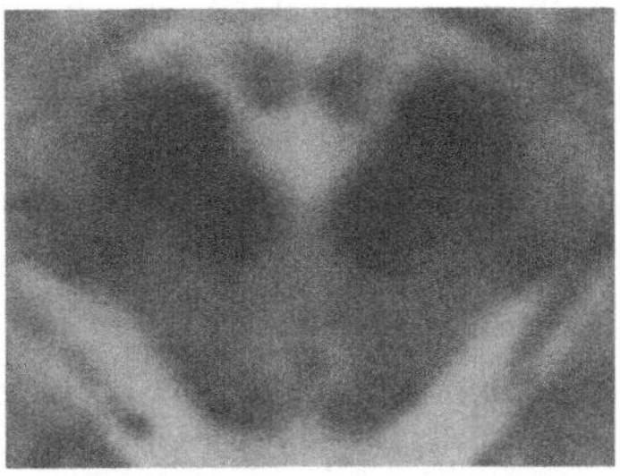

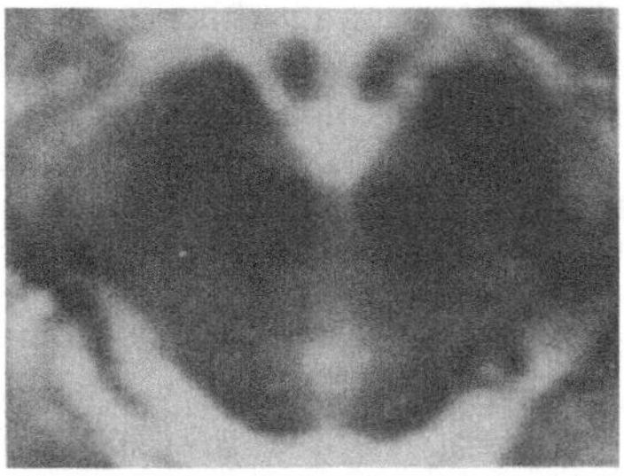

Parkinson's disease
(52-year-old male)

Control
(41-year-old male)

Fig.2. T2-weighted MRI at the level of the pars compacta

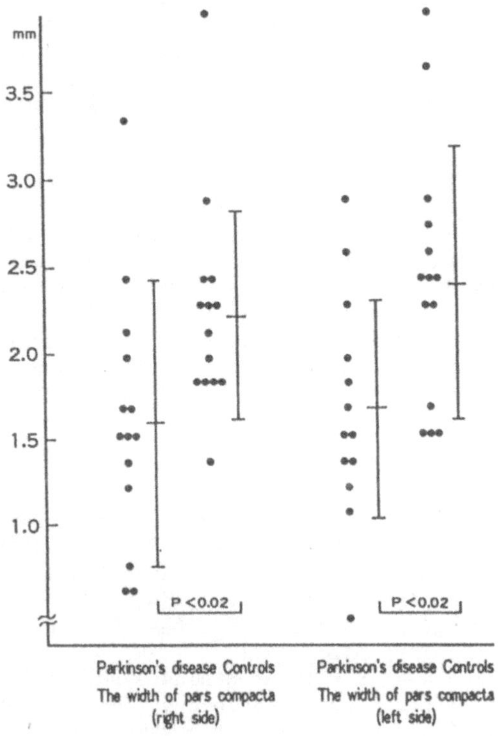

Fig.3. The width of the pars compacta

son's disease patients was statistically smaller than that of controls according to the chi-square test (t=2.41, p<0.02, on right side; t=2.74, p<0.02, on left side).

DISCUSSION

A. lenticulostriata is an end artery and makes an acute angle with A. cerebri media. It also branches against A. cerebri media (3), and supplies blood to the putamen. Erythrocytes tend to flow in the center of blood vessels; and platelets, near the wall of the vessels. Due to these anatomical features, A. lenticulostriata often causes insufficiency of blood flow, and there are many lacunae in the putamen in controls and Parkinson's disease. If the patients with parkinsonism have lacunae in the putamen on MRI, we can't consider them as the patients with arteriosclerotic parkinsonism but rather as having Parkinson's disease because there are many lacunae in the putamen in Parkinson's disease.

Drayer et al. (4)suggested that compared with that of controls, the signal intensity of the putamen in patients with multiple system atrophy was hypointensive on T2-weighted image MRI. But Rutledge et al.(5)described patients who were not parkinsonian but who showed hypointensity in the putamen. In Braffman's study the signal intensity of the putamen in Parkinson's disease was not always hypointensive and there was no statistical difference between Parkinson's disease and controls (1).In our study the signal intensity of the putamen on the T2-weighted image tended to demonstrate slight hypointensity in Parkinson's disease. Changes in the putamen detected by MRI are often recognized in the patients with parkinsonism, but it is rare in those with Parkinson's disease.

Drayer et al. (6)described the red nucleus and pars reticularis to be hypointensive on T2-weighted image MRI because of iron deposit in those regions. The cells of the pars compacta contain much melanin and little iron. Mukai (7)reported that the width of hypointensity of the substantia nigra was not enlarged on T2-weighted image MRI in Parkinson's disease and suggested little change in the pars reticularis. Braffman (1)measured width between the red nucleus and pars reticularis and defined it as the width of the pars compacta. In our study the significant changes in Parkinson's disease patients on T2-weighted MRI was narrowness and hazy border of the pars compacta. Interestingly, pathological changes of Parkinson's disease include degeneration and disappearance of pars compacta cells. The narrowness of the pars compacta in Parkinson's disease suggests atrophy of that region. These observed changes could reflect the primary pathology in Parkinson's disease . Further sophisticated studies on these changes should be performed with respect to the clinical course of Parkinson's disease.

REFERENCES

1) Braffman BH, Grossman RI : MR Imaging of Parkinson Disease with Spin-Echo and Gradient-Echo Sequences. AJR 152:159,1988
2) Duguid JR, Paz RDL : Magnetic Resonance Imaging of the midbrain in Parkinson's disease. Ann Neurol 20:744,1986
3) Katuki S, Tomonaga H : Angioarchitectual Specity of the Human Brain. Shinkei kenkyu no Shinpo 5:277,1961
4) Drayer Bp, Olanow W : Parkinson plus syndrome:Diagnosis using high field MR imaging of brain iron. Radiology 159:493,1986
5) Rutledge JN, Hilal SK : Study of movement disorders and brain iron by MR.AJNR 8:397,1987
6) Drayer BP, Burger P : High-resolusion magnetic resonance for mapping brain iron deposition(abstract). AJNR 6:466,1985
7) Mukai E,Makiko N : Magnetic resonance imaging of Parkinsonism. Clin Neurol 29:720,1989

COMPARISON OF LOCAL CEREBRAL GLUCOSE AND DOPAMINE METABOLISM BETWEEN PARKINSON'S DISEASE AND PARKINSONISM BY POSITRON EMISSION TOMOGRAPHY

Shinichi Hosokawa,[1] Ikuo Goto,[1] Motohiro Kato,[2] Makoto Otsuka,[3] Yasuo Kuwabara[3] and Yuichi Ichiya[3]

Departments of Neurology[1], Neurophysiology[2] and Radiology[3], Faculty of Medicine, Kyushu University Fukuoka, Japan

INTRODUCTION

The measurement of local cerebral energy and dopamine metabolism by positron emission tomography (PET) has been proved to be a potent tool to examine the pathophysiological mechanism of the movement disorders[1]. This study was performed to deterine by PET the similarities and differences in pathophysiological mechanism among Parkinson's disease (PD) and its related disorders. The findings on local cerebral glucose and dopamine metabolism were compared among Parkinson's disease, other Parkinsonism (PK) and progressive supranuclear palsy (PSP) patients.

SUBJECTS AND METHODS

PET was performed on 5 healthy controls (33-66 years old), 6 patients with PD, 6 patients with other PK (vascular PK, multiple system atrophy, and other atypical PK) and 3 patients with progressive supranuclear palsy (PSP).

The local cerebral glucose metabolism was measured by the ^{18}F-fluorodeoxyglucose (FDG) method developed by Phelps et al.,[2] and the local cerebral dopamine metabolism, by the ^{18}F-6-fluorodopa (F-dopa) method.[3] PET images were obtained with a SET-130W (HEADTOME III, Shimadzu). The full width at half maximum (FWHM) measure of spatial resolution was 8.2 mm in the medial resolution mode. We determined attenuation correction individually for each patient by obtaining a transmission scan using a ring source of activity (^{68}Ge). The values of local cerebral metabolic rate for glucose (LCMRglc) were calculated according to the equation developed by Phelps et al.[2] The F-dopa uptake in the striatum was evaluated by the ratio of concentration of fluorine-18 in the striatum to that in the cerebellum (striatum/cerebellum ratio) 120 min after F-dopa injection. The values of LCMRglc in the striatum in PD and PK were also compared between the patients in whom dopaminergic (DAergic) drugs were effective ("DAergic drug-effective group") and those in whom DAergic drugs were not effective ("DAergic drug-non-effective group).

Edited by T. Nagatsu *et al.*, Plenum Press, New York, 1990

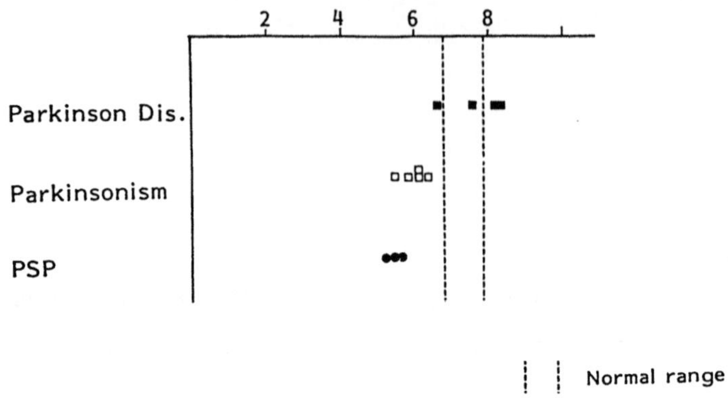

Fig. 1. LCMRglc in the striatum (mg/min/100 ml).

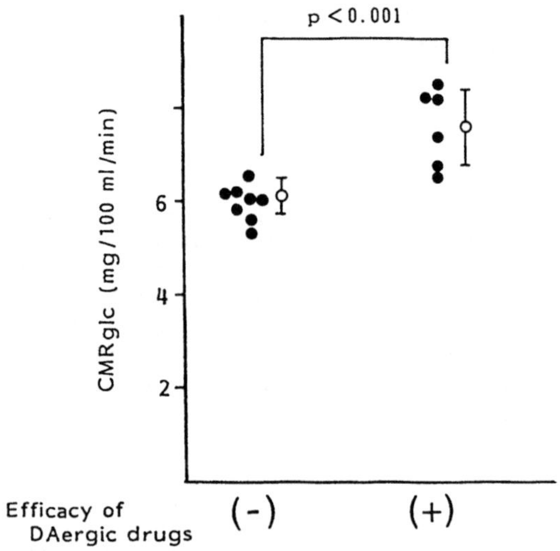

Fig. 2. Efficacy of DAergic drugs and LCMRglc in the striatum (Parkinson's disease and Parkinsonism).

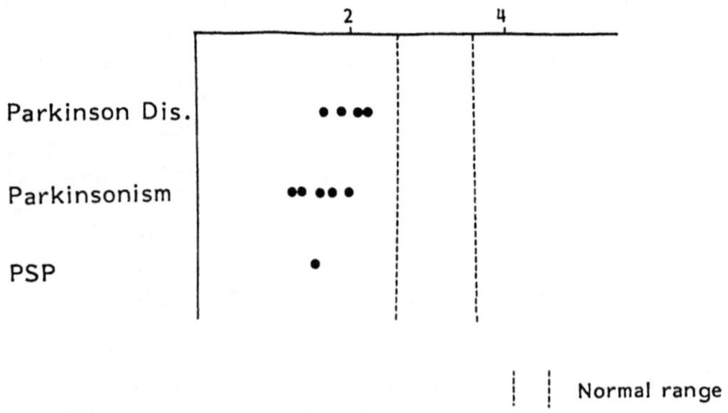

Fig. 3. ^{18}F-DOPA uptake in the striatum (striatum/cerebellum).

RESULTS

1. Cerebral glucose metabolism

In most PD patients, LCMRglc values were within normal range in all the structures including the striatum (Fig. 1) and the cerebral cortices. In other PK, there was a decrease in LCMRglc in the striatum and other cerebral structures including the cerebral cortices (Fig. 1). In PSP, there was also a decreased LCMRglc in the striatum and cerebral cortices, with the decrease being predominant in the frontal cortex.

In PD and PK patients, the LCMRglc in the stratum was significantly lower in the "DAergic drug-non-effective group" than in the "DAergic drug-effective group" (Fig. 2).

2. Dopamine metabolism

In PD, other PK and PSP, there was a common decrease in ^{18}F-dopa uptake in the striatum, compared with the control uptake, as revealed by the striatum/cerebellum ratio determined 120 min after ^{18}F-dopa injection (Fig. 3).

DISCUSSION

It has been shown that F-dopa uptake in the striatum reflects the dopamine storage capacity in the terminals of nigro-striatal DAergic neurons.[3] The observed decrease in F-dopa uptake in the striatum in PD and PSP patients in this study supports the previously reported findings of a reduction in striatal dopamine storage in these patients.[3]

This paper is the first to show a common dysfunction of the striatal dopamine storage in other PK including vascular PK, multiple system atrophy, and other atypical PK.

This study also provides the evidence of a difference in PET findings on glucose metabolism among PD, PK, and PSP patients. In PD, the LCMRglc was normal in all the structures including the striatum. On the other hand, in PK and PSP, there was a decrease in LCMRglc in the striatum and cerebral cortices. Since it has been assumed that LCMRglc reflects the total sum of neuronal activity in that structures, the decrease in LCMRglc in the striatum in PK and PSP may reflect the dysfunction of postsynaptic receptor of nigrostriatal DAergic neurons or dysfunction of other neurons in the striatum.

It was also noted that LCMRglc in the striatum was significantly lower in the "DAergic drug-non-effective group", compared with the "DAergic drug-effective group" in PD and PK patients. This finding suggests that the difference in glucose metabolism may underlie the difference in the efficacy of DAergic drugs.

In conclusion, these results indicate that the main dysfunction of the striatum in PD is restricted in the nerve terminals of the nigro-striatal DAergic neurons. Whereas, the dysfunction of the striatum in PK and PSP is localized not only in the nerve terminals of the nigro-striatal DAergic neurons, but also in the postsynaptic DAergic receptors or other neurons in the striatum.

REFERENCES

1. D.J. Brooks and R.S.J. Frackowiak, PET and movement disorders, J. Neurol. Neurosurg. & Psychiat. Suppl.:68 (1989).

2. M.E. Phelps, S.C. Huang, E.J. Hoffman et al., Tomographic measurement of local cerebral glucose metabolic rate in humans with (F-18)2-fluoro-2-deoxy-D-glucose. Validation of method. Ann. Neurol. 6:371 (1979).
3. K.L. Leenders, A.J. Palmer, N. Quinn et al., Brain dopamine metabolism in patients with Parkinson's disease measured with positron emission tomography. J. Neurol. Neurosurg. & Psychiat. 49: 853 (1986).

STRIATAL DOPAMINE D2 RECEPTORS IN PARKINSON'S DISEASE AND STRIATO-NIGRAL DEGENERATION DETERMINED BY POSITRON EMISSION TOMOGRAPHY

Hitoshi Shinotoh,[1,2] Akiyo Aotsuka,[1,2] Hisashi Yonezawa,[1] Hiroshi Fukuda,[1] Osamu Inoue,[1] Toshiro Yamasaki,[1] Yukio Tateno,[1] and Keizo Hirayama[2]

[1]Division of Clinical Research, National Institute of Radiological Sciences, Chiba City; and [2]Department of Neurology, School of Medicine, Chiba University, Chiba City, Japan

INTRODUCTION

Most patients with Parkinson's disease (PD) respond well, at least temporarily, to L-dopa therapy. However, patients with striato-nigral degeneration (SND) usually fail to respond to this therapy. Although this difference in therapeutic response in both disorders may be attributed to several factors, the most plausible explanation seems to be a loss of striatal dopamine receptors in patients with SND.

To examine this probability, we investigated the striatal dopamine D2 receptors in patients with PD and SND by positron emission tomography (PET) and carbon-11-labelled N-methylspiperone (C-11 NMSP), in order to elucidate the pathological mechanism underlying the different therapeutic responses to L-dopa therapy in the two disorders.

MATERIALS AND METHODS

Seven patients with PD (mean age±SD, 65±7 yrs) were investigated (Table 1). All patients showed typical parkinsonian tremor and responded to anti-parkinsonian drugs. Clinical disability was scored according to Hoehn and Yahr stage. The CT scan showed no abnormal findings.

Four patients with clinically diagnosed SND (mean age±SD, 61±9 yrs) were also investigated (Table 1). The diagnosis of SND was based on the following: 1) the predominance of rigidity and akinesia with absent or minimal tremor; 2) poor response to anti-parkinsonian drugs; 3) rapid disability compared with PD patients; 4) mild to moderate ponto-cerebellar atrophy on CT scan in cases that may correspond to "combined SND-OPCD" (Adams and Salam-Adams, 1986); and 5) parkinsonism was the predominant symptom from the onset of disease until the current study was conducted.

Table 1. Clinical Profiles

Case No.	Age/ Sex	Disease Duration (yrs)	Akinesia	Rigidity	Tremor	Stage (Hoehn & Yahr)	Pontocerebellar Atrophy (CT scan)	Clinical Diagnosis
1.	54/F	2	+	+	+	II	-	PD
2.	61/F	3	+	++	+++	II	-	PD
3.	66/M	2	+	++	+	III	-	PD
4.	68/F	3	+	+	+	III	-	PD
5.	74/F	12	+	++	++	III	-	PD
6.	71/F	4	++	++	++	IV	-	PD
7.	61/F	11	+++	+++	+	V	-	PD
8.	48/F	2	++	++	+	IV	+	SND
9.	60/M	3	++	++	-	IV	+	SND
10.	68/F	3	++	++	+	IV	+	SND
11.	66/F	8	+++	+++	+	V	++	SND

-: none, +: mild, ++: moderate, +++: severe

All neuropsychiatric drugs were withdrawn 2 days before the PET study. Seventeen volunteers (11 males and 6 females; age, 22-77 yrs; mean age, 51 yrs), free from any neurological disorders, were used as control subjects. Informed written consent was obtained from all patients and volunteers.

C-11 NMSP was prepared by N-alkylation of spiperone with carbon-11-labelled methyliodide. The specific activity of C-11 NMSP at the time of injection was 27±13 GBq/umol and a dose of 381±88 MBq, 7.7±5.5 ug of C-11 NMSP, was injected into each of the subjects.

A three-ring PET system was used to follow radioactivity in five sections of the brain covering an axial distance of 72 mm of the brain (Takami et al., 1983). The spatial resolution of the reconstructed images was 10 mm full width at half maximum (FWHM), and the axial thickness was also 10 mm.

K3, the rate constant of the ligand from the exchangeable tissue pool in the striatum to the receptors, was obtained by the following equation:

$$\frac{Ast(T)}{Acb(T)} = 1 + K3 \frac{\int_0^T Acb(T)dt}{Acb(T)}$$

in which Ast represents striatal radioactivity, and Acb, cerebellar radioactivity.

A newly developed, high resolution PET was also used for imaging of one control and one SND patient (case 10). The brain was imaged by 9 simultaneous image planes having a 3.6-mm FWHM resolution and 6-mm FWHM axial thickness.

RESULTS

The caudate and putamen were clearly visible 20-90 min following C-11 NMSP injection (Fig. 1). There was a significant age-related decline of K3 in controls (Fig. 2).

The distribution of C-11 radioactivity in the brain of PD patients was similar to that seen in controls. Three of the seven PD patients showed somewhat high K3 but the rest of them had normal ones (Fig. 2).

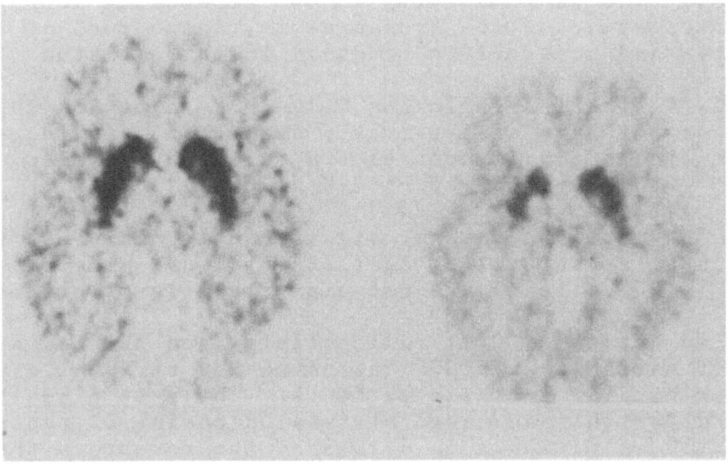

Fig. 1. Left image is a PET image of a control (61-year-old male) and right image is a PET image of a patient with SND (case 10).

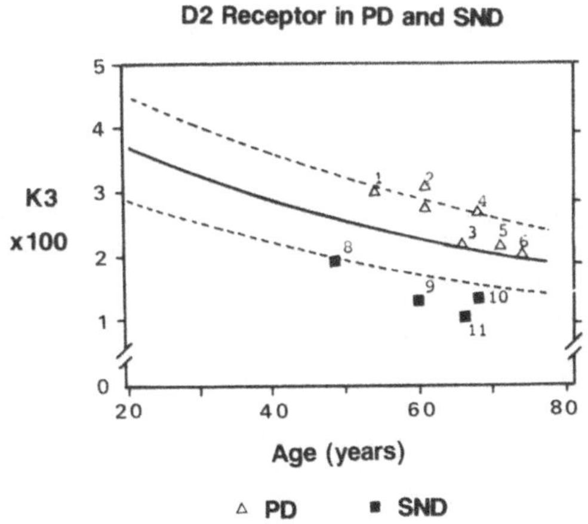

Fig. 2. K3 in PD (△) and SND (■). The solid line shows the exponential fit, and the dashed lines, the SD, for K3 in controls.

The distribution of C-11 radioactivity in the brain of SND patients was much different from that in controls and PD patients. The striatal accumulation of C-11 radioactivity in SND patients was much lower than that in controls and PD and images of SND patients by the high resolution PET showed that the decrease in C-11 radioactivity accumulation in the striatum was more prominent in the posterior putamen than in the anterior putamen and caudate nucleus (Fig. 1). K3 of SND patients was below the normal range (Fig. 2).

DISCUSSION

We assessed the striatal dopamine D2 receptors using the Patlak plot. The radioactivity in the cerebellum is regarded as a representative of the exchangeable tissue pool of the ligand (free and nonspecific binding) in the striatum.

The C-11 kinetics following C-11 NMSP injection in controls was the same as previously described, and the age-related decline of K3 in controls in our study is also consistent with previous reports (Wong et al., 1984).

K3 in PD patients was not significantly different from that in controls, suggesting that striatal dopamine D2 receptors in PD patients are not significantly altered.

The low uptake of C-11 radioactivity and the low K3 in the striatum of patients with SND indicates low striatal dopamine D2 receptor binding in these patients. The most plausible explanation for the decreased striatal dopamine D2 receptor binding in these patients is a loss of the receptors themselves in association with striatal neuron degeneration. The distribution of the decreased striatal dopamine D2 receptors in our study corresponds to the known distribution of striatal neuronal loss in SND.

Our results indicate that the inefficiency of anti-parkinsonian drug therapy in SND should be attributed, at least in part, to the loss of dopamine D2 receptors.

REFERENCES

Adams, R. D., Adams, M. S., 1986, Striatonigral degeneration, in: "Handbook of Clinical Neurology", Vol. 5(49), Vinken, P. J., Bruyn G. W., and Klawans, H. L., eds., Elsevier Science Publishers B. V., Amsterdam.

Wong, D. F., Wagner, H. N., Dannals, R. F., Links, J. M., Frost, J. J., Ravert, H. T., Wilson, A. A., Rosenbaum, A. E., Gjedde, A., Douglass, K. H., Petronis, J. D., Folstein, M. F., Toung, J. K. T., Burns, H. D., and Kuhar, M. J., 1984, Effects of age on dopamine and serotonin receptors measured by positron emission tomography in the living human brain, Science, 226: 1393.

ALZHEIMER TYPE DEMENTIA

- ONE OR SEVERAL DISORDERS?

C.G. Gottfries, Kaj Blennow, Björn Regland, and Anders Wallin

Department of Psychiatry and Neurochemistry
Gothenburg University
St. Jörgen´s Hospital, S-422 03 HISINGS BACKA, Sweden

INTRODUCTION

When Alzheimer and Kraepelin in the beginning of this century described the dementia disorders that thereafter were named Alzheimer´s disease, they had a presenile characteristic disease of the cerebral cortex in mind. Later on in the European textbooks Alzheimer´s disease was considered a form of presenile primary dementia. However, as neuropathologists have found Alzheimer lesions in brains from patients who have late onset dementia (senile dementia), these forms have been named senile dementia of the Alzheimer type (SDAT). The two forms have been combined into one, the Alzheimer type dementia (AD/SDAT).

The main reasons for combining AD and SDAT are the following: the disorders have the same clinical phenomenology, there are Alzheimer lesions present in both disorders, and there are significant correlations between Alzheimer lesions and behavioural disturbances. None of these reasons, however, are valid for looking upon the two orders as one entity. Although it is evident that there are similarities in symptomatology between the two disorders, there are also differences. It is also evident that senile plaques and fibrillary tangles are found both in AD and in SDAT, the only difference between the two disorders being quantitative. However, it has not been documented to what extent Alzheimer lesions delimit a homogeneous group from an etiological or clinical point of view. It is well known that Alzheimer lesions are found in normal aging, Down´s syndrome and other types of dementia, for instance vascular dementia, parkinsonism with dementia and dementia pugulistica. Thus, Alzheimer lesions delimit an Alzheimer encephalopathy which, however, may be the common pathway for different degenerative disorders. It is also true that statistically significant correlations between Alzheimer lesions and behavioural disturbances have been demonstrated (Blessed et al. 1968), but from these statistical correlations, it is not possible to draw any conclusions about a possible causal relationship.

The aim of this paper is to discuss differences between the presenile form of AD and the senile form SDAT.

Basic, Clinical, and Therapeutic Aspects of Alzheimer's and Parkinson's Diseases, Volume 2
Edited by T. Nagatsu *et al.*, Plenum Press, New York, 1990

DIFFERENCES IN SYMPTOMATOLOGY

It is usually assumed that the symptomatology in AD is similar to that in SDAT, although the SDAT symptomatology is somewhat milder. In some studies (Gustafson and Nilsson 1982, Bråne et al. 1989), differences were found in the amount of irritability, anxiety, confusion and restlessness. These kinds of symptoms dominated in the SDAT group, possibly indicating subcortical involvement in this group.

In a prospective study at our institute, preliminary results indicate that aphasia, apraxia, visual agnosia and more severe spatial disability are more frequent in AD than in SDAT. In SDAT patients with equal severity of the disease, there are more non-regional symptoms, such as reduced orientation, calculating ability and logical/analytical ability, and also a more pronounced tendency to confusion, perceptual disturbances, anxiety and increased or diminished motor activity. As is evident, this symptomatology indicates a focal parieto-temporal disturbance in AD, while the disturbance in SDAT is more general and also involves subcortical parts of the brain (see Fig. 1).

BRAIN IMAGING TECHNIQUES

Brain imaging techniques have improved greatly in recent years, and the localization and the degree of morphological changes in the brain, when present, can be relatively clearly described. In a study by Wallin et al. (1989a), morphological changes on CT scans are described. White matter low attenuation (WMLA) was found in a higher frequency in the SDAT group than in the AD group. In fact, the SDAT group did not differ significantly from the group of vascular dementias. The frequency of central atrophy in the SDAT group was also significantly higher than in the AD group. There were no significant differences between the AD and SDAT groups with regard to cortical atrophy. The lack of subcortical findings on CT in AD patients, suggests that AD is mainly a cortical disorder, while in SDAT there is also an involvement of subcortical regions in the degenerative process.

LABORATORY FINDINGS

With regard to peripheral markers, some differences have also emerged in the diagnosis of AD and SDAT. Regland et al. (1988) studied a group of patients with dementia disorders. The concentration of vitamin B12 in serum and the activity of MAO in platelets were investigated. The vitamin B12 levels were normal in AD patients, patients with vascular dementia and controls, while 23% in the SDAT group were found to have significantly reduced vitamin B12 concentrations (below 130 pmol/l)(Fig. 2). The investigation also showed that the demented patients with low vitamin B12 levels had significantly increased concentrations of MAO in platelets, indicating the presence of immature platelets, possibly because of mild peripheral deficiency of vitamin B12. None of the patients were anaemic.

POSTMORTEM HUMAN BRAIN INVESTIGATIONS

Neurochemical differences between AD and SDAT have also been stressed in postmortem human brain investigations (Rossor et al. 1984, Gottfries 1985). Usually the neurotransmitter disturbances seen in early onset dementia are more severe than those seen in late onset dementia. Postmortem studies of white matter changes (Brun and Gustafson 1976, Gottfries et al. 1985, Svennerholm et al. 1988) have also shown that the

	AD	SDAT
APHASIA	++	+
APRAXIA	++	+
VISUAL AGNOSIA	++	+
SPATIAL DISABILITY	++	+
REDUCED ORIENTATION	+	++
REDUCED CALCULATING ABILITY	+	++
REDUCED LOGICAL ANALYTIC ABILITY	+	++
CONFUSION	+	++
"FRONTAL LOBE" SYMPTOMS	(+)	+
"SUBCORTICAL" SYMPTOMS	(+)	+

Fig. 1. Symptomatology in patients with Alzheimer's disease (AD) and senile dementia of the Alzheimer type (SDAT)

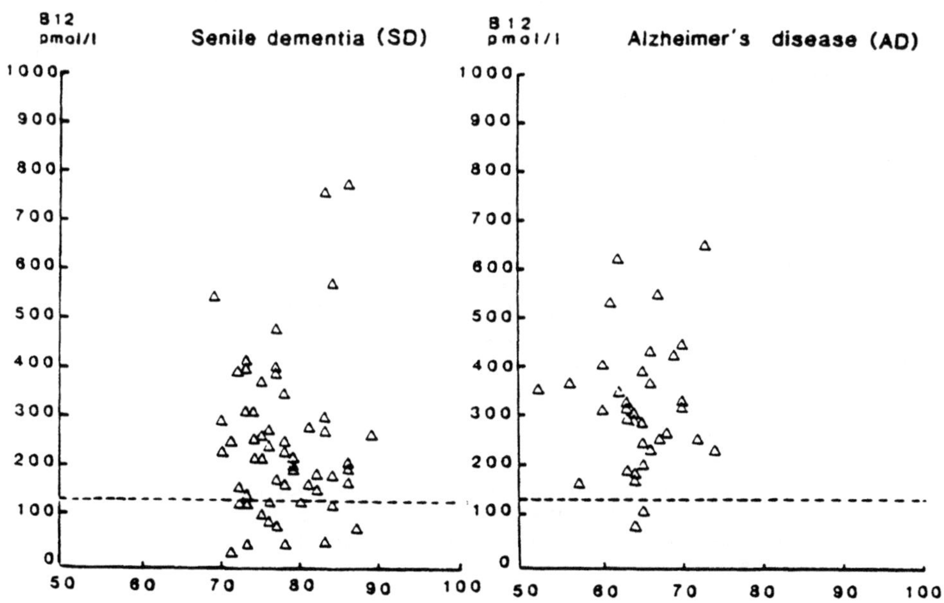

Fig. 2. Vitamin B12/serum in patients with dementia disorders and controls

recorded white matter changes are more severe in the group of demented patients with late onset of the disorder, which is in line with the CT scan findings mentioned above.

The gangliosides have also been shown to be reduced in patients with Alzheimer-type dementia. In as yet unpublished studies at our institute, preliminary data indicate that gangliosides are more severely reduced in AD than in SDAT, and the presence of gangliosides in CSF seems to be a marker of AD.

EPIDEMIOLOGICAL INVESTIGATIONS

According to the epidemiological investigations by Sjögren et al. (1952) and Sourander and Sjögren (1970), AD and SDAT behave like two from each other different disorders. There is a fourfold increased risk of SDAT in siblings of SDAT patients. No instances of AD were found in the 2.675 family members investigated. In other studies of familial aggregation in the AD/SDAT complex, Heston (1981) found that relatives of AD patients ran a markedly increased risk of getting AD. The highest risk of getting dementia among relatives was found in siblings of autopsied Alzheimer patients with comparatively early onset of the disorder. In this group, the cumulative risk was 50:50. The lowest risk was found in siblings of patients who had developed a dementia disorder after the age of 70.

IS AD A HOMOGENEOUS GROUP?

It is natural to assume that patients with the rather malignant early onset AD constitute a small homogeneous group. However, in a study by Mayeux et al. (1985) it was shown that patients with AD (n=110) could be subdivided into four groups: patients with extrapyramidal symptoms, patients with myoclonic symptoms, patients with a benign form of dementia, and patients with the typical form of AD. The subdividing of the dementias was not related to the age at onset. Extrapyramidal symptoms were found in 21% of the individuals investigated, and these patients had a higher prevalence of psychotic behaviour and greater mental impairment. According to Kaye et al. (1988), a subgroup of AD with extrapyramidal features also have reduced concentrations of HVA in CSF to a greater extent than other AD patients. In the study by Mayeux et al., myoclonus was found in 9.9%. Seven of the 110 patients had a benign form of dementia, yet those patients met the criteria for dementia at the assessment.

IS SDAT A HOMOGENEOUS GROUP?

SDAT is delimited by the age at onset and the absence of evident factors that cause the cognitive impairment. Since it is an exclusion diagnosis, SDAT patients may very well be a heterogeneous group. Some subgroups of SDAT are suggested below.

Normal aging: Several investigations have shown that there is an age-related decrease in the activity of neurotransmitter systems in normal aging (Gottfries 1985). This decrease seems to have its onset in the 60s but, because of the reserve capacities of the human brain, behavioural disturbances do usually not become evident until the age of 80. In higher age, however, it can be assumed that the aging process, together with negative environmental factors, may cause such cognitive impairment that a dementing disorder is diagnosed. From a clinical point of view, this type of high-age dementia cannot be separated from SDAT. It can also be questioned whether it is possible to differentiate these disorders by histological investigations of the brain, as the amount of senile plaques and fibrillary tangles increase with age.

Secondary dementias: According to investigations by Regland et al. (1988), 23% of patients with SDAT have reduced concentrations of vitamin B12 and significantly increased platelet MAO activity compared with controls. Regland et al. (unpublished data) also have data indicating that patients with SDAT and low vitamin B12 levels suffer from atrophic gastritis, as marked by reduced pepsinogen I concentrations in serum. The high MAO activity in platelet was significantly reduced in the patients after substitution of vitamin B12. As the cobalamin molecule is actively transported over membranes, the reduced concentrations of vitamin B12 in the group of late onset dementias can possibly be considered a marker of reduced membrane transport capacity. It can be assumed that other essential nutrients, such as folic acid, vitamin B6, tryptophane, tyrosine, zinc, selenium etc. are also transported insufficiently over membranes, thus giving rise to brain dysfunctions. If these findings have importance for the etiology of the dementia disorders, then the disorders of these patients, at present diagnosed as SDAT, should be diagnosed as secondary dementias attributable to deficiency of essential nutrients.

Confusion states: In DSM-III-R it is stated that a diagnosis of dementia should be made only in the absence of severe confusion. Ratings of patients with SDAT have shown that mild confusion is a very common symptom (Andersson 1989, unpublished data). The level of confusion is fluctuating during the day and the patients often show the "sundown syndrome". Slowly the level of confusion increases more and more, but the chronic non-dramatic appearance of the symptoms makes the differential diagnosis to SDAT difficult.

Vascular non-MID patients: According to Wallin et al. (1989b), some patients with vascular risk factors have dementia of a non-MID type, and this group is named probable vascular dementia (PVD). In patients with this type of disorder, the onset of the disorder is insidious and the progression is slow. Sometimes patients with PVD may have had stroke attacks, but if the infarct is small with mild symptoms and if brain imaging is not done, these attacks may remain unidentified. Erroneous diagnoses of SDAT cannot be excluded for these cases.

CONCLUSIONS

At present, cumulative data indicate that the presenile form of Alzheimer´s disease should be separated from senile dementia of the Alzheimer type.

The presenile form of Alzheimer´s disease seems to be a relatively homogeneous group, although some clinical features indicate subgroups. The hereditary pattern may also argue for the presence of a familiar and a non-familiar form.

SDAT is an exclusion diagnosis; the concept is used for primary degenerative disorders with late onset and when no other possible explanation of the dementia can be given. The group is assumed to include several subgroups. High age and negative environmental factors may go with cognitive impairment that is difficult to separate from SDAT. The deficiency of essential nutrients caused by degenerative changes in the mucus membrane of the gut may cause a brain dysfunction that is not possible to differentiate from SDAT. Mild confusion states and vascular dementias may also contaminate the SDAT group.

The combining of AD and SDAT into one large group, Alzheimer-type dementia, and naming it "the disease of the century" has, indeed, created great interest in dementing illness. Politicians, as well as doctors and researchers, have displayed a keen interest. As a consequence, the status of patients and their relatives has risen substantially. However, from a scientific point of view, the sampling of AD and SDAT has not been to advantage. On the contrary, it has been a step back. AD and SDAT must be distinguished. At least SDAT, which is an exclusion diagnosis, can be assumed to be a heterogeneous group, including several different forms of dementing illness.

The question whether Alzheimer lesions are specific or only of secondary nature must be further investigated. It would certainly be of great value to find antemortem biological markers for the diagnosis of dementia disorders. The antemortem clinical diagnosis should then be related not only to neuropathological findings but also to neurochemical changes and disturbances of the calcium homeostasis and glucose metabolism.

REFERENCES

Andersson M. (1989) Institutionalization of the elderly and characteristics of nursing home patients. Compr. Gerontol., accepted for publication.

Blessed G., Tomlinson B.E., Roth M. (1968) The association between quantitative measures of dementia and of senile change in the cerebral grey matter of elderly subjects. Br. J. Psychiatry 114:797-811.

Bråne G., Gottfries C.G., Blennow K., Karlsson I., Lekman A., Parnetti L., Svennerholm L., Wallin A. (1989) Monoamine metabolites in cerebrospinal fluid and behavioural ratings in patients with early and late onset of Alzheimer dementia. Alz. Dis. Assoc. Disord. 3:148-156.

Brun A., Gustafson L. (1976) Distribution of cerebral degeneration in Alzheimer´s disease. A clinicopathological study. Arch. Psychiatr. Nervenkr. 223:15-33.

Gottfries C.G. (1985) Definition of normal aging, senile dementia and Alzheimer´s disease; in: "Normal aging, Alzheimer´s disease and senile dementia. Aspects on etiology, pathogenesis, diagnosis and treatment". Proceedings of two symposia held at The C.I.N.P. 14th Congress, June 22-23, 1984, Florence, Italy. Gottfries C.G. (ed). L´Université de Bruxelles, Brussels, pp. 11-17.

Gustafson L., Nilsson L. (1982) Differential diagnosis of presenile dementia on clinical grounds. Acta Psychiatr. Scand. 65:194-209.

Heston L.L., Mastri A.R., Anderson E., White J. (1981) Dementia of the Alzheimer type: clinical genetics natural history and associated conditions. Arch. Gen. Psychiatry 38:1085-1090.

Kaye,J.A., May C., Daly E., Atack J.R., Sweeney D.J., Luxemberg J.S., Kay A.D., Kaufman S., Milstien S., Friedland R.P. Rapport S.I. (1988) Cerebrospinal fluid monoamine markers are decreased in dementia of the Alzheimer type with extrapyramidal features. Neurology 38:554-557.

Mayeux R., Stern Y., Spanton S. (1985) Heterogeneity in dementia of the Alzheimer type. Evidence of subgroups. Neurology 35:453-461.

Regland B., Gottfries C.G., Oreland L., Svennerholm L. (1988) Low B12 levels relatied to high activity of platelet MAO in patients with dementia disorders. A retrospective study. Acta Psychiatr. Scand 78:451-457.

Rossor M.N., Iversen L.L., Reynolds G.P., Mountjoy C.Q., Roth M. (1984) Neurochemical characteristics of early and late onset types of Alzheimer´s disease. Br. Med. J. 288:961-964.

Sjögren T., Sjögren H., Lindgren A.G.H. (1952) Morbus Alzheimer and morbus Pick. A genetic, clinical and patho-anatomical study. Acta Psychiatr. Neurol. Scand. suppl. 82.

Sourander P., Sjögren H. (1970) The concept of Alzheimer´s disease and its clinical implications; in: "Alzheimer´s disease and related conditions", Wolstenholme F.E.W., O´Connor M. (eds), Churchill, London, pp. 11-36.

Svennerholm L., Gottfries C.G., Karlsson I. (1988) Neurochemical changes in white matter of patients with Alzheimer´s disease; in: "A multidisciplinary approach to myelin disease", Serlupi Crescenzi G. (ed), Plenum, New York, pp. 319-328.

Wallin A., Blennow K., Gottfries C.G., Långström G., Uhlemann C. (1989a) White matter low attenuation on computed tomography in Alzheimer´s disease and vascular dementia. Acta Neurol. Scand., in press.

Wallin A., Alafuzoff I., Carlsson A., Eckernäs S.Å., Gottfries C.G., Karlsson I., Svennerholm L., Winblad B. (1989b) Neurotransmitter deficits in a non-multi-infarct category of vascular dementia. Acta Neurol. Scand. 79:397-406.

QUANTITATIVE TOPOGRAPHIC ANALYSIS OF EEG IN ALZHEIMER'S DISEASE

Gastone G. Celesia, Mitchell Brigell, and Joseph Mayer

Department of Neurology, Loyola University Chicago
Stritch School of Medicine, Maywood, Il. USA

INTRODUCTION

EEG abnormalities in Alzheimer's disease (AD) consist of accentuation of changes seen in normal aging (Soininen et al., 1982, Pentilla et al., 1985) with slowing of alpha rhythm and appearance of diffuse delta and theta activity. Coben et al. (1985 and 1989) longitudinally studied a group of patients with AD and a group of matched elderly controls. They noted that the AD group had higher percentage delta and theta, but lower mean frequency and percentage alpha and beta activity. Duffy et al. (1984) applied topographic mapping of EEG to AD patients and reported right temporal slow activity in younger AD patients and frontal slow activity in older AD patients. Most of these studies used spectrum analysis (Fourier transform) of EEG and concluded that quantitative analysis of EEG was feasible in AD patients and superior to conventional EEG analysis. Yet, none of these studies assessed the reliability and sensitivity of quantitative topographic analysis of EEG (QTA-EEG) compared to conventional EEG interpretation. In the present study, we look at this issue in 24 AD patients to determine the validity and the role of QTA-EEG in Alzheimer's disease.

METHOD

Twenty four patients (13 males and 11 females) meeting the criteria of probable Alzheimer's disease according to the definition of the NINCDS-ADRDA (McKhann et al., 1984) were studied. The patients' age varied from 57 to 90 with a mean age of 74.2 ± 8.2. Each patient underwent a complete physical and neurological examination. Serum vitamin B12, serum folate, VDRL, blood chemistry, WBC, hematocrit, hemoglobin, urine analysis, and thyroid function tests were normal in every patient. CT scans and/or head MRI were performed in every patient and were either normal or showed moderate cortical atrophy. The Hachinski ischemia score was less than or equal to 4. Each patient was evaluated with the Global Deterioration Scale, the Mini-Mental State Test and the Hamilton Depression Scale. The patients' Mini-Mental score varied from 12 to 20. Patients with history or signs of focal ischemic neurological disorders and/or depression were excluded from the study. EEG recordings were obtained with a Nicolet BEAM System. Silver-silver chloride electrodes were applied to the scalp with collodion in accordance with the 10-20 international system. Electrode impedance was equal or below 3,000 ohms. Band-pass filters were 1 to 250 Hz. The BEAM Nicolet System provided simultaneous output of conventional EEG and digitized EEG for further quantitative analysis.

Basic, Clinical, and Therapeutic Aspects of Alzheimer's and Parkinson's Diseases, Volume 2
Edited by T. Nagatsu *et al.*, Plenum Press, New York, 1990

The digitized EEG was Fast Fourier Transformed and the amplitude spectrum produced at each recording site. Quantitative topographic color maps of the entire scalp surface were produced by linear interpolation of the amplitude of the 4 nearest electrodes weighted by distance (Duffy et al., 1984, Duffy, 1986). Maps of absolute amplitude, relative amplitude and interhemispheric asymmetry were provided for each frequency band. Z-statistics between each individual patients' map and the age and sex matched normal groups' map were produced. One of the authors (J.M.) read conventional EEGs, blind to the patient's clinical data and to the QAT-EEG data. Similarly the interpretation of QAT-EEG was done by M.B. blind to the conventional EEG data.

RESULTS

Conventional analysis of EEG tracings was normal in eight patients. Fifteen patients had diffuse slowing in their EEG, and four of these patients had additional focal slowing over the temporal regions. One patient had focal right temporal slowing as a sole abnormality by conventional EEG analysis. Table 1 compares the results of conventional EEG with QTA-EEG.

Table 1

CLASSIC EEG	QUANTITATIVE TOPOGRAPHIC EEG: NORMAL	DIFFUSE SLOW	DIFFUSE + FOCAL	FOCAL	TOTAL
NORMAL	6	0	0	2	8
DIFFUSE SLOW	1	2	4	5	12
DIFFUSE + FOCAL	0	0	1	2	3
FOCAL	0	0	0	1	1
TOTAL	7	2	5	10	24

Quantitative analysis of EEG showed a greater frequency of focal abnormalities than was evident by conventional techniques. Seven patients had normal QTA-EEG. Seven showed diffuse slowing of EEG, and five of these patients also had areas of focal slowing. The remaining 10 patients showed focal slowing. Focal slowing was most frequent in posterior head regions (figure 1).

Bilateral parietal slowing (figure 2) was the most frequent (30%) focal abnormality followed closely by bioccipital (25%) and unilateral temporal slowing (15%). All patients with normal EEGs had milder dementia with mini-mental deficit in the range of 18 to 20.

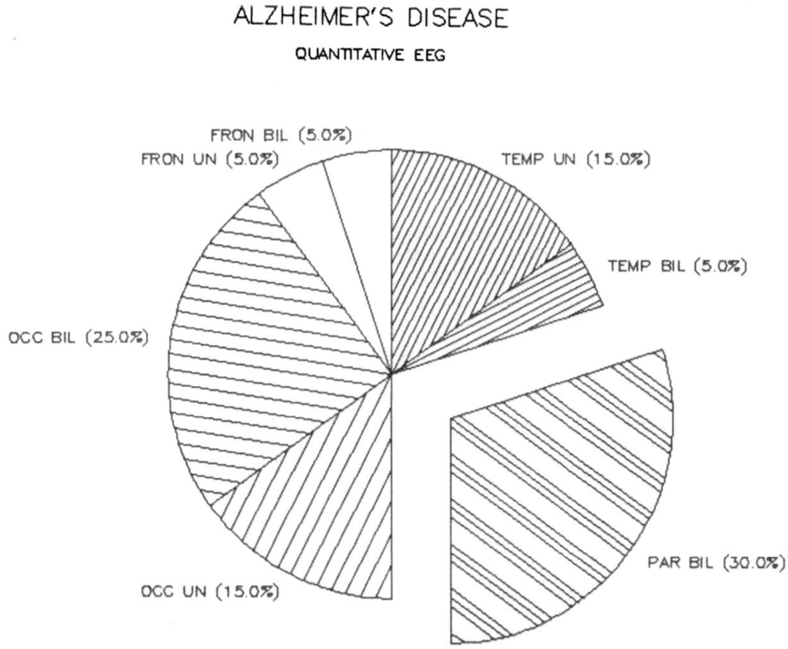

Figure 1

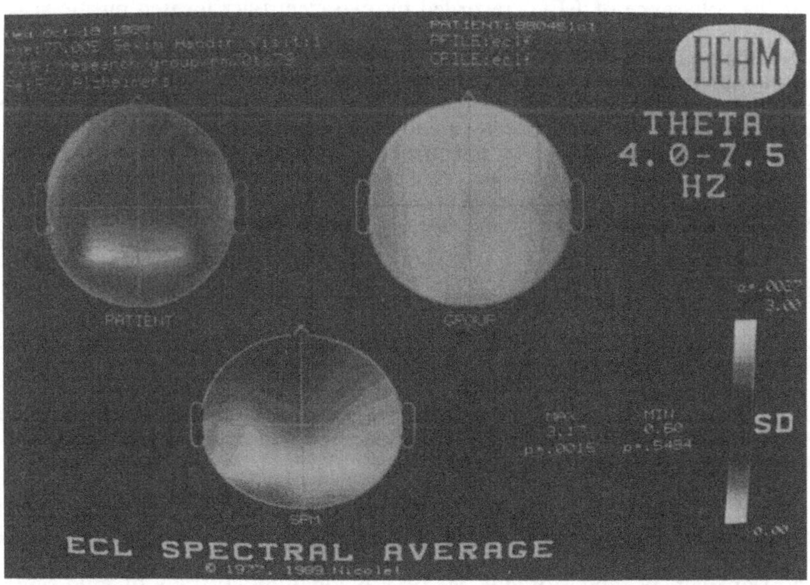

Figure 2. Alzheimer's patient with bilateral parietal theta. SPM refers to statistically significant changes in theta activity of our patient compared to the group data.

DISCUSSION

Conventional EEG and quantitative topographic analysis of EEG had similar yield of normal and/or generalized slow background activity. However, QTA-EEG was more sensitive in detecting focal abnormalities in AD patients than conventional EEG. Conventional EEG and computerized spectral EEG analysis were compared in 35 patients with AD by Brenner et al., (1988). They found that computer EEG analysis was not more sensitive than eyeballing of EEG in the identification of AD patients. This discrepancy between our results and Brenner et al., (1988) findings may be related to their computerized analysis that was limited to only four channels, therefore precluding the detection of possible regional EEG abnormalities. Our data were obtained using 21 electrodes, Z-statistics and amplitude asymmetry. Abnormalities were defined as a deviation equal or greater than 2.5 standard deviation from the data of age-matched controls. The utilization of multiple EEG electrode recordings is the crucial factor that permitted the detection of focal abnormalities. QTA-EEG using mapping of spectral activity has been applied to Alzheimer's disease by other authors (Duffy et al., 1984, Breslau et al., 1989). These studies report increased slow activity, particularly in the delta range in AD patients compared to elderly controls. Duffy et al., (1984) noted that the slow activity was mostly frontal and right temporal while Breslau et al., (1989) reported slow activity in the delta frequency range in frontal regions. Our study confirmed the importance of regional EEG changes in AD but noted a greater theta and delta activity in parieto-occipito-temporal regions. Our demonstration of bilateral parieto-temporo occipital slow EEG activity and hence of decreased alpha activity in the same regions relates well with the reported decreased regional metabolic rate studied by positron emission tomography in temporo-parietal areas of AD patients (Dura et al., 1986, Sheridan et al., 1988). Riekkinen et al. (1989) in a study of 23 AD patients showed that low level of CSF acetylcholinesterase (AChE) correlated with the presence of EEG delta activity. They further showed in a pilot autopsy study that AD patients with the lowest level of cortical ChAT activity and decreased cell density in the nucleus basalis of Meynert had the highest amount of delta slowing. They concluded that the cholinergic deficit was at least partly responsible for the EEG slowing in AD patients.

We conclude that quantitative topographic analysis of EEG is a valid and sensitive method to analyze EEG in AD. QTA-EEG may further allow the objective monitoring of therapeutic trials in the treatment of AD. Recently, Bolcioni and Baggio (1989) have applied the analysis of coherence to EEG in the evaluation of the probability that two stochastic signals arise from a common generator within a specific frequency band. These authors suggest that the coherence of EEG "recorded by two electrodes located on the scalp tends to increase when the underlying brain areas are involved simultaneously in a cognitive process." Bolcioni and Baggio (1989) have developed a graphic representation of coherence and phase differences of interhemispheric and intrahemispheric regions that allows the display of 64 possible electrode couples in sets of triangular correlation matrices. In collaboration with these authors, we are in the process of evaluating AD patients with this new method of EEG analysis. Preliminary data show a decrease in coherence and phase opposition between frontal and occipital regions and a decrease of intrahemispheric coherence in parieto-temporal regions. Further studies are needed, however, before the reliability of this methodology of EEG analysis can confidently be demonstrated.

REFERENCES

Bolcioni, G., and Baggio, C., 1989, EAGLE, Entwined Arrays Graphic Layout for EEG, FIDIA Biomedical Information, 2/3:1.

Brenner, R.P., Reynolds, C.F., and Ulrich, R.F., 1988, Diagnostic efficacy of computerized spectral versus visual EEG in elderly normal, demented and depressed subjects, Electroencephalogr. Clin. Neurophysiol., 69:110.

Breslau, J., Starr, A., Sicotte, N., Higa, J., and Buchsbaum, M.S., 1989, Topographic EEG changes with normal aging and SDTA, Electroencephalogr. Clin. Neurophysiol., 72:281.

Coben, L.A., Danziger, W., and Storandt, M., 1985, A longitudinal study of mild senile dementia of Alzheimer's type: Changes at 1 year and at 2.5 years, Electroencephalogr. Clin. Neurophysiol., 61:101.

Coben, L.A., Chi, D., Snyder, A.Z., and Storandt, M., 1989, Replication of a study of frequency analysis of the resting awake EEG in mild probable Alzheimer's disease, Electroencephalogr. Clin. Neurophysiol., (in press).

Duara, R., Grady, C., Haxby, J., Sundaram, M., Cutler, N.R., Heston, L., Moore, A., Schlagater, N., Larson, S., and Rapaport, S.I., 1986, Positron emission tomography in Alzheimer's disease, Neurology, 46:879.

Duffy, F.H., Albert, M.S., and McAnulty, G., 1984, Brain eletrical activity in patients with presenile and senile dementia of the Alzheimer's type, Ann. Neurol., 16:439.

Duffy, F.H., 1986, "Topographic Mapping of Brain Electrical Activity," Butterworth, Boston.

McKhann, G., Drachman, D., Folstein, M., Katzman, R., Price, D., and Stadlan, E.M., 1984, Clinical diagnosis of Alzheimer's disease: A report of NINCDS-ADRDA work group under auspices of Department of Health and Human Services task force on Alzheimer's disease, Neurology, 34:939.

Pentilla, M., Partanen, V.J., Soininen, H., and Riekkinen, P.J., 1985, Quantitative analysis of occipital EEG in different stages of Alzheimer's disease, Electroencephalogr. Clin. Neurophysiol., 60:1.

Riekkinen, P., Riekkinen P., Jr., Paljarvi, L., Soininen, H., Halonen, T., Reinikainen, K., and Partanen, J., 1989, Relationship beween cholinergic pathology and EEG slowing. In: Rierderen, P., and Beckman, H., (eds.), Key Topics in Brain Research. Alzheimer's disease: Epidiemology, neuropathology, neurochemistry, clinics. Springer-Verlag, Berlin (in press).

Sheridan, P.H., Sato, S., Foster, N., Bruno, G., Cox, C., Fedio, P., and Chase, J.N., 1988, Relation of EEG alpha background to parietal lobe function in Alzheimer's disease as measured by positron emission tomography and psychometry, Neurology, 38:747.

Soininen, H., Partanen, V.J., Helkala, E-L., and Riekinen, P.J., 1982, EEG findings in senile dementia and normal aging, Acta. Neurol. Scand., 65:59.

THE CONCEPT OF SUBCORTICAL DEMENTIA (SCD): ARE THERE CLINICAL ANATOMICAL, BIOCHEMICAL AND NEUROPHYSIOLOGICAL BASIS ?

Hélène Ollat

Association pour la Neuro-Psycho-Pharmacologie
16, Boulevard de Charonne
75020 PARIS - FRANCE

The term "Subcortical Dementia" (SCD) was introduced 15 years ago by ALBERT et al. (1) to describe the behavioral changes observed in patients with progressive supranuclear palsy (PSP). Later other conditions have been labelled as SCD including Huntington's disease (HD), Parkinson's disease (PD), multi-infarct dementia, depressive pseudo dementia...
In each case the clinical syndrome of SCD is opposed to this of cortical dementia (CD), whose model is the Alzheimer's disease (AD). And in the same way, SCD and CD are opposed at anatomical, functional and neurobiochemical levels (table 1).

For these 15 years the concept of SCD has generated controversy related to its validity, both as a clinical and an anatomical entity (2,3,4) : number of clinical signs are observed both in SCD and in CD ; the patterns of cognitive impairment in AD, HD, PD and PSP share many features in common (for review, see 5) ; cortical pathology may be associated with the clinical features of SCD and subcortical lesions are common during AD ; lastly there is also an overlap concerning neurobiochemical deficits.

In fact, it seems that only one conclusion is possible : the terms of CD and SCD are not appropriate and dichotomy between them can be considered as oversimplified (6).
But this semantic controversy should not hide two essential points : First, mainly or exclusively subcortical lesions may be accompanied by intellectual impairment ; Second, the pattern of this impairment is distinct from this of AD.
Intellectual deficit of AD is global ; in SCD there is a double dissociation : fundamental functions (motivation, attention,..) are more impaired than cognitive functions ; concerning cognitive functions global processes (memory, planning, solving problems, strategies) are more impaired than specific processes (language, praxias, gnosias) ; in other words, the SCD looks like an inability to use remaining abilities (3).
In SCD personality and mood disorders are more pronounced than in CD.
When fine neuropsychological analysis is performed, differences appear between CD and SCD. For example, some studies have examined specific aspects of memory function in AD, PD and HD. The evidence suggests that the primary cause of memory impairment in PD and HD is a retrieval deficit, concerning access to stored material and/or its mobilisation ; in addition encoding strategies may be also less efficient than normal, particularly in HD. In contrast, patients with AD, while being impaired at retrieval, are also impaired at the encoding stage. Other differences in memory function appear with abilities to improve an immediate span (for review, see 5).

Table 1
Features that distinguish cortical and subcortical dementias
from CUMMINGS and BENSON (Arch. Neurol., 1984, 41, 874–879)

CHARACTERISTIC	SUBCORTICAL DEMENTIA	CORTICAL DEMENTIA
MENTAL STATUS		
LANGUAGE	NO APHASIA	APHASIA
MEMORY	FORGETFUL (difficulty retrieving learned material)	AMNESIA (difficulty learning new maerial)
COGNITION	IMPAIRED (slowness, forgetfulness, impaired strategy and planning)	SEVERELY DISTURBED (agnosia, aphasia amnesia, apraxia and acalculia)
	SLOW PROCESSING TIME	RESPONSE TIME RELATIVELY NORMAL
PERSONALITY	APATHETIC	UNCONCERNED OR EUPHORIC
MOOD	AFFECTIVE DISORDER COMMON	NORMAL
MOTOR SYSTEM		
SPEECH	DYSARTHRIC	NORMAL
POSTURE	ABNORMAL	NORMAL *
GAIT	ABNORMAL	NORMAL *
MOTOR SPEED	SLOW	NORMAL *
MOVEMENT DISORDER	COMMON	ABSENT
		* except in the late stages
ANATOMY		
CORTEX	LARGELY SPARED	INVOLVED
BASAL GANGLIA, THALAMUS, MESENCEPHALON	INVOLVED	LARGELY SPARED
METABOLISM (FDG PET-Scan)	SUBCORTICAL HYPOMETABOLISM	CORTICAL HYPOMETABOLISM
NEUROTRANSMITTERS PREFERENTIALLY INVOLVED	HD : GABA PD : DOPAMINE	AD : ACETYLCHOLINE SOMATOSTATINE

Lastly, there is an essential characteristic of cognitive deficits in SCD, that is their **similarity to those observed after bilateral lesions of frontal lobes.**
In both the cases, registration and storage abilities are intact, as specific activities ; and retrieval abilities are impaired as all activities which need analysis of the current situation and its comparison to past ones, elaboration of programs and control of their execution, adaptation to shifting, sticking to a set-concept and ignoring the others.
In addition in both the cases there are emotional disorders and personality troubles, and clinical features of frontal syndrome may be observed in SCD.
This frontal dysfunction is common and severe in PSP, less severe in PD, and observed mainly during the early stages of HD, before the global intellectual deterioration (7). It is also evidenced by neuro functional imaging : cerebral blood flow and metabolism are reduced in frontal cortex during HD and PSP (8,9,10) ; this reduction is more diffuse during PD (11,12).
So perhaps today, the good question is not "does SCD really exist" ? but rather **why lesions in subcortical structures may be accompanied by frontal dysfunction signs ?**

This frontal dysfunction cannot be the result of cortical lesions.
In PSP frontal signs are pronounced, but cortical lesions (neurofibrillary tangles) are very rare and discrete.
Histological lesions of Alzheimer type (neurofibrillary tangles, granulo-vacuolar degeneration and senile plaques) are present in significant quantities in cerebral cortex and hippocampus of about 50% of parkinsonians (14). Yet they cannot explain the neuropsychological syndrome, characterized by the absence of instrumental troubles (aphasia, apraxia, agnosia) and the presence of frontal signs which are commonly absent or discrete in AD (15). Even if neuropsychological differences between AD and PD dementia are not admitted, we have to admit that the pattern of cerebral atrophy is different in PD dementia (subcortical), in PD with AD (cortical and subcortical) and in AD (cortical) as it has been recently shown by a morphometric analysis (16).
In HD, cortical neuronal loss is severe and diffuse ; but this atrophy occurs in late stages, with global intellectual deterioration, and frontal signs are observed mainly at the beginning of evolution. Moreover the severity of cognitive troubles is much more well correlated with the severity of caudate atrophy than with the severity of frontal atrophy (17).
On the other hand, frontal symptoms have been reported during caudate (18,19,20) or pallidal hemorrhages (21) sparing cortex (nor vascular, nor compressive lesions).

Therefore it appears clearly that **frontal dysfunction of SCD is the result of the lesions of subcortical structures.** This conclusion fits very well with data concerning functional relations between frontal lobes and subcortical areas.
With neurophysiological and anatomical approaches, it has been shown that there are several basal ganglia - thalamus - cortex circuits which are organized in parallel and which achieve segregation of information reaching the frontal lobe (22). Prefrontal cortex closes the loop of two of them, the dorso-lateral prefrontal and the lateral orbito-frontal circuits (Figure 1).

In animals selective lesions of striatum induce behavioural changes similar to those induced by lesions in the linked area of prefrontal cortex : behavioural suppression mechanisms are disturbed after lesions of orbito-frontal cortex or ventro-medial part of head caudate ; delayed alternance tasks are impaired after lesions of dorso-lateral frontal cortex or dorso-lateral part of head caudate (23).

Clinical observations recently reported (20) are also evidences for the physiological reality of these loops : patients with hemorrhagic lesions predominant in dorso-lateral caudate have the same behavioural troubles as patients with lesions in dorso-lateral convexity of frontal lobes (inertia, apathy, loss of interest) ; in contrast patients with hemorrhagic lesions predominant in ventro-medial caudate have behavioural troubles similar to those of patients with lesions in orbito-frontal areas (desinhibition, euphoria, impulsivity).

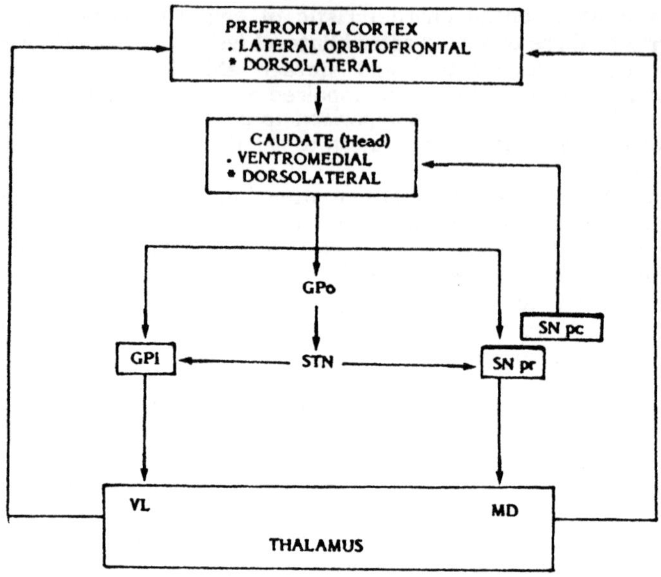

Figure 1
The dorsolateral prefrontal and lateral orbito-frontal circuits

(GPi : Globus Pallidus inner ; GPo : Globus Pallidus outer ; SN pc : Substantia Nigra, pars compacta ; SN pr : Substantia Nigra, pars reticulata ; VL : Ventro Lateral n. of thalamus ; MD : Mediodorsal n. of thalamus)

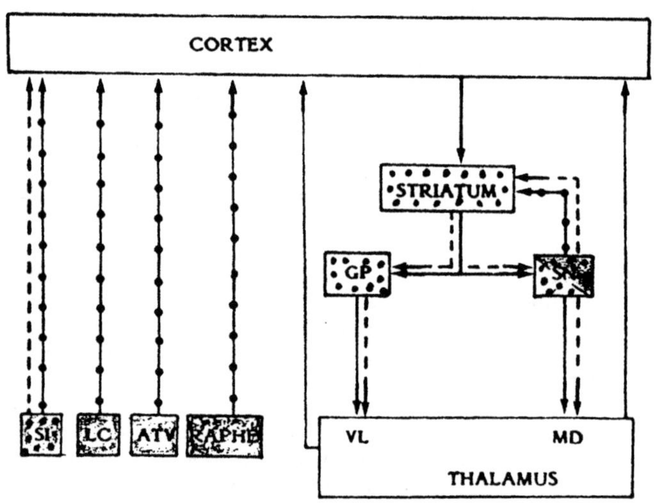

Figure 2
Cortical Deafferentation in PD and PSP
(Adapted from AGID et al)

areas [dotted] and pathways - - - with lesions in PSP
areas [shaded] and pathways ◆◆ with lesions in PD

GP	Globus Pallidus
SN	Substantia Nigra
VL	Ventrolateral n. of thalamus
MD	Mediodorsal n. of thalamus
SI	Substantia Innominata
LC	Locus Coeruleus
ATV	Tegmental Ventral Area

HD is characterized by a severe neuronal loss in striatum (caudate more than putamen) ; in PSP lesions (neurofibrillary tangles) are mainly observed in basal ganglia ; in both the cases the main frontal afferences from subcortical areas are destroyed (Figure 2). In other words frontal cortex is deafferented (7).
In PD the main lesion concerns substantia nigra with a severe denervation in the nigro-striatal pathway ; but other meso-cortical pathways, involved in modulation of cortical programs, are also destroyed (Figure 2).

If the frontal dysfunction observed in the so called SCD results from frontal cortex deafferentation, then it would be logical to try to re-set the disrupted connexions with an appropiated pharmacological treatment. Of course, the preliminary step is identification of the liable neuro-transmitter deficit(s).
In this field PD is the best known.
A lot of **evidences suggest that defective dopaminergic transmission affects neuropsychological functions** in PD : 1) some authors have reported a significant correlation between bradykinesia, a dopamine dependent symptom, and impaired performance on tests assessing visuo-spatial reasoning, psychomotor speed or conceptual ability (24,25) ; 2) in hemiparkinsonism, tasks mediated by dominant hemisphere are more impaired in patients with right hemiparkinsonism, (e.g. with presumably greater involvement of the left basal ganglia) ; in contrast tasks mediated by non-dominant hemisphere are more impaired in patients with left hemiparkinsonism (26,27,28) ; 3) patients with MPTP-induced parkinsonism, that is a quite pure dopaminergic condition, have cognitive disorders similar to those of patients with idiopathic PD ; 4) Levo-Dopa have been reported as improving the performance of PD patients on neuropsychological tests (30) ; 5) lesions of the mesocorticolimbic dopaminergic system interfere with behavior in animals and induce a severe dementia in human (31) ; 6) there has been reported a correlation between neuronal loss in the most medial part of SN (which projects mainly to caudate) and impairment of mental functions in PD (neuronal loss in lateral part of SN, which projects mainly to putamen was correlated to motor signs - 32) ; 7) bradyphrenia and depression may be, at least partially, explained by an hypoactivity of the mesocortico-limbic dopaminergic system (33).
Yet the dopaminergic deficit does not explain the cognitive disorders in all the cases of PD (34) and lesions in non-dopaminergic systems, as those of ascending cholinergic pathways, play probably a role (35).

In conclusion the concept of SCD has been very fruitful and has improved our knowledge of mental functioning. If it is preferable to give up this term, we should remind that **desorganization of fronto-subcortical circuits induces a clinical syndrome, with memory disorders and signs of frontal dysfunction, well distinct from the clinical signs of a desorganization in cortical circuits** (36).
This terminology, proposed by ALBERT himself, has numerous advantages : 1) it takes notice only of the functional aspect and does not implicate a strict anatomical correlation ; 2) it is a symptomatic categorization in which all forms of association or transition are possible ; 3) it allows inclusion of other pathologies where neuropsychological troubles are close to those of dysfunction in fronto-subcortical circuits (brain aging, pseudo-dementia depression,...) ; 4) it gives some therapeutical hopes : appropriate pharmacological treatments could restore the frontal cortex afferentation.
From this point of view, there are already results : dopaminergic agonists work in neglect syndromes induced by, at least partially, lesions of dopaminergic ascending pathways (37,38).

REFERENCES

1. M.L. Albert, R.G. Feldman and A.L. Willis, The "subcortical dementia" of progressive supranuclear palsy, J.Neurol.Neurosurg.Psych., 37:121-130 (1974)
2. R. Mayeux, Y. Stern, J. Rosen et al., Is "subcortical dementia" a recognizable clinical entity ? Ann.Neurol., 14:278-283 (1983)
3. J.L. Cummings and D.F. Benson, Subcortical dementia. Review of an emerging concept, Arch.Neurol., 41:874-879 (1984)

4. P.J. Whitehouse, The concept of subcortical and cortical dementia. Another look, Ann.Neurol., 19:1-6 (1986)

5. R.G. Brown and C.D. Marsden, Subcortical dementia : the neuropsychological evidence. Neurosc., 25:363-387 (1988)

6. H. Chang Hui, Dementia. A review emphasizing clinicopathologic correlation and brain-behavior relationships, Arch.Neurol., 46:806-814 (1989)

7. Y. Agid, M. Ruberg, B. Dubois and B. Pillon, Anatomo-clinical and biochemical concepts of subcortical dementia, in : Stahl SM, Iversen SD, Goodman EC, eds. "Cognitive neurochemistry". Oxford, UK ; Oxford Science Publications, 248-271 (1987)

8. N. Tanahashi, J.S. Meyer, Y. Ishikawa, P. Kansula and K.F. Mortel, Cerebral blood flow and cognitive testing correlate in Huntington's disease, Arch.Neurol, 42:1169-1175 (1985)

9. R. d'Antona, T. Baron, Y. Samson et al., Subcortical dementia : frontal cortex hypometabolism detected by positron emission tomography in patients with progressive supranuclear palsy, Brain, 108:785-799 (1985)

10. A.M. Goffinet, A.G. De Volder, C. Gillain, Positron tomography demonstrates frontal lobe hypometabolism in progressive supranuclear palsy, Ann.Neurol., 25:131-159 (1989)

11. D.E. Kuhl, M.E. Phelps, C.H. Markham et al., Cerebral metabolism and atrophy in Huntington's disease determined by ^{18}FDG and computed tomographic scan, Ann.Neurol., 12:425-434 (1982)

12. M. Globus, B. Mildworf, E. Melamed, Cerebral blood flow and cognitive impairment in Parkinson's disease, Neurology, 35:1135-1139 (1985)

13. K. Jellinger, P. Riederer, M. Tomonga, Progressive supranuclear palsy : clinicopathological and biochemical studies, J.Neural Transm., S16:111-128 (1980)

14. M. Hakim, G. Mathieson, Dementia in Parkinson's disease : a neuropathological study, Neurology, 29:1209-1214 (1979)

15. B. Pillon, B. Dubois, F. Lhermitte et al., Heterogeneity of cognitive impairment in progressive supranuclear palsy, Parkinson's disease, and Alzheimer's disease, Neurology, 36:1179-1185 (1986)

16. S.M. De La Monte, S.E. Wells, E.T. Hedley-Whyte et al., Neuropathological distinction between Parkinson's dementia and Parkinson's plus Alzheimer's disease, Ann.Neurol., 26:309-320 (1989)

17. K.A. Bamford, E.D. Caine, D.K. Kido et al., Clinical-pathologic correlation in Huntington's disease. A neuropsychological and computed tomography study, Neurology, 39:796-801 (1989)

18. J. Cambier, D. Elghozi, R. Strube, Hemorragie de la tête du noyau caudé gauche , Rev.Neurol. (Paris), 135:763-774 (1979)

19. E.K. Richfield, R. Twyman, S. Berent, Neurological syndrome following bilateral damage to the head of the caudate nuclei, Ann.Neurol., 22:768-771 (1987)

20. M.F. Mendez, N.L. Adams, K.S. Lewandowski, Neurobehavioral changes associated with caudate lesions, Neurology, 39:349-354 (1989)

21. R.L. Strub, Frontal lobe syndrome in a patient with bilateral globus pallidus lesions, Arch.Neurol., 46:1024-1027 (1989)

22. G.E. Alexander, M.R. De Long, P.L. Strick, Parallel organization of functionally segregated circuits linking basal ganglion and cortex, Ann.Rev.Neurosc., 9:357-381 (1986)

23. I. Divac, H.E. Rosvold, M.K. Swarcbart, Behavioral effects of selective ablation of the caudate nucleus, J.Comp.Physiol.Psychol., 63:184-190 (1967)

24. J.A. Mortimer, F.J. Pirozzollo, E.C. Hansch et al., Relationship of motor symptoms to intellectual deficits in Parkinson's disease, Neurology, 32:133-137 (1982)

25. D.G. Lichter, A.J. Corbett, G.H. Fitzgibbon et al., Cognitive and motor dysfunction in Parkinson's disease, Arch.Neurol., 45:854-860 (1988)

26. A.E. Taylor, J.A. Saint Cyr, A.E. Lang, Frontal lobe dysfunction in Parkinson's disease - the cortical focus of neostriatal outflow, Brain, 109:845-883 (1986)

27. S. Starkstein, R. Leiguarda, O. Gershanik, Neuropsychological disturbances in hemiparkinson's disease, Neurology, 37:1762-1764 (1987)

28. K.B. Spicer, R.J. Roberts, P.A. Le Witt, Neuropsychological performance in lateralized parkinsonism, Arch.Neurol., 45:429-432 (1988)

29. Y. Stern, J.W. Langston, Intellectual changes in patients with MPTP-induced parkinsonism, Neurology, 35:1506-1509 (1985)

30. D. Delis, L. Direnfeld, M.P. Alexander, E. Haplan, Cognitive fluctuations associated with on-off phenomenon in Parkinson's disease, Neurology, 32:1049-1052 (1982)

31. R.M. Torack, J.C. Morris, The association of ventral tegmental area histopathology with adult dementia, Arch.Neurol., 45:497-501 (1988)

32. J.O. Rinne, J. Rummukainen, L. Paljarvi, U.K. Rinne, Dementia in Parkinson's disease is related to neuronal loss in the medial substantia nigra, Ann.Neurol., 26:47-50 (1989)

33. Y. Agid, M. Ruberg, B. Dubois, F. Javoy-Agid, Biochemical substrates of mental disturbances in Parkinson's disease, Advances in Neurology, 40:211-218 (1984)

34. B. Pillon, B. Dubois, A.M. Bonnet, Cognitive slowing in Parkinson's disease fails to respond to levo-dopa treatment - the 15 objects test, Neurology, 39:762-768 (1989)

35. B. Dubois, M. Ruberg, F. Javoy-Agid, A. Ploska, Y. Agid, Degeneration of a subcortico-cortical cholinergic system in Parkinson's disease, Brain Res., 288:213-218 (1983)

36. M. Freedman, M.L. Albert, Subcortical dementia, in :"Handbook of Clinical Neurology", vol.2 (46) Neurobehavioral disorders, J.A.M. Frederick, ed., Elsevier Publishers, 311-316 (1985)

37. E. Ross, R.M. Stewart, Akinetic mutism from hypothalamic damage : successful treatment with dopamine agonists, Neurology, 31:1435-1439 (1981)

38. W. Fleet, E. Valenstein, R.T. Watson, K.M. Heilman, Dopamine agonist therapy for neglect in humans, Neurology, 37:1765-1770 (1987)

25. G.G. [illegible], [illegible], [illegible] et al., Cognitive and motor dysfunction in Parkinson's disease, Arch Neurol, [illegible] (198[illegible]).

26. A.E. Taylor, J.A. Saint-Cyr, A.E. Lang, Frontal lobe dysfunction in Parkinson's disease. The cortical focus of neostriatal outflow, Brain, [illegible] (1986).

27. S. Starkstein, [illegible], [illegible], [illegible] in Parkinson's disease, Neurology, [illegible]

28. [illegible], Roberts, [illegible]

29. [illegible]

VASCULAR DEMENTIA (VD) AND PRIMARY DEGENERATIVE DEMENTIA (PDD): CRITERIA FOR DIFFERENTIAL DIAGNOSIS

M. Nussbaum, I. Reider and A.D. Korczyn

Sackler School of Medicine, Tel-Aviv University
Ramat Aviv, Israel

INTRODUCTION

The differential diagnosis between vascular dementia (VD) and primary degenerative dementia (PDD) has been an issue of much debate. Attempts to differentiate between the two usually use the onset and progression of these entities as the major discriminative factors: it is commonly accepted that PDD has an insidious onset with relatively slow progression over several years. On the other hand, VD is characterized by an abrupt onset with a fluctuating course, the cumulative effect resulting from several strokes. The DSM-III criteria (1980) are based primarily on these clinical parameters.

Hachinski et al. (1975) devised a differentiating scale (HIS), taking into consideration not only the onset and progression of the diseases, but also the clinical and neurological outcome as well as risk factors and personality manifestations. The HIS is comprised of 13 items; each of them, in a somewhat arbitrary fashion, was assigned a point value of 1 or 2, and the summation results in a total score. A score of 4 or less is believed to be consistent with PDD while a score of 7 or more is thought to be indicative of VD.

Despite the wide use of the HIS, some unresolved questions persist, the most significant being the nature of the intermediate-score group, namely, patients with a HIS of 5 or 6, and the question of the relative contribution of the various elements of the disease to the total score. Some of the difficulties emanate from the fact that the existence of infarcts can only be judged indirectly by their clinical and neurological outcome. However, it is now believed that cognitive decline could be the primary, if not the only, manifestation of small lacunar strokes, otherwise being asymptomatic. The use of CT as a major non-invasive diagnostic tool can be helpful to clarify this issue (Erkinjuntti et al., 1987; Radue et al., 1978). Moreover, in view of an insufficient number of histopathologically-verified cases and the importance of an antemortem diagnosis, the role of CT in the study of dementia can not be overemphasized.

Basic, Clinical, and Therapeutic Aspects of Alzheimer's and Parkinson's Diseases, Volume 2
Edited by T. Nagatsu *et al.*, Plenum Press, New York, 1990

Thus, in the present study we have used a battery of clinical and neurological manifestations of PDD and VD together with the results of CT scans of the brain in an attempt to refine the differential diagnosis of VD from PDD.

PATIENTS AND METHODS

A series of 98 consecutively admitted mild to moderately demented outpatients (MMSE = 10-24) were assessed. Reversible types of dementia were ruled out by means of clinical and Hamilton depression scale examinations as well as tests for B_{12} and T_4 levels and VDRL. The patients were subjected to an extensive battery of neurological and neuropsychological examinations and complete blood chemistry and hematology analysis. An EEG and a CT scan of the brain were also obtained. A combination of four features was used as a differential diagnostic tool: 1) A history of stroke(s) preceding the onset of dementia; 2) The existence of focal neurological signs; 3) The existence of focal neurological symptoms; 4) A CT scan of the brain revealing at least one low-attenuation area indicative of stroke. Fulfilling at least one of the above criteria was sufficient for a patient to be diagnosed as having VD; otherwise, a diagnosis of PDD was made. This method was compared to the one of Hachinski et al. (1975) and the contribution of each of the HIS items to the differential diagnosis was evaluated.

RESULTS AND DISCUSSION

According to our differentiating method, 55 patients were diagnosed as having VD and 43 as having PDD. A HIS of ≥ 5 was calculated for 76% and 11.5% of the patients in these groups, respectively. Within the VD group, 54% of the patients had a history of stroke(s), 44% had focal neurological symptoms and 62% had focal neurological signs. The CT scan of 30 patients (55%) revealed at least one infarct. By definition, none of the patients in the PDD group had any of these signs (Fig. 1).

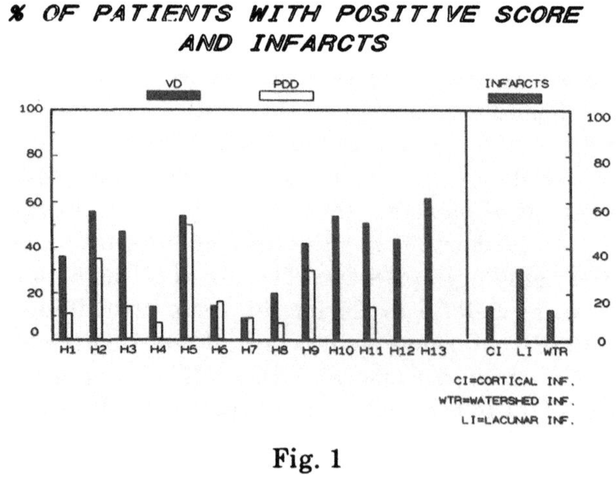

Fig. 1

These results suggest that in many cases there is a discrepancy between the diagnosis according to our criteria, basically parallel those employed by the DSM-III, and the one used by Hachinski et al. (1975).

In order to find out whether these results are biased due to the inherent nature of the design of our method, we also analyzed the various items of the HIS in terms of their individual discriminative power. The results of this analysis confirmed the strong diagnostic power of our test, namely: a history of stroke(s), focal neurological symptoms and focal neurological signs (H10, H12 and H13, respectively), were the most discriminative items of the HIS, regardless of the cut-off value (Figs. 2 and 3). On the other hand, items such as relative preservation of personality, depression, somatic complaints or emotional incontinence (H5-H8 of the HIS) had low diagnostic value due to lack of discriminative power, low frequency, or a combination of the two. It is therefore suggested that these items could be omitted from the HIS.

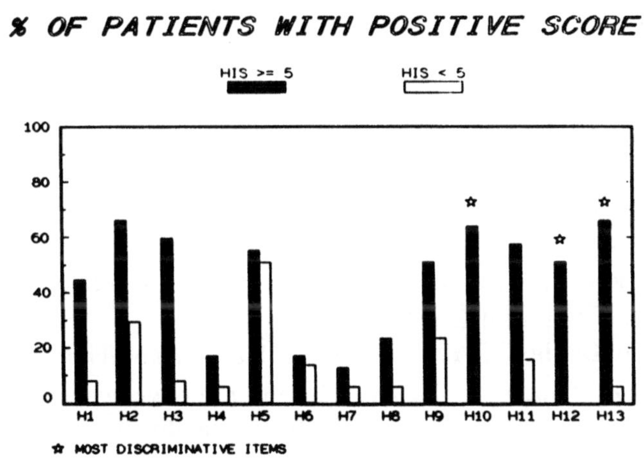

Fig. 2

In this context, it is worth mentioning that the existence of risk factors known to predispose to strokes, such as diabetes mellitus, polycythemia and smoking should also be evaluated in term of their contribution to the diagnosis of VD.

The results of our study also suggest that the inclusion of CT brain analysis can compensate for the lack of diagnostic power of certain items in the HIS. This is especially relevant in cases where the only manifestation of a patient with CT-verified strokes is cognitive decline.

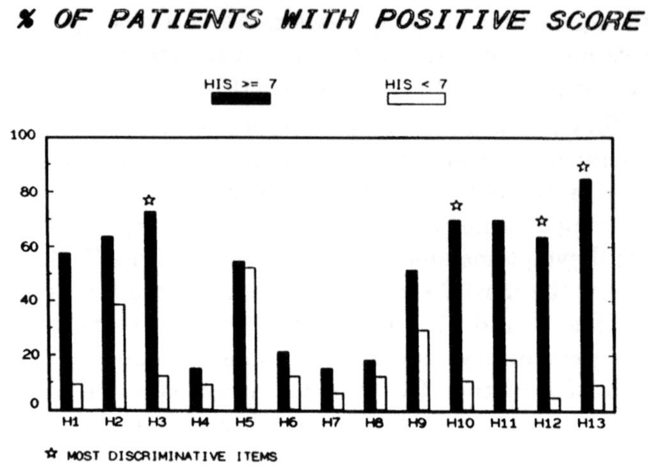

Fig. 3

REFERENCES

American Psychiatric Association: Diagnostic and Statistical Manual of Mental Disorders. 3rd ed. Washington DC: American Psychiatric Association, 1980.

Erkinjuntti, T., Ketonen, L., Sulkava, R., Vuorialho, M., and Palo, J., 1987, CT in the differential diagnosis between alzheimer's disease and vascular dementia, Acta Neurol. Scand., 75: 262.

Hachinski, V. C., Iliff, L., Du Boulay, G. H., McAllister, V., Marshall, J., Ross Russell, R. W., and Symon, L., 1975, Cerebral blood flow in dementia, Arch. Neurol., 32: 632.

Radue, L. W., Du Boulay, G. H., Harrison, M. J. G., and Thomas, D. J., 1978, Comparison of angiographic and CT findings between patients with multi-infarct dementia and those with dementia due to primary neuronal degeneration, Neuroradiology, 16: 113.

COGNITIVE FUNCTIONS IN SENILE DEMENTIA OF THE ALZHEIMER TYPE:

INTERACTION IN THE COGNITIVE MEASURES

M. Roudier, P. Marcie, C. Payan, J. Fermanian, and N. Podrabinek

Hôpital Charles Richet
95400, Villiers-le-Bel, France

INTRODUCTION

In dementia, language impairment is probably intricated with memory deficit and perceptive and attentional deficits can modify the measurment of other functions . Relationships among cognitive functions could be due to the nature of the tests which measure in fact several functions . We studied a sample of patients submitted to a test battery exploring language, memory, gnosia, praxia, ability of reasonning ; our aim was : firstly to determine what is measured by the various neuropsychological tests in demented patients, secondly to show relationships among tests in order to eliminate redundancy in tests .

METHODS AND PATIENTS

104 Patients (90 females, 14 males) fulfiling NINCDS (1) and DSM III R (2) criteria for clinical diagnostic of probable Alzheimer's disease were included in the study ; they were institutionalized in a geriatric long care unit in the Parisian area ; their mean age was 83.4 $\pm$ 5,9 (range 62 - 98) . Severity of SDAT was determined by MMSE value : mild category of severity when MMSE value was superior to 20 (20 patients), moderate when MMSE value was between 10 and 20 (70 patients), and severe when MMSE value was inferior to 10 (14 patients).

Neuropsychological explorations included MMSE (3), Wechsler Memory Scale (WM), Raven's coloured Progressive Matrice (PM-C); a detailed higher functions examination was described previously (4); language was explored using tests established from Irigaray (5), and examination of praxia and gnosia using tests established from Boller and Hécaen (6). A previous work has shown high reliability of this battery in demented patients (7).

STATISTICAL METHODS

32 Quantitative variables were described (mean, standard deviation, minimum and maximum values) . Patient's scores were submitted to Principal Component Analysis and the factors extracted from the correlation matrix were submitted to a varimax rotation (BMDP 4M)(table 1- table 2).

Basic, Clinical, and Therapeutic Aspects of Alzheimer's and Parkinson's Diseases, Volume 2
Edited by T. Nagatsu *et al.*, Plenum Press, New York, 1990

RESULTS AND DISCUSSION

Factor 1 (21% of explained variance) was characterized by high loadings for "mental abstract operation" tests : word spelling, mental calculation, evocation of antonyms, subtest III of Wechsler Memory Scale, grammatical transformations of sentences, generation of sentences, subtest V of Wechsler Memory Scale, writing and verbal fluency .

Factor 2 (17% of explained variance) was characterized by high loadings for number reading, word and sentence reading, color gnosia, number writing, phonemic discrimination, object naming, and writing .This factor was called "visual perceptive" because all the tests implied a visuoperceptive component . Naming impairment in dementia may be linked to a deficit altering visual recognition sphere (8) .

Factor 3 (12% of explained variance) was characterized by high loadings for subtests I, II, IV and VII of Wechsler Memory Scale, MMSE, and Famous Faces test. This factor represents "Memory" . Grober and Buschke distinguished in dementia patients between genuine memory deficit due to impairment of specific memory processes and apparent memory deficit due to impairment of other cognitive processes and this factor could match Grober and Buschke's hypothesis (9) .

Factor 4 (8% of explained variance), called "visuo-constructive factor" included constructive praxia, Raven's Matrices and Subtests IV and VI of Wechsler Memory Scale . The representation of drawing could interfer with the measure of visual memory in WMS or with intellectual ability measure in P.M .

In factor 5, the association relating object naming, picture naming, and oral verbal repetition could correspond to tasks altered in conduction aphasia . It may represent a disorder of "production" as Shallice and Warrington stated the basic disorder of conduction aphasia (10) . On another hand, here we observed the same dissociation of repetition from other language skills as suggested by Holland et al (11) . As for the Famous Faces test, an alteration of "production" could explain, in part, the difficulty to perform the test .

Some Tests were included in 2 factors simultaneously : object naming in factor 2 and factor 5 . One may envisage that perception and production participate in the deficit to name objects in dementia . The presence of writing in factor 1 and 2 suggests that perception and mental operation interact in the writing deficits occuring in dementia . The presence of Famous Faces test, in factors 3 and 5, may imply a conjunction of memory and production deficits in carrying out these tasks .

Thus, possibly several tests in dementia have not the same significance as in normal elderly subjects (Raven's Matrices) or (Word Comprehension, Clock Reading, localisation on a Map) do not measure the same functions because of the interference of deficit in other functions. The progressive and global nature of degenerative dementia requires a different approach in designing the tests : the tests commonly used with aphasics in classical neuropsychology do not fit evenly in case of dementia . The tests which were not correlated to the factors are without any interest for exploring these factors ; they could explore other functions not envisaged in this study .

Our study could help define specific tests devoted to describe the 5 functions above mentioned . This study will be continued with another sample of patients in order to confirm the results and to determine with cluster analysis if subtypes of subjects can be defined .

TABLE 1 - STATISTICS FOR EACH VARIABLE

TEST	MEAN	SD	RANGE
MMSE	13.34	4.57	5 - 25
RAVEN'S MATRICES	10.32	6.42	0 - 28
WECHSLER MEMORY SCALE :			
W1	2.70	1.68	0 - 6
W2	0.50	1.30	0 - 5
W3	2.07	2.49	0 - 9
W4	1.80	1.99	0 - 11
W5	6.43	2.38	0 - 11
W6	0.76	1.38	0 - 8
W7	2.52	2.87	0 - 18
VERBAL REPETITION	14.78	9.85	0 - 40
GRAMMATICAL TRANSFORMATION	6.50	5.45	0 - 15
SENTENCE JUDGMENT	2.10	2.43	0 - 8
EVOCATION OF ANTONYM	3.78	3.19	0 - 10
SENTENCE GENERATION	2.19	1.66	0 - 4
CONFRONTATION NAMING	1.18	1.85	0 - 10
VERBAL FLUENCY	7.81	7.31	0 - 28
AUDITORY COMPREHENSION	13.35	12.35	0 - 45
AUDITORY DISCRIMINATION	2.53	3.32	0 - 17
ORAL READING	1.86	2.67	0 - 15
WRITING	4.08	3.37	0 - 10
ORAL SPELLING	3.69	3.14	0 - 10
ORAL READING OF NUMBERS	1.75	2.63	0 - 10
WRITING OF NUMBERS	2.75	3.23	0 - 10
MENTAL CALCULATION	4.73	2.60	0 - 9
RIGHT-LEFT ORIENTATION	1.98	2.48	0 - 9
FINGER GNOSIA	5.24	3.53	0 - 11
PICTURE GNOSIA	4.26	3.38	0 - 15
COLOR GNOSIA	1.42	2.70	0 - 14
MAP LOCALIZATION	5.00	2.48	0 - 7
CLOCK READING	2.60	1.82	0 - 5
FAMOUS FACE TEST	12.48	2.29	0 - 14
CONSTRUCTIVE PRAXIA	26.03	14.79	0 - 43

TABLE 2 — PRINCIPAL COMPONENT ANALYSIS :

SORTED ROTATED FACTOR LOADINGS AFTER VARIMAX

	FACTOR 1 (21%)	FACTOR 2 (17%)	FACTOR 3 (12%)	FACTOR 4 (8%)	FACTOR 5 (7%)
ORAL SPELLING	.803				
MENTAL CALCULATION	.753				
EVOCATION OF ANTONYM	.728				
WECHSLER (W3)	.690				
GRAMMATICAL TRANSFORMATION	.637				
SENTENCE GENERATION	.624				
SENTENCE JUDGMENT	.618	(.405)			
WECHSLER (W5)	.616				
VERBAL FLUENCY	.577				
READING OF NUMBERS		.770			
ORAL READING		.759			
COLOR GNOSIA		.714			
WRITING OF NUMBERS		.689			
AUDITORY DISCRIMINATION		.670			
CONFRONTATION NAMING		.598			.536
WRITTING TO DICTATION	.558	.576			
WECHSLER (W2)			.831		
WECHSLER (W1)			.691		
MMSE	(.440)		.640		
WECHSLER (W4)			.639	.544	
WECHSLER (W7)	(.428)		.596		
WECHSLER (W6)				.745	
CONSTRUCTIVE PRAXIA	(.486)			.549	
RAVEN'S MATRICES				.542	
VERBAL REPETITION					.672
FAMOUS FACES TEST			.520		.631
PICTURE GNOSIA		(.454)			.591
MAP LOCALIZATION	(.445)			(.476)	
FINGER GNOSIA	(.417)				
AUDITORY COMPREHENSION	(.459)	(.418)			
RIGHT-LEFT ORIENTATION	(.434)	(.425)			
CLOCK READING			(.443)		

(Values enclosed within parentheses stand immediately below .5 retained arbitrarily as a critical level)

REFERENCES

1. G. Mc Kahnn, D. Drachmann, M. Folstein, R. Katzman, D. Price, E. Stadlan, Clinical diagnosis of Alzheimer's disease. Neurology 1984:34 ; 939-944 .
2. DSM III R, Diagnostic and Statistical Manual on Mental Disorders 3rd Edition revised (Edition Française, publiée sous la direction de P. PICHOT, Masson, Paris 1989). Americain Psychiatric Association, Washington DC 1987 .

3. M.F. Folstein, S.E. Folstein, P.R. Mc Hugh "Mini Mental State" : A practical method for grading the cognitive state of patients for the clinician . Journal of Psychiatric Research 1975: 12; 189-198.
4. M. Roudier, P. Marcie,N. Podrabinek, Y. Lamour, P. Davous : Etude neuropsychologique quantifiée dans la Démence Sénile de type Alzheimer ; Encéphale (1989) XV, 397-403 .
5. L. Irigaray, Le langage des déments . Paris, Mouton 1973.
6. F. Boller, H. Hecaen . L'évolution des fonctions neuropsychologiques : examen standard de l'unité de recherches neuropsychologiques et neurolinguistiques . Revue de Psychologie Appliquée 1979 : 29 ; 247-266.
7. M. Roudier, P. Marcie, N. Podrabinek, P. Davous, Y. Lamour, Neuropsychological study of senile dementia of Alzheimer's type . In : Bes A. et Al. Eds Senile Dementia Early Detection. Paris, London : Libbey, 1986 ; 614-618.
8. G. Rochford, A study of memory errors in dysphasic and in demented patients . Neuropsychologia 1971 : 90 ; 437-443.
9. E. Grober, H. Buschke, Genuine memory deficits in dementia. Developmental Neuropsychology, 1987 : 3 ; 13-36.
10. T. Shallice, E. Warrington, Auditory verbal short term memory impairment and conduction aphasia . Brain and Language 1977: 4 ; 479-491.
11. A.L. Holland, F. Boller, M. Bourgeois , Repetition in Alzheimer's disease . A longitudinal study. Journal of Neurolinguistics 1986 : 2 ; 162-177.

ACKNOWLEDGMENTS

This paper was supported in part by Contrat de Recherche Externe INSERM. CNAMTS N° 86-2.35.6.E. and Fondation Pierre Louis .

IS IT USEFUL TO MANAGE ALZHEIMER'S DISEASE AS TWO CLINICAL SUBTYPES: EARLY ONSET AND LATE ONSET SUBTYPES?

Kazuhiko Kono, Hidetoshi Endo, Takayuki Yamamoto and Fumio Kuzuya

Department of Geriatrics
Nagoya University School of Medicine
Nagoya, Japan

INTRODUCTION

The traditional distinction between early- and late-onset of Alzheimer's disease (AD) has been questioned in recent years. Although neurochemical and clinical differences between the two subtypes of AD have been observed, pathological examination of the brains of AD patients has disclosed no qualitative differences. We compared the two subtypes in terms of age at onset, clinical manifestations, degree of brain atrophy, and presence of human lumphocyte antigen (HLA) types A, B, and C. From the results, we attempted to determine whether or not it is useful to manage the two AD subtypes differently.

SUBJECTS AND METHODS

We studied the following five parameters:
I. Age at onset : Data were compiled from 258 AD patients (77 males and 181 females) seen as inpatients or outpatients during a recent 5-year period. Their ages on admission ranged from 46 to 94 years, and the duration of symptoms from 1 month to 20 years. On admission or at the first visit, their scores on Hasegawa's rating scale for dementia (HRSD) ranged from 0 to 29.5. The highest score on the HRSD is 32.5, and it has been established that scores on this scale are correlated with those on the Mini-mental state examination well.
II. Progression of ventricular dilatation: Brain CT scans were used to measure the maximum width of the third ventricle (MWTV) and obtain the cella media index (CMI) in 69 AD patients and 25 elderly controls. Since there were very few males, we analyzed only the data of female patients and controls. All subjects underwent brain CT scanning three or more times during a recent 5-year period. The controls included 15 elderly persons who lived in nursing homes, nine elderly mental patients, and an elderly nurse (average age, 77.4 years). All controls scored 25 or higher on the HRSD. The patients included 20 with early-onset and 24 with late-onset AD (average ages, 68.7 and 80.1 years, respectively). There was no significant difference between the average ages of the late-onset AD patients and the controls. The CMI is the ratio (%) of the minimum width of the body of the lateral ventricle to the minimum width of the external margin of the skull. CT measurements were taken over a 2.5-year period in AD patients and over a 2-year period in controls.

III. Change in HRSD score: We administered the HRSD to AD patients on admission and monthly thereafter for 5 months. First, we compared the HRSD scores of 16 early-onset and 14 late-onset AD patients. Each group included two males. The average ages of the early- and late-onset patients on admission were 61.2 and 74.8 years, respectively. The average duration of symptoms was the same in the two groups (3.5 years). Next, we compared the HRSD scores of 10 early-and eight late-onset AD patients. Each group included two males. The average age of early- and late-onset patients on admission were 63.2 and 73.3 years, respectively, and the average duration of symptoms 3.3 and 3.2 years, respectively (both age and duration data were matched, as were the average HRSD scores on admission). The patients were admitted on June 1, 1989 and were hospitalized for 5 or more months. All patients underwent psychiatric rehabilitation and received drugs for brain metabolism.

IV. Clinical assessment: The HRSD scores and modified Gottfries-Brâne-Steen (GBS) scale scores of the first group of patients studied in part III were compared. The HRSD assesses intellectual functions, including memory, orientation, calculation ability, and general knowledge. The modified GBS scale includes 14 items on intellectual impairment, five items on reduced activity, seven items on reduced emotion, five items on impaired motor function, and seven items on other symptoms associated with dementia.

V. Presence of HLA-A, B, and C: HLA typing was carried out for 124 early-onset AD patients (29 males), 219 late-onset patients (52 males) and 105 controls (28 males). Those with AD were inpatients during a recent period of 6 years and 5 months. Their ages on admission ranged from 46 to 94 years and the duration of symptoms from 1 month to 20 years. The controls included hospital and nursing home employees and elderly persons without dementia living in nursing homes. Their ages ranged from 23 to 100 years.

RESULTS

I. The overall peak age of onset of AD was 70-74 years. Among males, two peaks were observed, but these were too few males for this result to be significant (Fig. 1).

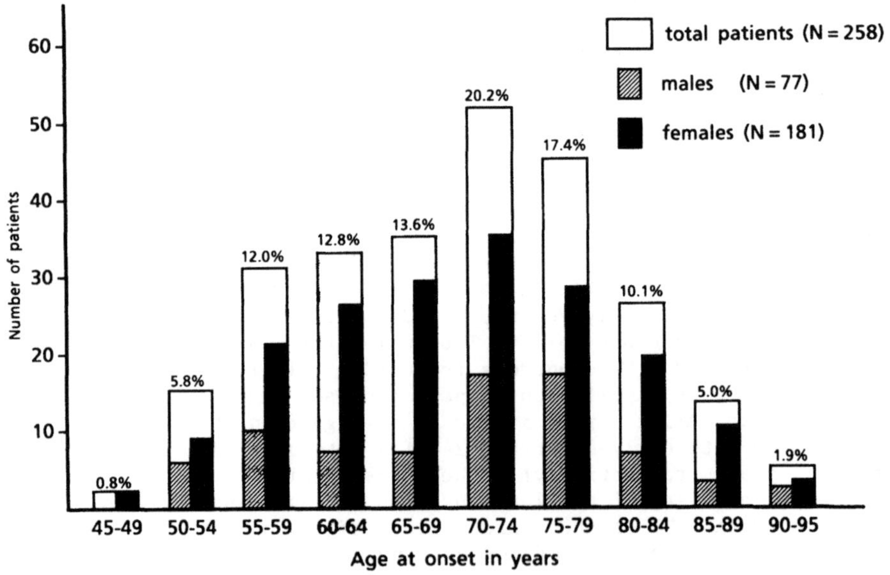

Fig. 1 Distribution of Alzheimer's Disease Patients by Age at Onset

II. Patients of both AD subtypes showed significantly greater ventricular dilatation than did controls. There were no significant differences between the two subtypes until 12 and 24 months after admission, when the CMI was significantly lower in the late-onset patients.
III. In the first group studied, HRSD scores of late-onset AD patients significantly improved at 1 and 2 months after admission, whereas the scores of early-onset patients significantly deteriorated within 5 months of admission. The mean HRSD scores of the early-onset patients were significantly lower at all time points. In the second group studied, the early-onset patients scored significantly lower at 2 and 3 months after admission.
IV. Using the scores of patients who were matched in terms of the duration of symptoms, we analyzed six items on the HRSD and 38 on the modified GBS scale. The early-onset patients scored significantly worse on 29 items (65.9%), while there were no items on which the late-onset patients scored worse. The items on which the differences between the two groups were particularly

Table 1 Intellectual, Interpersonal, Emotional, and Motor Functions at 2 Months after Admission

	index	early onset (N = 16)	late onset (N = 14)
HRSD	worst average	3.9 ± 3.3	12.0 ± 7.6**
	best average	8.2 ± 5.6	19.9 ± 8.4***
	memory for distant events	3.2 ± 1.9	5.2 ± 1.4**
	memory for recent events	1.7 ± 1.9	6.0 ± 3.0***
	orientation in space	1.1 ± 0.9	1.6 ± 1.0
	orientation in time	0.4 ± 0.9	1.3 ± 0.9*
	calcuration	0.3 ± 1.0	2.0 ± 1.8**
	knowledge	0.2 ± 0.8	1.8 ± 1.9**
modified GBS scale	A. Impaired intelligence	47.1 ± 15.0	24.4 ± 12.5***
	1. Impaired orientation in space	4.3 ± 1.8	2.6 ± 1.1**
	2. Impaired orientation in time	5.0 ± 1.6	3.7 ± 1.7*
	3. Impaired personal orientation	2.4 ± 1.9	1.3 ± 1.0
	4. Impaired memory for recent events	4.4 ± 1.7	3.4 ± 1.6
	5. Impaired memory for distant events	3.9 ± 1.2	2.6 ± 1.1**
	6. Impaired wakefulness	1.4 ± 1.5	0.8 ± 1.6
	7. Impaired concentration	4.1 ± 1.2	1.9 ± 1.5***
	8. Inability to increase tempo	3.3 ± 1.9	1.5 ± 1.5*
	9. Long-windedness	3.6 ± 1.7	2.4 ± 1.6
	10. Impaired oral comprehension	3.3 ± 1.7	1.6 ± 1.3**
	11. Impaired oral expression	3.6 ± 1.8	1.6 ± 1.4**
	12. Lack of knowledge	4.8 ± 1.6	2.6 ± 1.3***
	13. Impairement in abstract thinking	5.1 ± 1.6	3.1 ± 1.8**
	14. Impaired judgement	5.1 ± 1.8	3.1 ± 2.0*
	B. Impaired activity	47.7 ± 25.5	22.6 ± 18.7**
	1. Deficiency of spontaneous activity	2.6 ± 1.9	1.4 ± 1.6
	2. Reduced expression in wants and will	3.4 ± 2.0	1.5 ± 1.8*
	3. Reduced interest in acquaintances and amusements	3.3 ± 2.1	1.7 ± 1.5*
	4. Reduced responces to approaches given	1.9 ± 2.0	0.3 ± 0.5**
	5. Reduced motivation	3.2 ± 1.6	1.9 ± 1.2*
	C. Reduced emotion	33.9 ± 15.3	14.0 ± 13.8**
	1. Poor expression	2.3 ± 1.5	0.9 ± 1.4*
	2. Emotional blunting	3.1 ± 1.6	0.9 ± 1.6***
	3. Emotional lability	2.7 ± 1.6	0.8 ± 1.4**
	4. Reduced mood	1.9 ± 1.4	0.8 ± 0.9*
	5. Irritability	1.8 ± 1.5	1.0 ± 1.0
	6. Anxiety	1.4 ± 1.5	1.1 ± 0.9
	7. Bad humor	1.0 ± 1.0	0.4 ± 0.7
	D. Others	28.9 ± 13.4	17.2 ± 12.2*
	1. Sleep disturbance	2.3 ± 1.8	0.8 ± 1.1*
	2. Confusion	2.3 ± 1.8	1.4 ± 1.3
	3. Restlessness	3.7 ± 1.8	1.5 ± 1.5**
	4. Hallucination, Delusion	1.7 ± 1.6	1.6 ± 1.3
	5. Delirium	0.3 ± 0.7	0.4 ± 0.8
	6. Confabulation	1.2 ± 1.3	0.9 ± 0.6
	7. Talkativeness	1.1 ± 1.5	0.6 ± 1.1
	E. Impaired motor function	39.8 ± 27.3	11.2 ± 16.5**
	1. Motor incoordination in undressing and dressing	3.2 ± 2.0	1.0 ± 1.3**
	2. Motor incoordination in taking food	2.3 ± 2.0	0.0 ***
	3. Motor incoordination in walking	0.1 ± 0.3	0.4 ± 1.3
	4. Motor incoordination in managing personal hygiene	3.1 ± 2.4	1.0 ± 1.2**
	5. Inability to control bladder and bowel	3.1 ± 2.5	1.0 ± 1.8*

Student's t-test : *p<0.05 ; **p<0.01 ; ***p<0.001

striking ($p<0.001$) were "memory for recent events" on the HRSD, and "impaired concentration," "lack of knowledge," "emotional blunting," "motor incoordination while eating" on the modified GBS scale (Table 1).
V. Comparison of HLA studies showed HLA-B16 to be more frequent in AD patients than in controls and HLA-B35 to be more frequent in controls than in AD patients. There were no significant differences between the two AD subtypes, in frequency of specific antigens, but HLA-A10 in early-onset AD and HLA-B16 in late-onset AD were significantly more frequent than in controls.

DISCUSSION

Regarding the age of onset of AD, we did not observe two peaks. The presence of two peaks, representing presenile and senile phages, would have suggested that early- and late-onset AD may be different diseases. Mayeux et al. (1985) and Huff et al. (1987) found two peaks, whereas Froda et al. (1988) and Ortof and Crystal (1989) did not. In our study, unfortunately, the possibility of two peaks among males could not be verified because of the small number of patients.

In terms of ventricular dilatation, which we analyzed only in female patients and controls, there was a difference between early- and late-onset AD. CMI increased significantly more rapidly in early-onset patients than in late-onset patients and controls.

We compared changes in HRSD scores between early- and late-onset patients in two separate evaluations. In both cases, the results suggested a poor prognosis for early-onset AD and the possibility of improvement with late-onset AD.

Clinical assessment with the HRSD and modified GBS scales disclosed significant differences between the two subtypes on several items. This indicates to us that early- and late-onset AD should be managed differently.

The results of HLA analysis revealed no differences between the two AD subtypes, but there were differences between the AD patients and the controls. The frequency of HLA-B16 in all AD patients was significantly higher than that in controls, which agrees with the result of our previous study (Endo et al., 1986).

SUMMARY AND CONCLUSION

On the basis of age at onset alone, AD does not appear to comprise two distinct subtypes. However, we found several differences between early- and late-onset AD in terms of the rate at which ventricular dilatation progressed, changes in HRSD scores, and clinical features. We conclude, therefore, that consideration of the two subtypes is relevant to the management of AD.

REFERENCES

Endo H., Yamamoto T., and Kuzuya F., 1986, HLA system in senile dementia of Alzheimer type and multi-infarct dementia in Japan, Arch.Gerontol. Geriatr., 5:51.

Froda S., Gauvreau D., and DeBraekeleer M., 1988, Age at onset in Alzheimer's disease, J.Am.Geriatr.Soc., 36:179.

Huff F.J., Growdon J.H., Corkin S., and Rosen T.J., 1987, Age at onset and rate of progression of Alzheimer's disease, J.Am.Geriatr.Soc., 35:27.

Mayeux R., Stern Y., and Spanton S., 1985, Heterogeneity in dementia of the Alzheimer type: Evidence of subgroups, Neurology, 35: 453.

Ortof E., and Crystal H.A., 1989, Rate of progression of Alzheimer's disease, J.Am.Geriatr.Soc., 37: 511.

A MEMORY CLINIC LOCATED IN A GENERAL HOSPITAL: FIRST RESULTS

Florence Mahieux, Gilles Fénelon, Denyse Michelet,
and Alain Guillard

Clinique neurologique, Hôpital Tenon, Paris, France

Memory complaints can be related to a large spectrum of diseases. Nevertheless, it is a common belief that intellectual performances decline with age. This fact leads to an overlook of these diseases. The early detection of curable disorders with such an expression is at its dawn. On the other hand, the understanding and the spontaneous evolution of the complaints of memory loss without specific disease states have generated a growing debate (1). Memory clinics have been developed since the last 7 years (2) in order to set up suitable screening and new therapeutical ways.

METHODS

1. Course of the Consultation

The clinic works as an outpatients department. The board consists of a neurologist and a psychologist. Each patient is expected with an accompanient relative. Standard laboratory tests have to be done before consultation. They include full blood count, sedimentation rate, blood urea and electrolytes, serum calcium, glucose level, cholesterol and triglyceride levels, liver function tests, TPHA and VDRL serologies. The patients first fulfil a self-assessment including Zung depression scale (3), Zung anxiety scale (4) and a multi-example memory failure list derived from Mc Nair and Kahn self-assessment scale (5). The neurologist conducts an interview of the patient and accompanient and a full medical examination with Hachinski score and a Mini-Mental-State (6). The psychologist performs Raven's matrices, an evaluation of verbal fluency, the Wechsler Memory Scale (WMS) and Rey figure. The consultation is concluded by a diagnosis with ensued therapeutic plan, or by the decision of requiring further investigations.

2. Retained Criteria for Main Diagnosis

A dementia (Dem) is diagnosed according to the DSMIII-R criteria (7) with Mini-Mental-State score (MMS) $\leqslant 23$ and a Zung depression score (ZD)<50.

An associated disorder (dementia/depression) is diagnosed when there is evidence of dysphoric mood with possible organic mental disorder (cerebral atrophy at tomodensitometry, MMS$\leqslant 23$).

An Age-Associated Memory Impairment (AAMI) is diagnosed when the syndrome meets NIMH proposed criteria (1).

The diagnosis of possible AAMI is proposed when the patient complains

Basic, Clinical, and Therapeutic Aspects of Alzheimer's and Parkinson's Diseases, Volume 2
Edited by T. Nagatsu *et al.*, Plenum Press, New York, 1990

Table 1. Main parameters (a) of the described population.

	Dem	Assoc	Dep	OPD	AAMI	OBS	possible AAMI
Number	11	4	11	6	11	4	4
Mean age	74.8 (62-87)	69.0 (63-77)	53.7 (35-71)	54.5 (38-66)	69.9 (63-80)	44.8 (27-80)	50.3 (42-58)
Male/female	5/6	1/3	6/5	1/5	5/6	1/3	2/2
Provenance (b) GP	5 (45%)	3 (75%)	6 (55%)	4 (67%)	6 (55%)	1 (25%)	4 (100%)
HP	6 (55%)	1 (25%)	3 (27%)	2 (33%)	4 (36%)	3 (75%)	-
Own	-	-	2 (18%)	-	1 (9%)	-	-
Psychotropic Treatment	5 (55%)	2 (50%)	10 (91%)	6 (100%)	4 (36%)	1 (25%)	2 (50%)

(a) : data are expressed in number of patients with percentage in brackets ; except for mean age (years) with extreme values in brackets.
(b) : GP : general practitioners, HP : hospital physicians, Own:on their own.

of everyday problems resulting in reminding strategies and/or involvement of relatives with gradual onset , but when symptoms don't meet NIMH criteria for age (<50 years) and/or normal memory tests performances for WMS.

Others diagnosis can include psychiatric disorders (OPD) according to DSMIII-R criteria and organic brain syndromes (OBS) different from dementia.

RESULTS

1. Population

Since the first 18 months 51 outpatients consulted. 10 of them suffered for other diagnosis than the five described above. The 6 psychiatric patients suffered for somatoform disorders 4 times, anxiety disorder once and dissociative disorder once. All were under appropriate treatment. Four patients suffered for organic brain disease : cerebral hematoma (due to arterio-venous malformation), cerebral lymphoma and alcoholic intoxication (all 3 under appropriate treatment), the fourth patient suffered of an hydrocephalus which required investigations and treatment. Main parameters for all patients are resumed in table 1.

2. Psychological Evaluation

In consideration of the low size and the heterogeneity of the two "other" categories, we furnish detailed results for the three main diagnosis only with the addition of "possible" AAMI in respect to the potential interest of this discussed category. The mean score of Zung self-rating scales, two tests of verbal fluency and the 3 tests of secondary memory of WMS are detailed in table 2.

Table 2. Mood and psychological evaluation (a).

	Dem	Dep	AAMI	possible AAMI
ZD score	37.4 (28-44)	55.6 (51-63)	35.3 (23-47	43.0 (40-47)
ZA score	32.8 (24-52)	49.8 (37-61)	33.6 (23-54)	34.0 (30-37)
Verbal (b) fluency (1)	12.0 (6-18)	17.3 (9-24)	18.8 (11-29)	22.5 (19-28)
(2)	7.4 (1-14)	9.8 (3-18)	15.4 (4-29)	10.8 (7-14)
WMS Visual	5.3 (3-7)	9.2 (2-13)	7.6 (4-9)	13.8 (12-15)
Associate L.	5.4 (1.5-9.5)	16.3 (9.5-21)	14.2 (8.5-20)	19.6 (19-20.5)
Logical M.	5.7 (2.5-11.5)	10.5 (6.5-16.5)	8.7 (3.5-15.5)	13.9 (10-18)

(a) : data are expressed in mean score with extreme values in brackets
(b)1: Set test (animals) 2: Controlled oral word association, first letter.

3. Memory Complaints

The first pattern of the memory-failure list consists of 28 examples of difficulties which the patient have to score between 1 (none or rarely) to 4 (nearly all the time). Much of these troubles are scored as major complaints by a great majority of patients whatever the etiology of the disorders. We present some results concerning the complaints which show a tendency to vary according to the etiologies (table 3). Moreover we were interested in the ratio of personal complaint to relatives complaint.

DISCUSSION

The first memory clinics (2) were founded in order to detect early dementia. From that standpoint, it seems to be a success. We have seen 11 dementia, of different origin, but all living in the community and benefiting by precocious relations with familial associations. The frequency of familial rather than personal complaint reflects the need of early psychological support for relatives.

On another hand none of the depressed patients were diagnosed as an affective disorder. The clinic was of great benefit for them ! The nature of their memory complaints seems peculiar. A better assessment of these could be of great help for the understanding of mechanisms of memory.

At least, the existence of AAMI is warmly discussed. However, we established that it does exist some patients with a cluster of memory troubles in the everyday life and in some secondary memory tests, without evidence of begining dementia. It is of interest to point that it can exist discrepancies between the 3 secondary memory tests from WMS. Thus an only test issued from WMS cannot furnish sufficient accuracy for this diagnosis (1). We look forward to the evolution of these patients, especially our "possible" AAMI category.

Table 3. Main discriminating memory complaints.

	Dem	Dep	AAMI	possible AAMI
Main complainant				
Patient/Relative	3/8	10/1	11/0	1/3
Troubles in remembering (a)				
. Telephone numbers	2.3	1.5	2.7	1.0
. The day in the week	1.5	1.0	0.8	0
. Names of friends/relatives	1.0	2.0	0.8	1.0
. Train of cooking	0.5	1.3	0.3	0.5
Difficulties in recalling (a)				
. What one's said just before	4.0	3.0	0.6	1.0
. To call up again when promised	2.3	1.5	0.7	0.3
. Where are everyday objects	1.0	4.0	2.7	1.0
. The last television emission	1.0	2.3	1.2	1.0
. What one's asked just before	1.0	0.8	0.6	0.5
Forgetting (a)				
. To fulfil important papers	3.0	0.4	0	0
. Ingredients when cooking	1.0	0.3	0.1	0.5

(a) : data are expressed in ratio : major complaint (most or all the time) to minor complaint (none or a little of the time).

REFERENCES

1. T. Crook, R. T. Bartus, S. H. Ferris, P. Whitehouse, G. D. Cohen and S. Gershon, Age-Associated Memory Impairment : proposed diagnostic criteria and measures of clinical change-Report of National Institute of Mental Health Work Group, Dev. Neuropsychol., 2 : 261-276 (1986).
2. T. J. M. Van der Cammen, J. M. Simpson, R. M. Fraser, A. S. Preker and A. N. Exton-Smith, The memory clinic. A new approach to the detection of dementia, Br. J. Psychiatry, 150 : 359-364 (1987).
3. W. W. K. Zung, A self-assessing depression scale, Arch. Gen. Psychiatry, 12 : 63-70 (1965).
4. W. W. K. Zung, A self-rating instrument for anxiety disorders, Psychosomatics, 12 : 371-379 (1971).
5. D. M. Mc Nair and R. J. Kahn, Self-assessment of cognitive deficits, in : Assessment in geriatric pharmacology, T. Crook et al., ed., Mark Powley, New Canaan (Connecticut), pp : 137-143 (1984).
6. M. F. Folstein, S. E. Folstein and P. R. Mc Hugh, Mini Mental State. A practical method for grading the cognitive state for the clinician, Psychiatric Res., 12 : 189-198 (1975).
7. American Psychiatric Association, The diagnostic and statistical manual of mental disorders, third edition, revised, Washington (DC) (1987).

VISUAL COGNITIVE DISTURBANCE IN ALZHEIMER'S DISEASE —A STUDY BY MEANS OF VISION ANALYZER AS A NEW METHOD FOR EYE MOVEMENT

Mitsuru Fujii,[1] Ryo Fukatsu,[1] Yuji Aizawa,[1] Naohiko Takahata,[1]
Mitsuho Yamada,[2] Tadahiko Fukuda,[2] and Shinji Murakami[3]

[1]Department of Neuropsychiatry, Sapporo Medical College
Sapporo; [2]NHK Science and Technical Research Laboratories
Tokyo; and [3]Sapporo Yamanoue Hospital, Sapporo, Japan

INTRODUCTION

Dementia has been called the "disease of the century". And Alzheimer's disease (AD) is one of the most common forms of dementia. Neuropsychological dysfunction, such as constructional apraxia, spatial agnosia, and dysgraphia, are observed frequently in the early stage of AD. We assumed that visual cognitive dysfunction may be involved in the pathophysiology of these symptoms. Therefore, we investigated the visual cognitive function of patients with AD by means of newly developed technology, the vision analyzer.

SUBJECTS

We chose 18 right-handed cases as shown in Table 1. The evaluation of patients included history; physical, neurological, and neuropsychological examinations; computed tomographic scan, magnetic resonance imaging, electroencephalogram, single photon emission computed tomographic scan, chemical screening, blood count, and thyroid function. All the patients in the early stage of AD(Sjögren's stage II) showed some language difficulties, dysgraphia, constructional apraxia, and some gnosic disturbances in the neuropsychological examination.[1,2] General neurological examination was negative especially with respect to motor and sensory functions, the Hachinski ischemic scores were below 4, and the results were normal in the neuroophthalmological examination. No visual field defect nor nystagmus could be confirmed. Total IQ's of Wechsler Adult Intelligence Scale of each subject, including the 5 patients with AD, were 66-88.

METHOD

To detect the gazing point of focus, we used a vision analyzer. A block diagram of the device is shown in Fig.1. Photo transistors were set up on an eye camera, and eye movements detected as a combination of horizontal and vertical movements. The detector used is an eye-movement monitor manufactured by Takei Co. For data processing purposes, this detector was connected to a microcomputer, so that we could obtain the fixed and motion components of the gazing point.[3] Using this vision analyzer, we carried out 5 experiments. They were divided into two groups: the first was mainly for the lower oculomotor functions; and the second, for the

Basic, Clinical, and Therapeutic Aspects of Alzheimer's and Parkinson's Diseases, Volume 2
Edited by T. Nagatsu *et al.*, Plenum Press, New York, 1990

Table 1. Subjects

Case 1.	J. M.	26 M	Healthy control, young	
Case 2.	Y. A.	33 M	〃	
Case 3.	T. T.	56 M	Healthy control, aged	
Case 4.	H. E.	68 F	〃	
Case 5.	H. N.	69 F	〃	
Case 6.	M.F.	58 M	Multi infarct dementia (MID)	rt. paresis
Case 7.	E. F.	55 M	〃	paresis, dysarthria
Case 8.	H. O.	45 F	〃	cheiro-oral syndrome
Case 9.	S. M.	72 F	〃	(parietal infarction)
Case 10.	T. S.	92 F	〃	
Case 11.	S. H.	54 F	Parkinson's disease	I～II (Yahr's grade)
Case 12.	S. K.	77 M	〃	II (〃)
Case 13.	S. M.	58 M	〃	IV (〃)
Case 14.	K. O.	60 M	Alzheimer's disease	II (Sjögren's grade)
Case 15.	Y. S.	58 F	〃	II (〃)
Case 16.	A. K.	60 F	〃	II (〃)
Case 17.	T. Y.	55 M	〃	II (〃)
Case 18.	T. U.	60 M	〃	II (〃)

higher oculo-motor functions. The five experiments were as follows:
Experiment I: for the lower oculomotor functions such as 1) fixation, 2) saccadic eye movement, and 3) vestibulo-oculomotor reflex.
Experiment II: for the higher oculomotor functions such as 1) tracking a circle and following a line, and 2) copying and overwriting figures. For Exp. I-1 each subject was asked to look at a circle (1.3-degree radius) in the center of a 29-inch TV monitor with a random dot moving right and left for a period of 15 seconds at a speed of 3.7 degrees per second. We evaluated the distribution and the number of the gazing points during a 10-second. In Exp. I-2, each subject was instructed to look at the lamp (1.5V) lit among a group of 9 lamps set up horizontally on the TV monitor. We evaluated the positions of the gazing points and the velocities of movements of each eye. In Exp. I-3, each subject on a revolving chair was instructed to look at a circle with a 1.3-degree radius on the TV monitor. The speed of revolution was from 1.8 to 12.5 degrees per second. In experiment I, we used a 29-inch TV monitor for visual stimuli. The stimuli were stored beforehand on video tapes. Viewing distance from the eyes of the subjects to the TV monitor was about 1.3 m. The experimental room was semidark. Head movements were minimized by chin rest and head supports.

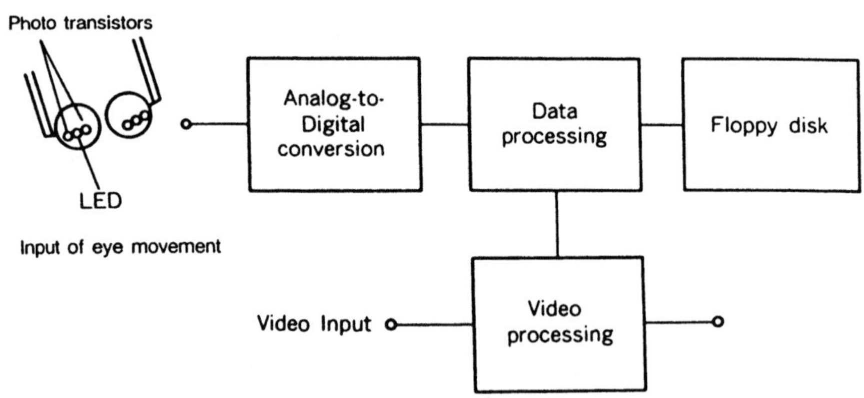

Fig. 1. Vision analyzer. The detector output was first analog-to-digital converted, and then divided into gazing points and motion components of the viewing points. These data were stored on a floppy disk. The entire system is referred to as the vision analyzer.

Experiment II involved examinations for the higher oculomotor functions. In Exp. II-1), each subject was instructed to find and track the circles and was instructed to follow the line from the left arrow to the right arrow of the geometrical figure with his eyes. We evaluated the locations, gazing times, and the number of gazing point. In Exp. II-2), each subject was instructed to copy on the right side of the screen triangle, square, cross, cube, flower, and house from the original figures on the left side. Also each subject was instructed to overwrite the figures with a pencil. In this experiment, a micro camera was put on their foreheads, and viewing distance from eye to the piece of paper was about 30 cm. The calibration was made before and during experiments, as necessary.

RESULTS AND DISCUSSION

1. Exp. I-1: Fixation. In young healthy controls, gazing points were almost always seen on the circle, but in the elderly groups they were seen out of the target. Although the increased number of short saccades and the variation in gazing times were notable, few significant differences between the patients with AD and other elderly groups were seen. These results may suggest that in the patients with AD, the function to look at the object in the center of the visual field is retained and also the comprehension of the meaning of instructions is good. 2, Exp. I-2: Saccadic eye movement. In the controls, gazing points moved to the lit lamp rapidly and correctly, although the number of correcting short saccades increased in the elderly. In the patients with AD, saccadic reaction times were extremely longer than those of the other subjects, and gazing points had a tendency to return to the center lamp. Four patients with AD were unable to fix on the end lamp on the left side of the screen, and no gazing point was seen on the lamp. These findings were characteristic in the patients with AD and might suggest the presence of neglect syndrome, which was not obvious clinically. 3, Exp. I-3: VOR. In the young healthy controls, the gazing points moved reflectively to the fixed target. On the other hand, the gazing points of the elderly subjects showed difficulty in reaching the target when the target approached the edge of their visual field. But in the patients with AD, the vestibulo-oculomotor reflex was well retained. These results obtained in our these experiments indicate that the lower oculomotor functions may be well preserved in patients with AD in the early stage. 4, Exp, II-1: Tracking the circle and following the line. Gazing points of the elderly healthy controls as well as of the young healthy controls covered all over the figure and hit all of the circle when the subjects tracked the circles. The strategy, that is to say, the intention to minimize their energy to search for the circles was recognizable in their eye movements. But the gazing points in the patients with AD increased very much and showed a tendency to be localized on a part of the whole figure. And also gazing times were prolonged. The same results were seen when the patients with AD were instructed to follow the line. In the patients with AD, the number of gazing point increased at the turning points to the vertical direction, and many short saccadic eye movements were seen. 5, Exp, II-2): Copying and overwriting the figure. In young and elderly healthy controls, the gazing points were distributed on the models and on the copied figures equally and went back and forth from the models to the copied figures. However, the number of the gazing points of the patients with AD increased, and the gazing points had a tendency to be localized away from the copied and original figures. The "closing in" phenomenon, which is a well known phenomenon in AD patients, was seen frequently during this experiment. The patients with AD failed to gaze back and forth from the model to the copied figures when they were copying during this phenomenon, and there was no gazing point on the model (Fig, 2). On the other hand, the patients with AD could overwrite the figures, although the number of gazing point was increased when they instructed to overwrite the figures twice with a pencil.

The results of Exp. II-1,2 are similar to "Seerenlahmung des Shauens"[4] or "locking of fixation"[5] and "visual inattention" of Bálint's syndrome.[2,4] Common higher brain mechanisms may be involved in AD patients, even in the early stage,

Alzheimer's disease T.Y.

Fig. 2. Copying a flower. " Closing in" phenomenon was seen. There was no gazing point of focus on the model.

and in the patients with Bàlint's syndrome. We assume that these findings might be related to the praxic and gnostic disorders in the patients with AD.

SUMMARY

The visual information processing in the patients with early onset AD was evaluated in an early stage of the disease by means of a vision analyzer. The results are summarized as follows:

1) Lower oculomotor functions, such as fixation, reflective saccade, and VOR were almost intact.

2) But some characteristic findings were noted from examination of the higher oculomotor functions. These findings in AD patients, which were obtained objectively for the first time by means of our vision analyzer, were similar to those of Bálint's syndrome. The objective assessment of higher brain functions by use of this new method, i. e., the vision analyzer, might be useful for better understanding of the pathomechanism of the dementia seen in AD.

REFERENCES

1. M. Fujii, R. Fukatsu, N. Takahata, T. Takahashi, and K. Morita, Neuropsychological evaluation and MRI in Alzheimer's disease and multi-infarct dementia, Seishin Shinkeigaku Zassi 89(9) : 713 (1987).
2. M. Fujii, R. Fukatsu, N. Takahata, M.Yamada, T. Fukuda, S.Murakami A neuropsychological study in Alzheimer's disease -analysis for the visual cognitive processing by using vision analyzer-, Seishin Shinkeigaku Zassi 89 (10) : 760 (1989).
3. M. Yamada and T. Fukuda, Quantitative evaluation of eye movements, SMPTE Journal 95(12) : 1230 (1986).
4. R. Bálint, Seelenlähmung des "Schauens", optische Ataxie, räumliche Störung der Aufmerksamkeit, Monatschr. Psychiat. Neurol. 25:51 (1909).
5. H. Hécaen and J. de Ajuriaguerra, Bálint syndrome (psychic paralysis of visual fixation) and its minor forms, Brain LXXVII : 373 (1954).

ELECTROPHYSIOLOGICAL COMPARISON BETWEEN PARKINSON'S DISEASE AND ALZHEIMER'S DISEASE

Hisao Tachibana, Masanaka Takeda, Makoto Izuta, Kyoko Takeda, Tomohiro Kaku, Masato Tsuchiyama and Minoru Sugita

The Fifth Department of Internal Medicine
Hyogo College of Medicine, Nishinomiya, JAPAN

INTRODUCTION

Parkinsonism appears frequently in patients with Alzheimer's disease. Dementia, a uniform finding in Alzheimer's disease, is commonly present in patients with Parkinson's disease. Some authors[1] have attributed the dementia of Parkinson's disease to concomitant Alzheimer's disease. However, Alzheimer's disease and demented Parkinson's disease have many neuropathological differences in the location of lesions and in the degree of neural degeneration[2].

The purpose of the present study was to investigate the short-latency somatosensory evoked potentials (SSEPs) and brainstem auditory evoked potentials (BAEPs) in patients with Alzheimer's disease or Parkinson's disease with dementia and to evaluate the electrophysiological differences in response between these two demented groups.

PATIENTS AND METHODS

The study was performed on 14 patients with Parkinson's disease with dementia, 16 patients with Alzheimer's disease, and 25 age-matched normal subjects without any history of neurological disease or of drug abuse. In addition, patients with Parkinson's disease had at least two of its cardinal features (bradykinesia, tremor, and rigidity), each of which had developed insidiously. Patients had some of the signs peculiar to Parkinson's disease, involving masked face, stooped posture, pulsion phenomenon, and Myerson's sign. Furthermore, patients were excluded who had Parkinsonism that was due to other causes or was part of a more widespread neurological illness. Patients with Alzheimer's disease met the criteria for primary degenerative dementia according to the Diagnostic and Statistical Manual of Mental Disorders of the American Psychiatric Association (DSM-III-R) and the National Institute of Neurological and Communication Disorders and Stroke-Alzheimer's Disease and Related Disorders Association (NINCDS-ADRDA) criteria for probable Alzheimer's disease. All had a progressive, nonstepwise, clinical course, and no other causes for their altered mental functions were revealed by laboratory evaluation. In addition, all had EEGs that were normal or demonstrated mild to moderate slowing, and all had CT scans that demonstrated mild to moderate cortical atrophy. The Alzheimer group contained only patients with ischemic scores of 3 or less. Both Parkinson's and Alzheimer's paitents met the criteria for dementia assigned in DSM-III-R.

Basic, Clinical, and Therapeutic Aspects of Alzheimer's and Parkinson's Diseases, Volume 2
Edited by T. Nagatsu *et al.*, Plenum Press, New York, 1990

Dementia was further classified as mild, moderate, or severe according to DSM-III-R criteria. Patients data are summarized in Table 1. There were no significant differences between the Parkinson's disease and the Alzheimer's disease groups in patient age, duration of dementia, or severity of dementia.

All recordings of SSEPs and BAEPs were made with a Neuropack II MEB-5100 (Nihon-Kohden). SSEPs were simultaneously recorded from the second cervical vertebra (C_2) and the contralateral primary somatosensory cortex (7 cm lateral to, and 2 cm posterior from the vertex) with a midfrontal reference (Fpz). The median nerve at the wrist was stimulated at a rate of 5 Hz and a duration of 0.2 ms with bipolar surface electrodes (cathode proximal). The stimulation intensity was adjusted to produce a small thumb twitch and was set at the same intensity bilaterally for each patient. The responses were amplified 100,000 times with filter settings of 20-3000 Hz. Averaged potentials obtained from 256 pulses were obtained. The values analyzed were peak latencies (N13, N20) and the central conduction time (CCT, the interpeak latency between N13 and N20). BAEPs were recorded between a vertex (Cz) reference and linked ear lobes (A_1, A_2) using click stimuli delivered at a rate of 10 Hz and 0.1 ms pulse duration monoaurally via earphones at 60 dB above hearing threshold. Masking white noise (-20 dB) was delivered to the unstimulated ear. The responses were amplified 100,000 times with filter settings of 100-3000 Hz. Averaged potentials to 2048 clicks were obtained. The latencies of the most stable components (waves I, III and V) and interpeak latencies (IPLs, waves I-V, I-III, and III-V) were studied. Mean values of SSEPs or BAEPs to left and right side stimulation were used for comparisons among groups. The results obtained were expressed as means and standard deviations (SD). All statistical procedures were performed using Student's t test or analysis of variance.

RESULTS

Table 2 shows the peak latencies and the interpeak latencies of SSEPs in patients with demented Parkinson's disease, patients with Alzheimer's disease and normal subjects. As indicated, there were no significant differences in peak latencies or CCTs between demented Parkinson's disease patients, those with Alzheimer's disease and normal subjects. Table 3 presents group means and SDs of BAEP peak latencies and IPLs. The mean peak latency of the V wave in patients with Alzheimer's disease were significantly increased than that of normal subjects ($p<0.05$). However, there was no significant difference in peak latency between patients with demented Parkinson's disease and normal subjects. Both patients with Parkinson's

Table 1. Case Series

	Parkinson's Disease with Dementia	Alzheimer's Disease	Normals
Number of Cases	14	16	25
Age (years)	71.1± 6.3	70.6± 7.0	66.6± 9.5
Duration of Dementia (months)	15.0± 10.3 (69.1± 51.5)*	13.8± 7.8	
Severity of Dementia			
mild	2	3	
moderate	6	4	
severe	6	9	

* Duration of illness

Table 2. Peak Latencies and Interpeak Latencies of SSEPs in Patients with Parkinson's Disease and Patients with Alzheimer's Disease and Normal Subjects

	Parkinson's Disease with Dementia	Alzheimer's Disease	Normals	F Value
N13	12.70± 0.92	12.85± 1.07	12.52± 0.79	0.12
N20	19.09± 1.03	19.33± 1.31	18.65± 0.92	2.05
CCT	6.39± 0.42	6.48± 0.67	6.13± 0.47	2.55

disease and those with Alzheimer's disease showed significantly increased I-V IPLs compared with normal subjects ($p<0.001$ and $p<0.005$, respectively). Furthermore, both patients with Parkinson's disease and those with Alzheimer's disease had significantly prolonged III-V IPLs ($p<0.01$ and $p<0.005$, respectively) although there were no significant differences in I-III IPLs between these three groups. The Alzheimer's and Parkinson's groups did not differ with regard to IPLs and peak latencies.

DISCUSSION

The present study showed normal peak latencies and CCTs in patients with demented Parkinson's disease or Alzheimer's disease. These patients might have been expected to have had alterations in potentials, as they showed mild to moderate changes in cortical electrical activity as measured by EEG. Our results, though, accorded with those in previous reports[3,4]: neither Parkinson's disease with dementia nor Alzheimer's disease appear to modify scalp-recorded SSEPs significantly.

However, both patients with demented Parkinson's disease and those with Alzheimer's disease had significant prolongation of I-V IPLs and III-V IPLs. These results suggest that their brainstem lesions were located in the auditory pathways. There are conflicting reports regarding BAEPs in Parkinson's disease and Alzheimer's disease[4-7]. Although some authors have reported normal BAEPs in patients with Parkinson's disease[5] or Alzheimer's disease[6], others have reported prolonged peak latencies or IPLs[7,8]. These inconsistent results may be due partly to differences in severity or

Table 3. Peak Latencies and Interpeak Latencies of BAEPs in Patients with Parkinson's Disease and Patients with Alzheimer's Disease and Normal Subjects

	Parkinson's Disease with Dementia	Alzheimer's Disease	Normals	F Value
I	1.75± 0.19	1.76± 0.10	1.80± 0.18	0.62
III	3.97± 0.26	3.97± 0.19	3.95± 0.20	0.04
V	5.97± 0.27	6.01± 0.28	5.82± 0.20	3.38*
I-III IPL	2.22± 0.22	2.21± 0.16	2.15± 0.15	0.93
III-V IPL	2.00± 0.15	2.05± 0.18	1.87± 0.11	8.30****
I - V IPL	4.22± 0.16	4.26± 0.27	4.02± 0.16	8.95****

* p< .05 **** p< .001

duration of disease. Another possible explanation is that Parkinson's disease[9] and Alzheimer's disease[10] are heterogeneous syndromes composed of several clinical subtypes. There is some evidence for the impairment of brainstem structures and systems other than the nigralstriatal pathways known to be impaired in Parkinson's disease[11]. Similarly, some brainstem or midbrain structures, such as locus coeruleus, the dorsal raphe nuclei, and pontine nuclei, are also known to be abnormal in patients with Alzheimer's disease[12]. These evidence may be consistent with our present demonstration of a significant prolongation of III-V IPLs because III-V IPL is considered to be the auditory transmission time between superior olivary nucleus and inferior colliculus. On the basis of these reports, the prolonged I-V IPLs (particularly III-V IPLs) occurring in patients with Parkinson's disease with dementia or Alzheimer's disease obtained in this study might be explained by a more general involvement of these disease proceses in brainstem pathway impairment and/or dysfunctions in some neurotransmitter systems.

The present study did not demonstrate any significant differences in BAEPs and SSEPs between patients with Alzheimer's disease and those with Parkinson's disease with dementia, although the two diseases had affected cognitive function differently in our series of patients. We therefore, conclude that the discrimination between patients with demented Parkinson's disease and those with Alzheimer's disease may be difficult if based upon SSEP and BAEP measurements alone.

REFERENCES

1. F. Boller, T. Mizutani, U. Roessmann, and P. Gambetti, Parkinson's disease, dementia, and Alzheimer's disease: Clinicopathological correlations, Ann. Neurol., 7:329 (1980).
2. S. M. de la Monte, S. E. Wells, E. T. Hedley-Whyte, and J. H. Growdon, Neuropathological distinction between Parkinson's dementia and Parkinson's plus Alzheimer's disease, Ann. Neurol., 26:309 (1989).
3. G. Abbruzzese, L. Reni, L. Cocito, S. Ratto, M. Abbruzzese, and E. Fovare, Short-latency somatosensory evoked potentials in degenerative and vascular dementia, J. Neurol. Neurosurg. Psychiatry, 47:1034 (1984).
4. H. Tachibana, M. Takeda, and M. Sugita, Short-latency somatosensory and brainstem auditory evoked potentials in patients with Parkinson's disease, Int. J. Neurosci., 44:321 (1989).
5. K. Chiappa, "Evoked potentials in clinical medicine", Raven Press, New York (1983).
6. A. M. Grimes, C. L. Grady, and A. Pikus, Auditory evoked potentials in patients with dementia of the Alzheimer type, Ear Hearing, 8:157 (1987).
7. S. W. Harkins, Effects of presenile dementia of the Alzheimer's type on brainstem transmission time, Int. J. Neurosci., 15:165 (1981).
8. M. Fradis, A. Samet, J. Ben-David, L. Podoshin, B. Sharf, J. Wajsbort, M. Zellinger, and H. Pratt, Brainstem auditory evoked potentials to different stimulus rates in Parkinsonian patients, Eur. Neurol., 28:181 (1988).
9. J. A. Mortimer, F. J. Pirozzolo, E. C. Hansch, and D. D. Webster, Relationship of motor symptoms to intellectual deficits in Parkinson's disease, Neurology, 32:133 (1982).
10. H. C. Chui, E. L. Teng, V. W. Henderson, and A. C. Moy, Clinical subtypes of dementia of the Alzheimer type, Neurology, 35:544 (1985).
11. K. Jellinger, The pedunculopontine nucleus in Parkinson's disease, progressive supranuclear palsy and Alzheimer's disease, J. Neurol. Neurosurg. Psychiatry, 51:540 (1988).
12. E. Iseki, M. Matsushita, K. Kosaka, H. Kondo, T. Ishii, and N. Amano, Distribution and morphology of brain stem plaques in Alzheimer's disease, Acta Neuropathol., 78:131 (1989).

A CLINICAL EEG STUDY OF ALZHEIMER-TYPE DEMENTIA

Makoto Uchiyama, Kunihiro Isse, Kuniaki Tanaka, Akinori Kuroda, Hisae Komazaki[1], Makoto Hamamoto[2], Hironaka Igarashi[2], and Tokuzo Miyazaki[2]

Departments of Psychiatry, [1]Clinical Pathology, and [2]Neurology
Tokyo Metropolitan Tama Geriatric Hospital
Higashi-Murayama, Tokyo, Japan

INTRODUCTION

A number of papers have reported EEG alterations in Alzheimer-type dementia (ATD), which included decreased alpha waves, accentuated theta waves, and increased delta waves (Giannitrapani et al.,1988). However, the relationship between clinical symptoms and EEG parameters has not been studied in detail. The difficulty affecting this kind of attempt may exist not in the techniques of EEG analysis, but in the systematization of symptomatology.

Investigators have recently applied radionuclide techniques to ATD patients, such as positron emission tomography (PET), and single photon emission computed tomography (SPECT) to document the distribution of brain pathology in the living patient. Several authors (Fraser,1987; Sharma et al.,1988) have suggested that various symptoms in the different clinical stages could be classified into three groups based on the possible localization of impairments: temporal symptoms, parietal symptoms, and frontal symptoms. Thus recent studies in the clinical symptomatology of ATD have been oriented to define the degree of dementia more accurately in order to represent the brain pathology.

We have tried to evaluate the severity of ATD by establishing a new simple grading scale (Isse et al.,1989). Using this scale, we compared EEG pathology in different stages of ATD with the clinical severity.

SUBJECTS AND METHODS

Thirty inpatients (mean age=75 years) with a clinical diagnosis of ATD were studied. All patients selected in this study were diagnosed by the criteria of the DSM-3R, and coincided with a diagnosis of primary degenerative dementia of Alzheimer type.

The clinical course of ATD is best characterized by progressive deterioration, which starts with memory impairment and ultimately compromises all central nervous system functions. Thus 3 core symptoms, representing the severity of deterioration, were selected based on

Table 1. Tama ATD Scale

Severity	Symptom				
Normal					
Mild	Memory Disturbance				
Moderate	Memory Disturbance	+	Higher Cortical Dysfunctions		
Severe	Memory Disturbance	+	Higher Cortical Dysfunctions	+	Personality Deterioration

clinical experience of the patient's natural history. The first symptom concerned memory disturbance, consisting of recent memory loss, remote memory loss and/or forgetfulness of words. The second symptom was higher cortical dysfunctions, manifested in behavioral disability, for example that in dressing, brushing teeth, or handling of chopsticks, and in disorientation or the degree of receptive dysphasia. The third symptom consisted of personality deterioration, which means that the patients lose their premorbid personality, manifesting animal-like primitive behavioral reactions and becoming totally incoherent. Based on these core symptoms our 4-point grading scale for ATD (Tama ATD Scale) was established (Table 1).

The mild group is characterized by memory disturbance. The moderate group has additional higher cortical dysfunction. The severe group progresses to manifest personality deterioration. According to the severity which was determined with the Tama ATD Scale, the patients were divided into 3 groups: mild (n=10), moderate (n=10), and severe (n=10). Ten healthy aged volunteers (mean age=70 years) served as a control group.

EEG activity from 21 channels (10-20 convention) was simultaneously recorded on paper and magnetic tape. After an inspective judgment by three EEG specialists (M.U., K.I. and K.T.), artifact-free resting records from F3, C3, O1 (ipsilateral ear reference) were quantized using FFT. EEG epochs (1024 points, sampling interval 5 msec.) were average-accumulated 10 times using a Signal Processor (San-Ei, 7T18). The mean values of absolute power for 6 frequency bands (delta, 2.0-3.8Hz; theta 1, 4.0-5.8Hz; theta 2, 6.0-7.8Hz; alpha 1, 8.0-9.8Hz; alpha 2, 10.0-12.8Hz; and beta 1, 13.0-19.8Hz) were compared among the groups using Wilcoxon T test.

RESULTS

1)EEG findings by inspection

Several characteristic findings were inspected in this investigation. Among the mild group (n=10), 6 cases showed scattered theta waves predominantly in the central region, and 1 case showed poor alpha activity. With respect to the moderate group (n=10), scattered theta waves appeared in 7 cases, and poor alpha activity was observed in 8. EEG tracings of all the severe cases (n=10) lacked alpha activity and consisted of scattered theta waves. Nine of the severe group had another characteristic finding of frontal delta waves. Hemispheric asymmetry of EEG activity was not recognized in any group (Table 2).

Table 2. Summary of EEG Findings by Inspection

Severity		EEG Findings: Scattered Theta	Poor Alpha	Random Delta
Mild	(n=10)	6	1	0
Moderate	(n=10)	7	8	1
Severe	(n=10)	10	10	9
Total		23	19	10

2)Quantitative analysis

Compared with the control group, the mild group was characterized by higher theta 1,2 power values in F3, C3, and O1 (Table 3). The moderate group had lower alpha 1 power values than the mild group at O1 (Table 4). As for the severe group, compared with the moderate group it had lower alpha 1,2 power values at O1, and a higher theta 1 power value at F3, and it also had a higher delta power value than the moderate group at F3 and C3 (Table 5).

COMMENTS

Our results can be summarized as follows: 1) Theta 1,2 waves increased in the mild group (vs. the control group), 2) Alpha 1 waves decreased in the moderate group (vs. the mild group) and in the severe

Table 3. Mild Group (vs. Control Group)

	DELTA	THETA1	THETA2	ALPHA1	ALPHA2	BETA1
F3		↑↑	↑			
C3		↑	↑			
01		↑↑	↑			

WILCOXON T TEST (↑ : $p<0.05$, ↑↑ : $p<0.01$)

Table 4. Moderate Group (vs. Mild Group)

	DELTA	THETA1	THETA2	ALPHA1	ALPHA2	BETA1
F3						
C3						
01				↓		

WILCOXON T TEST (↓ : $p<0.05$)

Table 5. Severe Group (vs. Moderate Group)

	DELTA	THETA1	THETA2	ALPHA1	ALPHA2	BETA1
F3	↑↑	↑				
C3	↑					
01				↓	↓	

WILCOXON T TEST (↑ or ↓ : $p<0.05$, ↑↑ : $p<0.01$)

Table 6. Scheme of EEG Alteration in ATD

Severity	EEG Alterations
Normal	
Mild	Theta ↑
Moderate	Theta ↑ + Alpha ↓
Severe	Theta ↑ + Alpha ↓ + Delta ↑

group (vs. the moderate group), 3) Delta waves increased in the severe group (vs. the moderate group). Therefore, with respect to EEG alterations, ATD was considered to have three different processes in accordance with the severity. The findings are schematized in Table 6.

In this study we performed clinical assessment of severity employing the Tama ATD scale, where the severity was represented as an accumulation of three core symptoms (Table 1). Taking these core symptoms into consideration, we speculate the following on the relationship between the core symptoms and the EEG parameters: The increased theta waves in the mild stage may be associated with memory disturbance, the decrease in alpha waves in the moderate stage may correspond to higher cortical dysfunction, and the appearance of delta waves in the severe stage may be related to personality deterioration.

REFERENCES

American Psychiatric Association, 1987, Diagnostic and Statistical Manual of Mental Disorder (DSM-3R), 3rd ed. revised. APA.

Diannitrapani, D. and Collins, J., 1988, EEG Differentiation between Alzheimer's and Non-Alzheimer's Dementias, in: "The EEG of Mental Activities", pp26-41, Karger, Basel.

Fraser, M., 1987, Diagnosis, in: "Dementia: Its Nature and Management", pp96-146, Wiley Medical Publication, London.

Isse, K., Uchiyama, M. and Tanaka,K., 1989, Drug Therapy for Alzheimer Type Dementia (in Japanese), Clinical Neuroscience 7:68.

Sharma, R. and Gierl, B. L., 1988, Dementia, in: " Psychiatry: Diagnosis & Therapy (Lange Clinical Manual)", pp153-165, Appleton & Lange, San Metro.

A CASE OF ALZHEIMERISM INDUCED BY TRIHEXYPHENIDYL

Naofumi Kajimura, Masatomo Suetsugi, Tsuyoshi Imai,
Shusaku Kai, Yasushi Mizuki, and Michio Yamada

Department of Neuropsychiatry, Yamaguchi University School
of Medicine, 1144 Kogushi, Ube 755, Japan

INTRODUCTION

Alzheimer's disease (AD) is a multisymptom disease that may be mediated by multiple neurochemical abnormalities in the brain. Especially, cholinergic impairment is the most consistent and pervasive abnormality, and the degree of the damage correlates closely with severity of dementia. It is well known that anticholinergic agents impair intellectual performance in both normal humans and patients with AD (Collerton, 1986) and that cholinergic neurotransmission is involved in human memory processes (Drachman and Leavitt, 1974). However, whether cognitive functions are mediated cholinergically has not been definitely established. We describe a 55-year-old male patient with Alzheimerism, which was dominated by memory loss and cognitive impairments, induced by trihexyphenidyl medication, an anticholinergic drug, given over a long period.

CASE REPORT

The patient was noted to have essential tremor at 18 years of age, but he was given no treatment. He was diagnosed as having parkinsonism at 31 years old, and was thereafter administered 20 mg/day of trihexyphenidyl for 23 years. He began to complain of memory loss when he reached 54 years old and was admitted to our hospital at age 55. He was ordered to cancel the drugs, and several psychometric tests were performed to evaluate his mental state: Wechsler Adult Intelligence Scale, Hasegawa's Dementia Rating Scale, Mini-Mental State, Benton's Visual Retention Test, and Memory Test (related antonyms). The results of these tests revealed coexistence of visuospatial agnosia and constructive and ideational apraxia with memory disturbance. CT of his head showed mild cortical atrophy, and his resting EEG's exhibited slow and irregular patterns. His symptoms were not improved for about two months, and organic dementia was suspected. However, we discovered that he had been secretly taking the same doses of trihexyphenidyl as in the past, and complete withdrawal of the drugs was done with his consent. Consequently, both cognitive impairments and memory disturbances were rapidly ameliorated, and the scores of all psychometric tests improved markedly (Table 1). The resting EEG's also showed improvement, and α rhythm began to appear after the withdrawal of the drugs. Monoamines and their metabolites in plasma (DA, NA, 5-HT, HVA, MHPG, and 5-HIAA) were measured several times during the admission; however, remarkable changes were not observed.

Table 1. Scores of Psychometric Tests before and after the Withdrawal of Trihexyphenidyl

Psychometric Tests	Before	After
Hasegawa's Dementia Rating Scale (32.5)	24.5	32.5
Mini-Mental State (30)	25	30
Memory Test <related antonyms> (10)	5.5	9.5
Wechsler Adult Intelligence Scale		
Total IQ	71	107
Benton's Visual Retention Test		
Correct (10)	2	8
Error	17	3

The scores of psychometric tests before and after the withdrawal of trihexyphenidyl are shown. Each test was administered right after the admission and before the discharge.

DISCUSSION

It is well known that administration of anticholinergic drugs (e.g., atropine, scopolamine, trihexyphenidyl) leads to deficits in recent memory (Collerton, 1986). However, it has not been clearly elucidated whether these drugs contribute directly to cognitive impairments such as visuospatial agnosia. In our case, the patient had been taking overdoses of trihexyphenidyl for a long period, which resulted in Alzheimerism. Thus, we infer that anticholinergic agents can induce not only memory disturbances but also cognitive impairments and that cholinergic neurotransmission might be concerned with these functions. Mesulam et al. (1983) pointed out that the effects of anticholinergic drugs on learning and recent memory are probably mediated by cholinergic neurons in the basal forebrain, particularly those arising from the medial septal nuclei or the diagonal band of Broca and projecting to the hippocampus. In addition, Nukina et al. (1984) reported an increased number of muscarinic receptors in the hippocampus, but no changes in the striatum or cerebral cortex, when rats were given chronic intraperitoneal medication of trihexyphenidyl. Therefore, both memory disturbances and cognitive impairments might be caused by the lesion in the basal forebrain innervating the hippocampus. By the withdrawal of the drugs, the symptoms of the patient were rapidly ameliorated, and the resting EEG's showed remarkable improvement, while the concentrations of the monoamines and their metabolites in plasma were not changed. These results suggest that anticholinergic drugs might produce cognitive impairments and that cholinergic systems in the basal forebrain might be responsible for these dysfunctions.

REFERENCES

Collerton, D., 1986, Cholinergic function and intellectual decline in Alzheimer's disease, Neuroscience, 19: 1.

Drachman, D. A., and Leavitt, J., 1974, Human memory and the cholinergic system, Arch Neurol., 30: 113.

Mesulam, M. M., Mufson, E. J., Levey, A. I., and Wainer, B. H., 1983, Cholinergic innervation of cortex by the basal forebrain: Cytochemistry and cortical connections of the septal area, diagonal band nucleus, nucleus basalis (substantia innominata), and hypothalamus in the rhesus monkey, J Comp Neurol., 214: 170.

Nukina, I., Ogawa, N., Kuroda, H., and Ota, Z., 1984, Altered muscarinic acetylcholine receptors in rat brain following chronic trihexyphenidyl treatment, Jpn J Neuropsychopharmacology., 6: 57. (in Japanese)

PROBABLE ALZHEIMER'S DISEASE WITH APPEARANCE OF PERIODIC SYNCHRONOUS DISCHARGES FOR A LONG PERIOD

Yasushi Mizuki, Yuichiro Furukawa, Masatomo Suetsugi, Shusaku Kai, Naofumi Kajimura, and Michio Yamada

Department of Neuropsychiatry, Yamaguchi University School of Medicine, 1144 Kogushi, Ube 755, Japan

INTRODUCTION

Periodic synchronous discharge (PSD) seen in Creutzfeld-Jakob disease (CJD) is a useful index for discrimination between Alzheimer's disease (AD) and CJD. Recently, it has been reported that PSD also appears in AD; however, its appearance in this disease is limited to a short period within one year. We describe herein a female patient with probable AD who displayed PSD for more than 3 years and in whom a close relationship was seen between its appearance and brain atrophy.

CASE REPORT

There were no noteworthy findings in the patient's personal and family history. She graduated from high school with an average record, was married at 23 years of age, and had three children. At 53 years of age, her family noticed that she was frequently forgetful and depressive. In her 55th year, she suddenly complained to the family, "Someone is here and is setting fire to my house". Her lapses of memory gradually intensified and at age 57 she began to forget her daily chores. The patient became unable to leave the house by herself, and she was admitted to the hospital at 60 years of age. At admission she presented with disorientation, poor memory, loss of calculating ability, and disorders of attentiveness and understanding, in addition to a wandering behavior and euphoria. Laboratory examinations performed at admission, such as hematological and electrolytic examinations, lipid and vitamin determinations, liver function tests, kidney tests, endocrine tests, urinalysis, and Wassermann's reaction test, did not reveal any abnormalities in the patient. Laboratory examinations using cerebrospinal fluid also showed no abnormalities. After admission, her symptoms gradually intensified and at 61 years of age she became bedridden. In November of the next year (62 years of age) she suddenly underwent epileptic seizures with generalized convulsion. Thereafter, she rapidly lost her mental ability and she began to show muscle rigidity, Babinski's sign, myoclonus and akinetic mutism in July of the next year (63 years of age).

EEG AND CT FINDINGS

At admission, a CT scan of the patient's head revealed an enlargement

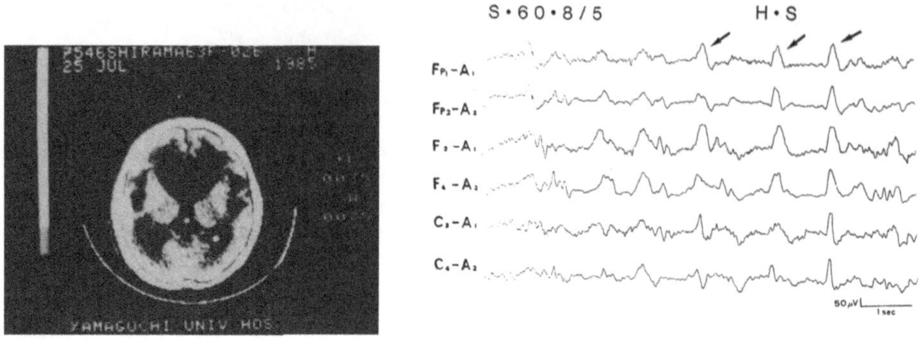

Figure 1. EEG's (right side) and a CT scan (left side) of the patient at 60 years of age. The arrows indicate PSD's.

of the cerebral ventricles and a widening of the cerebral sulci. A visual analysis indicated the presence of a moderate cerebral atrophy. In her EEG's, slow-wave activities with a low amplitude were observed in all regions. In August of the 63th year (9 months after epileptic seizures), PSD with an amplitude of 50 to 200 μV and a periodicity of 1 sec appeared as shown in the Figure. At the same time, a CT scan showed severe cerebral atrophy. PSD's were observed in all regions at the initial period; however, at age 64 they began to appear predominantly in the frontal and central areas and to show a higher amplitude in the right hemisphere. The continuity of the background activities was relatively intact, but the regularity was more damaged in the right hemisphere. At 67 years of age, the configuration and periodicity of PSD's and the laterality of background activities became unclear. At the same period, another CT scan showed a slight progression of cerebral atrophy.

DISCUSSION

According to the criteria of McKhann et al. (1984), the patient was diagnosed as having probable AD from her clinical course and symptoms, neurological signs, EEG and CT findings, and results of the laboratory examinations. In general, PSD in CJD appears at the initial period of the disease and often shows laterality (Au et al., 1980). The laterality of PSD disappears as CJD progresses, and then its configuration becomes irregular and unclear and the background activities are flattened. On the other hand, PSD's in this patient had some different characteristics; that is, the laterality appeared as the disease developed, the configuration was duller than that in CJD, and the background activities still remained even when PSD's became unclear. These results suggest that the nature of PSD in AD is different from that in CJD. As for the generating mechanism of PSD, Gloor et al. (1968) speculated that PSD appears when both the cerebral cortex and the gray-white matter in subcortical regions such as thalamus and brainstem are strongly impaired. Therefore, we assume that PSD's in this patient were caused by impairment in these regions. The laterality probably appeared due to the differential speed of damage development between the right and the left hemisphere, but the reason for its configuration and periodicity becoming unclear at the final stage of the disease is not clear.

REFERENCES

Au, W. J., Gabor, A. J., Vijayan, N., and Markand, O. N., 1980, Periodic lateralized epileptiform discharges (PLEDs) in Creutzfeld-Jakob disease, Neurology, 30: 611.

Gloor, P., Kalabay, O., and Giard, N., 1968, The electroencephalogram in diffuse encephalopathies: electroencephalographic correlates of grey and white matter lesions, Brain, 91: 779.

McKhann, G., Drachman, D., Folstein, M., Katzman, R., Price, D., and Stadlan, E. M., 1984, Clinical diagnosis of Alzheimer's disease, Neurology, 34: 939.

INCIDENCE AND PREVALENCE OF ALZHEIMER-TYPE DEMENTIA: EPIDEMIOLOGICAL INVESTIGATIONS IN 1987, 1988, AND 1989

Isao Fukunishi, Taketoshi Utsumi, Toshiyuki Hayabara, and Kiyoshi Hosokawa

Department of Neuropsychiatry, Kagawa Medical School
1750−1 Miki-cho, Kita-gun, Kagawa 761−07, Japan

INTRODUCTION

Epidemiological investigations of senile dementia have often been reported in various countries.[1-2] In most cases, however, the studies were usually only concerned with calculations of the morbidity prevalence rate of senile dementia, while the incidence rate has not been researched satisfactorily. According to reports on the prevalence rate, senile dementia was investigated on a large scale, that is, the subjects examined numbered occasionally greater than 1,000. However, all the past reports[3-6] on the incidence rate were made based on a subject population in subjects of less than 1,000. When the number of subjects is 1,000 or less, the occurrence of senile dementia is approximately 10 cases per year. If the number of patients with senile dementia increases or decreases even by one, the incidence differs to a great extent. In respect to incidence research, therefore, it is desirable that the subjects should be more than 1,000 in number.

In the present study, an epidemiological investigation was made on all aged persons at home (more than 3,000 persos) in Miki Town, Kagawa Prefecture, Japan, for 3 years (1987, 1988, and 1989). The incidence research in addition to the morbidity prevalence rate of senile dementia were examined on the basis of the above-mentioned preconditions.

MATERIALS AND METHODS

The subjects all consisted of the aged at home (3,754 persons 65 years of age or older) in Miki Town, Kagawa Prefecture, Japan, as of June 1, 1987. Of these 3,754 subjects, 1,569 were males and 2,185 were females. At the time of investigation, the total population in Miki Town was given by census at 26,801.

The reasons why Miki Town was selected as the district for epidemiological investigation were as follws: (1) The composition of the population in Miki Town is typical of many towns in Kagawa Prefecture and Japan in general. Thus, it was possible to carry out the investigation reflecting the situation in Kagawa Prefecture as being representative of Japan at large. (2) Miki Town has the largest population among the towns or villages in Kagawa Prefecture. (3) Statistically, urban areas, rural areas, and semi urban/rural areas are distributed evenly in relation to the size in Miki Town. (4) Owing to comprehensive district medical care and health guidance, it was possible to carry out an accurate epidemiological investigation on all the aged at home.

A two-step selection procedure consisting of primary and secondary investigations was used for all three years, 1987–89. In the primary investigation, four well-experienced neuropsychiatrists visited the aged persons directly, making individual diagnostic interviews. The cases of suspected dementia were then selecled for secondary investigation in more detail. In the secondary investigation, as in the primary investigation, the neuropsychiatrists visited the aged persons individually, making neurological and psychiatric examinations. Further, the examinations of GBS scale developed by Gottfries, C.G., Brane, G. and Steen, G. in 1980,[6] Hasegawa's dementia scale, and Hachinski's ischemic score were further used to determine if the cases of suspected dementia in the primary investigation were in fact demen tia or not. The total duration of examination was 60 to 90 minutes per case. The period lasted two months, from June to August.

The criteria for diagnosis of senile dementia and its severity classification (mild, moderate, and severe) were made according to DSM-III-R. The types of dementia were also classified into 3 groups according to DSM-III-R: primary degenerative dementia of the Alzheimer type (PDD), muti-infarct dementia (MID), and senile dementia not otherwise specified. For statistical analysis, One-Way ANOVA, Student's t–test and χ^2–test were used.

RESULTS

In the primary investigation comprising a total of 3,754 subjects, 313 (8.3 %) of the subjects were judged as suspected dementia. In the secondary investigation in 1987 the total prevalence rate was 4.1 %, and the proportion of PDD to MID was 0.98. The individual morbidity prevalence rates according to age group (5-year unit group) increased (data not shown). There were no significant differences in the prevalence rate between males and females in any of the age groups. However, the prevalence rate of males was higher than that of females in the first half (the ages of 65 to 75 years) but lower in the second half (the ages of more than 75 years) of the age range. No significant

Table 1. Prevalence Rate in 1987

Parameters		Prevalence Rate (%)
Total	Morbidity Prevalence	4.1 % (153/3,754)
Sex	Male	3.8 % (59/1,569)
	Female	4.3 % (94/2,185)
Types of Dementia		
	Primary Degenerative Dementia of the Alzheimer Type (PDD)	1.6 % (59/3,754)
	Multi-Infarct Dementia (MID)	1.6 % (60/3,754)
	Senile Dementia Not Otherwise Specified	0.9 % (34/3,754)
	PDD/MID	0.98 (59/ 60)
District		
	Urban Area	3.8 % (48/1,272)
	Rural Area	4.4 % (73/1,646)
	Semi Urban/Rural Area	3.8 % (32/ 836)

differences in the prevalence rate were noted between the types of dementia of the age groups. However, the prevalence rate of MID was higher than that of PDD in the first half but lower in the second half. The yealy incidence rate was 1.01 %. According to the age group the incidence rose as the age increased. Although no significacant differences in the incidence were noted between males and famales in a given age group, the incidence rate of males was higher than that of females in the first half but lower in the second half. Although no significant differences in the incidence rate were noted between the types of dementia of the age groups, the incidence rate of MID was higher than that of PDD in the first half but lower in the second half of the age range. The data obtained in the second and third years are not shown, because they were almost equivalent to those of the first year.

Table 2. Incidence Rate in 1988

Parameters		Incidence Rate	(%)
Yearly Incidence Rate		1.01 %	(38/3,754)
Sex	Male	1.02 %	(16/1,569)
	Female	1.01 %	(22/2,185)
Types of Dementia			
	Primary Degenerative Dementia of the Alzheimer Type (PDD)	0.40 %	(15/3,754)
	Multi-Infarct Dementia (MID)	0.40 %	(15/3,754)
	Senile Dementia Not Otherwise Specified	0.21 %	(8/3,754)
	PDD/MID	1.00	(15/ 15)
District			
	Urban Area	1.02 %	(13/1,272)
	Rural Area	1.03 %	(17/1,646)
	Semi Urban/Rural Area	0.96 %	(8/ 836)

Table 3. Incidence rates of the present and previous studies

Investigator (year)		Number of Subjects	Number of Dementia patients	Incidence Rate (%)
O. Hagnell	(1966), Sweden	455	7	1.54 %
K. Bergmann	(1969), U. K.	760	8	1.05 %
M. Roth	(1977), U. S. A.	309	4	1.29 %
I. Fukunishi et al	(1989), Japan	3,754	38	1.01 %

DISCUSSION

We also researched the morbidity prevalence rate of senile dementia. As in our case, most othey investigations were carried out on a large scale such as involving 1,000 subjects or more. As shown in Table 3, concerning the incidence rate, the sample populations of other investigators[3-5] were small in number (less than 1,000), because these investigations were carried out under difficult conditions such as involving the same subjects, the same methods, and the same district for two consecutive years. If we suppose that the incidence were to be approximately 1 %, then it follows that dementia would be noted yearly in no more than one out of one hundred senile subjects. Assumably, therefore, it is necessary that the subjects should bc at least 1,000 in number for a clear investigation of the incidence also. According to the other reports[3-5] of the incidence (Table 3), the subjects were 455, 760, and 309 in number, and the numbers of dementia patients detected were only 7, 8, and 4, respectively. It is easy to understand that a difference in subject number also leads to a great difference in the incidence. Thus, it is necessary that the number of subjects should be more than 3,000 in one year for a comprehensive epidemiological investigation of the incidence of dementia. Accordingly, the value obtained by the present study is a better representation of the true picture as it was obtained by a more accurate method accompanied by less errors.

ACKNOWLEDGIMENT

We are grateful to H. Izumi, and E. Morioka who helped with this investigation.

REFERENCES

1. D. W. K. Kay, P. Beamish and M. Roth, Old age mental disorders in Newcastle upon Tyne: a study of prevalence, Brit. J. Psychiatry. 110: 146 (1964).
2. H. O. Akesson, A population study of senile and arteriosclerotic psychoses, Hum. Hered. 19: 546 (1969).
3. O. Hagnell, A Prospective Study of the Incidence of Mental Disorder, Scandinavian Univ. Books, Norsteds, Lund (1966).
4. K. Bergmann, The epidemiology of senile dementia, Brit. J. Hosp. Med. 2: 727 (1969)
5. M. Roth, Epidemiological studies, in: "Alzheimer's disease", Terry and Bick, ed., Raven Press (1977).
6. C. G. Gottfries, G. Brane and G. Steen, A new rating scale for dementia syndromes, Arch. Gerontol. Geriatr. 1: 311 (1980).

DIFFERENCES IN MEMORY IMPAIRMENT IN PATIENTS WITH ALZHEIMER'S DISEASE AND PARKINSON'S DISEASE WITH AND WITHOUT DEMENTIA

Eeva-Liisa Helkala, Raija Soikkeli, Päivi Hartikainen,
Veikko Laulumaa, Hilkka Soininen, and Paavo Riekkinen

Department of Neurology
University Central Hospital, Kuopio
Finland

INTRODUCTION

In recent years, it has become evident that the pattern of memory impairment in different types of dementia differ[1-2]. Various memory deficits are commonly seen in idiopathic Parkinson's disease (PD). The pathology of hippocampus is believed to underly the severe memory deficit of Alzheimer's disease (AD)[3]. Because such a pathology has not been shown in PD, it is possible that in spite of impaired learning, PD patients can better store information. In PD factors such as duration of disease, age or medication may influence the learning.

The purpose of this study was to investigate the effect of age and dementia on learning and retention ability and on retrieval from semantic memory in Parkinson's disease patients compared to Alzheimer's disease patients and controls.

MATERIALS AND METHODS

Patients

The Parkinsonian study group included 16 young patients (PDY) and 18 old patients (PDO) with idiopathic Parkinson's disease and 18 demented (PDD) patients with idiopathic Parkinson's disease. All PD patients were consecutive referrals to the Department of Neurology of University Central Hospital in 1987, suitable to the research. The PD disability was assessed by the staging scale of Hoehn and Yahr[5], and severity of dementia was assessed using the dementia staging scale (CDR) of Hughes et al.[5]. We compared the PD groups with 32 patients with probable AD, as defined by the criteria of the NINCDS-ADRDA Work Groups[6]. Table 1 shows the characteristics of the patient groups. The controls comprised 23 subjects: 12 were healthy relatives of staff members, and 11 subjects were those who had been checked for minor neurological symptoms. The mean age of the control group was 68$\pm$7 years, range 56-81 years, 18 were women and 5 were men.

Table 1. Characteristics of young (PDY), old (PDO), and demented (PDD) patients with Parkinson's disease and Alzheimer's disease patients (AD)

	PDY	PDO	PDD	AD
Number of patients	16	18	18	32
Male/female	7/9	9/9	7/11	21/11
Age (mean±SD)	51±6	72±7	70±8	68±6
Duration of PD (mean±SD)	2.7±1.7	4.1±3.3	6.9±4.1	
Disability of PD[4]	2.1±1.0	2.6±1.0	3.0±1.0	
Duration of dementia			1–5 years	1–3 years
Severity of dementia[5]				
mild			8	18
moderate			6	13
severe			4	1

[4]According to Hughes et al.
[5]According to Hoehn and Yahr

Table 2. Medication of young (PDY) old (PDO) and demented (PDD) patients with Parkinson's disease and Alzheimer's disease patients (AD)

	PDY	PDO	PDD	AD
Anticholinergics	4	1	4	–
Amantadin	1	–	1	–
L-dopa	4	7	15	–
Bromocriptine	4	4	6	–
L-deprenyl	2	2	6	–
Neuroleptics	–	–	4	3
Antidepressants	–	2	4	–
Codergocrine	–	–	–	1

Methods and Memory Assessment

Learning was examined using a list-learning test based on Buschke's selective reminding method as modified by Fuld[7]. A yes/no recognition of words from a 20-word shopping list was requested after a 30-minute delay filled with other psychometric tests. Ten of the 20 recognition test words were drawn from the learned shopping list (target stimuli), whereas the other ten had not been presented earlier (distractor stimuli). To calculate the specifity of the learned items, we substrated half of the sum of the learned items and false positive errors from the total of learned items and added a constant of 8. The retrieval from semantic memory was examined by category fluency task[1].

RESULTS

One-day analysis of variance revealed a significant group effect on list-learning test between PDY, PDO, PDD, AD and C groups (F = 24.4,

Table 3. Scores of learning and recognition of list learning and catecory naming tests in young (PDY), old (PDO), and demented (PDD) patients with Parkinson's and Alzheimer's disease patients (AD) and controls (C).

	PDY	PDO	PDD	AD	C
Learning	34.4±9.9	22.6±11.1	15.5±10.7	12.5± 9.0	35.5±6.0
Recognition	96.7±4.7	82.2±10.7	84.7±12.2	68.2±16.4	99.2±2.5
Category	15.1±3.1	11.7± 3.0	7.4± 5.2	7.1± 4.3	13.5±4.3

df = 3,65, $p < 0.001$). Post hoc analysis (Scheffe's method, $p < .02$) showed that PDD, PDO, and AD groups were equally impaired and significantly impaired as compared to PDY or C groups. The PDY and C groups produced more words on category naming test than PDD or AD groups.

One-way analysis of variançe revealed also a significant group effect on delayed recognition of the words of learned words ($F = 15.0$, df = 3,65, $p < .001$). Post hoc analysis showed that PDY, PDD, PDO and C groups recognized more words than AD group.

To investigate the interaction effects of age, duration of the PD and anticholinergic medication on learning and retention in Parkinson's disease with and without dementia, we compared the scores of list learning test in PDD group and combined PDY and PDO groups. The four-way analysis of variance revealed a significant main effect of age on list learning test ($F = 11.67$, df = 1,38, $p < .02$). The younger patients learned more words. Age and anticholinergic medication had an interaction effect on recognition ($F = 5.00$, df = 1,38, $p < .05$). The old age together with anticholinergic medication impaired the recognition of the words.

DISCUSSION

In the list learning test, the all PD patients showed superior delayed recognition memory capability as compared to the AD patients, while only the PDY group's acquisition capability was superior to AD, PDO or PDD groups. The result suggest that PD patients are able to bind information better into long-term storage than AD patients. This may be due to better functioning of the entorhinal cortex and hippocampus. In AD, severe memory impairment has been suggested to be due to the pathology of hippocampus[3]. Such a pathology has not been shown in PD.

Only the PDD and AD groups were impaired on retrieval from semantic memory. Although general retrieval problems has been suggested to be the underlying cause of memory problems of other diseases of basal ganglia, i.e. Huntington's disease, this study does not support that the general retrieval problems might be an underlying cause of memory deficit in PD. However, the impaired recall memory compared to better recognition memory may suggest similar underlying processes in memory deficiencies of PD with dementia.

The patients who were impaired on learning test tended to be older than those who were not. This might reflect greater cell loss in nucleus basalis or greater incidence of cortical pathology in older parkinsonian patients. The levels of dopamine in monkeys have been shown to decrease dramatically with age both in the caudate nucleus and in prefrontal cortex[8]. Thus, the old parkinsonian patients may reach rapidly the critical level for production of cognitive symptoms. The duration of the disease tended to be higher in older patients in our study.

The anticholinergic medication and age had an interaction effect on recognition memory. The explanation may be similar as previously: in old patients the anticholinergic medication reach easily the critical level for production of cognitive symptoms. It has also been shown that there is the cholinergic dysfunction with advancing age[9-10].

In conclusion, the results suggest that the better retention ability on PD patients may be due to the better functioning entorhinal cortex and hippocampus. The advancing age seem to impair learning ability in PD more than in normal aging, and the anticholinergic medication together with aging process impairs retention ability.

REFERENCES

1. N. Butters, P. Milliotis and M. Albert, Memory assessment evidence of the heterogeneity of amnesic symptoms, in: "Clinical Neuropsychology". Vol I G. Goldstein (Ed.) pp. 127-159. Plenum Publishing Corporation. New York (1984).
2. E-L. Helkala, V. Laulumaa, H. Soininen, P. J. Riekkinen. Recall and recognition memory in patients with Alzheimer's and Parkinson's diseases, Ann Neurol 24:214-217 (1988).
3. B. T. Hyman, G. W. Hoesen, and L. J. Kromer, Perforant pathway changes and the memory impairment of Alzheimer's disease, Ann Neurol 20:472-481 (1986).
4. M. M. Hoehn, and M. D. Yahr, Parkinsonism: onset, progression and mortality, Neurology (Minneap.) 17:427-442 (1967).
5. C. P. Hughes, L. Berg, and W. L. Danziger, A new clinical scale for the staging of dementia, Br J Psychiatry, 140:566-572 (1982).
6. G. McKhann, D. Drachman and M. Folstein, Clinical diagnosis of Alzheimer's disease: report of the NINCDS-ADRDA Work Group under the auspices of Department of Health and Human Services Task Force on Alzheimer's disease, Neurology 34:939-944 (1984).
7. P. A. Fuld, Guaranteed stimulus-processing in the evaluation of memory and learning, Cortex 16:255-271 (1980).
8. P. S. Goldman-Rakic, and R. M. Brown, Regional changes of monoamines in cerebral cortex and subcortical structure of aging rhesus monkeys, Neuroscience 6:177-187 (1981).
9. J. Sirviö, J. O. Rinne, A. Valjakka, Different forms of brain acetylcholinesterase and muscarinic binding in Parkinson's disease, J Neurol Sci 90:23-32 (1989).
10. G. L. Neuk, D. J. Pierce and R. G. Struble, Age related changes in multiple neurotransmitter systems in the monkey brain, Neurobiol Aging 10:11-19 (1989).

DEMENTIA IN PARKINSON'S DISEASE

Amos D. Korczyn

Departments of Physiology and Pharmacology
and Neurology, Sackler Faculty of Medicine
Tel Aviv University, Ramat Aviv 69978, ISRAEL

James Parkinson[1], in his famous essay describing paralysis agitans, was of the opinion that "the senses and intellect are unaffected", and this view was maintained for many decades. It is now well recognized however, that in its later stages mental changes do occur. It is therefore interesting to note that Parkinson did remark that eventually a delirious state may develop. This observation is particularly noteworthy because in these later stages of the disease it may be quite difficult to communicate with the patient, particularly since speech is impaired. As treatment became available for Parkinsonism another difficulty arose, namely to differentiate endogenous from drug-induced confusion. The frequent coexistence of depression with Parkinson's Disease (PD) necessitates the exclusion of "pseudodementia". However, true cognitive decline is part and parcel of advanced PD, occuring in 20-80% of cases (depending on the population studied and criteria for diagnosing dimentia)[2-4]. Our own data (Fig 1) demonstrate a clear-cut cognitive decline in about 30% of cases, with a higher frequency in advanced PD.

The immediate question arising is what is the nature and which are the causes of this phenomenon. Mental deterioration depends not only on the disease but also upon the age of the patient. However, it is interesting that after correction for age and motor disability, the disease duration does not affect the cognitive status[5]. It is therefore possible that some patients may be more susceptible to degenerative processess in more than one system, e.g. nigrostriatal loss and Alzheimer changes.

One way to try and answer this question is by examination for possible risk-factors for dementia in PD. In two such studies recently reported by us[5,6] we have shown that demented and non-demented PD patients differ in that the demented have more frequent cardiovascular risk factors than non-demented subjects. This may indicate that in some

Basic, Clinical, and Therapeutic Aspects of Alzheimer's and Parkinson's Diseases, Volume 2
Edited by T. Nagatsu *et al.*, Plenum Press, New York, 1990

cases the dementia was at least partly caused by vascular damage to the brain (although patients with frank strokes were excluded). Another interesting difference is that demented patients were more likely to have been through the holocaust in Europe during the Second World War, thus suggesting that occurrences at early age may contribute to the later development of dementia[6]. It is still an open question whether the factors involved are physical or psychological.

We have tried to find out neuro-radiologic characteristics of PD patients with dementia[7]. Towards this goal we have measured several parameters on brain CT of patients with PD. Cerebral atrophy correlated with age, and less strongly (but still statistically significantly) with the degree of motor disablility, and least with cognitive decline. Since these three parameters are interdependent, it was difficult to isolate the factor of dementia. However cognitive decline was mostly related to subcortical radiologic changes such as ventricular dilation. Interestingly, no correlation existed with the cortical parameters, suggesting a subcortical origin for the dementia.

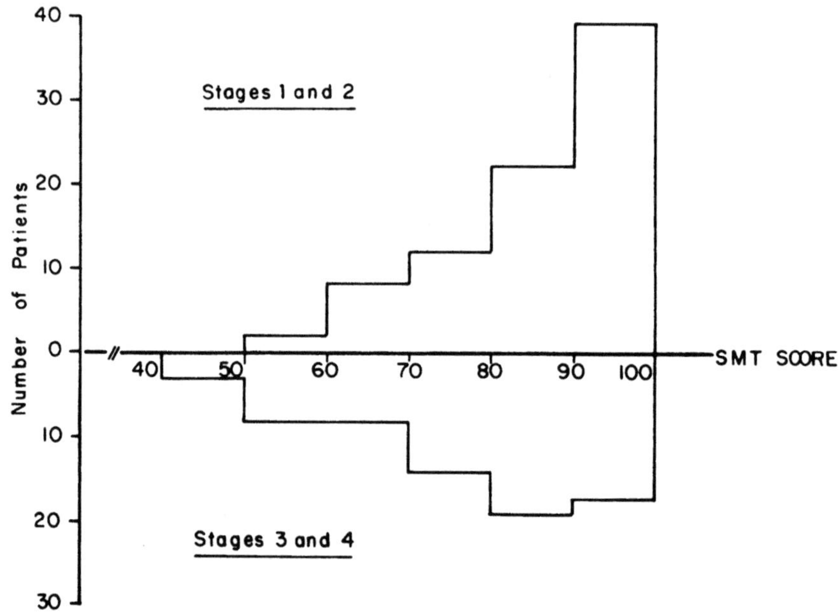

Fig 1. Results of short mental test in a consecutive series of 190 patients with Parkinson's disease. Note that abnormal abnormal results occur in about 20% of subjects at stage I or Stage II (Hoehn Yahr) as opposed to over 30% in more advanced stages (Treves and Korczyn, in preparation, 1990).

In an attempt to further substantiate these findings, we have also looked at the EEG of PD patients and related the changes to the existence of dementia. We have found that demented patients were more likely to show EEG abnormalities, particularly slowing. However caution must be exercised in the interpretation of these results since demented patients were more likely to be at an advanced motor disability[8]. Therefore another study was undertaken in which the EEG of PD patients, matched for age and motor disability but discordant for dementia, was evaluated. This study included computerized BEAM analysis and revealed only small differences, particularly decreased power in the alpha range and slightly increased power in lower frequencies (Blumen, Neufeld and Korczyn, in preparation).

The neuroradiologic findings mentioned above are reminiscent of those in Alzheimer's disease (AD). It was therefore of interest to examine whether PD patients are also sensitive to cholinergic blockade. To this effect we have performed neuropsychologic evaluation of PD patients with and without dementia and compared their performance to that of subjects with AD, multi-infarct dementia (MID), and normal age-matched controls. We have found that demented PD patients form an intermediate group in their senstivity to atropine, falling between AD and MID patients (Fig. 2).

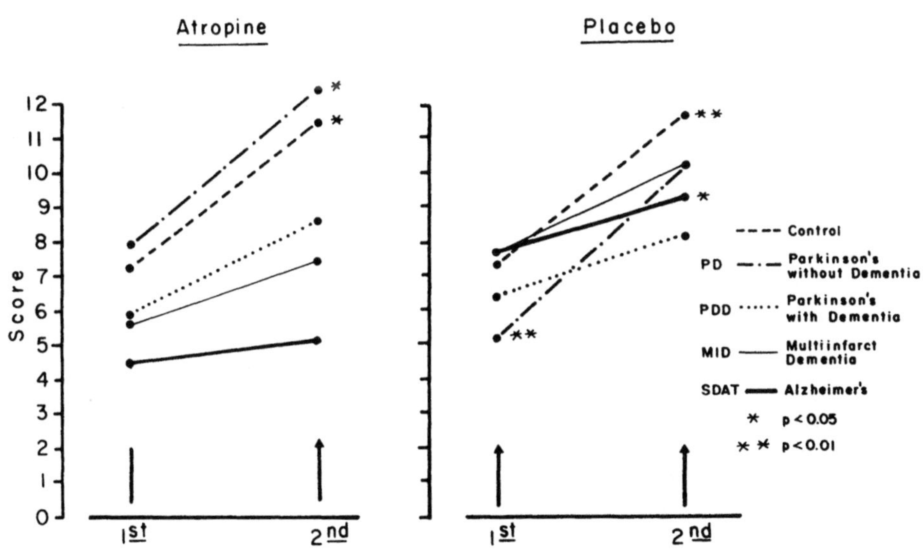

Fig 2. Effect of 1 mg atropine (or placebo) given i.v. on subsequent performance. Note that on retest there was improvement in logical memory function signifying learning. In PD patients the effect of atropine was insignificant.

These results demonstrate that dementia in PD is not a single nosologic entity. Rather, it probably is a heterogenous group. Several patients have coexistent Alzheimer's disease[9] or other dieseases, e.g. diffuse Lewi body disease[10], or sucbortical vascular damage[11]. In a large proportion, however, the basic pathological process of PD affects sucbcortical structures responsible for cognition such as the nucleus basalis of Meynert[12]. This complex pathophysiology may explain the relatively normal EEG pattern, the predominant subcortical damage seen radiologically and the limited neuropharmacological similarity to AD. These results have important implications regarding the prevention and treatment of dementia in PD, since it is not to be expected that a single measure would suffice.

REFERENCES

1. J. Parkinson, "An Essay on the Shaking Palsy, 1817". Medical Classics 2:964-997 (1938).
2. M. Hietman, and H. Teravainen, Cognitive performance in early Parkinson's disease. Acta Neurol. Scand. 73: 151-159 (1956).
3. R.H.S. Mindham, S.W.A. Ahmed, and C.G. Clough, A controlled study of dementia in Parkinson's disease. J. Neurol. Neurosurg. Psychiat. 45:969-974 (1982).
4. A.D. Korczyn, R. Inzelberg, T. Treves, M. Neufeld, I. Reider, and M. Rabey, Dementia of Parkinson's disease. Adv. Neurol. 45:399-403 (1988).
5. A.D. Korczyn, R. Inzelberg, T. Treves, M. Neufeld, I. Reider, and M. Rabey. Dementia in Parkinson's disease. in: "Alzheimer's and Parkinson's Diseases". Plenum Press, NY. pp 177-183 (1986).
6. A.D. Korczyn, and I. Salganik, Risk factors for dementia in Parkinson's disease. Neurology 38 (Suppl 1):321 (1988).
7. R. Inzelberg, T. Treves, I. Reider, I. Gerlenter, and A.D. Korczyn, Computed tomograpy brain changes in Parkinsonian dementia. Neuroradiol. 29:535-539 (1987).
8. M.Y. Neufeld, R. Inzelberg, and A.D. Korczyn, EEG in demented and non-demented Parkinsonian patients. Acta Neurol. Scand. 78:1-5 (1988).
9. H.C. Chui, J.A. Mortimer, U. Slager, C. Zarow, W. Bondareff, and D.D. Webster, Pathologic correlates of dementia in Parkinson's disease. Arch. Neurol. 43:991-995 (1986).
10. A.W. Clark, and J. Lehmann, Dementia with widespread Lewy bodies: Studies of the neocortical cholinergic system. Canadian Association of Neuropahtologists. (1983).
11. N.M. Bornstein, I. Reider, and A.D. Korczyn, Clinical correlates of lacunar infarcts in the basal ganglia. Neurology 39 (Suppl 1):303 (1989).
12. P.J. Whitehouse, J.C. Hedreen, C.L. White, and D.L. Price, Basal forebrain neurons in the dementia of Parkinson's disease. Ann. Neurol. 13:243-248 (1983).

JUVENILE OR EARLY-STARTING PARKINSONISM AS A PROTOTYPE OF NIGROSTRIATAL DOPAMINE DEFICIENCY

Hirotaro Narabayashi

Neurological Clinic
Tokyo, Japan

INTRODUCTION

Juvenile or early-starting parkinsonism (JP as defined by Yokochi[1,2] and Narabayashi et al.[3]) occupies about 7 to 10% of the total parkinsonism population seen in the clinic in Japan. Analysis of these cases from both clinical and basic approaches seems important and may lead to elucidation of the underlying mechanism causing the disease.

CHARACTERISTICS OF CLINICAL FEATURES

JP has been described as slower in progression and presenting mostly rigidity and akinesia with much less tremor at rest, when compared with the idiopathic Parkinson's disease (PD). Responses to levodopa therapy are generally more marked than in PD, often drastic; but reverse effects, such as levodopa-induced dyskinesia (DID) and up-down oscillation-producing effects, are also marked, which sometimes makes continuous medication impossible. Lowering of mental activity or autonomic symptoms is usually not observed.

RESULTS OF LONG-TERM TREATMENT

When cases of JP are successfully treated over a long period with pharmacological agents such as the combination of levodopa and/or dopamine (DA) agonist without production of reverse effects, the patients can maintain almost normalized daily living (including work) for as long as 10 to 20 years. Fig. 1 shows the results obtained by Yokochi[4] of long-term pharmacological treatment in 31 cases of JP for 10 to 12 years since start of medication. He found that the cases of grade A, i.e., the group of cases with normalized social activity (including housework or job) occupied more than half of the cases studied.

On the other hand, in PD the majority of cases tend to deteriorate steadily within several to ten years, even when medical therapy is smoothly applied as in JP. Fig. 2 gives the results of observation of 191 PD cases under long-term levodopa treatment for 12 years,[5] as obtained by a team of 27 neurological centers in Japan. The total clinical picture of the patients gradually deteriorated, with appearance of pronounced postural

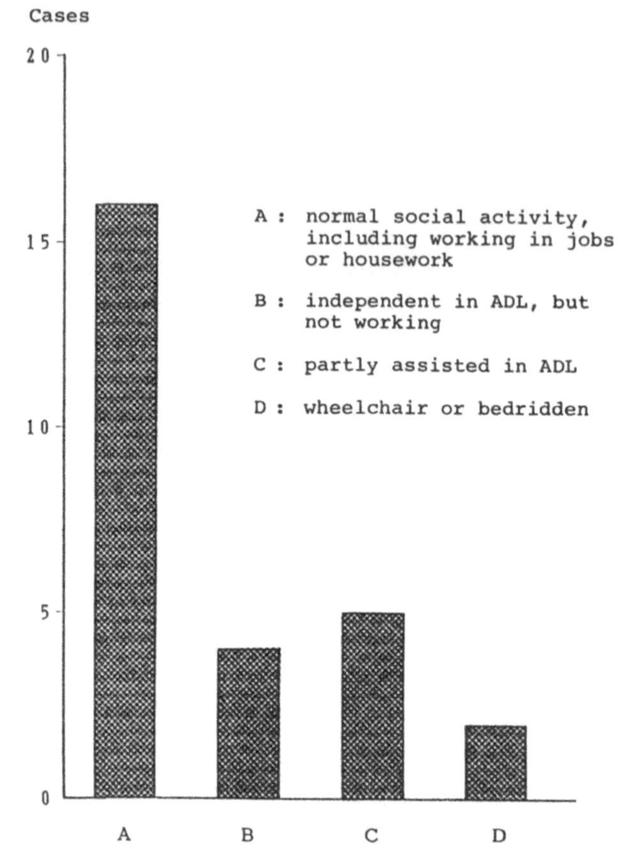

Fig. 1. Status of ADL (activity in daily life) in follow-up study for 10–12 years in 31 cases. (from Yokochi[4])

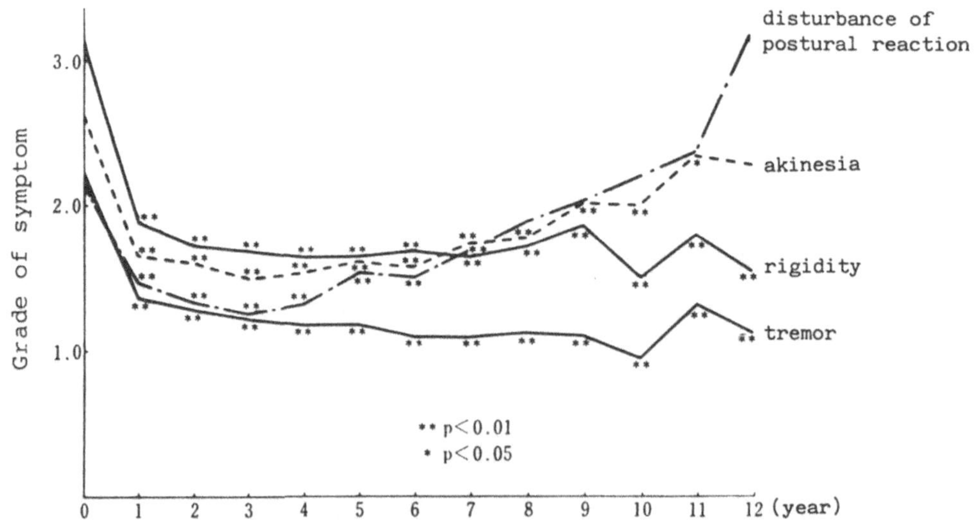

Fig. 2. Course of clinical symptoms of parkinsonism during long-term levodopa treatment for twelve years. (from Kase et al.[5])

Table 1. Improvement in Cases Operated-on Bilaterally

	A (drug free)	B	Total
Tremor	9 (1)	5	14
Juvenile parkinsonism	8 (2)	3	11
Total	17	8	25

A: Almost normal in job & social life

B: Independent at home, normal housework

instability and akinesia, including emotional changes, although both rigidity and tremor were well controlled by the medication.

Some cases of JP are treated by stereotaxic thalamotomy using a microelectrode recording technique because of an insufficient drug effect, gastrointestinal side effect, or severe and uncontrollable ballistic dyskinesia of the extremities. When JP cases are operated on unilaterally or, exceptionally, bilaterally with several years' interval between the operations on the two sides, the patients become free from rigidity and tremor with less akinesia and often require no or only minimum medicine. Table 1 indicates the clinical improvement in daily living of bilaterally operated-on cases followed postoperatively for three to eight years.[6,7] All eleven cases of JP were observed to maintain an almost normalized condition in daily living in jobs or housework under the same, less, or no medicine as used preoperatively.

Fig. 3 indicates the anatomical projection pathway from the nigrostriatum that might operate in the production of various symptoms of parkinsonism. By surgical interruption of the terminus of the pallidothalamic tract, which tract is known as a main output of the extrapyramidal impulses, at the ventrolateral nucleus (VL) of the thalamus, rigidity is totally abolished, and by addition of a lesion to the ventral intermediate nucleus (Vim), tremor disappears.

If most of the clinical symptoms such as rigidity, tremor and hypokinesia in JP are markedly and sustainedly improved by surgical intervention unilaterally or bilaterally resulting in the normalized social and daily living under minimum dose of medicine, the difficulties in these cases are interpreted as solely due to pathophysiology within the striopallidum secondary to the nigrostriatal pathology. Other neural structures that are considered responsible to produce other symptoms, such as so-called akinesia in the broad sense, difficulty of postural balancing, autonomic symptoms, or psychological changes, seem not to be affected in JP.

These experiences in and analysis of both pharmacological and neurosurgical treatment may indicate that the main difficulty in JP cases is due to the nigrostriatal DA deficiency. Also morphological and chemical pathology is expected to be limited to the nigrostriatum, and this expectation held true in sections made of the midbrain of the case described below. If such pathological mechanism is treated properly by medicine and/or surgery, the patient can maintain almost normalized social living for many years.

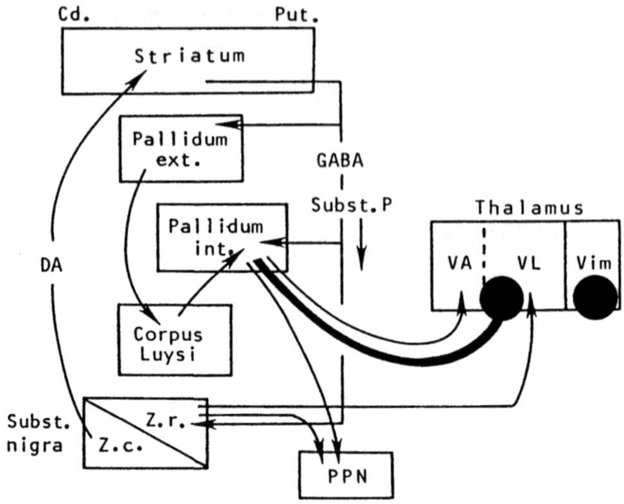

Fig. 3. Schema of projection within the extrapyramidal system producing rigidity (pallidum internum to VL) and tremor (Vim). Black circles indicate the placement of surgical lesions; see text.

PATHOLOGICAL STUDY OF ONE CASE

One of the female JP cases, a woman who died at the age of 39 due to acute ileus, was neuropathologically verified.[3,8] This case started as dystonia of the feet at the end of age six and became progressively worse very slowly. About ten years later dystonia and rigidity spread to both sides of the extremities and trunk, resulting in the bedridden state. Unilateral (left) thalamotomy was done at the age of 24, but did not lead to sufficient improvement in ADL (activity in daily life).

When levodopa was administrated to the patient at the age of 30, she responded markedly and improved to almost normal mobility enabling her to stand and walk alone, but a year later severe ballistic DID started in the extremities of the unoperated side and trunk, which practically made further medication impossible. Thalamotomy was then performed on the right side thalamus attempting to control the dyskinesia, when the patient was 38. After this operation, DID did not appear on either side of the body, and as a result, levodopa therapy again became applicable. A minimum dose of 0.3g per day without decarboxylase inhibitor was found to be sufficient to bring her back to normal mobility for daily living, such as strolling around the countryside, writing poems, and shopping by herself.

On macroscopy of the midbrain, the nigral area showed no pigmentation, as in other parkinsonian cases. However, by microscopic examination, neurons within the compact zone (SNc) were found to be relatively well preserved, but with almost no melanin pigment; although some reduction in number of neurons was observed. Reduction in neuron number in the part of the SNc projecting to the putamen was more severe than in the part projecting to the caudate, as pointed out by Gibb et al.[9] Such regional difference in the grade of neuronal degeneration within the SNc is interpreted as a characteristic neuropathological feature of PD, and it is quite significant that such a younger-starting JP case presented the same or very similar pattern of regional changes within the SNc. Other structures such as the locus ceruleus (LC), caudate and putamen, pallidum, and hypothalamic nucleus were neuropathologically almost intact. Therefore, the only structure morphologically involved in pathology in this case was the SNc. On the other hand, marked loss of tyrosine hydroxylase (TH) activity was seen in the striatum.

CASES OF PD AND THE ROLE OF NOREPINEPHRINE DEFICIENCY

As described briefly in the previous paragraph, in classical PD, the whole clinical picture tends to worsen progressively even after diminution of rigidity and tremor by medical or surgical treatment. The symptoms that start to appear and become more manifest in the later stage of the disease under long-term treatment are postural instability, difficulty of gait, emotional deterioration, and autonomic symptoms. I have postulated that progressive deficiency in norepinephrine (NE) biosynthesis in addition to that of DA might be the cause of these later appearing difficulties. L-threo-3,4-dihydroxyphenylserine (L-DOPS), the industrial precursor of NE, was then applied to these later-stage patients with relative and mild improvement of these symptoms.[10,11] Therefore, metabolic disturbances in these longstanding PD cases are assumed to be a consequence of the combined deficiency of DA and NE, although in the initial few years, DA deficiency may predominate.

In contrast, DA deficiency is the main disturbance for many years in the younger-starting cases. Such differences between the two conditions, JP and PD, are assumed to be related to the difference in biological aging of the brain.

In experiments in the monkey, 1-methyl-4-phenyl-1,2,3,6-tetrahydropyridine (MPTP) produces selective neuronal degeneration in the SNc. When animals of older age are used, the changes observed are more severe and widespread in the SNc and LC, whereas the changes in the younger animals are limited mostly to within the SNc.[12] If the hypothesis that the human parkinsonism is the consequence of an endogenically produced substance affecting the brain, such as tetrahydroisoquinoline (TIQ) as proposed by Nagatsu and Yoshida et al. (both in this volume) or some similar substance(s), morphological and chemical pathology and clinical manifestation could be slightly different depending on the grade of aging of the brain affected. If it affects the younger brain, changes might be confined to the nigrostriatum. When the older brain is exposed, pathology might be more severe and widespread, reaching other neuronal systems, such as the LC, etc. Such a hypothesis seems to coincide with and explain the difference in clinical pictures and courses of the disease between JP and PD.

REFERENCES

1. M. Yokochi, Juvenile Parkinson's disease, Pt. 1: Clinical aspects, Adv. Neurol. Sci. (Tokyo) 23: 1048 (1979).
2. M. Yokochi, Juvenile Parkinson's disease, Pt. 2: Pharmacokinetic study, Adv. Neurol. Sci. (Tokyo) 23: 1060 (1979).
3. H. Narabayashi, M. Yokochi, R. Iizuka, and T. Nagatsu, Juvenile parkinsonism, in: "Handbook of Clinical Neurology," Vol. 5, P. J. Vinken, G. W. Bruyn and H. L. Klawans, eds., Elsevier Science Publishers, Amsterdam (1986).
4. M. Yokochi, unpublished data, personal communication.
5. M. Kase, J. Atarashi, K. Ando, K. Ito, T. Ohmoto, and H. Kowa, Twelve years' course of Parkinson disease with levodopa treatment, Adv. Neurol. Sci. (Tokyo) 28: 681 (1984).

6. H. Narabayashi, Surgical treatment in the era of levodopa and other pharmacological treatment, in: "Parkinson's Disease," G. Stern, ed., Chapman & Hall, London (in press).
7. H. Narabayashi, Lessons from stereotaxic surgery using microelectrode techniques in understanding parkinsonism, Mt. Sinai J. Med. 55: 50 (1988).
8. M. Yokochi, H. Narabayashi, R. Iizuka, and T. Nagatsu, Juvenile parkinsonism. Some clinical, pharmacological and neuropathological aspects, in: "Advances in Neurology," Vol. 40, R. G. Hassler and J. F. Christ, eds., Raven Press, New York (1984).
9. W. R. G. Gibb, H. Narabayashi, M. Yokochi, and R. Iizuka, Additional pathological observations in juvenile onset parkinsonism with dystonia, Neurology 39 (Suppl. 1): 139 (1989).
10. H. Narabayashi, T. Kondo, A. Hayashi, T. Suzuki, and T. Nagatsu, L-threo-3,4-dihydroxyphenylserine treatment for akinesia and freezing of parkinsonism, Proc. Japan Acad. 57: 351 (1981).
11. H. Narabayashi, T. Kondo, F. Yokochi, and T. Nagatsu, Clinical effects of L-threo-3,4-dihydroxyphenylserine in cases of parkinsonism and pure akinesia, in: "Advances in Neurology," Vol. 45, M. D. Yahr and K. J. Bergmann, eds., Raven Press, New York (1986).
12. L. S. Forno, J. W. Langston, L. E. DeLanney, I. Irwin, and G. A. Ricaurte, Locus ceruleus lesions and eosinophilic inclusions in MPTP-treated monkeys, Ann. Neurol. 20: 449 (1986).

CLINICAL FEATURES OF AUTOSOMAL RECESSIVE TYPE JUVENILE PARKINSONISM AND IMPROVEMENT OF SYMPTOMS BY SMOKING

Atsushi Ishikawa and Tadashi Miyatake

Department of Neurology
Brain Research Institute
Niigata University
Niigata 951, Japan

INTRODUCTION

Juvenile parkinsonism (JP) was defined as parkinsonism starting before the age of 40.[1] One type of JP is autosomal recessive type juvenile parkinsonism (AR-JP), first reported by Yamamura et al.[2]

We had the opportunity to examine 27 patients with AR-JP, 15 of which comprised 9 families. The present report describes the clinical features of AR-JP, efficacy of smoking for symptom improvement which was observed in 5 patients with AR-JP, and catecholamine metabolism in rat brain after nicotine administration.

CLINICAL FEATURES

AR-JP has characteristic symptoms compared with those of Parkinson's disease (PD). We show these characteristic symptoms of AR-JP (Table 1) and the results of a few clinical examinations studied in our 27 AR-JP patients.

There was no sex preponderance (male/female: 14/13) in our study. Inheritance was usually autosomal recessive, although a rare autosomal dominant case has been reported. There were also sporadic cases who had the same clinical findings as the autosomal recessive type, but did not have any inheritance trait. The average age at onset was 27.3 years in our study. The clinical course was usually very slowly progressive. In our study the average clinical observation period after onset was 19.4 years, but the patients were still able to perform daily routines by themselves. As an initial symptom AR-JP patients tended to complain of gait disturbance. Foot dystonia such as inversion and plantar flexion at the ankle, plantar flexion of the toes, extension of the great toe, and plantar arching was found as one of the characteristic findings. In our study we found foot dystonia in 64% of the patients. This symptom disappeared after treatment or in on-period. Dystonic posture was usually restricted to the feet. Another characteristic finding was that the patients experienced improvement of movement disability following sleep. Severity of clinical symptoms was reduced for a period ranging from a half to two hours after the patients

Basic, Clinical, and Therapeutic Aspects of Alzheimer's and Parkinson's Diseases, Volume 2
Edited by T. Nagatsu *et al.*, Plenum Press, New York, 1990

Table 1. Clinical Features of Autosomal Recessive Type Juvenile Parkinsonism

Sex	M=F (M/F, 14/13)
Heredity	autosomal recessive
	consanguinity: 8/9 families
	13/15 patients
Age at onset	27.3 ± 8.2 y.o.
Clinical course	very slowly progressive
	(clinical observation period: 19.4 ± 9.6 y.)
Initial symptom	gait dist.: 52%, tremor: 33%, others: 15%
Foot dystonia	64%
Improvement after sleep	84%
Smoking efficacy	45% (5/11 smoker patients)
Levodopa efficacy	markedly, long lasting
Wearing-off phenomenon	96%
Dyskinesia	
in the extremities	81%
in the orolingual area	30%
Tremor	fine/coarse, 69%/31%
Frozen gait	56%
Hyperreflexia	59%

woke up in the morning, and an afternoon nap could also lead to a temporary reduction, but simple rest without sleeping had no effect. Eighty-four percent of the patients experienced this phenomenon in our study. After the initiation of treatment this improvement after sleep became obscure due to the occurrence of the wearing-off phenomenon. We also found smoking to be efficacious in the patients with AR-JP. Levodopa therapy was markedly effective, but the wearing-off phenomenon and dopa-induced dyskinesia usually occurred during it. Dyskinesia often appeared in the extremities and trunk, but it was not usually observed in the orolingual area. The wearing-off phenomenon and dyskinesia were important clinical factors for medical control in the treatment of AR-JP. Although tremor, rigidity, gait disturbance, and impaired postural reflex are cardinal symptoms in parkinsonism, AR-JP had clinical differences from PD in regard to these symptoms. Tremor was usually postural and fine. Rigidity sometimes showed a characteristic distribution, with that in the ankle joint being often intense. The gait pattern was sometimes limping or the heel-up type, and frozen gait also occurred. An impaired postural reflex was usually evident even in the patients whose other neurological disabilities were not so severe. Deep tendon reflexes were often exaggerated. Involvement of the autonomic nervous system such as orthostatic hypotension was seldom found. We could not detect dementia on examination, although the patients sometimes complained of mild memory disturbance.

To clarify the pathophysiology of AR-JP, we examined several biochemical markers including HVA and 5HIAA in the CSF of AR-JP patients without medication (Fig. 1). HVA levels of AR-JP were lower than those of controls, but there was no significant difference between HVA levels of AR-JP and those of PD. 5HIAA levels of AR-JP were also lower than those of controls, but the difference in the levels between AR-JP and PD was much milder than that in HVA levels.

Since the patients tend to easily develop the wearing-off phenomenon, we examined the levels of levodopa, dopamine (DA), and dihydroxyphenylacetic

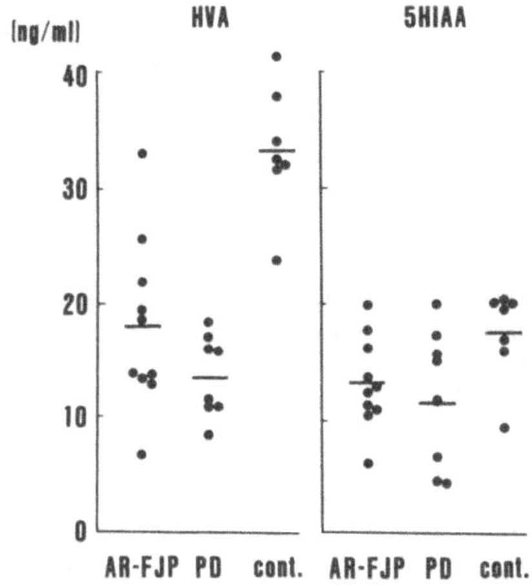

Fig. 1. HVA and 5HIAA levels in the patients with AR-JP and PD compared with those in controls.

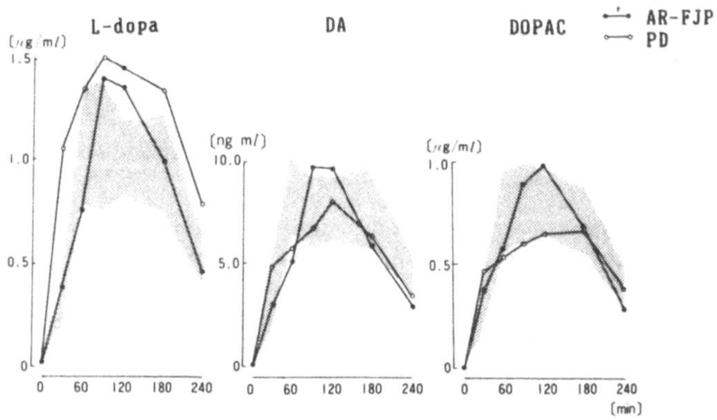

Fig. 2. Changes in L-dopa, DA, DOPAC levels in the serum after oral loading of 400 mg of L-dopa.

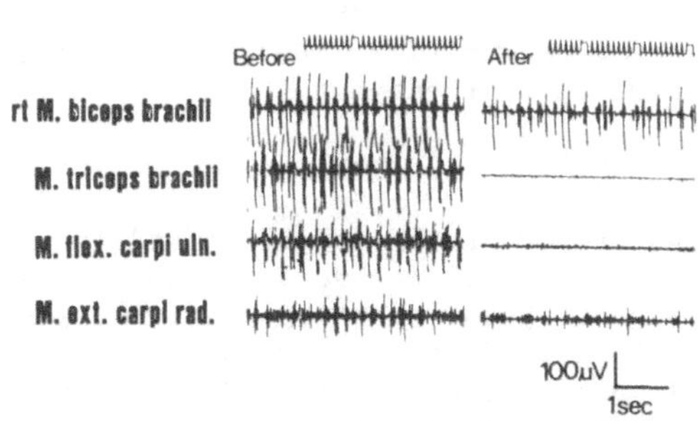

Fig. 3. Surface EMG showing improvement after cigarette smoking in a patient with AR-JP.

acid (DOPAC) in the serum after oral loading of levodopa. Fig. 2 shows that patterns of absorption and metabolism of levodopa in AR-JP did not differ from those in PD.

A typical patient with AR-JP is presented as follows.

Case Report

He developed fine tremor and difficulty in writing at the age of 24. These symptoms gradually increased, and 3 years later he showed heel-up gait and bradykinesia. He was aware of improvement of such symptoms for a half or one hour following sleep. Marriage of his parents was consanguineous and one elder brother of 3 siblings suffered from the same disease.

In July 1982, he was admitted to our hospital. He showed a masked face and monotonous speech. Fine tremor was observed more easily as the postural type than as the rest type. There were also rigidity, bradykinesia, impaired postural reflex, adiadochokinesia, and gait disturbance with frozen gait. Deep tendon reflexes were slightly exaggerated. He also showed hand deformities and foot dystonia such as equinus foot and extension of the great toe. Only a few days after the administration of 600 mg of levodopa, he showed improvement in his symptoms. After the dose of levodopa was increased, he began to notice a mild wearing-off phenomenon and foot dyskinesia. He has been well controlled for 7 years by the administration of 800 mg of levodopa in combination with decarboxylase inhibitor in 7 to 8 divided daily doses, and he is able to work as a businessman without almost any difficulties.

SMOKING EFFICACY

We experienced improvement of parkinsonian symptoms following cigarette smoking in AR-JP. Eleven patients with AR-JP had experienced cigarette smoking in the past or were still smoking cigarettes. Five of the 11 ex-smoker and smoker patients showed smoking efficacy in terms of parkinsonian symptoms. Eight patients were ex-smokers or smokers out of 13 patients with juvenile onset PD, and 16 patients were ex-smokers or smokers out of 52 patients with PD. However, none of them showed any smoking efficacy.

The 5 patients showing a good effect of smoking consisted of 4 males and 1 female, ranging in age from 33 to 58 years, with a mean age of 47 years. Their ages at onset ranged from 14 to 29 with a mean of 22 years. The parents of two cases had a consanguineous marriage, and the sibling of one case suffered from the same disease. They showed tremor, rigidity, bradykinesia, impaired postural reflex, adiadochokinesia, and gait disturbance including frozen gait. Efficacy with levodopa was significant in all cases, and they all displayed the wearing-off phenomenon and dopa-induced dyskinesia in the extremities. Four cases had improvement following sleep before the initiation of treatment.

Case Report of One Patient

He developed hand tremor at the age of 31 and later he began to limp in the left leg. He was relieved by levodopa therapy but dyskinetic movement in the neck and wearing-off phenomenon appeared. He was admitted to our hospital at 51 years of age. He showed tremor, rigidity, bradykinesia, and adiadochokinesia. His gait was small stepped and limping. He sometimes showed frozen gait and festination. Impaired postural reflex was eminent. HVA level in the CSF was decreased to 6.6 ng/ml (normal, 33.2 ± 5.1 ng/ml) without medication. His parents were not consanguineous, and none of his family suffered from the same disease.

Table 2. Efficacy of Smoking of One Cigarette in 4 Patients with AR-JP

	Case 1	Case 2	Case 3	Case 4
Duration of efficacy (min.)	5-20	5-15	15	5-30
Tremor	↓	↑	↓	-
Rigidity	↓	→	↓	↓
Bradykinesia	↓	↓	↓	↓
Gait disturbance	↓	↓	↓	↓
Occurrence of dyskinesia	+	+	+	-

↓: decreased, ↑: increased, →: unchanged.

He was not aware of the smoking efficacy. He used to smoke before getting out of his car to improve his movement disorders. Efficacy was greater in the off-period, especially after waking up in the morning. At first he felt relaxation in the rigid extremities. Tremor, rigidity, and bradykinesia were improved (Fig. 3, Table 2). The improvement in his gait was remarkable. He became able to walk faster with wide steps. However, adiadochokinesia, handwriting, drawing of circles, and finger tapping were not improved. Mild dyskinesia that was the same as the dopa-induced type appeared in the right arm. Efficacy appeared quickly after several puffs and lasted for about 10 minutes. We made him try to smoke again after this efficacy disappeared rapidly. He could obtain the same efficacy by smoking the second cigarette, but the subsequent third smoking had no efficacy and tremor became slightly worse. Nicotine chewing-gum (Nicorette®) revealed similar but milder efficacy, and its efficacy lasted about 30 minutes.

Summary of Smoking Efficacy

We examined the efficacy of one cigarette containing 0.9 mg to 1.1 mg nicotine in 4 AR-JP patients (Table 2). Improvement of symptoms appeared soon after several inhalations, and lasted for 5 to 30 minutes. Thereafter it disappeared rapidly. The patients noticed relaxation in the rigid extremities following smoking. The efficacy was observed in tremor, rigidity, and bradykinesia. Improvement of gait disturbance was more remarkable. Discrete movement of the hand was not improved. On the contrary, tremor was increased slightly in one case. Dyskinesia similar to the dopa-induced type appeared mildly in the extremities in 3 cases. We were unable to find this efficacy with the subsequent second or third cigarette. We also tried the efficacy of nicotine chewing-gum (containing 2 mg nicotine) in 14 AR-JP patients. Five smoking-effective patients revealed milder effects than those achieved by cigarette smoking, and 4 other patients with AR-JP had much milder effects in parkinsonian symptoms by nicotine chewing-gum. This indicates that smoking efficacy might be based on a nicotinic pharmacological action in the CNS.

CATECHOLAMINE METABOLISM IN THE RAT STRIATUM AFTER NICOTINE ADMINISTRATION

To clarify the mechanism of smoking efficacy observed in AR-JP, we examined, using the method of brain microdialysis, catecholamine metabolism in the rat striatum after nicotine administration.

The probe was inserted by means of a stereotaxic apparatus into the striatum of Wistar male rats under pentobarbital anesthesia. Ringer's solution was perfused at a flow rate of 2 μl/min and the sample solution was obtained. After the metabolites had reached a constant level, 0.5 mg/kg nicotine was subcutaneously injected. Samples were collected thereafter every 20 minutes for 4 hours. The same volume of saline was injected subcutaneously as the control. To examine the interaction with nicotinic acetylcholine receptor (nAChR) antagonist, hexamethonium or mecamylamine was administered before the injection of nicotine. We measured levels of DOPAC, homovanillic acid (HVA), DA, and 5-hydroxyindoleacetic acid (5-HIAA) by HPLC.

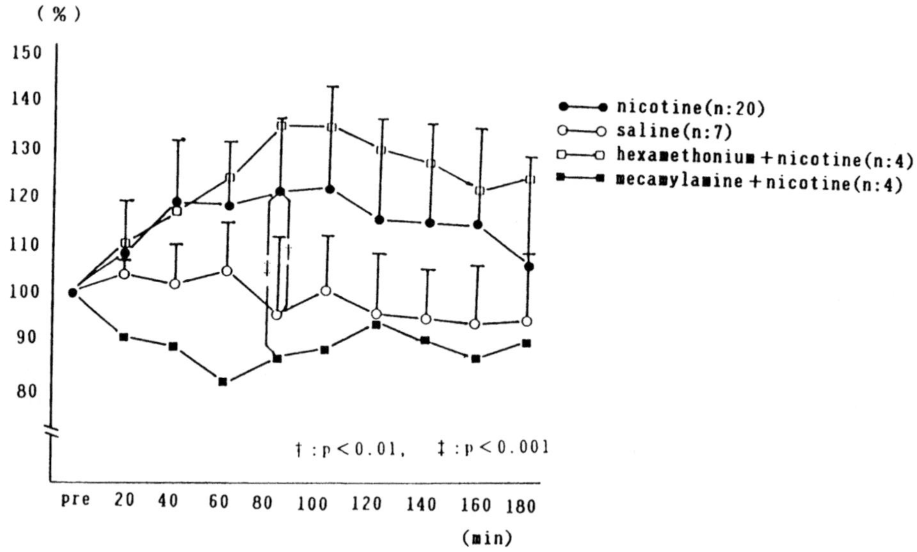

Fig. 4. Changes in striatal DOPAC levels following administration of nicotine, saline, hexamethonium and nicotine, or mecamylamine and nicotine.

Fig. 4 shows the changes in DOPAC levels following nicotine injection, and interaction with nAChR antagonists. After the nicotine injection, the DOPAC levels were obviously increased. The maximum level was found to be 119%. The central-active nAChR blocker mecamylamine (5 mg/kg, i.p.) suppressed this increase in DOPAC level following nicotine injection. In contrast, the peripheral-active nAChR blocker hexamethonium (10 mg/kg, i.p.) did not affect this increase. These data indicate the increase in DOPAC levels following nicotine injection was due to the central action of nicotine.

HVA levels following nicotine injection were also increased, and the increase was also suppressed by mecamylamine but not by hexamethonium. DA levels did not change following nicotine injection. The pattern of 5HIAA levels revealed a similar change to those of DOPAC and HVA, but the change was not statistically significant.

DISCUSSION

Several clinical and pathological cases of JP have been reported.[1-10] In JP, an autosomal recessive type with minor dystonia and improvement following sleep was found mainly in Japan.[2,9] The cases reported by Yamamura et al.[2] in 1973, and the cases reported by Sunohara et al.[9] as idiopathic dystonia-parkinsonism in 1985, are considered to be the same disease as our cases of AR-JP based on inheritance type and clinical symptoms. But minor differences were observed. In the cases of Yamamura et al., moderate dystonia was found in the extremities and trunk, and in idiopathic dystonia-parkinsonism of Sunohara et al., neck dystonia was observed in 3 cases of 5. AR-JP is considered to be classified in the Ia group of JP reported by Narabayashi et al.[10] Ota et al.[4] in 1958 reported a case similar to ours from neuropathological observation, and they found neuronal loss in the substantia nigra without description about the existence of Lewy bodies. In 1984, Yokochi et al.[1] reported an autopsied case that showed a decrease in melanin pigment content of nigral neurons with normal cell population. But this case seems to differ from ours in the clinical features and the age at onset.

In our patients with AR-JP we found smoking efficacy. There were several clinical reports on the influence of smoking on neurological symptoms, such as increase in ataxia in patients with spinocerebellar degeneration or multiple sclerosis,[11] decrease in hemiballism,[12] and decrease in parkinsonian symptoms in post-encephalitic parkinsonism.[13] As for tremor there were two different reports: Shiffmann et al.[14] observed an increasing effect; and Marshall et al.,[15] a decreasing one. Locomotion activity in MPTP-treated mice was found to be increased by nicotine.[16] But this is the first time to present the efficacy of smoking in several cases with AR-JP.

After smoking, the blood concentration of nicotine was reported to increase to the maximum level in 10 minutes,[17] and nicotine was found in the CNS in 7 seconds.[18] Nicotine was found to activate the nigrostriatal dopaminergic system and thus release dopamine in the striatum,[19] and the existence of a nicotinic presynaptic receptor on the terminal of the dopaminergic neuron was considered.[20] Nicotine induced firing of neurons in the compact zone of the substantia nigra,[21] and increased dopamine turnover in the striatum.[22] Recently the increase in DA in the nucleus accumbens by local administration of nicotine was reported.[23]

Nicotinic action in the CNS has not been well studied previously. Our smoking and nicotine chewing-gum efficacy in AR-JP and the data of increases in catecholamine metabolites obtained by brain microdialysis <u>in vivo</u> revealed that nicotine might have a stimulant effect on the catecholamine system in the CNS. The nicotinic action in the CNS may differ from muscarinic action even with respect to AChR. This nicotinic effect may not be restricted to AR-JP, but AR-JP may be easily influenced by nicotine.

Distribution of nAChR in the CNS was reported and there were large numbers of nAChRs in the striatum and substantia nigra.[24] Recently it was reported that nAChR was decreased in the frontal cortex in PD and Alzheimer's disease.[25]

As another interesting finding, many epidemiologic studies showed a negative correlation between smoking rate and incidence of PD.[26] Such studies may have some correlation to the nicotinic action in PD.

Our present study and previous reports indicate that the nicotinic action in the CNS in parkinsonism is worthy of our attention. These studies may contribute to a better understanding of the pathophysiology of AR-JP and PD and help lead to a new approach for the treatment of parkinsonism.

REFERENCES

1. M. Yokochi, H. Narabayashi, R. Iizuka, and T. Nagatsu, Juvenile parkinsonism - Some clinical, pharmacological and neuropathological aspects, in: "Advances in Neurology," R.G. Hassler and J.F. Christ, ed., pp. 407-413, Raven Press, New York (1984).
2. Y. Yamamura, I. Sobue, K. Ando, M. Iida, T. Yanagi, and C. Kono, Paralysis agitans of early onset with marked diurnal fluctuation of symptoms, Neurology 23: 239-244 (1973).
3. J. R. Hunt, Progressive atrophy of the globus pallidus (Primary atrophy of the pallial system), Brain 40: 58-148 (1917).
4. Y. Ota, S. Miyoshi, O. Ueda, T. Mukai, and A. Maeda, Familial paralysis agitans juvenilis: A clinical, anatomical and genetic study, Folia Psychiatr. Neurol. Jpn. 12: 112-121 (1958).
5. W. E. Martin, J. A. Resch, and A. B. Baker, Juvenile parkinsonism, Arch. Neurol. 25: 494-500 (1971).
6. K. K. Sachdev, N. Singh, and M.S. Krishnamoorthy, Juvenile parkinsonism treated with levodopa, Arch. Neurol. 34: 244-245 (1977).
7. G. Carlier, and J. M. Dubru, Familial juvenile parkinsonism, Acta Paediatr. Belg. 32: 123-127 (1979).
8. C. G. Clough, M. Mendoza, and M. D. Yahr, A case of sporadic juvenile Parkinson's disease, Arch. Neurol. 38: 730-731 (1981).
9. N. Sunohara, Y. Mano, K. Ando, and E. Satoyoshi, Idiopathic dystonia-parkinsonism with marked diurnal fluctuation of symptoms, Ann. Neurol. 17: 39-45 (1985).
10. H. Narabayashi, M. Yokochi, R. Iizuka, and T. Nagatsu, Juvenile parkinsonism, in: "Handbook of Clinical Neurology," P.J. Vinken, G.W. Bruyn, and H.L. Klawans, eds., pp. 153-165, Elsevier, Amsterdam (1986).
11. J. D. Spillane, The effect of nicotine on spinocerebellar ataxia, Br. Med. J. 2: 1345-1351 (1955).
12. D. Grigoresco, and S. Axente, Syndrome luysien: Influence du tabac sur les grands mouvements d'hemiballisme, Rev. Neurol. 1: 361-363 (1931).
13. H. Moll, The treatment of post-encephalitic parkinsonism by nicotine, Br. Med. J. 1: 1079-1081 (1926).
14. S. M. Shiffmann, E. R. Gritz, J. Maltese, M.A. Lee, N.G. Schneider, and M.E. Jarvik, Effects of cigarette smoking and oral nicotine on hand tremor, Clin. Pharmacol. Ther. 33: 800-805 (1983).
15. J. Marshall, and H. Schnieden, Effect of adrenaline, noradrenaline, atropine, and nicotine on some types of human tremor, J. Neurol. Neurosurg. Psychiatry 29: 214-218 (1966).
16. H. Sershen, A. Hashim, and A. Lajtha, Behavioral and biochemical effects of nicotine in an MPTP-induced mouse model of Parkinson's disease, Pharmacol. Biochem. Behav. 28: 299-303 (1987).
17. K. S. Rotenberg, R. P. Miller, and J. Adir, Pharmacokinetics of nicotine in rats after single cigarette smoke inhalation, J. Pharmacol. Sci. 69: 1087-1090 (1980).
18. W. A. Check, New knowledge about nicotine effects, JAMA 247: 2333-2338 (1982).
19. T. C. Westfall, Effect of nicotine and other drugs on the release of ^{3}H-norepinephrine and ^{3}H-dopamine from rat brain slices, Neuropharmacology 13: 693-700 (1974).
20. M. F. Giorguieff, M. L. LeHoc'h, T. C. Westfall, J. Glowinski, and M.J. Besson, Nicotinic effect of acetylcholine on the release of newly synthesized [^{3}H]dopamine in rat striatal slices and cat caudate nucleus, Brain Res. 106: 117-131 (1976).
21. W. Lichtensteiger, D. Felix, R. Lienhart, and F. Hefti, A quantitative correlation between single unit activity and fluorescence intensity of dopamine neurons in zona compacta of substantia nigra,

as demonstrated under the influence of nicotine and physostigmine, Brain Res. 117: 85-103 (1976).

22. K. Andersson, K. Fuxe, L. F. Agnati, and P. Eneroth, Effects of acute central and peripheral administration of nicotine on ascending dopamine pathways in the male rat brain: Evidence for nicotine induced increase of dopamine turnover in various telencephalic dopamine nerve terminal systems, Med. Biol. 59: 170-176 (1981).

23. J. C. Mifsud, L. Hernandez, and B. G. Hoebel, Nicotine infused into the nucleus accumbens increases synaptic dopamine as measured by in vivo microdialysis, Brain Res. 478: 365-367 (1989).

24. P. B. S. Clarke, C. B. Pert, and A. Pert, Autoradiographic distribution of nicotine receptors in rat brain, Brain Res. 323: 390-395 (1984).

25. P. J. Whitehouse, A. M. Martino, M. V. Wagster, D. L. Price, R. Mayeux, J. R. Atack, and K. J. Kellar, Reductions in [^{3}H]nicotinic acetylcholine binding in Alzheimer's disease and Parkinson's disease: An autoradiographic study, Neurology 38: 720-723 (1988).

26. M. D. Nefzger, F. A. Quadfasel, and V. C. Karl, A retrospective study of smoking in Parkinson's disease, Am. J. Epidemiol. 88: 149-158 (1968).

THE PROGNOSTIC COMPARISON BETWEEN THE PROGRESSIVE CLINICAL COURSE AND THE STABLE COURSE OF PARKINSON'S DISEASE OVER 10 YEARS

Takehiko Ogawa, Masao Kinoshita, and Fuminori Segawa

The Fourth Department of Internal Medicine, Toho University
2-17-6 Ohashi, Meguro-ku, Tokyo, Japan

Introduction

Parkinon's disease usually has a progressive clinical course, though the speed of progression is variable according to the cases. After dopamine treatment has been induced, patients' prognosis has been dramatically improved. But according to the recent literature, 81% of 191 patients are classified above stage III by Yahr scale, and they need some help in their daily activities of living when the duration lasts over 10 years. The data collected here from these patients offered an opportunity to explore the difference between patients in milder stages or advanced stages of illness.

Subjects

We have studied 23 patients who had Parkinson's disease(PD) over 10 years and had been followed at Toho University, Ohashi Hospital. They were classified into 2 groups by Yahr scale. Group A consisted of 13 patients of stages I and II; group B consisted of 10 patients of stage III and above. Each patient recieved dopamine treatment, and other antiparkinsonism medication, icluding amantadine hydrochloride or anticholinergic drugs. In recent years, dopamine agonists was added 2 of the 23 patients.

Methods

We evaluated the duration of illness prior to dopamine treatment, and checked levodopa dosage at 5 and 10 years after onset. When levodopa was given with peripheral acting dopa inhibitor, levodopa dosage is multiplied by 5 to convert to a common base. Eleven symptoms and signs of PD (tremors, rigidity, akinesia, postural instability, shuffling gait, start-hesitation, falling, masked face, constipation, dementia, depression) were observed by physicians. We analysed the incidences of clinical fluctuations (wearing-off and on-off phenomena) and side effects of levodopa (dopa-induced dyskinesia, psychiatric side effects). And four questions about their behaviours at the daily living were answered by the patients, or their families (physiotherapy at home, having hobbies, doing house work and going out for a walk). Each of the items was graded 0 for absence, 1 for presence and was multiplied in a numeric score.

Edited by T. Nagatsu *et al.*, Plenum Press, New York, 1990

Table 1

	Group A	Group B	p value
Samples			
Sample size	13	10	
Male:female	5:8	3:7	
Mean Yahr stage at study	1.85±0.38	3.48±0.70	
Mean age at study	65.8±9.6	69.5±7.5	NS
Therapeutic Course			
Mean age at onset	52.8±10.5	57.0±7.8	NS
Duration of PD (yrs)	12.9±2.8	12.5±3.2	NS
Duration of illness before levodopa tx (yrs)	4.23±3.37	1.50±1.84	<0.05
Symptoms and Side effects			
Mean score of symptoms	5.46±2.70	6.30±1.89	NS
Incidence of main symptoms			
Akinesia	46.2%	60.0%	
Rigidity	53.8%	70.0%	
Tremor	92.3%	100%	
Postural instability	38.5%	60.0%	
On-off	23.1%	80.0%	
Wearing-off	23.1%	80.0%	
Psychiatric side effect	23.1%	20.0%	
Dopa induced dyskinesia	76.9%	60.0%	
Behabiour at daily living	2.38±1.19	1.17±0.75	<0.05

Table 2

	Group A	Group B	p value
Dosage at 5 years (mg)	1480±1549	2600±1410	NS
At 10 years	2231±1305	3325±1041	<0.05
At the time of the study	2058±1279	3373±1062	<0.02

Results

Table 1 shows no significant differences among the two groups with respect to the mean age of onset, nor the mean duration of illness. Two groups were similar with regard to the mean scores of variable symptoms. As to four main symptoms of PD (akinesia, rigidity, tremor and postural instability), the incidence of postural instability tends to be higher in Group B than those of Group A. The patients of Group B showed a high rate of clinical fluctuations. With regard to the behavior at the daily living, the mean score of group A was significantly higher.

In Table 2 mean dosasage of levodopa was summarized and Group B had significantly higher dose than that of Group A at 10 years and at the time of the study, whereas they showed no significant differences at 5 years after disease onset.

Discussion

In long-term follow up studies of levodopa treatment, Markham et al recommended early levodopa treatment, whereas Lesser et al suggested that the treatment should be delayed as much as possible. Our data supported the benefits of beginning dopamine treatment as late as possible. During long term treatment of levodopa, wearing-off and on-off phenomena were common in patients whose final level of ADL was rated to be poor. Among the main symptoms of PD, the patients who had more progressive course of illness showed high incidence of postural instability. The dopa dosage was increased after 5 years of clinical course when they reached progressive stages. This suggests a poor response to levodopa therapy in those advanced patients. The behaviour at the daily living is one of the most important factors influencing the therapeutic course.

References

1. Kase M, Atarashi J, et al: Twelve year's course of Parkinson disiease with levodopa treatment. Advances in Neurological Sciences 28: 681-691, 1984.
2. Grnerus A-K, Carlsson A, Svanborg A: The aging neuron-influence on symptomatology and therapeutic response in Parkinson's syndrome. Advances in Neurology 24: 327-334, 1978.
3. Barbeau A: Treatment of Parkinson's disease with L-dopa and Ro 4-4602: Review and present status. Advances in Neurology 2: 173 -198, 1973.
4. Lesser R-P, Fahn S, Snider S-R, Cote L-J, et al: Analysis of the clinical problems in Parkinsonism and complications of long-term levodopa therapy. Neurology 29: 1253-1260, 1979.
5. Markham C-H, Diamond S-G: Evidence to support early levodopa therapy in Parkinson disease. Neurology 31: 125-131, 1981.

FAMILIAL PARKINSON'S DISEASE AND FAMILIAL MANIC-DEPRESSIVE ILLNESS

Akira Takei, Shigeru Chiba, Yuzuru Sato, and Tsutomu Miyagishi

Department of Psychiatry and Neurology, Asahikawa Medical College, Asahikawa, 078 Japan

INTRODUCTION

There are only a few reports of familial Parkinson's disease (PD) and familial manic-depressive illness (MDI) being found in an individual family.[1-3] The present study suggests a genetic association between PD and MDI.

REPORT OF CASES

The family pedigree is illustrated in Figure 1. Two grandparents of the patients (Generation I) were all from the same isolated farm village, where marriage in the neighborhood had been common. In Generation II, there were two couples of cousin marriage, and out of the nine siblings (Generation III) from one couple, there were two that developed PD (D and E) and two, MDI (B and C).

The clinical features of the four cases are summarized in Table 1. None of them showed both PD and MDI. The two patients with MDI showed a good response to drug treatment, and in those with PD, L-dopa was effective but the neurological symptoms slowly progressed. At present, C and D are being treated in our hospital, B is in another mental hospital, and E is attending a medical college.

In addition to the four cases, A and F had previous episodes that suggested psychiatric disorders. Case A developed several episodes with delusion of persecution (e.g. "Somebody is going to kill me") and psychomotor excitement, and was placed under house detention in her adulthood. Case F was particular about keeping clean(e.g., washing hands many times). Although A appears to have been psychotic, and F, mysophobic, more details of the psychiatric symptoms in these two cases were not clear.

DISCUSSION

In the present study, two of the nine siblings from a couple of cousin marriage were diagnosed as having PD, and the other two as having MDI. A few occurrences of PD and MDI in multiple members of a family have been reported. Taguchi et al.[1] described two cases of PD and one case of depression in two generations of a family, and Kaiya et al.[2]

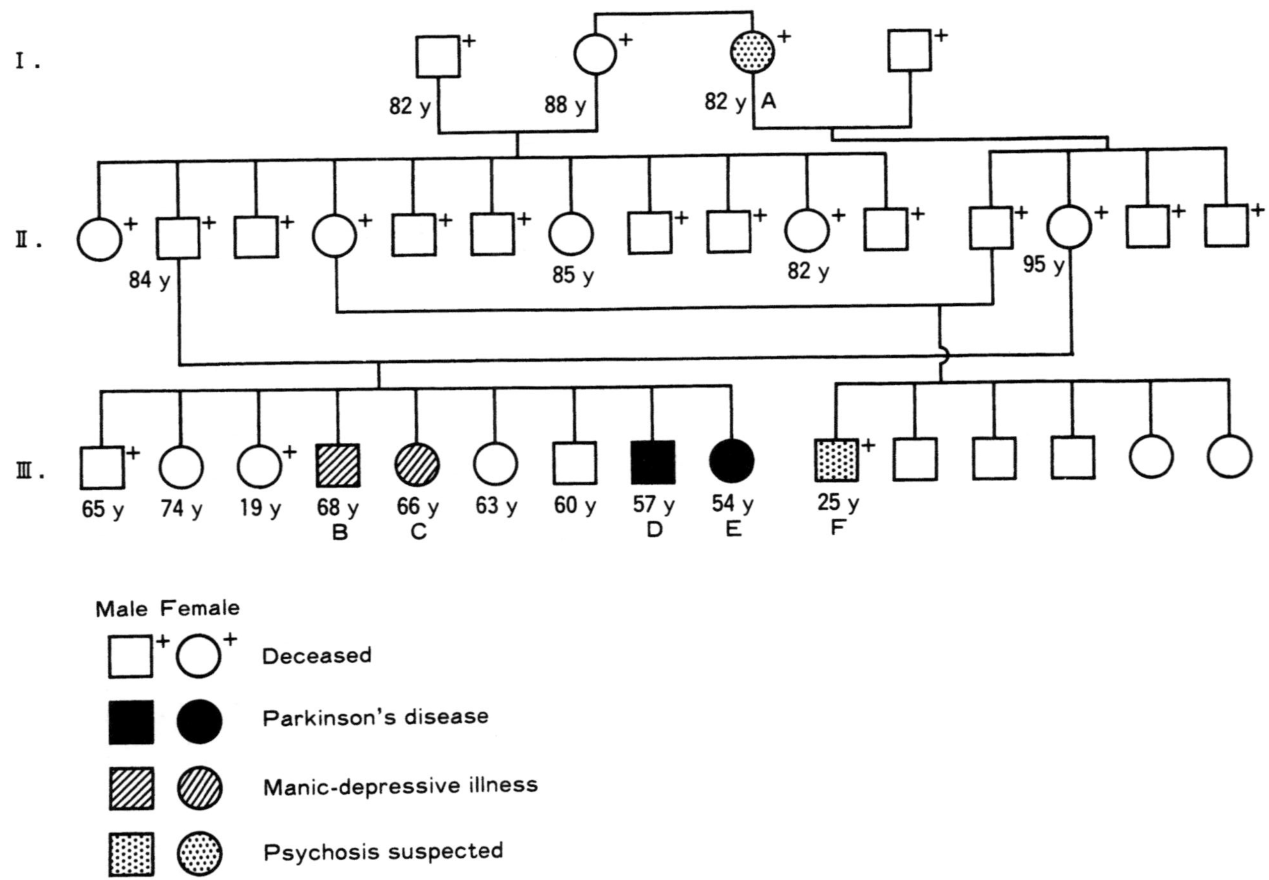

Figure 1. Family pedigree.

Table 1. Clinical features of four patients (two with PD and two with MDI).

	Case B	Case C	Case D	Case E
age/sex	68/M	66/F	57/M	54/F
diagnosis	MDI	MDI	PD (Yahr III)	PD (Yahr II)
age at onset	45	55	31	25
disease duration (yrs.)	23	11	26	29
history of depressive/ manic phases	depressive (3 times), manic (4 times)	depressive only (5 times)	negative	negative
neurologic symptoms	negative	negative	Parkinsonian symptoms, L-dopa-induced dyskinesia	Parkinsonian symptoms, L-dopa-induced dyskinesia
drugs used for therapy	lithium carbonate, haloperidol, tricyclic antidepressants	tricyclic antidepressants	L-dopa, anticholinergic drugs	L-dopa, anticholinergic drugs

also reported a family in which mother was diagnosed as having PD and her son, as having MDI. Tune et al. [3] found a family showing three patients with PD and MDI, one with PD only, and one with MDI only in three generations. These studies and the present study suggest a genetic association between these two diseases.

Characteristics of the padigree of this family were that the parents of the patients were cousins and that marriage was repeated in the neighborhood of an isolated farm village, thought the generation of the parents. Such consanguine marriages did not occur in other reported cases of PD and MDI, but it may be important for investigating the causes of the multiple occurrence of PD and MDI in this family to consider involvement of genetic factors in the pathogenesis ofthe two diseases.

From the neurochemical viewpoint, disturbances of catecholamine and serotonin metabolism are also considered to play an important role in the pathogenesis of MDI. [4] In PD, on the other hand, a decrease in dopamine metabolism in the striatum and a wide variety of disorders in catecholamine and serotonin metabolism have been reported. [5] Therefore, the neurochemical features of amine metabolism are considered to overlap partially between the two diseases . [1-3]

From these findings, PD and MDI may be thought to have a common feature in both neurochemistry and genetics.

REFERENCES

1. H. Kaiya, S. Adachi, H. Yoshida, and M. Namba, A family consisting of a Parkinsonian mother and a manic-depressive son with dyschromatopsia, Japanese J. of Clin. Psychiatry 8:105(1979).
2. K. Taguchi, S. Watanabe, Y. Nakashima, and S. Otsuki, A family with two cases of Parkinsonism and one case of depression in two generations, Seishin Igaku 15:45(1973).

3. L. E. Tune, M. Folsein, P. Rabinson, G. Jayaram, and P. McHugh, Familial manic-depressive illness and familial Parkinson's disease: A case report, Johns Hopkins Med. J. 151:65(1982).
4. J. J. Schildkraut, The catecholamine hypothesis of affective disorders: A review of supporting evidence, Am. J. Psychiatry 122: 509(1965).
5. O. Hornykiewicz and S. J. Kish, Biochemical pathophysiology of Parkinson's disease, in: "Advances in Neurology, Vol. 45, Parkinson's disease," M. D. Yahr and K. J. Bergmann, eds., Raven Press, New York(1986).

NEUROPSYCHOLOGICAL DISTURBANCES IN PARKINSON'S DISEASE

Yasuo Iwasaki, Masao Kinoshita, Ken Ikeda, and
Kiyoshi Takamiya

The Fourth Department of Medicine
Toho University, Tokyo, Japan

INTRODUCTION

Idiopathic Parkinson's disease (PD) is heterogenous in many respects. Mental changes may also range from a cluster of deficits to global dementia. Disease severity has been identified as a factor predictive of cognitive dysfunction in PD. Mortimer et al. measured symptoms severity and found that bradykinesia was correlated with performance of visuospatial tasks. However,the strength of the association was not impressive, leading those authors to speculate that many other factors could contribute to patients scores. The present study was undertaken to elucidate the nature and frequency of cognitive impairment in PD by comparing group of PD patients against age and education matched normal counterparts on a extensive range of neuropsychological abilities. We also examined the relationship between severity of the motor disabilities and intellectual impairment in PD.

METHODS

Thirty patients with idiopathic PD participated. At the time of testing all patients were receiving levodopa. None of the patients met DSM-III criteria for dementia. PD patients had a mean age of 68.7 years and had completed a mean of 11.9 years of education. The mean duration of illness was 5.9 years. Thirty normal controls also participated. None of the controls had evidence of neurologic disease or a history of alcoholism. All subjects gave informed consent.

PROCEDURE

NEUROPSYCHOLOGICAL ASSESSMENT

1.Mini-mental state examination (MMS)
2.Zung self-rating depression scale (SDS)
3.Memory-learing of ten words for five trials,ten minute delayed verbal recalls.

Basic, Clinical, and Therapeutic Aspects of Alzheimer's and Parkinson's Diseases, Volume 2
Edited by T. Nagatsu *et al.*, Plenum Press, New York, 1990

NEUROLOGICAL ASSESSMENT

Each patients with PD was examined by the complete clincal scale by Yahr. Major parkinsonian signs,bradykinesia, tremor and rigidity,were also rated on a scale from 0 to 3,with 0 indicating the absence of a symptoms and 3 indicating severity.

RESULTS

No statistically significant difference was found between the PD patients and the controls with respect to age and education. The mean MMS score for the patients was 26.1 and 29.0 for the controls. This difference was significant. The mean SDS score of the patients was higher than that of the ocntrols. The mean index of memory performance was significantly lower in the PD patients group compared to the control group.

RELATIONSHIP BETWEEN COGNITIVE DYSFUNCTION AND MOTOR DISABILITY IN PATIENTS WITH PD

A significant correlation between MMS score and bradykinesia existed in the patients. There was a siginificant negative correlation between rigidity and memory impairment in patients with PD.

DISCUSSION

There appears to be a distinct relationship between intellectual impairment in patients with PD and classic triad of motor symptoms: bradykinesia,tremor,and rigidity. Previous research found significant relationship between severity of bradykinesia and intellectual impairment. In our study,severity of bradykinesia was associated with results of MMS,and a significant relationship between severity of rigidity and memory impairments was found. Damage to the dopamine-containg mesocorticolombic pathway is not known to occur in patients with PD. The neuropsychological abnormalities observed in this study might be caused by the destruction of this pathway,and a comparable loss of dopaminergic neurons in the neighboring nigrostriatal bundle could explain the reported correlation between motor deficits and intellectual decline.

CONCLUSIONS

The performance of patients with Parkinson's disease (PD) on selective neuropsychological tests was examined in regard to the applicability of such examinations to diagnosis. Thirty patients with PD,and 30 age-and education-matched controls were given a battery of tests designed to assess motor and intellectual functions. The PD group displayed significantly lower scores on all tests than those in the control group. Correlation anlyses on the several motor and neuropsychological results in PD group revealed that there was a significant negative correlation between bradykinesia and Mini-mental state score,and rigidity and memory tests. Results are discussed with respect to the nivolvement of dopaminergic pathways in the central nervous system.

REFERENCES

Bowen FP,Hoehn MM,Yahr MD:Alterations in spatial orientation as determined by a route walking test.
Neuropsychologia 10: 335-361,1972.

Folstein MF,Folstein SE,McHugh PR:Mini-mental state:A practial method for grading the mental state of patients for the clinician. J Psychiatr Res 12: 189-198,1975.

Galassi R,Montagna P,Ciardull C,Lorusso S,Mussuto V,Stracciar A: Cognitive impairment in motor neuron disease.
Acta Neurol Scand 71:480-484,1985.

Mortimer JA,Pirozzolo FJ,Hansch EC,Webster DD:Relation of motor symptoms to intellectual deficits in Parkinson's disease.
Neurology 31: 133-137,1982.

Zung WWK:A self-rating depression scale.
Arch General Psychiatry 12: 63-70,1965.

STUDY ON INCIDENCE OF CANCER AND SMOKING RATE IN PARKINSON'S DISEASE

Ken Ikeda, Masao Kinoshita, Toshiya Shiojima, Yasuo Iwasaki, and Kiyoshi Takamiya

The Fourth Department of Internal Medicine
Toho University Ohashi Hospital, Tokyo, Japan

INTRODUCTION

Parkinson's disease is a degenerating disorder of the central nervous system characterized by akinesia, rigidity and tremor. The disease usually begins after the 6th decades of the age, however, a low cancer rate in this disease has been reported(1,2). Jansson et al(2) reported the cancer rate in the parkinsonian patients was about one-third compared to the general populations. In Japan, epidemiologic research for Parkinson's disease has been conducting not so mach and there is only one report by Harada and colleagues(3). As the treatment in Parkinson's disease being progressed so, the survival period may be expected more prolongly. Accordingly, the cancer incidence may be changing because the geriatric populations are increasing. From above mentioned documents it is important to investigate the cancer rate in Parkinson's disease. In the present study, we reported the cancer and smoking rate in the patients with Parkinson's disease.

MATERIALS and METHODS

We studied the cancer and smoking rate among 133 patient swith Parkinson's disease who admitted in our hospital between 1976 and 1988. The age ranged between 41 and 80 years old. The average age was 64.1±10.8(64.2±11.1 for men, 64.8±10.7 for women). At the last follow-up the duration of Parkinson's disease was an average of 6.3 4.5 years and the range varied one to 15 years. (6.2±4.5 for men, 6.4±4.5 for women). The sex ratio (male/female) was 52:81. We studied the following variables:
Age and sex-specific cancer rate;
smoking status at the onset of Parkinson's disease;
relationship between smoking and cancer;
correlation between during the treatment period and cancer. Furthermore, These results were compared with the cancer rate of Japanese statistics and information department, minister's secretariat and smoking rate of national investigation in the general population by two side test.

RESULTS

Age and sex-specific cancer rate

The results are summerized in Table 1.
The relative ratio was defined as parkinsonian cancer rate divided by cancer rate of the general populations. The cancer rate and relative ratio of parkinsonian group were 6.01% and 0.60,respectively(3.84% and 0.40 for men, 7.40% and 0.72 for women.)

The sex ratio was 0.51. These results were not significantly different from those for the general populations,but seemed to be lower.

As far as age-specific cancer rate,there were no patients associated with cancer for the ages from 35 to 54,and between 75 and 79 years.
The cancer rate and relative ratio,aged between 55 and 64 years,were 6.06% and 0.48,respectively. However,the age between 65 and 69 years,the cancer rate and relative ratio were 12.50% and 1.01,respectively and showed almost the same cancer rate. The cancer rate and relative ratio for the ages between 70 and 74 years were 8.75% and 0.74, respectively. For the ages between 80 and 84,the cancer rate was 16.66% and relative ratio was 2.50. There was no significant differences,but this was considered to be due to the low number of patients.

As regards for age and sex-specific cancer rate, only 2 female patients whose ages were between 55 and 64 years, were associated with cancer. Then the cancer rate and relative ratio were 9.09% and 0.68,respectively. For the ages between 65 and 69 years,the cancer rates were 14.28% for men and 11.1% for women. Each rate was almost the same figure to those of the general populations.
The cancer rate and relative ratio aged between 70 and 74 years were 8.33%,0.58 for men and 8.69%,0.99 for women, respectively. For the ages more than 80 years old,the cancer rate and relative ratio for women were 16.66% and 3.57,respectively.

Smoking rate

Among 133 patients with Parkinson's disease smokers were only 6 males. The smoking rate was 4.51% in parkinsonian patients(11.53% for men and 0% for women). Accordingly the smoking rate in Parkinson's disease was considerably lower than the rate for the general community.

Relationship between smoking and cancer

Eight patients with Parkinson's disease were having with cancer. However,these all patients were non-smokers. No correlations could be obtained between smoking and cancer in this study.

Correlations between during the treatment period and cancer

Eight patients(2 males,6 females) developed the cancer during survey,in whom the primary lesions were variable; uterus,ovary,mamma,stomach,colon,liver and pancreas. Within these 6 women,3 cases were originated from the genital cancer. However,there was no relationship between duration of the treatment in Parkinson's disease and the development of cancer,because 3 patients affected the cancer before the onset of Parkinson's disease.

Table 1.Age and sex-specific cancer rate for Parkinson's disease and control subjects

Age group	34~44			45~54			55~64			65~69			70~74			75~79			80~84			34~84		
	No. of total	M*	F**	No. of total	M	F	No. of total	M	F	No. of total	M	F	No. of total	M	F	No. of total	M	F	No. of total	M	F	No. of total	M	F
Case	5	2	3	21	10	11	33	11	22	16	7	9	35	12	23	17	10	7	6	0	6	133	52	81
Cancer																								
No.	0	0	0	0	0	0	2	0	2	2	1	1	3	1	2	0	0	0	1	0	1	8	2	6
%	0	0	0	0	0	0	6.06	0	9.09	12.50	14.28	11.11	8.57	8.33	8.69	0	0	0	16.66	0	16.66	6.01	3.84	7.40
Control																								
%	7.14	3.65	12.64	9.41	6.44	14.30	12.50	12.03	13.09	12.34	14.22	10.53	11.14	14.23	8.76	8.91	12.37	6.69	6.65	10.51	4.66	9.86	9.53	10.20
RR***							0.48		0.68	1.01	1.00	1.05	0.76	0.58	0.99				2.50		3.57	0.60	0.40	0.72

*M:Male, **F:Female, ***RR:Relative ratio=parkinsonian cancer rate/control cancer rate

DISCUSSION

Doshay(4) reprted the first low cancer rate in the patients with Parkinson's disease. Subsequently a low cancer incidence has been contradicted by others(5). Recently, Jansson et al(2) reported that the cancer rate was about one-third compared to the general populations, but the cancer risk increased during the treatment period. It is well known that there are no obesity, the low smoking rate and high total body potassium levels among parkinsonian patients(6). These factors might protect the patients against cancer. Although low cancer rates in Parkinson's disease has been pointed out by many investigators, it is possible that the cancer rate could increase due to prolongation of survival span in parkinsonian patients.

In this study, we examined the cancer and smoking rate among 133 patients in Parkinson's disease.
The cancer rate and relative ratio in Parkinson's disease were 6.01%and 0.60, respectively. The sex ratio was 0.51 and the cancer rate for women was about two times that of men. The cancer rate was lower than that for the general pupulations. Nevertheless, these results were not significantly different and this might influence the low number of parkinsonian patients.

However, there was the same cancer rate in the patients for the ages between 65 and 69 years. Furthermore, the ages between 80 and 84 years, the cancer rate increased about 2.5 times compared to the general populations.
The primary organs of cancer were varied, but within the femal patients the genital organs were half the originated cancer. In Parkinson's disease, the smoking rate was 4.51% and this was significantly lower than that of the general populations. There were no correlations among cancer, smoking, duration of the treatment and onset age of Parkinson's disease.

Our results contribute to the incidence and rate of cancer during follow up in the parkinsonian patients. Further studies including autopsy findings were needed to established the more exact cancer rate in Parkinson's disease.

REFERENCES

1)Barbeau A, Joly JG: Parkinson et cancer. Union Med Can 92:169-174, 1963
2)Jansson B, Jankovic J: Low cancer rates among patients with Parkinson's disease. Ann Neurol 17:505-509, 1985
3)Harada H, Nishikawa S, Takahashi K: Epidemiology of Parkinson's disease in a Japanese city. Arch Neurol 40: 151-154, 1983
4)Doshay LJ: Problem situations in the treatment of paralysis agitans. JAMA 156:680-684, 1954
5)Westlund K, Hougen A: Cancer as a cause of death among patients with other chronic disease. JAMA 162:1003, 1956
6)Ellis KJ, Shukla KK, Cohn SH, Pierson RN: A predictor for total body potassium in man based on height, weight, sex, and age: applications in metabolic disorders. J Lab Cli Med 83:716-727, 1974

HIGH SPEED MEMORY SCANNING IN PARKINSON'S DISEASE: ADVERSE EFFECT OF L-DOPA

W. Poewe, Th. Benke, W. Berger, and L. Schelosky

University Hospital for Neurology
A-6020 Innsbruck, Anichsstrasse 35, Austria

INTRODUCTION

Cognitive slowing or bradyphrenia has been claimed to form a characteristic part of the mental changes found in patients with idiopathic Parkinson's disease [1] and it has been linked to meso-cortico-limbic dopamine deficiency [2]. Pure mental slowness is difficult to prove in these patients because of interfering motor retardation and previous studies with different reaction time paradigms have yielded conflicting results [3,4]. In the present study the Sternberg paradigm [5] was used to assess high speed memory scanning in a group of patients with fluctuating Parkinson's disease and healthy controls.

PATIENTS AND METHODS

12 Patients with idiopathic Parkinson's disease (PD) and dose-dependent, predictable motor oscillations in response to oral L-Dopa were included in the study (4 females, 8 males; mean age 57 yrs.; mean duration of PD 10.5 yrs.; mean duration of L-Dopa 8.5 yrs.; mean L-Dopa dose 1211 mg/d; mean Hoehn & Yahr stage 1.3 when 'on' and 3.7 when 'off'). 13 age-, sex- and IQ-matched healthy individuals volunteered as controls.

High speed memory scanning was studied by a modified version of the Sternberg paradigm, which is designed to measure the rate at which short term memory is scanned mentally. Subjects are asked to remember sets of digits (up to a set-size of four for the present study) presented on a screen at a frequency of one digity every 2 seconds. After each set and following an acoustic cue they are then presented with a single test digit and required to decide as fast as possible whether or not this number was a member of the previous set or not by pressing a 'yes' or 'no' button held in the right ('yes') and left ('no') hand. In healthy volunteers Sternberg (1966) has shown a linear correlation between set-size and reaction time (RT) with an RT-increase of approximately 30msec per additional item in the memory set. When RT's are plotted against set-sizes the slope of the linear curve is a measure of the purely cognitive speed of memory scanning while its extrapolated-intercept represents the motor part of the measured RT.

In the present study patients were first tested at 8 a.m. following a drug free interval of 12 hours, when their parkinsonian signs were fully expressed ('off'), and a second session was performed following onset of clinical benefit from an individually optimal dose of L-Dopa taken immediately after completion of the first test. Controls were tested twice on one morning (8 a.m. and 10 a.m.).

RESULTS

When tested in the 'off'-condition the patients had a markedly prolonged motor component of their RT's compared to the first test session of controls. As expected this normalised following L-Dopa (table 1).

Memory scanning speed was not different between controls in session one and patients when 'off' medication but values changed significantly in both groups when retested: patients showed significant slowing of memory scanning after L-Dopa while controls had become faster (table 2). Accordingly patients' cognitive processing speed when 'on' was significantly delayed compared with that of controls in the corresponding test session.

Error rates are summarized in table 3 for all sessions of patients and controls. As can be seen patients produced more incorrect responses than controls under both test conditions. Controls improved their response accuracy upon retesting and a similar, but less clear tendency was also present in patients when retested following L-Dopa.

DISCUSSION

The present study has failed to detect evidence for slowing of high speed memory scanning in a group of patients with Parkinson's disease after temporary withdrawal of L-Dopa. This is in contrast to previous findings in an elderly subgroup of patients [6], but confirms the results obtained Rafal and co-workers [7]. In line with several other studies employing choice reaction time paradigms [4,8] it does not lend further support to the concept of generalized cognitive slowing as part of the mental changes of Parkinson's disease. Discrepant results concerning the presence of absence of mental slowing in Parkinson's disease may reflect differences in patient populations studied as well as in methodology or domains of cognitive functions tested [9,10].

More importantly this study has revealed clear effects of L-Dopa on the speed of memory scanning in patients with Parkinson's disease which significantly delayed mental processing. That this was not due to unspecific performance deterioration induced by dopaminergic effects on arousal and attention is suggested by the fact that the patients' error rate improved simultaneously and to similar degrees as that of controls when retested. These findings are in contrast to some earlier observations suggesting bradyphrenia could be ameliorated by L-Dopa [11,12] and does not support the hypothesis that cognitive slowing could result from meso-cortico-limbic dopamine deficiency [2]. The dissociation of dopaminergic effects on motor and cognitive speed in Parkinson patients as observed in this study would rather suggest that both are mediated via different neurochemical systems.

Performance deterioration on certain frontal lobe tests following L-Dopa has recently been described in a group of patients with fluctuating Parkinson's disease [13]. It was suggested that such negative effects of L-Dopa might result from 'overstimulation' of dopamine receptors occuring in the caudate nucleus when therapeutic dopamine levels are reached in the more severely denervated putamen and that this might disrupt function in fronto-caudate neuronal loops subserving cognitive processes [13]. Whether such mechanisms could be responsible for the decreased speed of memory scanning following L-Dopa in this study has to remain speculative, but it ist clear that drug effects have to be carefully controlled when studying cognitive functions in patients with Parkinson's disease.

table 1. Motor results

Group Session	PD	Co
1 ('off'for PD)	960.2 (+/-459.1)	725.4 (+/-299.9)
2 ('on' for PD)	688.7+ (+/-380.0)	726.3 (+/-246.0)

+...Comparison within groups, $p < 0.001$

table 2. Cognitive results

Group Session	PD	Co
1 ('off'for PD)	80.6 (+/-59.4)	82.4 (+/-33.5)
2 ('on' for PD)	117.4+ (+/-92.8)	46.3*♣ (+/-29.2)

+...Comparison within groups, $p < 0.002$
*...Comparison within groups, $p < 0.05$
♣...Comparison between groups, $p < 0.05$

table 3. Errors

Group Session	PD	Co
1 ('off'for PD)	2.16 (+/-2.5)	0.84♣ (+/-1.3)
2 ('on' for PD)	1.75 (+/-0.8)	0.46♠ (+/-0.6)

♣...Comparison between groups, $p < 0.05$
♠...Comparison between groups, $p < 0.002$

REFERENCES

1. D. Rogers, Bradyphrenia in parkinsonism: a historical review Psych. Med. 16:257 (1986).

2. Y. Agid, M. Ruberg, B. Dubois, and F. Javoy-Agid, Biochemical Substrates of Mental Disturbances in Parkinson's Disease, in: "Advances in Neurology," R.G. Hassler et al., eds., Raven Press, New York (1984).

3. D. Rogers, A. J. Lees, E. Smith, M. Trimble, and G. M. Stern, Bradyphrenia in Parkinson's disease and psychomotor retardation in depressive illness. Brain 110:761 (1987).

4. E. V. Evarts, H. Träväinen, and D. B. Calne, Reaction Time in Parkinson's Disease. Brain 104:167 (1981).

5. S. Sternberg, High-Speed Scanning in Human Memory, Science 153:652 (1966).

6. R. S. Wilson, A. W. Kaszniak, H. L. Klawans, and D. C. Garron, High speed memory scanning in Parkinsonsm. Cortex 16:67 (1980).

7. R. D. Rafal, M. I. Posner, J. A. Walker, and F. J. Friedrich, Cognition and the Basal Ganglia. Brain 107:1083 (1984).

8. B. Dubois, B. Pillon, F. Legault, Y. Agid, and F. Lhermitte, Slowing of Cognitive Processing in Progressive Supranuclear Palsy. Arch Neurol 45:1194 (1988).

9. E. C. Hansch, K. Syndulko, S. N. Cohen, Z. I. Goldberg, A. R. Potvin, and W. W. Tourtellotte, Cognition in Parkinson's Disease: An Event-Related Potential Perspective. Ann Neurol 11:599 (1982).

10. B. Pillon, B. Dubois, A-M. Bonnet, M. Esteguy, J. Guimaraes, J-M. Vigouret, F. Lhermitte, and Y. Agid, Cognitive slowing in Parkinson's disease fails to respond to levodopa treatment: The 15-objects test. Neurology 39:762 (1989).

11. J. de Ajuriaguerra, Etude psychopathologique des parkinsoniens, in: "Monoamines Noyaux Gris Centraux Et Syndrome De Parkinson, "J. de Ajuriaguerra et al., eds., Librairie de 1'Universite, Georg et Cie S. A., Geneve (1971).

12. D. A. Drachman, and S. Stahl, Extrapyramidal Dementia and Levodopa Lancet i:809 (1975).

13. A. M. Gotham, R. G. Brown, and C. D. Marsden, 'Frontal' Cognitive Function in Patients with Parkinson's Disease 'on' and 'off' Levodopa. Brain 111. 299 (1988).

WECHSLER ADULT INTELLIGENCE SCALE AND OTHER RELATED FACTORS FOR PARKINSON'S DISEASE PATIENTS, MAIN FOCUS: BRAIN FUNCTION

Norio Taniguchi, Hideki Kondo, Keiko Toyonaka, Goichi Makinaga
Masashi Inui, Hide-aki Kameda, and Tsuyoshi Nishimura

Department of Neuropsychiatry
Osaka University Medical School, Osaka, Japan

PATIENTS AND METHOD

Using Wechsler Adult Intelligence Scale(WAIS), we studied 17 out-patients with Parkinson's disease(PD). Secondary parkinsonism cases including cerebrovascular disease were excluded. We checked these patients by WAIS, EEG, and brain-CT. We analyzed basic activity of EEG by averaging the bilateral occipital wave sequence during 30 seconds of resting. Three indexes were used in the analysis of brain-CT. The first is VAI(Ventricle Area Index), which was described by Synek et al.[1] and which shows the ratio of the ventricle to the brain. The second is BAI(Brain Atrophy Index), which was proposed by Kitani et al.[2] and indicates the ratio of the brain to the whole intracranial space. The third is CMI(Cella Media Index). VAI and BAI were measured at the level of the Foramen of Monro. Figure 1 presents the method for measurement of VAI and BAI; and Figure 2, that of CMI.

RESULTS

Tables 1-3 present the means and standard deviations for the 17 subjects on demograhic factors, disease variables, WAIS, EEG, and brain-CT data. Regarding the WAIS data of the patients, the verbal IQ(VIQ) scores were superior to those of the performance IQ(PIQ).

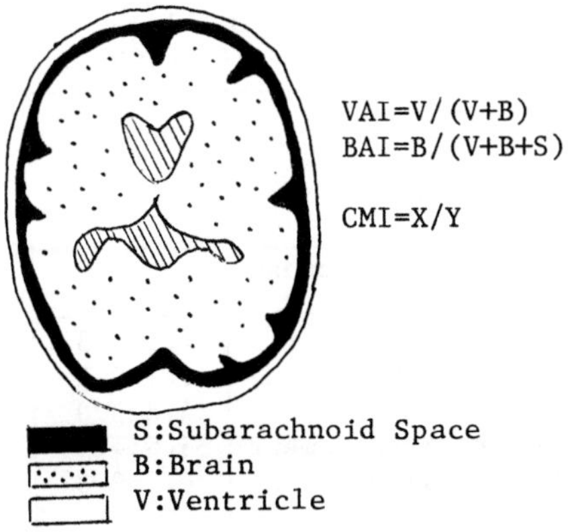

Fig. 1. Measurement of VAI and BAI

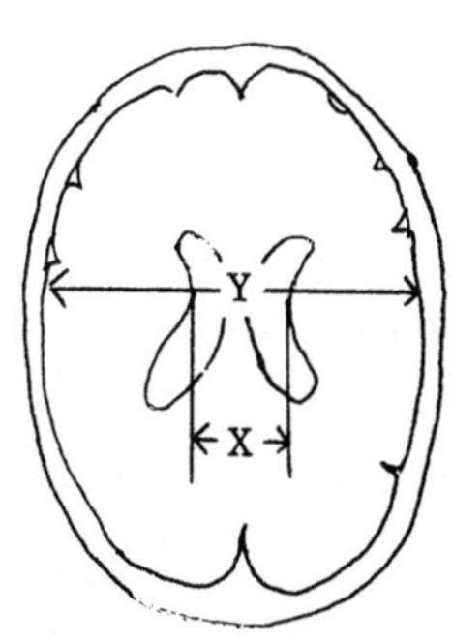

Fig. 2. Measurement of CMI

Table 1. Background Information

Age, y	71.9±5.9	
M/F	8/9	
Education, y	9.5±3.1	
Onset of PD, y	65.8±7.1	
Symptoms Duration, y	6.1±3.7	
Parkinsonian stages according to Hohen and Yahr	Stage1	1
	Stage2	6
	Stage3	7
	Stage4	3

Table 2. WAIS Data

FSIQ	84.0±16.2
VIQ	88.9±17.4
PIQ	82.1±15.6

Table 3. EEG and CT Data

EEG	8.26±1.49	Hz
BAI	88.7±5.5	
VAI	8.21±3.37	
CMI	0.286±0.063	

We checked the correlation coefficients among full-scale IQ(FSIQ), VIQ, PIQ, EEG, VAI, BAI, CMI, Age, and duration of symptoms(Table 4). Mean frequencies of EEG showed significant correlation with FSIQ, VIQ, PIQ, and CMI. VAI, BAI, and CMI showed no significant correlation with FSIQ, PIQ, and VIQ. VIQ and PIQ had significant intercorrelation. Table 5 presents the mean scores with SD of the WAIS subtests in all 17 subjects. All averaged scores in the verbal subtests were superior to any of the averaged scores of the performance subtests. Table 6 shows the correlation coefficients among the WAIS subtests. Among the 6 verbal subtests, all but the Arithmetic scores were significantly correlated with each other. Regarding the 5 performance subtests, all were significantly interrelated. Information and Similarities verbal subtests showed significant correlations with all performance subtests But Comprehension, Arithmetic, and Digit Span showed no significant correlation with any performance subtests.

Table 4. Correlation Coefficients among WAIS, EEG, CT, Indexes and Duration of Symptoms

	FSIQ	VIQ	PIQ	EEG	VAI	BAI	CMI	AGE	SYM.DUR.
FSIQ	1								
VIQ	.91***	1							
PIQ	.88***	.65**	1						
EEG	.66**	.64**	.66**	1					
VAI	-.03	-.17	-.11	-.37	1				
BAI	.06	.04	.05	.29	-.91***	1			
CMI	-.45	-.45	-.37	-.56*	.64**	-.62**	1		
AGE	-.23	-.17	-.30	.03	.01	-.07	.01	1	
SYM.DUR.	.01	-.07	.04	-.21	.43	-.37	-.10	-.02	1

* :p< .05
** :p< .01
*** :p< .001

Table 5. Scores of the 11 Subtests

Verbal Subtests		Performance Subtests	
Information	7.4±2.9	Digit Symbol	4.5±1.3
Comprehension	9.1±4.1	Picture Completion	6.1±2.2
Arithmetic	6.8±2.3	Block Design	4.8±3.0
Similarities	7.4±3.5	Picture Arrangement	5.5±1.7
Digit Span	6.5±2.2	Object Assembly	6.2±2.8
Vocabulary	6.3±5.2		

Table 6. Correlation Coefficients among the WAIS Subtests

TEST	VIQ	I	C	A	S	D	V	PIQ	DS	PC	BD	PA	OA
VIQ	1												
I	.79***	1											
C	.79***	.49*	1										
A	.64**	.43	.55*	1									
S	.78***	.69**	.49*	.20	1								
D	.82***	.51*	.58*	.62**	.52*	1							
V	.90***	.71**	.71**	.63**	.59*	.80***	1						
PIQ	.65**	.72**	.38	.36	.68**	.41	.58*	1					
DS	.45	.53*	.30	.19	.54*	.15	.30	.84***	1				
PC	.50*	.70**	.28	.28	.50*	.33	.55*	.82***	.59*	1			
BD	.60*	.58*	.45	.40	.55*	.41	.48	.91***	.81***	.66**	1		
PA	.65**	.63**	.31	.37	.60*	.45	.61**	.80***	.59*	.56*	.67**	1	
OA	.54*	.58*	.25	.15	.68**	.32	.51*	.86***	.67**	.63**	.73***	.62**	1

*   :p<.05
** :p<.01
***:p<.001

VIQ:Verbal IQ, I:Information, C:Comprehension, A:Arithmetic, S:Similarities
D:Digit Span, V:Vocabulary, PIQ:Performance IQ, DS:Digit Symbol,
PC:Picture Completion, BD:Block Design, PA:Picture Arrangement,
OA:Object Assembly

Table 7 gives the results of the WAIS subtests compared statistically with EEG, VAI, BAI, CMI, Age, duration of symptoms, and education. In the 6 verbal subtests, the scores of Arithmetic, Digit Span, and Vocabulary were significantly interrelated with EEG. Similarities showed significant correlation with CMI. Also, Information showed a significant correlation with education. In the 5 performance subtests, the scores of Block Design and Picture Arrangement were highly interrelated with EEG. Picture Completion was significantly correlated with education.

DISCUSSION

The studies of the intellectual function of PD based on WAIS are controversial. Some authors reported that the intellectual functions of PD were impaired in comparison with those of the controls.[3-5] Another reported that they were the same as those of the controls.[6] Among the former opinions, two articles stated that both the verbal and performance subtests scores in PD were lower than those in the controls,[3,4] while the other stated that only the performance scores of PD were lower than those of the controls.[5]

Table 7. Correlation Coefficients among the WAIS Subtests, EEG, CT Indexes Age, Duration of Symptoms, and Education

* :p<.05 ** :p<.01 ***:p<.001	Verbal Subtests						Performnce Subtests				
	I	C	A	S	D	V	DS	PC	BD	PA	OA
EEG	.41	.41	.55*	.41	.61*	.56*	.43	.45	.73***	.74***	.42
VAI	.06	-.13	.06	-.13	-.09	-.01	-.30	.05	-.25	-.26	.16
BAI	-.17	-.04	-.27	.07	.01	.01	.21	-.14	.10	.30	-.07
CMI	-.17	-.30	-.05	-.51*	-.24	-.23	-.38	.00	-.38	-.45	-.31
Age	-.10	-.15	-.07	-.28	-.06	-.07	-.35	.10	-.26	-.39	-.34
Duration	.23	-.16	-.22	-.22	-.33	-.11	-.16	.04	-.07	.03	.22
Education	.56*	.40	.20	.31	.34	.38	.43	.52*	.43	.33	.22

In our patients, the performance subtests scores(82.1±15.6) were lower than the verbal subtests scores (88.9±17.4). The correlation coefficient between the performance scores and the verbal ones was 0.65 and significant($p<.01$). Regarding the relation between EEG and WAIS, the WAIS scores (FSIQ, VIQ, PIQ) decreased as the frequencies of EEG decreased. The correlation coefficient between EEG and the verbal scores was 0.64 and significant ($p<.01$), and that between EEG and performance scores, 0.66, was also significant ($p<.01$).

PD is said to be accompanied by brain atrophy.[7] VAI and BAI, the two-dimensional measurements of the ventricle size are reported to be significantly correlated with age.[2] Both the verbal and performance scores showed no significant relation to VAI, BAI, CMI, age or duration of symptoms, although they showed significant relation to EEG. These results suggest that the frequencies of EEG reflects the intellectual functions measured in the WAIS more than the brain atrophy observed by a CT-scan.

Among the WAIS subtests, the scores of Block Design and Picture Arrangement showed a strongly significant relation to EEG, with the correlation coefficients being 0.73 ($p<.001$) and 0.74 ($p<.001$) respectively. In the verbal tests, the Arithmetic, Digit Span, and Vocabulary subtests showed significant relation to EEG, and their coefficients were 0.55 ($p<.05$), 0.61 ($p<.05$), and 0.56 ($p<.05$), respectively. In the verbal subtests, all but the Arithmetic test demonstrated significant relation to each other. In the performance tests, all tests showed a significant relation to each other.

Patients with PD are said to have deficits in visuospatial ability as well as in handling abstractions.[8] In the WAIS subtests, Block Design is said to be the best cognitive discriminator between patients with PD and control subjects. And Picture Arrangement seems to need a greater degree of "active" intelligence. Accordingly, the fact that the Block Design and Picture Arrangement tests showed a highly significant relation to the frequencies of EEG in occipital basic activity is very interesting.

In summary, we studied 17 outpatients with Parkinson's disease by Wechsler Adult Intelligence Scale, EEG, Brain-CT, and clinical background factors. In these 17 patients, every averaged verbal subtest score was numerically higher than any averaged performance subtests score. Some subtests showed statistically significant relationship to EEG results and educational years, but not to brain atrophy, age, or duration of symptoms. The scores of Block Design test and Picture Arrangement test were highly interrelated with the EEG results.

REFERENCES

1. V. Synek, J.R. Reuben, G.H. Du Boulay, Compairing Evans index and computerized axial tomography in assessing relationship of ventricular size to brain size, Neurology 26:231 (1976).
2. M. Kitani, S. Kobayashi, S. Yamaguchi, T. Katsube, K. Okada, A. Murata, T. Tsunematsu, CT measurement of age-related brain atrophy by two-dimensional measurement with a digitizer, Progress in Computerized Tomography 6:659 (1984).
3. A. W. Loranger, H. Goodell, F. H. Mcdowell, J. E. Lee, and R. D. Sweet, Intellectual impairment in Parkinson's syndrome, Brain 95:405 (1972).
4. R. Portin and U. K. Rinne, Neuropsychological responses of parkinsonian patients to long-term levodopa treatment, in: "Parkinson's disease," U. K. Rinne, M. Klinger, and G. Stamm, eds., Elsevier/North-Holland Biomedical Press, Amsterdam, New York:271 (1980).
5. C. G. Matthews and K. Y. Haaland, The effect of symptoms duration on cognitive and motor performance in parkinsonism, Neurology 29:951 (1979).

6. A. J. Lees and E. Smith, Cognitive deficits in the early stages of Parkinson's disease , *Brain* 106:257 (1983).
7. J. Horiguchi, M. nakagawa, Y. Inami, T. Shoda, D. shimizu, M. shudou, and Y. Kakimoto, Correlation of intelligence, cerebral atrophy, and psychotic symptoms in idiopathic Parkinsonian patients under the treatment of l-Dopa, *Japanese Journal of Clinical Psychiatry* 12:623 (1983).
8. J. L. Cummings, The dementia of Parkinson's disease, *Eur. Neurol.* 28:15 (1988).

STUDY ON RELATIONSHIP BETWEEN COMPUTED TOMOGRAPHY BRAIN CHANGES AND MOTOR FUNCTION IN PARKINSON'S DISEASE

Kiyoshi Takamiya, Masao Kinoshita, Yasuo Iwasaki, Ken Ikeda and Toshiya Shiojima

The Fourth Department of Internal Medicine, Toho University School of Medicine, Tokyo, Japan

PURPOSE

It has been reported that there is a significant relationship between the motor disability and cerebral atrophy in patients with Parkinson's disease. There has not been reported the relationship between the Parkinsonian triad(tremor, rigidity, bradykinesia) and degree of brain atrophy. Computed tomography(CT) is useful tool to determine the cerebral atrophy. The aim of this study was to evaluate the relationship between the sign of Parkinsonian triad and the degree of brain atrophy by the evaluation of CT changes.

MATERIALS AND METHODS

Fifty patients with idiopathic Parkinson's disease(PD) were included in this study. Patients with normal pressure hydrocephalus, cerebral vascular events, alcoholism, cerebral tumor, cranial trauma were excluded. None of the patient met DSM-III criteria for dementia. The age range of the patients was 45-78 years, mean 62.6 ± 10.6(SD).

Each patients was examined by the complete clinical scale by Hoehn and Yahr. Major Parkinsonian signs, tremor, rigidity, and bradykinesia were also rated on a scale from 0 to 3, with 0 indicating the abscence of the sign and 3 indicating severity.

The statistical analysis was performed in Student's t-test and Chi^2-test.

Table 1. Subject Profile

Number(Male:Female)	50(22:28)
Mean age (±SD)	62.6 ± 10.6
Duration of illness;ys (±SD)	6.2 ± 3.0
Stage of disease;Hoeha and Yahr score (±SD)	2.5 ± 0.9
Tremor (±SD)	1.3 ± 0.9
Rigidity (±SD)	1.8 ± 0.5
Bradykinesia (±SD)	2.2 ± 0.5

SD: standard deviation

The CT studies were performed using a TOSHIBA TCT-70A-30, and the cuts were parallel to the orbito-mental line, with slice thickness of 13mm and distance of 10mm between contiguous levels. The different parameters used for the evaluation of the CT images are as follows:

F : Distance between the tips of frontal horns.

C : Transverse distance between the caudate nuclei, at midportion level.

SC : Maximum septum-caudate distance.

V_3 : Maximum width of third ventricle.

CM : Minimal with of lateral ventricles, in the cella media region.

CBR: Cistern Brainstem Ratio = (width of prepontine cistern) / (distance from posterior clinoid to floor of fourth ventricle).

RESULTS

Table 2. Neuroradiological parameters of 50 PD patients.

	mean ± SD
F (mm)	37.2 ± 3.0
C (mm)	15.6 ± 1.7
SC (mm)	5.6 ± 1.6
V_3 (mm)	5.0 ± 1.4
CM (mm)	32.0 ± 2.9
CBR	0.19 ± 0.08

Table 3. Relationship between neuroradiological parameters and Parkinsonian triad, age, duration of illness, and Hoehn and Yahr score.

Neuro-radiological parameters	Tremor	Rigidity	Brady-kinesia	Age	Duration	H-Y score
F (mm)	NS	NS	NS	NS	NS	NS
C (mm)	NS	NS	NS	$P<0.05$	NS	NS
SC (mm)	NS	NS	NS	$P<0.05$	NS	NS
V_3 (mm)	NS	NS	NS	$P<0.01$	$P<0.05$	$P<0.05$
CM (mm)	NS	NS	NS	$P<0.01$	$P<0.05$	NS
CBR	NS	NS	NS	NS	NS	NS

NS: no significant differences

DISCUSSION

There is a controversy between the degree of brain atrophy and motor disability in patients with PD. Inzelberg et al. evaluated the relationship between brain atrophy and the motor and cognitive function in 132 consecutive patients. They stressed that age as a most important factor relating to brain atrophy and the motor and cognitive decline were interdependent but they affected similar CT parameters after correction for age. The effect of motor impairment was reported to be the stronger than the cognitive decline and to be the only one whitch correlated with cortical atrophy indicated with the mesurment of the distance of frontal horns, caudate, septum-caudate, cella media, and the third ventricle. In our study, there was no significant differences between Parkinsonian triad and the neuroradiological parameters. Age, duration of illness and Hoehn-Yahr score had relationship to the neuroradiological parameters, however age was found to affect most parameters, e.g. the ventricular, or cortical or both.

In conclusion, our study highlighted the effect of age on CT parameters, and it was suggested that age as a most important factor relating to the brain atrophy.

REFERENCES

1. Steiner I, Gomori JM, Melamed E. Features of brain atrophy in Parkinson's disease, a CT scan study. Neuroradiology 27: 158, 1985.
2. Inzelberg R, Treves T, Reider I, Gerlenter I, Korczyn AD. Computed tomography brain changes in Parkinsonian dementia. Neuroradiology 29: 535, 1987.
3. Iwasaki Y, Kinoshita M, Ikeda K, Takamiya K. Cognitive function in Parkinson's disease: in relation to motor symptoms. Intern. J. Neuroscience 47: 295,1989.

A PATIENT PRESENTING MAINLY DYSTONIA IN A FAMILY WITH JUVENILE PARKINSONISM

Atsushi Ishikawa, Keiko Tanaka, Akira Koyama,
and Tadashi Miyatake

Department of Neurology
Brain Research Institute
Niigata University
Niigata 951, Japan

INTRODUCTION

Juvenile parkinsonism (JP) consists of several clinical phenotypes. Yamamura et al.[1] and Sunohara et al.[2] reported juvenile parkinsonian patients who showed diurnal fluctuation of symptoms that was relieved markedly with levodopa. The patients of this group also had minor dystonia in the feet, and in the neck in some cases, as one of the characteristic findings. Klawans et al.[3] and Katchen et al.[4] reported a group of patients who presented initially with dystonia and subsequently developed parkinsonian symptoms in which the dystonia was worsened with levodopa. Segawa et al.[5] reported other variants showing mainly dystonia with diurnal fluctuation that was relieved with levodopa.

We report herein a case presenting mainly dystonia in a family of JP. The proband showed dystonic movement in the neck and extremities that was worsened with levodopa, and mild rigidity in the ankles. Cardinal symptoms of the other 6 patients of this family were parkinsonism and mild dystonic posture in the neck and feet that were relieved with levodopa.

The proband and the other members of this family provide important information for us to understand the clinical variations of JP and the correlation between parkinsonism and dystonia.

CASE REPORTS

Case 1 (IV-9), the Proband

A 34-year-old man first noticed stiffness in the right knee on walking at the age of 25. This symptom spread to the left knee. Then he felt difficulty in walking and picking up something with the right fingers because of stiffness. His arm was forced to draw backwards when he started to walk. He also felt difficulty in writing. Two years later the dystonic movement spread to the neck. He also felt stiffness in both knees, ankles, and left inguinal area. He was admitted to our hospital at the age of 27.

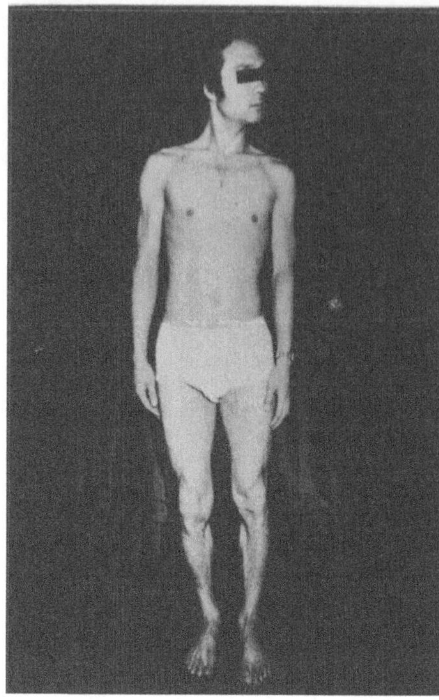

Fig. 1. Photograph of the proband.

His mental activity was normal. Dystonic movement was found especially on walking. Even at rest, his face turned to the left and the right forearm was supinated with finger flexion. He felt difficulty in turning his head to the right and dull pain in the nape. His steps looked irregular and bizarre, because the right arm was elevated backwards, the right leg was lifted up, and the knee joints did not stretch fully on walking. The left arm swing was poor in contrast. He showed neither small-stepped nor frozen gait. There was mild rigidity in both ankle joints. Deep tendon reflexes were decreased and horizontal gaze nystagmus was observed. He did not show bradykinesia, resting or postural tremor, impaired postural reflex, weakness, nor cerebellar ataxia, but did show mild plantar flexion at the ankle. The Kayser-Fleischer ring was not found. His symptoms were relieved after sleep for about 20 minutes. Routine blood examination showed no specific abnormalities. Serum copper and ceruloplasmin levels were normal. HVA level in the CSF was slightly decreased to 20.2 ng/ml (normal, 33.2 ± 5.1 ng/ml), but his 5HIAA level was normal. Brain CT and EEG revealed no abnormal findings.

His symptoms were not relieved with any dose of levodopa from small (10 mg/day) to large (3000 mg/day). With levodopa administration neck rotation to the left became more severe, and the right arm was continuously elevated backwards. The knee joints were semi-flexed, feet were inverted, and the voluntary movement apparently became slow and neither heel reached to the floor on walking. Trihexyphenidyl treatment relieved his symptoms, and diazepam was also effective, so he was treated with trihexyphenidyl (12 mg/day) and diazepam (6 mg/day) for the past 7 years. The dystonic movement has gradually decreased, and at present he does not frequently feel stiffness in the nape, knee, inguinal, or ankle area. He still shows mild rigidity of the ankle joints, but other parkinsonian symptoms, such as bradykinesia, tremor, impaired postural reflex, and frozen gait, have not appeared over these 7 years. He can work without much difficulty.

Case 2 (III-4)

A 64-year-old man, an uncle of Case 1, was healthy until the age of 32 when he noticed gait disturbance. On walking he lifted the left leg high and swung the left arm inside to outside. Three years later he felt weakness in the knee joints and difficulty in associated movements of the hip, knee, and ankle joints on walking. At the age of 40, he noticed hand tremor. Such symptoms were relieved after sleep for about 15 minutes. He was admitted to our hospital at the age of 45.

He showed marked rigidity, bradykinesia, and impaired postural reflex. On walking he lifted both legs high, but the step was not wide with poor arm swing. Sometimes frozen gait appeared. Postural fine tremor was also found.

Levodopa was very effective for these symptoms and he was discharged, being instructed to take levodopa daily, 3000 mg/day. But 2 months later choreic dyskinesia appeared, especially in the extremities. Later he was treated with levodopa in combination with decarboxylase inhibitor (carbidopa) and bromocriptine. The daily dose of levodopa/carbidopa was about 500 mg divided into 10 to 12 parts taken per day, and bromocriptine was 15 mg/day.

In February, 1989, he was readmitted to our hospital for the control of severe wearing-off phenomenon and dyskinesia. He showed marked bradykinesia in the off-period, and his speech was slurred. Marked rigidity was found in both extremities and the neck. He also showed adiadochokinesia and impaired postural reflex. Fine postural tremor was found in the jaw and fingers. On walking he lifted both legs high, looking as if he were hopping, and displayed a small-stepped gait on turning. Arm swing was absent. He also showed numbness in the left upper extremity and in both lower extremities, especially in the off-period, and mild horizontal nystagmus. Deep tendon reflexes were normal without pathological reflexes. Dyskinesia could be found in the on-period in the extremities, especially in the distal portion. Brain CT revealed no pathological findings except for a small low-density area in the cerebellum probably due to an infarction there.

Pergolide was effective to control the severe wearing-off phenomenon and dyskinesia in the extremities. He was treated with 1.2 mg pergolide, and the levodopa/carbidopa was decreased to 200 mg daily, divided into 8 portions.

In this patient it seemed that dystonia had been replaced by parkinsonism.

Case 3 (V-3)

A 37-year-old man, the son of Case 2, noticed pronation movement of the right forearm on holding some object at the age of 15, and fine tremor on movement at the age of 20. When he was 22 years old his head started to tilt backwards, and he had mild gait disturbance. Then he noticed that he could not flex the right knee joint easily on walking, and showed mild bradykinesia.

He was admitted to our hospital when he was 25 years old. He showed mild bradykinesia, rigidity, impaired postural reflex, and adiadochokinesia. Fine postural tremor was found, and deep tendon reflexes were exaggerated without pathological reflexes. On walking the right knee flexion was incomplete and the right arm swing was absent. Sometimes frozen gait was found especially on turning. There was mild plantar flexion of the left ankle. These symptoms were improved after sleep for about 1 hour.

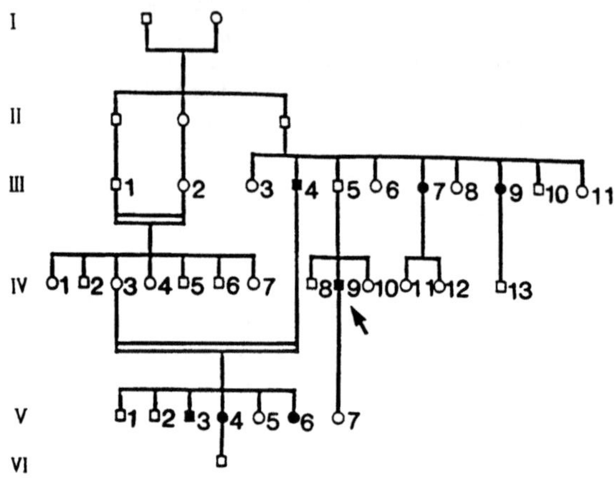

Fig. 2. Family pedigree.

Table 1. Clinical Features of the Patients

	III-4 (Case 2)	III-7	III-9	IV-9 (Case 1)	V-3 (Case 3)	V-4	V-6
Age/Sex	64/M	56/F	52/F	34/M	37/M	34/F	29/F
Age at onset	32	31	28	25	15	19	13
Dystonic movement	+	−	−	++	−	−	−
Dystonic posture	−	+	+	+	+	+	+
Area of dystonia	foot	foot	neck	neck	neck	neck	neck
Tremor (fine)	+	+	+	−	+	+	−
Rigidity	+	+	+	+*	+	+	+*
Bradykinesia	+	+	+	−	+	+	−
Frozen gait	+	+	+	−	+	+	−
Deep tendon reflex	→	↑	↑	↓	↑	↑	→
Improve. after sleep	+	+	+	+	+	+	−
Levodopa effect							
for parkinsonism	++	++	++	−	+	++	n.t.
for dystonic move.	?			−**			
for dystonic post.		+	+	?	+	+	n.t.
Wearing-off phenom.	++	+	+	−	−	+	n.t.
Dopa-induced dyskinesia	++	+	+	−	+	+	n.t.
HVA levels in CSF (normal, 33.2±5.1 ng/ml)				20.2	29.7		24.5

*: ankle only, **: worsened, n.t.: not tried

He was administered trihexyphenidyl with moderate effect. He also benefited from levodopa/carbidopa treatment, but at the dose of 300 mg/day he presented mild foot dyskinesia. He is now being treated with 10 mg of trihexyphenidyl and 150 mg of levodopa/carbidopa a day.

SUMMARY OF THIS FAMILY

We presented 3 patients (IV-9, III-4, V-3) having different clinical features in a family of JP. The cardinal symptom of the proband, Case 1 (IV-9), was dystonic movement. Dystonia was found in the neck and the extremities, predominantly expressed on the right. Dystonia was increased with even a small dose of levodopa, but was decreased by trihexyphenidyl and diazepam. He showed mild rigidity in the ankles, decreased HVA level in the CSF, and improvement after sleep before the initiation of treatment, but other parkinsonian signs did not appear.

Cardinal symptoms of Case 2 (III-4) and Case 3 (V-3) as well as Cases III-7, III-9, V-4, V-6 presented in Table 1 were parkinsonian ones and were relieved with levodopa. Case 2 (III-4) had clinical similarity of dystonic movement and nystagmus to Case 1 (IV-9). Dystonic movement in Case 2, however, was gradually replaced by parkinsonian symptoms. Case 1 (IV-9) also seemed to resemble Case V-6 who had dystonia without parkinsonian symptoms except for ankle rigidity.

A high incidence of neck dystonia (3 with torticollis and 3 with retrocollis) is a characteristic feature of this family. These symptoms were relieved with levodopa except for Case 1. The initial symptom of 4 cases (III-9, IV-9, V-3, V-6) seemed to be symptoms due to dystonia.

Patients of the third generation, Cases 2 (III-4), III-7, and III-9, had rather late onset (average age at onset, 30.3 years old) with the favorable effect of levodopa, and there were wearing-off phenomenon and dopa-induced dyskinesia in the extremities. They also had numbness in the feet and lower extremities, especially during the off-period. In contrast, the patients of the fifth generation, Cases 3 (V-3), V-4, and V-6, had earlier onset (average age at onset, 15.7 years old), and their symptoms were relatively mild. Cases 3 (V-3) and V-4 sometimes noticed mild dopa-induced dyskinesia and Case 3 did not show the wearing-off phenomenon. Progression of their symptoms was very slow, and in Case 1 (IV-9) symptoms did not change for 7 years. The disorder in this family is thought to be inherited as an autosomal dominant trait.

DISCUSSION

Dystonia has been recognized as both an initial and a late feature of Parkinson's disease.[6–8] During levodopa therapy, dystonia is observed during the off-period or in the early morning,[9] usually in the feet, and at peak-dose period in the face and neck, and is assumed to be due to striatal dopaminergic receptor supersensitivity.

Dystonia is also seen in untreated juvenile parkinsonian patients. In 1917, Hunt[10] already described foot dystonia in parkinsonian patients who showed diurnal fluctuation. Yamamura et al.[1] and Sunohara et al.[2] reported a group of patients as idiopathic dystonia-parkinsonism[2] who showed minor dystonic posture and parkinsonian symptoms with marked diurnal fluctuation. These dystonia and parkinsonian symptoms were markedly improved with levodopa.

The patients reported by Segawa et al.[5] as hereditary progressive dystonia (HPD) had marked dystonic posture in the extremities with marked response to levodopa, but the age at onset was younger, before the age of 10. Nygaard and Duvoisin[11] reported a case similar to HPD, with autosomal dominant inheritance, age at onset before 10, dystonia followed by parkinsonism, dramatic response to levodopa, and minimal progression; however, this disorder included neck dystonia, making it different from HPD.

On the other hand, Klawans et al.[3] and Katchen et al.[4] reported a group of patients as dystonia-parkinson syndrome[3] showing dystonia followed by parkinsonism, whose dystonia was worsened with levodopa but whose parkinsonism was relieved.

Recently Fahn et al.[12] and Lee et al.[13] reported X-linked recessive dystonia and parkinsonism in Filipino males. The patients showed generalized dystonic movements and parkinsonism, and no response to levodopa.

All the patients except for Case 1 (IV-9) in our family had minor dystonic posture in the neck and feet together with other parkinsonian symptoms showing diurnal fluctuations, marked efficacy of levodopa, wearing-off phenomenon, and dopa-induced dyskinesia. They resemble the cases of those reported by Yamamura et al.,[1] Sunohara et al.,[2] and are also similar to the case of Nygaard and Duvoisin[11] except for the age at onset. However, our Case 1 resembles those cases reported by Klawans et al.[3] and Katchen et al.,[4] though ours has not shown parkinsonism yet. The latter group was reported as sporadic and the response to levodopa differed. It is conceivable that our Case 1 has the same genetic background as other patients of his family.

As the pharmacological basis of dystonia is not fully understood, it is important to study these patients with familial JP for better understanding of the pathophysiology of dystonia.

REFERENCES

1. Y. Yamamura, I. Sobue, K. Ando, M. Iida, T. Yanagi, and C. Kono, Paralysis agitans of early onset with marked diurnal fluctuation of symptoms, Neurology 23:239 (1973).
2. N. Sunohara, Y. Mano, K. Ando, and E. Satoyoshi, Idiopathic dystonia-parkinsonism with marked diurnal fluctuation of symptoms, Ann. Neurol. 17:39 (1985).
3. H. L. Klawans, and N. Paleologos, Dystonia-parkinson syndrome: Differential effects of levodopa and dopamine agonists, Clin. Neuropharmacol. 9:298 (1986).
4. M. Katchen, and R. C. Duvoisin, Parkinsonism following dystonia in three patients, Movement Disorders 1:151 (1986).
5. M. Segawa, A. Hosaka, F. Miyagawa, Y. Nomura, and H. Imai, Hereditary progressive dystonia with marked diurnal fluctuation, in: "Advances in Neurology," R. Eldridge and S. Fahn, eds., pp. 215-233, Raven Press, New York (1976).
6. P. A. Nausieda, W. J. Weiner, and H. L. Klawans, Dystonic foot response of parkinsonism, Arch. Neurol. 37:132 (1980).
7. D. Kidron, and E. Melamed, Forms of dystonia in patients with Parkinson's disease, Neurology 37:1009 (1987).
8. W. H. Poewe, A. J. Lees, and G. M. Steru, Dystonia in Parkinson's disease: Clinical and pharmacological features, Ann. Neurol. 23: 73 (1988).
9. E. Melamed, Early-morning dystonia: A late side effect of long-term levodopa therapy in Parkinson's disease, Arch. Neurol. 36:308 (1979).

10. J. R. Hunt, Progressive atrophy of the globus pallidus (Primary atrophy of the pallidal system), Brain 40:58 (1917).
11. T. G. Nygaard and R. C. Duvoisin, Hereditary dystonia-parkinsonism syndrome of juvenile onset, Neurology 36:1424 (1986).
12. S. Fahn and C. Moskowitz, X-linked recessive dystonia and parkinsonism in Filipino males, Ann. Neurol. 24:179 (1988).
13. L. V. Lee, F. M. Pascasio, F. D. Fuentes, and G. H. Viterbo, Torsion dystonia in Panay, Philippines, in: "Advances in Neurology," R. Eldridge and S. Fahn, eds., pp. 137-151, Raven Press, New York (1976).

HEREDITARY DYSTONIA-PARKINSONISM SYNDROME OF JUVENILE ONSET WITH MARKED DIURNAL FLUCTUATIONS

Alain Destee*, Luc Defebvre*, Brigitte Delisse*, and Jean-Daniel Guieu**

Departments of *Neurology and **Clinical Neurophysiology
Chu Lille, France

We have studied a family with dystonia-parkinsonism of juvenile onset with marked diurnal fluctuations. As always reported anticholinergics and levodopa were responsible of a clear improvement.

Proband

We examined this woman for the first time in 1987. She was born in 1929. Pregnancy, labor, delivery and developmental milestones were normal. She regularly attended school at 5. At age 7 inversion of the feet, more pronounced on the left side, responsible for poor balance, was noted when walking or standing up for a long time in the evening. The symptoms quickly worsened and few weeks later an equinovarus posture, flexion of the legs and increased lumbar lordosis soon appeared after the beginning of the walk. Few months later there was also general bradykinesia, rest, postural and intention tremor in the upper extremities, increased muscle tone, mainly in the lower limbs. In fact dystonic and parkinsonian symptoms were fluctuating in the day time : much less severe when she awoke from sleep, they progressively worsened, mainly after exercise ; in the evening she was unable to stand up and to walk. After a nap there was an improvement, less important than after the night's rest. Sometimes freezing with "feet glued to the floor" appeared after emotion.

At age 21 the reflexes were judged to be hyperactive in the four limbs ; the plantar responses were flexor. Anticholinergics (Diethazine) dramatically improved all the extrapyramidal symptoms : 3 days later the patient was able to stand up and to walk. Five years later, at age 26, Diethazine was replaced by profenamide (Parsidol, 100 mg/d) with better results.

At age 58 she was examined by us few days after the treatment was stopped. There was equinovarus posture of the feet more pronounced on the left side, worsened by voluntary movement of the other limbs. The left big toe was tonically extended and the other toes flexed. Unable to walk alone, she moved with someone's help with increased lumbar lordosis and a tendency to fell backwards ; her legs being extended she "waddled" along. These dystonic symptoms worsened the day through. There were also parkinsonian features (without diurnal fluctuations) : facial hypomimia, general bradykinesia, increased tone with cogwheel rigidity in the limbs and the neck, mild rest and postural tremor of the upper extremities, abnormal glabellar response. The reflexes were judged to be hyperactive in the limbs. Sometimes bucco linguo facial dyskinesias were observed. The general examination revealed angor pectoris.

All laboratory studies, including serum ceruleoplasmin, copper and amino acid, were normal. MRI revealed some hyperintensities in the white matter and hypointensities in the both pallidum (fig), without significance because of her age. There was some improvement with trihexyphenidyl, 12 mg/d, but this drug had to be stopped because of headaches and accomodation disturbances.

The response to levodopa (150 mg/d) with benserazide (37,5 mg/d) was quick and "dramatic". The patient was able to stand up and walk alone keeping her balance. The examination revealed the improvement of both dystonic and parkinsonian features. There was no diurnal fluctuation of the therapeutic effect. Eighteen months later the improvement persisted.

PATIENT 2

The eldest brother of the proband was born in 1912. Pregnancy, labor, delivery and developmental milestones were normal. At age 13, he tended to turn his left foot in after a 2 miles walk in the evening. Gradual progression of the gait disturbance developed few months later: he was free of symptom on the morning but in the evening an equinovarus posture of the feet, much more pronounced on the left side appeared after a 100 yards walk forcing him to stop, then disappearing after a few minutes. These symptoms were not observed when he ran, he went upstairs or rode his bicycle. Two years later there was a postural tremor of the right upper extremity. Theses symptoms with diurnal fluctuations remained unchanged until he was 40 years old and then spontaneously disappeared. During 17 years he was free of symptom. At age 57 micrographia, general and bradykinesia bradyphrenia appeared. Two years later there was a mild rest tremor of the upper extremities, increased tone of the limbs then excess of saliva. There was no diurnal fluctuations. Anticholinergics (trihexyphenidyl, 6 mg/d) alone then with piribedil (100 mg/d) levodopa (300 mg/d) and benserazide (75 mg/d) clearly improved his condition. At age 75, after 12 hours without medication the symptoms increased with "marche à petits pas", sometimes freezing...

The score obtained with part III of the UPDRS* was 26 ; 60 mn after 100 mg levodopa and 25 mg benserazide intake it was 11. All laboratory studies, including serum ceruleoplasmin, copper and amino acid, were normal. The human leucocytes antigènes were A^{28} A^{192} B^{7} B^{21}. MRI showed some hyperintensities in the white matter.

PATIENT 3

This elder brother of the proband was born in 1919. Pregnancy, labor, delivery and developmental milestones were normal. At age 15 a "dystonic claudication" appeared : the gait was normal in the morning but in the afternoon an equinovarus posture of the right foot appeared after a 300 yards walk ; this symptom disappeared when the patient had a rest, was not observed when the patient ran, went upstairs or rode his bicycle. At age 44 during surgical therapy for lumbar disk herniation there was a lesion of the L5 and S1 nerve roots : the dystonic symptom disappeared because of motor deficit.

---

*Fahn S., Elton R.L. and members of the UPDRS Development Committee (1987)
Unified Parkinson's Disease Rating Scale.
In Recent Developments in Parkinson's disease
Fahn S., Marsden C.D., Calne D., Goldstein M. Eds., pp 153-163.
Mac Millan Healthcare Information, USA.

For 16 years the patient was free of symptom. At age 61 slowing of his pace, hypomimia, excessive saliva were the first symptoms of a parkinsonian syndrome. At age 68 there was general bradykinesia, active glabellar response, cogwheel rigidity of the limbs, postural tremor of the upper extremities and rest tremor of the left one ; postural stability was impaired with tendency to fall backwards ; there was some freezing. There was not diurnal fluctuations. The UPDRS* score (part III) was 43 (max = 108) spontaneously and 18 after 100 mg levodopa and 25 mg benserazide. Treatment with levodopa (200 mg/d) and benserazide (50 mg/d) clearly improved the patient : the Schwab and England score (modified Schwab and England ADL score : UPDRS, part IV) was 70 after and 30 before. Eighteen months later the therapeutic benefit was the same.

All laboratory studies, including serum ceruleoplasmin, copper and amino acid, were normal. The human leucocytes antigenes were A^2 A^{192} B^{18} B^{21}.

PATIENT 4

This elder sister of the proband was born in 1923. Pregnancy, labor, delivery and developmental milestones were normal. At age 8 poor imbalance with tendency to fall backwards appeared progressively the day through, due to inversion of the left foot when walking.

At age 15 equinovarus posture of the left foot was maintained through the day but worsened with exercise, improved with sleep. When walking the left shoulder went up. This symptomatology remained unchanged until age 62. Then parkinsonian symptoms appeared : rest tremor of the upper extremities and postural tremor of the left one, cogwheel rigidity, general bradykinesia, excessive saliva. Diethazine yielded little benefit.

At age 66 UPDRS* score (part III) was 27 (max = 108) without therapeutic and 9 after levodopa (200 mg) and benserazide (50 mg) which also clearly improved the dystonic foot. The rest and postural tremor frequency was 5/sec (see fig). All laboratory studies, including serum ceruleoplasmin, copper and amino acid, were normal. Human leucocytes antigenes were A^1 A^2 B^{18} B^{15}. Levodopa (150mg/d) and benserazide (37,5 mg/d) obviously improved the patient.

COMMENTS

Though they have been examined late in their life, after more than 50 or 60 years evolution, these patients can be considered similar to that studied in 1988 by Nygaard, Marsden and Duvoisin[1] and present "Dopa-responsive dystonia".

This disorder is characterized by :

1. Onset with dystonia in childhood or adolescence;
2. The concurrent or subsequent development of "parkinsonian" signs;
3. A dramatic therapeutic response to L - Dopa and sometimes other medications as anticholinergics;
4. Frequently there is diurnal fluctuations (worsening of the dystonia on exercise and improvement with sleep) and a family history suggesting autosomal dominant inheritance.

As Deonna[2] pointed to, the dystonia always started in the lower extremities by the twisting of one foot, toe-walking and gait difficulties. Others dystonic symptoms are less frequent and pronounced : increased lumbar lordosis, scoliosis, occasionally retrocollis or torticollis. As reflexes are brisked in the legs, a dystonic extension of the big toe observed spontaneously or after a

---

* See Case n°2.

plantar stimulation is often misinterpreted as a Babinski's sign. Parkinsonian features occured simultaneously or later : postural instability, bradykinesia, hypomimia, cogwheel rigidity, rest and/or postural tremor.

Segawa[3] emphasised diurnal fluctuations and sleep benefit which were observed in 67% cases[1].

Some cases (as well as our patients) don't show these symptoms in the morning and are unable to stand erect in the evening with mainly kinesiogenic exacerbation of the dystonia.

In every case the dystonic and parkinsonian symptoms are quickly and dramatically improved by levodopa alone or in combination with IDC : after a few days or weeks the patients have nearly normal function. After the drug is discontinued reemergence of symptoms quickly occured.

According to Segawa[4] "Dopa responsive dystonia" is probably related to functional lesions restricted to the terminal of the nigrostriatal DA neurons with perhaps the lowering of the content or the activity of tyrosine hydroxylase at the terminal. A biopterin deficiency could be another explanation[5].

Our family confirms some clinical features already pointed to[1] : the age of onset is earlier in girls who are generally more severely involved than boys[2], a good improvement can be obtained not only by levodopa but also by anticholinergic drugs[6].

The originality of this family is due to the clinical evolution of male patients: dystonia limited to the lower limbs appeared in adolescence and disappeared spontaneously (at least in one case) about twenty seven years later, thereafter there was no symptom and later parkinsonian syndrome foretelling a common Parkinson's disease appeared. Though this evolution is less easily observed in the two female patients the dystonic symptoms get less pronounced, progressively disappear being replaced by parkinsonian syndrome. These data suggest that dystonic and parkinsonian signs may be due to different specific mechanisms.

REFERENCES

1. NYGAARD T.G., MARSDEN C.D., DUVOISIN R.C. Dopa-Responsive Dystonia. Adv Neurol 1988 ; 50 : 377-384.
2. DEONNA T. Dopa-sensitive progressive dystonia of chilhood with fluctuations of symptoms - Segawa's syndrome and possible variants. Neuropediatrics 1986 ; 17 : 81-85.
3. SEGAWA M., HOSAKA A., MIYAGAWA F, et al. Hereditary progressive dystonia with marked diurnal variation. Adv Neurol 1976 ; 14 : 215-233.
4. SEGAWA M., NOMURA Y., TANAKA S. et al. Hereditary Progressive Dystonia with marked diurnal fluctuations. Consideration on its pathophysiology based on the characteristics of clinical and polysomnographical findings. Adv Neurol 1988 ; 50 : 367-376.
5. FINK J.K., BARTON N., COHEN W. et al. Dystonia with marked diurnal variation associated with biopterin deficiency. Neurology 1988 ; 38 : 707-711.
6. CORNER B.D. Dystonia musculorum deformans in siblings : treated with Artane (trihexyphenidyl). Proc Soc Med 1952 ; 45 : 451-452.

DYSTONIA-PARKINSONISM SYNDROME SHOWING GOOD RESPONSE TO AN ANTI-CHOLINERGIC DRUG BUT NOT TO L-DOPA

Masahiro Nomoto, Koichi Machigashira, Hiroko Atsuchi, and Mitsuhiro Osame

Third Dept. of Internal Medicine, Kagoshima University School of Medicine, Kagoshima 890, Japan

INTRODUCTION

A number of the symptomatic dystonias with widespread basal ganglia pathology such as Wilson's disease, gangliosidosis, Lesch-Nyhan syndrome, Hallervoden-Spatz disease, neuroacanthocytosis, CO intoxication,encephalitis can cause parkinsonian syndrome combined with dystonia. On the other hand, DOPA-responsive dystonias such as hereditary progressive dystonia (Segawa's disease), idiopathic dystonia-Parkinsonism, Hereditary dystonia-Parkinsonism of Juvenile onset, and some types of Juvenile Parkinsonism are also conditions of parkinsonism syndrome combined with dystonia, and that shows dramatic response to L-DOPA therapy. We present 2 cases of dystonia-Parkinsonism syndrome that can not be classified in either these two groups of disorders and that show good response to an anti-cholinergic drug.

CASE REPORTS

Patient 1

A 42-year-old female had been well until the age of 35 years, when she developed a slight left-sided limp. Two years later she developed abnormal posture in the left shoulder with elevation, in the left arm with abduction, and in the head with rotation(spasmodic torticollis). The symptoms gradually worsened over the next few years and she developed blepharospasm at the age of 42 years. On examination she was a well-developed and well-nourished woman 162cm in height and 60 kg in weight. She was alert and cooperative. No Kayser-Fleischer rings were detected. Her face was masked and she had rigidity of the neck and the limbs. Blepharospasm and torticollis were prominent, and were worsened by voluntary movements. She showed stooped posture and dystonia of the trunk on walking. Glabellar tap gave an abnormal result. Fine finger tremor was found in the postural position, and was dominant in the right hand.

Investigations. Findings from the following studies were normal: full blood count and red-cell morphology; blood Wasserman test, serum electrolytes, urea, calcium, phosphorous, magnesium, alkaline phosphatase, glutamic oxaloacetate transaminase, glutamic pyruvate transaminase, ceruloplasmin, and copper;

Basic, Clinical, and Therapeutic Aspects of Alzheimer's and Parkinson's Diseases, Volume 2
Edited by T. Nagatsu *et al.*, Plenum Press, New York, 1990

Table 1. CSF Levels of HVA and 5HIAA.

Metabolite	10 a.m.	8 p.m.
HVA		
Patient 1	17.1	-
Patient 2	9.2	32.6
5HIAA		
Patient 1	10.4	-
Patient 2	18.1	13.2

Control level at 10 a.m. (n=16)
HVA:27-54
5HIAA:16-39 (ng/ml)

urinary and serum amino acids, the content of oligosaccharide in urine; cerebrospinal fluid cell count, glucose, and protein levels; skull and chest roentgenograms; electroencephalogram; CT scan and MRI. The cerebrospinal fluid was sampled at 10:00 a.m. without any treatment.

HVA and 5HIAA levels are given in the table. The level of HVA was low. EMG with superficial electrodes revealed dystonic discharge at bilateral orbicularis oculi, left trapezius, and the extensor of the left forearm.

Therapy. The L-DOPA/carbidopa was commenced at a dosage of 100/10 mg daily and increased to 300/30 mg daily. The rigidity of the limbs decreased slightly; however, dystonia of the arms, spasmodic torticollis, and blepharospasm are worsened greatly. Therapy was then changed to trihexyphenidyl given at a dosage of 4 mg daily. The rigidity and akinesia then decreased and blepharospasm almost completely abated. Spasmodic torticollis also improved and appeared only on tension.

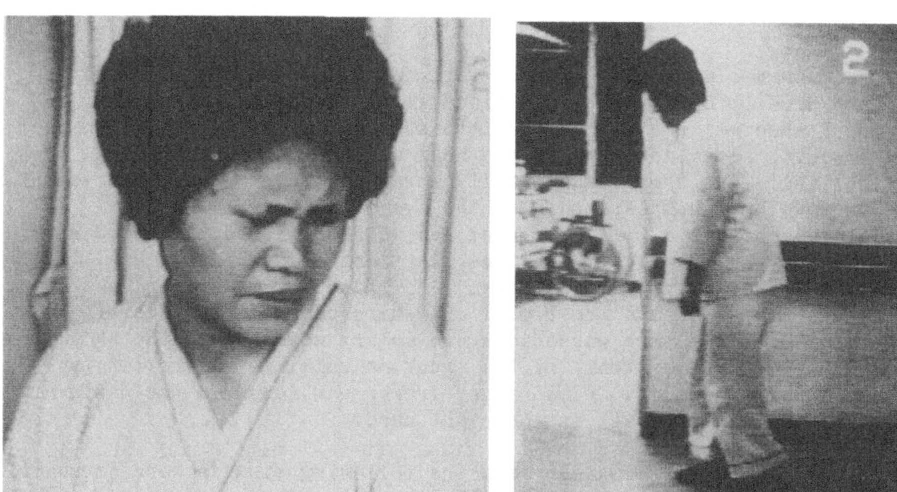

Fig. 1. Blepharospasm and torticollis (left) and stooped posture (right) in Patient 1.

Patient 2

A 37-year old male had been well until the age of 29 years, when he developed a right-sided limp. He also began to develop "bowing" movement on every step, which became more noticeable toward the end of the day. He did not show muscle weakness. The symptoms gradually worsened over the next few years. There were no family consanguinity or history of neurological disorders. On examination he was a well-nourished man 168 cm in height and 69 kg in weight. His intelligence was normal. No Kayser-Fleischer rings were detected. His face was mildly masked, and a glabellar tap yielded an abnormal result. There was no muscle weakness or wasting. Plastic rigidity was present in the limbs, and fine postural tremor was found in his fingers. On standing he developed dystonia at the trunk and bowed every 2 or 3 seconds. On walking his dystonia at the trunk was more prominent, making him bow on every step. He also showed propulsion and stooped posture. There were no cerebellar signs and sensory examination was normal. The plantar responses were flexor.

Investigations. Findings from the following studies were normal: full blood count and red-cell morphology; blood Wasserman test, serum electrolyte, urea, phosphorous, magnesium, alkaline phosphatase, glutamic oxaloacetate transaminase, glutamic pyruvate transaminase, ceruloplasmin and copper; urinary and serum amino acids, the content of oligosaccharide in urine; cerebrospinal fluid cell count, glucose, and protein level; skull and chest roentgenograms; electroencephalogram; CT scan and MRI. The cerebrospinal fluid was sampled at 10:00 a.m. and 8:00 p.m. on the same day without any treatment. HVA and 5HIAA levels are given in the table. The HVA level was low: however, its diurnal fluctuation was preserved.

Therapy. Initially, the patient received L-DOPA/ carbidopa at a dose of up to 750/75 mg daily. The rigidity was fairly abated, but there was no improvement in walking, dystonia at the trunk, akinesia, and propulsion. He also received bromocriptine without any benefit. Therapy was then changed to trihexyphenidyl at a dose of 4 mg daily, which was increased over the next 2 weeks to 15 mg daily. By this therapy the dystonia at trunk and the limbs disappeared and the rigidity was abated. Also the akinesia improved excellently, and he was able to run again. Thus the patient was discharged from the hospital and now works on his farm.

DISCUSSION

Flexion dystonia seen in Parkinson's disease, described by Denny-Brown(1), consists of a slowly progressive, persistent flexed posture of the limbs. The dystonia presented in these 2 patients, however, was not flexion dystonia but dystonic movement at the trunk, spasmodic torticollis, and blepharospasm. Dystonic phenomena are relatively frequent in Parkinson's disease on L-DOPA therapy and usually disappear if L-DOPA therapy is withdrawn. In DOPA-treated patients, foot dystonia is usually seen as an off-period phenomenon, particularly likely to occur in the early morning before the first dose of L-DOPA is taken and L-DOPA led to definite improvement or abolition of foot dystonia. Dystonia has often been regarded as a side effect of L-DOPA therapy(2). None of our patients had received neuroleptic medications or anti-parkinsonian drugs before they developed dystonia. The development of dystonia and Parkinsonism has been noted in DOPA-responsive fluctuating dystonia or in similar syndromes of young adult onset(3-8); such patients develop dystonia of the legs (with "turning in") and show a dramatic response to a small dose of L-DOPA. The precise nosological position of these syndromes remains uncertain; however, they differ from the 2 cases presented here. Our cases developed dystonia in the upper half of the body; and treatment with trihexyphenidyl, but not L-DOPA, gave considerable relief of their signs.

The dystonia and parkinsonism appeared at the same time in Patient 2 and within 2 years in Patient 1. No independent etiology for dystonia can be demonstrated in any of these cases. Duvoisin and his colleague(9) reported 3 cases of parkinsonism following dystonia and concluded that dystonia is one of the manifestations of Parkinson's disease that is sometimes alleviated, but more often is activated or exacerbated, by dopaminergic therapy and that may infrequently appear as the initial manifestaion. Similar cases of dystonia-Parkinsonism have been reported(10-12).

In our 2 cases, there was no evidence of involvement of the cerebellar, pyramidal, oculomotor, or other neuronal systems. The patients showed no dysautonimia or dementia. The clinical features observed suggest that there is the selective lesion in the basal ganglia in these patients and are compatible with Parkinson's disease except for the dystonia. This kind of dystonia is, however, rare in Parkinson's disease and the nature of underlying disease process in these 2 cases cannot be established without postmortem study. In any event, the dystonia-Parkinsonism syndrome presented here may be one of the degenerative movement disorders characterized by the combination of dystonia and Parkinsonism without involvement of other neuronal systems. One more important clinical feature in this dystonia-Parkinsonism syndrome is the excellent response to trihexyphenidyl but not to L-DOPA.

REFERENCES

1. D. Denny-Brown, Clinical symptomatology of diseases of the basal ganglia. in: "Handbook of Clinical Neurology,Vol.6, Disease of basal ganglia", P.J. Vinken and G. W. Bruyn, eds., North Holland, Oxford(1975).
2. W. Poewe, A.J. Lees, D. Steiger:357 and G.M. Stern: Foot dystonia in Parkinson's disease: Clinical Phenomenology and neuropharmacology, Adv. Neurol. 45(1986).
3. M. Segawa, A. Hosaka, E. Miyagawa, Y. Nomura and H. Imai:Hereditary progressive dystonia with marked diurnal fluctuation, Adv. Neurol. 14:215(1976).
4. M. Yokochi, H. Narabayashi, R. Iizuka and T. Nagatsu: Juvenile Parkinsonism--some clinical, pharmacological and neurological aspects, Adv. Neurol. 40:407(1984).
5. I. Kumamoto, M. Nomoto H. Yoshidome, M. Osame and A. Igata: Five cases of dystonia with marked diurnal fluctuation and special reference to homovanilic acid in CSF, Clini. Neurol. 24:697 (1984).
6. R.A. Quvier: Progressive dystonia with marked diurnal fluctuation, Ann. Neurol. 4:412(1978).
7. T. Nygaard and R.C. Duvoison: Hereditary juvenile onset dystonia-Parkinson syndrome, Neurology 36:1424(1896).
8. N.S. Sunohara, Y.Mano, K. Ando and E. Satoyoshi: Idiopathic dystonia-parkinsonism with marked diurnal fluctuation of symptoms, Ann. Neurol. 17:39(1985).
9. M. Katchen and R.C. Duvoisin: Parkinsonism following dystonia in three patients, Movement Disorders 1:151(1986).

10. A.J. Lees, R.J. Hadie and G.M. Stern: Kinesiogenic foot dystonia as a presenting feature of Parkinson's disease, J. Neurol. Neurosurg. Psychiat. 47:885(1984).
11. P.A. Lewitt, R.S. Burns and R.P. Newman: Dystonia in untreated Parkinsonism. Clin. Neuropharm. 9:293(1986).
12. H. N. Klawans, and N. Paleologos: Dystonia-Parkinson syndrome: Differential effects of levodopa and dopamine agonists. Clin. Neuropharm. 9:298(1986).

NEUROLEPTIC MALIGNANT SYNDROME ASSOCIATED WITH WITHDRAWAL OF ANTIPARKINSONIAN DRUGS

Sadako Kuno, Osamu Komure, Eiji Mizuta,
Shunzou Yamazaki and Hiroshi Nishitani

Department of Neurology, Utano National Hospital
Kyoto, Japan

INTRODUCTION

Neuroleptic malignant syndrome (NMS) occurs rarely but can be a fatal disorder induced by adverse effects of neuroleptic medication.[1] This syndrome is characterized by muscular rigidity and/or tremor, hyperthermia, autonomic dysfunction, and altered consciousness. Also, a syndrome similar to NMS has been reported to be induced by a withdrawal of antiparkinsonian drugs.[2] We analyzed 15 patients with NMS associated with withdrawal of antiparkinsonian drugs (NMS-APD) to elucidate characteristics of this syndrome.

CASE REPORT

A 56-year-old man had suffered from Parkinson's disease for 9 years and had been treated daily during the last 4 years with L-dopa (700 mg) plus benserazide (175 mg), amantadine (200 mg), and trihexyphenidyl (6 mg). He complained of visual and auditory hallucinations and had a delusion that his son had been arrested by the police. Two days after an increase of L-dopa medication to 750 mg/day with carbidopa (75 mg/day), he complained that these drugs were toxic and refused to take antiparkinsonian drugs. Four days after cessation of these drugs, he was admitted to the neurology ward of Utano National Hospital on October 20, 1980, because of unresponsiveness, dehydration, and hyperthermia (Figure). When he was hospitalized, the patient could not move, eat, and drink anything. He showed high fever, severe dehydration, and delirium. After admission, all the symptoms including high fever (40 degrees centigrade), diaphoresis, skin pallor, deterioration of consciousness, severe rigidity, and akinesia continued, in spite of saline infusion with glucose, antibiotics, and vitamins. Three days after admission, he fell into a preshock state following GI tract bleeding. The patient finally recovered by treatment with intravenous infusion of saline and glucose as well as L-dopa and adrenocorticosteroid hormones for a period of 30 days. A high serum creatine kinase level of 1,690 IU (20 < normal < 220 IU) on the day of admission declined with amelioration of clinical manifestations.

Basic, Clinical, and Therapeutic Aspects of Alzheimer's and Parkinson's Diseases, Volume 2
Edited by T. Nagatsu *et al.*, Plenum Press, New York, 1990

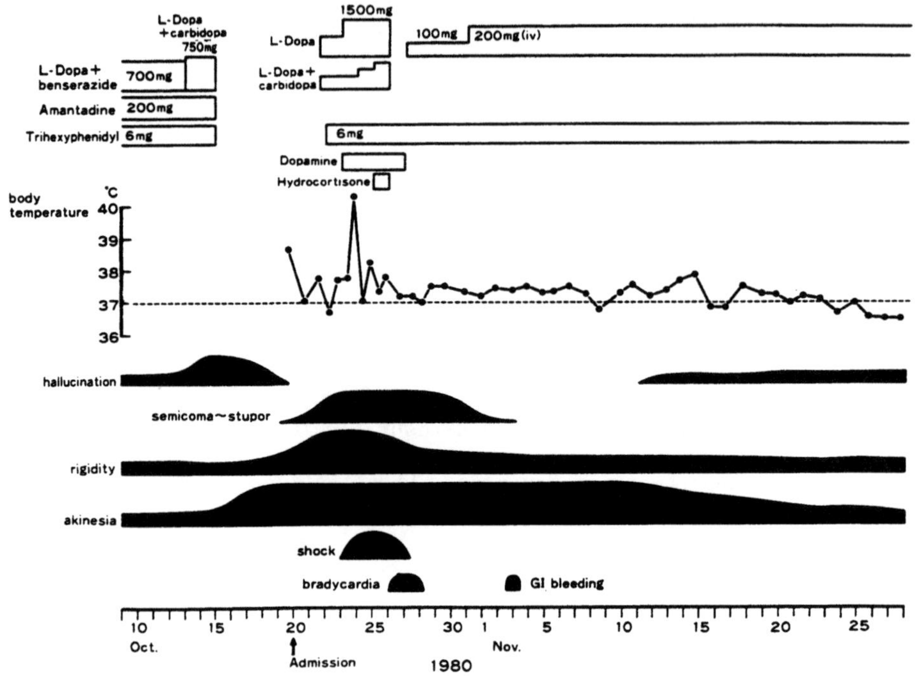

Figure 1. Temporal relationship between drug therapy and clinical manifestations of a 56 year-old man who suffered from Parkinson's disease. Adapted from Fujitake et al. (1984)[3].

CLINICAL CHARACTERISTICS OF 15 CASES OF NMS-APD

During the past 13 years, we have had an opportunity of seeing other NMS-APD patients who showed symptoms similar to those described above. This allowed us to analyze clinical characteristics manifested before and at the onset of NMS-APD in 15 patients among about 900 patients who had been treated with antiparkinsonian drugs for many years. The frequency of occurrence of NMS-APD was 1.67%. In order to investigate the susceptibility of NMS-APD, we analyzed the clinical background before the onset of NMS-APD, using criteria summarized in Table 1. The salient findings are listed as follows:

1) Thirteen cases (Cases 1 - 13) suffered from Parkinson's disease, and the 2 other cases had OPCA and PNLA, respectively. These patients included 5 males and 10 females. The age at the onset of NMS-APD ranged from 32 to 74 years old (mean age, 61 ± 11). The disease duration varied from 2.7 to 21.6 years (mean duration, 9.0 ± 4.4 years). The period of treatment with L-dopa ranged from 0.6 to 15.0 years (mean duration, 6.4 ± 3.4 years). The severity of parkinsonism by Hoehn and Yahr grade before the onset of NMS-APD was 3 or 4 (mean grade, 3.6 ± 0.5).
2) All 15 patients (100 %) showed adverse psychiatric effects such as hallucinations and illusions before the onset of NMS-APD.
3) Slight to moderate mental deterioration was recognized in 9 patients (60 %).
4) Seven (47 %) were tremor-dominant cases, and 9 (60 %) displayed the wearing-off phenomena or end-of-dose deterioration.

Table 1. Clinical Background prior to Onset of NMS

No.	age/sex	Dx.	Yahr's stage	tremor	wearing off	psychosis	dementia	duration (disease/dopa)
1.	32/M	PD	3	++	+++	+++	+	(6.7/ 6.3)
2.	44/F	PD	4	++	+++	++	-	(7.0/ 5.0)
3.	59/F	PD	3	++	+++	++	-	(7.6/ 7.4)
4.	60/F	PD	4	++	+++	+++	-	(21.6/15.0)
5.	62/M	PD	3	+++	+++	+++	-	(10.0/ 9.0)
6.	63/F	PD	4	+	++	+	+	(9.0/ 6.0)
7.	66/M	PD	3	+++	+	+++	+	(9.0/ 5.0)
8.	56/M	PD	4	+	-	+++	+	(9.0/ 6.0)
9.	59/F	PD	4	+++	-	+++	+	(15.0/11.0)
10.	61/F	PD	3	++	-	+++	++	(9.0/ 7.0)
11.	64/F	PD	4	+++	-	++	+	(4.0/ 3.0)
12.	69/F	PD	3	++	-	+	-	(11.0/ 8.0)
13.	75/F	PD	4	+	-	++	-	(6.0/ 2.0)
14.	69/M	OPCA	(4)	+	-	+++	-	(2.7/ 0.6)
15.	74/F	PNLA	(4)	+	-	+	+	(8.0/ 5.0)

PD: Parkinson's disease, OPCA: Olivo-ponto-cerebellar atrophy, PNLA: Pallido-nigro-luysian atrophy.

Table 2. Clinical Features at Onset of NMS

No.	labile B.P.	tachy-cardia	disturbed consciousness	Yahr's stage	CK (IU/L)	ESR	WBC
1.	++	+	+++	5	500	14/36	10700
2.	-	+	+++	5	1107	43/32	6800
3.	++	++	+++	5	700	45/79	7600
4.	-	+	-	5	764	4/15	5500
5.	+++	++	++	5	387	23/43	5500
6.	++	-	+++	5	308	10/22	9700
7.	++	+	++	5	NT	21/53	5900
8.	+	+	++	5	NT	10/24	12800
9.	+	-	+	5	630	40/71	8200
10.	+	+	+++	5	1690	53/98	9800
11.	-	++	+	5	980	24/57	5000
12.	++	+++	+	5	24	32/71	11000
13.	+	-	+++	5	1100	31/61	10400
14.	-	++	++	(5)	358	4/18	8300
15.	++	++	+++	(5)	NT	15/37	22800

B.P.: blood pressure, CK: creatine kinase, ESR: erythrocyte sedimentation rate, WBC: white blood cells, NT: not tested.

The clinical characteristics and laboratory findings at the onset of NMS-APD are shown in Table 2. The following points were notable:

1) All fifteen patients (100 %) had hyperthermia over 38 degrees centigrade; 11 (73 %) had an abnormally high perturbation in blood pressure, and 12 (80 %) had tachycardia, being indicative of autonomic nervous disturbances.
2) Fourteen patients (93 %) had altered consciousness (7, stupor; 7, coma).

3) All fifteen patients (100 %) showed marked rigidity and tremor, and their mean Hoehn and Yahr's severity grade elevated from 3.6 to 5.0.
4) Eleven patients (92 %) showed abnormally high serum creatine kinase levels, and 8 (53 %) had high erythrocyte sedimentation rates, whereas 7 (47 %) had leukocytosis.

Our NMS-APD cases were not fatal, and all 15 patients recovered by treatment with saline infusion including glucose, L-dopa, some vitamins, and adrenocorticosteroid hormones.

DISCUSSION

The NMS in patients with schizophrenia has been considered to result from toxic adverse effects caused by neuroleptic medications. The NMS-APD cases presented in this paper were induced apparently by a withdrawal of antiparkinsonian drugs in patients with dopamine deficiency disorder, particularly with adverse effects of psychosis by dopaminergic drugs. Recently, it has been suggested that dopaminergic receptor activity in the brain may be increased in some type of patients with schizophrenia, since ^{3}H-spiperone binding to a particular type of schizophrenic's nerve cell membrane homogenate is higher in Bmax than the binding in controls.[4] A similar result was obtained in vivo by PET studies.[5] In Parkinson's disease, the nigrostriatal dopaminergic deficiency is the main cause of disabilities in the patients. Therefore, the state of the dopaminergic system in these two diseases appears to be opposite. One might then speculate that suppression of the brain dopaminergic system by neuroleptic medication may lead to the expression of the NMS in schizophrenic patients. In the case of Parkinsonian patients, continued treatment with L-dopa may enhance the brain dopaminergic system, reaching high levels at which psychiatric disorder such as schizophrenia may appear. A sudden withdrawal of L-dopa treatment would suppress the brain dopaminergic system, thereby inducing NMS-APD. A major difference between NMS and NMS-APD is a high mortality rate in the former compared with the latter. The reason for this difference is not clear.

REFERENCES

1. J. Delay and P. Deniker, Sur quelques erreurs de prescription des medicaments psychiatriques, Bull. Mem. Soc. Med. Hop. Paris 116: 487 (1965).
2. M. Toru, O. Matsuda, K. Makiguchi K. Sugano, Neuroleptic malignant syndrome like state following a withdrawal of antiparkinsonian drugs, J. Nerv. Ment. Dis. 169: 324 (1981).
3. J. Fujitake, S. Kuno and H. Nishitani, Neuroleptic malignant syndrome-like state in eight patients with parkinsonism, Clin. Neurol. 24: 371 (1984).
4. T. Lee and P. Seeman, Elevation of brain neuroleptic/dopamine receptors in schizophrenia, Amer. J. Psychiatry 137: 191 (1980).
5. D. F. Wong, H. N. Wagner Jr., L. E. Tune, R. F. Dannals, G. D. Pearlson, J. M. Links, C. A. Tamminga, E. P. Broussolle, H. T. Ravert, A. A. Wilson, J. K. T. Toung, J. Malat, J. A. Williams, L. A. O'Tuama, S. Snyder, M. J. Kuhar, and A. Gjedde, Positron emission tomography reveals elevated D-2 dopamine receptors in drug-naive schizophrenics, Science 234: 1558 (1986).

FREEZING OF GAIT AND RESISTANCE TO DOPAMINERGIC THERAPY

Gilles Fénelon (1), Marc Ziégler (2), Florence Mahieux (1), and Alain Guillard (1)

(1) Clinique neurologique, Hôpital Tenon, Paris, France
(2) Service de neurologie, Hôpital Sainte-Anne, Paris, France

Freezing of the gait is a sudden, involuntary, and transient episode of immobility. The patient remains "riveted to the spot" (1) or, sometimes, falls. Frequently encountered in the course of Parkinson's disease (PD) it sometimes occurs in other parkinsonian syndromes. We report three such cases where freezing was a prominent symptom.

CASE REPORTS

Case 1. This 77 year-old woman complained, at age 68, of unsteadiness, micrographia and weight loss. She had no history of hypertension. The disease has been slowly progressive and, after a couple of years, is characterized by freezing of the gait. The freezing episodes are numerous and impair the gait markedly. They occur indoors or outdoors, at anytime of the day, and are frequent on initiation, when turning and crossing the street. They also are triggered by obstacles and emotions. To try to overcome the freezing, she pulls a leash tied up to her right foot. This unusual strategy is inconstantly successful. She can climb stairs and step over obstacles. The gait is otherwise normal. She has a slight hypomimia and loss of speech expression, and complains of hypersalivation. Turning in bed, wiping her feet on a door-mat is difficult but she has no upper limb akinesia at all, no rigitidy, no tremor. The rest of the neurological examination, including oculomotor function, and the mental status are normal. Unified Parkinson's Disease Rating Scale (UPDRS) score for activities of daily living (ADL) is 11 and for motor examination is 6. A computed tomographic (CT) scan of the brain was normal. T2-weighted cerebral magnetic resonance imaging (MRI) disclosed a few hyperintense foci of the white matter of both hemispheres. Neither levodopa associated to benserazide (1000 mg/d) nor subcutaneous apomorphine (single shot injections, from 1 to 4 mg) improved the freezing phenomenon.

Case 2. At age 43, this man progressively suffered micrographia and right leg stiffness. A postural tremor of the hands had appeared 10 years beforehand. At age 46 the examination revealed, besides the postural tremor, an akinetic rigid syndrome most prominent on the right limbs. The gait appeared "mechanical", with numerous freezing episodes. There was no oculomotor abnormality, no intellectual impairment. The neurological status was otherwise normal. The UPDRS score for ADL was 12 and for motor examination was 13. A CT scan of the brain was normal. Dopaminergic drugs (levodopa plus benserazide, 600 mg/d, amantadine, bromocriptine) provided no benefit.

Basic, Clinical, and Therapeutic Aspects of Alzheimer's and Parkinson's Diseases, Volume 2
Edited by T. Nagatsu *et al.*, Plenum Press, New York, 1990

Case 3. This 77 year-old man has had an akineto-hypertonic syndrome for the past 7 years. The progression of the disease has been slow. His gait is characterized by intermittent freezing, particularly when turning, and is otherwise normal. The examination discloses severe dysarthria and signs of axial akinesia. He has only slight upper limbs akinesia, rigidity and postural tremor of the hands. Voluntary downgaze is impaired. He shows a moderate intellectual decline. UPDRS score for ADL is 16 and for motor examination is 21. MRI of the brain showed only a moderate cerebral and brainstem atrophy. During the first years, levodopa and benserazide improved his condition. At present, his neurological status is no longer changed by doparminergic drugs (last trial : levodopa associated to carbidopa : 1000 mg/d).

DISCUSSION

In our three cases, the freezing of the gait was the major symptom, nearly isolated in case 1, associated with slight parkinsonian symptoms, without rest tremor, in the two other cases. No improvement was observed with levodopa and dopaminergic agonists (bromocriptine, apomorphine). In the three cases, freezing was disabling and responsible for falls. In case 1, the neurological status is similar to "pure akinesia". In this rare condition, "difficulty in initiation of movements" with freezing of gait is the dominant symptom (2, 3). The patients display neither rigidity nor tremor and respond poorly to levodopa therapy. However, in a recently reported case, levodopa did alleviate the symptoms. In this case, postmortem examination of the brain showed the classical changes of Lewy body PD (4). Besides, a "lower body parkinsonism" has also been individualized. The patients had a marked gait disturbance with freezing, no or minimal upper limb involvement, frequent hypertension and poor response to levodopa (5). Because CT or MRI scans showed deep subcortical lesions the authors assumed that this syndrome had an ischemic etiology. However, a vascular origin is not likely in our case 1 as there is no history of hypertension or stroke-like episodes, and the MRI changes are not specific (6).

Freezing of the gait was first described in a thesis about postencephalitic parkinsonism by Gabrielle Lévy in 1922 (7). it was later found to occur in the course of idiopathic PD (8) and named "enrayage cinétique" (kinetic "jamming"), or "bégaiement de la mise en route" ("start-stuttering"). After the advent of levodopa, freezing episodes attracted attention again, since they often appear or reappear under treatment. They have been described together with the other fluctuations in performance in patients on chronic levodopa therapy (1, 9, 10, 11). Such episodes can occur spontaneously but they are often triggered by situations such as starting to walk, turning, crossing the street or passing through a doorway. Patients often develop strategies to overcome the freezing such as stepping over a real or imaginary obstacle. Sometimes the stimulus is an obstacle, a sudden noise, an emotion. They usually occur at anytime of the day.

The underlying mechanisms of freezing of the gait remain unknown. It is not a mere global enhancement of akinesia, as akinesia does not occur simultaneously in upper limbs (10). Its relation to triggering stimuli and the shortness of episodes clearly individualize them from "on-off" effect which is a fluctuation in performance specifically seen in PD patients on chronic levodopa therapy. Freezing is not a complication of levodopa therapy since it occurs in the natural history of PD and in other parkinsonsian syndromes, and it is usually not related to timing of dosage in treated PD (1). A prevailing dopaminergic mechanism is unlikely because in most cases (in PD and other cases like ours) dopaminergic drugs do not influence neither the frequency nor the characteristics of the freezing episodes. Selegiline adjunction has no efficacy on random freezing (12). Early treatment with bromocriptine, alone or in combination with levodopa, does not prevent significantly the occurence of freezing in PD (13). However, freezing of the gait was present in 6 out of 7 patients with MPTP-induced parkinsonism, a condition assumed to be due to selective lesions to the nigrostriatal dopaminergic system (14). DL-threo-3,4-dihydroxyphenylserine was tried in the hypothesis of a norepinephrine deficiency. Its beneficial effect appeared inconstant (4, 15). Which neurotransmitters ever might be involved in freezing, we are struck by the discrepancy between the low intensity of the triggering stimulus and the severity of the motor response. It implies a strong influence of sensory or associative inputs on motor pathways. Such a convergence has been demonstrated anatomically at the pallidal level (16). Moreover, electrophysiological studies of pallidal neurons activity have suggestef that, in monkeys rendered parkinsonian by MPTP, the basal ganglia respond to noise excessively and are unable to select the appropriate input signals (17). This abnormal response and lack of selectivity could lead to the freezing phenomenon even for a low intensity stimulus.

Acknowledgment : The authors wish to thank M. M. Pomès for preparing the manuscript.

REFERENCES

1. C. D. Marsden and J. D. Parkes, "On-off" effects in patients with Parkinson's disease on chronic levodopa therapy, Lancet, 1:292-296 (1976).
2. H. Narabayashi, H. Imai, M. Yokochi, K. Hirayama and R. Nakamura, Cases of pure akinesia without rigidity and tremor and with no effect by L-dopatherapy, in: "Advances in Parkinsonism", W. Birkmayer and O. Hornykiewicz, ed., Roche, Basle, pp:335-342 (1976).
3. H. Imai, H. Narabayashi and E. Sakata, "Pure akinesia" and the later added supranuclear ophtalmoplegia, Adv. Neurol., 45:207-212 (1986).
4. N. P. Quinn, P. Luthert, M. Honavar and C. D. Marsden, Pure akinesia due to Lewy body Parkinson's disease : a case with pathology, Movement Disorders, 4:85-89 (1989).
5. P. M. Fitzgerald and J. Jankovic, Lower body parkinsonism : evidence for vascular etiology, Movement Disorders, 4:249-260 (1989).
6. A. Kertesz, S. E. Black, G. Tokar, T. Benke, T. Carr and L. Nicholson, Periventricular and subcortical hyperintensities on magnetic resonance imaging, Arch. Neurol., 45:404-408 (1988).
7. G. Lévy, Contributions à l'étude des manifestations tardives de l'encéphalite épidémique (formes prolongées et reprises tardives). Mémoire et thèse, Paris (1922).
8. R. Melaragno Filho, Sur quelques aspects des troubles de la dénervation volontaire. Enrayage cinétique, persévération tonique, "bégaiement de la mise en route du mouvement", piétinement observés dans les syndromes extrapyramidaux et sur leur relation avec quelques autres phénomènes analogues observés en neuropathologie, Arq. Neuropsiquiatr., 6:334-351 (1948).
9. A. Barbeau, Long term side-effects of levodopa, Lancet, 1:395 (1971).
10. G. Boudin, B. Pépin, A. Guillard, S. Godlewski, J. M. Fabiani and M. Haguenau, Résultats après trois à cinq ans de traitement de 219 cas de syndrome parkinsonien traité par la L-Dopa et l'association L-Dopa inhibiteur de la décarboxylase, Rev. Neurol. (Paris), 130:273-284 (1974).
11. J. L. Ribadeau Dumas, P. Rondot and P. Cardon, Développement des études sur le traitement des syndromes parkinsoniens par la lévodopa. II. Effets à long terme de la L-Dopa seule ou en association avec un inhibiteur de la décarboxylase, Thérapie (Paris), 32:161-172 (1977).
12. U. K. Rinne, R-(-)-Deprenyl as an adjuvant to levodopa in the treatment of Parkinson's disease, J. Neural. Transm. (suppl) 25:149-155 (1987).
13. U. K. Rinne, Early dopamine agonist therapy in Parkinson's disease. Movement Disorders, 4, Suppl 1:586-594 (1989).
14. P. A. Ballard, J. W. Tetrud and J. W. Langston, Permanent human parkin-

sonism due to 1-methyl-4-phenyl-1,2,3,6-tetrahydropyridine (MPTP) : seven cases, Neurology, 35:949-956 (1985).
15. H. Narabayashi, T. Kondo, F. Yokochi and T. Nagatsu, Clinical effects of L-threo-3,4-Dihydroxyphenylserine in cases of parkinsonism and pure akinesia, Adv. Neurol., 45:593-602 (1986).
16. G. Percheron, J. Yelnik and C. François, A Golgi analysis of the primate globus pallidus. III. Spatial organization of the striato-pallidal complex, J. Comp. Neurol., 227:214-227 (1984).
17. M. Filion, L. Tremblay and P. J. Bédard, Abnormal influences of passive limb movement on the activity of globus pallidus neurons parkinsonian monkeys, Brain Research, 444:165-176 (1988).

NORMAL THYROID FUNCTION IN 50 PATIENTS

WITH PARKINSON'S DISEASE

Toshiya Shiojima, Masao Kinoshita, Yasuo Iwasaki,
Ken Ikeda and Kiyoshi Takamiya

The Fourth Department of Internal Medicine
Toho University Ohashi Hospital
2-17-6 Ohashi, Meguro-ku, Tokyo, JAPAN

PURPOSE

The clinical picture of patients with Parkinson's disease(PD) has a certain resemblance to hypothyroidism. Many of the patients in both groups are characterized by general slowness and diminution of movements, monotonous voice and an attitude of inactivity, apathy or depression. In these cases the disorder cannot be mistaken for hypothyroidism, and usually there is no difficulty in differentiating between the two diseases. Hypothyroidism might be overlooked if it occurs with PD. Furthermore, in the study of Berger and Kelly 3 of 46 Parkinson patients were hypothyroid, hence evaluation of thyroid function was recommended. The purpose of this study was to evaluate thyroid function in patients with PD in order to diagnose masked hypothyroidism.

MATERIALS AND METHODS

Since 1969 the diagnosis of PD was established in our depertment in more than hundred patients by history, clinical and laboratory examination. Thyroid function tests were determined in 50(24 men and 26 women aged 46 to 84 years(means 70.4 years)) of these patients. All patients were receiving treatment with L-dopa, carbidopa/L-dopa, or benserazide/L-dopa. The control group consisted of 50 non-neurological patients. In all patients the serum concentration of triiodothyronine(T_3), thyroxine(T_4), thyroid-stimulating hormone(TSH), and free T_4 were measured by the radioimmunoassay. All sera were also tested for the presence of antithyroid antibody activities(antithyroglobulin(Tab), antimicrosomal antibody(Mab)) by the commmercial hemagglutination tests. No patients were receiving any medication known to influence thyroid function assessment. With statistical variable as T_3, T_4, TSH, free T_4, Tab, and Mab, the following tests were performed: calculation of mean values(+ standard deviation), examination for the normal range and Student's t-test. The limit of significance was $p<0.05$.

Edited by T. Nagatsu *et al.*, Plenum Press, New York, 1990

RESULTS AND DISCUSSION

The results of thyroid function tests and the frequency of the circulating antithyroid antibody titers were summarized in Table 1. The T_3, T_4, TSH and free T_4 levels were found to be within the normal range. There were no statistically significant difference between patients with PD and the normal subjects. Thyroid antibodies were not detectable in the sera of all patients with PD. No patients exhibited the thyroid enlargement. PD is primarily a disease of the aged; 84% of cases occur in patients aged 65 years and older. Hypothyroidism can be a difficult diagnosis to make in the aged, particularly when it coexists with PD. In one study of geriatric inpatients, only 13 of 46 hypothyroid patients had classical signs and symptoms of myxedema. The most common presentation was apathy and general deterioration of mobility and health. The most common psychiatric manifestation was depression. Concurrence of these nonspecific features with the bradykinesia, rigidity, and facial hypomimia of PD could be overlooked. The main results of our study were the normal thyroid parameters found in all the Parkinson patients. Berger and Kelly reported 3 of 46 PD patients were hypothyroidism. Johannessen et al. studied 52 patients with PD and found normal thyroid values and TSH-levels in all patients except one, who turned out to have an incipient hypothyroidism. Administration of L-dopa has been found to alter some endocrine functions. The basal T_3, T_4, TSH, and free T_4 concentrations in our study were normal, and no affection of long time L-dopa treatment on T_3, T_4, TSH, and free T_4 were found. Unfortunately there were no opportunity to accomplish the study by scintilating thyroid scan and a suppression test to give additional proof of noncoexisting thyroid autonomy. In conclusion,we found normal thyroid values in patients with PD and PD could notbe related to thyroid disorders.

Table 1
Tyroid function in patients with Parkinson´s disease

Thyroid function test	Normal range	PD Mean±SD
T_3, ng/dl	70 ～ 190	121.7±27.7
T_4, μg/dl	5.0 ～ 13.0	9.1±1.3
TSH, μU/dl	～ 10	4.2±2.3
free T_4, ng/dl	1.1 ～ 2.0	1.4±0.1
TBG	14 ～ 29.4	24.3±1.5
Antithyroid antibody titers		
Tab-positive		none
Mab-positive		none

T_3 : triiodothyronine, T_4 : thyroxine, TSH : thyroid-stimulating hormone, TBG : thyroxine-binding globulin, Tab : antithyroglobulin antibody, Mab : antimicrosomal antibody, PD : Parkinson´s disease, SD : standard deviation

REFERENCES

1. Berger JR, Kelley RE. Thyroid function in Parkinson disease. Neurology(NY) 31; 93, 1981.
2. Iwasaki Y, Kinoshita M. Thyroid function in patients with multiple sclerosis. Acta Neurol Scand 77; 269, 1988.
3. Iwasaki Y, Kinoshita M. Normal thyroid function in 32 patients with amyotrophic lateral sclerosis. Jpn J Med 28; 309, 1989.
4. Iwasaki Y, Kinoshita M, Ikeda K, Takamiya K, Shiojima T. Amyotrophic lateral sclerosis and thyroid function. J Neurol 236; 373, 1989.

RISK FACTORS TO PARKINSON'S DISEASE;

HERITABILITY AND LIFE STYLES

Kiyotaro Kondo

Department of Public Health, Hokkaido University School of Medicine, Sapporo, Japan

INTRODUCTION

Reviewing current biometrical data including those of our own studies, this report proposes that idiopathic Parkinson's disease (PD) is a final consequence of a long-standing preclinical process influenced by multiple genes and environmental factors, especially life styles.

NATURE / NURTURE IN PARKINSON'S DISEASE

Major results are as follows.

Genetic Analyses

Frequency of PD differs in populations in which individuals share common genes to different extents. In monozygotic twins the concordance rate is rather low, however (Table 1). Even if the etiology is not elucidated, mechanism of familial aggregations can be tested biometrically, because statistical characteristics reflect the mechanism, and inversely the latter can be tested by comparing observed patterns with the expected patterns derived from possible bypotheses (Table 2).

A multifactorial theory has been accepted in this way (Kondo et al, 1973), proposing that genes and environmental factors collaborate to determine underlying liability, and that if its value crosses a certain threshold, signs and symptoms develop which clinicians call PD.

Table 1. Frequencies of Parkinson's Disease in Various Caucasian Populations

Populations	Frequencies(%)
General populations	less than 1
Siblings	5.7
Parents	5.7
Dizygotic twins	6.67
Monozygotic twins	6.25

These are pooled data of those aged over 60 from the most quoted reports and, frequency is expressed as the prevalence rate for the general populations, as the recurrence rate for siblings and parents, and as the concordance rate for twins.

Risk Factors and Case - Control Studies

Factors predating clinical onset of a disease are called risk factors to the disease, when individuals with the factors have elevated probability of being affected. Risk factors are not necessarily the cause. In obscure diseases they are useful mainly in three ways : 1) as clues for the cause, 2) to identify high-risk individuals, and 3) when eliminated, to prevent the disease.

Established risk factors to PD are advanced age, less smoking, rigid preclinical personality and affection of the relatives. Possible risk factors include head injuries, low blood pressure, habitual constipation, reduced physical activity, etc (Kondo, 1986).

Table 2. Expected and Observed Family Patterns, Various Genetic Hypotheses, Parkinson's Disease

Expected Patterns	Compatibility with the Observed Patterns
A. Single-gene inheritances :	
X-linked recessive. Parents normal;1/2 sons but no daughter affected ; no paternal males affected	no
X-linked dominant. All daughters but no son of affected fathers affected	no
autosomal recessive. No parent affected ; 1/4 children affected	no
autosomal dominant. One parent affected ; 1/2 children affected ;few sporadic mutants	no
B. Multifactorial inheritance with a threshold effect :	
sib recurrence rate is square root of prevalence rate	yes
risk decreases in remote relatives	yes
concordance rate larger in monozygotic than dizygotic twins	yes?
risk increases with younger age at onset	yes
risk increases with more sibmates already affected	yes

Adapted from Kondo et al (1973) and other reports.

Despite some methodological inadequacies, a case-control study is useful for a quick screening of candidate risk factors. It is also useful to test hypotheses proposed from other areas of investigation. Few studies were not particularly rewarding however, that evaluated histories of various diseases prior to PD.

Our study asked 271 questions to 166 cases with PD and two each of age-and-sex matched controls, to evaluate factors including lifestyles. PD-prone profiles were ; 1) non- Japanese diet, 2) less smoking, 3) less drinking, 4) less hobbies and if any indoor, non-sporting hobbies requiring no partner, 5) narrow interests only in business, own surroundings, limited human relations etc, 6) more habitual constipation, 7) less circulatory disturbances, 8) less physical exercises, and 9) rigid preclinical personality (Kondo, 1986).

Twins and the Heritability Estimates

In theory, twins are useful to evaluate the nature / nurture problem in a given trait, but in reality, the results are often perplexing mainly because of difficulties of collection of sufficient data without biases and errors.

The essential point with the twin data is to compare groups of individuals who share genes to known different extents. The same purpose can also be achieved based on other types of populations which are larger and more accessible in epidemiological surveys. Table 1 shows that frequency of PD grows as the extent becomes larger to which the members share genes.

Heritability gives a rough estimate of the degree of genetic determination. Reported values are 79.2 % for Minnesota, U.S.A. and 90.8 % for Sweden, based on the incidence among the siblings and in the general population, as calculated by Falconer's method (Kondo et al, 1973). It was about 65 % based on the data in Table 1, calculated by the present author by Smith's nomogram. Parenthetically, a value of 22 % was reported by the author (Kondo, 1986), based on the twins of Ward et al (1983), but this value should be modified in light of the new twin data that have been added.

PD is probably a threshold consequence of a long-standing interaction of genes and environmental factors, where genes are slightly more decisive than the environmental factors. Otherwise stated, postnatal influences modify a rather strong inherited inclination toward PD, as the host ages.

<u>Difficulties of Adequate Etiological Studies</u>

Causal relation can easily be detected when a short-time exposure to a single specific agent rapidly causes a specific syndrome. Acute MPTP experiments represent such a situation. Action of the environmental factors in PD probably is not so simple, and it may be that cumulative effects occur after a long-time exposure to small amounts of various agents interacting with age-dependent predispositions influenced by plural loci, and superimposing themselves on the aging process. The agents may be plural, each within unharmful amounts but jointly they become substantially harmful. These situations are proven in experimental carcinogenesis, and unlike occupational cancers due to heavy exposures to single or few identifiable carcinogens, most cancers are caused in these ways depending on different lifestyles.

PROPOSED STUDY OF THE MIGRANTS

Above are the highlights of biometrical information we have now, which have apparent limitations. Even in these situations, however, migrants may give a breakthrough for better understanding about the nature / nurture problem of PD. For example, in the following three groups,

1. Japanese in Japan showing a prevalence rate of PD, 50-60 per 100,000
2. Japanese-Americans born in U.S.A. to purely Japanese parents, and
3. Caucasians born in U.S.A., with the rate, 100-150 per 100,000

the second group involves those who are genetically identical to the first group, but subject to the almost same environment as the third group. Frequency of PD may show a gradient among the three groups, and a new case-control study utilizing the cases as index cases who emerged from these groups may give explanations for such a gradient, and significances of genes and the environment in the causation of the disease.

The case-control study must use the same questionnaire involving highly selected questions, in all three populations in order to compare the risk factors between the Japanese and the US Caucasians. Factors in the lifestyle are manageable and useful for primary prevention, or postponement of clinical onsets.

Food preferences are important possible risk factors. At least in acute experiments in squirrel monkeys, MPTP-related tetrahydroisoquinoline produces a condition highly resembling human PD. Quoting and amplifying the data on amounts of MPTP-related substances in various foods (Niwa et al, 1989), one should be able to estimate lifelong exposures to such substances in three groups stated above. In each group, a dose-response relation may be analyzed to establish whether PD is associated with increased exposures to such substances or other food constituents.

Knowledge about the genes for tyrosine hydroxylases is increasing, on the other hand. These genes affect dopamine synthesis and therefore influence host predisposition to environmental neurotoxins. Interracial comparisons in these aspects must also be developed and correlated with the proposed case-control studies.

REFERENCES

Falconer, D. S., 1965, The inheritance of liability to certain diseases, estimated from the incidence among the relatives, Ann hum Genet, 29:51.

Kondo, K, Kurland, L. T., Schull, W. T., 1973, Parkinson's disease, genetic analysis and evidence of a multifactorial etiology, Mayo Clin Proc, 48:465.

Kondo, K,1987, Epidemiological evaluation of risk factors in Parkinson's disease, in " Parkinson's Disease " Yahr, M. D. and Bergman, K. J. eds., Raven Press, New York.

Niwa, T., Yoshizumi, H., Tatematsu, A., Matsuura, S., Nagatsu, T., 1989, Presence of tetrahydroisoquinoline, a parkinsonism-related compounds, in food, J Chromatogr, 493:347.

Smith, C., 1970, Heritability of liability and concordance in monogygotic twins, Ann hum Genet, 34:85.

Ward, C. D., Duvoisin, R. C., Ince, S. E., Nutt, J. D., Eldridge, R., Calne, D. B., 1983, Parkinson's disease in 65 pairs of twins and in a set of quadruplets, Neurology, 33:815.

PREVALENCE OF PARKINSON'S DISEASE IN IZUMO CITY, JAPAN

Kazunori Okada, Shotai Kobayashi, and Tokugoro Tsunematsu

The 3rd Division of Internal Medicine
Shimane Medical University
Izumo, Shimane, 693, Japan

INTRODUCTION

Parkinson's disease is caused by the degeneration of dopaminergic neurons in the substantia nigra, and develops mainly later in life. It is a major public health problem especially in geriatric societies. Recently, the discovery of MPTP-induced parkinsonism has led to a renewed interest in the epidemiology of Parkinson's Disease.[1] Campanella reported that pesticides are just one of the potential environmental neurotoxins that might be important in the etiology of PD.[2] There is an urgent need for accurate information on its prevalence in different communities world-wide. Japan has become an industrial and gerontrogical society very rapidly in recent years. Therefore, as it is an important problem whether the prevalence of Parkinson's disease is increasing or not, we studied the prevalence ratio in the area of Japan having the largest population of elderly persons.

SUBJECTS AND METHODS

An epidemiological survey was carried out to screen a community of 80,639 people in Izumo city of Shimane Prefecture (this prefecture has the highest elderly population in Japan) for Parkinson's disease(PD). This city is located in western Japan and covers an area of 175.44 sq Km. The elderly population in this city were mainly engaging in agriculture. The direct screening questionnaire including questions about tremor, bradykinesia, gait disturbance, and ADL disturbances was sent to all members of the adult population (56,869 people over the age of 20) and from these questionnaires we tabulated a list of persons suspected of having parkinsonian symptoms. Next, investigations were carried out in hospitals equipped for CT scanning. For the population who was not under medical treatment, a door-to-door survey was carried out by medical students. In addition to these surveys, we referred to the list of registered PD patients who received public financial assistance. To exclude the parkinsonian syndrome by known causes, questionable cases were examined by neurologists.

Basic, Clinical, and Therapeutic Aspects of Alzheimer's and Parkinson's Diseases, Volume 2
Edited by T. Nagatsu *et al.*, Plenum Press, New York, 1990

A diagnosis of PD was made if two or more of the following clinical features were present: postural instability, bradykinesia, rigidity, and tremor at rest. The frequency of PD is expressed as a point prevalence ratio (prevalence day, October 1, 1985).

At the same time, to assess the prevalence of dementia in PD, we asked each patient's family about his or her mental state: memory disturbance, nocturnal delirium, and activity of daily living. The suspected cases were examined by neurologists. The diagnosis of dementia was based upon DSM III criteria and apparent and severe cases were tabulated.

RESULTS

The direct screening questionnaire was returned by 62% of all who had received it. There were 439 people who had suspected symptoms of parkinsonism based on the direct questionnaire. The medical information from hospitals was collected in 109 cases and door-to-door survey was carried out in 330 cases. Thirty questionable cases were examined by neurologists. Of the 439, 66 patients (mean age of 67.0) were concluded to be suffering from PD. The names of only 11 these patients appeared on the public registration list. Fig.1 shows the age distribution of Parkinson's disease. There was one family with cases of juvenile PD; their ages were 37, 35 and 32 y. The age-adjusted prevalence ratio was 83.1 per 100,000 population. When the prevalence ratios were adjusted to Shimane Prefecture and the total Japanese population, their adjusted ratios were 94.0 and 68.3, respectively.(Fig.2) Their mean duration of illness was 7.0 years. There were 18 males and 48 females, with the age-adjusted prevalence ratio for women (131.9 per 100,000) exceeding that for men (53.7 per 100,000). There was a 1.4 preponderance of females in the city. When the sex ratio was equalized, the preponderance of PD in women was twice that in men. Seven cases who were positive for parkinsonism as judged from the direct questionnaire died before they could receive a neurological examination.

To assess the prevalence of dementia in PD, 41 PD cases were given a mental examination and head CT examination. The diagnosis of dementia was based upon DSM III criteria. To rule out depression and the side effects of anti-parkinson drugs, only apparent and severely demented cases were tabulated. Of these patients, 17.1% showed apparent dementia (Table.1). Demented cases were older and were at a more advanced stage in Yahr's classification than non-demented ones.

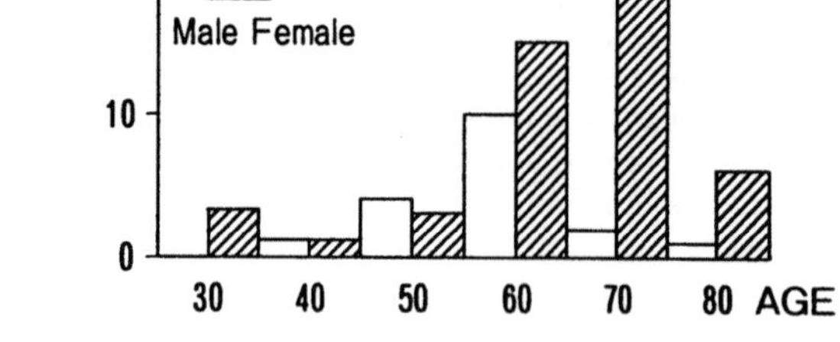

Fig.1. Age distribution of Parkinson's disease

Table.1. Dementia in Parkinson's Disease(N=41)

AGE	30	40	50	60	70	80
STAGE						
I	1		3	2	4	1
II		1	1	7	1	2(1)
III	2			4	3(1)	1(1)
IV				2(1)	1	
V			1		3(2)	1(1)

Figures in parentheses indicate numbers of PD patients with dementia.

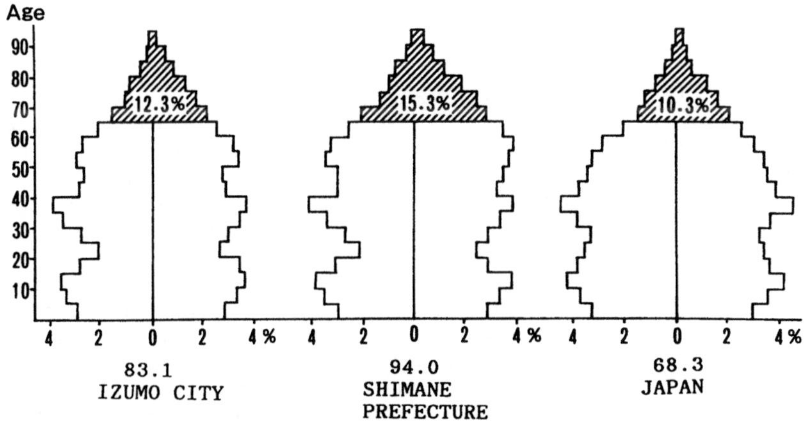

Fig.2. Populational distribution and age-adjusted prevalence ratio

Table 2. Japanese prevalence reports of Parkinson's Disease

Author	Year	Area	Population (1,000)	Prevalence rate (/100,000)
Nishitani	1978	Kyoto	26	34.3
Takahashi	1978	Yonago	120	38.1(35.9)
Igata	1980	Kagoshima	1767	37.0(29.6)
Takahashi	1979	Aichi	25	55.0
Harada	1983	Yonago	125	80.6(65.4)
This report	1985	Izumo	81	83.1(68.3)

():adjusted by Japanese population

DISCUSSION

In Japan there are only a few reports about the prevalence of PD. These reports described the prevalence rate of PD in Japan to be 34.3-55.0/100,000 people(Table 2). Harada reported that the prevalence rate of PD in Yonago City, which is located near Izumo City was 80.6 / 100,000 people in 1983.[3] In these studies except Harada's, the population sample was too small or a limited hospital-based group was reported. In this study, we carried out community-based research to assess the prevalence rate of PD. As a result, we tabulated 66 PD patients representing an age-adjusted prevalence ratio of 83.1 / 100,000, which is the highest ratio in Japan. When this prevalence ratio was adjusted for Shimane Prefecture, it became 94.0. In the Japanese national survey of 1985, the percentage of aged people over 65 was 12.3% in Izumo City, 15.3% in Shimane Prefecture and 10.3% in Japanese population.

Recently, the discovery of MPTP-induced parkinsonism has led to a renewed interest in the epidemiology of PD. Campanella and colleagues studied the incidence of PD in Quebec province and reported that there was strong correlation between the incidence of the disease and the level of pesticide use. He suggested that pesticides are just one of the potential environmental neurotoxins that might be important in the etiology of PD.

The finding of an increasing prevalence ratio of PD with increasing age is consistent with the hypothesis that an environmental neurotoxin may play a role in the cause of PD. People in the old age groups could have been exposed for a longer period to putative environmental factors and would therefore be more likely to sustain PD. From a clinical viewpoint, Marttila described the possibility of the effect of levodopa treatment to increase the prevalence. Since measures diminishing mortality and ensuring longer survival necessarily are reflected in an increase in the number of persons with Parkinson's disease, the prevalence of Parkinson's disease will increase by 1.4-1.8 times in the near future.[4]

There are controversies over the sex difference of the prevalence rate. Marttila et al[5], Jenkins[6] and Harada et al[3] reported female preponderance, just as reported here. But Rosati et al[7] and Kessler[8] obtained opposite results.

There are many reports about the prevalence rate of dementia in Parkinson's disease. Brown and Marsden[9] summarized the prevalence of dementia in PD in major studies published over the past 60 years and reported that an average 35.1% were thought to have some degree of dementia. Recent populational surveys reported the prevalence rate to vary from 20% to 31%. Mayeux et al[10] reported 37 demented patients(10.9%) out of 339 patients with idiopathic Parkinson's disease, which population comprised both hospitalized patients and outpatients. The demented patients were older at onset and showed more rapid progression than non-demented patients. Our results are consistent with those of Mayeux's and coworkers.

Our results in this study suggest a chronological increasing proportion of Parkinson's disease in Japan. The problem of increasing Parkinson's disease and associated dementia is not only a medical issue but also a social one. It is necessary to institute a nationwide prevalence study of this disease.

REFERENCES

1. J. W. Langston, P. Ballard, J. W. Tetrud, and I. Irwin : Chronic parkinsonism in humans due to a product of meperidine-analog synthesis, Science 219:979 (1983).
2. G. Campanella, M. Roy: A case-control study of Parkinson's disease in South Quebec: exposure to metals and pesticide. Proceedings of the Ninth International Symposium on Parkinson's disease, 57 (1988).
3. H. Harada, S. Nishikawa, K. Takahashi: Epidemiology of Parkinson's disease in a Japanese city, Arch. Neurol. 40:151 (1983).
4. R. J. Marttila, U. K. Rinne: Epidemiology of Parkinson's disease-An overview, J. Neural. Transmission. 51:135 (1981).
5. R. J. Marttila, U. K. Rinne: Epidemiology of Parkinson's disease in Finland, Acta Neurol.Scand. 53:81(1976).
6. A. C. Jenkins: Epidemiology of Parkinsonism in Victoria, Med.J.Aust. 2:496(1966).
7. G. Rosati, E. Granieri, L. Pinna, et al: The risk of Parkinson's disease In Mediterranean people, Neurology 30:250(1980).
8. I. I. Kessler: Epidemiological studies of Parkinson's disease:III.A community-based survey, Am. J. Epidemiol. 1972;96:242-254.
9. G. Brown, D. Marsden: How common is dementia in Parkinson's disease? Lancet 2:1262(1984).
10. R. Mayeux, Y. Stern, R. Rosenstein, K. Marder: An estimate of prevalence of dementia in idiopathic Parkinson's disease, Arch. Neurol. 45:260(1988).

CLUSTERING OF PARKINSON'S DISEASE IN SOUTHERN ISRAEL

AN ENDEMIC PARKINSONISM-IMPAIRED VIBRATION SENSE SYNDROME

Y.O. Herishanu* and J.R. Goldsmith**

*Department of Neurology and **Epidemiology and Health Services Evaluation Unit, Faculty of Health Sciences Ben-Gurion Univ. of the Negev, POB 151, Beer-Sheva, Isreal

INTRODUCTION

Parkinson's disease (PD) is a common neurological condition of unknown origin afflicting mostly elderly people. In recent years much research has been done to elucidate the factors involved in its etiology. Prominent scientists in the field believe that environmental factors are involved in the etiology of PD, acting on a background of slow, sustained dopaminergic neuronal loss due to advancing age (1)

Detection of clusters of PD patients may help to focus on some environmental etiological factors. The aim of this project is to study an unusual clustering of PD in several kibbutzim in a limited area of the Negev (Southern Israel).

METHODS

A review of the medical records of the Neurology Outpatient Clinic of Soroka University Medical Center (serving all 250,000 inhabitants of the Negev) revealed 156 PD patients alive in October 1988. Thirteen PD patients live in three adjacent kibbutzim (4-5 Km apart) in the North-Western Negev. Kibbutz A and B were settled in 1941-42 and C in 1948. Kibbutz A and B received their water supply from the same well situated in another kibbutz, and later had their own wells using water from the same aquifer. Kibbutz A is situated higher (100 m) than the other two and is separated from B by a gully. These kibbutzim have plantations of oranges and vegetables with a plastics factory in A, a small metallurgic factory in B, and frozen food and ceramics in C.

Step 1

A questionnaire was administered to the PD patients in the cluster, to 13 others matched for age in the cluster area, and to 32 PD patients and 32 senile and essential tremor patients in the Negev. The questionnaire included data on birthdate, birthplace and its size, water supply, date of immigration to Israel, incarceration in a labor or extermination camp, occupations, contact with herbicides, pesticides, manganese, mercury, etc., hobbies, diseases, admissions to hospital, chronic drug treatment, family conditions, habits: smoking, alcohol and coffee drinking, use of salad oil, etc.

Basic, Clinical, and Therapeutic Aspects of Alzheimer's and Parkinson's Diseases, Volume 2
Edited by T. Nagatsu *et al.*, Plenum Press, New York, 1990

Step 2

Eighty randomly-chosen, neurologically healthy people, aged above 40, living in kibbutz A and B had a neurological examination and filled in the same questionnaire as in Step 1. Webster Score (2,3) was used for assessment of 8 PD signs: rigidity, posture, upper extremity swing, gait, tremor, facies, seborrhea and presence of glabella tap reflex. The value rating is 0-3 for each item: 0=no involvement and 1,2,3 are equated to early, moderate and severe disease, respectively. Any score between 1-8 is accepted as signs of early PD.

Data on population composition and PD were obtained from 14 kibbutzim and from 8 moshavim (villages) in the Negev.

RESULTS

This study revealed a low prevalence rate of PD in the Negev of 62.1/100,000.

Six PD patients live in kibbutz A, four in B and three in C. Their ages range from 51 to 82 years. Seven out of 13 are female. The standardized age specific prevalence (SPR) in these three kibbutzim is approximately 5 times higher than in the Negev population, indicating a cluster of PD (Fig. 1).

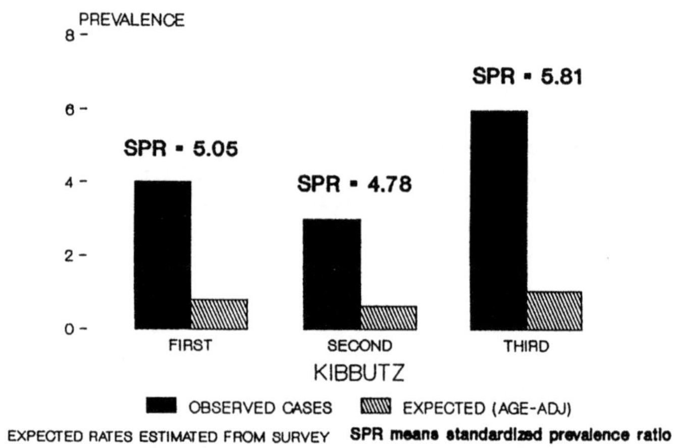

Fig. 1 Observed and expected prevalence of Parkinson's disease in three kibbutzim

All the PD patients have lived for more than 20 years in these kibbutzim and all were born in Central and Eastern Europe, Brazil and Argentina. Eleven out of 13 PD patients lived before their immigration, for at least 15 years, in small towns where the water supply was from wells. The mean age at onset of PD was 68 years. Twelve patients have had symptoms of PD for 1-6 years. Five PD patients had akinesia, rest tremor and rigidity, 5 had akinesia and rest tremor and 3 only resting tremor. Four PD patients smoked in the past. Four PD patients worked as teachers, clerks, etc., 6 in agriculture, 3 raising cows and sheep and in the local industry. These kibbutzim utilize various pesticides, herbicides and fungicides, including Maneb, and the metallurgic factory utilizes Mn, Zn and Ni. Only 3 PD patients were detected in all other kibbutzim and moshavim mentioned above. Of the neurologically "healthy" people (28 females), 48 presented signs of PD, and of these 29 lived in kibbutz A (Fig. 2).

The age of 37 people was between 60-79 and of 9 between 40-59. The age of the 32 people (17 females) with a score of zero was in the same range. The people presenting signs of PD were living in these kibbutzim for more than 25 years. A score of 1,2 and 3 was found in twelve persons, each. Three had a score of 6, two a score of 7 and one a score of 8. The most frequent sign was that of neck and/or limb rigidity (in 15) and the next sign was glabella tap reflex (in 12). Half of the persons with PD signs never smoked. Vibration sense was impaired in 44 persons (in 27 out of 48 persons with PD signs). In people from kibbutz A, vibration sense was found impaired in 27 out of 46 (Fig. 2).

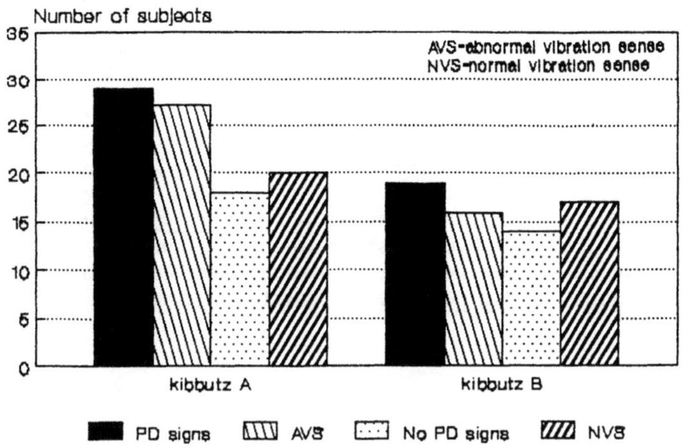

Fig. 2 PD signs and impaired vibration sense in two kibbutz populations

DISCUSSION

Thirteen PD patients were found in three adjacent kibbutzim, and the age specific prevalence in these kibbutzim is substantially higher than that observed in the Negev population, pointing to a cluster of PD (3). Very few cases were found scattered between other tens of kibbutzim and moshavim in the Negev. Most of the PD patients, and those presenting signs of PD, consumed well water before inhabiting the kibbutzim and have continued to use well water along with other residents of these kibbutzim.

A significant percentage of the people considered neurologically healthy in two of these kibbutzim show signs of PD, together with impaired vibration sense. The people presenting signs of PD did not differ from those free of symptoms in relation to age, sex, country of origin, occupation and habits. We do not yet know what is the significance of the early PD signs for further development of PD in the future. We suppose that the impaired vibration sense may represent indirect evidence for the activity of a/some neurotoxic agents on the population of the cluster. These neurotoxic agents might be also involved in the etiology of PD. We suggest that these factors begin their influence after the age of 20, since most of our patients and other kibbutz residents immigrated from different countries after this age, and the subclinical changes took several decades until the onset of clinical symptoms. These kibbutzim utilize a wide variety of herbicides, pesticides and fungicides which are spread mostly from airplanes and could contaminate the water supply. One of the latter is Maneb (manganese ethylene-bis-dithiocarbamate) which was recently incriminated in development of cog-wheel rigidity, memory disturbances and sleepiness (4). The population of these kibbutzim is also exposed to Paraquat (somewhat similar to MPTP) and to Carbon disulfide. The latter, a fumigant, was incriminated in production of an atypical parkinsonism, peripheral sensory changes, hearing loss and cerebellar signs (5).

Since not only these kibbutzim use these compounds, we suppose that these, and presumably other, factors present in the environment act in conjunction with the natural ageing process to induce the pathological changes of Parkinson's disease.

REFERENCES

1. D.B. Calne and J.W. Langston: Aetiology of Parkinson's disease. Lancet II: 1457 (1983).
2. D.D. Webster: Clinical analysis of the disability in Parkinson's disease. Mod. Treat. 5:257 (1968).
3. Y.O. Herishanu, J.R. Goldsmith, J.M. Abarbanel and Z Weinbaum: Clustering of Parkinson's disease in Southern Israel. Can.J.Neurol.Sci. in press.
4. H.B. Ferraz, P.H.F. Bertolucci, J.S. Pereira, J.B.C. Lima and L.A.F. Andrade: Chronic exposure to the fungicide maneb may produce symptoms and signs of CNS manganese intoxication. Neurology 38:550 (1988).
5. H.A. Peters, R.L. Levine, C.G. Mattews and L.J. Chapman: Extrapyramidal and other neurologic manifestations associated with carbon disulfide fumigant exposure. Arch. Neurol. 45:537 (1988).

THE EXTEROCEPTIVE CONTROL OF VOLUNTARY MUSCLE CONTRACTION IN PARKINSONISM

Fuminori Segawa, Teturou Ishida, and Masao Kinoshita

The Fourth Department of Internal Medicine
Toho University
2-17-6, Oohasi, Meguroku, Tokyo, Japan

Introduction

The electromyographic silent period (SP), first described by Hoffmann in 1919, can be conveniently defined as "a transitionary relative or absolute decrease in EMG activity, evoked in the midst of an otherwise sustained muscle contraction." The silent period can be evoked by electrical stimulation of the muscle nerve, by tendon percussion, or by sudden release of tension applied to the muscle tendon. This phenomenon may have several mechanisms of origin, as indicated by the variety of methods employed. A number of studies have described an abnormal latency of the silent period in patients with a variety of diseases affecting the motor system and in patients with involuntary movements. This may be a way to show an exteroceptive control of voluntary movement.

Both prolonged and normal silent periods have been reported in patients with parkinsonism. In this disease, Angel et al. (1966) and Higgins et al. (1971) found normal silent periods. On the other hand, McLellan (1972) found an abnormal silent period latency in untreated parkinsonian patients. The mean silent period was significantly shortened for all patients after treatment with levodopa. The present study was undertaken in order to further investigate this problem by using double stimulation, and we have also analyzed the enhancement response to voluntary contraction (CE) after the silent period.

Subjects

Ten normal subjects (age 37.3 ± 15.7), twenty patients with Parkinson's disease (age 58.8 ± 7.5), and five patients with essential tremor (age 41.5 ± 10.5) were tested. All subjects gave informed consent. All the Parkinson's disease patients were receiving levodopa therapy.

Method

Two surface recording electrodes were placed over the hypothenar muscle. For stimulation of the ulnar nerve, surface electrodes were placed longitudinally over the nerve at the elbow (stimulus duration 0.2 ms, rate 0.8 Hz, constant current source). Stimulus intensity was selected to be well above the threshold for motor fibers of the ulnar nerve, as judged by the EMG response and the contraction of the hypothenar muscle. The EMG signal was filtered (corner frequencies 5 and 1000 Hz), using the onset of stimulation as the trigger point.

The full-wave rectified surface EMGs were digitized on separate channels of a 7T17 signal processor (sampling rate 2500 Hz) and stored on hard disk for analysis with the averaging program. The subject was seated comfortably, with the forearm extended horizontally forward to grasp a strain gauge. The subjects were instructed to sustain the voluntary contraction under isometric conditions with the maximal voluntary force they could maintain during the recording period. The contractile force was displayed on a strain gauge for visual feedback for the subjects.

The test was repeated over 80 to 160 trials with each subject to get an averaging wave. The averaging wave was calculated as a percentage compared to the mean muscle potential during the 40 ms before stimulation. Double stimulations with intervals of 0 to 150 ms were adopted, and the silent periods after the second stimulation were analyzed.

Results

After a stimulation of supramaximal intensity to the ulnar nerve, suppression of the EMG lasting for about 100 ms can be recorded, until the potential of the voluntary contraction (VC) returns (Fig. 1). In the middle of the silent period (about 50 ms) a volitional potential (VP) can be obtained. VP has also been called V2 (Upton et al., 1971) or the C response (Sutton and Mayer, 1974), and has been discussed as a long loop reflex (cortical reflex). On the averaging wave an enhancement of the voluntary contraction (CE) can be obtained (during 100 to 200 ms, with peak time 150 ms in the controls) (Fig. 2). Following a small suppression and small enhancement, the EMG amplitude returns to the previous (prestimulation) level. The results of calculating the suppression and enhancement rates are presented in Table 1.

In the cases of Parkinson's disease, a small VP and a small enhancement of the voluntary contraction were obtained (Fig. 3). In the cases of essential tremor, a larger VP and larger second enhancement (CE2) were obtained (Fig. 4)

Analyzing of the silent period showed no significant differences among Parkinson's disease, essential tremor, and normal subjects. In Parkinson's disease, patients with postural instability showed longer silent periods than patients without it. Upon analyzing the silent periods after 30 to 120 ms interval double stimulations, it became quite obvious that the silent period following the second stimulus was shortened to about 70% of its original value and the voluntary contraction returned earlier. With 20 to 40 ms interval double stimulations, VP was also suppressed and hardly found. With 60 to 120 ms interval double stimulations, VP can be distinguished from the potential of the voluntary contraction, and the second VP showed a larger amplitude than the VP after the first stimulation. The silent periods after double stimulation were not significantly different between the cases of essential tremor and the normal controls, but in the Parkinson's disease cases the shortening of the silent period after the 30 to 120 ms interval stimulations was hardly noticeable (Fig. 5).

Discussion

The VP and CE waves were evoked with a shorter latency after stimulation to the nerve at the elbow than at the wrist, indicating a conduction pathway involving the spinal cord. The long central delay of the VP and CE waves suggest a long conduction pathway in the spinal cord.

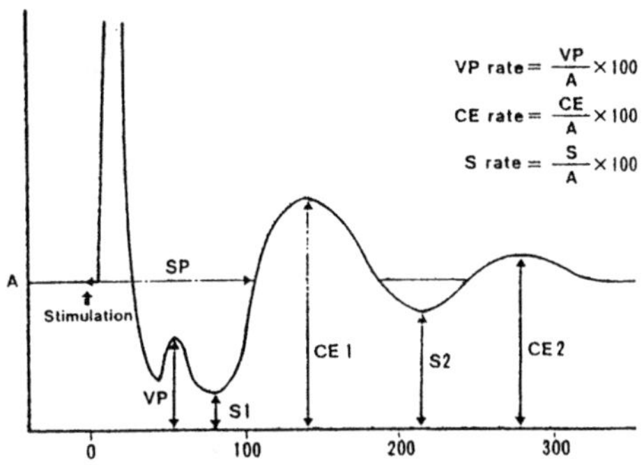

Fig. 1

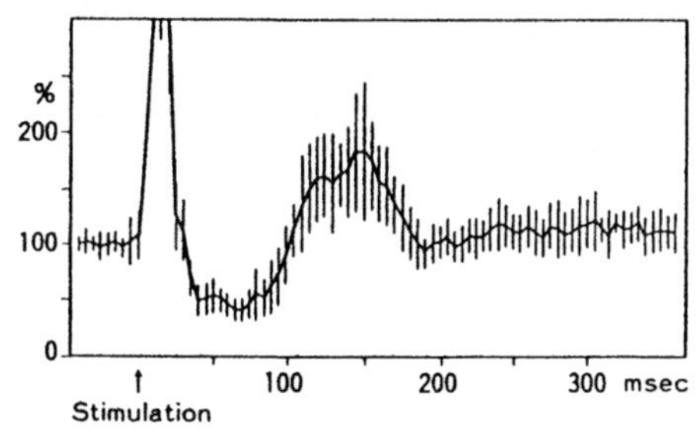

Fig. 2. Normal control.

Table 1

	Control	Parkinson s disease	Essential tremor
VP rate (%)	55.38±10.29	52.43±16.37 (NS)	95.33±21.24 (p<0.01)
S1 rate (%)	37.23± 8.95	33.43± 6.85 (NS)	27.85± 9.50 (NS)
CE1 rate (%)	201.00±49.57	143.52±27.33 (p<0.01)	210.77±59.25 (NS)
S2 rate (%)	82.38± 9.78	78.76±12.81 (NS)	56.08± 8.90 (p<0.01)
CE2 rate (%)	144.00±18.60	112.29±27.54 (p<0.01)	174.38±37.17 (p<0.1)
SP (ms)	100.67±10.49	109.50±14.96 (NS)	107.00±13.21 (NS)
postural instability (+)		120.71±11.95	
postural instability (−)		100.00± 9.83	

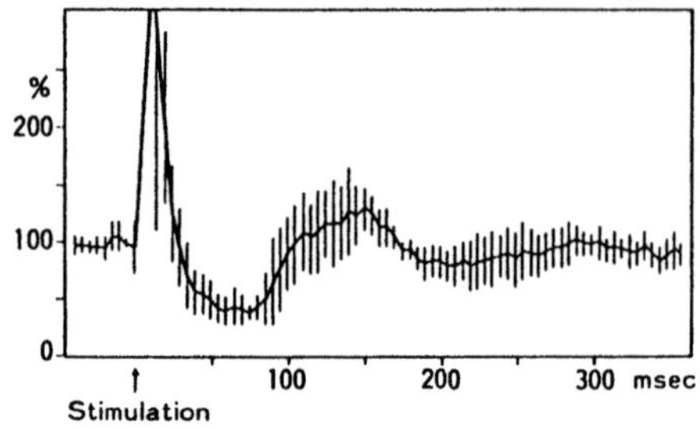

Fig. 3. Parkinson's disease

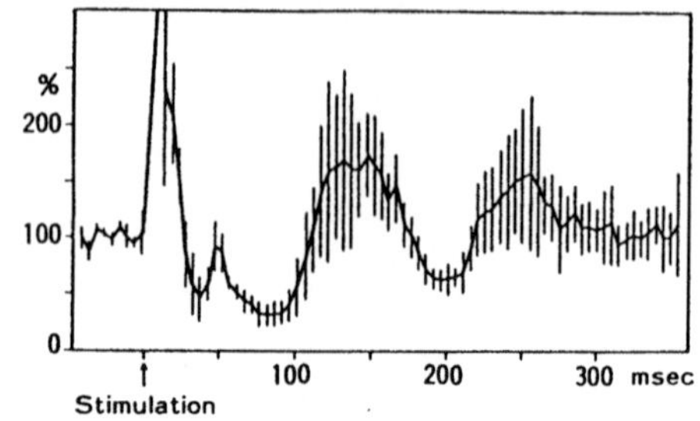

Fig. 4. Essential tremor

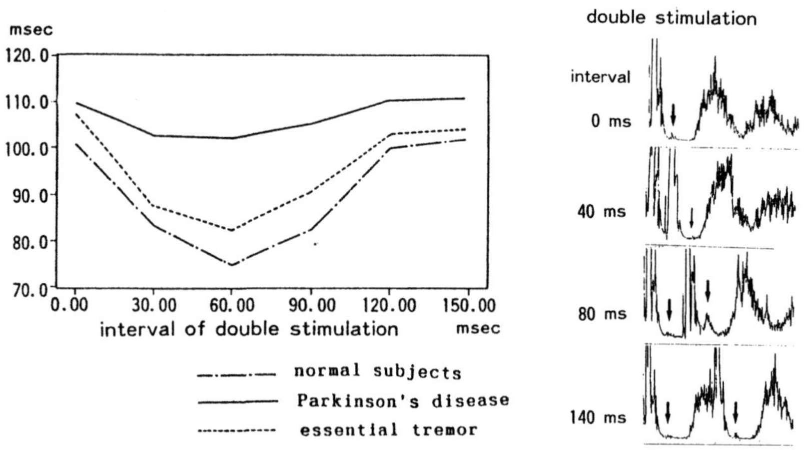

Fig. 5. Silent period after second stimulation

VP reflex pathways involving transmission longitudinally within the spinal cord have been described. Shimamura and Livingston (1963) gave evidence for a spino-bulbo-spinal reflex pathway, and Marsden et al. (1973) have proposed a spino-cortical pathway. Whether such pathways are involved in the genesis of the VP reflex remains to be determined.

Lee and Stein (1981) reported the difference between essential tremor and parkinsonian tremor by resetting of tremor by mechanical perturbations. They revealed that the phase of essential tremor could be readily reset by external perturbation, and the oscillations within spinal reflex loops play a major role in the generation of essential tremor. In our study essential tremor showed large VP and good enhancement effect of VC (CE1 and 2). This may be the result of such a peripheral loop control, and the reflex inputs are capable of modifying essential tremor. This is one of the differences between parkinsonian tremor and essential tremor.

The silent period has been described by several mechanisms; the result of the spindle pause, the discharge of the Golgi tendon organs, an antidromic inhibition involving the Renshow feedback mechanism. But abnormal silent periods have been reported in patients with cerebellar disease and CNS lesions. The silent period needs some inhibitory system of CNS lesions. The silent period was reported to be prolonged in patients with parkinsonism by Pinelli (1955) and later by Hofschmidt (1959). On the other hand, Angel et al. (1966) and more recently Higgins et al. (1971) found a normal silent period in this disease. Our results show a normal silent period in the mean value, but patients with postural instability have long silent periods. Such an abnormality of exteroceptive control may play a role in postural instability. Further study may be necessary on this point.

Hofschmidt and Linke (1976) demonstrated the silent period in the masseter muscles of parkinsonian patients and showed a longer silent period and a long silent period even after double stimulation. The origin of the shortening of the silent period after short double stimulations in normal subject is unknown, but convincing evidence supports the concept that the enhancement effect (during 100 to 200 ms after the first stimulation) may cancel the suppression effect of the second stimulation. As the enhancement response is small in parkinsonian patients, the shortening of the silent period is also small.

References

1. Angel RW, Hofmann WW, Eppler W: Neurology, 16: 529-532, 1966.
2. Dietrichson P: Acta Neuro Scand, 47: 183-193, 1971.
3. Higgins DC, Haidri NH, Wilbourn J: J Neurol Neurosurg Psychiatry, 34: 508-511, 1971.
4. Hoffmann P: Zeitschrift für Biologie, 70: 515-524, 1919.
5. Hofmann WW: J Neurol Neurosurg Psychiatry, 25: 203-207, 1962.
6. Hofschmidt HJ: Deutsche Zeitschrift für Nervenheilkunde, 179: 298-308, 1959.
7. Hofschmidt HJ, Linke D: J Neurol Neurosurg Psychiatry, 39: 536-537, 1976.
8. Lee RG, Stein RB: Ann Neurol, 10: 523-531, 1981.
9. Marsden CD, Merton PA, Morton HB: Lancet, 1: 759-761, 1973.
10. McLellan DL: J Neurol Neurosurg Psychiatry, 35: 373-378, 1972.
11. Pinelli, P: Bolletino della Società Medico-Chirurgica, Pavia, 69: 877-901, 1955.
12. Shimamura M and Livingston RB: J Neurophysiol, 26: 258-272, 1963.
13. Stanley EF: J Neurol Neurosurg Psychiatry, 41: 1016-1023, 1978.
14. Sutton GG, Mayer RF: J Neurol Neurosurg Psychiatry, 37: 207-217, 1974.
15. Upton ARM, McComas AJ, Sica REP: J Neurol Neurosurg Psychiatry, 25: 699-711, 1971.

EXTEROCEPTIVE SUPPRESSION OF THE MASSETER AND TEMPORALIS MUSCLES IN PATIENTS WITH PARKINSON'S DISEASE

Kenji Nakashima, Kazuro Takahashi, Toshio Azumi and Gen Ishida

Division of Neurology, Institute of Neurological Sciences
Tottori University School of Medicine
86 Nishimachi, Yonago 683, Japan

INTRODUCTION

Although failure of facial expression is one of the characteristic symptoms in patients with Parkinson's disease (PD), the mechanism is still unclear. Electrical stimulation of the mental nerve to suppress voluntary contraction of the masseter and temporalis muscles is known as exteroceptive suppression.[1-3] This suppression consists of two components: early (ES1) and late (ES2) components. In the present study, exteroceptive suppression was evaluated in patients with PD.

MATERIALS AND METHODS

Twenty-three patients with idiopathic PD (mean age$\pm$SE, 57.1$\pm$2.1 years) and 20 normal subjects (53.3$\pm$3.9 years) were examined. According to the Hoehn and Yahr Scale,[4] three patients were rated stage II; sixteen, stage III; and four, stage IV. All patients were receiving medication (levodopa-carbidopa, trihexyphenidyl, amantadine, bromocriptine). Before the experiments, the consent of subjects was obtained after they had been informed fully about the study.

In all experiments, the electromyogram (EMG) was recorded from surface electrodes placed over the unilateral temporalis and masseter muscles. Thirty-two EMG signals (bandwidth -3dB at 53 Hz to 3 kHz) were rectified

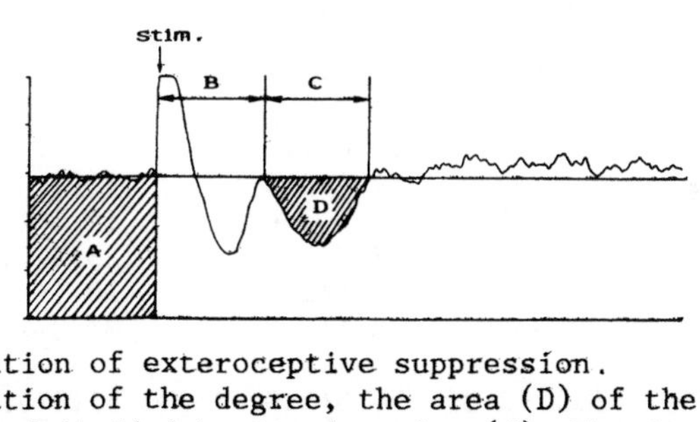

Fig.1. The evaluation of exteroceptive suppression.
For estimation of the degree, the area (D) of the suppression was measured and divided by its duration (C). The degree was expressed relative to the background EMG activity (A/50): 100 x (D/C)/(A/50).
B: latency.

Basic, Clinical, and Therapeutic Aspects of Alzheimer's and Parkinson's Diseases, Volume 2
Edited by T. Nagatsu *et al.*, Plenum Press, New York, 1990

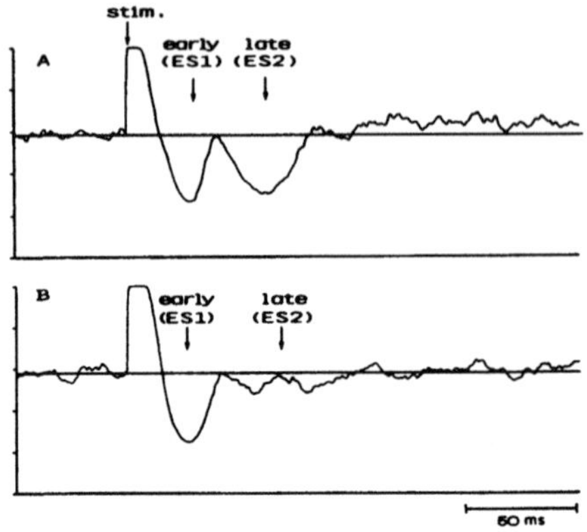

Fig.2. Representative EMGs of two subjects used in the study.
A: normal 30-year-old male. B: 63-year-old female with PD.

and averaged. Percutaneous electrical stimulation of 0.5 ms duration was delivered to the unilateral mental nerve during voluntary contraction. Moderate and strong intensities of stimulation were used: three (3ST) or five times (5ST) the sensory threshold. An intensity of 5ST was not used in some subjects because of the pain sensation. Maximal contractions were serially repeated after 5-10 s of rest. The mean background EMG activity recorded during the 50 ms prior to the stimulus was estimated. The latency and duration of suppression were evaluated at the point where the trace crossed the baseline indicating the background EMG activity. The area was measured and divided by its duration. The degree of suppression was expressed relative to the background EMG activity (Fig.1). In the present study, the early and late components were estimated, although stimulus artifact appeared to merge into the early component. Statistical significance was evaluated with the two-tailed Wilcoxon rank-sum test.

RESULTS

Background EMG activity of the masseter and temporalis muscles showed no difference between normal subjects and patients with PD (Table 1). The latency, duration, and degree of the early component in patients with PD were also not different from those in the normal subjects (Fig.2), except

Table 1. Background EMG, Early Component and the Latency of Late Component

		masseter		temporalis	
		normal	Parkinson	normal	Parkinson
Background EMG (μV)		79.1±9.5	82.7±7.7	78.6±6.5	86.5±6.9
Early component					
Latency (ms)	3ST	18.8±1.1	20.4±1.1	16.8±1.0	17.8±0.9
	5ST	17.5±0.9	19.4±0.5	17.1±0.7	16.3±0.3
Duration (ms)	3ST	13.6±1.6	13.1±1.6	18.5±1.7	17.5±1.6
	5ST	23.4±1.6	17.8±1.4*	23.8±0.9	23.7±1.3
Degree (%)	3ST	17.4±3.4	11.6±2.6	20.1±2.9	13.5±2.7
	5ST	28.9±2.8	25.1±2.5	31.3±2.3	27.9±2.2
Late component					
Latency (ms)	3ST	51.5±3.3	56.5±2.9	51.3±3.2	55.6±2.9
	5ST	46.6±1.5	48.8±2.3	43.4±0.7	50.4±2.5

mean ± SE * $p<0.05$

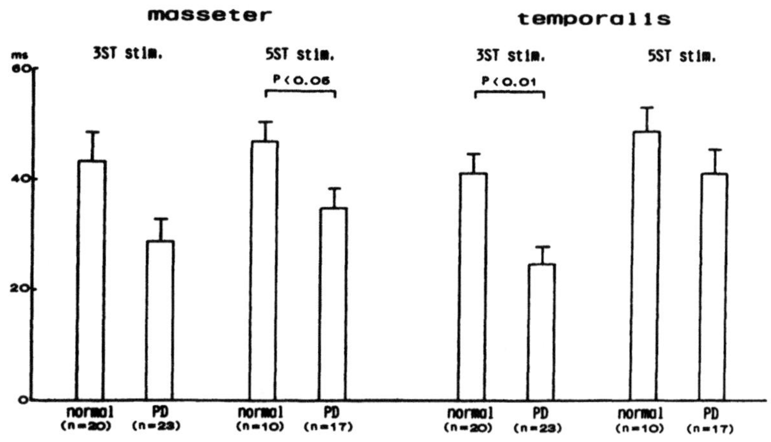

Fig.3. Duration of the late component.

for one instance in which the duration in the masseter muscle after stimulation with 5ST was shortened (Table 1). The latency of the late component did not show a significant difference between normal subjects and patients with PD (Table 1). However, the duration and degree of the late component were lower in patients with PD than in normal subjects (Figs.3, 4).

DISCUSSION

The precise neural networks of exteroceptive suppression remain unknown. The afferent and efferent fibers may pass through the trigeminal nerve. The fact that background EMG activities were the same as in normal subjects may indicate that the excitability of the motoneurons themselves was equal in the normal subject and PD groups. The early and late components are thought to be mediated by separate neural networks.[1] The early component may be mediated oligosynaptically and the late component multi-

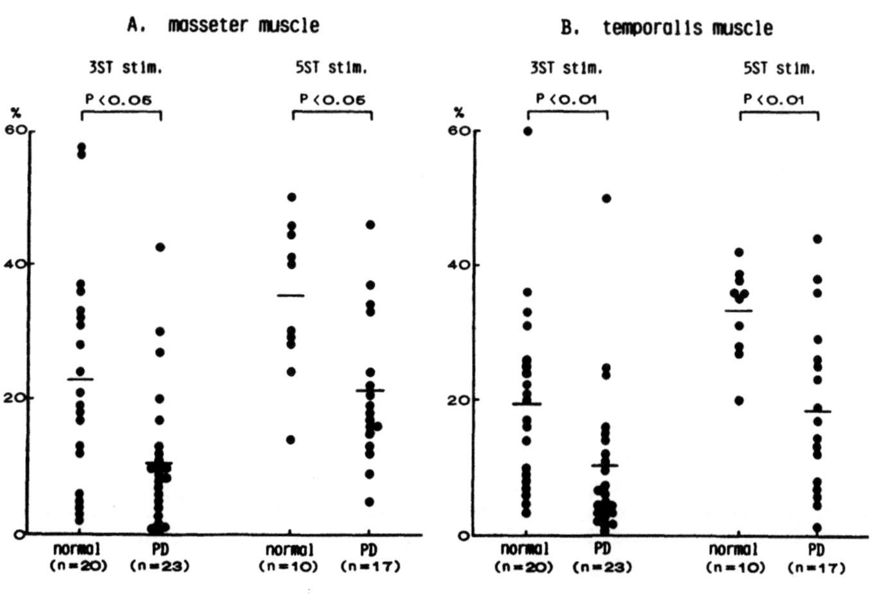

Fig.4. Degree of the late component.

synaptically.[2] A phenomenon accounting for the differences in the late component between normal subjects and patients with PD may be the disturbed function of interneuronal activity contributing to the late component of exteroceptive suppression. The function of the interneurons mediating the early component may be mostly preserved in PD.

As decreased exteroceptive suppression has also been observed in patients with idiopathic dystonia[5,6] and chronic headache[3,7], the reduction of the late component may not be specific in PD. Inhibitory interneurons may be modulated by input from other neuronal networks.[3] Berardelli et al.[5] showed that exteroceptive suppression was absent in some patients with blepharospasm and oromandibular dystonia. Exteroceptive suppression in the sternocleidomastoid muscle was reduced in patients with spasmodic torticollis.[6] Dysfunction of the basal ganglia might exist in patients with dystonia.[5,6] The reduction of the late component in patients with PD might be produced by disturbed interneurons in the brainstem, which might be modulated by descending volleys from the basal ganglia.

The silent periods in the muscles of the hand could be normal[8] or prolonged[9] in patients with PD. These silent periods might be related to spindles and the Golgi tendon organs.[10] Exteroceptive suppression of the jaw-closing muscles may differ from the silent periods associated with muscle twitches in the hand muscles.[2] Cutaneous afferents may contribute to exteroceptive suppression of the jaw-closing muscles.[2]

In conclusion, we have described the reduction of exteroceptive suppression in the parkinsonian face. This abnormality might result from the disturbed function of the basal ganglia.

REFERENCES

1. G. Cruccu, R. Agostino, M. Fornarelli, M. Inghilleri, and M. Manfredi, Recovery cycle of the masseter inhibitory reflex in man, Neurosci. Lett. 49:63 (1984).
2. E. Godaux, J. E. Desmedt, Exteroceptive suppression and motor control of the masseter and temporalis muscles in normal man, Brain Res. 85:447 (1975).
3. Y. Schoenen, B. Jamart, P. Gerard, P. Lenarduzzi, and P. J. Delwaide, Exteroceptive suppression of temporalis muscle activity in chronic headache, Neurology 37:1834 (1987).
4. M. M. Hoehn, M. D. Yahr, Parkinsonism: onset, progrssion, and mortality, Neurology 17:427 (1967).
5. A. Berardelli, J. C. Rothwell, B. L. Day and C. D. Marsden, Pathophysiology of blepharospasm and oromandibular dystonia, Brain 108:593 (1985).
6. K. Nakashima, P. D. Thompson, J. C. Rothwell, B. L. Day, R. Stell, and C. D. Marsden, An exteroceptive reflex in the sternocleidomastoid muscle produced by electrical stimulation of the supraorbital nerve in normal subjects and patients with spasmodic torticollis, Neurology 39:1354 (1989).
7. K. Nakashima, T. Azumi, G. Ishida, and K. Takahashi, Exteroceptive suppression produced by mental nerve stimulation in patients with chronic headache, Clin.Neurol. (Tokyo), in press.
8. D. C. Higgins, N. H. Haidri, and A. J. Wilbourn, Muscle silent period in Parkinson's disease, J.Neurol.Neurosurg.Psychiatry 34:508(1971).
9. D. L. McLellan, Levodopa in Parkinsonism: Reduction in the electromyographic silent period and its relationship with tremor, J.Neurol. Neurosurg.Psychiatry 35:373 (1972).
10. D. C. Higgins and J. S. Lieberman, The muscle silent period: Variability in normal man, Electroenceph.Clin.Neurophysiol. 24:17 (1968).

MECHANISM OF THE DIFFICULTY IN SET SHIFTING TASK IN PARKINSON'S DISEASE

Fuyuhiko Tamaru[1] and Nobuo Yanagisawa[2]

1 Department of Neurology, Kakeyu Hospital, Maruko, and
2 Department of Medicine (Neurology), Shinshu University School of Medicine, Matsumoto, Japan

INTRODUCTION

Recent studies on mental function in Parkinson's disease have suggested that patients with Parkinson's disease (PD) have a deficit in frontal lobe function. The most common method used, and the one which brough about this idea, is the Wisconsin Card Sorting Test (WCST). However, there is still debate on what aspects of mental function are estimated with this test. The results of WCST tend to be interpreted differently depending on the opinion of the authors: concept fromation, set shifting, perseveration, and so on (Bowen et al., 1975; Cools et al., 1984; Lees and Smith, 1983). However, it may safely be said that PD patients have difficulty in dealing with multiple sets. In an attempt to elucidate the deficient mental process in dealing with multiple sets in the light of information processing, we have devised a new test called the 'criterion shift task.'

SUBJECTS AND METHODS

Thirty-one patients with Parkinson's disease (PD) were examined. Twenty-three control subjects (NC) were selected from persons with non mental deterioration or neurological deficits. There was no significant difference in age between the control subjects (mean 61.4; SD, 10.0) and PD (mean 62.0; SD, 11.2).

'Criterion Shift Test'

The equipment consisted of a personal computer to which were attached a timer-counter board and a parallel I/O board, and a pulse generator which emitted single pulses when the button switch, held in the subject's hand, was pressed. Test stimuli appeared in the center of a 14-inch CRT one by one. Each stimulus was held on the screen for 0.75 seconds, and then disappeared. Two different types of stimuli were presented, one a circle and the other a square. The order of appearance of stimulus type was random. In the first stage of the experiment, subjects were told to press the button in their hand as quickly as they could whenever a stimulus appeared. The mean reaction time was calculated (simple reaction time=SRT). In the second stage of the experiment, the subjects were told to press the button in their hand when the test stimulus was the same as the immediately preceding one. In this case, the present test stimulus was the conditioning stimulus for the next trial. In this task, subjects had to shift successively the criterion for pushing the button according to the conditioning stimulus of each trial (Shift task). Mean reaction time (shift reaction time) was calculated from 60 trials. Shift processing time (Sh-PT) was obtained by subtracting the simple reaction time from the shift reaction time in order to eliminate any slowness of motor performance. There are two possible deficits detected by this task: 1) difficulty in shifting between two alternative criteria and 2) difficulty in maintaining shifted criteria. In order to decide which of the two possible deficits is critical, both simple reaction task and shift task were performed under two different inter-trial intervals: a longer interval (from 5 to 7.5 seconds, randomized) and a shorter interval (from 1.5 to 2.25 seconds, randomized).

Other Tests

"Raven's colored matrices' was employed to evaluate intelligence without any effect due to motor deficit. And the 'Wisconsin Card Sorting Test' was given for comparison with criterion shift test.

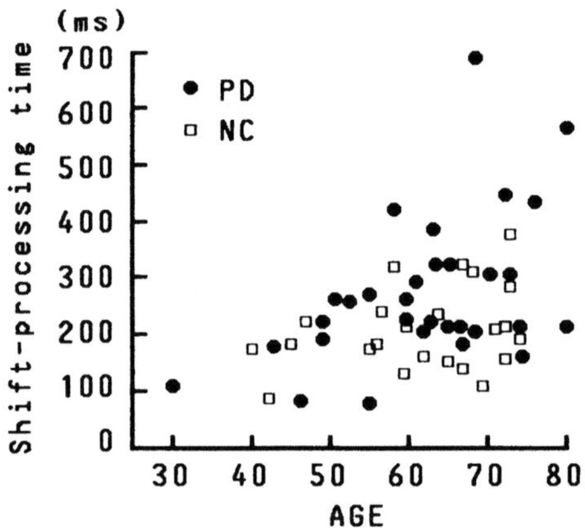

Fig. 1. Shift processing time vs. age. PD: Parkinson's disease, NC:controls.

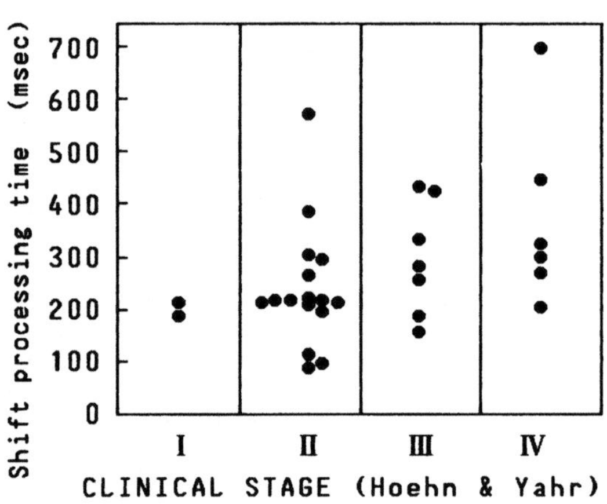

Fig. 2. Correlation between the shift processing time and the clinical stage of illness.

RESULTS

Fig. 1 shows the Sh-PT for each subject. PD patients seemed to show longer Sh-PT than controls. At the same time, Sh-PT increased as the age of subject became higher. In order to examine the effect of aging and illness, both PD and controls were divided into two age groups separated at age 65. An ANOVA revealed significant main effects for age ($F(1,50)=4.758$, $P<0.05$), and for illness ($F(1,50)=4.857$, $P<0.05$). No interaction between age and illness was observed.

Correlation between Sh-PT and the clinical stage of illness (Hoehn and Yahr) was significant (Spearman's rank correlation RHO=0.463, $P<0.001$). Sh-PT became longer as the stage of illness advanced (Fig. 2). Sh-PT showed a mild but significant correlation with perseverative error of the WCST (Spearman's RHO=0.354, $P<0.02$). On the other hand, correlation between Sh-PT and Raven's colored matrices was not significant (Spearman's RHO=0.210, n.s.).

Fig. 3-A shows the effect of change of the inter-trial interval. In PD patients (Hatched bars), Sh-PT became shorter as the inter-trial interval was shortened ($t=3.266$, $df=15$, $P<0.01$). Such an effect was not observed in controls (open bars) ($t=1.264$, $df=8$, $0.1<P<0.3$). The change in SRT by the change of interval (Fig. 3-B), however, was not significant in PD patients ($t=1.571$, $df=15$, $0.1<P<0.2$) or in controls ($t=1.962$, $df=8$, $0.05<P<0.1$).

DISCUSSION

In the criterion shift test, PD patients showed more errors and longer Sh-PT than control subjects. The deficit was independent of aging. And seemed to be characteristic to PD because Sh-PT correlated with stage of illness. The deficit could not be attributed to dementia. Significant correlation with WCST means that the deficit may be caused by some frontal lobe dysfunction. Sh-PT improved when inter-trial interval was shortened. If patients with PD have a difficulty in shifting criteria, Sh-PT would be

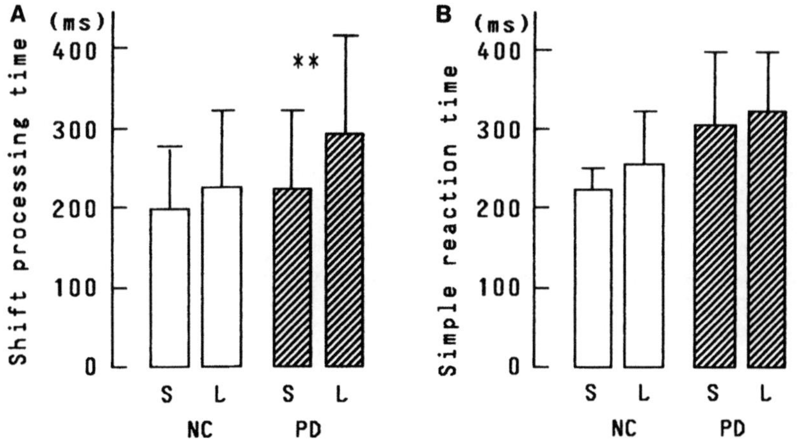

Fig. 3 Mean and SD bars of (A) shift processing time and (B) simple reaction time for patients with Parkinson's disease (PD) and controls (NC) at shorter (S) and longer (L) inter-trial intervals. **$P<0.01$.

expected to worsen as the preparation period for the next trial (=inter-trial interval) is shortened. As the reverse was the case, the deficit is not the result of difficulty in shifting criteria. The result that Sh-PT improved as inter-trial interval became shorter makes us speculate that there is a deficit in memory storage. It is true that such an improvement may have occurred as the arousal level was raised. However an arousal level change can be recognized by a change in SRT, but in the present study, no change in SRT was observed. Therefore, change in arousal level as the result of alternation of the inter-trial interval was excluded.

Flowers and Robertson (1985) pointed out that difficulty in card sorting tasks arises from the difficulty in maintaining mental sets. The card sorting task can be regarded as a kind of delayed response. PD patients are known to show deterioration in delayed response (Freedman and Oscar-Berman, 1986). Our result showed the attenuation of stored criterion occurred within only a few seconds. Such a rapid attenuation means that the rehearsal mechanism, which enables one to maintain a stored memory, doesn't work properly in PD patients.

According to the hypothesis of 'working memory,' there are several chambers for short-term memory. When we carry out a task that requires us to deal with multiple sets, each set is registered in a separate chamber and these chambers are organized by the central executive. If we postulate dysfunction in the mechanism to maintain a set held in one of these chambers under some condition, the result of the present study can be explained. The task to deal with multiple sets simultaneously may cause such a difficulty in the working memory system in PD. It is still unclear whether the difficulty in maintaining short-term memory contents is task dependent or not in PD. Hereafter, we must study what type of task leads to difficulty in maintenance of memory contents.

In conclusion, the results obtained with our newly devised test have disclosed that there is no difficulty in shifting criteria in Parkinson's disease and deficits in cognitive function, as shown in the present study, may be attributed to the failure of the memory system.

REFERENCES

Bowen, F. B., Kamienny, R., Burns, M. M., and Yahr, M., 1975, Parkinsonism: Effect of levodopa treatment on concept formation, Neurology, 25: 701.

Cools, A. R., Bercken, J. H. L. VAN DEN, Horstink, M. W. I., Spaendonck, K. P. M. VAN, and Berger, H. J. C., 1984, Cognitive and motor shifting attitude disorder in Parkinson's disease, J.Neurol, Neurosurg. Psychiat., 47: 443.

Flowers, K. A. and Robertson, C., 1985, The effect of Parkinson's disease on the ability to maintain a mental set, J.Neurol, Neurosurg.Psychiat., 48: 517.

Freedman, M. and Oscar-Berman, M., 1986, Selective delayed response in Parkinson's and Alzheimer's disease, Arch.Neurol., 43: 886.

Lees, A. J. and Smith, E., 1983, Cognitive deficits in the early stages of Parkinson's disease, Brain, 106: 257.

CEREBRAL BLOOD FLOW, EVOKED POTENTIAL AND REACTION TIME

IN PARKINSON'S DISEASE

Hisao Tachibana, Kazuo Toda, Keita Kawabata, Hiroyuki Nishimura, Naomi Yokota, Minoru Sugita and Kenzo Konishi*

The Fifth Department of Internal Medicine and Department of Science of Behavior*, Hyogo College of Medicine, Nishinomiya Japan

INTRODUCTION

Although cerebral blood flow (CBF) decreases in patients with Parkinson's disease[1-4], particularly in those with dementia[3,4], the underlying mechanism of CBF reduction is not fully understood. The previous results on the relationship between neuropsychological tests and CBF are controversial[2,4]. The P300, a long-latency positive component of the scalp -recorded event related potential (ERP) associated with cognitive processing increases in some patients with Parkinson's disease[5,6]. Likewise, reaction time (RT) also increases in patients with Parkinson's disease[7] and is influenced by the presence of dementia[8]. To further study the underlying mechanism of CBF reduction in Parkinson's disease, we investigated the relationship between CBF and P300 or RT.

PATIENTS AND METHODS

The study was performed on 21 patients with Parkinson's disease (mean age, 66.5 ± 7.2 years). For comparison with the Parkinson's disease group, the regional CBF (rCBF) measurement was done for 15 age-matched normal subjects (mean age, 68.4 ± 5.7 years). Furthermore, ERP and RT were also measured for another 15 age-matched normal subjects (mean age, 65.4 ± 11.8 years). Patients with Parkinson's disease had at least two of its cardinal features (bradycardia, tremor, and rigidity), which had developed insidiously. Patients had some of the signs peculiar to Parkinson's disease such as masked face, stooped posture, pulsion phenomenon, and Myerson's sign. Furthermore, patients were excluded who had Parkinsonism that was due to other causes or was part of a more widespread neurological illness. None of the patients had overt dementia.

The rCBF was measured by the ^{133}Xe inhalation method described by Obrist et al[9]. The Aloka RRG 526 rCBF measurement system was used. The investigation was carried out with subjects lying at rest on a bed in a quiet room with the eyes closed. The subjects inhaled a concentration of 1.25mCi/l ^{133}Xe mixed with room air and oxygen through a close-fitting face mask with a one-way valve for a period of one minute. The radioactivity emerging from the head was monitored for the next 10 minutes by NaI collimated scintilation detectors set around homologous regions of both hemispheres (seven over each hemisphere) and with the head related to the detectors in a standard position. Each probe was incorporated into an on-line computerized system.

Air drawn directly from the face mask by a thin catheter was continuously monitored by a separate detector in order to record the ^{133}Xe concentration in the exhaled air. The end-tidal values of this air curve were used to correct the "head" clearance curves for the recirculation of the radioisotopes. The fast component (F1) of two-compartmental analysis described by Obrist et al.[9] was calculated by on-line computer. Arterial blood pressure was measured by auscultation immediately before and after the study. $PECO_2$ and PEO_2 were measured continuously with the exhaled gas monitor (Nihon Denki Sanei; IH21A).

ERP and RT data were recorded in a sound-attenuated electrically shielded chamber. EEG was recorded using Ag/AgCl electrodes placed on the scalp sites Fz, Cz, Pz (International 10-20 system) and referred to linked earlobes. An electrode to measure eye movements (EOG) was attached below the left inferior orbita. Electrode impedance was maintained below 5K Ω. EEG and ocular activity were amplified (filter 0.02-50 Hz) and stored together with event markers on harddisc after A-D conversion for off-line analyses. Three kinds of stimuli were presented on a TV screen: rare target, rare non-target and frequent non-target, which were presented in a random order with the probabilities of 19, 19 and 62%, respectively. Stimulus duration was 66 msec and the interstimulus interval was 1240 msec (offset to onset). The rare target stimuli were designated as a target to be responded to by the subject, while the remaining 81% were background to be ignored. A total of 256 to 320 visual stimuli were generated by the computer and displayed on a TV screen. The subjects were instructed to respond with a button, pressed as quickly and accurately as possible to rare targets presented. Subjects used their dominant forefinger for response. RT was defined as the interval from the onset of the target stimuli to the button press. To obtain ERP, EEGs free from ocular or other artifacts were averaged separately for the three kinds of stimuli. The averaging epoch began from 200 msec before stimulus onset to 1200 msec after stimulus onset. The P300 component was defined as the largest positive-going potential on the Pz site occurring between 0 and 1200 msec after stimulus onset. Latency was defined as time in milliseconds from stimulus onset presentation to identified component. Amplitudes were measured relative to the prestimulus baseline. The results obtained were expressed as means± standard deviations (SD). All statistical procedures were performed using Student's t-test to compare different groups, and Pearson's r was calculated for correlations.

RESULTS

Table 1 shows the mean hemispheric F1 values in patients with Parkinson's disease and age-matched normal subjects. As indicated, there were significant differences in mean F1 values between these two groups ($p<0.02$). No significant differences were observed in mean arterial blood pressure (MABP) and $PECO_2$.

Table 1. Comparison of Mean Hemispheric F1 Values in Patients with Parkinson's Disease and Age-Matched Normal Subjects

Group	Number of Cases	Age [yrs]	Mean F1 Values [ml/100g brain/min]	MABP [mmHg]	$PECO_2$ [mmHg]
Parkinson's Disease	21	66.5 ± 7.2	63.0* ± 13.6	97.3 ± 9.4	39.1 ± 4.9
Normal Subjects	15	68.4 ± 5.7	73.1* ± 9.6	98.4 ± 11.7	40.1 ± 3.2

* $p < 0.02$

Table 2. Comparison of P300 Components and Reaction Time in Patients with Parkinson's Disease and Age-Matched Normal Subjects

Group	Number of Cases	Age [yrs]	P300		Reaction Time [msec]
			Latency [msec]	Amplitude [μ V]	
Parkinson's Disease	21	66.5 ± 7.2	454.8 ± 55.6	19.6 ± 6.7	508.0* ± 119.1
Normal Subjects	15	65.4 ± 11.8	437.3 ± 31.1	17.9 ± 7.0	431.3* ± 56.2

* $p < 0.05$

Table 2 compares the P300 components and RT of Parkinson's disease patients and normal subjects. As shown in Table 2, there were significant differences in RT between the two groups ($p<0.05$). Furthermore, there was a negative correlation between the mean F1 values and RT in patients with Parkinson's disease ($p<0.01$) as demonstrated in Figure 1. On the other hand, no significant differences were noted in either latency or amplitude of the P300 between the Parkinsonian group and normal subjects. There was no significant correlation between the mean F1 values and P300 latency (Figure 2).

DISCUSSION

In the present study, mean F1 values were found to be significantly lower in patients with Parkinson's disease than in age-matched controls. This is consistent with the findings of previous studies[2,3] which have shown CBF reduction in patients with nondemented Parkinson's disease.

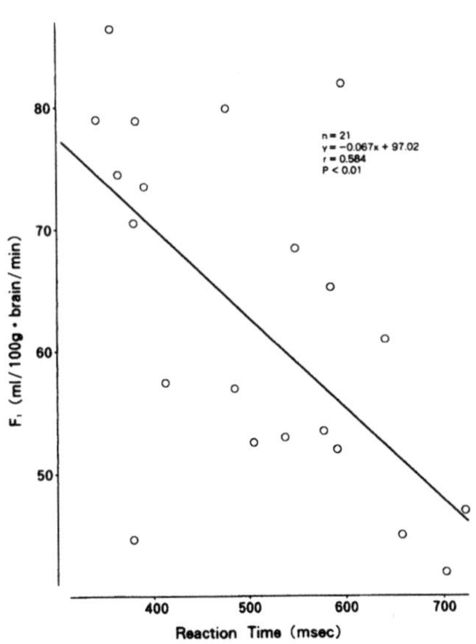

Fig.1 Relationship between mean hemispheric F1 values and RT in patients with Parkinson's disease

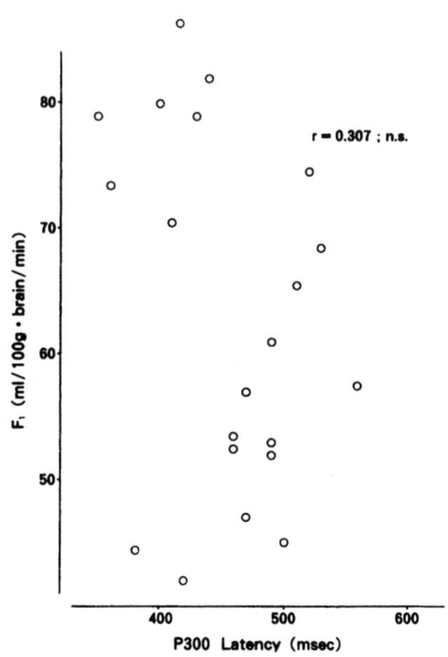

Fig.2 Relationship between mean hemispheric F1 values and P300 latency in patients with Parkinson's disease

The present study also showed that P300 latency was neither increased nor correlated with the mean F1 values. RT in the same patient was increased and inversely correlated with mean F1 values. P300 latency increases both with age and with dementia[6]. However, as reported by Goodin et al.[6], and verified in the present study, a normal P300 latency is found in patients with nondemented Parkinson's disease, while patients with Parkinson's disease with dementia have a prolonged P300 latency. Therefore, the present study's results indicate that the reduction in mean F1 values found in Parkinson's disease occurs without any significant cognitive impairment, as judged by prolongation of P300 latency. These results may be consistent with those of Globus et al.[2], who found no correlation between the magnitude of rCBF reduction and the presence or severity of intellectual deterioration in patients with Parkinson's disease.

RT prolongation has also been reported to occur with Parkinson's disease and is considered a consequence of a defect in the "autoevoked" arousal system[7]. Although a possible role of an altered epinephrine or dopamine mechanism prolonging RT has been suggested[10], the relationship between deficits in other neurotransmitter system and RT prolongation has not yet been clarified. Bes et al.[1] suggested that the decrease of CBF in the prefrontal region of Parkinson's patients may be related to the degeneration of ascending dopamine pathways. Pizzolato et al.[4] found a low parietal rCBF in patients chronically treated with anticholinergic drugs which was unrelated to the degree of mental deterioration. Together with dopamine deficiency, changes in other neurotransmitter systems including the adrenergic, cholinergic, GABAergic and peptidergic systems, appear to occur in Parkinson's disease. Therefore, the CBF reduction in Parkinson's disease patients may be regarded as an integrated result of an imbalance of multiple neurotransmitters.

REFERENCES

1. A. Bès, A. Güell, N. Fabre, Ph. Dupui, G. Victor, and G. Géraud, Cerebral blood flow studied by Xenon-133 inhalation technique in Parkinsonism: Loss of hyperfrontal pattern, J. Cereb. Blood Flow Metab., 3:33 (1983).
2. M. Globus, B. Mildworf, and E. Melamed, Cerebral blood flow and cognitive impairment in Parkinson's disease, Neurology, 35:1135(1985).
3. H. Tachibana, J. S. Meyer, Y. Kitagawa, N. Tanahashi, P. Kandula, and R. L. Rogers, Xenon contrast CT-CBF measurements in Parkinsonism and normal aging, J. Am. Geriatr. Soc., 33:413 (1985).
4. G. Pizzolato, M. Dam, N. Borsato, B. Saitta, C. Da. Col, N. Perlotto, P. Zanco, G. Ferlin, and L. Battistin, [^{99}Tc]-HM-PAO SPECT in Parkinson's disease, J. Cereb. Blood Flow Metab., 8:s101 (1988).
5. B. F. O'Donnell and N. K. Squires, Evoked potential changes and neuropsychological performance in Parkinson's disease, Biol. Psychiatr., 24:23 (1987).
6. D. S. Goodin and M. J. Aminoff, Electrophysiological differences between demented and nondemented patients with Parkinson's disease, Ann. Neurol., 21:90 (1987).
7. K. M. Heilman, D. Bowers, R. T. Watson, and M. Greer, Reaction times in Parkinson disease, Arch. Neurol., 33:139 (1975).
8. G. Ladurner, M. Tschinkel, H. Klebl, and H. Lytwin, Reaction time in cerebrovascular disease, Arch. Gerontol. Geriatr., 4:373 (1985).
9. W. D. Obrist, H. K. Thompson, H. S. Wang, and W. E. Wilkinson, Regional cerebral blood flow estimated by ^{133}Xenon inhalation, Stroke, 6:245 (1975).
10. Y. Stern, R. Mayeux, and L. Côté, Reaction time and vigilance in Parkinson's disease. Possible role of altered norepinephrine metabolism, Arch. Neurol., 41:1096 (1984).

TOPOGRAPHY OF AUDITORY EVENT-RELATED POTENTIALS IN PARKINSON'S DISEASE

Kenji Kani, Hiroko Yamamoto, and Koji Ohashi

Dept. of Neurology and Clinical Laboratory Research Center
Fujita Health University School of Medicine, Toyoake, Aichi
470-11 Japan

INTRODUCTION

Recently, attention has been paid to cognitive impairment in Parkinson's disease (PD) in connection with subcortical dementia. But clinical evaluations of cognitive deficits in PD may be complicated by motor disabilities or drug administration.

The event-related potential (ERP) evoked by a detected rare signal contains a late positive wave that has been called P3 or P300. Since Sutton et al.[1] first described that P300 reflected information processing-related activities and evaluated cognitive function independent of specific motor responses, several reports have indicated that in dementing illnesses, c. g., Alzheimer's disease, Huntington's chorea, and chronic alcholism, prolonged P3 latencies are found. However, the scalp distribution and topography of P3 in those demented states have been less often reported. Accordingly, we examined the scalp distribution and topography of P3 and the slow wave (SW), which appears following P3, for the purpose of electrophysiological identification of cognitive changes in PD.

Table 1. Characteristics of Subjects.

	PD	Young adults	Elderly
Number	10	10	11
Mean age (year)	61.2 ± 7.8	20.1 ± 0.3	66 ± 7.0
Men/Women	3/7	1/9	4/7
Yahr disability score			
I	1		
II	5		
III	4		
Duration of illness(y)	0.5-9		
	(4.5 + 2.6)		
L-Dopa administration	+ = 6		
	- = 4		

SUBJECTS

We examined 10 patients with PD, 11 normal elderly subjects and 10 young adults. The PD Patients were selected for no clinically detectable intellectual and normal psychiatric disturbances. The elderly subjects, having no neurological or other diseases, were relatives of our inpatients or outpatients. The young subjects were volunteer students of our university. Table 1 presents characteristics of the subjects. A mini mental state was required for all patients and all elderly subjects, and those scoring less than 24 were excluded.

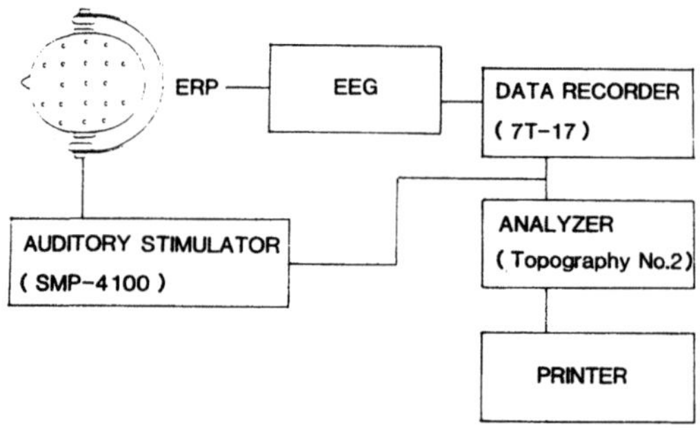

Figure 1. Schema of experimental design.

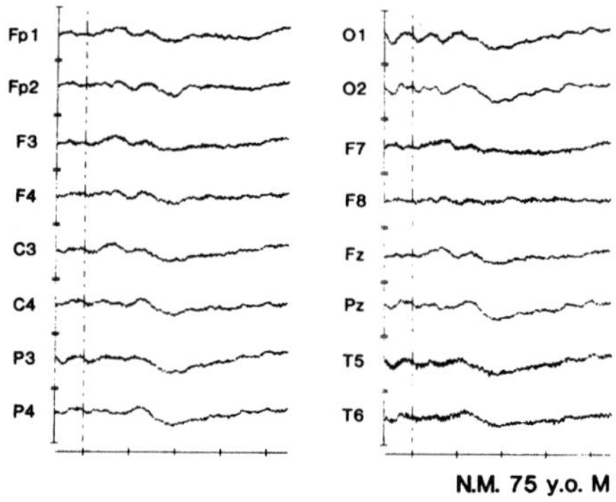

Figure 2. Averaged ERP wave form from a single subject (a 75-year-old normal elderly male).

METHODS

Figure 1 shows the apparatus used in our experiment. An auditory oddball stimulus paradigm was employed, in which subjects kept a running mental count of target tone pips of 2,000 Hz scattered among non-target pips of 1,000 Hz. The ratio of target to non-target stimuli was 1:4. A total of 160 tone pips of 0.5 Hz random stimulus rate and 5 msec. rise and fall time were produced, and from the average of 32 artifact-free trials the ERP components were obtained. A time constant of 0.3 sec was used for the EEG and the ERP. The EEG was recorded from 16 scalp electrodes placed according to the international 10-20 system, using linking electrodes placed on left and right ear lobes as a reference.

Figure 2 shows the wave forms oftained from a single subject. Amplitudes were measured from the peaks to their bases. The base line voltage was calculated from the mean voltage over the 16-msec. period prior to the tone pip stimulus. ERP scalp topography was determined from the amplitudes of each component, which were analyzed with the "Topography No.2" program. We also made an overall mean ERP map for the patients with PD, elderly, and young groups, by analysis of the overall mean peak voltage of each component.

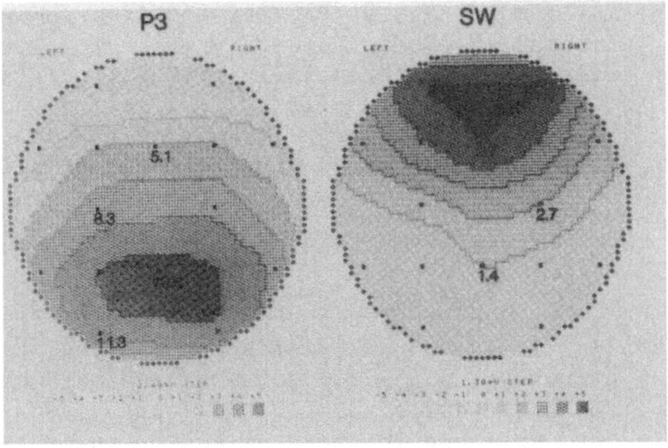

Figure 3. P3 and SW topography in the young adult group.

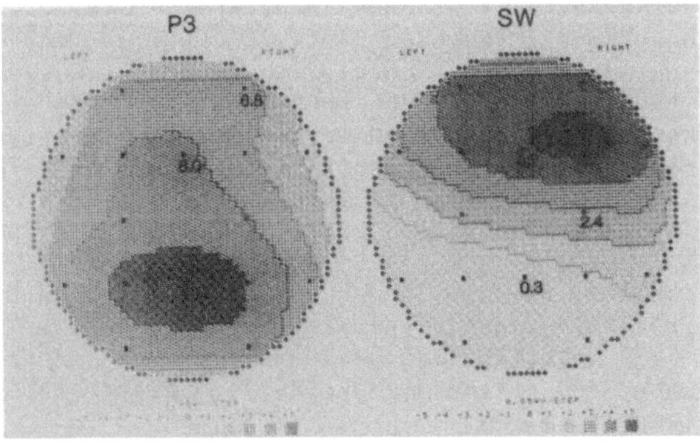

Figure 4. P3 and SW topography in the elderly group.

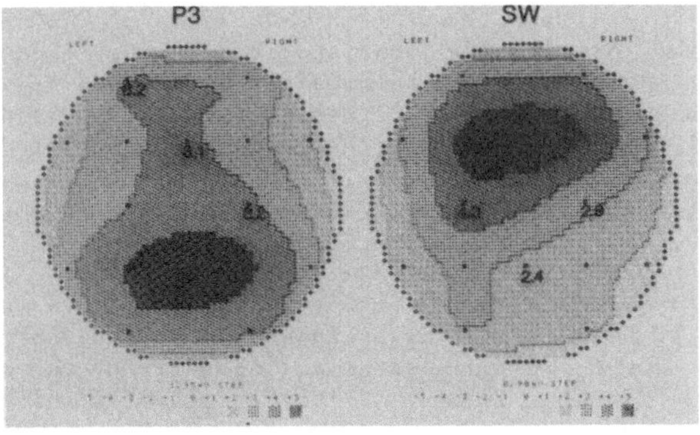

Figure 5. P3 and SW topography in the patients with PD.

RESULTS

Figure 3 shows representative P3 and SW topography in the young adult group. The maximum voltage of P3 was concentrated in the Pz region and voltage declined laterally. Regarding SW, the maximum voltage was in the prefronal region and also declined laterally in regular form. For the elderly group (Fig.4), the maximum voltage of P3 had a diffuse distribution extending from the parietal region to the frontal vertex, while that of SW was shifted posteriorly in comparison with the SW of the young group. Figure 5 shows the waves in patients with PD. The maximum voltage of P3 extended from the parietal region to the most prefrontal region. Concerning P3, in young subjects, P3 distribution extended circumferentially from Pz, but in both elderly and PD groups it was much broader. In PD, the maximum voltage of SW is shifted posteriorly in comparison with that of the young group and had a much broader distribution compared with that of the eldery.

DISCUSSION

We found that the topography of the P3 potential evoked by an auditory oddball stimulus paradigm differed between the young group and the other two groups. The maximum voltage area extended further forward from Pz in PD and elderly than in the young group. There was no difference in the amplitude in the Pz area for the three groups, but the amplitude was higher in the frontal region in the PD and elderly.

Several authors have stated that P3 amplitude decreases with age, but others have reported that only latency is altered. Our results agree with those of Pfefferbaum et al., who reported that P3 was relatively larger at the mid-frontal area than at the vertex in older subjects. He thought this was due to the age-related decrease of the overlapping frontal negative slow wave. Our SW topography showed a decrease in amplitude in the frontal region in the elderly and PD groups, and in PD, the SW was distributed more broadly than that of elderly, that is, it extended from Pz to the area farthr back.

In conclusion, P3 amplitude and distribution might reflect information processing activites in humans, and SW distribution contributes to P3 distribution. The difference in pattern of SW topography between PD patients and elderly might reflect cognitive differences between them.

REFERENCES

1. S. Sutton, M. Braren, J. John, and E.R. John, Evoked potential correlates with stimulus uncertainly, Sceience 150: 1187 (1965).
2. A. Pfefferbaum, J.M. Ford, B.G. Wenegrat, W.T. Roth,and B.S. Kopell, Clinical application of the P3 component of event-related potentials I. Normal aging, Electroencephal. Clin. Neurophysiol. 59: 85 (1984).
3. T.W. Picton, D.T. Stuss, S.C. Champagne, and R.F. Nelson, The effects of age on human event-related potentials,Psychophysiology 21:312 (1984).
4. G. Barret, R. Neshige, and H. Shibasaki, Human auditory and somato-sensory event-related potentials: effect of response condition and age, Electroencephal. Clin. Neurophysiol. 66:409 (1987).

SLEEP-WAKE COMPLAINTS IN PARKINSON'S DISEASE

Jun Horiguchi, Yasushi Inami, Toshihiko In'nami, Ohichi Nishimatsu, Yoshinari Yamamoto, and Yasuo Kakimoto

Dept. of Neuropsychiatry, Ehime Univ. School of Med. Shigenobu-cho, Onsen-gun, Ehime 791-02, Japan

INTRODUCTION

Parkinsonian patients frequently claim various kinds of sleep-wake disturbance. Nevertheless, there is no study based on large numbers of patients and control subjects that proved this clinical impression. The purpose of the present study was to investigate the frequencies of the sleep-wake complaints, especially restless legs symptoms, nocturnal myoclonus, and sleep apnea in Parkinson's disease.

MATERIALS AND METHODS

A questionnaire asking current sleep-wake complaints was sent to 4,090 parkinsonian patients and 3,601 control subjects. The patients who had been diagnosed as being Parkinson's disease were chosen from a list of members of the Japanese Parkinson's Disease Association. The control subjects were office workers and the aged living at home. Shift workers were excluded from the sample population.

The following parameters were investigated for both groups: age, sex, inability to initiate sleep, frequent awakenings, early morning awakening, daytime sleepiness, parasomnias, restless legs symptoms, nocturnal myoclonus, snoring, apneic episodes, and dyspnea during sleep. Answers were made by "Yes" or "No". From the parkinsonian patients, additional information was also gathered concerning duration of Parkinson's disease, present and past history of drug treatment, physical disturbance based on Hoehn and Yahr functional classification,[1] hallucinosis and/or hallucinosis with confusion, and night terrors. Questionnaires were completed by the household member if the patient could not write himself or herself. Chi-square method was used to analyze the results.

RESULTS

The replies of 1,575 parkinsonian patients and 2,509 control subjects were used for the analysis. The parkinsonian

Table 1 Prevalence of Sleep-Wake Complaints

Question	Parkinsonian N (%)	Control N (%)	Difference (chi-Square)
Inability to Initiate Sleep	549 (36.6 %)	683 (32.8 %)	P < 0.05
Frequent Awakenings	1329 (84.8 %)	1490 (70.3 %)	P < 0.01
Early Morning Awakening	630 (43.3 %)	521 (25.6 %)	P < 0.01
Daytime Sleepiness	1154 (77.2 %)	985 (46.8 %)	P < 0.01
Parasomnias	340 (23.8 %)	49 (2.5 %)	P < 0.01
Restless Legs Symptoms	372 (26.0 %)	185 (9.6 %)	P < 0.01
Nocturnal Myoclonus	437 (30.3 %)	150 (7.7 %)	P < 0.01
Snoring	792 (59.8 %)	866 (46.8 %)	P < 0.01
Apneic Episodes	323 (16.2 %)	110 (6.0 %)	P < 0.01
Dyspnea during Sleep	291 (20.0 %)	150 (6.0 %)	P < 0.01

patients had an average age of about 65.6 years; and the control subjects, about 60.2 years. The male/female ratio was 0.97 in the parkinsonian patients and 1.28 in the control subjects. The mean duration of Parkinson's disease was about 11 years. Of the parkinsonian patients, 90.2% had been receiving antiparkinsonian drugs. The duration of drug treatment was less than 5 years in 35.3% of the patients, from 5 to 10 years in 39.8%, from 10 to 15 years in 23.7%, and more than 15 years in 1.2%. Classification of physical disturbance based on Hoehn and Yahr functional classification and the percentage in each category were I (11.3%), (11.7%), (34.8%), (26.0%), and (16.2%).

Table 1 shows the results of the present study about sleep-wake complaints. The prevalences of inability to initiate sleep, frequent awakenings, early morning awakening, daytime sleepiness, parasomnias, restless legs symptoms, nocturnal myoclonus, snoring, apneic episodes, and dyspnea during sleep were statistically higher in the parkinsonian patients than in the control subjects.

Figure 1a shows the incidences of restless legs symptoms in the parkinsonian patients and control subjects classified by age. The incidence at the sixth decade was the highest in the both groups. Figure 1b shows the results for nocturnal myoclonus. The incidence of the sixth decade in the parkinsonian patients was the highest, while the parcentage for the control subjects had a tendency to increase with age.

The prevalence of restless legs symptoms according to the grade of physical disturbance based on Hoehn and Yahr functional classification was the following: I(12.2%), (23.4%), (25.1%), (34.1%), and (34.1%). Similarly, the prevalence of nocturnal myoclonus was as follows: I(19.5%), (21.1%), (27.0%), (37.0%), and (42.3%). The prevalence of both symptoms increased significantly with the progression of

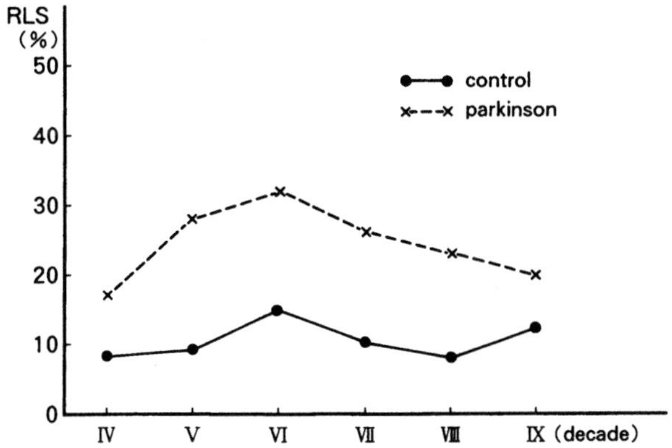

Fig. 1a. Incidence of restless legs symptoms (RLS)

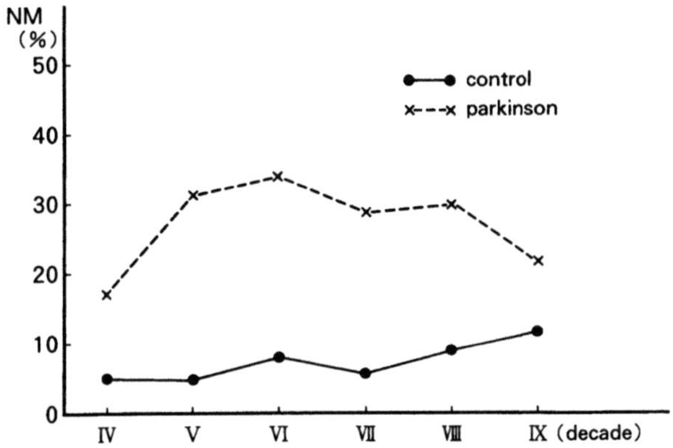

Fig. 1b. Incidence of nocturnal myoclonus (NM)

the disease according to Yahr's classification ($P<0.01$, chi-square value=32.247; $P<0.01$, chi-square value=41.468, respectively).

The prevalences of hallucinosis and/or hallucinosis with confusion and night terrors in the parkinsonian patients were 31.3% and 46.5%, respectively.

DISCUSSION

The prevalence of various sleep-wake complaints, e.g., inability to initiate sleep, frequent awakenings, early morning awakening, and daytime sleepiness, were statistically more frequent in the parkinsonian patients than in the control subjects. The following three factors may induce this sleep disturbance:

(1) Sleep disruption due to motor disturbance of Parkinson's disease. The daily hypoactivity of the parkinsonian patients may cause daytime sleepiness and insomnia at night. Various involuntary movements such as early morning dystonia[2] may cause early morning awakening.

(2) Higher prevalence of restless legs symptoms, nocturnal myoclonus, and sleep apnea syndrome such as snoring, apneic episodes, and dyspnea during sleep in the parkinsonian patients. The main restless legs symptom is a creeping or crawling sensation between knee and ankle, and it inhibits to initiate sleep. We[3] reported earlier that amantadine induced this syndrome in our parkinsonian patients.

Nocturnal myoclonus syndrome[4] also causes sleep-wake disturbance due to periodic leg movements during sleep, and it is frequently combined with restless legs symptoms. In the present study, 16.2% of the parkinsonian patients reported both symptoms. Telestad et al.[5] found a higher prevalence of restless legs symptoms among the middle-aged population, as we found in our control subjects and parkinsonian patients. Coleman[4] reported that the prevalence of nocturnal myoclonus gradually increased with age in the middle-aged population, as we found in our control subjects. The parkinsonian patients, however, showed a higher prevalence of nocturnal myoclonus among the middle-aged population. Progression of Parkinson's disease may inhibit the myoclonus. Sleep apnea syndrome[6] is also a cause of sleep disruption and accompanies daytime sleepiness.

(3) Significantly higher prevalence of parasomnias in the parkinsonian patients than in the control subjects. The phenomenon of parasomnias may include night terrors, nightmares, somnambulism, and hallucinosis and/or hallucinosis with confusion. These phenomena tend to occur at night, and cause sleep-wake disturbances. It should be noted that 90.2% of the parkinsonian patients were on antiparkinsonian drugs at the time of examination in the present study. The possibility that some of thee sleep disturbances were facilitated by antiparkinsonian drugs is not excluded.

REFERENCES

1. M.M. Hoehn and M.D. Yahr, Parkinsonism: onset, progression and mortality, Neurology 17:427(1967).
2. E. Melamed, Early-morning dystonia: A rate side effect of long term levodopa therapy in Parkinson's disease, Arch. Neurol. 36:308(1979).
3. J. Horiguchi and Y. Inami, Restless legs syndrome in four Parkinsonian patients treated with amantadine, Clin. Neurol.(Tokyo) 25:153(1985).
4. R.M. Coleman, Periodic movements in sleep(nocturnal myoclonus) and restless legs syndrome, in:"Sleeping and Waking Disorders", C. Guilleminault ed., Addison-Wesley, Menlo-Park, pp.265-95(1982).
5. W. Telestad, Treatment of the restless legs syndrome with carbamazepine: a double blind study, Brit. Med. J. 288:444(1984).
6. C. Guilleminault, F.L. Eldridge, W.C. Dement, Insomnia with sleep apnea: a new syndrome, Science 188:856(1973).

ESTIMATION OF BIOPTERIN AND NEOPTERIN IN CEREBROSPINAL FLUID IN PARKINSON'S DISEASE

Kenichiro Fujishiro, Masako Hagihara[1], Akira Takahashi, and Toshiharu Nagatsu[1]

Departments of Neurology and [1]Biochemistry
Nagoya University School of Medicine, Nagoya, Japan

INTRODUCTION

Biopterin and neopterin are natural pteridines and exist in body fluids. Recently changes in their levels have been observed in certain diseases. Tetrahydrobiopterin (BH_4) has the role of a cofactor for phenylalanine hydroxylase and tyrosine hydroxylase, which are enzymes essential for the biosynthesis of catecholamines.[1] Neopterin is a metabolite of D-*erythro*-7,8-dihydroneopterin triphosphate, an intermediate of tetrahydrobiopterin. In Parkinson's disease, disorders of the dopamine synthesis system and decreased biopterin levels in the brain,[2] serum,[3] and cerebrospinal fluid (CSF)[4] were reported. On the other hand, neopterin has not been examined in CSF of parkinsonian patients. Neopterin was shown to be elevated in body fluids in various disease conditions involving activation of cell-mediated immunity, and has been regarded as a biochemical marker for the *in vivo* activation state of cell-mediated immunity.[5] Recently antibodies against nervous system elements in parkinsonian patients have been reported,[6-8] suggesting some changes in the immune system of Parkinson patients. In the present study, we measured neopterin and biopterin in CSF by using high-performance liquid chromatography with fluorescence detection (HPLC-FD) to examine whether CSF neopterin and biopterin changes exist in parkinsonian patients.

MATERIALS AND METHODS

CSF samples from 17 controls (age:16-60 years) and from 18 patients with Parkinson's disease (age:52-78 years, stage: II-IV[9]) were obtained in the morning at about 10 a.m.; samples were either used immediately or stored at -80°C. The control CSF samples were obtained from patients who had no neurological diseases or other diseases with altered immunity. Informed consent was obtained from each patient. The cell counts and protein or glucose concentrations in CSF were within normal ranges.

Total biopterin and neopterin concentrations in CSF were measured by HPLC by a modification of the method of Fukushima and Nixon.[10] One ml of CSF was mixed with 3.5 ml of 0.1 N phosphoric acid and then 0.5 ml of 2 N trichloroacetic acid (TCA) was added. Next, 0.5 ml of an iodine solution (0.5% I_2/0.9% KI in 0.2 N TCA) was added. The mixture was allowed to stand 1 h at room temperature under reduced light, and then the sample was

Edited by T. Nagatsu *et al.*, Plenum Press, New York, 1990

centrifuged at 700 g for 10 min. To the supernatant ascorbic acid (1% solution, 0.1 ml) was added and then the mixture was applied to a Dowex 50 mini column (200 to 400 mesh ; H^+form) with a column bed volume of 0.2 ml. After the column had been washed twice with 5 ml of water, pteridines were eluted with 2.0 ml of 1 N NH_4OH. The eluate was lyophilized and then to the residue was added 0.2 ml buffer (0.02M disodium phosphate and 0.14 M citric acid, pH 3.0).[11] A 50-µl aliquot of the solution was injected into an HPLC apparatus equipped with a Shimadzu LC-5A pump and an LC injector SIL-1A (Shimadzu, Kyoto, Japan). The analytical column (4.0 mm I.D. x 250 mm) was a Whatman partisil A 10 ODS. An RF-535 fluorescence detector (Shimadzu, Kyoto, Japan) was used for detection of pteridines (excitation at 350 nm, emission at 450 nm). The mobile phase consisted of 5% methanol, 0.02 M disodium phosphate, and 0.04 M citric acid (pH 3.0). Flow rate was 1 ml/min. Statistic analysis of the data was carried out by Student's *t* test.

Table 1. Neopterin and biopterin contents of CSF for normal controls and patients with Parkinson's disease

Subjects	Age(yrs.) mean (range)	Number of subjects	Pteridine content (pmol/ml CSF) Neopterin	Biopterin	Neopterin/ Biopterin ratio
Controls	35(16-64)	17	25.18±8.88	22.09±7.86	1.24±0.46
Parkinson's disease	61(51-78)	18	13.98±6.72[a]	7.7 ±3.51[a]	2.01±1.13[b]
Stage II*	58(52-61)	4	12.36±4.37	10.92±3.50	1.16±0.29
Stage III*	60(54-77)	9	13.71±6.84	7.75±2.72	1.67±0.63
Stage IV*	66(54-78)	5	15.70±7.05	5.35±2.10	3.18±1.11

Mean ± S.D.

*The stages of Parkinson's disease were evaluated by the criteria reported by Hoehn and Yahr.[9]

[a]p<0.001 for controls

[b]p<0.05 for controls

RESULTS

As shown in Table 1, the biopterin and neopterin levels in CSF of patients with Parkinson's disease were low ($p<0.001$) but neopterin/biopterin ratios were high ($p<0.05$) as compared with those of normal controls. The biopterin levels were lower and the neopterin/biopterin ratio tended to be higher in the patients who had more severe symptoms as evaluated by Hoehn and Yahr classification.[9]

DISCUSSION

(BH_4) is the cofactor for phenylalanine, tyrosine and tryptophan hydroxylases, enzymes that play important roles in the biosynthesis of catecholamines and indolamines. BH_4 is oxidized to quinonoid 7,8-dihydrobiopterin, 7,8-dihydrobiopterin, and biopterin. A decreased level of total biopterin in the brain[3] and CSF[4] of parkinsonian patients has been reported, which may reflect the degeneration of the nigrostriatal dopaminergic neurons. Neopterin is derived from 7,8-dihydroneopterin triphosphate, which is an intermediate in the biosynthesis of BH_4 in the dopaminergic neurons. Therefore, decreased biopterin may be accompanied by a reduction in the neopterin level. The formation of 7,8-dihydroneopterin triphosphate from guanosine triphosphate (GTP), the first and probably rate-limiting step of BH_4 biosynthesis, is catalyzed by GTP cyclohydrolase I. In the brain of parkinsonian patients, low activity of GTP cyclohydolase I was reported.[12] The present results on low levels, of both neopterin and biopterin agree with this previous report. In the present study, we found decreased neopterin, and biopterin levels and an increased neopterin/biopterin ratio in CSF in Parkinson's disease. Levels of biopterin tended to be lower, but those of neopterin tended to be higher in patients having more severe parkinsonian symptoms than in those with milder symptoms, which indicates the possibility of formation and release of neopterin into the CSF from cells different from dopaminergic neurons. Neopterin has recently been regarded as an *in vivo* marker for activation of cell-mediated immunity, for *in vitro* experiments have demonstrated a correlatin between neopterin release from monocytes/macrophages and the induction of these cells by interferon-gamma.[13] In addition, antibodies against nervous system components have been detected in the CSF of parkinsonian patients. Therefore we speculate that the tendency of an increased neopterin and neopterin/biopterin ratio in patients who have more severe parkinsonian symptoms reflects increasing involvement of the immune system in the course of Parkinson's disease.

CONCLUSION

Biopterin and neopterin in CSF of eighteen patients with Parkinson's disease were measured by HPLC-FD. The CSF biopterin and neopterin levels of patients with Parkinson's disease were lower, but their neopterin/biopterin ratios were higher, than those of normal controls. Also the level of biopterin became lower, while the neopterin/biopterin ratio increased as Parkinson's disease became advanced. Our results suggest influence of the immune system on the clinical evolution of Parkinson's disease.

REFERENCES

1. T. Nagatsu, S. Matsuura, and T. Sugimoto, Physiological and clinical chemistry of biopterin, Med. Res. Reviews 9: 25 (1982).
2. T. Nagatsu, T. Yamaguchi, T. Kato, T. Sugimoto, S. Matsuura, M. Akino, I. Nagatsu, R. Iizuka, and H. Narabayashi, Biopterin in brain and urine from controls and parkinsonian patients: application of a new radioimmunoassay, Clin. Chem. Acta 109: 305 (1981).
3. W. Lovemberg, R. A. Levine, D. S. Robinson, M. Ebert, A. C. Williams, and D. B. Calne, Hydroxylase cofactor activity in cerebrospinal fluid of normal subjects and patients with Parkinson's disease, Science 204: 624 (1979).

4. T. Yamaguchi, T. Nagatsu, T. Sugimoto, S. Matsuura, T. Kondo, R. Iizuka, and H. Narabayashi, Effects of tyrosine administration on serum biopterin in normal controls and patients with Parkinson's disease, Science 219: 75 (1983).
5. A. Hasen, D. Fuchs, G. Reibnegger, E. R. Werner, and H. Wachter, Neopterin in clinical use, Pteridines 1: 3 (1989).
6. A. McRae-Degueurce, L. Rosengren, K. Haglid, S. Booj, C.-G. Gottfries, A. C. Granerus, and A. Dahlstrom, Immunocytochemical investigations on the presence of neuron-specific antibodies in the CSF of Parkinson's disease cases, Neurochem. Res 13: 679 (1988).
7. A. Pouplard, J. Emile, F. Pouplard, and D. Hurez, Parkinsonism and autoimmunity: antibody against human sympathetic ganglion cells in Parkinson's disease, Adv. Neur 24: 321 (1979).
8. G. Husby, L. Li, L. E. Davis, E. Wedege, E. Kokmen, and R. C. Williams, Jr., Antibodies to human caudate nucleus neurons in Huntington's chorea, J. Clin. Invest 59: 922 (1979).
9. M. M. Hoehn and M. D. Yahr, Parkinsonism: Onset, progression, and mortality, Neurology 17: 427 (1967).
10. T. Fukushima, and C. Nixon, Analysis of reduced forms of biopterin in biological tissues and fluids, Analyt. Biochem 102: 176 (1980).
11. J. L. Dhondt, G. Forzy, J. M. Hayte, J. P. Farriaux, and C. Largilliere, Pterins and monoamine metabolites in cerebrospinal fluid, in: "Unconjugated Pterins and Related Biogenic Amines," L. Walter and R. A. Levine, eds. Taylor & Francis, London, New York and Philadelphia (1987).
12. T. Nagatsu, T. Horikoshi, M. Sawada, I. Nagatsu, T. Kondo, R. Iizuka, and H. Narabayashi, Biosynthesis of Tetrahydrobiopterin in Parkinsonian human brain, Adv. Neur 45: 223 (1986).
13. C. Huber, J. R. Batchelor, D. Fuchs, A. Hausen, A. Lang, D. Niederwieser, G. Reibnegger, P. Swetly, J. Troppmair, and H. Wachter, Immune response-associated production of neopterin, J. Exp. Med 160: 310 (1984).

QUANTITATIVE EVALUATION OF VOCAL CHANGES IN PARKINSON'S DISEASE: INTERCORRELATION AMONG VOICE PARAMETERS AND MONOAMINE SYSTEMS

Kazutaka Tanabe[1], Yoshiaki Furukawa[1], Tomoyoshi Kondo[1], Yoshikuni Mizuno[1], Kiyonori Endo[2], and Hirotaro Narabayashi[2]

[1]Department of Neurology, Juntendo University School of Medicine, [2]Neurological Clinic
Tokyo, Japan

INTRODUCTION

The characteristic abnormality in the speech of patients with Parkinson's disease (PD) is the "hypokinetic dysarthria" described by Darley et al. (1969) (1). However, vocal disorder of this disease remains unclarified, and factors contributing to the worsening of phonation are still unknown. The purpose of this study is to analyze quantitatively the voice of PD, and to investigate the correlation between the vocal changes and the neurochemical parameters of ventricular CSF. We developed a method for the quantitative evaluation of vocal changes in PD. Total biopterin contents and ventricular CSF monoamine metabolites were assayed in the same patients. Of particular interest is whether specific vocal changes have any correlation with the parameters of the monoamine metabolism.

MATERIALS AND METHODS

Eighteen patients with PD were studied. The mean age of the patients was 56.6 ± 10.7 years old, and the duration of illness was 7.0 ± 5.0 years. Patients were asked to vocalize loudly three vowels. 'A', 'E', and 'O' for approximately 2 second-period for each with 1 second interval between respective vowels. They were recorded by a dynamic microphone (SONY F-U5T) into a high quality tape recorder (SONY TC-200). These vowels were sampled at 10 KHz, and quantized to 14 bits as digital signals which were processed using an adaptive inverse filter and the linear prediction algorithm according to the method of Atal (1971), Itakura (1972), and Nakajima (1978) (2,3,4). Two kinds of quantitative voice parameters were evaluated. The vocal tract from the vocal cord to the mouth was divided equally into 12 cross sectional areas. Upon vocalization of each vowel, the area of each cross section was calculated as the proportion of the section which included the glottis (CSG). The sum of the difference in each one of the 12 cross sectional areas was calculated as an index of the vocal tract movement (E-O: difference in the cross sectional area between 'E' and 'O'). The regularity of the vocal cord vibration (P.A.C. : Pitch Auto-Correlation) was estimated by a calculating the auto-correlation of pitch intervals in phonating the vowel 'A'. This value was 100% when the pitch intervals were completely constant, and 0% if it is quite random. Samples of the ventricular CSF were obtained during stereotaxic surgeries, which were performed under the local anesthesia, and they were frozen immediately and stored in darkness at -70°C until neurochemical analyses. Levels of homovanillic acid (HVA), 3-methoxy-4-hydroxyphenylethylene glycol (MHPG), and 5-hydroxyindoleacetic acid (5HIAA) were measured by liquid chromatography. Total biopterin (T-BP) content was assayed by HLPC (Yanaco L-2000) with fluorometric detection (Shimadzu RF 530: Ex 360 nm; Em 445 nm) according to the method of Fukusima and Nixon (1980) (5) with a slight modification.

RESULTS

The ventricular HVA level was 218.6 ± 16.6 ng/ml (mean ± SE), MHPG 9.05 ± 0.86 ng/ml, and 5HIAA 89.67 ± 6.52 ng/ml. The T-BP level was 15.58 ± 0.89 pmol/ml. Mean percentile of P.A.C. was 39.15 ± 3.33 %, and E-0 28.6 ± 5.0 CSG. A positive correlation was observed between the two voice parameters E-0 and P.A.C. ($r=0.75$, $p<0.0005$; Fig.1). Significant positive correlations were found between the T-BP level and the voice parameters for P.A.C. ($r=0.77$, $p<0.001$; Fig.2) and E-0 ($r=0.67$, $p<0.005$; Fig.3). The proportion of P.A.C. was significantly correlated with MHPG ($r=0.52$, $p<0.05$) and 5HIAA ($r=0.49$, $p<0.05$), and E-0 with HVA ($r=0.47$, $p<0.05$) and 5HIAA ($r=0.46$, $p<0.06$). No significant correlations were seen between the age and the neurochemical or the voice parameters.

DISCUSSION

The reduction in the parameter E-0 implies the decrease in the vocal tract movement during the phonation of two vowels studied which is mainly accomplished by the orofacial muscle activities. Hypokinesia and rigidity of the orofacial muscles during articulation were observed in previous studies with surface EMG or mechanical devices (6,7). Abnormalities in the vocal cord vibration were frequently observed in PD patients (8), and can be

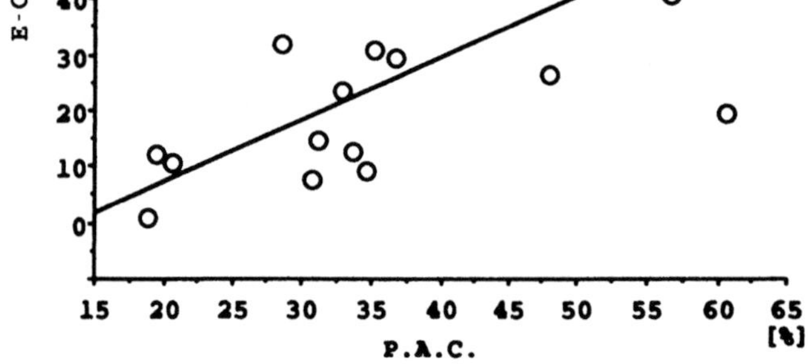

Fig. 1. Correlation between two voice parameters ; E-0 : Difference in the cross sectional area between 'E' and 'O' (vocal tract parameter), P.A.C. : Pitch Auto-Correlation (vocal cord parameter). CSG : Cross Section of Glottis (used as an unit).

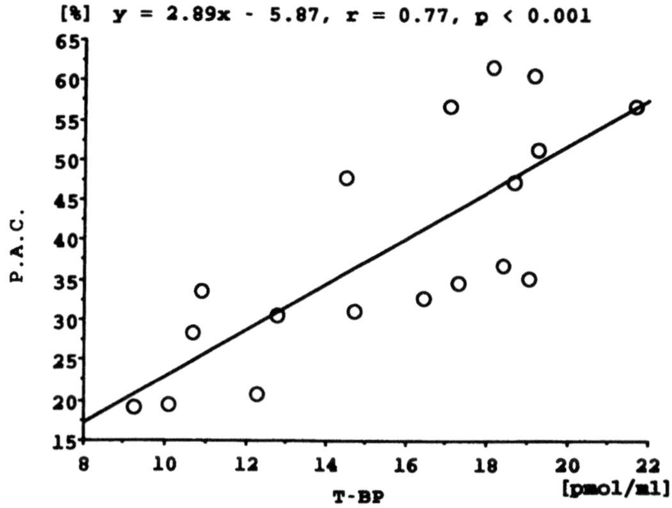

Fig. 2. Correlation between vocal cord parameter (P.A.C.) and total biopterin contents in ventricular CSF (T-BP).

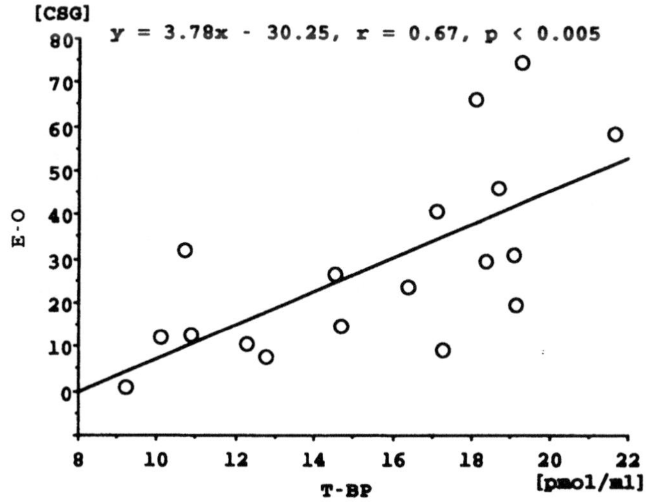

Fig. 3. Correlation between vocal tract parameter (E-O) and total biopterin contents in ventricular CSF (T-BP).

expressed as the changes in the parameter of P.A.C.. Our study revealed a strong positive correlation between these two kinds of vocal parameters, and our data are considered as a quantitative expression of the hypokinetic dysarthria described by Darley et al. (1969) (1). In patients with PD, diminished dopamine metabolism in the nigro-striatal nerve terminals is well known (9), however, to our knowledge, no one has yet correlated the loss of dopamine metabolism quantitatively with the clinical disabilities evaluated by a quantitative method. The present results demonstrate that certain combinations of vocal and neurochemical parameters show significant correlations. Particularly, each one of the vocal parameters showed a highly significant positive correlation with the T-BP level. As the T-BP level is considered to be the best indicator of the state of the monoamine metabolism (9), our observation suggests that the severity of dysarthria as studied by our method seems to represent a quantitative expression of the damage to the metabolism of monoamine transmitters.

REFERENCES

1. F. L. Darley, A. E. Aronson and J. B. Brown. Differential diagnostic patterns of dysarthria. J Speech & Hear Res. Vol. 12: pp. 246-269, (1969).
2. B. S. Atal, and S. L. Hanauer. Speech Analysis and Synthesis by Linear Prediciton of the Speech Wave. J Acoust Soc Am. Vol. 50 (2): pp. 637-655, (1971).
3. F. Itakura and S. Saito. An audio response unit based on partial correlation. IEEE Trans. Comm. COM-20(4): pp. 792-797, (1972).
4. T. Nakajima, T. Suzuki, H. Oomura, S.Ishizaki and K. Tanaka Estimation of Vocal Tract Area Function by Adaptive Deconvolution and Adaptive Speech Analysis System. J Acoust Soc Japan. Vol. 34, No. 3: pp. 157-166, (1978).
5. T. Fukushima, and J. Nixon. Analysis of the reduced form of biopterin in biological tissues and fluids. Ann Biochem. Vol. 102: pp. 176-188, (1980).
6. M. P. Caligiuri, Labial kinetics during speech in patients with parkinsonian rigidity. Brain. Vol. 110: pp. 1033-1044. (1987).
7. C. J. Hunker, J. H. Abbs, and S. M. Barlow. The relationship between parkinsonian rigidity and hypokinesia in the orofacial system: A quantitative analysis. Neurology (Ny) Vol. 32: pp. 755-761, (1982).
8. J. Logemann, B. Boshes and H. Fisher. The steps in the degeneration of speech and voice control in Parkinson's disease. In :"Parkinson's disease." J. Siegfried, (Ed.) Vol.2, pp. 101-112, Hans Huber Bern, (1973).
9. T. Nagatsu, S. Matsuura and T. Sugimoto. Physiological and Clinical Chemistry of Biopterin. Med Res Rev. Vol. 9 (No. 1): pp. 25-44, (1989).

PERIODIC MOVEMENTS DURING SLEEP IN PATIENTS WITH PARKINSON'S DISEASE

Akira Iwasaki, Yohsuke Narabayashi, Mamoru Takakusagi,
Shinichi Iwasaki, and Katsuhiko Hamaguchi

Department of Neurology, Saitama Medical School
Moroyama-machi, Iruma-gun, Saitama, Japan

INTRODUCTION

It is well known that various involuntary movements appear in the course of Parkinson's disease (PD) during long-term treatment with L-DOPA.[1,2] However, periodic movements during sleep (PMS) have rarely been reported in patients with PD, and little has been clarified concerning its mechanism. With polysomnography we studied whether patients with a long history of PD would have PMS. In patients with PMS, in order to approach the pathomechanism of this phenomenon, we investigated the influence of L-DOPA on PMS and the relation between PMS and the clinical background of the patient.

PATIENTS AND METHODS

Ten patients who had been suffering from Parkinson's disease for longer than five years were chosen for this study (Table 1). There were six males and four females, and their average age was 62.7±7.7 years old(49 to 76). Duration of illness was 10.4 ±3.5 years(5 to 16). According to the Hoehn & Yahr scale, 5 cases were in stage III; and the other 5, in stage IV. They had been prescribed L-DOPA plus a decarboxylase inhibitor with a mean dose of 412.5±209.2 mg/day(100 to 700). Duration of L-DOPA therapy was 6.5±3.4 years(1 to 13). L-DOPA was stopped on the day of the examination. Polysomnography was done at 2:00 p.m. for three to six hours under natural sleep or controlled sleep induced with 10 or 15 mg of pentobarbital. EEG was recorded at C_3 and C_4. EOG was recorded at the right and left outer canthi. EMG was recorded from submentalis, bilateral quadriceps, hamstrings, tibalis anterior, and gastrocnemius muscles. Involuntary movements were monitored and recorded with a videocamera. In cases showing PMS, 100 or 200 mg of L-DOPA plus decarboxylase inhibitor was orally administered just before polysomnography in order to study the effects of L-DOPA on PMS. Relation of PMS to clinical backgrounds was also investigated.

RESULTS

We observed PMS in two (cases 2 and 7) out of patients with Parkinson's disease who had been treated long-term with L-DOPA.

Table 1. Clinical Data from Ten Patients with Parkinson's Disease

Cases	Sex Age	Duration of PD(yrs)	Hoehn & Yahr scale	Dose of L-DOPA(mg)	Duration of L-DOPA therapy(yrs)	W-O	P
1	M 49	14	III	500	7	−	−
2	M 57	8	III	700	7	+	−
3	F 58	7	III	600	7	−	−
4	M 61	10	IV	400	6	−	−
5	M 62	14	IV	125	3	−	−
6	F 62	8	III	200	4	−	−
7	M 63	10	IV	600	10	+	+
8	M 66	5	III	100	1	−	−
9	F 73	16	IV	500	13	−	+
10	F 76	12	IV	400	7	−	−

W-O : wearing-off, P : peak-dose dyskinesia

Case 2

He is a 57-year-old merchant who has been suffering from PD since October 1980. His speech was slightly monotonous and he had a masked face and mild cog-wheel rigidity in the neck and in the left upper extremity. Postural finger tremor was mild on both sides. His gait was short stepping, he had a lumbar-stooped posture, and his arm swing was poor bilaterally. Postural reflexes showed propulsion and retropulsion. Trihexyphenidyl and amantadine were administered till 1981, when it was replaced by L-DOPA plus carbidopa. Parkinsonian signs had been well controlled for three years, but since 1984 the effect of a single dose of L-DOPA continued no longer than 2 hours. Bromocriptine added in 1986 didn't ameliorate the wearing-off phenomenon. In October in that year when he was drowsy, he noticed jerky movements in the feet. Polysomnography on November 22 in 1988 demonstrated simple or polyclonic contractions in the right tibialis anterior muscle at sleep stage 1 with irregular intervals(3 to 159 seconds). At sleep stage 2, periodic polyclonic contractions of duration between 0.5 and 2.0 seconds were observed with an interval of 16 to 21 seconds in the right tibialis anterior muscle for 20 minutes. Polysomnography on May 9 in 1989 showed disappearance of the polyclonic muscle contractions at both sleep stages 1 and 2 under a dose of 100 mg of L-DOPA plus carbidopa.

Case 7

This 63-year-old man had a 10-year history of PD. He had a mask-like waxy face, a small unclear voice, and sometimes stuttering and micrographia. Gait was lumbar-stooped with short steppage, and he had mild or moderate rigidity in the neck, arms, and forearms. Tremor was induced in bilateral upper extremities on positioning. Postural reflexes showed propulsion and retropulsion. Since 1978 he had been well treated with L-DOPA plus benserazide till October in 1985,when diurnal fluctuation of motor performance was noticed. During on-periods, he was able to walk normally, but he could not speak or write as well as when he had been in good health. When he was in off-periods, his gait became slower, he could neither dress without support nor fold a newspaper. Polysomnography on November 29, 1988, showed involuntary movements with irregular intervals of 5 to 61 seconds in the left tibialis anterior and hamsrtrings at sleep stage 1. At sleep stage 2, these polyclonic contractions with a duration of 1 to 4.7 seconds recurred periodically with an interval of 15 to 19 seconds and also occurred on the opposite side; they were sustained for 18 minutes. They were stereotyped

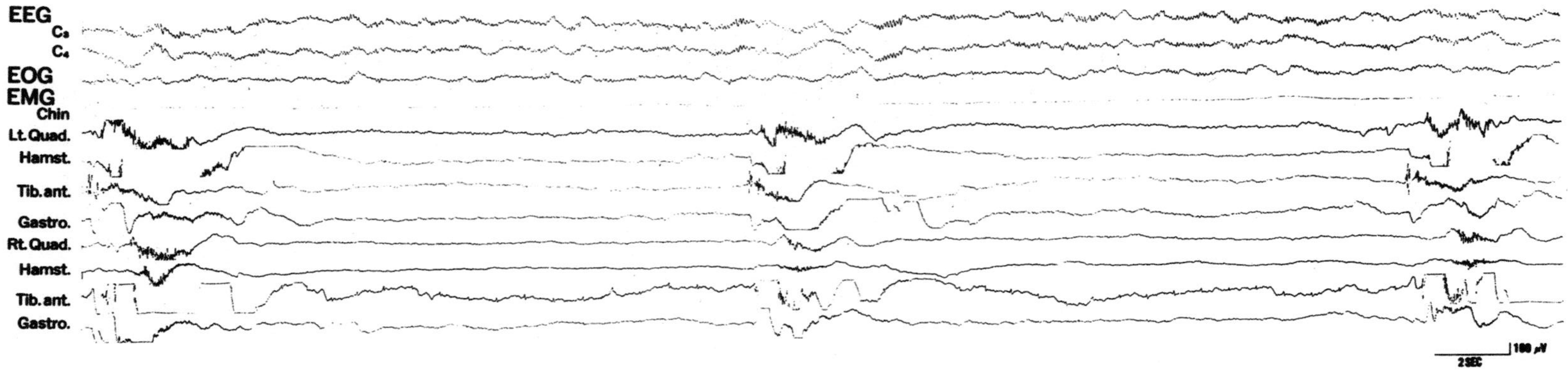

Fig. 1. Polysomnogram of Case 7 recorded during sleep stage 2 with L-DOPA. EOG: electrooculogram, Quad.: quadriceps femoris, Tib. ant.: tibialis anterior, Gastro.: gastrocnemius.

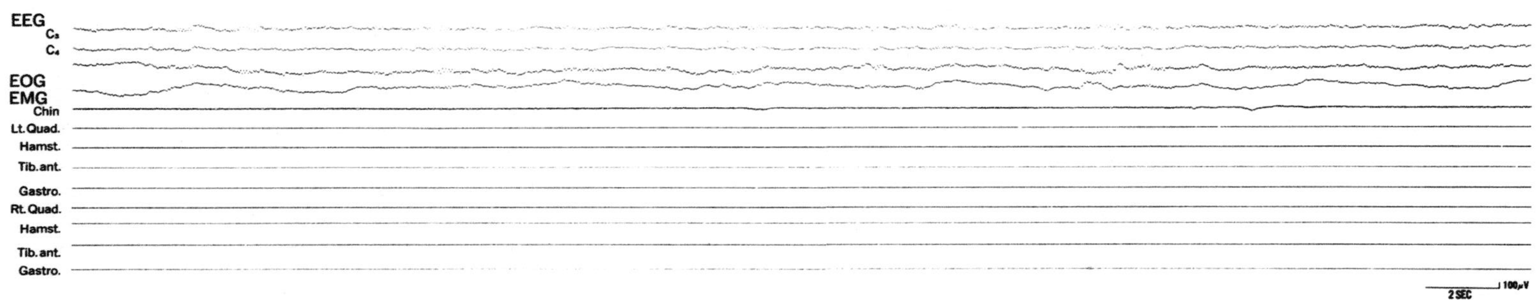

Fig. 2. Polysomnogram of the same case as in Fig. 1 during sleep stage 2 with L-DOPA.

polyclonic contractions initiating in the left tibialis anterior muscle, and then spreading to the hamstrings, then to the quadriceps, and terminating in the gastrocneminus muscle. On the opposite side, the polyclonic contractions manifested themselves in the same way with a delay of 0.5 second. These muscle activities were observed periodically with an interval of 17 seconds at sleep stage 2 (Fig. 1). Polysomnography after oral dosing of 200 mg of L-DOPA plus benserazide on April 25, 1989, showed suppression of involuntary movements at sleep stage 2 (Fig. 2).

DISCUSSION

One of features observed in our cases is that involuntary movements appeared during sleep or at the drowsy state. Case 2 showed a dorsiflexion of the right foot, and case 7 showed a triple flexion consisting of dorsiflexion of foot, knee, and hip joint. Another feature is that each muscle contraction was a polyclonic contraction whose duration was within 5 seconds and which occurred at an interval of 16 to 21 seconds in Case 2, and of 15 to 19 seconds in Case 7. These involuntary movements are considered periodic movements during sleep (PMS) as proposed by Lugaresi.[3] While PMS appears in a variety of diseases, it has rarely been reported in Parkinson's disease. It is notable that PMS in our study was inhibited by oral dosing of L-DOPA plus decarboxylase inhibitor. Askenasy et al.[4] reported PMS in three patients with PD. L-DOPA plus decarboxylase inhibitor suppressed PMS in two cases, and trihexyphenidyl and a dopaminergic potentiator were also effective in one case. Thus, our present study is in agreement with that of Askenasy and colleagues. It is reported that L-DOPA was effective on PMS in restless legs syndrome[5] and that L-DOPA plus carbidopa was also effective on PMS in elderly subjects free of restless legs syndrome or other neurological disorders.[6] Recently it was showen that PD patients have an increased incidence of pathological changes other than basal ganglia. From the facts that PMS appears not only in PD, but also in other diseases, and that PMS in both pathological conditions is suppressed with antiparkinson drugs, it is implied that dopaminergic neurons other than those of striatonigral system may be involved in the pathomechanism of PMS.

REFERENCES

1. A. Friedman, Levodopa-induced dyskinesia;clinical observations, J. Neurology. 232:29(1985).
2. H. L. Klawans, C. Goetz, and D. Bergen, Levodopa-induced myoclonus, Arch. Neurology. 32:331(1975).
3. E. Lugaresi, F. Cirignotta, G. Coccagna, and P. Montagna, Nocturnal myoclonus and restless legs syndrome, in "Advances in Neurology, vol. 43, Myoclonus", S. Fahn et al., Raven Press, New York (1986).
4. J. M. Askenasy, E. D. Weitzman, and M. D. Yahr, Are periodic movements in sleep a basal ganglia dysfunction? J. Neural Trans. 70:337(1987).
5. J. Montplaisir, R. Godbout, G. Poirier, and M. A. Bedard, Restless legs syndrome and periodic movements in sleep; pathophysiology and treatment with L-dopa, Clin. Neuropharmacol. 9:456(1986).
6. C. Guilleminault, S. Mondini, J. Montplaisir, J. Mancuso, D. Cobasco, and W. C. Dement, Periodic leg movement, L-dopa, 5-hydroxytryptophan, and L-tryptophan, Sleep. 10(4):393(1987).

(±)-CIS-METHYL-SPIRO(1,3-OXATHIOLANE-5,3')QUINUCLIDINE (AF102B): A NEW M1 AGONIST AS A RATIONAL TREATMENT STRATEGY IN ALZHEIMER'S DISEASE - AN OVERVIEW

ABRAHAM FISHER and ELIAHU HELDMAN

Israel Institute for Biological Research, P.O.Box 19
Ness-Ziona 70450, Israel

INTRODUCTION

Alzheimer's disease (AD) is a widespread and progressive neurodegenerative disorder associated with aging. This clinical syndrome is evidenced by reduced intelectual functioning with impairment of cognitive functions, memory and learning, and visuospatial skills; short-term working memory loss and a decrease in alertness, attention span and motivation are also characteristic features of AD.

Morphological and neurochemical studies indicate a select degeneration of cholinergic cells in the basal forebrain and of their projections to cortex and hippocampus (reviews 1-3). This degeneration is reflected by a marked reduction of presynaptic cholinergic indices in these brain regions involved in mnemonic processes (1-3). These regions may be involved in mnemonic processes and hence the loss of cognitive functions.

There is also evidence that noradrenergic, serotonergic, and somatostatin-like immunoreactivity are abnormal in AD, though to lesser extent than the cholinergic hypofunction (3-5). The contribution of the cholinergic dysfunction to the cognitive symptoms in AD is reflected by a direct correlation between loss of presynaptic cholinergic markers and the severity of memory deficits (1, 2).

Whilst the presynaptic cholinergic deficits are well established, somewhat conflicting results are reported regarding muscarinic receptors in AD. A loss of M2 muscarinic receptors (low-affinity for pirenzepine), located presynaptically and regulating acetylcholine (ACh) release, is reported in a few studies (6-8 but see also 9, 10) while postsynaptic muscarinic receptors (mainly M1; high-affinity for pirenzepine), facilitating cellular excitation, are relatively unchanged (6-9), decreased (10) or even upregulated (11).

Given presynaptic but not postsynaptic losses in AD, the cognitive dysfunction might be restored by treatment of the cholinergic hypofunction. In principle, an increase in central cholinergic activity can be achieved via presynaptic or postsynaptic modulation. The greatest efforts to date has been directed toward a so-called "replacement therapy" which should enhance cholinergic activity (for reviews see 12, 13). Some of these therapeutic strategies evaluated in clinical studies are summarized in Table 1. The outcome of the clinical trials ranked from lack of significant success to modest improvement (12, 13). The possible reasons for lack of a more significant effect in AD patients with the currently available cholinergic compounds could be attributed to:

a) the properties of the drugs tested such as
   - inadequate pharmacokinetic profile (e.g. short duration of action, poor absorption);
   - minimal selectivity for brain regions involved in memory processes leading to adverse central and peripheral side-effects;
   - insufficient penetration of the blood-brain barrier;
   - narrow therapeutic index.

b) the special problems regarding drug development in AD such as
   - unknown etiology of the disorder;
   - no-well defined neurochemical hypotheses;
   - no reference compounds with demonstrated clinical efficacy;
   - lack of perfect animal models capable of mimicking all aspects of this disorder;
   - homogeneity and/or heterogeneity of AD patients and relevant diagnostic difficulties;
   - the demonstration of clinical efficacy and existence of various types of protocols each with its pros and cons.

c) presynaptic facilitation of cholinergic activity would require at least a limited number of presynaptic neurons capable of synthesizing and releasing ACh thus limiting, a priori, the potential use of compounds acting through such a mechanism (e.g. cholinesterase inhibitors) to only mildly or moderately affected AD patients.

In light of these problems, but especially since the most widely accepted neurochemical hypothesis for the cognitive deficit in AD involves still the cholinergic system, the following type of drugs can be suggested as a rational treatment strategy: a. M1 agonists (2, 12, 14), b) M2 antagonists (2, 12, 14); c. mixed M1 agonistic and M2 antagonistic activity in the same compound. It is important that such candidate drugs should have all or at least most of the following properties: a. selectivity for synapses involved in cognitive functions; b. positive effects on mnemonic processes; c. lack of peripheral and central side-effects; d. wide therapeutic index; e. passage through the blood-brain barrier; f. adequate pharmacokinetic profile and activity when taken orally; g. long duration of action; h. lack of, or minimal, tolerance when administered chronically.

Since ACh is a highly flexible molecule capable of attaining a number of conformations with the same energy, one can speculate that it is this conformational flexibility which enables this neurotransmitter to interact with all types of cholinergic receptors (e.g. M1, M2, and further subtypes and nicotinic). In this context, we have hypothesized that a rational drug design of an M1 selective agonist could be via rigid analogs of ACh. This is based on the premise that utmost rigidity limits the ability of ligands to adapt to minor differences in receptor substructure and thereby provides selectivity towards a limited population of receptors. ($\pm$)-Cis-2-methyl-spiro(1,3-oxathiolane-5,3)

Table 1. Potential drugs with some activity on the cholinergic system evaluated in AD patients

Drug tested	Mechanism of Action
Choline	Precursor of ACh
Lecithin	Precursor of ACh
4-Aminopyridine	Enhancer of ACh release
Piracetam	*
Aniracetam	*
Oxiracetam	*
Physotigmine	Inhibitor of acetylcholinesterase
Tetrahydro-aminoacridine	Inhibitor of acetylcholinesterase
Arecoline	Muscarinic agonist (mixed M1 and M2)
Bethanecol	Muscarinic agonist (mixed M1 and M2)
Oxotremorine	Muscarinic agonist (mainly M2)
RS86**	Muscarinic agonist (mixed M1 and M2)
Pilocarpine**	Muscarinic agonist (mixed M1 and M2)

* Mechanism not clear; these nootropic compounds are believed to enhance cholinergic stimulation;

** RS86 and pilocarpine show predominant M1 agonistic activities in certain systems but still significant M2 (and other type) agonistic activities prevail in a number of tests (18, 19, 42-44).

Table 2. Potential animal models for AD

1. Aged models (rats, monkeys)

2. Anoxia/hypoxia in rats

3. Drug-induced deficit models (anticholinergics; hemicholinium; benzodiazepine)

4. Genetic models (hippocampally deficient mice; senescence-accelerated mouse)

5. Aluminum-induced lesions in experimental animals

6. CNS lesions:
   - electrolytic (nucleus basalis)
   - excitotoxins (nucleus basalis)
   - AF64A (hippocampus; nucleus basalis)

quinuclidine (AF102B), (Figure1), a closely related analog of ACh embodying the "muscarinic pharmacophore" in a framework of utmost rigidity, is such a selective M1 agonist (14-16). The aim of this paper is to overview some of the properties of AF102B in vitro and in vivo and to show that this compound fulfills most of the conditions required from a candidate drug aimed to treat AD.

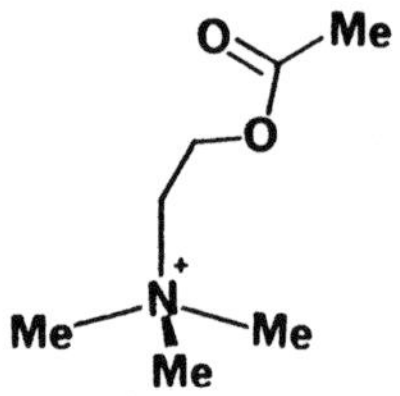

ACh

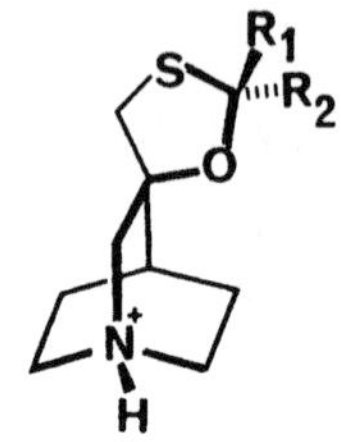

	R_1	R_2	Code
<u>cis</u> -	CH_3	H	AF102B
<u>trans</u> -	H	CH_3	AF102A

Figure1. Solid lines represent the "muscarinic pharmacophore" in these structures enclosed in the spiro-quinuclidine-oxathiolane framework.

NEUROCHEMICAL, PHYSIOLOGICAL AND PHARMACOLOGICAL FEATURES OF AF102B

AF102B is a full muscarinic agonist as shown on the isolated guinea-pig ileum and trachea, being 87 and 1.3 fold less potent than ACh (16). One of the remarkable features of AF102B is its M1 selectivity. This was shown in a number of studies including:

Binding studies. These experiments were done with selective and non-selective radiolabelled ligands (in vitro, rat brain regions) and in comparison with muscarinic agonists such as: oxotremorine and carbachol (CCh) (mainly M2 agonists), McN-A-343 (a prototype M1 ganglionic stimulant), <u>cis</u>-AF30 and <u>trans</u>-AF30 (relatively selective ganglionic and central muscarinic agonists), (17-19). The relative M1 selectivity was assayed using the radiolabelled ligands: ^{3}H-pirenzepine (^{3}H-PZ) an M1 selective antagonist (20); (-)-^{3}H-QNB (a mixed M1 and M2 antagonist), (21, 22); and (+)-^{3}H-<u>cis</u>-dioxolane, a mixed M1 and M2 agonist with some preference for M2 receptors (23). In these studies, on rat forebrain or frontal cortex homogenates, AF102B showed preference for PZ-sensitive binding sites (16). In the same line, when evaluated on rat forebrain (rich in M1 receptors) vs. cerebellum homogenates (predominantly M2 receptors; see refs. 21, 24) respectively, AF102B had higher affinity for the first, as expected from a selective M1 probe (unpublished results). In fact, when compared with the other agonists in this study, AF102B is most selective for M1 receptor sites (15).

Neurochemical and electrophysiological studies. All these experiments are supporting the M1 selectivity of this compound. Thus in contrast to CCh, AF102B did not potentiate phosphoinositides (PI) hydrolysis nor did it inhibit adenylate cyclase (AC) activity in rat cerebral cortex in vitro; however, AF102B blocked CCh-induced activation of PI hydrolysis without altering CCh-induced inhibition of AC (15). This would imply that AF102B, as an M1 probe, is apparently more selective than PZ, since the later shows only a 15 fold preference for M1 sites (e.g. inhibition of CCh-induced stimulation of PI turnover) vs M2 sites (e.g. inhibition of CCh-induced inhibition of AC activity), (25). The remarkable M1 agonistic activity of AF102B was shown in the following studies: i) Both AF102B and McN-A-343 induced only PZ-sensitive depolarization of isolated rabbit superior cervical ganglion (26); ii) AF102B had preferential (40 fold) agonistic activity on potassium-evoked release of ^{3}H-dopamine (possibly M1 mediated) as compared to release ^{3}H-ACh (possibly M2 mediated) in rat striatal and hippocampal synaptosomes, respectively (27); iii) in rat hippocampal slices, AF102B acts primarily to block Ca^{2+}-dependent potassium current underlying the slow after hyperpolarization (AHP). Pirenzepine blocks the effects of AF102B on AHP in a selective pattern indicating that the blockade of AHP induced by AF102B is mediated by M1 receptors (unpublished results). Our results are in agreement with those reported for the M1/M2 receptor classification in hippocampal slices (28).

In line with the above mentioned findings, whole body autoradiography revealed that ^{3}H-AF102B (mice; 2.3 mg/kg, iv) concentrates preferentially in brain regions rich in M1 receptors such as the cerebrum but significantly less in the cerebellum (predominantly M2 receptors). These findings together with pharmacokinetic studies reveal that, regardless of the route of administration, AF102B is capable of crossing the blood-brain barrier at a very short time after peripheral administration and has a rapid plasma absorption (29).

Unlike other well-known cholinergic agonists such as oxotremorine or RS86 (18), AF102B (5 and 50 mg/kg, ip) did not increase significantly the steady-state concentration of ACh and choline (Ch) in mice brain. On the other hand, these doses of AF102B did antagonize the increase in ACh (but not Ch) levels induced by oxotremorine in mice brain (unpublished results). The increased levels of ACh induced by oxotremorine, mainly an M2 agonist, can be attributed to activation of presynaptic M2 autoreceptors which inhibit ACh release (30). Albeit the mechanism of action by which AF102B modulates ACh levels is not clear as yet, these effects of AF102B might be beneficial in the treatment of AD patients in which inhibition of ACh release (by an M2 agonist, for example) might further diminish the activity of hypofunctioning cholinergic system. One possibility that cannot be ignored at this stage is that AF102B exerts its effect on ACh levels indirectly via presynaptic M1 receptors located on cholinergic neurons, or terminals. This mechanism can be inferred in light of the recent suggestion of functional presynaptic M1 receptors on brain cholinergic neurons (31).

Since the etiology of AD is unknown, no homologous animal model capable of modeling all aspects of this neurological disorder has been developed as yet. Under such limitations the evaluation of new drugs for the treatment of AD would require approximate animal models that are mimicking different aspects of this disease. Table 2 summarizes some of the most widely used models and the pros and cons of these models are critically evaluated in a number of reviews (12, 32-34). The effects of AF102B on cognitive functions was evaluated, inter alia, in the following animal models:

Ethylcholine aziridinium (AF64A)-induced cholinotoxicity (3 nmole/ 2 ul/ side, icv) in rats (33). AF64A induces a long-term presynaptic cholinergic hypofunction confined mainly to the hippocampus. In this respect, this model mimicks the cholinergic dysfunction in AD. In this model, AF02B (1-5 mg/kg, ip or po) consistently restored cognitive impairments (inter alia) in a step-through passive avoidance (PA), a Morris water maze (MWM) and an 8-arm radial maze (RAM) task. AF102A (the <u>trans</u> isomer (1 mg/kg,ip) reversed the cognitive impairments in the PA and MWM, but not in the RAM, whereas physostigmine (0.06 and 0.1 mg/kg, ip, for PA and MWM, RAM, respectively), for example, was beneficial only in the PA test (15, 16, 35). Similarly, in the same animal model AF102B is ameliorating memory deficits in a PA test at 0.1-1 mg/kg,ip and 1-5 mg/kg, po (36). In addition, AF102B significantly improved the performance of AF64A-treated rats in a T-maze and RAM task (37).

Scopolamine-induced amnesia in rodents. In this context, cognitive dysfunctions induced by scopolamine in young human volunteers mimic some clinical manifestations of AD patients (38). In rats, AF102B (3-5 mg/kg, ip) restored scopolamine.HBr (0.5 mg/kg, sc)-induced cognitive impairments in a PA task (29). Similar results in a PA task were recently reported in scopolamine-induced amnesia in mice (37).

Aged rats. Aged rats can serve as a useful model for the study of behavioral aspects of brain aging in the human (39). AF102B (1 mg/kg, ip) attenuated aged (19-26 old months) rats associated cognitive impairments in MWM (reference memory) and RAM (working memory) task, respectively (40).

Basal forebrain electrolytic lesion. This animal model mimics to some extent the reported loss in AD of cholinergic cell bodies in the nucleus basalis of Meynert which project to the cortex (2, 12). In this animal model, AF102B (1-5 mg, po) reversed memory impairments in a 2-way shuttle avoidance task (41).

In the animal models tested, AF102B shows a wide therapeutic index. This is reflected by restoration of cognitive functions at much lower (< 1/50-100) doses than required to produce other unwanted central (e.g. tremors, hypothermia, but especially purposeless chewing) or peripheral (e.g. salivation, lacrimation, diarrhea) cholinomimetic effects (Ref.16 and unpublished results). Interestingly, when compared with pilocarpine (10 mg/kg, ip or po) which shows a relative M1 selectivity (42), but induces strong purposeless chewing behavior in rats, AF102B (5, 10 and 50 mg/kg, po) did not induce such a behavior. In this regard, purposeless chewing behavior was suggested as a pharmacological model of dystonic disorders (43). The implication of these results is that AF102B is not expected to produce dystonic reactions in AD patients. On the other hand, cholinergic drugs such as pilocarpine, for example, might show a propensity to induce dystonic reactions in AD patients and this can be considered a serious side-effect.

Taken together, all these results would indicate that it is the M1 selectivity of AF102B which is responsible for the effect of AF102B on learning and memory processes. On the other hand, central effects such as tremors, hypothermia and purposeless chewing, albeit cholinergic, are associated with modulation of muscarinic receptors other than M1, most probably M2 type or subtype (18, 19, 43, 44).

A prerequisite for a drug that has to be administered chronically is that it should not produce tolerance to its beneficial effects. In this regard, repetitive administration of AF102B (0.2 mg/kg/day, ip for three weeks) restored cognitive impairments in AF64A-treated rats in the MWM (working memory paradigm) without a diminution of the effect of the agonist (unpublished results). Similarly, repetitive administration of AF102B in the same animal model was beneficial in RAM and T-maze (37). This might be in line with the findings that, although AF102B is a full agonist on the guinea-pig ileum, no down-regulation of muscarinic receptors (assayed from Bmax and Kd using ^{3}H-QNB as the radioligand) from rat forebrain occured upon chronic daily administration of AF102B (po, 5-100 mg/kg; 14 and 90 days studies), (29). The absence of desensitization or tolerance in both binding and behavioral studies could be attributed to lack of stimulation of AF102B on PI turnover in the rat cerebral cortex. Two studies demonstrating a direct correlation between desensitization and PI turnover in rat brain are described briefly in order to support our hypothesis. Thus, PI turnover induced by full agonists (maximal stimulation of PI turnover; class A) was associated with the desensitization of electrophysiological responses following prolonged ejection periods of the compounds on hippocampal pyramidal cells. On the other hand, partial agonists (poor stimulation of PI turnover; class B) produced stimulatory responses upon prolonged ejection without causing a desensitization (45). A recent study showed that muscarinic desensitization of cell firing in rat hippocampus is related to large increases in PI hydrolysis. All muscarinic agonists tested stimulated firing to a similar extent, but the efficacy for stimulation of PI turnover varied greatly. Desensitization occured only for class A agonists (46). Based on these reports and our results with AF102B, we can hypothesize that a rational drug design of M1 muscarinic agonists should focus especially on those compounds which are poor stimulators of PI turnover, since they are not expected to induce desensitization.

A number of basic questions with regard to the mechanism of action of AF102B both in vitro and in vivo, need further clarifications. Thus the selectivity of AF102B to M1 receptors in vivo could emerge also from distributional and metabolic factors, the existence of different receptor reserves and from variations in efficacy (18). The results in vitro strongly indicate that AF102B is an M1 agonist; therefore the distributional and metabolic factors alone cannot explain the results in vivo. It is still possible that due to different receptor reserves AF102B might show full agonistic activity on certain M1 receptors (perhaps those modulating learning and memory processes) and partial agonistic or even antagonistic activity on other type of functions mediated by peripheral or central muscarinic (M2?) receptors. Albeit this last explanation appears plausible, it still cannot explain why AF102B does not induce mainly M2 agonistic effects (evidenced by effects such as tremors, hypothermia, purposeless chewing). In fact, the receptor reserves for M2 receptors are significantly larger than for M1 receptors (25). A most reasonable explanation would be that AF102B is a unique M1 agonist both in vitro and in vivo, and its structure represents the exact or at least an approximate M1 agonistic pharmacophore.

In summary, based on the results accumulated so far on AF102B, this compound is considered as a promising candidate drug for the treatment of AD and is presently developed towards this challenging goal.

Acknowledgement: This work was supported by Snow Brand Milk Products, Tokyo, Japan.

REFERENCES

1. Bartus, R.T., Dean, R.L., Beer, B. and Lippa, A.S., The cholinergic hypothesis of geriatric memory dysfunction, Science 217:408 (1982).
2. Whitehouse, P.J., Neuronal loss and neurotransmitter receptor alterations in Alzheimer's disease, in: "Alzheimer's and Parkinson's Diseases: Strategies for Research and Development," (A. Fisher, I. Hanin, and C. Lachman, eds.) Plenum Press, New York (1986).
3. Whitehouse, P. J. and Unnerstall, J.R., Neurochemistry of dementia, Eur. Neurol. 28 (suppl.): 36 (1988).
4. Rossor, M. N., and Iversen, L. L., Non-cholinergic neurotransmitter abnormalities in Alzheimer's disease, Brit. Med. Bull. 42:70 (1986).
5. Sims, N. R., Bowen, D. M., Allen, S. J., Smith , C. C. T., Neary, D., Thomas, D.J., and Davison, A.N., Presynaptic cholinergic dysfunction in patients with dementia, J. Neurochem. 40:503 (1983).
6. Mash, D.C., Flynn, D.D. and Potter, L.T., Loss of M2 muscarine receptors in the cerebral cortex in Alzheimer's disease and experimental cholinergic denervation, Science 228:115 (1985).
7. Araujo, D. M., Lapchak, P. A., Robitaille, Y., Gauthier, S., and Quirion, R., Differential alteration of various cholinergic markers in cortical and subcortical regions of human brain in Alzheimer's disease, J. Neurochem. 50:1914 (1988).
8. Rinn e, J. O., Lonnberg, P., Marjamaki, P. and Rinne, U. K., Brain muscarinic receptor subtypes are differently affected in Alzheimer's disease and Parkinson's disease, Brain Res. 483:402 (1989).
9. Probst, A., Cortes, R., Ulrich, J., and Palacios, J. M., Differential modification of muscarinic cholinergic receptors in the hippocampus of patients with Alzheimer's disease: an autoradiographic study, Brain Res. 450:190 (1988).
10. Smith, C. J., Perry, E. K., Perry, R. H., Candy, J. M., Johnson, M., Bonham, J. R., Dick, D.J., Fairbairn, A., Blessed, G., and Birdsall, N. J. M., Muscarinic cholinergic receptor subtypes in hippocampus in human cognitive disorders, J. Neurochem. 50:847 (1988).
11. Waller, S.B., Ball, M.J., Reynolds, M.A. and London, E.D., Muscarinic binding and choline acetyltransferase in postmortem brains of demented patients, Canad. J. Neurol. Sci. 53:528 (1986).
12. Moos, W. H., Davis, R. E., Schwarcz, R.D. and Gamzu, E.R., Cognition activators, Med. Res. Rev. 8:353 (1988).
13. Becker, R. E. and Giacobini, E., Mechanisms of cholinesterase inhibition in senile dementia of the Alzheimer type: clinical, pharmacological, and therapeutic aspects, Drug Dev. Res. 12: 163 (1988).
14. Fisher, A., Karton, I., Heldman, I., Levy, A. and Grunfeld, I., Derivatives of quinuclidine, Israel Patent (1986).
15. Fisher, A., Brandeis, R., Pittel, Z., Karton, I., Sapir, M., Dachir, S., Levy, A., Mizobe, F. and Heldman, E., AF102B: A new M1 agonist with potential application in Alzheimer's disease, Soc. Neurosci. (Abstr.) 13:657 (1987) .
16. Fisher, A., Brandeis, R., Pittel Z., Karton I., Sapir, M., Dachir, S., Levy, A. and Heldman E., (±)Cis-2- methyl-spiro(1,3-oxathiolane-5,3')quinuclidine (AF102B): a new M1

agonist attenuates cognitive dysfunctions in AF64A-treated rats, Neurosci. Lett. 102:325 (1989).

17. Fisher, A., Weinstock, M., Gitter, S. and Cohen, S, A new probe for heterogeneity in muscarine receptors: 2-Methyl-spiro(1,3-dioxolane-4,3')-quinuclidine, Eur. J. Pharmacol. 37:329 (1976).

18. Palacios, J.M., Bolliger, G., Closse, A., Enz, A., Gmelin, G. and Malanowski, J., The pharmacological assessment of RS 86 (2-ethyl-8-methyl-2,8-diazaspiro-[4,5]-decan-1,3-dion hydrobromide). A potent, specific muscarinic acetylcholine receptor agonist, Eur. J. Pharmacol. 125:45 (1986).

19. Pazos, A., Wiederhold, K.H. and Palacios, J.M., Central pressor effects induced by muscarinic receptor agonists: evidence for a predominant role of M2 receptor subtype, Eur. J. Pharmacol. 125:63 (1986).

20. Hammer, R., Berrie, C.P., Birdsall, N.J.M., Burger, A.S.V. and Hulme, E.C., Pirenzepine distinguishes between different subclasses of muscarinic receptors, Nature 283:90 (1980).

21. Tonnaer, J.A.D.M., Van Vugt, M.A., DeBoer, T.H. and DeGraaf, J.S., Differential interactions of muscarinic drugs with binding sites of [^{3}H]pirenzepine and [^{3}H]-quinuclidinyl benzilate in rat brain tissue, Life Sci. 40:1981 (1987).

22. Watson, M., Yamamura, H.I. and Roeske, W.R., [^{3}H]Pirenzepine and (-)-[^{3}H]quinuclidinyl benzilate binding to rat cerebral cortical and cardiac muscarinic cholinergic sites. I. Characterization and regulation of agonist binding to putative muscarinic subtypes, J. Pharmacol. Exptl. Therap. 237:411 (1986).

23. Closse, A., Bittiger, H., Langenegger, D. and Wanner, A., Binding studies with [^{3}H]cis-methyldioxolane in different tissues, N. S. Arch. Pharmacol. 333:372 (1987).

24. Potter, L.T., Flynn, D.D., Hanchett, H.E. and Kalinski, D.L.D., Independent M1 and M2 receptors: ligands, autoradiography and functions, Trends Pharmacol.Sci. (suppl.) 22 (1984).

25. Gil, D.W. and Wolfe, B.B., Pirenzepine distinguishes between muscarinic receptor-mediated phosphoinositide breakdown and inhibitor of adenylate cyclase, J. Pharmacol. Exptl. Therap. 232:608 (1985).

26. Mochida, S., Mizobe, F., Fisher, A., Kawanishi, G. and Kobayashi, H., Selective M1 muscarinic agonists McN-A-343 and AF102B cause dual effects on superior cervical ganglia of rabbits, Brain Res. 45:9 (1988).

27. Ono, S., Saito, Y., Ohgane, N., Kawanishi, G., and Mizobe, F., Heterogeneity of muscarinic autoreceptors and heteroreceptors in the rat brain: effects of a novel M1 agonist, AF102B, Eur. J.Pharmacol. 155:77 (1988).

28. Dutar P. and Nicoll R.A., Classification of muscarinic responses in hippocampus in terms of receptor subtypes and second-messenger systems: electrophysiological studies in vitro, J. Neurosci. 8:4214 (1988).

29. Fisher, A., Brandeis, R., Pittel Z., Karton I., Sapir, M.,Dachir, S., Levy, A. and Heldman E., AF102B: a potential drug for the treatment of Alzheimer's disease; further characterization, Soc. Neurosci. (Abstr.) 14:905 (1988).

30. Consolo, S., Ladinsky, M., Vinci, R., Palazzi E. and Wang J.X., An in vivo pharmacological study on muscarinic receptor subtypes regulating cholinergic neurotransmission in rat striatum, Biochem. Pharmacol. 36:3075 (1987).

31. Pittel, Z., Heldman, E., Rubinstein, R. and Cohen S, Synaptosomal acetylcholine release is modulated by M1 and M2 mACh receptors, The Fourth International Symposium on Subtypes of Muscarinic Receptors. Weisbaden, W. Germany, July 20-22, Abstr. # 29 (1989).
32. Fisher, A. and Hanin, I., Potential animal models for senile dementia of Alzheimer's type with particular emphasis on AF64A-induced cholinotoxicity. Ann. Rev. Pharmacol. Toxicol. 26:161 (1986).
33. Fisher, A. and Hanin, I., The AF64A-treated rat as an experimental model for Alzheimer's disease: a critical evaluation, in: "Alzheimer's and Parkinson's Disease: Strategies for Research and Development," (A. Fisher, I. Hanin and C. Lachman, eds.) Plenum Press, New York (1986).
34. Smith, G., Animal models of Alzheimer's disease: experimental cholinergic denervation, Brain Res. Rev. 13: 103 (1988).
35. Brandeis, R., Pittel, Z., Lachman, C., Heldman, E., Luz, S., Dachir, S., Levy, A., Hanin, I. and Fisher, A., AF64A-induced cholinotoxicity: behavioral and biochemical correlates, in: "Alzheimer's and Parkinson's Diseases: Strategies for Research and Development" (A. Fisher, I. Hanin, and C. Lachman, eds) Plenum Press, New York (1986).
36. Nakahara, N., Iga, Y., Mizobe, F., and Kawanishi G., Amelioration of experimental amnesia (passive avoidance failure) in rodents by the selective M1 agonist AF102B, Japan. J. Pharmacol. 48:502 (1988).
37. Nakahara, N., Iga, Y., Saito, Y., Mizobe, F. and Kawanishi, G., Beneficial effects of FKS-508 (AF102B), a selective M1 muscarinic agonist, on the impaired working memory in AF64A-treated rats, Japan. J. Pharmacol. 51:539 (1989).
38. Drachman, D.A. and Leavitt, J.L., Human memory and the cholinergic system. A relationship to aging? Arch. Neurol. 30: 113 (1974).
39. Schuurman, T., Horvath, E., Spencer, D.G., Jr. and Traber, J., Old rats: an animal model for senile dementia in: Senile Dementias: Early Detection (A. Bes, J. Cahn, S. Hoyer, J.P. Marc-Vergnes, and H.M. Wisniewski, eds) John Libbey Eurotext, London, Paris (1986).
40. Brandeis, R., Dachir, S., Sapir, M., Levy, A. and Fisher, A., Reversal of age-related cognitive impairments by an M1 cholinergic agonist, AF102B, Pharmacol. Biochem. Behav., accepted for publication.
41. Iga, Y., Nakahara, N., Mizobe, F., Ohgane, N., Sawai, T. and Kawanishi, G., SND-5008 (AF102B) a novel muscarinic agonist improves experimental cognitive dysfunction in rodent, Soc. Neurosci. (Abstr.) 14:57 (1988).
42. McCormick, D.A. and Prince, D.A., Two types of muscarinic response to acetylcholine in mammalian cortical neurons, Proc. Natl. Acad. Sci., USA 82:6344 (1985).
43. Stewart, B.R., Jenner, P. and Marsden, C.D., The pharmacological characterization of pilocarpine-induced purposeless chewing behavior in the rat, Psychopharmacol. 96:55 (1988).
44. Stewart, B.R., Jenner, P. and Marsden, C.D., Assessment of the muscarinic receptor subtype involved in the mediation of pilocarpine-induced purposeless chewing behaviour, Psycharmacol. 97:228 (1989).
45. Lippa, A.S., Crichett, D.J. and Joseph, J.A., Desensitization of muscarinic acetylcholine receptors: a possible relation to receptor heterogeneity and phosphoinositides, Brain Res. 366: 98 (1986).

46. Crews, F.T., Pontzer, N.J. and Chandler, L.J., Muscarinic cholinergic neurotransmission: receptor subtypes and signal transduction, First Ann. Suncoast Workshop on the Neurobiology of Aging, St. Petersburg, Fl., USA, Feb 26-March 1 (1989).

BEHAVIORAL AND NEUROCHEMICAL EFFECTS OF AF64A IN YOUNG AND OLD FISCHER 344 MALE RATS

C.E. Tedford, S.A. Lorens, J.C. Corey, D. Lokhorst, G. Kindel, E. Wülfert[1], and I. Hanin

Department of Pharmacology, Loyola University of Chicago
2160 South First Avenue, Maywood, IL 60153
[1]UCB-Pharmaceutical Sector, Chemin du Foriest, B-1420
Braine-l'Alleud, Belgium

INTRODUCTION

Normal senescence in humans and in animals is accompanied by losses in cognitive function, and by learning and memory deficits (3, 10, 16, 38, 48). In addition, increasing age has been associated with decreases in locomotor and exploratory activity (17, 52).

The cholinergic hypothesis of geriatric memory dysfunction (3) has implicated the cholinergic system in age-related impairments in cognitive function. Results from several studies have demonstrated that anticholinergic drugs, such as scopolamine, induce learning and memory impairments in animals (1), as well as in humans (41, 42). Furthermore, pharmacological enhancement of cholinergic neurotransmission can restore impaired cholinergic function and reverse cognitive deficits in animals (1, 14, 15). However, as reviewed by Decker (9), the relationship between age-related dysfunctions in learning and memory, on the one hand, and cholinergic hypofunction, on the other, is not entirely clear.

Cholinergic hypofunction has been implicated not only in age-related cognitive impairments (3), but also in diseases that result in losses in learning and memory capability (28). In senile dementia of the Alzheimer type (SDAT), one of the most consistent findings is a reduction in choline acetyltransferase (ChAT) activity in the neocortex and hippocampus (5, 8). Numerous studies have demonstrated, moreover, that presynaptic cholinergic markers are reduced in SDAT, indicating that cholinergic neurotransmission is impaired (6, 8, 37). SDAT also is associated with a loss of cholinergic neurons that project to the neocortex and hippocampus, areas that play an important role in cognitive function (22, 28, 51).

Our laboratory has been interested in the neurochemical and behavioral effects produced by the cholinotoxin, 1-ethyl-1-(2-hydroxyethyl) aziridinium chloride (AF64A; 12, 18). The bilateral i.c.v. administration of AF64A reduces hippocampal ACh content, ChAT activity, acetylcholinesterase (AChE) activity, high affinity choline transport

Basic, Clinical, and Therapeutic Aspects of Alzheimer's and Parkinson's Diseases, Volume 2
Edited by T. Nagatsu *et al.*, Plenum Press, New York, 1990

(HAChT) (11, 13, 20, 21, 25, 27), as well as acetylcholine (ACh) release from hippocampal slices (24). Moreover, rats treated with AF64A show deficits in the acquisition of a passive avoidance response, and in radial-arm and water maze performance (4, 7, 32), as well as elevations in locomotor and exploratory activity (2, 50). These results suggest that the AF64A-treated animal may provide a model for studying drugs which may reverse the cholinergic dysfunction and cognitive deficits seen in SDAT.

All of the AF64A studies cited above have been conducted in young adult rats. In the present studies we investigated the effects of AF64A on aged and young animals. Comparisons were made between the age-related behavioral deficits associated with normal senescence, and the behavioral deficits induced by the i.c.v. administration of AF64A. We also examined the neurochemical cholinergic changes seen with normal senescence and those produced by AF64A.

METHODS

ANIMALS

Young (4-6 mo) and aged (20-22 mo) F344 male rats (Harlan, Indianapolis), weighing 343-420 g at the time of surgery, were used. The animals were housed individually, and maintained on a 12 hr light/dark cycle. Food and water were available <u>ad libitum</u>.

SURGERY

All animals were anesthetized with pentobarbital sodium (50 mg/kg, i.p.), treated with ampicillin (50 mg/kg, i.m.), then placed in a Kopf stereotaxic instrument with the incisor bar set 3.3 mm below the interaural plane. AF64A (0.5, 1.0 and 1.5 nmole) or vehicle (1.5 ul), was prepared daily (12). Vehicle consisted of pH-adjusted H_2O, prepared in the same manner as AF64A. AF64A or vehicle was injected into each lateral ventricle using the following coordinates: 0.8 mm caudal, $\pm$ 1.5 mm lateral, and 3.6 mm ventral to bregma (34). AF64A was infused (1.0 ul/min) via 28 gauge cannulae connected to a CMA/100 microsyringe pump (Carnegie Medicin AB; Solna, Sweden). Following completion of the drug or vehicle infusions, the cannulae were left <u>in situ</u> for 5.0 min before being removed.

Bilateral electrolytic lesions (n=9) of the cingulum and cingulate gyrus were produced by passing 3.0 mA DC for 20 sec between an intracranial cathode (a 26 stainless steel needle insulated with Epoxylite except for 2.0 mm at its tip) and an anode clipped to the wound margin. The drugs and stereotaxic coordinates employed were the same as for the i.c.v. AF64A injections. Sham operated rats (n=7) were treated in the same manner as the lesioned rats, except that an electrode was not lowered intracranially.

EXPERIMENTAL PROTOCOL

Two experiments were performed. In the first, the dose-dependent behavioral, neurochemical and histological effects of AF64A were examined using both young and aged F344 rats. In the second, the behavioral and histological effects of electrolytic cingulate lesions were assessed using only young F344 rats.

Behavioral testing was initiated 28 days postoperatively, and following 5-7 days of handling (4-5 min/day). This postsurgical interval was chosen since, by this time period, AF64A-induced cholinergic deficits are maximal and relatively stable, whereas transient noncholinergic effects are minimal (20, 21). The AF64A treated animals were tested using three behavioral paradigms, in order to evaluate lesion effects on locomotor activity, exploratory behavior, and spatial learning capability. The electrolytic cingulate lesioned rats were examined using only the latter two paradigms. Behavioral testing was conducted over a 5 day period. The locomotor and exploratory test were conducted during the morning and afternoon, respectively, on day 1. One-way avoidance acquisition and retention were examined on days 2 and 5, respectively.

One week following completion of behavioral testing the AF64A and vehicle treated animals were sacrificed instantaneously, using a guillotine. The neostriatum, medial frontal cortex and left hippocampus were obtained (19) and stored at -80°C prior to biochemical analysis. The right hippocampus was immediately homogenized and utilized for HAChT determinations. Coronal blocks (4.0 mm thick) of brain tissue immediately surrounding the i.c.v. injection sites were fixed in buffered formalin for subsequent histological evaluation. The electrolytic cingulate lesion and sham operated rats were anesthetized with pentobarbital sodium (65 mg/kg, i.p.) and perfused transcardially with 50 ml isotonic saline followed by 100 ml buffered formalin.

NEUROCHEMICAL ANALYSIS

High Affinity Choline Transport. HAChT was determined according to Yamamura and Snyder (53). It was defined as the amount of choline transported into tissue in the presence of Na^+, minus that accumulated in the absence of Na^+ (25, 53).

Choline Acetyltransferase and Acetylcholinesterase Activities. The hippocampal, cortical and neostriatal ChAT and AChE activities were measured as reported by Leventer et al. (25).

BEHAVIORAL TESTING

Locomotor Activity. Activity level was measured in enclosed cylindrical photocell chambers with wire mesh floors as described previously (33). The animals were placed individually in a chamber for 3.0 h, and their activity scores recorded.

Open Field Activity. The open field measured 50 cm x 50 cm x 40 cm high. The floor was painted flat white and divided into nine squares (16.7 x 16.7 cm) by thin black lines. Three equidistantly spaced holes (3.8 cm diameter) were located in the center square. The open field was located in a sound attenuated dark room and was illuminated by a 15 W bulb positioned 100 cm above the center of the chamber, as well as a 20 W fluorescent tube located under the floor of the chamber. The animals were placed in the chamber for 21 min and their behavior videotaped for subsequent analysis. The number of squares entered, rears (rat stands on hind legs and sniffs), and nose pokes (rat inserts snout at least 1.0 cm into one of the holes) were counted.

One-Way Active Avoidance. Acquisition of an unsignalled (5.0 sec shock delay) one-way avoidance task was tested as detailed previously (26).

Testing was continued until criterion (10 consecutive avoidance responses) was attained. Escape responses and returns to the "shock" compartment were scored as "errors". All rats achieved the learning criterion in a single session, 30-45 min. in duration. Retention of the one-way avoidance task was examined 72 h later. The rats were treated in the same manner as during the acquisition phase, and tested until they made 10 consecutive avoidance responses.

HISTOLOGY

The blocks of brain tissue containing the i.c.v. injections and electrolytic lesions were fixed in buffered formalin for at least two weeks prior to sectioning. Frozen coronal sections (50 um) were cut using a sliding microtome. Every fifth section was float mounted onto gelatin-chrom alum-coated slides, and stained by the cresyl violet acetate method. The histological damage produced by the vehicle and AF64A injections and by the electrolytic lesions was assessed microscopically.

STATISTICS

The data presented in the figures and tables are expressed as group means $\pm$ the standard errors of the means. The AF64A lesion data were analyzed using a two-way analysis of variance. Duncan's multiple range test was used for post hoc group comparisons when significant differences were observed. The cingulate lesion data were analyzed using Student's t-test (two-tailed) for independent means. Group differences were considered significant if they reached the 0.05 level of confidence.

RESULTS

EXPERIMENT I: AF64A STUDY

The animals lost weight immediately post-operatively. However, by the time behavioral testing was initiated all rats appeared healthy and had regained their preoperative weights. At the time of sacrifice the rats weighed 354-395 g.

NEUROCHEMISTRY. No age-related differences in HAChT, ChAT or AChE were observed between the young and old vehicle treated rats in any of the brain regions assayed (Table 1). The AF64A-treated young and old rats showed a dose-dependent reduction in hippocampal HAChT [$F(3,59)=19.39$, $p < 0.001$], ChAT activity [$F(3,59)=31.48$, $p < 0.001$], and AChE activity [$F(3,59)=4.08$, $p < 0.001$]. The magnitudes of the reductions (44-60%) in hippocampal cholinergic markers were similar in both the young and aged AF64A-treated rats.

Effects of AF64A administration on cholinergic parameters in other brain regions were either small or non-significant (Table 1). No significant reductions in ChAT or AChE activity were seen in the neostriatum. Small but significant reductions in medial frontal cortical ChAT and AChE were seen. However, the magnitude of these reductions was much less in the cortex than in the hippocampus. These results suggest that, in both young and old F344 rats, administration of AF64A primarily affects hippocampal cholinergic function.

HISTOLOGY. The vehicle injection left a 0.25 mm diameter glial scar through the cingulate gyrus, cingulum, and corpus callosum. Increasing

TABLE 1

DOSE-DEPENDENT EFFECTS OF AF64A ON CHOLINERGIC PARAMETERS IN YOUNG AND OLD FISCHER 344 RATS

	HAChT HIPPOCAMPUS	ChAT HIPPOCAMPUS	ChAT STRIATUM	ChAT FRONTAL CORTEX	AChE HIPPOCAMPUS	AChE STRIATUM	AChE FRONTAL CORTEX
YOUNG-VEH	5.4 ± 0.2	19.3 ± 0.9	58.6 ± 2.7	13.0 ± 0.4	1317 ± 38	6468 ± 199	956 ± 33
YOUNG-0.5	4.6 ± 0.3	16.3 ± 1.1	56.1 ± 2.5	12.0 ± 0.4	1134 ± 59	6083 ± 382	827 ± 51
YOUNG-1.0	4.5 ± 0.3	12.8 ± 1.2*	60.0 ± 2.2	11.7 ± 0.3	937 ± 86*	6992 ± 401	772 ± 35*
YOUNG-1.5	3.0 ± 0.4*	8.1 ± 0.4*	55.9 ± 1.9	11.4 ± 0.3*	524 ± 31*	6064 ± 191	707 ± 46*
OLD-VEH	5.4 ± 0.2	19.6 ± 0.8	53.0 ± 2.2	12.6 ± 0.4	1283 ± 29	5806 ± 157	811 ± 35
OLD-0.5	4.7 ± 0.3	16.4 ± 1.0	50.7 ± 2.7	12.2 ± 0.4	1109 ± 56	6216 ± 279	750 ± 37
OLD-1.0	4.5 ± 0.4	12.5 ± 1.0*	47.9 ± 2.2	11.8 ± 0.6	799 ± 84*	6021 ± 216	656 ± 58*
OLD-1.5	3.0 ± 0.4*	11.9 ± 1.2*	57.2 ± 3.8	11.4 ± 0.4	739 ± 86*	6241 ± 380	716 ± 57

* Indicates significant difference from age-matched vehicle-treated group, $p < 0.05$.

n = 6 - 8 animals/group.

HAChT = pmoles/mg protein/8.0 min.

ChAT and AChE = nmoles/mg protein/30 min.

VEH = vehicle-treated rats. AF64A (0.5, 1.0 and 1.5 nmoles/ventricle) was bilaterally administered i.c.v.

doses of AF64A led to ventricular enlargement. Some non-specific tissue damage marked by cortical cavitation and transection of the cingulum and corpus callosum occurred at the highest dose used. The maximal extent of the histological damage approximated 1.5 mm in the anteroposterior plane and 2.0 mm dorsoventrally. The septum, caudate-putamen, and dorsal hippocampus were <u>not</u> affected. No age-related differences were apparent.

LOCOMOTOR ACTIVITY (Figure 1). The aged control rats were hypoactive in comparison to the young control rats [$F(1,60)=46.75$, $p < 0.001$]. Furthermore, the young rats which received 1.5 nmole/side of AF64A, showed a significant increase in locomotor activity which was not observed in the old animals [$F(3,60)=5.58$, $p < 0.002$].

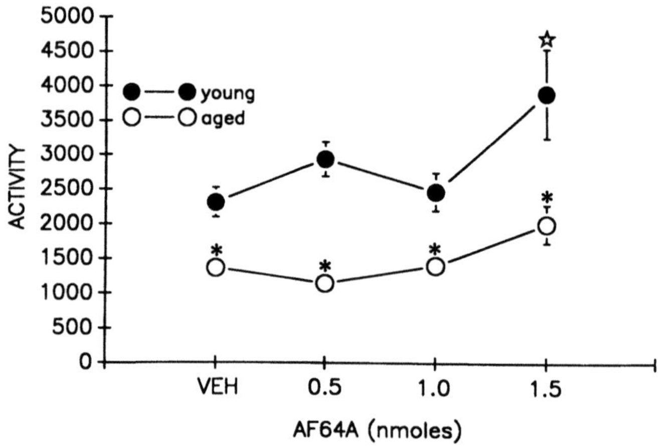

Figure 1

Effect of AF64A (i.c.v., bilateral) on locomotor activity in photocell chambers. Total activity was recorded for 3.0 h. Doses refer to concentrations of AF64A administered per lateral ventricle. (n) = 6 - 8 animals/group. (*) Indicates a significant ($p < 0.05$) age difference between similar treatment groups. Open star indicates a significant ($p < 0.05$) AF64A-induced difference from age-matched controls.

OPEN FIELD/HOLE BOARD ACTIVITY. Young and old vehicle-treated rats did not differ with respect to the number of squares entered, rears, and nose pokes (Table 2). The young and old rats differed, however, in their response to AF64A. Young AF64A-treated rats demonstrated significant dose-dependent increases in the number of squares entered, rears, and nose pokes [$F(3,52)=4.22$, $p < 0.01$; $F(3,52)=7.06$, $p < 0.001$; $F(3,52)=11.13$, $p < 0.001$; respectively). In contrast, only the old animals treated with 1.5 nmoles/ventricle of AF64A showed a significant increase in the total number of nose pokes and rears.

ONE-WAY ACTIVE AVOIDANCE CONDITIONING AND RETENTION (Figure 2). In comparison to the young rats, the old vehicle-treated animals were deficient in the acquisition of the one-way active avoidance response. The old vehicle-treated rats required more trials before emitting their first avoidance response [$F(1,52)=32.48$, $p < 0.001$], made more errors [$F(1,52)=30.82$, $p < 0.001$], and required a greater number of trials to reach criterion [$F(1,52)=5.60$, $p < 0.02$]. AF64A treatment produced a

TABLE 2

DOSE-DEPENDENT EFFECTS OF AF64A ON OPEN FIELD/HOLE BOARD ACTIVITY

	TOTAL SQUARES	REARS	NOSE POKES
YOUNG-VEH	36.4 ± 4.8	15.9 ± 3.0	8.3 ± 1.8
YOUNG-0.5	56.5 ± 3.4	49.5 ± 9.2*	15.9 ± 3.0
YOUNG-1.0	90.0 ± 10.4*	50.1 ± 5.0*	21.8 ± 2.9*
YOUNG-1.5	111.6 ± 26.6*	55.0 ± 10.0*	29.4 ± 4.5*
OLD-VEH	27.3 ± 4.7	22.3 ± 3.5	3.9 ± 0.8
OLD-0.5	30.2 ± 5.1	24.1 ± 4.1	5.9 ± 1.3
OLD-1.0	12.3 ± 3.2	15.5 ± 1.3	4.3 ± 1.5
OLD-1.5	32.0 ± 7.9	45.7 ± 8.0*	13.3 ± 2.1*

* Indicates significant difference from age-matched vehicle-treated group, $p < 0.05$.

n = 6 - 8 animals/group.

VEH = vehicle-treated group. The different doses (0.5, 1.0, and 1.5 nmoles) of AF64A were administered i.c.v. bilaterally.

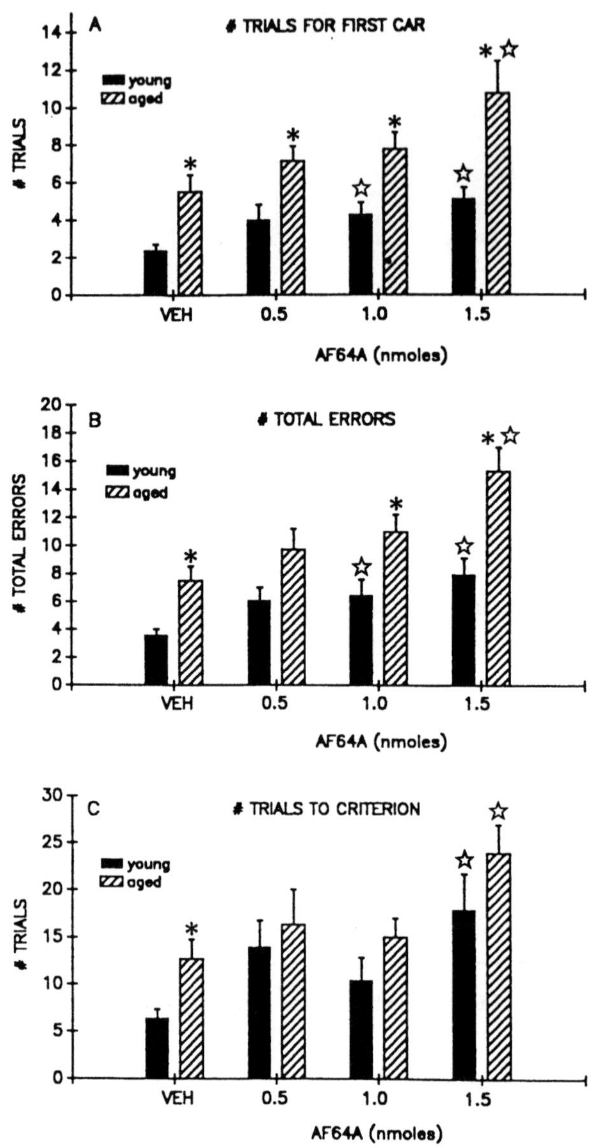

Figure 2

Effect of AF64A (i.c.v., bilateral) on one-way active avoidance acquisition in young and aged F344 rats.

A: Number of trials required before first conditioned avoidance response (CAR) emitted.

B: Total number of errors (escape responses plus returns to the "shock" compartment).

C: Number of trials to criterion.

Doses refer to concentrations of AF64A administered per lateral ventricle.

(n) = 6 - 8 animals/group.

(*) Indicates a significant ($p < 0.05$) age difference between similar treatment groups.

Open star indicates a significant ($p < 0.05$) AF64A-induced difference from age-matched controls.

dose-dependent impairment in the acquisition of the conditioned avoidance response in both the young and old rats. No differences were found between young and old vehicle and AF64A treated rats during retention testing conducted 72 h later, at which time all rats reached criterion after making only 1-4 errors.

EXPERIMENT II: ELECTROLYTIC CINGULATE LESION STUDY

Aspiration of the cingulate gyrus has been reported to impair the acquisition of a two-way conditioned avoidance response as well as spatial learning and memory (43, 45). Thus, this experiment was performed using young F344 male rats in order to determine whether the non-specific damage to the cingulate gyrus, cingulum, and corpus callosum associated with the higher i.c.v. doses of AF64A contributed to the AF64A-induced behavioral deficits described above. The electrolytic lesions produced cavitation and glial scarring in the cingulate gyrus, cingulum transection, and damage to the corpus callosum which was similar to, or greater in extent than that observed following the higher doses of AF64A. The cingulate electrolytic lesions, however, did not significantly affect the young animals' open field behavior or acquisition of the one-way avoidance response (Table 3). These data indicate that the behavioral effects of i.c.v. administered AF64A are not due to non-specific cortical, cingular, or callosal damage.

TABLE 3

BEHAVIORAL EFFECTS OF ELECTROLYTIC CINGULATE LESIONS

OPEN FIELD/HOLE BOARD ACTIVITY

GROUP	(N)	TOTAL SQUARES	REARS	NOSE POKES
SHAM-OPERATED	(7)	57.0 ± 7.5	27.0 ± 3.0	9.0 ± 2.0
CINGULATE LESION	(9)	52.2 ± 12.0	31.1 ± 5.0	9.2 ± 2.2

ONE-WAY ACTIVE AVOIDANCE ACQUISITION

GROUP	(N)	1ST TRIAL ESCAPE LAT (SEC)	# TRIALS 1ST CONDITIONED AVOIDANCE-RESPONSE	TOTAL ERRORS	TRIALS TO CRITERION
SHAM-OPERATED	(7)	4.5 ± 2.2	2.7 ± 0.3	3.1 ± 0.3	5.1 ± 1.5
CINGULATE LESION	(9)	5.7 ± 1.3	4.0 ± 0.5	4.8 ± 0.8	6.0 ± 1.6

DISCUSSION

The present studies utilized young and old F344 male rats, and compared their susceptibility to the neurochemical and behavioral effects of bilateral AF64A administration i.c.v. The cholinergic parameters assayed in the hippocampus, neostriatum and frontal cortex did not reveal any differences between the young and old rats. These findings suggest that no dramatic biochemical changes occur in forebrain cholinergic neurons during normal aging in F344 rats. The dose-dependent reductions in hippocampal cholinergic neurochemical parameters seen after i.c.v. AF64A administration were equivalent in both age groups (Table I). These results indicate that, from a neurochemical perspective, the old F344 rat is not any more susceptible to the cholinotoxic effects of AF64A than the young F344 rat.

The cholinotoxic effects of AF64A were restricted to the hippocampus. Significant cholinergic changes were not observed in the neostriatum. Reductions in ChAT and AChE activities were observed in the medial frontal cortex within the same groups. However, the magnitude of the cortical reduction was never as substantial as in the hippocampus. These results are in agreement with earlier reports (7, 11, 12, 20, 21, 24, 25, 27) which indicate that i.c.v. administration of AF64A leads primarily to cholinergic deficits in the hippocampus.

The young vehicle-treated animals were more active in the photocell chambers than the old vehicle-treated rats. This observation is in agreement with previous reports (17, 52) indicating reductions in locomotor activity with increasing age. In contrast, no difference in the open field behavior of the young and old control rats was discerned. This apparent discrepancy may be due to the length of the test session (3.0 h exposure to the photocell chambers, versus the 21 min exposure to the open field). In agreement with earlier reports (17, 52), we have recently found that old (21-22 mo) F344 male rats exhibit less exploration than young rats (6-7 mo) in a larger more brightly lit open field.

AF64A enhanced the locomotor and exploratory behavior of the young animals in a dose-dependent manner. Increased activity in the photocell chambers was observed after the 1.5 nmole/ventricle dose (Fig. 1), and significantly increased ambulation (squares entered), rearing, and investigatory behavior (nose poking) in the open field was seen after the 1.0 nmole/ventricle dose of AF64A (Table 2). In contrast, AF64A failed to alter the locomotor activity of the old rats as measured in the photocell chambers, and enhanced rearing and nose poking in the open field/hole board only after the 1.5 nmole/ventricle dose. These observations suggest that old F344 male rats are less sensitive to the behavioral effects of AF64A than young animals.

The old vehicle-treated F344 rats acquired the one-way conditioned avoidance response (CAR) more slowly than the young animals (Fig. 2). This finding is consistent with the reports of Doty (10) and Goodrick (16) who demonstrated that aged rats are impaired in the acquisition of a discriminated avoidance and a T-maze task. These investigators also found that once the aged animals had acquired the task, long term retention was not impaired. No differences were observed between the young and old F344 rats during the retention phase of the one-way avoidance conditioning paradigm.

AF64A (1.5 nmole/ventricle dose) not only impaired one-way conditioned avoidance response (CAR) acquisition in the young animals, but produced additional deficits in the old rats. This impairment was manifested by an increase in the number of trials before the first CAR was emitted, by an increase in the number of trials before the criterion of 10 consecutive CARs was reached, and by an increase in the total number of errors committed (Fig. 2). Interestingly, as with the age-related deficits, none of the AF64A-treated animals evidenced impaired retention upon retesting 72h later.

Comparing both the neurochemical and behavioral studies, our findings indicate that the age-related and AF64A-induced behavioral effects appear to be independent. If one were to consider only the behavioral data, one might infer that AF64A produces a deficit in CAR acquisition that is similar to that seen with normal aging. Furthermore, if one were to use only young animals treated with AF64A, the reduction in hippocampal biochemical cholinergic markers in young rats would be observed to parallel the increase in acquisition deficits seen. This interpretation would be consistent with Bartus et al. (3), and the cholinergic hypothesis of geriatric memory dysfunction. Neurochemical differences, however, were not observed between the young and old untreated F344 rats, and the hippocampal cholinergic deficits observed following AF64A administration were equivalent in the young and old rats. Thus, one might conclude that the behavioral deficits seen in the old control animals are independent of hippocampal cholinergic hypofunction. It must be stated, however, that not all indices of cholinergic function were assessed in the present study. Changes in ACh turnover and muscarinic receptor densities have been observed during normal senescence (30, 35, 36, 38, 39, 40, 46). Thus, we cannot exclude the possibility that the behavioral deficits observed in the old control rats are due to an impairment in cholinergic neurotransmission. Deficits in receptor density or decreased in ACh release may represent a more subtle loss in the control of cholinergic dynamics and may be more representative of normal aging. Moreover, the loss in biochemical markers seen with AF64A may represent destruction and loss of cholinergic neurons, as seen in Alzheimer's disease.

Histological analyses also were conducted in order to define the specific effects of AF64A, and to exclude effects due to non-specific tissue damage. We found that the necrotic effects of AF64A were confined to the corpus callosum, cingulum and cingulate gyrus, and that they appeared to be due to spread of the drug dorsally along the cannula track. Aspiration of the entire cingulate gyrus has been reported to impair the acquisition of a two-way conditioned avoidance response, as well as spatial learning and memory (43, 45). We thus assessed whether the non-specific damage to the cingulate gyrus, cingulum, and corpus callosum which we observed with the higher i.c.v. doses of AF64A might have contributed to the AF64A-induced behavioral deficits described above. Electrolytic cingulate lesions in young F344 rats, which produced histological damage similar to, or greater than that associated with the higher doses of i.c.v. AF64A administration, did not affect open field behavior or one-way avoidance conditioning. Thus, the AF64A induced behavioral changes observed in the present study cannot be ascribed to non-selective cingulate damage. Importantly, non-selective tissue damage was <u>not observed</u> in these studies in the dorsal hippocampal formation, posterior septal area, or neostriatum following AF64A administration.

SUMMARY

Several key observations can be drawn from the present results:

1) Cholinergic deficits in both old and young rats may underlie the impaired one-way (spatial) CAR acquisition induced by AF64A.

2) Reductions in hippocampal cholinergic parameters following AF64A administration are comparable in both old and young rats, indicating no differential susceptibility with age to neurochemical changes resulting from neurotoxic insult.

3) Old, control animals function less efficiently than their young counterparts, in behavioral tests of locomotor activity and one-way active avoidance conditioning.

4) AF64A administration in young rats induces significant changes in CAR acquisition, locomotor and exploratory behavior. In old rats, on the other hand, locomotor and exploratory behavior exhibit little change following AF64A administration. These observations indicate differential susceptibility, with age, to behavioral changes induced by neurotoxic insult to the cholinergic system.

The combined data suggest that some behavioral and neurochemical changes produced by i.c.v. AF64A administration are distinct from those associated with the normal aging process. Thus, the aged AF64A-treated animal might serve as a useful animal model by which to study the behavioral, neurochemical and histological consequences of superimposing cholinergic hypofunction on the normal aging process.

ACKNOWLEDGEMENTS

The work reported in this paper has been supported by NIMH Grant #MH42572, and by the UCB s.a. Pharmaceutical Sector, Brussels.

REFERENCES

1. Aigner, T. G. and M. Mishkin. Behavioral and Neural Biology 45: 81-87, 1986.
2. Bailey, E. L., D. H. Overstreet and A. D. Crocker. Behavioral and Neural Biology 45: 263-274, 1986.
3. Bartus, R. T., R.L. Dean III, B. Beer and A. S. Lippa. Science 217: 408-417, 1982.
4. Blaker, W. D. and S. D. Goodwin. Pharmacology Biochemistry and Behavior 28: 157-163, 1987.
5. Bowen, D. M., C. B. Smith, P. White and A. N. Davison. Brain 99: 459-496, 1976.
6. Bowen, D. M., N. R. Sims, J. S. Benton, G. Curzon and A. N. Davison. New England Journal of Medicine 305: 1016, 1981.
7. Chrobak, J. J., I. Hanin, D. E. Schmechel and T. S. Walsh. Brain Research 463: 107-117, 1988.
8. Davies, P. and A. J. F. Maloney. The Lancet Dec.25: 1403, 1976.
9. Decker, M. W. Brain Research Reviews 12: 423-438, 1987.
10. Doty, B. A. Journal of Gerontology 21: 287-290, 1966.
11. Eva, C., M. Fabrazzo and E. Costa. Journal of Pharmacology and Experimental Therapeutics 241 (1): 181-186, 1987.
12. Fisher, A. and I. Hanin. Ann. Rev. Pharmacol. Toxicol. 26: 161-181, 1986.
13. Fisher, A., C. R. Mantione, D. J. Abraham and I. Hanin. Journal of Pharmacology and Experimental Therapeutics 222: 140-145, 1982.
14. Flood, J. F., G. E. Smith and A. Cherkin. Pharmacology Biochemistry and Behavior 21: 169-173, 1984.
15. Flood, J. F., G. E. Smith and A. Cherkin. Psychopharmacology 86: 61-67, 1985.
16. Goodrick, C. L. Journal of Gerontology 23: 298-304, 1968.
17. Goodrick, C. L. Journal of Gerontology 26: 58-62, 1971.

18. Hanin, I., A. Fisher, H. Hörtnagl, S. M. Leventer, P. E. Potter and T. J. Walsh. In: Psychopharmacology: The Third Generation of Progress, Meltzer H.Y. (Ed.), Raven Press, N.Y., 1987, pp.341-349.
19. Heffner, T., J. A. Hartman and L. S. Seiden. Pharmacology Biochemistry and Behavior 13: 453 - 456, 1980.
20. Hörtnagl, H., P. E. Potter and I. Hanin. Neuroscience 22: 203-213, 1987.
21. Hörtnagl, H., P. E. Potter and I. Hanin. Brain Research 421: 75-84, 1987.
22. Hyman, B. T., G. W. VanHoesen, A. R. Damasio and C. L. Barnes. Science 225: 1168-1170, 1984.
23. Kozlowski, M. R. and R. E. Arbogast. Brain Research 372: 45 - 54, 1986.
24. Leventer, S., D. McKeag, M. Clancy, E. Wülfert and I. Hanin. Neuropharmacology 24: 453-459, 1985.
25. Leventer, S. M., E. Wülfert and I. Hanin. Neuropharmacology 26:361-365, 1987.
26. Lorens, S. A., H. C. Guldberg, K. Hole, C. Kohler and B. Srebro. Brain Research 108: 97-113, 1976.
27. Mantione, C. R., A. Fisher and I. Hanin. Life Sciences 35: 33-41, 1984.
28. McGeer, P. L., E. G. McGeer, J. Suzuki, C. E. Dolman and T. Nagai. Neurology 34: 741-745, 1984.
29. McGurk, S. E., S. L. Hartgraves, P. H. Kelly, M. N. Gordon and L. L. Butcher. Neuroscience 22: 215-224, 1987.
30. Meyer, E. M., E. St. Onge and F. T. Crews. Neurobiology of Aging 5: 315-317, 1984.
31. Mesulam, M. M. and VanHoesen, G. W. Brain Research 109: 152-157, 1976.
32. Moos, W. H., R. E. Davis, J. J. Kinsora, R. D. Schwarz, M. E. Smith and C. J. Spencer. (In press).
33. Paris, J. P. and S. A. Lorens. Behavioral Brain Research 26: 139-151, 1987.
34. Paxinos, G. and C. Watson. In: The Rat Brain in Stereotaxic Coordinates. 2nd Edition, Academic Press, Australia, 1986.
35. Pedigo Jr., N. W., D. M. Lerna and T. N. Krumrei. Neurobiology of Aging 5: 227-233, 1984.
36. Rinne, J. O. Brain Research 404: 162-168, 1987.
37. Rylett, R. J., M. J. Ball and E. H. Colhoun. Brain Research 289: 169-175, 1983.
38. Schuurman, T., E. Horvath, D. G. Spencer Jr. and J. Traber. In: Senile Dementia: Early Detection, A. Bes et. al.,(Eds.) J. Libby Eurotext, 1986.
39. Stein, D. G. In: Aging of the Brain, D. Samuel (Ed.), Raven Press, New York, 1983.
40. Strong, R., J. C. Waymire, T. Samorajski and Z. Gottesfeld. Neurochemical Research 9 (11): 1641-1651, 1984.
41. Sunderland, T., P. Tariot, D. L. Murphy, H. Weingartner, E. A. Mueller and R. M. Cohen. Psychopharmacology 87: 247-249, 1985.
42. Sunderland, T., P. N. Tariot, E. A. Mueller, D. L. Murphy, H. Weingartner and R. M. Cohen. Psychopharmacology Bulletin 21: 676-679, 1985.
43. Sutherland, R. J., I. Q. Whislaw and B. Kolb. Journal of Neuroscience 8: 1863 - 1872, 1988.
44. Tateishi, N., Y. Takano, K. Honda, K. Yamada, Y. Kamiya and H. Kamiya. Clinical and Experimental Pharmacology and Physiology 14: 611-618, 1987.
45. Thomas, G. J. and L. S. Otis. Journal of Comparative and Physiological Psychology 51: 130 - 134, 1958.
46. Thompson, J. M., C. L. Makino, J. R. Whitaker and J. A. Joseph. Brain Research 299: 169-173, 1984.

47. Tonnaer, J. A. D. M., A. J. J. C. Lammers, J. H. Wieringa and H. W. M. Steinbusch. Brain Research 370: 200-203, 1986.
48. Wallace, J. E., E. E. Krauter and B. A. Campbell. Journal of Gerontology 35: 355-363, 1980.
49. Waller, S. B. and E. D. London. Experimental Gerontology 18: 419-425, 1983.
50. Walsh, T. J. and I. Hanin. Advances in Behavioral Biology 29: 461-167, 1986.
51. Whitehouse, P. J., D. L. Price, R. G. Struble, A. W. Clark, J. T. Coyle and M. R. DeLong. Science 215: 1237-1239, 1982.
52. Willig, F., A. Palacios, P. Monmaur, M. M'Harzi. J. Laurent and J. Delacour. Neurobiology of Aging 8:393-402, 1987.
53. Yamamura, H. I. and S. H. Snyder. Journal of Neurochemisty 21: 1355-1374, 1973.

SELECTION OF CHOLINERGIC AGENTS WHICH MAY IMPROVE MEMORY

Lincoln T. Potter, Bradley D. Pearce and Abraham Fisher

Department of Pharmacology
University of Miami School of Medicine
P.O. Box 016189
Miami, Florida 33101, USA

Summary

In Alzheimer's disease (AD) there is extensive cholinergic denervation of the cerebral cortex and hippocampus. The muscarine receptors which are normally linked to cerebral excitation (m1 and probably m3) remain and appear capable of responding to exogenous agonists. The receptors which promote cerebral inhibition (m2 and probably m4) also remain functional and are suitable targets for selective antagonists. A selective and efficacious m1 agonist is necessary to test the idea that cerebral excitation alone can improve memory in humans. The best way to screen new m1 agonists is by competition with 1 nM ^{3}H-pirenzepine for hippocampal m1 receptors, since the ratio of agonist affinities for the low and high affinity states of these receptors correlates closely with agonist efficacy. Similar screening of each agonist at m2 - m4 receptors yields data which allow comparisons of relative agonist affinities and efficacies at different receptors. These data allow predictions of agonist selectivity. Recent studies indicate further that muscarine receptor sites are paired, either because of receptor dimers, or more likely, because of two sites per molecule. Only one site is regulated by GTP. It is not yet clear whether occupation of one site by an agonist or antagonist is sufficient to change the activation of receptors; simultaneous agonist occupation is probable during normal neurotransmission. Techniques have been developed to assess the binding of ligands to each site. New drugs may therefore be developed to modify receptor function by binding to either site.

Introduction

This brief paper discusses the basic science behind the idea that cholinergic agents may improve memory, especially in Alzheimer's disease (AD). Emphasis is given to information about muscarine receptors and cholinergic neurotransmission which helps to constrain the choice of suitable agonists and antagonists for cholinergic replacement therapy.

The idea that some form of cholinergic intervention may improve memory is based on two kinds of information. First, there is general agreement that cholinergic denervation of the cerebral cortex and hippocampus is one of the earliest, most severe and most characteristic features of AD (1, 2). There is a good correlation between the degree of loss of cholinergic cell bodies and enzymes, and the degree of loss of cognitive functions (1-3). Second, there is general agreement that normal cholinergic neurotransmission plays an important and possibly essential role in the formation and retrieval of memories, especially short-term memories which are severely disturbed in AD (4). Much of the scientific evidence is based upon the ability of cholinergic antagonists like scopolamine to impair cognition, and the ability of acetylcholinesterase inhibitors like physostigmine and tetrahydroaminoacridine to cause small improvements in the performance of memory tasks. Observations that esterase

Basic, Clinical, and Therapeutic Aspects of Alzheimer's and Parkinson's Diseases, Volume 2
Edited by T. Nagatsu *et al.*, Plenum Press, New York, 1990

inhibitors produce some beneficial effects in persons with AD (5) add to the idea that cholinergic agents may be useful. In addition, there is experimental evidence that cholinergic denervation of the hippocampus in rats causes memory deficits, and that reinnervation diminishes these deficits (6).

<u>How do cholinergic nerves in the cortex and hippocampus modify memory?</u>

At present there is no simple answer to this question. Both the cortex, which is presumably the major site of memory storage, and the hippocampus, which helps to direct and modify impulse traffic between different regions of the cortex, appear essential for normal memory (7). The effects of acetylcholine in these structures are exerted on muscarine and nicotine receptors, and are clearly complex. The receptors are distributed unevenly (8-10), and we have no assurance that all the physiological mechanisms which might underlie memory functions have been described. What does seem clear is that cholinergic neurotransmission causes: (a) brief inhibition of the cortex by activating muscarine receptors on GABAergic interneurons (11), (b) some increase in sensory input by activating nicotine receptors, and (c) more prolonged cerebral excitation by activating muscarine receptors on relay neurons (11). One interpretation of these three effects is that cholinergic neurotransmission is designed to accomplish three goals: to reduce ongoing cerebral activity briefly, to increase the input of new data, and to strongly facilitate the use of this new data (12). In each case the action of acetylcholine is modulatory on the primary excitatory actions of glutamate and aspartate, and the primary inhibitory actions of GABA. In a sense, cholinergic denervation leaves the cerebral computer intact, but severely under-activated and under-modulated.

<u>How can we best mimic cholinergic neurotransmission with exogenous drugs?</u>

Obviously this is a formidable task, and we cannot expect perfection. Nonetheless there is sufficient physiological and pharmacological information available to allow reasonable hypotheses and some selection of agents which may prove effective. It is useful to consider the following points together.

(a) *A priori*, cerebral excitation is more likely to improve memory than is cerebral inhibition. (b) McCormick and Prince (11) found that the ability of acetylcholine to excite cerebral cells was readily blocked by pirenzepine, indicating that excitation is mediated by pharmacologically-defined "M1" receptors. Inhibition was readily blocked by atropine but not by pirenzepine, indicating that inhibition is mediated by "M2" receptors. (c) About two-thirds of the cholinergic receptors in the cortex are M1 receptors (as defined pharmacologically), about one-third are M2 receptors, and only a few percent are nicotine receptors (12). The percentage of M1 receptors in the hippocampus is still higher. (d) There are at least five genes for muscarine receptors, of which four are known to be expressed. Bonner et al named these "m1 - m4" (13), and Peralta et al named the equivalent receptors "HM1, HM2, HM4 and HM3," respectively (14). (e) According to Buckley et al (15) pirenzepine binds with its highest affinity to m1 receptors, and with about 5, 11 and 57-fold lower affinity to m4, m3 and m2 receptors, respectively. (f) *In situ* hybridization experiments show that the cortex and hippocampus have considerably more messenger RNA for m1 and m3 receptors than for m2 and m4 receptors (13, 16). (g) Genetically-defined m1 and m3 receptors are linked to the activation of phospholipase C and to the closure of K^+ channels, which produces cellular excitation (17, 20); in contrast, m2 and m4 receptors are linked to the inhibition of adenylate cyclase or opening of K^+ channels, which often produces cellular inhibition (20-23). (h) Oxotremorine is known to be highly effective at m2 and m4 receptors, moderately effective at m3 receptors, and nearly ineffective at m1 receptors (21, 22, 24). It crosses the blood-brain barrier readily and produces depression in patients with AD (25), as might be expected from its inhibitory electrophysiological effects (11).

Taken together, the evidence above points to two conclusions: activation of the cerebral cortex and hippocampus is most likely to be enhanced with agonists which are selective for m1 and m3 receptors, while inhibition is to be expected from agonists which are selective for m2 and m4 receptors. The problem in attempting to make use of these conclusions is that we do not know whether cerebral excitation alone can improve memory. As a first step in answering this question the clearest choice is to develop and test a selective m1 agonist for its effects on memory in humans. Meanwhile more experimental data are necessary concerning the specific locations and functions of m1 and m3 receptors, and concerning the possibility that selective m2 and/or m4 antagonists (including weak agonists) may improve cerebral excitation. Ultimately the use of choline (to improve residual cholinergic nerve function in the same manner as L-DOPA helps residual dopaminergic nerves) plus the use of an M2 antagonist, could prove as satisfactory in early AD as the use of an m1 agonist.

Is the m1 receptor response mechanism sufficiently intact in AD to permit successful cholinergic replacement therapy?

The receptor mechanism for m1 and m3 receptors is believed to include receptors, an unidentified GTP-binding regulatory protein, phospholipase C, two second messengers (diacyl glycol and IP_3), kinases and K^+ channels (12, 19, 27). To our knowledge no one has shown that the activation of cholinergically-denervated M1 receptors in AD tissue leads to normal K^+ channel closure and cellular excitation. However there is general agreement that the total number of M1 receptors is little altered in AD (28, 29), and we have shown that GppNHp-induced shifts in agonist affinity are normal in AD tissue in which there is severe depletion of choline acetyltransferase (Pearce and Potter, unpublished). In addition, after experimental cholinergic denervation of hippocampal tissue it is clear that cholinergic agonists cause nearly normal excitation for at least two weeks (30), and that the activation of phospholipase C is only slightly diminished after denervation for more than a year (Pearce and Potter, unpublished). These data indicate that denervated m1 receptors should still be capable of responding to exogenous agonists. Moreover the partial effectiveness of esterase inhibitors in AD (5) implies the presence of functional response mechanisms.

What are the best methods for testing selective m1 agonists?

The selectivity of an agonist for one receptor over another is dependent upon the relative affinities of the agonist for each receptor, the efficacy of each agonist-receptor complex for producing cellular responses, and the numbers of spare receptors. The numbers of receptors in human brains are not under our control. In practice no agonist has been found with substantially higher affinity for m1 then for other muscarine receptors, although new agents may be found with this desirable characteristic. Large differences have been found, however, in agonist efficacies, which typically have favored m2 responses. For example, oxotremorine is particularly weak at m1 receptors (11, 24). Selective efficacy at m1 receptors is therefore a critical feature for new cholinergic agonists for AD.

Only electrophysiological and biochemical approaches can give direct information about the efficacy of an agonist at different receptors. Since cerebral excitation is the desired result, every promising agonist must be tested physiologically. No other technique can avoid assay artifacts due to receptor desensitization. For screening purposes, however, physiological tests are slow, affinity measurements require null methodology (31), and there are no ideal brain tissues for comparative studies which have pure m1, m3 and m4 receptors. Satisfactory studies of human brain tissue are nearly impossible. Biochemical measurements of phospholipase C activation (in terms of the turnover of phosphoinositides) in cortical and hippocampal tissue have provided the next-best picture of agonist effectiveness (32), although without differentiation between ml and m3 receptor responses. The use of cell lines expressing genetically pure receptor subtypes has yielded physiological and biochemical information about receptor subtypes, and is highly desirable. For example, Baumgold and White have compared the effectiveness of different agonists for eliciting phosphoinositide turnover in cell lines expressing m3 and m4 receptors (22). Unfortunately few laboratories have had access to cells with pure m1 receptors (18, 20, 23), and some reservations are appropriate concerning whether existing cell lines have the same G-proteins and other response proteins that are present in the brain.

At first glance it appears easy to compare the relative affinities of an agonist for different receptors. To accomplish this many groups of investigators have competed an agonist vs a ^{3}H-labelled antagonist (e.g. 22). However it is now clear that each muscarine receptor can exist in (at least) two affinity states (26, 33). To establish the selectivity of an agonist it is therefore optimum to compare the high affinity state of one receptor with the high affinity state of another receptor, and also to compare the two low affinity states.

Moreover the available data show that the ratio of low/high agonist affinity at each receptor correlates closely with physiological and biochemical data concerning efficacy. Simple competition curves between agonists and ^{3}H-antagonists for each receptor can therefore yield the information about relative affinities and efficacies necessary to predict the selectivity of new drugs. We believe that such binding assays are the best available screening tests for the development of new selective m1 agonists.

As an example of the binding assays noted, we have competed a wide variety of agonists against m1 receptors (labelled with 1 nM ^{3}H-pirenzepine) in m1-rich hippocampal membranes and against nearly pure m2 receptors (labelled with 0.1 nM ^{3}H-N-methyl-scopolamine) in brainstem membranes (24, 33). An attractive feature of these assays is that the receptors are coupled with their native G-proteins. Moreover the brain membranes used can be obtained from human, including AD, tissue. There is an excellent correlation between the ratio of low/high affinities at m1 receptors and the ability of each agonist to cause cerebral excitation and phosphoinositide turnover (24). Similarly there is an excellent correlation between the ratio of low/high affinities at m2 receptors with the ability of each agonist to inhibit adenylate cyclase (21, 33). These results indicate that the efficacy of different agonists is determined by the nature of the interaction between the agonist-receptor complex and a G-protein. In general, agonists with tertiary amine groups are more efficacious at m2 than m1 receptors. The extreme case is oxotremorine which shows K_L/K_H ratios of 1 and 200 at m1 and m2 receptors, respectively (24, 34). This difference alone demonstrates that the nature of the binding sites on m1 and m2 receptors is markedly different, and sufficiently different to permit the development of new selective m1 agonists. It is necessary to develop tertiary agonists which have selective m1 efficacy.

Evidence for paired muscarine receptors

A striking feature of competition curves between agonists and m1 receptors, and agonists and m2 receptors, is that the numbers of receptor sites showing high and low affinity are equal (24, 26, 33). To examine this phenomenon further, we varied not only the receptor subtype, but the agonist (including the irreversible agonist, acetylethylcholine mustard), the radioligand (including both agonists and antagonists), the receptor concentration, the temperature, and, under certain conditions, the presence of divalent cations, guanine nucleotides, N-ethyl maleimide and digitonin. The numbers of high and low affinity sites remained equal. Such a fixed ratio cannot be explained as the result of different receptors, different effector proteins, limited receptor-G protein ratios, fortuitous equilibrium conditions, or artifacts of receptor distribution. Instead muscarine receptors behave as if they are either dimeric or bivalent (34). Evidence for equal numbers of high and low affinity sites for agonists and antagonists on receptors in solution (34, 35), evidence for the binding of propylbenzilylcholine mustard to more than one aspartate residue near the active site of m1 receptors (36), and evidence for the cooperative binding of QNB under certain conditions (34, 37, 38), leads us to the hypothesis that there are normally two receptor sites per receptor molecule (34). We have called these sites H and L. This hypothesis requires a fundamental change in our thinking. In the past, high and low affinity were considered as two states of one receptor site. Now it appears that H and L sites are distinct, and that only H sites can cycle between high and low affinity, depending upon receptor-G protein interactions. Further data indicate that agonists normally occupy H sites before L sites, and that both are probably occupied simultaneously during cholinergic neurotransmission (34).

What is the meaning of paired H and L receptors for cholinergic replacement therapy?

It is evident that the presence of paired receptor sites brings up the point that two acetylcholine molecules may normally act as *coagonists*; hence exogenous agonists can be designed to act either in concert with acetylcholine or independently. The good correlation between agonist efficacy and GppNHp-sensitive phenomena (24) suggests that H sites must be activated for receptor function. However some evidence suggests that low affinity sites must (also) be occupied for receptor functions (39). This point is not trivial since low concentrations of exogenous agonists are most likely to occupy H sites preferentially. At the present time it seems likely that the presence of an antagonist on H sites is sufficient to block receptor function (40), but blockade at L sites alone may also be effective. Moreover we do not know whether the presence of a nonefficacious agonist on L sites behaves as a successful coagonist with acetylcholine or as an antagonist. These considerations bring up the intriguing possibility that ligands can be developed with L-site selectivity which act as m1 coagonists, even though they do not appear to have high efficacy alone. AF102B may be such an agent.

To screen ligands for their interactions with only one receptor site, we have developed techniques for the selective blockade of H sites with acetylethylcholine mustard, and for the selective blockade of L sites with propylbenzylylcholine mustard, which then permit binding studies of the free site (41). Studies of the efficacy of coagonists require physiological or biochemical measurements of the actions of mixed agonists.

What are the potential problems with cholinergic replacement therapy?

Cholinergic agents will not cure AD, and their use is only likely to be of benefit early in the disease before there is extensive loss of other neurons. An m2 antagonist would be of benefit only in those patients with considerable retention of cholinergic neurotrans-mission, or in patients receiving a nonselective m1 agonist. An m2 antagonist would give far more peripheral side effects than an m1 agonist, because of the large numbers of m2 receptors in cardiac and smooth muscle.

Some question may be raised as to whether the tonic application of an m1 agonist can successfully mimic the normal phasic action of acetylcholine. In this regard it is useful to remember that the m1 action of acetylcholine only modulates excitatory neurotransmission due to the phasic action of the neurotransmitters glutamate and aspartate, which are well preserved in early AD. As a parallel, it may also be noted that exogenous bromocryptine and pergolide successfully replace the modulatory action of dopamine after the degeneration of dopaminergic nerves in Parkinson's disease.

More serious concerns with an m1 agonist include the probability of causing or exacerbating Parkinsonian signs and symptoms because of the activation of striatal m1 receptors, and the probability of causing downregulation of normal m1 receptor mechanisms. We have suggested that the use of an *esterase-sensitive* m1 agonist could overcome these problems by targeting the action of the agonist to synapses which have lost acetylcholin-esterase (by reason of cholinergic denervation), and by preventing agonist actions in esterase-rich brain regions like the striatum (12).

References

1. JT Coyle, D Price & M DeLong. Science 219, 1184-1190, 1983.
2. PT Francis et al. New Eng J Med 313, 7-11, 1985.
3. LA Hansen et al. Neurology 38, 48-54, 1988.
4. RT Bartus et al. Ann. N.Y. Acad Sci 444, 332-358, 1985.
5. WK Summers et al. New Eng J Med 315, 1241-1245, 1986.
6. SB Dunnet et al. Brain Res 251, 335-348, 1982.
7. LR Squire. Memory and Brain, Oxford Univ Press, NY 1987.
8. DC Mash & LT Potter. Neuroscience 19, 551-564, 1986.
9. R Cortes, A Probst & JM Palacios. Neuroscience 20, 65-107, 1987.
10. PBS Clarke, CB Pert & A Pert. Brain Res 323, 390-395, 1984.
11. DA McCormick & DA Prince. Proc Nat Acad Sci 82, 6344-6348, 1985.
12. LT Potter. (In)"International Symposium on Muscarinic Cholinergic Mechanisms" (Eds. S. Cohen, M. Sokolovsky) pp 294-305 Freund Publishing Ltd., London, 1987.
13. TI Bonner et al. Science 237, 527-532, 1987.
14. EG Peralta et al. EMBO J 6, 3923-3929, 1987.
15. NJ Buckley et al. Mol Pharmacol 35, 469-476, 1989.
16. NJ Buckley, TI Bonner & MR Brann. J Neurosci 8, 4646-4652, 1988.
17. SVP Jones et al. Proc Nat Acad Sci 85, 4056-4060, 1988.
18. S Numa et al. Cold Spring Harbor Symp Quant Biol 53, 295-301, 1988.
19. P Dutar & RA Nicoll. J Neurosci 8, 4214-4224, 1988.
20. A Ashkenazi et al. Cold Spring Harbor Symp Quant Biol 53, 263-272, 1988.
21. FJ Ehlert. Adv Exp Med Biol 236, 265-276, 1988.
22. J Baumgold & T White. Biochem Pharmacol 38, 1605-1616, 1989.
23. A Yatani et al. Science 241, 828-831, 1988.

24. LT Potter & CA Ferrendelli. J Pharmacol 248, 974-978, 1989.
25. JL Davis et al. Amer J Psychiatry 144, 468-471, 1987.
26. LT Potter, CA Ferrendelli & HE Hanchett. Cell Molec Neurobiol 8, 181-191, 1988.
27. I Litosch. Biochem J 244, 35-40, 1987.
28. DC Mash, DD Flynn & LT Potter. Science 228, 1115-1117, 1985.
29. PJ Whitehouse & K Sing Au. Prog Neuro-psychopharmacol 10, 665-676, 1986.
30. SJ Bird & GK Aghajanian. Brain Res 100, 355-370, 1975.
31. RF Furchgott. Adv Drug Res 3, 21-55, 1966.
32. SK Fisher, JC Figueiredo & RT Bartus. J Neurochem 43, 1171-1179, 1984.
33. LT Potter, LA Ballesteros & HE Hanchett. Submitted for publication.
34. LT Potter, CA Ferrendelli, A Fisher & HE Hanchett. Submitted for publication.
35. GL Peterson et al. Proc Nat Acad Sci 81, 4993-4997, 1984.
36. CM Fraser et al. Mol Pharmacol (in press).
37. B Hedlund, M Grynfarb & T Bartfai. NS Arch Pharmacol 320, 3-13, 1982.
38. R Mattera et al. J Biol Chem 260, 7410-7421, 1985.
39. JC Venter et al. Biochem Pharmacol 38, 1197-1208, 1989.
40. SP Baker & P Posner. Mol Pharmacol 30, 411-418, 1986.
41. LT Potter, LA Ballesteros, A Fisher & R Zhang. Submitted for publication.

RATIONALE FOR THE PLANNED CLINICAL TRIALS WITH NERVE GROWTH FACTOR IN ALZHEIMER'S DISEASE

Franz F. Hefti and Lon S. Schneider

Andrus Gerontology Center and Department of Psychiatry
University of Southern California
Los Angeles, CA 90089

Approximately ten years ago, it was discovered that Alzheimer's disease is associated with a pronounced loss of cholinergic neurons of the basal forebrain, and that this population of neurons is trophically affected by nerve growth factor (NGF). It was speculated that lack of NGF may be responsible for the loss of cholinergic neurons in Alzheimer's disease (Hefti, 1983). Since the early observations a considerable research effort has substantiated that NGF selectively acts on cholinergic neurons and that NGF administration counteracts degenerative changes of these cells. During the same time the cholinergic cell loss has been overshadowed by recent discoveries which emphasize the role of the classical hallmarks, neuritic plaques and neurofibrillary tangles, in the disease process. Despite the relative diminution of the role of cholinergic neurons in Alzheimer's disease pathology, the cholinergic deficit seems responsible for part of the behavioral and cognitive deficits; and clinical trials with cholinergic drugs suggest that cholinergic hypertrophy may be beneficial in Alzheimer's disease.

FUNCTION OF BASAL FOREBRAIN CHOLINERGIC NEURONS IN COGNITIVE PROCESSES AND ALZHEIMER'S DISEASE

Cholinergic neurons of the basal forebrain are located in the medial septal nucleus, diagonal band of Broca and nucleus basalis of Meynert and provide a widespread and topographically organized innervation to the hippocampus and cortex. A large body of evidence supports the view that these cholinergic neurons are involved in functions related to cognition and memory (reviews: Russell, 1982; Wenk et al., 1987; Smith, 1988). In rodents and primates, lesions of cholinergic neurons in the septum and nucleus basalis or of their respective projections to the hippocampus and cortex result in deficits in learning and memory. The behavioral deficits are, at least in part, reversed by administration of cholinergic compounds. In unlesioned animals and humans, anticholinergic drugs impair performance in a variety of tasks, whereas agonistic compounds produce inconsistent effects.

The massive decline of the cholinergic basalo-cortical projection in Alzheimer's disease is well established. Early findings of reductions in the activity of cerbral cortical choline acetyltransferase (ChAT), the key enzyme in acetylcholine synthesis (Bowen et al., 1976; Davies and Maloney, 1976; Perry et al., 1977), were followed by the demonstration of a massive loss of cholinergic cell bodies in the nucleus basalis

(Whitehouse et al., 1982; Nagai et al., 1983; Arendt et al., 1983). Further studies then revealed correlations between cortical ChAT reduction or nucleus basalis cell reduction and plaque densities in cortical areas (Arendt et al., 1984,1985, Etienne et al., 1986). A tight topographical correlation between cholinergic input to the cortex and plaque density was postulated by Arendt and Bigl (1985), but disputed by Mesulam (1986). The cholinergic deficit was shown to correlate with the decline in mental scores (Perry et al., 1978). However, after the initial enthusiasm for the then-called "cholinergic hypothesis of Alzheimer's disease," it became clear that many other neuronal systems are affected besides the cholinergic ones (Katzman, 1986; Price, 1986; reviews). In particular, there is an equally pronounced decline in the noradrenergic projection from the locus coeruleus to the cortex. Furthermore, the emphasis of Alzheimer research has shifted back to the study of the classical neuropathological markers of Alzheimer's disease, neuritic plaques and neurofibrillary tangles (Muller-Hill, 1989; Selkoe, 1989, reviews). While clearly putting the cholinergic deficit in perspective and emphasizing that it is not an exclusive pathological change, these studies should not detract from the fact that the cholinergic deficit represents the most pronounced transmitter deficit and seems to represent one of the early events in the disease process (Francis et al., 1985).

An implication of the cholinergic hypothesis is that potentiation of central cholinergic function should lessen the cognitive impairment associated with Alzheimer's disease. If the cholinergic deficit is responsible for cognitive or behavioral changes typical of Alzheimer's disease, one would expect cholinergic agonistically acting drugs to be effective in its treatment. Thus far there have been three main cholinergic approaches to Alzheimer's treatment: the use of precursor loading, cholinesterase inhibition, and direct cholinergic agonists. Numerous trials with precursors have been unimpressive, failing to reliably improve cognitive function in Alzheimer patients. However, a meta-analytic review of the literature suggests that treatment with precursor (lecithin) may enhance the effects of an anticholinesterase inhibitor (Jorm, 1986). Cholinesterase inhibitors have been the most frequently used experimental treatment of Alzheimer's disease. Modest but reliable improvement in memory were produced by the acetylcholinesterase inhibitor physostigmine (e.g., Davis and Mohs, 1982; Thal and Fuld, 1983; Stern et al., 1987, Thal, et.al., 1986). Dramatic improvements have been claimed for another cholinesterase inhibitor, tacrine (THA; Summers et al., 1986), but these results await replication. Studies with direct cholinergic agonists produced mixed results (Penn et al., 1988; Caine, 1980; Davis et al., 1987). Cholinesterase inhibitors seem relatively more effective than direct agonists. It can be speculated that acetylcholinesterase inhibitors enhance the phasic cholinergic input typical for at least some of the basal forebrain cholinergic neurons (Dutar et al., 1986). Direct receptor agonists may be relatively ineffective because they provide a prolonged and tonic cholinergic stimulation of postsynaptic cells, or perhaps they actually inhibit acetycholine release through their action on presynaptic receptors.

TROPHIC ACTIONS OF NGF ON BASAL FOREBRAIN CHOLINERGIC NEURONS

NGF is necessary for normal development of peripheral sympathetic and sensory neurons and the cholinergic neurons of the basal forebrain (reviews: Thoenen et al., 1987; Whittemore and Seiger, 1987; Hefti et al., 1989). During adult life, forebrain cholinergic neurons remain responsive to NGF. mRNA coding for NGF receptors is present in rat basal forebrain (Buck et al., 1987; Higgins et al., 1989) and these receptors have been visualized using various histological methods. In rats, primates and humans, localization and morphology of the NGF receptor positive neurons is very similar to that of cholinergic neurons of the basal forebrain (Hefti 1986; Riopelle et al., 1987; Kordower et al., 1988; Schatteman et al., 1988; Yan and Johnson, 1988; Hefti and Mash, 1989).

Co-localization studies on primate and human brains revealed that the population of ChAT-positive cells which also express NGF receptors is >90% of the entire population of cholinergic neurons (Kordower et al., 1988,1989). A similar percentage was found in the basal forebrain of the adult rat (Eckenstein et al., 1988; Batchelor et al., 1989). Concentrations of NGF as well as mRNA coding for NGF are highest in the hippocampus and cortex, and hippocampal target cells of the cholinergic neurons have been shown to express NGF mRNA (Shelton and Reichardt, 1986, Rennert and Heinrich, 1986; Ayer-LeLievre et al., 1988).

The finding that adult cholinergic neurons express NGF receptors and that this population of cells degenerates in human Alzheimer's disease prompted several research groups to study the effects of NGF in animals with experimental lesions of forebrain cholinergic neurons. Intraventricular infusions of NGF were found to prevent degenerative changes and loss of cholinergic cell bodies in the septum of rats with transections of the septo-hippocampal pathway (Hefti, 1986; Williams et al., 1986; Kromer, 1987). Such infusions also stimulated expression of ChAT and NGF receptors by remaining cholinergic neurons (Hefti et al., 1984; Higgins et al., 1989). Intraventricular administration of NGF was also found to counteract reductions of cortical ChAT activities following ibotenic acid-induced lesions of the nucleus basalis (Haroutunian et al., 1986) but failed to attenuate reductions in ChAT activity induced by administration of the ethyl choline mustard aziridinium ion (Johnson et al., 1988). NGF treatment failed to prevent the loss of septo-hippocampal GABAergic neurons or dopaminergic nigrostriatal neurons (Montero and Hefti, 1988). These findings suggest that NGF prevents the lesion-induced loss of septal cholinergic neurons by directly acting on NGF receptors expressed by cholinergic cells and that NGF does not simply protect any neuron with an axonal lesion. NGF treatment was found to prevent the lesion-induced loss of septal cholinergic neurons in old rats as effectively as in young animals (Montero and Hefti, 1989).

It is still not entirely clear whether the loss of septal cholinergic neurons after fimbrial transections is due to the death of cholinergic neurons or, alternatively, to shrinkage associated with a complete loss of expression of cholinergic marker enzymes. Results obtained with Nissl staining in rats with fimbrial transections clearly indicate a loss of large cell bodies in the medial septal nucleus and diagonal band of Broca (Daitz and Powell, 1954; Gage et al., 1986; Hefti, 1986). However, Hagg et al. (1988) reported partial recovery of cholinergic neurons after delayed NGF treatment to rats with complete fimbrial transections. Besides saving cholinergic cell bodies from degenerative changes, NGF infusions into rats with fimbrial transections induce local sprouting of cholinergic fibers in the septal area (Williams et al., 1986). It is still unclear whether NGF is able to promote regeneration of cholinergic axons into denervated target areas. In animals with partial fimbrial transections which leave a tissue bridge between the septum and hippocampus, NGF administration during several months stimulate local growth of cholinergic fibers, they, however, do not seem to induce cholinergic axons to reestablish the previous input of the hippocampus (Hefti, unpublished observations). By stimulating ChAT synthesis in cholinergic cell bodies, these NGF infusions elevate ChAT activity in remaining axons and, in consequence, probably the effectiveness of cholinergic transmission (Hefti et al., 1984; Higgins et al., 1989). Chronic administration of NGF to rats with such partial fimbrial transections were found to accelerate the behavioral recovery of the animals from learning deficits manifested in a radial maze test (Will and Hefti, 1985) in support of the view that NGF administration results in behavioral improvement.

Only tentative evidence is available on the performance of NGF-related mechanisms in the aged brain. Levels of NGF and its mRNA were found diminished in the brain of aged rats (Lärkfors et al., 1987) and there is an age-related decrease in the intensity of

NGF receptor immunoreactivity in the basal forebrain of aged rats and humans suggesting a down-regulation of receptor synthesis (Koh and Loy, 1988; Hefti and Mash, 1989). Gage and collaborators selected behaviorally impaired individuals which performed poorly in a swimming maze paradigm from a group of aged rats. In the forebrain of the impaired animals, the size of cholinergic cell bodies was smaller than in non-impaired aged rats, again supporting the view that cholinergic neurons play an important role in cognitive functions. Intraventricular NGF infusions were found to counteract both the morphological atrophy of cholinergic neurons and the behavioral deficits (Fischer et al., 1987,1989; Mandel et al., 1989). According to these findings it seems likely that intraventricular infusions of NGF are able to counteract age-related atrophic changes of cholinergic neurons in the mammalian brain and improve those behavioral functions which are dependent upon an intact cholinergic system.

NGF AND ALZHEIMER'S DISEASE

So far, there is no direct evidence supporting a primary involvement of NGF in the pathogenesis of Alzheimer's disease. Normal levels of mRNA coding for NGF were obtained in cortical samples from Alzheimer patients (Goedert et al., 1986). NGF receptor mRNA levels in the basal forebrain were found to be normal (Goedert et al., 1989). However, as recently discussed by Kordower and Mufson (1989), these findings do not exclude the possibility that other NGF-related mechanisms are affected. Furthermore, it has to be pointed out that the finding of unaltered NGF receptor mRNA levels in Alzheimer forebrains reported by Goedert et al., (1989) probably has been interpreted erroneously as evidence against a change in Alzheimer's disease. Since NGF receptors are selectively expressed by cholinergic neurons in the adult human basal forebrain and since more than 50% of these cells are lost in Alzheimer's disease (Hefti and Mash, 1989; Kordower et al., 1989), unaltered prevalence of NGF receptor mRNA demonstrated suggests a twofold upregulation of NGF receptor expression by remaining cholinergic neurons. It seems important to check this surprising result at the cellular level using in situ hybridization techniques. If substantiated, this finding would suggest that upregulation of NGF receptor synthesis is a compensatory mechanism to counteract the decline in another NGF-related mechanism.

The originally published hypothesis (relating NGF to Alzheimer's disease) focussed on the possibility that NGF administration may counteract diminished function of one of the NGF-related mechanisms (Hefti, 1983). However, animal studies discussed above suggest that NGF administration will counteract cholinergic atrophy irrespective of the actual cause. Accordingly, administration of NGF to Alzheimer patients as planned now should not be perceived as replacement therapy but, rather, as a pharmacological attempt to induce hypertrophy of cholinergic neurons surviving in the Alzheimer brain. In experimental studies, NGF-induced hypertrophy is manifested by increases in the expression of structural and transmitter-specific proteins. According to studies on animals with experimental lesions, NGF-induced hypertrophy of basal forebrain cholinergic neurons increases their resistance to insults. Behavioral studies suggests that hypertrophy elevates the ability of cholinergic neurons to influence their postsynaptic neurons. Based on these considerations, it can be anticipated that NGF administration would attenuate the rate of degeneration of cholinergic neurons surviving in Alzheimer brains and their functional performance. Accordingly, NGF treatment may attenuate the deterioration of or, eventually, improve those aberrant behaviors which are a consequence of the cholinergic deficit. The potential use of NGF in the treatment of Alzheimer's disease was considered by a Study Group of the National Institute on Aging. This group concluded that there is a strong scientific rationale for clinical trials and gave detailed recommendations for the development towards the clinical use of NGF (Phelps et al., 1989a,b).

CONSIDERATIONS IN THE ORGANIZATION OF CLINICAL TRIALS WITH NGF

While it is possible to administer purified mouse NGF to the human brain, it seems more prudent to use recombinant human NGF to diminish the possibility of immunological responses. To be able to reach neuronal populations in the brain, NGF will have to be given intracerebrally since the protein does not cross the blood-brain barrier. In experimental animals, NGF has been injected through chronically implanted cannulas or has been chronically infused with the help of osmotic minipumps. In humans, NGF may be chronically infused into the brain with the help of similar mechanical pump devices. Subcutaneous pumps are available which deliver medication proteins through a small tube to the cerebral ventricles (Harbaugh, 1986, 1987). Pumps and the intraventricular delivery systems are well tolerated in animals (DeYebenes et al., 1987), and, apparently, are at least adequately tolerated in humans (Harbaugh, 1987) but are not without risk. If intraventricular infusion of NGF is proven to be clinically effective it may become worthwhile to attempt replacing the infusion pumps with other ways of NGF administration. Such alternative possibilities have been discussed in a recent review (Hefti et al., 1989).

The organizational design of NGF clinical trials will have to be carefully considered in accord with ethical, scientific and FDA guidelines. The work group convened by the National Institute on Aging recommended an initial, open toxicity study, followed by a three to four month therapeutic trial, and then by a controlled long-term study to assess differential deterioration (Phelps, et al., 1989a). Allowing a patient to first receive an intraventricular pump implantation and then to be randomized to receive placebo infusion raises serious ethical issues. One concern is that placebo-randomized patients may be deprived of an alternative, active medication such as tacrine during the long-term study. Another consideration is the amount of deterioration to be tolerated in patients during longterm studies. The conventional parallel-group study with a parametric comparison of mean deterioration scores after a fixed period of time may be inappropriate, and even unethical, if carried out for as long as a year. To address this issue, it has been suggested that a maximum level of deterioration be set as the study end point, and that a nonparametric survival analysis be used to determine efficacy and study end. Study subjects reaching this minimum detectable level of deterioration could be then placed on alternative potentially active medication.

Issues in sample selection involve age, duration, severity, and the clinical heterogeneity of the illness itself. A differential response among subtypes might be expected. This might be especially true among those with increased cerebral amyloid deposits or in the subset of patients who improve with cholinesterase inhibitors. It may be desirable to choose, or screen, potential subjects for NGF based on their clinical response to acetylcholinesterase inhibitors. More severely demented patients have more cholinergic impairment and have been considered less likely to respond to cholinesterase inhibitors. NGF, acting on surviving cholinergic neurons, may also be more effective at early stages of the disease. Early controlled phase II studies of NGF will provide estimates of NGF's effect on short-term clinical improvement. For studies attempting to demonstrate that NGF retards Alzheimer's disease progression, a larger sample size is likely to be needed for placebo-controlled studies (Overall, 1987). Sample size determination will depend, in part, on the sensitivity of the instrument used to measure change and the study duration. For example, the commonly used Alzheimer's Disease Assessment Scale (ADAS; Mohs, et al., 1983), detects a mean deterioration of about 7 points ($\pm$ 10 S.D.) over the course of one year in a cohort of untreated Alzheimer patients (Kramer-Ginsberg et al., 1988). If NGF is hypothesized to prevent this deterioration in a placebo-controlled study, then a sample size of 33 in each treatment group would be needed to conduct a trial with a power of 0.80. However, if NGF only attenuates worsening on the ADAS by 3.5 points, then a sample size of 129 in each group would be needed.

Objective proof of NGF's efficacy will likely have to be in the form of improvement in cognitive function rather than in behavior, functional activities, or survival (Leber, 1986). There is a wide choice of reliable and valid neuropsychological tests. Nevertheless, the ADAS is probably the most frequently used outcome measure in Alzheimer disease clinical trials (Mohs et al., 1983). The ADAS assesses memory, dysphasia, dyspraxia, and behavior. (If a cholinesterase inhibitor is determined to be effective in Alzheimer's disease, it will be on the basis of the ADAS, hence making it the "standard" response variable in Alzheimer clinical trials).

Detrimental effects of NGF are concerns when considering its clinical use in humans. The notion that NGF will be useful in Alzheimer's disease is based on substantial biochemical and anatomical evidence indicating trophic stimulation of forebrain cholinergic neurons and observations that the cholinergic hypertrophy is accompanied by behavioral improvement in animals. Based on these findings it seems unlikely that NGF-mediated cholinergic stimulation is functionally detrimental. However, long-term administration in humans may induce aberrant cholinergic sprouting as observed in lesioned rodents (Williams et al., 1986). NGF treatment may be associated with behavioral toxicity not previously recognized in animals, such as depression or psychosis. Since intraventricular NGF administration to neonatal animals elevates brain levels of amyloid precursor protein mRNA (Mobley et al., 1988), it has to be determined whether application of NGF to the adult brain elevates amyloid precursor protein levels and whether such elevations produce pathological changes typical for Alzheimer's disease. Furthermore, since some non-neuronal cells seem responsive to NGF (Hefti et al., 1989), the possibility of undesired proliferation has to be considered. NGF infusions may act on intracerebral sympathetic neurons and produce changes in local blood flow (Morgan, 1989).

Given these considerations, the institution of phase I toxicity studies in human pose certain problems. Since a significant surgical procedure needs to be performed in order to administer the NGF, and the longterm effects in humans are potentially harmful, initial toxicity studies probably should be done in patients who might be considered least at risk. These are likely to be patients with moderate to advanced disease, and those few surviving patients who have had intraventricular pumps implanted. Although it is in these patients that issues of free and informed consent become paramount.

While there seems to be a spectrum of levels of enthusiasm for clinical trials with NGF, neuroscientists being most cautious and neurosurgeons most enthusiastic, it has to be emphasized that such trials will require approval by the United States Food and Drug Administration or, in other countries, by a similar board. These agencies are responsible for weighing potential benefits against dangers and they will require exclusion of undesired effects by rigorous toxicology studies. As the first example of an entirely new class of compounds, NGF will probably be scrutinized more than other drugs.

CONCLUSIONS

Experimental studies on animals suggest that intraventricular administration of NGF to Alzheimer patients may attenuate the degeneration of cholinergic neurons and the behavioral consequences associated with the cholinergic deficits. Based on these findings clinical trials with NGF seem justified, if potential detrimental effects are ruled out. If intraventricular NGF administration is effective, it may be possible to develop alternative ways of NGF administration or to pharmacologically manipulate the synthesis of endogenous NGF.

Acknowledgements: The authors were supported by grants from NIH, NSF, the Alzheimer's Disease and Related Disorders Association and the National Parkinson Foundation.

REFERENCES

Arendt, T., Bigl, V., Tennstedt, A. and Arendt, A., 1985, Neuronal loss in different parts of the nucleus basalis is related to neuritic plaque formation in cortical target areas in Alzheimer's disease. Neuroscience 14:1-14.

Arendt, T., Bigl, V., Tennstedt, A. and Arendt, A., 1984, Correlation between cortical plaque count and neuronal loss in the nucleus basalis in Alzheimer's disease. Neurosci. Lett. 48:81-85.

Arendt, T., Bigl, V. and Tennstedt, A., 1983, Loss of neurons in the nucleus basalis of Meynert in Alzheimer's disease, paralysis agitans and Korsakoff's disease. Acta Neuropathol. 61:101-108.

Ayer-LeLievre, C., Olson, L., Ebendahl, T., Seiger, A. and Persson, H., 1988, Expression of the beta-nerve growth factor gene in hippocampal neurons. Science 240:1339-1341.

Batchelor, P.E., Armstrong, D.M., Blaker, S.N. and Gage, F.H., 1989, Nerve growth factor receptor and choline acetyltransferase co-localization in neurons within the rat forebrain: response to fimbria-fornix transection. Neuroscience, J. Comp. Neurol. 284:187-204.

Bowen, D.M., Smith, C.B., White, P. and Davison, A.N., 1976, Neurotransmitter-related enzymes and indices of hypoxia in senile dementia and other abiotrophies. Brain 99:459-496.

Buck, C.R., Martinez, H.J., Black, I.B. and Chao M.V., 1987, Developmentally regulated expression of the nerve growth factor receptor gene in the periphery and brain. Proc.Natl.Acad.Sci. U.S.A. 84:3060-3063.

Caine E.D., 1980, Cholinomimetic treatment fails to improve memory disorders. New Eng. J. Med. 303:585-586.

Daitz, H.M. and Powell, T.P.S., 1954, Studies on the connections of the fornix system. J. Neurol. Neurosurg. Psychiatry 17:75-82.

Davies, P. and Maloney, A.J. F., 1976, Selective loss of central cholinergic neurons in Alzheimer's disease. Lancet II:1403.

Davis K.L., Hollander E., Davidson M., et al., 1987, Induction of depression with oxotremorine in patients with Alzheimer's disease. Am. J. Psych. 144:468-471.

Davis K.L., Mohs R.C., 1982, Enhancement of memory processes in Alzheimer's disease with multiple-dose intravenous physostigmine. Am. J. Psych. 139:1421-1424.

DeYebenes, J.G., Fahn, S., Lovelle, S., Jackson-Lewis, V., Jorge, P., Mena, M.A., Reiriz, J., et al., 1987, Continuous intracerebroventricular infusion of dopamine and dopamine agonists through a totally implanted drug delivery system in animal models of Parkinson's disease. Mov. Dis. 2:143-158.

Dutar, P., Rascol, O., Jobert, A. and Lamour, Y., 1986, Comparison of septo-hippocampal with basalo-cortical projection neurons in the rat: an electrophysiological approach. Neurosci. Lett. 63:86-90.

Eckenstein, F., 1988, Transient expression of NGF-receptor-like immunoreactivity in postnatal rat brain and spinal cord. Brain Res. 446:149-154.

Etienne, P. Robitaille, Y., Wood, P., Gauthier, S., Nair, N.P.V. and Quirion, R., 1986, Nucleus basalis neuronal loss, neuritic plaques and choline acetyltransferase activity in advanced Alzheimer's disease. Neuroscience 19:1279-1291.

Fischer, W., Gage, F.H. and Bjorklund, A., 1989, Degenerative changes in forebrain cholinergic nuclei correlate with cognitive impairments in aged rats. Europ. J. Neurosci. 1:33-45.

Fischer, W., Wictorin, K., Bjorklund, A., Williams, L.R., Varon, S. and Gage, F.H., 1987, Amelioration of cholinergic neurons atrophy and spatial memory impairment in aged rats by nerve growth factor. Nature 329:65-68.

Francis, P.T., Palmer, A.M., Sioms, N.R., Bowen, D.M., Davison, A.N., Esiri, M.M.,Neary, D., 1985, Neurochemical studies of early-onset Alzheimer's disease. Possible influence on treatment. Smowden, J.S. and Wilcock G.K. Lancet, 4:7-11.

Gage, F.H., Wictorin, K., Fischer, W., Williams, L.R., Varon, S. and Bjorklund, A., 1986, Life and death of cholinergic neurons: in the septal and diagonal band region following complete fimbria fornix transections. Neuroscience 19:241-256.

Goedert, M., Fine, A., Dawbarn, D., Wilcock, G.K. and Chao, M.V., 1989, Nerve growth factor receptor mRNA distribution in human brain: normal levels in basal forebrain in Alzheimer's disease. Mol. Brain Res. 5:1-7.

Goedert, M., Fine, A., Hunt, S.P. and Ullrich, A., 1986, Nerve growth factor mRNA in peripheral and central rat tissues and in the human central nervous system: Lesion effects in the rat brain and levels in Alzheimer's disease. Mol. Brain Res. 1:85-92.

Hagg, T., Manthorpe, M., Vahlsing, H.L. and Varon, S., 1988, Delayed treatment with nerve growth factor reverses the apparent loss of cholinergic neurons after acute brain damage. Exp. Neurol. 101:303-312.

Harbaugh, R.E., 1987, Intracerebroventricular cholinergic drug administration in Alzheimer's disease: preliminary results of a double-blind study. J. Neural. Transm. [Suppl.] 24:271-277.

Harbaugh, R.E., 1986, Intracranial drug administration in Alzheimer's disease. Psychopharmacol. Bull. 22:106-109.
Haroutunian V., Kanof, P.D. and Davis, K.L., 1986, Partial reversal of lesion-induced deficits in cortical cholinergic markers by nerve growth factor. Brain Res. 386:397-399.
Hefti, F., Hartikka, J. and Knusel, B., 1989, Neurotrophic factors in the adult and aging brain and their potential use in the treatment of neurodegenerative diseases. Neurobiol. Aging 10:515-533.
Hefti, F. and Mash, D.C., 1989, Localization of nerve growth factor receptors in the normal human brain and in Alzheimer's disease. Neurobiol. Aging, 10:75-87.
Hefti, F., 1986, Nerve growth factor (NGF) promotes survival of septal cholinergic neurons after fimbrial transections. J. Neurosci. 6:2155-2162.
Hefti, F., David, A. and Hartikka, J., 1984, Chronic intraventricular injections of nerve growth factor elevate hippocampal choline acetyltransferase activity in adult rats with septo-hippocampal lesions. Brain Res. 293:305-311.
Hefti, F., 1983, Alzheimer's disease caused by a lack of nerve growth factor? Ann. Neurol. 13:109-110.
Higgins, G.A., Koh, S., Chen, K.S. and Gage, F.H., 1989, Nerve growth factor (NGF) induction of NGF receptor gene expression and cholinergic neuronal hypertrophy within the basal forebrain of the adult rat. Neuron 3:247-256.
Johnson, G.V.W., Simonato, M. and Jope, R.S., 1988, Dose- and time-dependent hippocampal cholinergic lesions induced by ethylcholine mustard aziridinium ion: effects of nerve growth factor, GM1 ganglioside, and vitamin E. Neurochem. Res. 13:685-692.
Jorm, A.F., 1986, Effects of cholinergic enhancement therapies on memory function in Alzheimer's disease: a meta-analysis of the literature. Aust. New Zeal. J. Psychiatr. 20:237-240.
Katzman, R., 1986, Alzheimer's disease. New Engl. J. Med. 314:964-973.
Koh, S and Loy, R., 1988, Age-related loss of nerve growth factor sensitivity in rat basal forebrain neurons. Brain Res. 440:396-401.
Kordower, J.H. and Mufson, E.J., 1989, NGF and Alzheimer's disease: unfulfilled promise and untapped potential. Neurobiol. Aging 10:543-544.
Kordower, J.H., Gash, D.M., Bothwell, M., Hersh, L. and Mufson, E.J., 1989, Nerve growth factor receptor and choline acetyltransferase remain colocalized in the nucleus basalis (Ch4) of Alzheimer's patients. Neurobiol. Aging 10:76-74.
Kordower, J.H., Bartus, R.T., Bothwell, M., Schatteman, G. and Gash, D.M., 1988, Nerve growth factor receptor immunoreactivity in the non-human primate (Cebus apella): distribution, morphology, and colocalization with cholinergic enzymes. J. Comp. Neurol. 277:465-486.
Kramer-Ginsberg E., Mohs R.C., Aryan M., et al., 1988, Clinical predictors of course for Alzheimer patients in a longitudinal study: a preliminary report. Psychopharmacology Bulletin 24:458-462.
Kromer, L.F., 1987, Nerve growth factor treatment after brain injury prevents neuronal death. Science 235:214-216.
Larkfors, L., Ebendal, T., Whittemore, S.R., Persson, H., Hoffer, B. and Olson L., 1987, Decreased level of nerve growth factor (NGF) and its messenger RNA in the aged rat brain. Molec. Brain Res. 3:55-60.
Leber P., 1986, Establishing the efficacy of drugs with psychogeriatric indications. In: Crook, T. et al (eds), Treatment Development Strategies for Alzheimer's Disease. Madison, CT: Mark Powley, 1986.
Mandel, R.J., Gage, F.H. and Thal, L.J., 1989, Spatial learning in rats: correlation with cortical choline acetyltransferase and improvement with NGF following NBM damage. Exp. Neurol. 104:208-217.
Mesulam, M.M., 1986, Alzheimer plaques and cortical innervation. Neuroscience 17:275-276.
Mobley, W.C., Neve, R.L., Prusiner, S.B. and McKinley, M.P., 1988, Nerve growth factor induces gene expression for prion- and Alzheimer's beta-amyloid proteins. Proc. Natl. Acad. Sci. USA 85:9811-9815.
Mohs R.C., Rosen W.G., Davis K.D., 1983, The Alzheimer's Disease Assessment Scale (ADAS): An instrument for assessing treatment efficacy. Psychopharmacol Bull 19:448-450.
Montero, C.N. and Hefti, F., 1989, Intraventricular nerve growth factor adiministration prevents lesion-induced loss of septal cholinergic neurons in aging rats. Neurobiol. Aging, in press.
Montero, C.N. and Hefti, F., 1988, Rescue of lesioned septal cholinergic neurons by nerve growth factor: specificity and requirement for chronic treatment. J. Neurosci. 8:2986-2999.
Morgan, D.G., 1989, Considerations in the treatment of neurological disorders with trophic factors. Neurobiol. Aging 10:547-548.
Muller-Hill, B. and Beyreuther K., 1989, Molecular biology of Alzheimer's disease. Ann. Rev. Biochem. 58:287-307.
Nagai, T., McGeer, P.L., Peng, J.H., McGeer, E.G. and Dolman, C.E., 1983, Choline acetyltransferase immunohistochemistry in brains of Alzheimer's disease patients and controls. Neurosci. Lett. 36:195-199.
Overall J.E., 1987, Estimating sample size for longitudinal studies of age-related cognitive decline. Journal of Gerontology 2:137-141.
Penn R.D., Martin E.M., Wilson R.S., Fox J.H., and Savoy S.M., 1988, Intraventricular bethanechol infusion for Alzheimer's disease: results of double-blind and escalating dose trials. Neurology. 39:219-222.
Perry, E.K., Tomlinson, B.E., Blessed, G., Bergman, K. and Gibson, P.H., 1978, Correlation of cholinergic abnormalities with senile plaques and mental scores in dementia. Br. Med. J. ii:147-159.

Perry, E.K., Perry, R.H. and Tomlinson, B.E., 1977, Necropsy evidence of central cholinergic deficits in senile dementia. Lancet i:189.
Phelps C.H., Gage F.H., Growdon J.H., et al., 1989a, Potential use of nerve growth factor to treat Alzheimer's disease. Science 243:1189.
Phelps C.H., Gage F.H., Growdon J.H., et al., 1989b, Potential use of nerve growth factor to treat Alzheimer's disease. Neurobiology of Aging, 10:205-07.
Price, D.L., 1986, New Perspectives on Alzheimer's Disease. Ann. Rev. Neurosci. 9:489-512.
Rennert, P.D. and Heinrich, G., 1986, Nerve growth factor mRNA in brain: Localization by in situ-hybridization. Biochem. Biophys. Res. Comm. 138:813-818.
Riopelle, R.J., Verge, V.M.K., Richardson, P.M., 1987, Properties of receptors for nerve growth factor in the mature rat nervous system. Mol. Brain Res. 3:345-353.
Russell, R.W., 1982, Cholinergic system in behavior: the search for mechanisms of action. Ann. Rev. Pharmacol. Toxicol. 22:435-463.
Schatteman, G.C., Gibbs, L., Lanahan, A.A., Claude, P. and Bothwell, M., 1988, Expression of NGF receptor in the developing and adult primate central nervous system. J. Neurosci. 8:860-873.
Selkoe, D.J., 1989, Molecular pathology of amyloidogenic proteins and the role of vascular amyloidosis in Alzheimer's disease. Neurobiol. Aging 10:387-395.
Shelton, D.L. and Reichardt, L.F., 1986, Studies on the expression of beta NGF gene in the central nervous system: level and regional distribution of NGF mRNA suggest that NGF functions as a trophic factor for several neuronal populations. Proc. Natl. Acad. Sci. USA 83:2714-2718.
Smith, G., 1988, Animal models of Alzheimer's disease: experimental cholinergic denervation. Brain Res. Revs. 13:103-118.
Stern, Y., Sano, M. and Mayeux, R., 1987, Effects of oral physostigmine in Alzheimer's disease. Ann. Neurol. 22:306-310.
Summers, W.K., Majovski L.V., Marsh G.M., Tachiki K., Kling A., 1986, Oral tetrahydroaminoacridine in long-term treatment of senile dementia, Alzheimer type. New Eng. J. Med. 315:1241-1245.
Thal L.J., Masur D.M., Sharpless N.S., Fuld P.A., Davies P., 1986, Acute and chronic effects of oral physostigmine and lecithin in Alzheimer's disease. Prog. Neuropsychopharmacol. Biol. Psychiatry 10:617-636.
Thal, L.J. and Fuld, P.A., 1983, Memory enhancement with oral physostigmine in Alzheimer's disease. New Engl. J. Med. 308:720.
Thoenen, H., Bandtlow, C. and Heumann, R., 1987, The physiological function of nerve growth factor in the central nervous system: comparison with the periphery. Rev. Physiol. Biochem. Pharmacol. 109:145-178.
Wenk, G., Hughey, D., Boundy, V., et al., 1987, Neurotransmitters and memory: Role of cholinergic, serotonergic, and noradrenergic systems. Behav. Neurosci. 101:325-332.
Whitehouse, P.J., Price, D.L., Struble, R.G., et al., 1982, Alzheimer's disease and senile dementia: loss of neurons in the basal forebrain. Science 215:1237-1239.
Whittemore, S.R. and Seiger, A., 1987, The expression, localization and functional significance of beta-nerve growth factor in the central nervous system. Brain Res. Revs. 12:439-464.
Will, B., Hefti, F., 1985, Behavioural and neurochemical effects of chronic intraventricular injections of nerve growth factor in adult rats with fimbria lesions. Behav. Brain. Res. 17:17-24.
Williams, L.R., Varon, S., Peterson, G.M., Wictorin, K., Fischer, W., Bjorklund, A. and Gage, F.H., 1986, Continuous infusion of nerve growth factor prevents basal forebrain neuronal death after fimbria fornix transection. Proc. Natl. Acad. Sci. USA 83:9231-9235.
Yan, Q. and Johnson, E.M., 1988, An immunohistochemical study of the nerve growth factor receptor in developing rats. J. Neurosci. 8:3481-3498.

NEUROCHEMICAL AND PSYCHOPHARMACOLOGICAL ASSESSMENT OF THE M1-SELECTIVE AGONIST AF102B (FKS-508)

Fumio Mizobe, Noboru Nakahara, Nobuo Ogane, Yasunari Saito, Masashi Ise, Yoshinori Iga, Kenji Fukui, Nobuaki Fujise, Tatsuya Haga*, and Gosei Kawanishi

Research Institute of Life Science, Snow Brand Milk Products Co. Ltd., Ishibashi, Tochigi; and *Tokyo University Faculty of Medicine, Hongo, Tokyo, Japan

INTRODUCTION

Senile dementia of the Alzheimer type (SDAT) is characterized by a progressive decline of memory and intellectual abilities. Although several neurotransmitter systems in the brain of SDAT patients are reported to be impaired, the abnormality in the cholinergic system is generally greater and more consistent, and is also found to correlate well with the cognitive deficits[1]. Moreover, a large body of evidence has suggested that the central cholinergic system plays an important role in memory and cognitive functions in humans as well as in animals[2]. In SDAT, cholinergic nerve terminals appear to be extensively degenerated while postsynaptic muscarinic receptors are relatively unchanged[3]. Accordingly, clinical trials with cholinomimetics, such as acetylcholinesterase inhibitors (physostigmine, tetrahydroaminoacridine) and muscarinic agonists (arecoline, oxotremorine, RS 86) have been conducted, intending to compensate for the central cholinergic defects. Although modest but significant clinical improvements have been documented, applicability is hampered by central and peripheral side effects, including chills, tremor, and hypersalivation[4,5].

It is generally accepted that there are at least two subtypes of muscarinic receptors, classified as M1 and M2, on the basis of their affinities for pirenzepine. Results indicating the heterogeneous localization of these receptor subtypes in peripheral tissues as well as in the central nervous system suggest that it may be possible to improve cholinergic therapy by selective activation of muscarinic receptor subtypes. More specifically, it is assumed that M1-selective agonists would compensate for aspects of cognitive dysfunctions in SDAT with less adverse side effects[5]. The observation that M1 site populations were well preserved in the brain of SDAT patients has also encouraged this therapeutic strategy[6].

In the present paper, we have overviewed our neurochemical and psychopharmacological studies on AF102B (FKS-508; cis-2-methylspiro(1,3-oxathiolane-5,3')quinuclidine hydrochloride hemihydrate), which have indicated that (i) AF102B is a potent muscarinic agonist (ii) with a high

Table 1. Potency of muscarinic agonists for displacement of [^{3}H]QNB binding to receptors in membrane preparations of rat cerebral cortex and to muscarinic receptors purified from porcine cerebrum

Agonist	$K_{0.5}$ (μM)	
	Membrane	Purified receptors
Oxotremorine	1.3	3.8
AF102B	5.1	10
Arecoline	15	37
AF102A (trans-isomer)	20	27

selectivity for M1 receptors and that (iii) it ameliorates experimental amnesia in rodents (iv) with less adverse central side effects.

AF102B IS A MUSCARINIC AGONIST

Binding to Muscarinic Receptors

Binding assay was carried out with synaptosomal membrane preparations of the rat cerebral cortex and with muscarinic receptors purified from the porcine cerebrum[7]. We found that AF102B potently inhibited binding of [^{3}H]QNB both to receptors in membrane preparations (Table 1) and to purified muscarinic receptors (Table 1, Fig. 1). The rank order of potency of AF102B and other muscarinic agonists to inhibit binding of [^{3}H]QNB to receptors in membrane preparations of the cerebral cortex was oxotremorine > AF102B ≒ pilocarpine > McN-A-343 > arecoline ≧ AF102A (the trans-isomer) > methacholine > bethanechol. Apparent Ki ($K_{0.5}$) values of AF102B and muscarinic agonists obtained in membrane preparations were comparable to those in the purified receptors (Table 1).

Contraction of the isolated guinea pig ileum

AF102B dose-dependently induced contraction of the smooth muscle in isolated guinea pig ileum. The effects of AF102B were antagonized by atropine. AF102B was far less potent than oxotremorine to induce the smooth muscle contraction (Fig. 2, Table 3). We found that AF102B also acts as an agonist in the isolated guinea pig urinary bladder.

AF102B IS HIGHLY SELECTIVE FOR M1 RECEPTORS

Presynaptic Muscarinic Receptors Regulating Transmitter Release

There is ample evidence to demonstrate that muscarinic receptors are located on not only cholinergic but also dopaminergic nerve endings in the rat brain (i.e., autoreceptors and heteroreceptors, respectively). Their stimulation by muscarinic agonists results in the inhibition of acetylcholine (ACh) release induced by depolarization and the potentiation of evoked dopamine (DA) release. It is suggested that the muscarinic heteroreceptors regulating DA release and the autoreceptors modulating ACh release are subdivided into the M1 and M2 subtypes, respectively[8]. Therefore, we examined the effects of AF102B on the presynaptic heteroreceptors potentiating the evoked release of DA from the rat striatum (an M1-mediated response) and on the autoreceptors

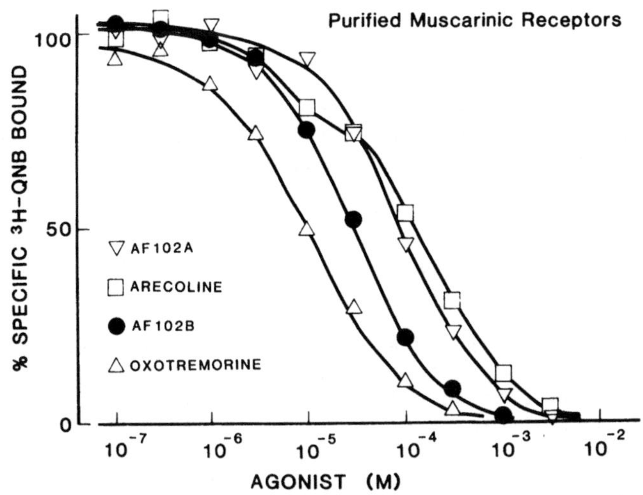

Fig. 1. Displacement by muscarinic agonists of [^{3}H]QNB binding to muscarinic receptors purified from the porcine cerebrum. Purification of muscarinic receptors and binding assay were carried out as described elsewhere[20,21].

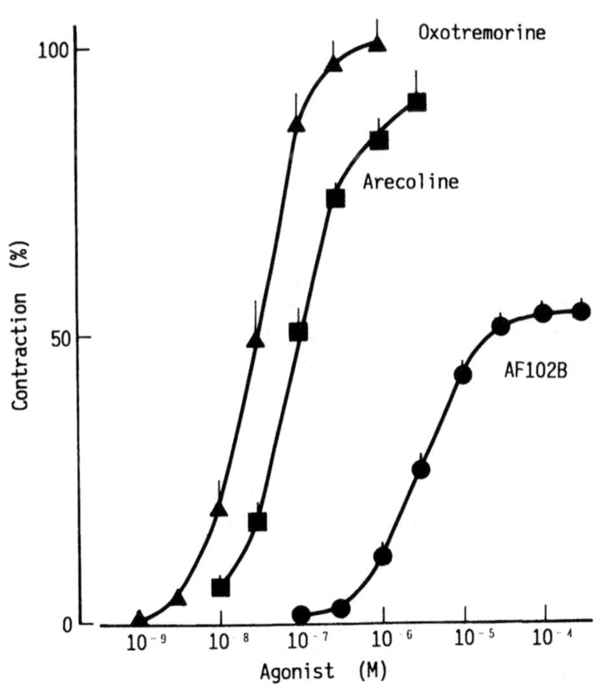

Fig. 2. Smooth muscle contraction induced by muscarinic agonists in the isolated guinea pig ileum. Points represent means ± S.E.M. (n=6-9).

inhibiting the evoked release of ACh from the hippocampus (an M2-mediated response) in comparison with oxotremorine, to elucidate the M1-selectivity of AF102B[9].

AF102B and oxotremorine potentiated the K^+-evoked release of [^{3}H]DA from rat striatal synaptosomes. The dose-response curves for AF102B and oxotremorine showed that AF102B (EC_{50}: 0.20 μM) was slightly more potent than oxotremorine (EC_{50}: 1.0 μM) in stimulating the hetero-receptors (M1 type). On the other hand, AF102B and oxotremorine inhibited the K^+-evoked release [^{3}H]ACh from rat hippocampal synaptosomes in a dose-dependent manner. AF102B (EC_{50}: 5.0 μM) was far less potent than oxotremorine (EC_{50}: 0.25 μM) in stimulating the auto-receptors (M2 type). These results suggest that AF102B is a muscarinic agonist with a high selectivity for M1 receptors (Table 3).

Recent pharmacological and electrophysiological studies have also provided evidence suggesting that AF102B is an M1-selective agonist. Fisher et al.[10] found that when the relative M1 selectivity was evaluated by the radioligand receptor binding assay in rat forebrain (mainly M1 sites) and cerebellum (mainly M2 sites) homogenates, AF102B was highly selective for M1 binding sites. Mochida and her colleagues reported that AF102B generated slow depolarization (an M1-mediated response) but did not induce hyperpolarization (an M2-mediated response)[11] and that it had no effect on M2-mediated Ca^{2+} entry during the action potential in rabbit superior cervical ganglia[12]. Xu et al.[13] demonstrated that AF102B produced a dose-dependent increase in striatal DA release, which was sensitive to pirenzepine (an M1 antagonist) but not to AF-DX 116 (an M2 antagonist).

AF102B AMELIORATES EXPERIMENTAL AMNESIA IN ANIMAL MODELS

Since the etiology of SDAT is not known, no conclusive animal models are yet available. Nonetheless, the efficacy of therapeutic drugs for SDAT should be evaluated in various animal models before final clinical trials. Based upon the consistent finding of a presynaptic cholinergic defect in SDAT, the cholinergic system was manipulated to produce animal models of SDAT: scopolamine treatment, AF64A injection, and basal forebrain (BF) lesioning. Psychopharmacological studies were made to evaluate the efficacy of AF102B for reversal of learning and memory deficits of animal models in several behavioral tasks using negative reinforcement (electric shock in passive and active avoidance tasks) and positive reinforcement (food reward in T-maze and radial-arm maze tasks) (Table 2).

Scopolamine-Treated Mice

Since scopolamine (a muscarinic antagonist) has been demonstrated to produce cognitive deficits in humans and animals, which are regarded to be similar to those observed in SDAT, scopolamine-induced experimental amnesia in mice is widely used as an animal model of SDAT[2].

Subcutaneous injection of scopolamine (1 mg/kg) impaired performance in a one-trial step-through passive avoidance task in mice. AF102B (1-10 mg/kg, i.p.) significantly ameliorated the scopolamine-induced memory deficits[14]. Peroral administration of AF102B also reversed the passive avoidance failure (Table 3).

AF64A-Treated Rats

Intracerebroventricular (i.c.v.) injection of a cholinotoxin, AF64A,

Table 2. Effects of AF102B on learning and memory deficits of animal models in behavioral tasks

Animal model	Task	Drug administration	Effect of AF102B on cognitive functions
SCP-treated mice	Passive avoidance	Single (i.p./p.o.)	Improved
AF64A-treated rats	Passive avoidance	Single (i.p./p.o.)	Improved
	T-maze	Repeated (i.p.)	Improved
	Radial-arm maze	Single (p.o.)	Improved
	Radial-arm maze	Repeated (p.o.)	Improved
BF-lesioned rats	Active avoidance	Repeated (p.o)	Improved

Abbreviations: SCP, scopolamine; BF, basal forebrain.

produces a cholinergic hypofunction in the hippocampus together with memory impairment, indicating that AF64A-treated rats (AF64A-rats) are a workable animal model for SDAT[2,15]. This animal model mimicks the septo-hippocampal cholinergic defects in SDAT.

AF64A injection (3 nmol/2 μl, each lateral ventricle) into rats brought about the impairment of performance in passive avoidance, T-maze, and radial-arm maze tasks, together with a selective decrease in hippocampal ChAT activity, a presynaptic cholinergic marker. AF102B (0.1-1 mg/kg, i.p. or 1-5 mg/kg, p.o.) ameliorated the retention latency in the one-trial passive avoidance task in AF64A-rats[14]. Single administration of AF102B (1-5 mg/kg, p.o.) significantly reduced the incorrect choices of AF64A-rats in the radial-arm maze task with 6-hr delay time, suggesting that AF102B has a beneficial effect on the impaired working memory. Repeated administrations of AF102B for 5 weeks ameliorated impaired performance of AF64A-rats in a delayed alternation task in the T-maze (5 mg/kg/day, i.p.) and acquisition failures in the radial-arm maze task (1-5 mg/kg/day, p.o.). These results indicate that pharmacological tolerance was not brought about during 5 weeks of treatment[16]. Neither abnormalities in general behavior nor adverse side effects were observed in rats treated with AF102B for 5 weeks. Fisher et al.[17] also reported beneficial effects of AF102B on cognitive deficits of AF64A-rats in passive avoidance and radial-arm maze tasks.

Basal Forebrain-Lesioned Rats

There is extensive cell loss in the nucleus basalis of Meynert (nbM) in SDAT, which is the origin of the major cholinergic input to the frontal, temporal, and parietal cortices. Rats with lesions in the corresponding nuclei (BF-lesioned rats) can serve as an animal model for some of the cholinergic and mnemonic deficits of SDAT[2].

Bilateral lesions of the rat basal forebrain were electrolytically produced. Repeated administration of AF102B (1 mg/kg/day, p.o.) for 4 weeks significantly enhanced the avoidance rate without increasing the unavailable response in a two-way shuttle active avoidance task, indicating that AF102B ameliorated cognitive deficits of BF-lesioned rats[18].

AF102B AMELIORATES AMNESIA WITH LESS CENTRAL SIDE EFFECTS

Stimulation of muscarinic receptors brings about various symptoms,

Table 3. Comparison of muscarinic activities of AF102B and oxotremorine in vivo and in vitro

(A) In Vivo

	Proposed receptor subtype	ED_{50} (mg/kg, p.o.) AF102B	Oxotremorine	Ratio (AF/Oxo)
Reversal of SCP-induced amnesia[a]	M1	2.9	0.76	3.8
Hypothermia	M2	33	0.19	170
Tremor[a]	M2	95	2.7	35

(B) In Vitro

	Proposed receptor subtype	EC_{50} (μM) AF102B	Oxotremorine	Ratio (AF/Oxo)
Transmitter release modulation				
[³H]DA release (rat striatum)	M1	0.20	1.0	0.20
[³H]ACh release (rat hippocampus)	M2	5.0	0.25	20
Guinea pig ileum contraction	M2	3.1	0.032	97

Abbreviations: AF, AF102B; Oxo, oxotremorine; SCP, scopolamine.
[a]ED_{50} values were estimated by use of the Probit equation.

including hypothermia and tremor as central effects and salivation and diarrhea as peripheral ones[4,5]. Although little is known about the relationship between muscarinic responses and receptor subtypes in the central nervous system, recent pharmacological studies have suggested that M1 sites could be involved in memory and cognitive functions while M2 sites might mediate hypothermia and tremor (see references in ref.19). If this is the case, the M1-selective and brain-entering agonist, AF102B, should highly differentiate between the central muscarinic effects. We examined muscarinic activities of AF102B on reversal of scopolamine-induced amnesia, hypothermia, and tremor in mice in comparison with those of the classical agonist oxotremorine[19].

Reversal of Scopolamine-Induced Amnesia

AF102B and oxotremorine dose-dependently ameliorated scopolamine-induced experimental amnesia in a one-trial step-through passive avoidance task[19]. ED_{50} values were estimated by use of the Probit equation, based on the number of mice that exhibited more than 150 sec of retention latency after drug administration. AF102B (ED_{50}: 2.9 mg/kg, p.o.) was less potent than oxotremorine (ED_{50}: 0.76 mg/kg, p.o., as sesquifumarate salt) in reversing scopolamine-induced amnesia (Table 3).

Hypothermia

Core body temperature was measured with a thermometer inserted into the rectum of mice. AF102B and oxotremorine dose-dependently produced hypothermia, however, AF102B was far less potent (170-fold) than oxotremorine (Table 3), and the maximum response induced by AF102B was 35 % of that induced by oxotremorine[19].

Tremor

AF102B and oxotremorine dose-dependently induced tremor[19]. Again, AF102B (ED_{50}: 95 mg/kg, p.o.) was far less potent than oxotremorine (ED_{50}: 2.7 mg/kg, p.o.) (Table 3).

These results show that AF102B discriminates between centrally mediated muscarinic effects, as it was potent in ameliorating amnesia but markedly weak in producing hypothermia and tremor, providing further evidence to suggest that M1 and M2 sites may be involved in memory and cognitive functions and in the expression of hypothermia and tremor, respectively.

In conclusion, our neurochemical and pharmacological studies indicate that the M1-selective agonist AF102B may be a potential therapeutic drug for SDAT.

REFERENCES

1. E. K. Perry, The cholinergic hypothesis: ten years on, Brit. Med. Bull. 42:63 (1986).
2. G. Smith, Animal models of Alzheimer's disease: experimental cholinergic denervation, Brain Res. Rev. 13:103 (1988).
3. R. T. Bartus, R. L. Dean, B. Beer, and A. S. Lippa, The cholinergic hypothesis of geriatric memory dysfunction, Science 217:408 (1982).
4. E. Hollander, M. Davidson, R. C. Mohs, T. B. Horvath, B. M. Davis, Z. Zemishlany, and K. L. Davis, RS 86 in the treatment of Alzheimer's disease: cognitive and biological effects, Biol. Psychiat. 22:1067 (1987).
5. E. Hollander, R. C. Mohs, and K. L. Davis, Cholinergic approaches to the treatment of Alzheimer's disease, Brit. Med. Bull. 42:97 (1986).
6. D. C. Mash, D. D. Flynn, and L. T. Potter, Loss of M2 muscarine receptors in the cerebral cortex in Alzheimer's disease and experimental cholinergic denervation, Science 228:1115 (1985).
7. F. Mizobe, S. Ono, N. Ohgane, K. Iwamoto, S. Katayama, and T. Haga, AF102B, a novel muscarinic M1-type agonist, stimulated presynaptic auto- and heteroreceptors in different potency, Trend. Pharmacol. Sci. 9 (Suppl.):81 (1988).
8. M. Raiteri, R. Leardi, and M. Marchi, Heterogeneity of presynaptic muscarinic receptors regulating neurotransmitter release in the rat brain, J. Pharmacol. Exp. Ther. 228:209 (1984).
9. S. Ono, Y. Saito, N. Ohgane, G. Kawanishi, and F. Mizobe, Heterogeneity of muscarinic autoreceptors and heteroreceptors in the rat brain: effects of a novel M1 agonist, AF102B, Eur. J. Pharmacol. 155:77 (1988).
10. A. Fisher, E. Heldman, R. Brandeis, I. Karton, Z. Pittel, S. Dachir, A. Levy, and F. Mizobe, (Cis)-2-methyl-spiro-(1,3-oxathiolane-5,3')-quinuclidine (AF102B): a novel M1 muscarinic agonist as a potential drug for the treatment of Alzheimer's disease, in: "Alzheimer's Disease: Advances in Basic Research and Therapies," R.J. Wurtman, S.H. Corkin, J.H. Growdon, eds., pp. 405-409, Center for Brain Sciences and Metabolism Charitable Trust, Cambridge (1987).
11. S. Mochida, F. Mizobe, A. Fisher, G. Kawanishi, and H. Kobayashi, Dual synaptic effects of activating M1-muscarinic receptors, in superior cervical ganglia of rabbits, Brain Res. 455:9 (1988).

12. S. Mochida and H. Kobayashi, Stimulation of M2-muscarinic receptor decreases calcium entry during action potential, Trend. Pharmacol. Sci. 9 (Suppl.):83 (1988).
13. M. Xu, F. Mizobe, T. Yamamoto, and T. Kato, Differential effects of M1- and M2-muscarinic drugs on striatal dopamine release and metabolism in freely moving rats, Brain Res. 495:232 (1989).
14. N. Nakahara, Y. Iga, F. Mizobe, and G. Kawanishi, Amelioration of experimental amnesia (passive avoidance failure) in rodents by the selective M1 agonist AF102B, Japan. J. Pharmacol. 48:502 (1988).
15. N. Nakahara, Y. Iga, F. Mizobe, and G. Kawanishi, Effects of intracerebroventricular injection of AF64A on learning behaviors in rats, Japan. J. Pharmacol. 48:121 (1988).
16. N. Nakahara, Y. Iga, Y. Saito, F. Mizobe, and G. Kawanishi, Beneficial effects of FKS-508 (AF102B), a selective M1 agonist, on the impaired working memory in AF64A-treated rats, Japan. J. Pharmacol. 51:539 (1989).
17. A. Fisher, R. Brandeis, Z. Pittel, I. Karton, M. Sapir, S. Dachir, A. Levy, and E. Heldman, (±)-Cis-2-methyl-spiro(1,3-oxathiolane-5,3') quinuclidine (AF102B): a new M1 agonist attenuates cognitive dysfunctions in AF64A-treated rats, Neurosci. Lett. 102:325 (1989).
18. Y. Iga, N. Nakahara, F. Mizobe, N. Ohgane, T. Sawai, I. Nakajima, and G. Kawanishi, SND-5008 (AF102B), a novel muscarinic agonist improves experimental cognitive dysfunctions in rodents, Soc. Neurosci. Abstr. 14:57 (1988).
19. N. Nakahara, N. Fujise, G. Kawanishi, and F. Mizobe, Central muscarinic activities of a M1-selective agonist: preferential effect on reversal of amnesia, Brain Res. 507:172 (1990).
20. K. Haga and T. Haga, Affinity chromatography of the muscarinic acetylcholine receptor, J. Biol. Chem. 258:13575 (1983).
21. K. Haga and T. Haga, Purification of the muscarinic acetylcholine receptor from porcine brain, J. Biol. Chem. 260:7927 (1985).

EFFECTS OF ACUTE AND CHRONIC ADMINISTRATION OF THE NEW MUSCARINIC AGONIST SDZ BOP-086 ON PHOSPHOINOSITIDE HYDROLYSIS

Shigeru Morinobu, Tomiko Kawanami, Norihiko Kuwayama, Shinobu Kawakatu, Katsuo Sagawa, and Shiro Totsuka

Department of Neuro-Psychiatry, Yamagata University, School of Medicine, Yamagata, Japan

INTRODUCTION

During the course of advances in neuropsychopharmacological analysis of Alzheimer's disease, a marked decrease in presynaptic cholinergic function has been found, and is presumed to be the main cause of the dementia. Neuropathological studies of Alzheimer's disease showed that the degeneration of the nucleus basalis of Meynert is closely related to the symptoms of the dementia. Therefore, the muscarinic receptor agonists or acetylcholine esterase inhibitors were developed and used for the treatment of the disease. Although many drugs that activate central cholinergic functions were developed, the effectiveness of these drugs has been rather controversial.

In the present study, we investigated the effects of acute and chronic administration of a newly developed muscarinic receptor agonist, SDZ BOP-086 (Sandoz Ltd.), on the carbachol-stimulated-phosphoinositide (PI) hydrolysis in rat cerebral cortical slices. [^{3}H]inositol-1-phosphate ([3H]IP1) was determined as an indicator of PI hydrolysis in the slices prelabelled with [^{3}H]myo-inositol.

MATERIALS and METHODS

Drug Treatment

In order to investigate the influence of chronic administration of SDZ BPO-086 (SDZ), we treated male Fischer rats (approx. 400g) with SDZ (0.3 mg/kg, i.p.) once daily for 14 days.

Tissue Preparation

Rats were decapitated, and their brains were rapidly removed. Cross-chopped slices (350 μm) were prepared from the rat frontal cerebral containing modified Krebs-Hensleit (K-H) buffer equilibrated with 95% O_2/5 % CO_2 at 37°C.

Assay of [^{3}H]IP1 Accumulation

After slices had been incubated for 60 min in K-H buffer, they were pipetted into vials containing 2 μCi of [^{3}H]myo-inositol and 150 mM LiCl in 240 μl of the same buffer. These slices were then incubated at 37°C for 90 min, after the buffer solution had been gassed with O_2 and CO_2. Then carbachol was added and the slices were incubated for a further 45 min. In the acute experiment, SDZ was added instead of carbachol to examine whether SDZ induces phosphoinositide hydrolysis. All reactions were stopped by addition of 150 μl of 10% trichloroacetic acid, the incubation mixture was centrifuged at 1800 rpm for 5 min, and 750 μl of the upper aqueous phase was collected. The aqueous phase was added to a column containing Dowex resin. After 8 ml of 5 mM sodium tetraborate containing 60 mM sodium formate was added to the column, 8 ml of 0.1 M formic acid/200 mM ammonium formate was added for extraction of [^{3}H]IP1. The samples of the extract were added to 8 ml of ACS-II and counted for radioactivity. All results are expressed as the percentage stimulation over the basal value of [^{3}H]IP1 formation in the absence of the agonist.

RESULTS

Agonist-stimulated-[^{3}H]IP1 Formation

Carbachol stimulated [^{3}H]IP1 formation in a concentration-related manner (Figure 1). The maximal response was obtained by 10^{-3}M carbachol and amounted to 516 % of the control.

Acute Experiment

Table 1 shows the effect of SDZ on the [^{3}H]IP1 forma-

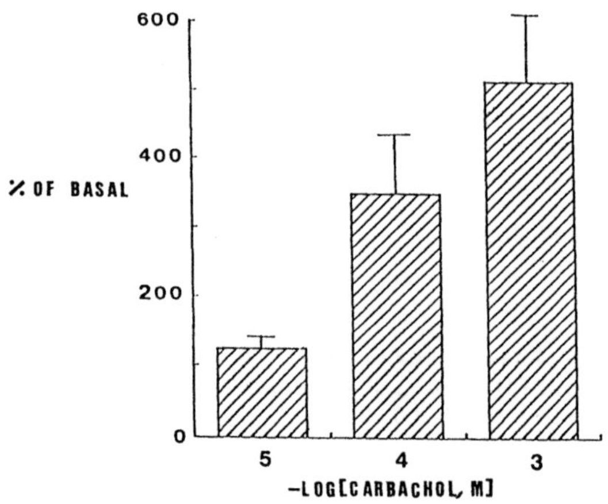

Fig 1. Carbachol-stimulated-[^{3}H]inositol-1-phosphate formation in rat cerebral cortex slices.

Table 1. Affinity and Relative Efficacy of SDZ BOP-086 in Rat Cerebral Cortex Slices.

	pD2	R.E.
SDZ BOP-086	5.0	41

pD2: -log of the dose producing half-maximal response
R.E.: Relative efficacy is expressed as percentage of the maximal response elicited by carbachol.

tion. These data suggest that SDZ stimulates phosphoinositide hydrolysis. The efficacy of SDZ was 41 % of that of carbachol.

Chronic Experiment

The effects of the chronic SDZ administration on carbachol-stimulated-[^{3}H]IP1 formation in rat frontal cortex slices are shown in Fig. 2. In the slices isolated from the animals treated chronically with SDZ, 10^{-5}M carbachol-stimulated-[^{3}H]IP1 formation was 160 %. In the control slices without the pretreatment with SDZ, 10^{-5}M carbachol-stimulated-[^{3}H]IP1 formation was 125 % (Figure 2). Thus there was no significant difference in formation between the two groups. On the other hand, [^{3}H]IP1 formation stimulated by 10^{-3}M carbachol in the SDZ-pretreated slices was 1310 %, and the carbachol-stimulated-[^{3}H]IP1 formation in the control slices was 516 % (Figure 2). These results indicate that the formation of [^{3}H]IP1 stimulated by 10^{-3}M carbachol was significantly enhanced by the chronic SDZ administration.

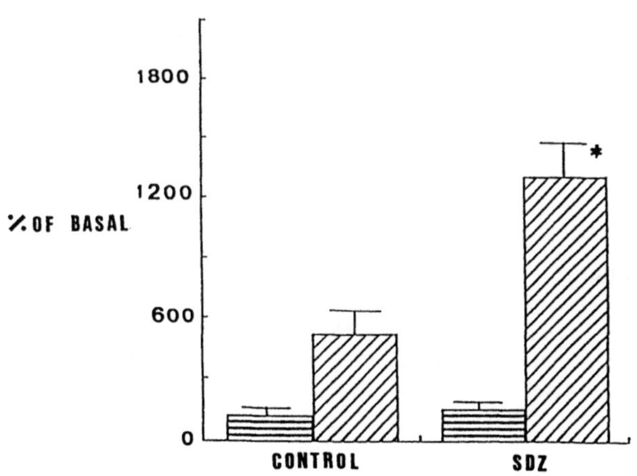

Fig 2. Carbachol (10^{-3}M, 10^{-5}M)-stimulated-[^{3}H]inositol-1-phosphate formation in rat cerebral cortex slices after chronic SDZ administration.
right: 10^{-3}M, left: 10^{-5}M.
*: $p < 0.05$ vs. control.

DISCUSSION

SDZ is a newly developed muscarinic cholinergic receptor agonist (with high affinity for and without selectivity between M1 and M2 receptor subtypes)[1]. SDZ dose not interact with other neurotransmitter receptors, such as monoamine neurotransmitter receptors. Therefore, SDZ is considered to be a selective and potent central and peripheral muscarinic cholinergic agonist.

It has been well recognized that the activation of muscarinic receptors is closely linked to the induction of phosphoinositide hydrolysis and increased IP1 formation in many tissues, including the cerebral cortex[2]. In the acute experiment SDZ stimulated IP1 formation and the maximal response of IP1 formation to SDZ was 41 % of that to carbachol. These results suggest that SDZ facilitates the central muscarinic cholinergic transduction, and may be effective in treatment of Alzheimer's disease.

In addition, the chronic consecutive SDZ administration markedly enhanced the carbachol-stimulated IP1 formation. These findings are unexpected and imply that the chronic SDZ administration may further facilitate the central muscarinic cholinergic transduction.

Therefore, it appears to be unlikely that the down-regulation of muscarinic receptors occurs during the course of chronic SDZ administration. The mechanism involved in the enhancement of carbachol-stimulated response by SDZ has not yet been unraveled. Since activation of muscarinic receptors is coupled to GTP-binding protein, facilitation of the coupling process by SDZ may be a potential mechanism involved. Investigation of the effects of SDZ on [^{3}H]GTP binding and its modification by cholera toxin and islet activating protein in rat cerebral slices may be relevant to clarify the mechanism.

In summary, acute administration SDZ stimulated IP1 formation in rat cerebral slices. The maximal response of IP1 formation to SDZ was 41 % of that to carbachol. The chronic SDZ administration significantly enhances the carbachol-stimulated IP1 formation in rat cerebral cortex slices. Therefore, SDZ may be effective in treatment of Alzheimer's disease.

ACKNOWLEDGMENTS

This study was supported in part by grants 63570500 from the Ministry of Education, Japan. SDZ BOP-086 was provided by Sandoz Ltd., Basel, Switzerland.

REFERENCES

1. Palacios, J.M., Pharmacology of SDZ BOP-086, (unpublished).

2. Brown, E. Kendall, D.A. and Nahorski, S.R., Inositol Phospholipid Hydrolysis in Rat Cerebral Cortical Slices, I: Receptor Characterization, J. Neurochem. 42:1379 1984.

EVALUATION OF "NOOTROPIC" DRUG IN RATS BY THE RADIAL MAZE TASK: PROBLEM OF ORDER IN ADMINISTERING SCOPOLAMINE AND TESTING DRUG AND EFFECT OF A SPECIFIC CHOLINERGIC AGONIST, SDZ-210-086

Hideko Kurata, Ken-ichi Fujimoto, Kunihiko Ikeguchi, and Mitsuo Yoshida

Department of Neurology, Jichi Medical School
Tochigi-ken, Japan

INTRODUCTION

Recently, the population of demented elderly has risen, and the lack of an effective treatment makes dementia a major medical and social problem (Brody, 1985). The most common cause of dementia is Alzheimer's disease (AD). Between 5 and 15% of the population in Japan over the age of 65 is estimated to be affected by this disorder. We are thus urged to develop an effective drug for dementia. In order to achieve this purpose, we must make an animal model of AD and establish a method to evaluate the effects of "nootropic" drugs.

As disturbance of cerebral cholinergic neurotransmission is a major feature in AD (Collerton, 1986) and scopolamine is a muscarinic cholinergic antagonist that disrupts cholinergic neurotransmissions, this drug can be used to make an animal model of AD. The radial maze task, which was devised by Olton and Samuelson (1976), has recently been used for screening nootropic drugs, and the performance of this task by rodents is impaired by scopolamine (Eckerman et al., 1980; Stevens, 1981; Watts et al., 1981). Therefore we tested a nootropic drug to see if it would reduce this impairment. Usually, the test drugs are administered before scopolamine injection; however, there is no agreement about the order in which to administer scopolamine and the test drug. So we administered a nootropic drug to rats both before and after the injection of scopolamine and tested the performance of the animals in an 8-arm radial maze. The nootropic drug we tried was (-)-(2S)-2,8-dimetyl-1,3-dioxa-8-azaspiro[4,5]decane hydrogen maleate (SDZ-210-086) (Sandoz Ltd., Basle, Switzerland), which is a specific muscarinic cholinergic agonist, and would be expected to reduce the impairment of spatial maze performance produced by scopolamine.

MATERIALS AND METHODS

Apparatus

The radial maze (Fig. 1), elevated 50 cm above the floor, consisted of an octagonal central platform (30 cm wide) surrounded by 8 equally spaced radial arms (60 cm long, 12 cm wide) with guillotine doors. Food wells, located 1 cm from the end of each arm, were 1.0 cm deep and 1.5 cm in diameter. The tests were carried out in a sound-attenuated room lit by 40 W fluorescent tubes. Several visually distinct objects (e.g., desk, shelf) were present in the room.

Procedure

Fifteen adult male Wistar rats, 8 weeks old at the start of experiments, were housed individually at 25±2 °C room temperature and on a 12 h-12 h light-dark schedule. Food deprivation was begun one week prior to testing. During this period, rats were handled daily for approximately ten min and fed a small quantity of sugar water on the platform. At the start of each session, 0.12 ml of 30% sugar water (food) was placed in each food well. Each rat was placed on the central platform and all guillotine doors were raised to allow access to any of the arms. If a rat chose 3 arms serially, we considered it not to use spatial memory but a pattern behavior, and all the guillotine doors were closed temporarily. When a rat entered an arm that it had not yet entered and ate the food, this was defined as a correct choice. However, if the rat re-entered this same arm, this was defined as a wrong choice or error. When the rat had eaten all the food or 10 min had elapsed, the session was terminated.

Training at the rate of one session per day continued until the rats achieved one of two criteria: (1) to choose 7 different arms with one or no error, and to repeat this performance 3 days in succession, or (2) to choose 7 different arms in the first 8 choices 5 days in succession, with 7 different arms being chosen first on the last day.

The next day, when each rat had met either criterion, scopolamine (0.5 mg/kg) was injected intraperitoneally 30 min prior to testing. After that, drug trials were performed. Before the next trial, each rat was required to choose 7 different arms in the first 8 choices, 2 days in succession. SDZ-210-086 was injected intraperitoneally 15 min after scopolamine. The dosages of SDZ-210-086 were 0.0625, 0.125, 0.25, 0.5,

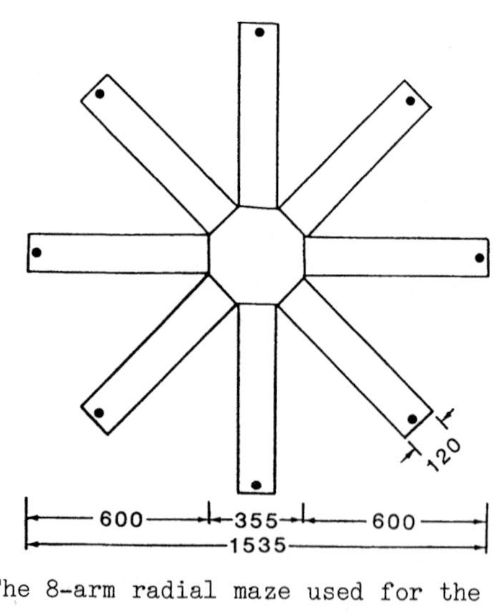

Fig. 1. The 8-arm radial maze used for the experiment.

and 1.0 mg/kg. Scopolamine was administered at 0.5 mg/kg. After this post-administration test of SDZ-210-086, scopolamine alone was tested. When the rate were ready again, SDZ-210-086 was injected 15 min before scopolamine. The dosages were 0.03125, 0.0625, 0.125, 0.25, 0.5, and 1.0 mg/kg. Scopolamine was administered at 0.5 mg/kg. After this pre-administration test of SDZ-210-086, scopolamine alone was tested again.

RESULTS

Data from 2 animals that did not run on the platform were excluded from statistical analysis. All the rest of the 13 rats which were used for the experiment met one or other of the criteria within 7.69±0.13 (mean±SE) days. On the day when each rat met a criterion, the mean number of correct choices out of the first 8 choices was 7.85±0.75 (mean±SE). On the next day, when scopolamine was injected 30 min prior to testing, the mean number of correct choices out of the first 8 choices was 5.46±0.29 (mean±SE). This decrease in the number of correct choices was statistically significant ($p<0.01$, Student's t-test).

Next, various doses of SDZ-210-086 were injected 15 min after scopolamine (post-administration trial). As shown in Table 1, so-called bell-shaped improvements of the spatial maze performance were observed with its dosage. The result of scopolamine only is the average of 3 trials, namely, before the post-administration of SDZ-210-086, between the post- and the pre-administration, and at the end of this test. A statistical difference was observed between the scopolamine only and the dosage of 0.25 mg/kg ($p<0.01$, Student's t-test). After another test of scopolamine only, various dosages of SDZ-210-086 were injected 15 min before scopolamine (pre-administration trial). As shown in Table 2, significant statistical improvement was observed with the dosages of 0.0625 mg/kg, 0.125 mg/kg, and 0.5 mg/kg ($p<0.05$, Student's t-test). With the dosage of 1.0 mg/kg, the mean number of correct choices was higher than the numbers with 0.0625 mg/kg and 0.5 mg/kg; however, there was no statistical significance because of the larger individual differences.

DISCUSSION

Until now, test drugs have usually been administered before scopolamine injection in tests to examine the effect of nootropic drugs on the spatial disorientation of rats mediated by scopolamine. When a cholinergic drug was administered before scopolamine, cholinergic receptors were occupied by the drug and afterwards scopolamine would be administered. However, in Alzheimer's disease, the cholinergic dysfunc-

Table 1. Post-Administration of SDZ-210-086

	Scopolamine only	Scopolamine and various dose of SDZ-210-086 .0625	0.125	0.25	0.5	1.0 (mg/kg)
Correct Choices	5.32	5.23	5.54	6.27	5.69	5.08 (mean)
out of First 8	0.41	0.43	0.45	0.30 **	0.33	0.47 (SE)

Correct choices in the first 8 selections in the radial maze task after the injection of scopolamine and the post-administration of various dosages of SDZ-210-086, 15 min after scopolamine injection.
**: $p<0.01$, Student's t-test

Table 2. Pre-Administration of SDZ-210-086

	Scopolamine only	Various dosage of SDZ-210-086 and Scopolamine .03125	.0625	0.125	0.25	0.5	1.0 (mg/kg)
Correct Choices	5.32	5.62	6.00	6.54	5.92	6.00	6.15 (mean)
out of First 8	0.41	0.37	0.29	0.29	0.40	0.41	0.37 (SE)
			*	*		*	

Correct choices in the first 8 selections in the radial maze task after the injection of scopolamine and the pre-administration of various doses of SDZ-210-086, 15 min before scopolamine injection. *: $p<0.05$, Student's t-test

tion is already present, and "nootropic drugs" are administered to the patients. Thus, when spatially disorientated rats are used as a model of Alzheimer's disease, scopolamine should theroretically be administered before testing drugs.

When SDZ-210-086 was administered after scopolamine, bell-shaped improvements of the spatial maze performance of rats were recognized with its dosage, and a statistical difference was recognized only with 0.25 mg/kg. When SDZ-210-086 was administered before scopolamine, improvement effects were shown over a wider dosage range; however, the individual differences were larger. In the pre-administration method, it seems to be easier to reduce the spatial disorientation mediated by scopolamine, but we possibly overestimate the effects of drugs by this method. From the results of our experiment, we conclude that the post-administration method should be used not only theoretically but also practically.

As for SDZ-210-086, it was disclosed that this potent specific muscarinic cholinergic agonist had the effect of reducing the impairment of spatial maze performance of rats mediated by scopolamine. It is especially important that the effect was confirmed by the post-administration method.

ACKNOWLEDGMENT

We thank Sandoz Ltd. for providing SDZ-210-086.

REFERENCES

Brody, J. A., 1985, Prospects for an aging population, Nature, 315: 463.

Collerton, D., 1986, Cholinergic function and intellectual decline in Alzheimer's disease, Neuroscience, 19:1.

Eckerman, D. A., Gordon, W. A., Edwards, J. D., MacPhail, R. C., and Gage, M. I., 1980, Effects of scopolamine, pentobarbital, and amphetamine on radial arm maze performance in the rat, Pharmac. Biochem. Behav., 12:595.

Olton, D. S., and Samuelson, R. J., 1976, Remembrance of places passed: Spatial memory in rats, J. Exp. Psychol. Anim. Behav. Proc., 2:97.

Stevens, R., 1981, Scopolamine impairs spatial maze performance in rats, Physiol. Behav., 27:285.

Watts, J., Stevens, R., and Robinson, C., 1981, Effects of scopolamine on radial maze performance in rats, Physiol. Behav., 26:845.

EFFECT OF DM-9384, A NEW PYRROLIDONE DERIVATIVE, ON PASSIVE AVOIDANCE AND CEREBRAL CHOLINE ACETYLTRANSFERASE ACTIVITY IN RATS

Hiroshi Kojima, Shinichi Kawajiri, Takeo Sakurai, and Terukiyo Yamasaki

Exploratory Research Laboratory II, Research Institute
Daiichi Pharmaceutical Co., Ltd.
16-13 Kitakasai 1-Chome Edogawa-ku, Tokyo 134, Japan

INTRODUCTION

It has been reported that DM-9384 (N-(2,6-dimethylphenyl)-2-(2-oxo-1-pyrrolidinyl) acetamide, Fig. 1) improves various types of amnesia such as the scopolamine-induced type and enhances acquisition of learning both negatively and positively reinforced (Sakurai et al., 1989). In the present study, we examined whether DM-9384 would improve one trial step-through passive avoidance failure in nucleus basalis magnocellularis (NBM)-lesioned rats, which is regarded as a reliable model of Alzheimer's disease. In addition, cerebral choline acetyltransferase (ChAT) activity in intact and NBM-lesioned rats was measured to clarify the possible mechanism of the cognition-enhancing effect of DM-9384.

MATERIALS AND METHODS

Animals

Male Wistar rats weighing 170-230 g were purchased from Japan SLC INC. The rats were housed in a room with controlled temperature (23±2°C), relative humidity (55±15%), and a 12-hour light-dark cycle. NBM were lesioned bilaterally with a lesion generator (RADIONICS, INC., 70°C, 20 sec) ; coordinates of the lesion site were A -1.5mm, L 2.7mm, H 7.5mm, according to Paxinos and Watson's atlas.

Compounds

DM-9384 and reference compounds, oxiracetam and bifemelane, were synthesized at our research institute. DM-9384 and oxiracetam were suspended in 0.5% sodium carboxymethyl cellulose solution, and bifemelane was dissolved in distilled water. These compounds were administered p.o. in a volume of 0.5 ml/100g body weight. Physostigmine sulfate (Merck) and

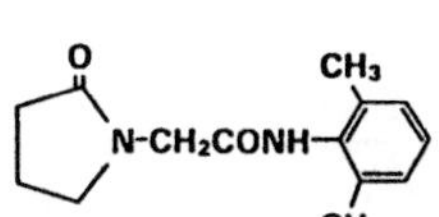

Fig. 1. Chemical structure of DM-9384

Basic, Clinical, and Therapeutic Aspects of Alzheimer's and Parkinson's Diseases, Volume 2
Edited by T. Nagatsu *et al.*, Plenum Press, New York, 1990

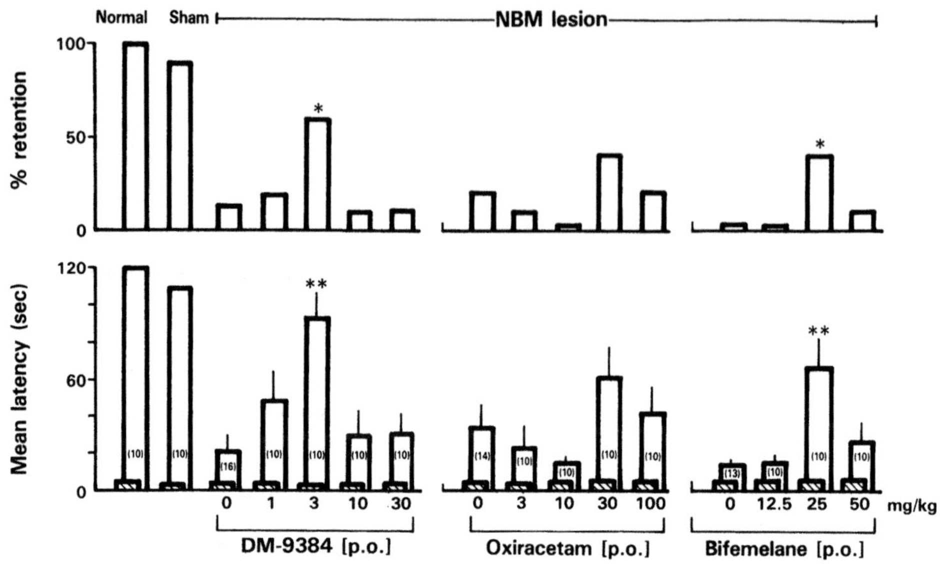

Fig. 2. Effects of DM-9384, oxiracetam and bifemelane on passive avoidance failure in NBM-lesioned rats. Drugs were administered 1 hour before training and test trial. % retention indicates the per cent of rats remaining in the light compartment for 120 sec in the retention test. Numbers in parenthesis indicate numbers of rats used. Shaded columns indicate the latency in training trial. Vertical bars represent SEM. * $P<0.05$, ** $P<0.01$ compared with value for vehicle control (0 mg/kg).

arecoline hydrochloride (Tokyo Chemical Industry) were dissolved in physiological saline and injected i.p. Their doses were expressed in terms of the salts.

Passive Avoidance Response

Light-dark passive avoidance apparatus of the step-through type (Sakurai et al., 1989) was used to test the amnesia.

Determination of Choline Acetyltransferase (ChAT) Activity

ChAT activity was assayed by the method of Fonnum (Fonnum, 1975). Protein content was determined by Lowry's method.

RESULTS AND DISCUSSION

Effects of DM-9384 and Reference Compounds on Passive Avoidance Failure in NBM-lesioned Rats

Results are shown in Fig. 2. Lesioning of the NBM produced marked failure of passive avoidance response. DM-9384 at a dose of 3 mg/kg,p.o., led to a significant improvement of this passive avoidance failure in NBM-lesioned rats. The dose-response curve was bell-shaped. The effects of oxiracetam and bifemelane were less potent than the effect of DM-9384. Cholinomimetics such as physostigmine and arecoline also showed potent ameriolating effects in this model (data not shown). These findings indicate that insufficiency of the cholinergic system may be involved in the passive avoidance deficit of this model.

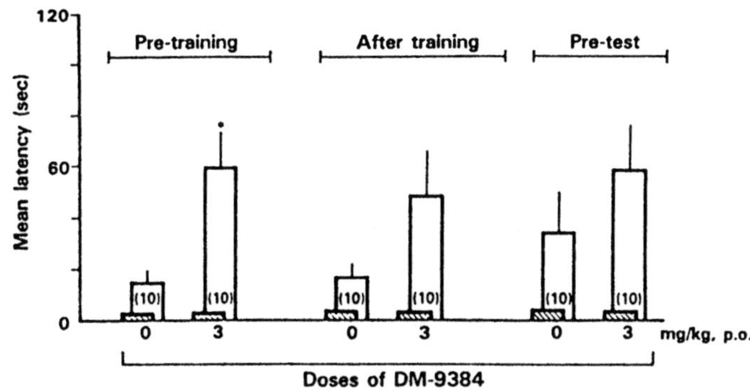

Fig. 3. Effects of administration time of DM-9384 on passive avoidance failure in NBM-lesioned rats. DM-9384 was administered singly 1 hour before training (Pre-training), immediately after training (After taining) or 1 hour prior to test trial (Pre-test), respectively. Other legends are the same as described in Fig. 2. * P<0.05 compared with value of vehicle control (0 mg/kg).

Influence of Administration Time of DM-9384 on Passive Avoidance Failure in NBM-lesioned Rats

Results are shown in Fig. 3. DM-9384 (3 mg/kg,p.o.) was significantly effective in ameriolating amnesia only when administered before training and was less effective given immediately after training or prior to the test trial, suggesting that DM-9384 may mainly affect those processes underlying memory acquisition and consolidation, effects which coincided with the results of a previous report (Sakurai et al., 1989).

Effect of DM-9384 on ChAT Activity in Normal Rat Brain

Result are shown in Table 1. DM-9384 at doses of 1, 3 and, 10 mg/kg, p.o./day for 14 days significantly increased ChAT activity in the cortex by approximately 7-9 % compared with the vehicle control value, whereas DM-9384 did not change the activity in the hippocampus.

Table 1. Effect of DM-9384 on ChAT Activity in Normal Rat Brain

		ChAT activity (pmol/mg protein/min)		
		DM-9384 (p.o., 2 weeks)		
Region	Control	1 mg/kg	3 mg/kg	10 mg/kg
Cortex	803 ± 9.7	873 ± 6.2 (9.4%)**	858 ±18.0 (6.9%)*	867 ±27.1 (8.0%)*
Hippocampus	1088 ±12.3	1122 ±32.9 (3.1%)	1143 ± 7.0 (5.1%)	1087 ±33.0 (0.0%)

Cortical tissue, excluding amygdala and olfactory bulb, was used in this experiment. DM-9384 was administered repeatedly for 14 days. Data indicate mean ± SEM (n=9). Percent increase in ChAT activity is given in parenthesis. * P<0.05, ** P<0.01 compared with control value (one-way analysis of variance and LSD).

Table 2. Effect of DM-9384 on ChAT Activity by Repeated Administration in NBM-lesioned Rat Brain

Treatment	ChAT activity (pmol/mg protein/min) Frontal cortex	Parietal cortex
Normal rat		
Control	1174 ± 49.1	1333 ± 51.3
NBM-lesioned rat		
Control	947 ± 35.7**	903 ± 50.6**
DM-9384 (3 mg/kg/day)	999 ± 43.8** (5.5%)	1028 ± 45.2** (13.9%)

NBM-lesioned rats were administered DM-9384 repeatedly for 15 days after surgery and decapitated 2 days later. Data indicate mean ± SEM (n=9). Percent increase in ChAT activity is given in parenthesis. ** $P < 0.01$ compared with normal control value (one-way analysis of variance and LSD).

Effect of DM-9384 on Decreased ChAT Activity in NBM-lesioned Rat Brain after Single and Repeated Administration

Single administration of DM-9384 at a dose of 3 mg/kg,p.o., failed to affect ChAT activity in the frontal cortex of normal rats. However, DM-9384 tended to potentiate the activity in the parietal cortex by 3.8 % compared with the control. NBM lesioning significantly lowered ChAT activity in both frontal and parietal cortices. DM-9384 did not influence the activity in the former cortex, but tended to cause recovery in the latter cortex, although the level of recovery was not statistically significant (data not shown). Furthermore, repeated administration of DM-9384 for 15 days recovered the decreased ChAT activity in both cortices by 5.5-13.9 % compared with NBM-lesioned control rats, although the recovery did not reach statistical significance (Table 2).

CONCLUSIONS

DM-9384, like cholinomimetics, showed an improvement of passive avoidance failure in NBM-lesioned rats, which animals are regarded as one of the reliable models of Alzheimer's disease. This drug was considered to affect selectively the processes underlying memory acquisition and consolidation. Decreased cortical ChAT activity may be one of the possible mechanisms whereby DM-9384 exerts its anti-amnesic effect in NBM-lesioned rats.

REFERENCES

Fonnum, F., 1975, A rapid radiochemical method for the determination choline acetyltransferase, *J. Neurochem.*, 24:407.

Sakurai, T., Ojima, H., Yamasaki, T., Kojima, H. and Akashi, A., 1989, Effects of N-(2,6-dimethylphenyl)-2-(2-oxo-1-pyrrolidinyl) acetamide (DM-9384) on learning and memory in rats, *Japan. J. Pharmacol.*, 50:47.

DM-9384, A PYRROLIDONE DERIVATIVE, AMELIORATES BASAL FOREBRAIN LESION-INDUCED AMNESIA AND INHIBITS CYCLOHEXIMIDE-INDUCED DECREASE IN THE NUMBER OF GABA AND ACETYLCHOLINE RECEPTORS

Tsutomu Kameyama, Toshitaka Nabeshima, Keiko Tohyama
Shin-ichi Ogawa, Kenshi Murase, and Sei-ichi Ishihara

Department of Chemical Pharmacology, Faculty of
Pharmaceutical Sciences, Meijo University, Nagoya 468
Japan

INTRODUCTION

DM-9384 [N-(2,6-dimethyl-phenyl)-2-(2-oxo-1-pyrrolidyl) acetamide] is a cyclic derivative of γ-aminobutyric acid (GABA), as is piracetam, which has already been marketed in several European countries as a nootropic drug. DM-9384 ameliorates several amnesic models including cycloheximide (CXM)-induced amnesia in animal experiments, but the mechanism of DM-9384 has not been defined well. We earlier reported that DM-9384 ameliorates GABA antagonist- and acethylcholine (ACh) antagonist-induced amnesia in mice (Nabeshima et al., 1987; Nabeshima et al., 1988b; Sakurai et al., 1989). It has also been indicated that CXM-induced amnesia is attenuated by DM-9384, and that the latter's antiamnesic effect is attenuated by treatment with a GABA antagonist such as bicuculline or picrotoxin. CXM-induced amnesia is known to be ameliorated by physostigmine, muscimol, and baclofen (Nabeshima et al., 1988a). These results suggest that the antiamnesic effect of DM-9384 may be developed via the GABAergic and the AChergic neuronal systems.

In the present experiments, to clarify the involvement of protein synthesis systems in the antiamnesic effect of DM-9384, we investigated whether DM-9384 ameliorates basal forebrain (BF) lesion-induced amnesia, whether DM-9384 protects the CXM-induced decrease in the number of binding sites for $GABA_A$ and muscarinic ACh (m-ACh), and whether DM-9384 displaces [^{3}H]muscimol and [^{3}H]QNB binding to $GABA_A$ and m-ACh receptors.

MATERIALS AND METHODS

Behavioral studies

Male Wistar rats (290-320 g) with a BF lesion were used for behavioral studies. An automated version of the 4-unit T-maze was used. Each rat had 1 training session per day for 10 days. The number of errors was recorded automatically. Step-through-type passive avoidance task was used. Retention test was carried out 24 hr after training. Step-through latency was recorded up to a maximum cutoff time of 300 sec.

Binding studies

Male ddY mice (30-35 g) were used for binding studies. Mice were administered DM-9384 (5 mg/kg, p.o.) 15 min before treatment with CXM (60 mg/kg, s.c.), and were sacrificed 24 hr after the administration of CXM. Then the whole brain excluding the cerebellum was rapidly removed and synaptic membranes (P_2B fractions) were prepared as described previously (Nabeshima et al., 1984). Incubations were carried out at 4 ℃ for 10 min (in the case of [^{3}H]muscimol binding) and at 25℃ for 30 min (in the case of [^{3}H]QNB binding). Non-specific binding was determined in the presence of 1 mM unlabeled GABA and 1 μM unlabeled atropine for [^{3}H]muscimol and [^{3}H]QNB binding, respectively. The binding reaction was stopped by rapid filtration through a Whatman GF/B filter, and then the filter was washed twice with 5 ml of ice-cold 50 mM Tris-HCl buffer (pH 7.4). Radioactivity was measured by liquid scintillation counting. Abilities of DM-9384 to displace the specific binding of [^{3}H]muscimol and [^{3}H]QNB were investigated in the concentration range of 10^{-3}～10^{-9} M. With regard to autoradiographic studies, frozen mouse brain was cut into 20-μm-thick coronal sections. The sections were thaw-mounted on gelatin-coated glass slides and incubated with [^{3}H]muscimol (at 4 ℃ for 30 min) or [^{3}H]QNB (at 25℃ for 120 min). After drying, the slide-mounted sections were juxtaposed with tritium-sensitive film and stored in a deep-freezer for 3 weeks.

Statistical Analysis

In the 4-unit T-maze, the effects of the drug were assessed by two-way analysis of variance (ANOVA). In the passive avoidance task, data were analyzed by the Kruskal-Wallis test followed by the Mann-Whitney's U-test. Scatchard plots were analyzed by the method of Rosenthal (1976) using a computer program to obtain the dissociation constant (Kd) and the maximum binding capacity (Bmax). The data were compared by Student's t-test. IC_{50} values were determined by a computer program for linear least-squares regression analysis from Hill plots of displacement curves.

RESULTS

Behavioral studies

BF-lesioned rats exhibited a larger number of errors compared with the sham-operated rats in 4-unit T-maze task ($F=96.87$, $P<0.01$). DM-9384 attenuated BF lesion-induced amnesia in terms of the number of errors (1 mg/kg: $F=23.57$, $P<0.01$; 3 mg/kg: $F=155.28$, $P<0.01$), as did aniracetam (3 mg/kg: $F=101.15$, $P<0.01$; 10 mg/kg: $F=43.18$, $P<0.01$, data not shown).

BF-lesioned rats showed an impaired retention on the test day compared with sham-operated rats in passive avoidanse task ($P<0.05$). Repeated administration of DM-9384 (3 mg/kg) over the 14 days significantly increased retention latency ($P<0.05$). Aniracetam (10 mg/kg) also significantly increased the retention latency ($P<0.05$, data not shown).

Table 1. Effects of DM-9384 on the CXM-induced Decrease in Number of Binding Sites of $GABA_A$ and m-ACh Receptors

Treatment	N	Kd (nM)	Bmax (fmol/mg protein)	Bmax % of control
[³H]muscimol binding				
Saline	3	13.1±1.19	991.5 ±47.9	100.0
CXM	3	11.0±0.70	700.0 ±27.3**	70.6
DM-9384 + CXM	3	12.3±1.39	860.3 ±47.6ᵃ	86.8
[³H]QNB binding				
Saline	3	0.44±0.04	195.6 ±5.4	100.0
DM-9384	3	0.44±0.03	271.3 ±9.9 **	138.7
muscimol	3	0.40±0.04	193.1 ±8.2	98.7
CXM	3	0.37±0.01	171.4 ±1.7 *	87.6
DM-9384 + CXM	3	0.54±0.03ᵃᵃ	233.1 ±7.1 *, ᵃ	119.2
muscimol + CXM	3	0.40±0.02	177.8 ±3.8	90.9

*$P<0.05$, **$P<0.01$ vs. saline, ᵃ $p<0.05$, ᵃᵃ$p<0.01$ vs. CXM

Binding studies

As shown in Table 1, CXM (60 mg/kg) significantly decreased Bmax of [^{3}H]muscimol binding sites to 70.6 % of the control group value, but did not affect the Kd value of the sites. Bmax of [^{3}H]QNB binding sites also significantly decreased, to 87.6 % of the control group, but the Kd value of the sites was not altered. Pretreatment with DM-9384 (5 mg/kg) protected both [^{3}H]muscimol and [^{3}H]QNB binding sites from the decreases in Bmax induced by CXM. Furthermore, DM-9384 itself increased the Bmax in [^{3}H]QNB binding sites.

DM-9384 strongly displaced the specific binding of [^{3}H]muscimol to $GABA_A$ binding sites, but not that of [^{3}H]QNB binding to muscarinic ACh binding sites (data not shown). The value of -log IC_{50} of DM-9384 was calculated as 8.07±1.88 M.

DISCUSSION

CXM, a protein synthesis inhibitor, is a well-known amnesic drug, and in fact is widely cited as a primary model for several types of dementia in humans. However, the mechanism of CXM affecting memory is still not clear. Probably CXM may inhibit the protein synthesis related to the cholinergic and GABAergic neuronal systems that play an important role in the memory process (Kameyama et al., 1986; Nabeshima et al., 1988a), because ACh and GABA agonists ameliorate the amnesia induced by CXM (Nabeshima et al., 1988a).

We have already shown that DM-9384 ameliorates GABA antagonist-, ACh antagonist-, and CXM-induced amnesia. These results suggest that DM-9384 is able to ameliorate the dysfunction in GABAergic and AChergic neuronal systems induced by CXM.

The results of the behavioral studies indicated that DM-9384 ameliorates the BF lesion-induced amnesia model, which is considered to be an animal model of Alzheimer's disease in both 4-unit T-maze and passive avoidance. The antiamnesic effects of DM-9384 were stronger than those of aniracetam.

To clarify the mechanism of DM-9384, we investigated how DM-9384 acts to protect GABA and ACh receptors from the decrease in Bmax induced by CXM. Our results indicated that DM-9384 not only protected against decreases in the number of $GABA_A$ and muscarinic ACh receptors, but also itself increased the number of muscarinic ACh receptors. Furthermore, DM-9384 displaced the specific binding of $[^3H]$muscimol to both synaptic membranes and brain slices, but not that of $[^3H]$QNB. Taking these results together, we conclude that DM-9384 protects against the CXM-induced decreases in the number of GABA and ACh receptors by interacting with $GABA_A$ receptors and increasing protein synthesis and that DM-9384 may have a preventative effect on Alzheimer's disease.

ACKNOWLEDGMENT

We are grateful to Daiichi Pharmaceutical Co. Ltd. for their supply of DM-9384.

REFERENCES

Kameyama, T., Nabeshima, T., and Noda, Y., 1986, Cholinergic modulation of memory for step-down type passive avoidance task in mice, Res. Commun. Psychol. Psychiatr. Behav., 11:193.

Lowry, O. H., Rosenbrough, N. J., Farr, A. L., and Randall, R. J., 1951, Protein measurement with the Folin phenol regent, J. Biol. Chem., 193:265.

Nabeshima, T., Noda, Y., and Kameyama, T., 1987, Effect of DM-9384, a pyrrolidone derivative, on amnesia model animal having GABAergic neuronal dysfunctions, Abstract of 10th Int. Natl. Cong. Pharmacol. (Sydney), p.301.

Nabeshima, T., Noda, Y., Itoh, K., and Kameyama, T., 1988a, Role of cholinergic and GABAergic neuronal systems in cycloheximide-induced amnesia in mice, Pharmacol. Biochem. Behav., 31:405.

Nabeshima, T., Noda, Y., Matsuno, K., Furukawa, H., and Kameyama, T., 1984, Phencyclidine interacts with $serotonin_2$ receptor, but not $serotonin_1$ receptor, Res. Commun. Subst. Abuse., 5:81.

Nabeshima, T., Noda, Y., Tohyama, K., Harrer, S., and Kameyama, T., 1988b, Antiamnesic effects of DM-9384, a pyrrolidone derivative, on drug-induced amnesia animal models, Psychopharmacology Suppl., 96:305.

Rosenthal, H. E., 1967, A graphic method for the determination and presentation of binding parameters in a complex system, Anal. Biochem., 20:525.

Sakurai, T., Ojima, H., Yamasaki, T., Kojima, H., and Akashi, A, 1989, Effects of N-(2,6-dimethyl-phenyl)-2-(2-oxo-1-pyrrolidinyl) acetamide (DM-9384) on learning and memory in rats, Japan. J. Pharmacol., 50:47.

THE EFFECT OF CHINESE HERBAL MEDICINE TJ-23 (TOKI-SYAKUYAKU-SAN) ON THE CHOLINE ACETYLTRANSFERASE ACTIVITY IN THE CORTEX OF RAT

Yaeko Nagatsuka, Ken-ichi Fujimoto, Kunihiko Ikeguchi, and Mitsuo Yoshida

Department of Neurology
Jichi Medical School, Tochigi-ken, Japan

INTRODUCTION

Recently, with the increase of the average span of human life, the population of demented elderly has risen. Within various causes of dementia, the most common cause of dementia is Alzheimer's disease (AD). The pathophysiology of AD has not been clarified yet and no effective treatment has been developed.

Some herbal medicines, which have been handed down in China, are known to improve psychiatric symptoms, however, as is usually the case of oriental medicine, the mechanism of action has not been clarified. Toki-Syakuyaku-San (TJ-23) is a mixture of these herbs. For the first step, we investigated the effect of TJ-23 on the presynaptic cholinergic function.

MATERIALS AND METHODS

Animals

Thirty senescent male Wistar rats, 20 months of age at the start of experiment, were used. Each rat was housed individually in a wire mesh cage 35(D) × 25(W) × 25(H)cm and fed a solid diet of CE-2 (Nippon Clea Co., Tokyo) with or without TJ-23, and tap water *ad libitum*. The animals were maintained at 25±2°C room temperature and on a 12h light-dark schedule (light period; 7:00-19:00, dark period; 19:00-7:00).

Administration of Medicine

Dried and powdered extract of TJ-23 was obtained by decocting a mixed compound of six herbal drugs listed in the Pharmacopeia of Japan No.11; i.e. peony root (4.0g), atractylodes lancea rhizome (4.0g), alisma rhizome (4.0g), hoelen (4.0g), cnidium rhizome (3.0g) and Japanese angelica root (3.0g); being supplied by Tsumura and Co., Tokyo. We prepared a special solid diet, CE-2 which contained 2.17% of dried and powdered extract of TJ-23. We used this CE-2 and control CE-2 without TJ-23. The experimental group of 20 rats was given the experimental CE-2 at a dosage level of 1g/kg/day of TJ-23. The control group was given CE-2 without TJ-23.

Basic, Clinical, and Therapeutic Aspects of Alzheimer's and Parkinson's Diseases, Volume 2
Edited by T. Nagatsu *et al.*, Plenum Press, New York, 1990

Table1. Effect of TJ-23 on Frontal Cortex of Rat

	Control	TJ-23 for 1W	TJ-23 for 2W
ChAT activity (μmol/h/g)	1.55±0.11	1.57±0.12 N.S.	1.80±0.24 **

The effect of chronic administration of "Toki-Syakuyaku-San" (TJ-23) on the choline acetyltransferase activity in the frontal cortex of senescent rat. mean±SD. N.S.: not significant. **: $p<0.01$, Student's t-test.

Assay of choline acetyltransferase activity

A total of 20 experimental rats and 10 control rats were treated over a period of 2 weeks. After 1 week of administration, 10 rats were selected randomly from the experimental group, anesthetized with ether, and sacrificed by decapitation. The brains were removed rapidly and stored at -80°C until assayed. After 2 weeks of administration, the remainder of the experimental group and the 10 control rats were treated similarly. Frozen sections of the frontal cortex were cut out with a razor blade and divided into 2 bilateral pieces. After determination of wet weight, the samples were homogenized with Sonifier in 0.2% Triton X-100 containing 10 mM EDTA, pH7.4. Choline acetyltransferase (ChAT) activities were assayed by radiochemical method (Fonnum, 1969) with an acetyl group donor, 1-(^{14}C)acetyl coenzyme A (New England Nuclear, Boston, MA.), of a specific activity of 51.3 mCi/mmol. The time of incubation was 15 min at 37°C. The ChAT activity was expressed as μmol/g wet weight/h.

RESULTS

As show in the Table, the ChAT activity in the frontal cortex of the control rats was 1.55±0.11 μmol/g wet weight/h (mean±SD, n=20). The administration of TJ-23 for 1 week, which showed no significant difference from the control group, was 1.57±0.12 μmol/g wet weight/h (n=20). However, the administration for 2 weeks did increase the ChAT activity to 1.80±0.24 μmol/g wet weight/h (n=20). The increase of ChAT activity was about 17% compared with that of the controls and statistically significant ($p<0.01$, Student's t-test).

DISCUSSION

From the 1960's, the adverse effects of anticholinergic drugs on memory function have been extensively studied (Collerton, 1986). In the mid 1970's, the marked decrease of the activity of ChAT, a synthetizing enzyme of acetylcholine, in the brains of patients dying with AD was discovered (Bowen et al., 1976; Davies and Maloney, 1976; Perry et al.,1977). It was also reported that the ChAT levels correlated with dementia rating scales (Perry et al., 1978) and pathological markers of AD, such as the number of plaques, tangles, or cell loss (Mountjoy, 1986). In 1981, selective loss of cholinergic neurons, whose cell bodies lie in the basal forebrain and provide widespread innervation of the cerebral cortex, was discovered (Whitehouse et al., 1981). Instead of this impairment of presynaptic cholinergic mechanism, muscarinic cholinergic receptors in the cortex are preserved (Bartus et al., 1982;

Collerton, 1986). Therefore, clinical trials to replace the decreased acetylcholine, such as choline precursor, ChAT activator, choline esterase inhibitor, and direct agonist of the cholinergic receptor, have been extensively made, but so far no sufficient effects have yet been obtained (Collerton, 1986).

Administration of TJ-23 for 1 week did not alter the ChAT activity of the frontal cortex of the senescent rat, but the administration for 2 weeks did increase the ChAT activity about 17%, compared with that of the controls. This suggest that TJ-23 has a chronic ChAT activating effect. Some drugs have already been reported to increase the ChAT activity in the brain of rats, however, the values show at most a 10% increase; therefore, in comparison with those drugs, the effect of TJ-23 to activate ChAT is strong. The fact that this herbal medicine, which has been handed down for use on persons suffering from psychiatric symptoms, has the effect of increasing ChAT activity deserves further investigation.

TJ-23 has now widely been used for sterile women and those with menopausal complications in Japan, and little side effects have been reported. The fact that TJ-23 increased the ChAT activity in the brain of the senescent rat could lead to the possibility that TJ-23 might increase the ChAT activity known to be significantly decreased in the brain of the patient with AD.

ACKNOWLEDGMENT

We thank the laboratory of Tsumura and Co., Ltd. for providing Toki-syakuyaku-san (TJ-23).

REFERENCES

Bartus, R. T., Dean, R. L., Beer, B., and Lippa, A. S., 1982, The cholinergic hypothesis of geriatric memory dysfunction, Science, 217:408.

Bowen, D. M., Smith, C. B., White, P., and Davison, A. N., 1976, Neurotransmitter-related enzymes and indices of hypoxia in senile dementia and other abiotrophies, Brain, 99:459.

Collerton, D., 1986, Cholinergic function and intellectual decline in Alzheimer's disease, Neuroscience, 19:1.

Davies, P., and Maloney, A. J., 1976, Selective loss of central cholinergic neurons in Alzheimer's disease, Lancet, 2:1403.

Fonnum, F., 1969, Radiochemical micro assays for the determination of choline acetyltransferase and acetylcholinesterase activities, Biochem. J., 115:465.

Mountjoy, C. Q., 1986, Correlations between neuropathological and neurochemical changes, Brit. Med. Bull., 42:81.

Perry, E. K., Perry, R. H., Blessed, G., and Tomlinson, B. E., 1977, Necropsy evidence of central cholinergic deficits in senile dementia, Lancet, 1:189.

Perry, E. K., Tomlinson, B. E., Blessed, G., Bergmann, K., Gibson, P. H., and Perry, R. H., 1978, Correlation of cholinergic abnormalities with senile plaques and mental test scores in senile dementia, Brit. Med. J., 2:1457.

Whitehouse, P. J., Price, D. L., Clark, A. W., Coyle, J. T., and DeLong, M. R., 1981, Alzheimer's disease: evidence for selective loss of cholinergic neurons in the nucleus basalis, Ann. Neurol., 10:122.

EFFECT OF TJ-23 ON SCOPOLAMINE-INDUCED DEFICIT OF LEARNING AND MEMORY IN RATS

Michihiro Fujiwara, Katsunori Iwasaki, and Tatsuhiko Ogata

Department of Physiology and Pharmacology, Faculty of Pharmaceutical Sciences, Fukuoka University
Fukuoka 814-01, Japan

INTRODUCTION

A traditional Kampo prescription, TSUMURA Toki-Shakuyaku-San(TJ-23), consisting of Anqelicae Radix(3g), Pasoniae Radix(4g), Cnigi Rhizoma(4g), Atractylodis Lanceae Rhizoma(4g), Alismatis Rhizoma(4g), and Hoelen(4g), has been used for the treatment of various clinical symptoms associated with gynecological disease. Pharmacological studies indicated that TJ-23 causes an increase in peripheral blood flow by its vasodilative action and diuretic action, etc. TJ-23 also ameliorates ovarian dysfunction by increasing estrogen through the hypothalamic-pituitary stimulation.[1,2]

Recently, TJ-23 was found to significantly increase the nicotine receptor numbers and noradrenaline content in the cerebral cortex. From animal studies, the cholinergic function has been shown to be very important for accurate cognitive performance. It is also known that cognitive function is impaired with muscarinic or nicotinic receptor blockers and this cognitive dysfunction is improved by cholinergic and/or noradrenergic drugs.[3] These reports suggest the possibility of a cognition enhancing effect of TJ-23. This experiment was thus undertaken to examine this possibility in an animal model of cognitive disruption.

ANIMALS AND METHODS

Animals

Male Wistar rats weighing 200-250g were used and housed under conditions of 22 ± 1 ℃ and reversed 12:12 light-dark cycles. The rats were allowed free access to water but were subjected to a restricted feeding schedule to maintain their body weight at 80-85% of the ad libtum level. Male ddY mice weighing 20-25g was used for the study of oxotremorine-induced tremor.

Apparatus

1)Radial-arm maze task. Spatial learning and memory were examined by use of an eight-arm radial maze similar to that described by Olton and Samuelson.[4] The maze consisted of an octagonal central platform(24cm distance between opposite sides) with 8 radial arms(48cm long and 10cm wide, with 30cm high side walls made of transparent plexiglass). A food pellet(10mg sugar chip) was placed at the distal end of each arm. In acquisition training, one trial per day, the rat was left in the maze either until he obtained all 8 pellets or until 10 min had elapsed. Correct choice was defined as entering an unchosen arm during that trial. Error was recorded when the rat reentered an in which no food remained. Scopolamine(i.p.) and TJ-23(p.o.) were administered 30 and 60 min respectively before the retention test in eight-arm radial maze task.

2)Passive avoidance task. A two-compartment step-through apparatus consisting of illuminated (25 × 10 × 25cm) and dark chamber (30 × 30 × 30cm) equipped with a grid floor was used. The two chambers were separated by a guillotine door (8 × 8cm). Since normal rats prefer dark to light, they enter the dark chamber within 15 sec. In the acquisition trial, they received an unavoidable foot-shock (AC:3 mA for 5 sec) through the grid floor. At 24 hour after the acquisition trial (retention test), the rat was placed in the illuminated chamber, and the latency to enter the dark chamber was measured for 600 sec at the maximum.

3)Assay of brain monoamines. The rat was sacrificed by decapitation and the brain was immediately removed. The frozen coronal section was punched out and regional samples were homogenized with an ultrasonic homogenizer. Following centrifugation, the supernatant was injected into HPLC-ECD.

4)Oxotremorine-induced tremor. TJ-23 was studied for its possible potentiating effect on oxotremorine-induced tremor which act directly on muscarinic acetylcholine receptors. Observation of tremor was made at 5min intervals starting 30 min after the oxotremorine treatment. The severity of tremor was scored in 5 grades; 0:absent, 1:mild, 2:moderate, 3:marked, and 4:severe.

RESULTS

Eight-arm radial maze task

Figure 1 shows the effect of TJ-23 on the number of correct choices (CC) and error choices(EC) in the first 8 choices. The control trained rats made approximately 8 correct choices and no errors in the first 8 choices (CC;7.8, EC;0.2 in Fig.1). Scopolamine at 0.5mg/kg,i.p., significantly decreased CC (5.8) and increased EC (8.0), indicating the disruption of spatial cognition. Oral administration of TJ-23 at 100 and 200 mg/kg, significantly reversed the decreased CC and increased EC in scopolamine- treated rats. At 500 mg/kg, however, The drug did not increase CC. Similar effects were also observed with so-called nootropic drugs such as tetrahydroaminoacridine, aniracetam and amantadine.

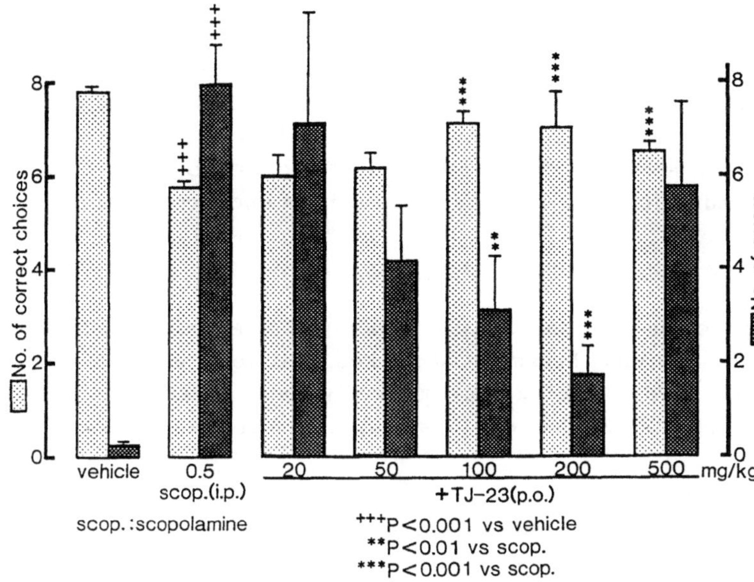

Fig.1 Effect of TJ-23 on the scopolamine-induced deficit of spatial cognition in rats

Figure 2 shows the changes in brain noradrenaline(NA) levels after TJ-23 (200mg/kg,p.o.) was given to scopolamine-treated rats. Brain NA levels in frontal cortex(FC), septal area(SEP), and dorsal hippocampus(DH) were decreased by scopolamine-treatment. TJ-23 increased to normal NA levels in these areas in accordance with improvement of spatial cognition.

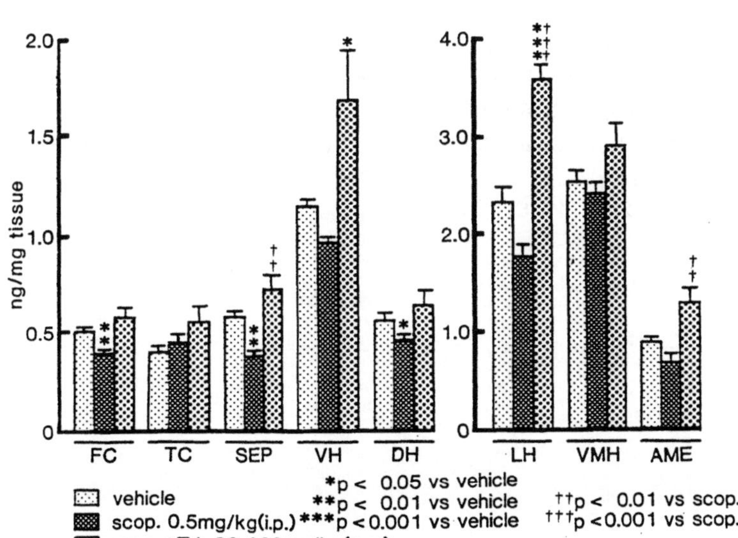

Fig.2 Effect of TJ-23 on brain NA levels in scopolamine-treated rats

Passive avoidance task

The mean latency to enter the dark chamber in the retention test of passive avoidance (24 hour after the acquisition test) was approximately 550 sec in normal rats. This latency was reduced to 60 sec in scopolamine (3 mg/kg,i.p.)-treated rats. TJ-23 (100 and 500 mg/kg,p.o.) prevented markedly the scopolamine-induced disruption of passive avoidance learning.

Oxotremorine-induced tremor potentiation

Oxotremorine (0.5mg/kg,i.p.)- induced tremor, indicating central acetylcholine activity, became most marked at 10 to 15 min and subsided with time. TJ-23 (100-500mg/kg,p.o.) markedly potentiated this tremor. This suggests that TJ-23 also has a stimulating effect on central cholinergic systems.

In conclusion, TJ-23 appears to improve the cognitive dysfunction caused by central cholinergic dysfunction, possibly by facilitating the cholinergic and/or noradrenergic mechanisms, which suggests the possibility of its usefulness in the treatment of Alzheimer's disease without any side effects.

REFERENCES

1. S. Usuki, Effect of Hachimijiogan, Tokishakuyakusan, and Keishi-bukuryou-gan on the estrogen and progesterone secretion by rat preovulatory follicles. Am.J.Chim.Med. 14:161 (1986).
2. N. Hagino, The role of maternal nicotine in the development of nicotine acetylcholine receptors and hypothalamic dopaminergic neurons in the fetal brain, in: "Developmental Neuroscience: Physiological, Pharmacological and Clinical Aspects" F. Caciagli, ed., Elsevier Science Publishers, B.V. (1984)
3. M. Fujiwara, K. Iwasaki, and Y. Matsumoto, Involvement of brain noradrenaline in spatial cognition in rats. Japan.J.Pharmacol. 46:18 (1988)
4. D. S. Olton and R. J. Samuelson, Remembrance of place passed: Spatial memory in rats. J.Exp.Psychol. 2:97 (1976)

ADVANTAGES OF A VISUAL RECOGNITION TASK TO DETECT COGNITIVE FACILITATION BY CHOLINOMIMETIC DRUGS IN PRIMATES

N.M.J. Rupniak and S.D. Iversen

Merck Sharp and Dohme Research Laboratories, Neuroscience Research Centre, Terlings Park, Eastwick Road, Harlow Essex, CM20 2QR, U.K.

Improvements in cognition in Alzheimer's disease have most reliably been obtained using cholinergic agonists such as physostigmine (see Kopelman, 1986). However, therapy with existing cholinergic agents has met with limited success owing to dose-limiting side-effects such as nausea, short duration of action and poor efficacy.

Preclinical screens for novel cholinomimetics in animals often rely on the ability of test compounds to reverse behavioural disruption induced by scopolamine. This strategy may be associated with several drawbacks. For example, in man scopolamine mimics some, but not all, aspects of cognitive disruption seen in dementia (Kopelman, 1986). Therefore, drugs which might be of benefit in Alzheimer's disease could appear inactive in a scopolamine-reversal paradigm. Such models also require the test compound to compete with and displace scopolamine from its cerebral binding sites. This pharmacological blockade clearly does not reflect the situation in Alzheimer's disease, where agents which might be unable to displace scopolamine could still gain access to postsynaptic muscarinic receptors. In addition, the presence of scopolamine would mask any adverse signs associated with muscarinic toxicity, and it would therefore not be possible to predict or compare the likely safety of different compounds.

One animal model in which it is not necessary to impair performance with scopolamine in order to reveal cognitive enhancement by physostigmine is a visual recognition task for primates (Aigner and Mishkin, 1986). In recent experiments (Rupniak et al, 1989) we have compared the ability of several cholinomimetic agents to improve performance in this test and to reverse the disruptive effects of scopolamine in a spatial task first described by Bartus (1978).

In the spatial delayed response task, the monkey sits in front of a touch-sensitive television screen upon which visual stimuli may be presented. At the start of a trial, a single white square is illuminated in one of nine possible spatial locations selected at random (see Figure). When the monkey touches this stimulus, the screen becomes blank and the retention interval ensues (either 2, 10 or 20 sec duration). At the end of this delay, a 3 x 3 choice matrix with all nine positions simultaneously occupied is presented. The

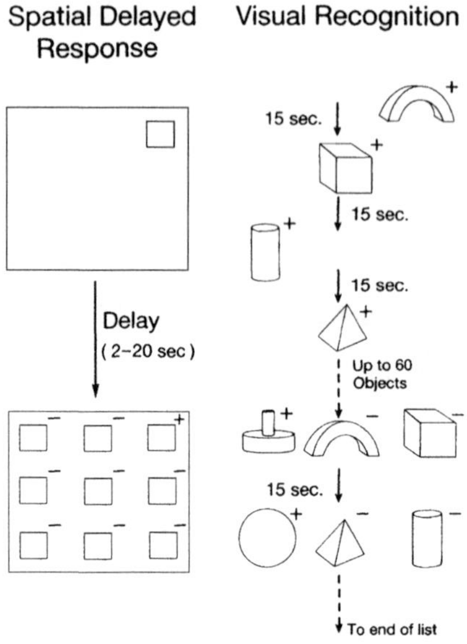

Figure 1

monkey is required to touch the square occupying the same location as the sample in order to obtain a food reward. Rhesus monkeys' performance normally approahces 100% correct, but following treatment with scopolamine choice accuracy after long (but not short) retention intervals falls close to chance levels (Bartus and Johnson, 1976).

The visual recognition task taxes the monkeys' memory by presenting a serial list of 3-dimensional junk objects, all of which are required to be remembered. During acquisition the monkey is shown a series of objects, one at a time every 15 sec, and displaces them in order to obtain a reward (see Figure). Once each member of the list has been seen, the retention test follows, in which the same objects are presented in the same serial order, two at a time, along with a novel object not presented before. The monkey is required to select the novel object to obtain a reward. By adjusting the number of objects presented in the list in accordance with the individual monkey's ability, it is possible to maintain performance in the undrugged state at around 60-70% correct, well above chance levels (33%), but with sufficient room for improvement by drug treatment.

In rhesus monkeys, physostigmine was able to fully reverse the effects of scopolamine on spatial delayed response performance. The minimum effective dose was 0.03mg/kg i.m. (coadministered with 0.03mg/kg of scopolamine). In the presence of scopolamine, no adverse signs were observed using up to 0.08mg/kg of physostigmine. In the visual recognition task, where physostigmine was administered alone, choice accuracy was improved from around 55 to 70% correct with a minimum effective dose of 0.001mg/kg (30 times lower than that required for scopolamine reversal; see Table). No overt adverse signs were observed, but two animals refused to be tested at a dose of 0.06mg/kg. Unlike physostigmine, pilocarpine was able to only partially reverse the scopolamine-induced impairment (minimum effective dose: 2.0mg/kg). At doses of 4 or 5mg/kg monkeys salivated and retched or vomited. Visual recognition performance was improved by a dose 16 times lower than that required to overcome the cognitive

Table 1

Comparison of Cognitive Facilitation in Two Primate Memory Tasks

Drug	Minimum Effective Dose (mg/kg i.m.) Scopolamine Reversal	Visual Recognition	Difference
Physostigmine	0.03	0.001	30 fold
Pilocarpine	2.0	0.125	16 fold
Arecoline	>1.8	0.1	>18 fold

effects of scopolamine (0.125mg/kg; Table). Salivation and emesis were induced by 0.5 and 0.75mg/kg of pilocarpine. Unlike these agents, doses of up to 1.8mg/kg of arecoline did not reverse cognitive disruption induced by scopolamine. Higher doses were not examined owing to the appearance of toxic signs (pallor, tremor or ataxia), particularly during the first 10 min after administration. However, in the absence of scopolamine, visual recognition was improved by 0.1mg/kg of arecoline in 3 out of 4 animals, who also exhibited salivation and emesis; the fourth animal refused to be tested (Table).

As expected, our findings show that the potency of muscarinic agonists to improve performance in the visual recognition paradigm is at least one order of magnitude lower than for reversal of the cognitive effects of scopolamine. This difference in potency appears to vary considerably for different compounds, suggesting that the therapeutic dose range might not be readily predicted from scopolamine reversal models. For example, arecoline appeared unable to reverse the effects of scopolamine at sub-toxic doses, presumably because of its relatively poor brain penetration. However, in other animal models which do not employ scopolamine (Ridley et al, 1987; present study) and in Alzheimer patients (Christie et al, 1981) arecoline does improve cognition.

The therapeutic use of cholinomimetics as cognitive enhancers is blighted by dose-limiting side-effects such as nausea. Allowing for possible differences between species and routes of administration, the visual recognition paradigm in primates may provide extremely valuable information about the likely active dose range and therapeutic index of new drug candidates. This task also enables detection of noncholinergic agents which might be of value as cognitive enhancers in Alzheimer's disease.

References

Aigner T.G., Mishkin M. (1986). The effects of physostigmine and scopolamine on recognition memory in monkeys. Behav. Neural. Biol. 45: 81-87.

Bartus R.T. (1978). Evidence for a direct cholinergic involvement in the scopolamine-induced amnesia in monkeys: effects of concurrent administration of physostigmine and methylphenidate with scopolamine. Pharmacol. Biochem. Behav. 9: 833-836.

Christie J.E., Shering A., Ferguson J., Glen A.I.M. (1981). Physostigmine and arecoline: effects of intravenous infusions in Alzheimer presenile dementia. Br. J. Psychiat. 138: 46-50.

Kopelman M.D. (1986). The cholinergic neurotransmitter system in human memory and dementia: a review. Q.J. Exp. Psychol. 38A: 535-573.

Ridley R.M., Baker H.F., Drewett B. (1987). Effects of arecoline and pilocarpine on learning ability in marmosets pretreated with hemicholinium-3. Psychopharmacology 91: 512-514.

Rupniak N.M.J., Steventon M.J., Field M.J., Jennings C.A., Iversen S.D. (1989). Comparison of the effects of four cholinomimetic agents on cognition in primates following disruption by scopolamine or by lists of objects. Psychopharmacology 99: 189-195.

TETRAHYDROAMINOACRIDINE AND SOME OF ITS ANALOGUES: EFFECTS ON THE CHOLINERGIC SYSTEM

A. Adem, A. Mohammed, A. Nordberg* and B. Winblad

Department of Geriatric Medicine, Karolinska Institute Stockholm and *Department of Pharmacology, University of Uppsala, Uppsala, Sweden

INTRODUCTION

Alzheimer's disease/senile dementia of Alzheimer type (AD/SDAT) is one of the most disabling conditions affecting the aged and is the major cause of dementia among the elderly people. AD/SDAT is characterized by changes in several transmitter systems of the brain (Gottfries, 1985; Hardy et al., 1985). One of the most pronounced changes has been found in the cholinergic system (Perry, 1986). The hypothesis that a defect in the cholinergic system is involved in AD/SDAT has wide acceptance in terms of a therapeutic strategy.
9-Amino-1,2,3,4-tetrahydroacridine (THA), a reversible acetylcholinesterase inhibitor, has been reported to improve performance in Alzheimer patients (Summers et al., 1986). In this study we have examined some properties of THA in-vitro (human and rat tissue) and in-vivo (rats) to define some of the biochemical and behavioral mechanisms by which THA might produce its therapeutic effects and report that THA has multiple mechanisms of action on the cholinergic system. We have also examined the interaction in-vitro of 20 analogues of THA with cholinergic enzymes (plasma cholinesterases) and receptor binding sites (muscarinic), in a search to find analogue(s) with better effect(s) on the cholinergic system.

METHODS

The AD/SDAT group consisted of 3 cases with a definitive diagnosis of AD/SDAT, verified at postmortem investigations. Histopathological investigations of silver stained sections revealed an abundance of plaques and tangles in cortex and hippocampus. The mean age was 76$\pm$ 1.27 years (age range 73-78 years). The time lapse between death and autopsy was 11.3$\pm$3.13 hrs (range 4-15 hrs). Human brains were also obtained at autopsy from 8 individuals without case histories of neuropsychiatric disorders. Frontal cortices were obtained from 4 cases (mean age 56.8$\pm$8.1, range 48-70 years). The mean time between death and autopsy was 11.5$\pm$5.12 hrs, range 5-17 hrs). Temporal cortices were obtained from 4 cases (mean age 72.5 +3.17 years, range 67-83 years) and the time lapse between death and autopsy was 17.5$\pm$0.83 hrs (range 15-19 hrs).
Frontal or temporal cortices, after removal of white matter were homogenized in 30 volumes of 0.32 M sucrose and centrifuged at 1000xg for

10 minutes. The supernatants were taken and centrifuged at 45000xg for 25 min. and the pellets (P2) were suspended in 0.32 M sucrose and used for binding assays (Adem, 1987).
Male Sprague Dawly rats (300-600 g) were decapitated and the cortices dissected. Homogenates were prepared as described above for human brains. In vitro displacement studies in which human or rat brain tissue homogenates were incubated in 50 mM Na-K phosphate buffer (pH=7.4) with increasing concentrations of THA or THA analogues (10^{-11}-10^{-4} M) and ^{3}H-quinuclidinyl benzilate (^{3}H-QNB, 0.22 nM, 39.4 Ci/mmol, NEN) or ^{3}H-pirenzepine (^{3}H-PZ, 5nM, 76 Ci/mmol, NEN) or ^{3}H-oxotremorine-M (^{3}H-OXO-M, 1 nM, 87 Ci/mmol) or ^{3}H-acetylcholine in the presence of cytisine or atropine (^{3}H-ACh, 10 nM, 81.7 Ci/mmol), Amersham) or 3H-nicotine (3H-Nic, 5 nM, 80 Ci/mmol, Amersham) were performed using filtration technique as described previously (Adem, 1987).
In saturation studies human cortical membranes were incubated with increasing concentrations of ^{3}H-QNB in the presence and absence of THA (10^{-6}M). Atropine (10^{-4}M) was used to measure non specific binding. Plasma cholinesterase activity was measured according to the method of Augustinsson et al(1978). Human plasma was diluted 100 times in 10 mM Na-K-phosphate buffer(pH=7.4) and aliquots (800ul) were taken for measurement of plasma cholinesterase activity using achetylthiocholine (ATC) or butyrylthiocholine (BTC) as substrate (Adem, 1987). Different concentrations of THA or its analogues (10^{-10}-10^{-4}M) were used to inhibit the plasma cholinesterase activity. The inhibition for each concentration of inhibitor was calculated as a percent of the plasma cholinesterase activity in the absence of inhibitor. IC_{50} values were determined from curves of percent inhibition versus inhibitor concentration.
Motor activity and cognitive function of THA treated and saline treated (controls) rats were tested during adulthood (6 months) and old age (22 months) as described previously (Mohammed et al, 1989).

RESULTS AND DISCUSSION

The ability of THA to displace a number of muscarinic radioligands from human brain cortical membranes was examined (Table 1). THA showed a similar affinity for M1 and M2 cortical receptors. Moreover, THA/^{3}H-QNB, THA/^{3}H-PZ and THA/^{3}H-OXO-M competition curves yielded Hill coefficients not significantly different from unity. THA inhibited the binding of low concentrations of ^{3}H-QNB, ^{3}H-PZ and ^{3}H-OXO-M indicating competition for primary sites.
To study if THA is competitive or non-competitive at the human muscarinic cholinergic receptor sites (frontal cortex), saturation studies with increasing concentrations of ^{3}H-QNB were performed in the absence or presence of THA (10^{-6}M) in the incubation media. Similar B_{maxes} of 828 and 839 pmol/g protein but different K_{ds} of 35 and 72 pM were obtained from Scatchard plots of the saturation curves in the absence and presence of THA respectively, thus indicating that THA has a competitive interaction with muscarinic receptors.

The ability of THA and 20 of its analogues to displace a number of muscarinic radioligands from muscarinic receptors in rat cortex was also examined (Table II). A few analogues showed slightly higher or similar affinity for the ^{3}H-PZ binding sites (THB 016, THB 019, THB 015, THB 013) or the ^{3}H-OXO-M binding sites (THB 018, THB 007, THB 016). THA-^{3}H-OXO-M competition curves in rat cortex yielded Hill coefficients significantly lower than unity (n_H=0.62), indicating that THA recognized in rat cortex two or more agonist (^{3}H-OXO-M) labelled sites with different affinities.

Table I. THA interferance with the binding of different muscarinic ligands. IC50 values are given in each case. Hill coefficients (n_H) from displacement curves are also given.

Ligand	Human cortex	
	IC_{50}(uM)	n_H
^{3}H-Oxotremorine	2.8+0.20	0.91
^{3}H-Pirenzepine	1.8+0.22	0.90
^{3}H-QNB	5.7+0.24	0.83

The computer assisted two site analysis yielded a significant resolution as compared with the one-site fit ($p<0.005$), revealing that THA distinguishes a high affinity agonist site (32%, K_H=0.02 uM) and a low affinity agonist site (62%, K_L= 1.7 uM) in the rat cortex.

In order to compare the affinity of THA for muscarinic and nicotinic receptors in human postmortal Alzheimer and non-Alzheimer (control) brains, different concentrations of THA were used to displace ^{3}H-QNB (muscarinic) and ^{3}H-ACh (nicotinic) from their binding sites in temporal cortices in these brains (Fig. 1A and B). No difference in the affinity of THA for muscarinic teceptors in temporal cortices from Alzheimer and control brains was observed. Likewise, the affinity of THA for nicotinic receptors in temporal cortices from Alzheimer and control brains were similar.

In-vitro, THA and 20 of its analogues inhibit human plasma cholinesterase (Table II). Our results indicate that THA at clinical levels (10^{-8}-10^{-6}M) is a potent inhibitor (IC_{50}= 0.03 uM) of human plasmacholinesterase. We have also observed that THA has about ten times lower affinity for rat brain acetylcholinesterase (IC_{50}=0.32 uM) and human erythrocyte acetylcholinesterase (IC_{50}= 0.30 uM). Although the clinical relevance of the potent inhibition plasmacholinesterase is difficult to speculate, the inhibition of brain acetylcholinesterase may be of clinical relevance since THA permits acetylcholine to diffuse from intact cholinergic synapses to cells which have been cholinergically denervated and thereby increase cholinergic activity. Some of the analogues (THB 008 and THB 009) showed similar affinity as THA to inhibit human plasmacholinesterase.However, THB 013 showed higher affinity to inhibit plasmacholinesterase when acetylthiocholine was used as substrate but not when butyrylthiocholine was used as substrate. It is possible that THB 013 distinguishes different forms of cholinesterases in plasma. Further biochemical as well as behavioural studies in animals are required to understand if some of these analogues might be of clinical use.

In-vivo, THA inhibits the general activity of rats in open field test system. Following injection of increasing doses (0.5, 1.0, 2.0, 4.0, 8 mg /kg body weight) of THA to rats, a dose dependent decrease in general activity was observed in THA treated rats compared to control rats at concentrations higher than 1 mg/kg body weight. However, at a concentration of 0.5 mg/kg body weight the general activity of the rats was increased compared to controls (Fig.2). The central mediation of motor stimulant effect of low doses of THA is at present unclear. However, the reduced motor acitivity following higher doses (>1 mg/kg body weight) of THA is consistent with the notion that acetylcholine suppresses motor activity, since THA inhibits acetylcholinesterase leading to availability of acetylcholine at the synaptic cleft.

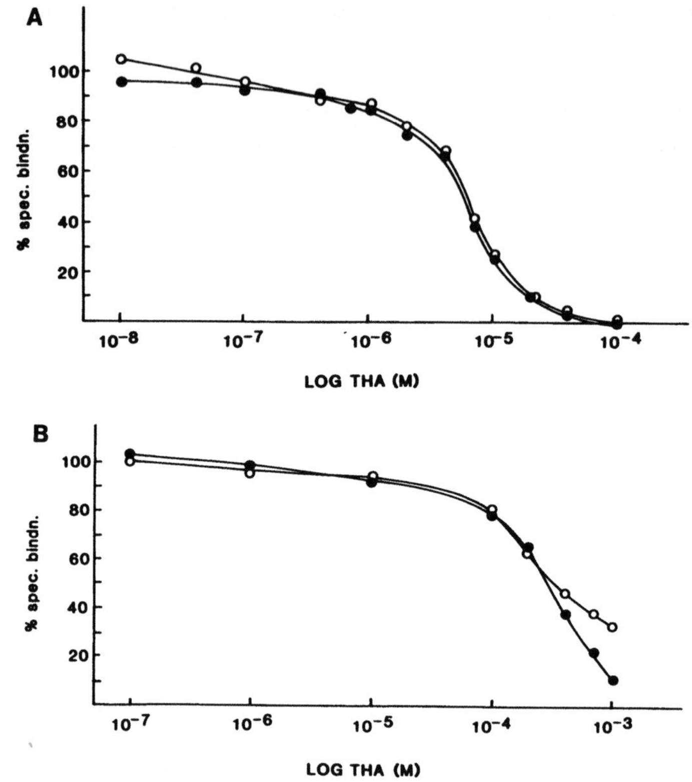

Fig 1.A. Inhibition of ^{3}H-QNB binding to temporal cortex muscarinic receptors from control (●-●, m=3) and Alzheimer (O-O, m=3) brains by THA. B. Inhibition of ^{3}H-ACh (in the presence of atropine) binding to temporal cortex nicotinic receptors from control (O-O,m=3) and Alzheimer (●-●,m=3) brains by THA.

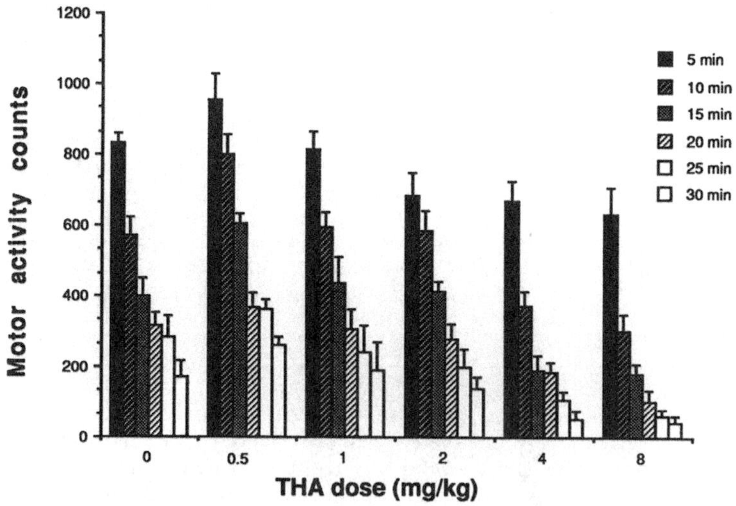

Fig 2. Total motor activity counts of saline and THA treated adult rats (6 months old). Statistical analysis revealed that 0.5 mg dose caused a significant increase of motor activity at 5,10,15, 20,25,30 ($p<0.05$). Results are presented as Mean ± SEM activity counts.

Table II. The affinities (IC_{50} values) of THA and 20 of its analogues to inhibit in-vitro the binding of muscarinic radioligands to rat cortical membranes and to inhibit human plasmacholinesterase activity when acetylthiocholine (ATC) or butyrylthiocholine (BTC) were used as substrates.

Substance	Muscarinic Receptors			Plasmacholinesterase	
	IC_{50} Values (uM)				
	^{3}H-PZ	^{3}H-OXO-M	^{3}H-Q NB	ATC	BTC
THB 001	10	17	75	>10	>10
THB 002	40	17	140	>10	>10
THB 003	40	65	90	>1000	>1000
TH 004	5	220	140	>1000	>1000
THB 005	10	7,5	17	3,0	4,0
THB 006	10	2,6	22	0,55	0,70
THB 007	5	0,8	16	7,0	2,0
THB 008	8	3,4	12	0,04	0,055
THB 009	5	17	5	0,03	0,045
THB 010	10	11	50	20	25
THB 011	20	24	100	20	30
THB 012	20	7	26	20	30
THB 013	4,5	2,3	3,4	0,008	0,01
THB 014	5	24	9,0	0,90	1,0
THB 015	4	4,4	1,2	0,35	0,30
THB 016	1,5	0,95	1,4	4,0	10
THB 017	5	9,5	5,0	2,0	2,0
THB 018	9	0,26	7,5	25	>10
THB 019	2	24	9,5	>10	>10
THB 020	>10	17	4,6	>1	>1
THA	3	0,6	6,7	0,03	0,04

Cognitive function of THA treated and saline treated rats was tested during adulthood (6 months old) and old age (22 months) in the Morris water maze. The latency to locate the platform was taken as a measure of performance (Fig. 3A and B). In the adult rats THA caused a slight improvement in acquisition of the Morris maze task, however, the difference in performance between the experimental and control (age matched) rats was not significant (p=4250). THA improved learning significantly in aged rats compared to age matched control rats. Facilitation of learning in aged rodents after treatment with THA have also been reported by others(Flood and Cherkin, 1988; Nabeshima et al., 1988; Mohammed et al., 1989). It would appear that aged rats with cognitive impairment are more likely to respond to the beneficial effect of THA than healthy young rats.

CONCLUSIONS

-The effect of THA may be partly mediated via direct effect on muscarinic receptors and/or nicotinic receptors.

- No difference in the affinity of THA for temporal cortices muscarinic receptors from control and Alzheimer brains was observed. Likewise, no difference in the affinity of THA for nicotinic receptors from these brains was observed.

-THA improved learning in aged rats.

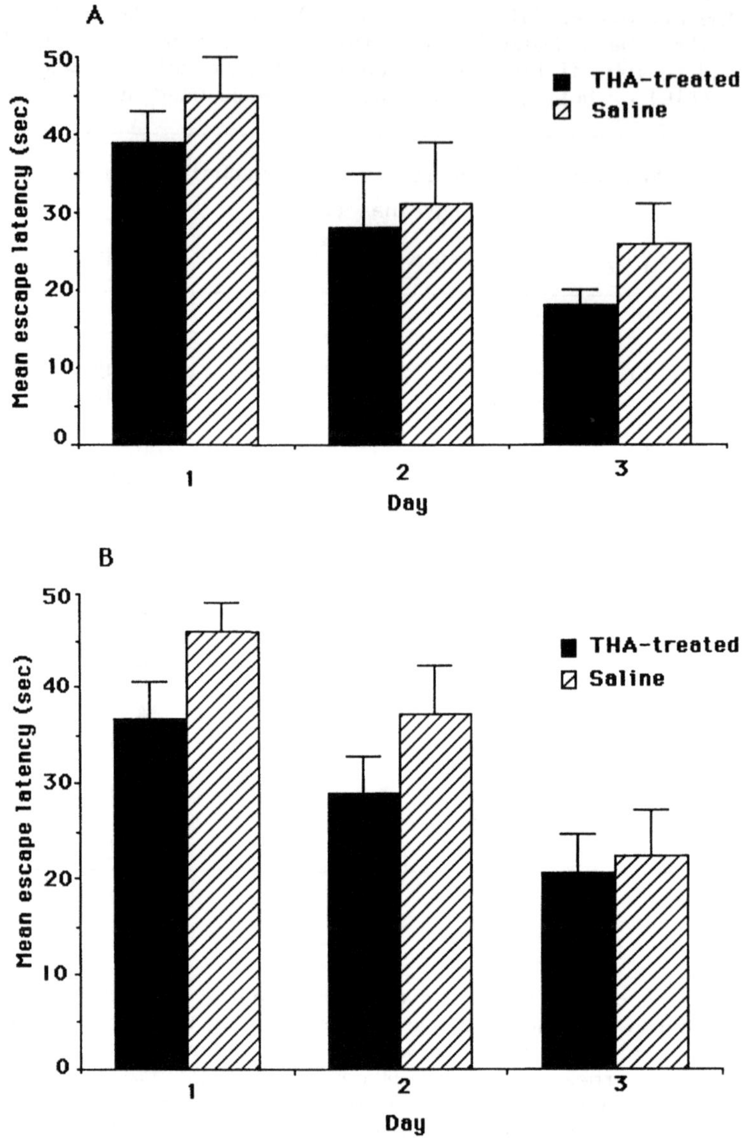

Fig 3. Effect of THA on Morris maze learning in (A) adult rats (6 months) and (B) aged rats (20 months). Results are presented as Mean + SEM latency to reach the platform.

-THA has a higher affinity for inhibiting plasmacholinesterase compared to membrane bound acetylcholinesterase.
-Some analogues showed higher affinity than THA for inhibiting cholinesterase activity while others compared to THA showed higher affinity for binding to muscarinic receptors. Studies are now in progress to investigate the effects of these analogues on learning.
Our findings indicate that the mechanisms of action of THA on the cholinergic system involves reversible inhibition of cholinesterases and reversible and competitive interaction with muscarinic cholinergic receptors. These effects might be of therapeutic value in the treatment of AD/SDAT.

REFERENCES

1. Adem, A., Characterization of muscarinic and nicotinic receptors in neural and non-neural tissue: changes in Alzheimerś disease, Acta Universitatis Upsaliensis, 32 (1987)4-61 (Doctoral thesis).

2. Augustinsson, K.B, Eriksson, H. and Faijersson, Y., A new approach to determining cholinesterase activities in samples of whole blood. Clin. Chim. Acta 89 (1978) 239-252.

3. Flood, J.F. and Cherkin A. Effect of acute arecoline, tacrine and arecoline + tacrine post-training administration on retention in old mice. Neurobiol. Aging 9:5-8, 1988.

4. Gottfries, C.G. Alzheimer´s disease and senile dementia: biochemical characteristics and aspects of treatment. Psychopharmacology 86: 245-252, 1985.

5. Hardy J.A., Adolfsson, R., Alafuzoff, I., Bucht, S., Marcusson,J., Nyberg, P., Perdahl, E., Wester, P., Winblad, B. Transmitter deficits in Alzheimer´s disease. Neurochem. Int. 7: 545-563, 1985.

6. Mohammed, A.K., Magnusson, O., Maehlen, J., Fonnum, F., Norrby, E., Schultzberg, M. and Kristensson, K. Behavioural deficits and serotonin depletion in adult rats after transient infant nasal viral infection. Neuroscience, in press.

7. Nabeshima, T., Yoshida, S., Kameyana, T. Effects of the novel compound NIK-247 on impairment of passive avoidance response in mice. Eur. J. Pharmacol. 154:263-269, 1988.

8. Perry,E.K. The cholinergic hypothesis: ten years on. Br. Med. Bull. 42:63-69, 1986.

9. Summers, W.K., Majovski, L.V., Marsh, G.M., Tachiki, K., Kling, A. Oral tetrahydroaminoacridine in long-term treatment of senile dementia, Alzheimer type. New Engl. J. Med. 315:1241-1245, 1986.

PHARMACOLOGICAL STUDIES OF CEREBROACTIVE DRUGS

Jyunji Yamamoto

Dept. of Pharmacology
Taiho Pharmaceutical Co, Ltd.
Tokushima, Japan

INTRODUCTION

It is difficult to clarify the pharmacological mechanisms of cerebroactive drugs (anti-dementia or Alzheimer's disease drugs, cognitive enhancers). Cortical and hippocampal EEG power spectra each have characteristic peaks, and the spectra after administration of psychotropic drugs were different from those obtained during normal behavioral states (Yamamoto, 1985). Antianoxic effects are often used to estimate the protective effects against brain damage. The present study was conducted to compare the EEG spectral effects of cerebroactive drugs with those of psychotropic drugs and to investigate the antianoxic effects of the cerebroactive drugs.

MATERIALS AND METHOD

1. EEG Power Spectra in Rabbits

Twenty Japanese White rabbits weighing 2.8-3.4 kg were used. The animals were anesthetized with pentobarbital sodium, and bipolar stainless steel wire electrodes (0.25-mm diameter, insulated except for the last 0.5 mm from the tips) were chronically implanted into the hippocampus (P:4, L:4, H:5) according to the brain atlas of Sawyer et al. (1954). Two stainless screw electrodes (1.0-mm diameter) were placed subdurally at an interval of 2-3 mm on the surface of the cortex (A:2, L:2). Each electrode and connector socket were fixed with dental cement. After at least 1 week for recovery, the animals were moved in a transparent plastic box (26x42x34 cm) to a sound-proof, shielded room. EEG's of the freely moving rabbits were recorded bipolarly on a polygraph. While recording the EEG's, we also performed power spectral analysis simultaneously for 15 min with a signal processor followed by FFT transformation.

2. Protective Effect against Cerebral Anoxia in Mice

Groups of 10 mice (ddY, male, 28-34g) were used for each dose of each drug. At 0.5 or 1 hr after i.p. or p.o. administration, all mice were placed into a transparent plastic box (13x13x16 cm) having intake and exhaust ports. They were exposed to a gas mixture of 4% oxygen and 96% nitrogen with a flow of 5 liters/min. The survival time, measured as the time from the induction of anoxia to the respiratory failure of the animals, was recorded. Statistic analyses were performed according to Student's t-test.

Basic, Clinical, and Therapeutic Aspects of Alzheimer's and Parkinson's Diseases, Volume 2
Edited by T. Nagatsu *et al.*, Plenum Press, New York, 1990

RESULTS

1. EEG Power Spectra in Rabbits

Tetrahydroaminoacridine (THA, 1.0, 3.0, 5.0 mg/kg,i.v.) dose dependently decreased delta wave peak power and total power of the cortex, and increased the theta wave peak power of the hippocampus. The heights of theta wave peaks after administration were higher than those of the wakefulness state. Physostigmine (0.05, 0.1, 0.2 mg/kg,i.v.), amantadine (10,

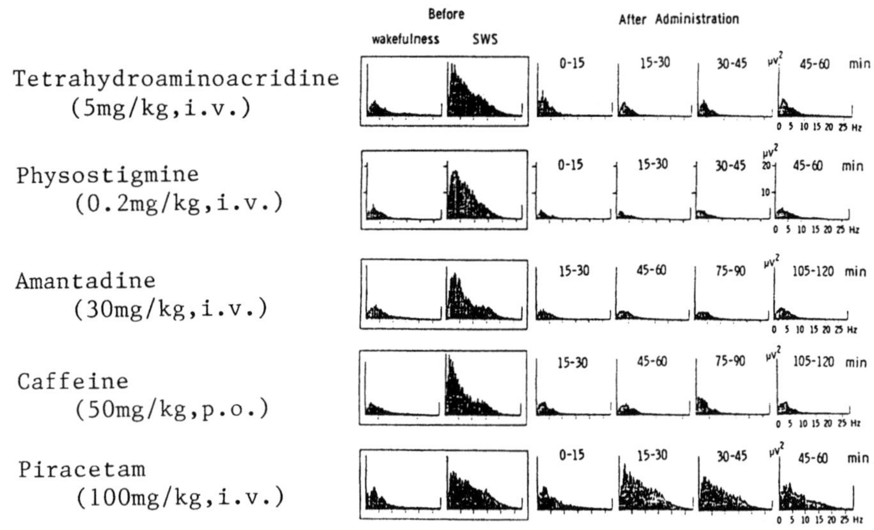

Fig. 1. Effects of cerebroactive drugs on cortical EEG power spectra in rabbits.

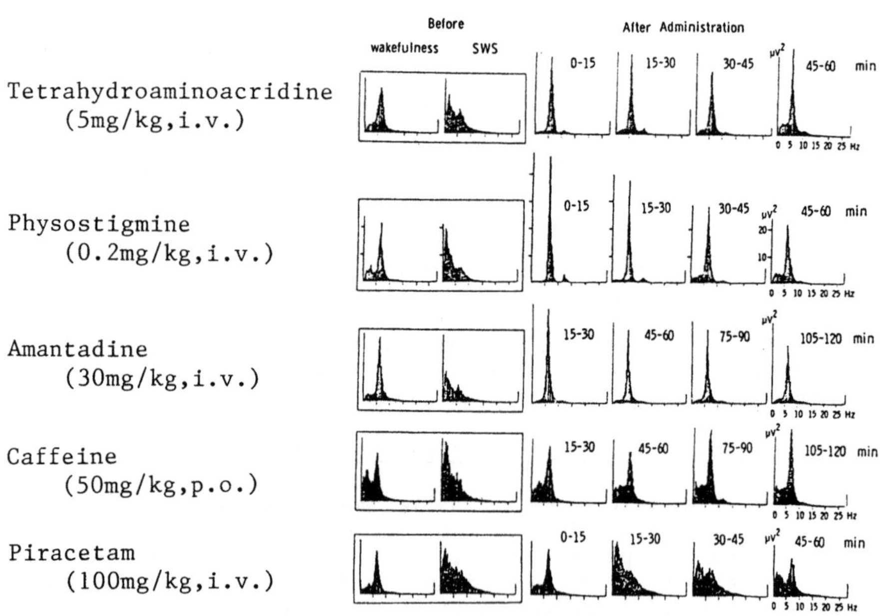

Fig. 2. Effects of cerebroactive drugs on hippocampal EEG power spectra in rabbits.

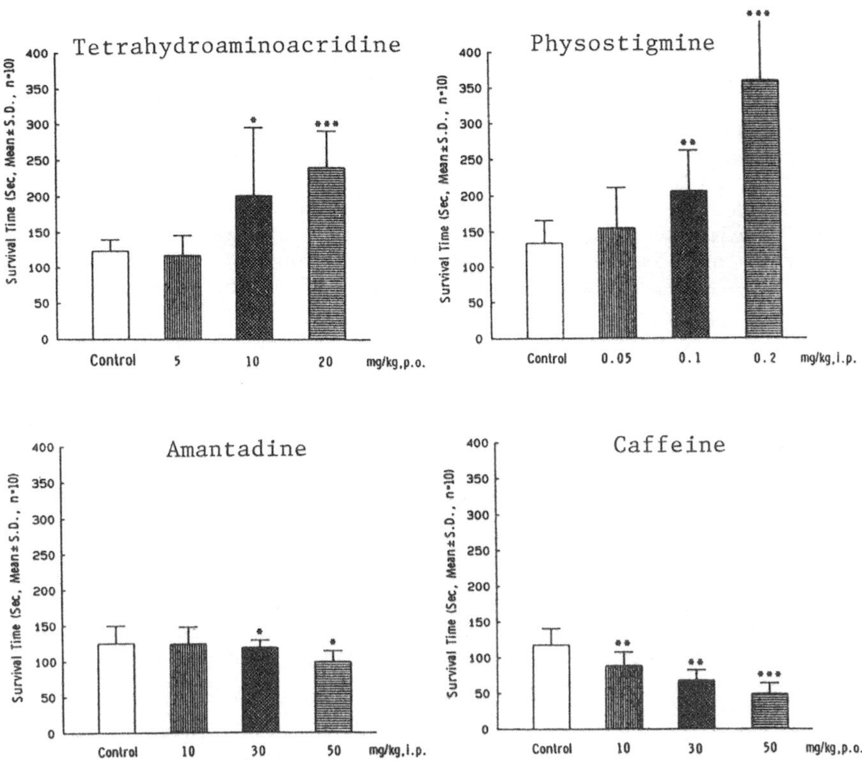

Fig. 3. Protective effect of cerebroactive drugs against cerebral anoxia in mice.
*: $p < 0.05$, **: $p < 0.01$, ***: $p < 0.001$

20, 30 mg/kg,i.v.), and caffeine (10, 30, 50 mg/kg,p.o.) produced similar spectral changes as THA. But, the effect of amantadine was longer acting, and the increase in theta wave peak height caused by caffeine was not remarkable. Piracetam administration (10, 50, 100 mg/kg,i.v.) resulted in no significant spectral changes (Figs. 1 and 2).

2. Protective effect against cerebral anoxia in mice

THA (10, 20mg/kg,p.o.) and physostigmine (0.1, 0.2 mg/kg,i.p.) increased the survival time of mice significantly. But, amantadine (30, 50 mg/kg,i.p.) and caffeine (10, 30, 50mg/kg,p.o.) decreased the survival time significantly (Fig. 3). Piracetam showed no effect.

DISCUSSION

Cerebroactive drugs, except piracetam (a nootropic), produced similar EEG power spectral changes; i.e., the delta wave peak of the cortex decreased and the theta wave peak of the hippocampus increased (arousal pattern). The spectral alterations caused by psychotropic drugs could be classified into three patterns:1) production of a new peak in cortical spectra, 2) shift of the theta wave peak of the hippocampus and 3) a change in the peak powers (Yamamoto, 1985). For example, diazepam produced a new peak at 14 Hz, and haloperidol shifted theta wave peak to lower frequencies. Methamphetamine shifted theta wave peak to higher frequencies. Therefore, the spectral changes caused by cerebroactive drugs, as seen in our present study, are different from those elicited by psychotropic drugs. The increase of theta wave peak seems to be important and common to cerebroactive drugs. The AChergic system was found to play an important role in the regulation of theta wave peak power (Yamamoto, 1988). However, the protective effects against cerebral anoxia did not parallel the EEG spectral changes. The different effects on EEG and anoxia may be due to different pharmacological action of the drugs.

These results suggest that the EEG spectral changes caused by cerebroactive drugs and psychotropic drugs are different and that the EEG arousal effect is not related to the antianoxic effect.

REFERENCES

Sawyer, H. W., Everret, J. W., and Green. J. D., 1954, The rabbit diencephalon in stereotaxic coordinates, J. Comp. Neurol., 101:801.

Yamamoto, J., 1985, Characteristics of cortical and hippocampal EEG power spectra of rabbits during normal behavioral states and after administration of CNS acting drugs, Japan.J. Pharmacol., 37:227.

Yamamoto, J., 1988, Roles of cholinergic, dopaminergic, noradrenergic, serotonergic and GABAergic systems in changes of the EEG power spectra and behavioral states in rabbits, Japan. J. Pharmacol., 47:123.

EFFECTS OF TETRAHYDROAMINOACRIDINE ON PASSIVE AVOIDANCE AND ENDOGENOUS NERVE GROWTH FACTOR LEVELS IN RATS

Shinichi Kawajiri, Takeo Sakurai, Hiroshi Kojima, and Terukiyo Yamasaki

Exploratory Laboratories II, Research Institute
Daiichi Pharmaceutical Co., Ltd.
1-16-13 Kitakasai, Edogawa-ku, Tokyo 134, Japan

INTRODUCTION

One of the most consistent changes in Alzheimer's disease is loss of cortical choline acetyltransferase (ChAT) activity. Recently, nerve growth factor (NGF) and its mRNA were reported to exist in the rat central nervous system (Korsching et al., 1985). According to the report, the NGF levels are relatively high in the regions innervated by the magnocellular cholinergic neurons of the basal forebrain (hippocampus, neocortex). Hatanaka et al. (1988) reported that cultured basal forebrain cholinergic neurons from postnatal rats retain the ChAT activity by the addition of NGF, whereas the neurons in culture for 7 days without NGF lose more than half of their ChAT activity. In addition, intraventricularly administered NGF elevated hippocampal ChAT activity in rats with septo-hippocampal lesions (Hefti et al., 1984); and NGF administered via either the lateral ventricle or the basal forebrain of nucleus basalis magnocellularis (NBM)-lesioned rats reduced the lesion-induced cortical cholinergic marker deficit (Haroutunian et al., 1989). Therefore, NGF is a highly probable candidate as a neurotrohic factor for cholinergic neurons in the brain.

In the present work, we confirmed the effect of tetrahydroaminoacridine (THA), an acetylcholinesterase inhibitor that has been reported to be beneficial in the treatment of Alzheimer's disease, on the step-through passive avoidance task in rats. Peripheral cholinergic signs induced by THA were also examined as side effects. In addition to these behavioral effects , we also determined whether the brain NGF levels were affected by the THA treatment or not, for there seems to be no report on the relation between THA and NGF.

MATERIALS AND METHODS

Male Wistar rats weighing 180-214 g were used. THA, purchased from Sigma Chemical Co. Ltd.(St. Louis, MO, USA), was dissolved in distilled water and administered p.o. in a constant volume of 0.5 ml/100 g body weight of the rat.

Behavioral observations were carried out on 7 groups of 3 rats each for 24 hr. After treatment, side effects were scored according to the multidimensional observation test of Irwin.

The NBM of rats under pentobarbital anesthesia (50 mg/kg, i.p.) was lesioned by a continuously monitored radio frequency from a Lesion Generator (RF-4a, Radionics Inc. Mass, USA). According to the atlas of Paxinos and Watson, the lesion coordinates were set as A-1.5 mm, L 2.7 mm, H 7.5 mm.

The bilaterally NBM-lesioned rats were trained in a conventional step-through passive avoidance apparatus 6 days after the operation. Passive avoidance was tested 24 hr after the training. THA was administered 60 min before each training and test trials. Retention time was measured up to 120 sec.

ChAT activity in the rat cortex was assayed by the method of Fonnum 14 days after unilateral NBM lesioning.

NGF was measured by the enzyme immunoassay (EIA) reported previously (Furukawa et al., 1983; Korsching et al., 1985; Larkfors and Ebendal, 1987). Ninety-six-well microtiter plate were coated with antiβ-mouse-(2.5s) NGF (Boehringer Mannheim Biochemica, West Germany) and blocked with 5% bovine serum albumin. β-NGF (Takara, Kyoto, Japan) was used as a standard. Frontal cortex or hippocampus was homogenized in 10% (w/v) of ice-cold buffer and centrifuged. The supernatants were used as samples. Standard solutions or samples were added to each well and incubated for 120 min at room temperature. After the wells had been washed repeatedly with Tris-buffered saline containing 0.5% Tween 20, β-galactosidase-conjugated antiβ-NGF (Boehringer) was then added. The enzyme reaction was started by the addition of 4-methylumbelliferyl-β-D-galactoside and stopped by the addition of glycine-NaOH buffer. The amounts of 4-methylumbelliferone formed were measured by fluorometry (excitation wavelength, 360 nm; emission, 450 nm).

RESULTS

Side Effects

No significant side effect was observed by the treatment with 1-3 mg/kg of THA. However, when 10-100 mg/kg was administered, a dose-dependent decrease in body temperature and increases in lacrimation, salivation, and miosis were induced. Twitches, exophthalmos, and a decrease in spontaneous behavior occurred with 30 mg/kg of THA. When 100 mg/kg of THA was given to rats, it produced diarrhea, tremor and convulsion, in addition to the above-mentioned peripheral cholinergic signs; and the animals died within 90 min.

Effects on Passive Avoidance

Passive avoidance was impaired in the rats with bilateral lesioning of the NBM. When rats without the lesion were tested by the same method, their average retention time was 120 sec, which was the cutoff time (data not shown in Fig. 1). THA increased the retention time in the test trial in a dose-dependent manner (Fig. 1A). Percent of rats whose retention time was 120 sec is shown in Fig. 1B. Ten mg/kg of THA significantly improved the performance of the NBM-lesioned rats in the passive avoidance task.

ChAT Activity in The Cortex after NBM Lesioning

ChAT activity significantly decreased by 36% in the anterior part of cortex 14 days unilateral lesioning of the NBM. The decrement in the middle or posterior part was small and insignificant.

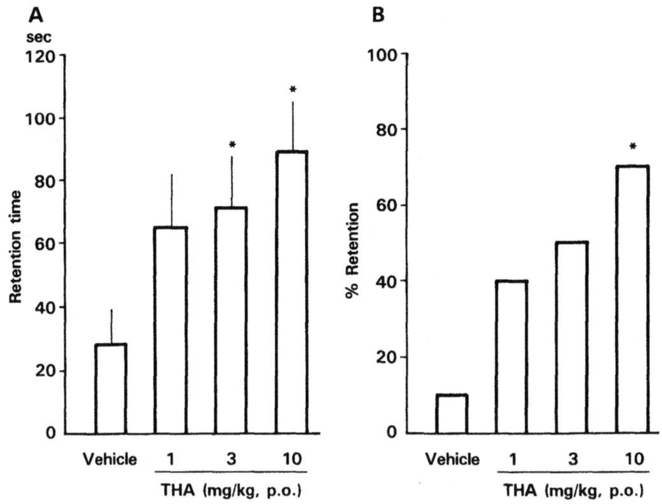

Fig. 1. Effect of THA on passive avoidance deficit of NBM-lesioned rats. Ordinate shows mean retention time + S.E.M. (n=10) (A) or percent of rats remaining in the light compartment for 120 sec (B) in the retention test.

Effect of THA on NGF Protein Level in The Brain

There was a close correlation between fluorescence intensity and the concentration of purified mouse β-NGF from 12.5 to 800 pg/ml. One or 5 mg/kg of THA was orally administered to naive rats 4 times every 2 days. Two days after the last administration, the NGF protein level in the

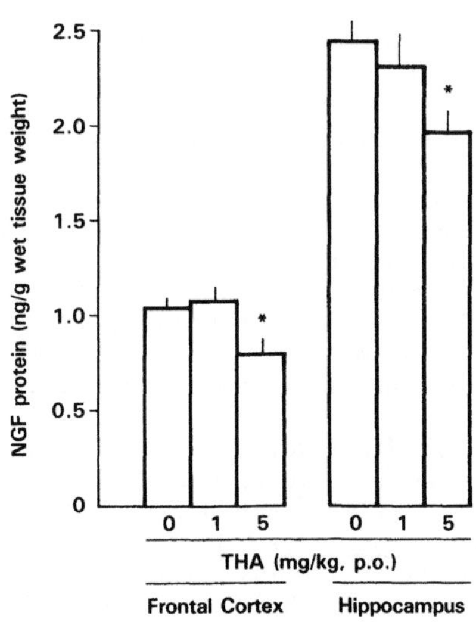

Fig. 2. Effect of THA on NGF protein levels in the frontal cortex and the hippocampus. Data are mean + S.E.M. (n=5). THA was administered to naive rats 4 times every 2 days. Rats were sacrificed 2 days after the last administration. * $p<0.05$ compared with control value (0 mg/kg).

frontal cortex and hippocampus were measured. NGF concentration in the hippocampus was approximately 2.3 times higher than that in the frontal cortex. Five mg/kg of THA reduced the NGF levels in the frontal cortex and the hippocampus by 24% and 19%, respectively (Fig. 2).

DISCUSSION

Orally administered THA significantly improved the performance of the NBM-lesioned rats in the passive avoidance task. But as the effective doses were close to those that induce undesirable behavioral actions, it is not easy to use THA as a memory enhancer without side effects. The passive avoidance deficit observed in this experiment probably stem from the damage to cholinergic neurons, since ChAT activity in the frontal cortex was decreased after NBM lesioning. After confirming the efficiency of THA on learning behavior, we determined that the NGF levels in the brain were decreased by THA treatment. Recently Kosasa et al. (1989) reported from experiments using a microdialysis method that THA increases extracellular acetylcholine (ACh). Considering the close relation between NGF and ACh in the central nervous system, we suppose that THA might decrease NGF via increasing the ACh concentration at synapses.

In summary, we confirmed that THA improves the impaired learning ability in rats and determined that THA decreases NGF levels in tha brain. It suggests that ACh elevated by acetylcholinesterase inhibition may reduce NGF levels in the central nervous system.

ACKNOWLEDGMENT

We thank Dr. S. Furukawa, National Institute of Neuroscience, National Center of Neurology and Psychiatry, for his helpful suggestions.

REFERENCES

Fonnum, F., 1975, A rapid radiochemical method for the determination of choline acetyltransferase, J. Neurochem., 24:407.

Furukawa, S., Kamo, I., Furukawa, Y., Akazawa, S., Satoyoshi, E., Itoh, K., and Hayashi, K., 1983, A highly sensitive immunoassay for mouse β-nerve growth factor, J. Neurochem., 40:734.

Haroutunian, V., Kanof, P. D., and Davis, K. L., 1989, Attenuation of nucleus basalis of Meynert lesion-induced cholinergic deficits by nerve growth factor, Brain Res., 487:200.

Hatanaka, H., Tsukui, H., and Nihonmatsu, I., 1988, Developmental change in the nerve growth factor action from induction of choline acetyltransferase to promotion of cell survival in cultured basal forebrain cholinergic neurons from postnatal rats, Dev. Brain Res., 39:85.

Hefti, F., Dravid, A., and Hartikka, J., 1984, Chronic intraventricular injections of nerve growth factor elevate hippocampal choline acetyltransferase activity in adult rats with partial septo-hippocampal lesions, Brain Res., 293:305.

Korsching, S., Auburger, G., Heumann, R., and Thoenen, H., 1985, levels of nerve growth factor and its mRNA in the central nervous system of the rat correlate with cholinergic innervation, EMBO J., 4:1389.

Kosasa, T.,Yamanishi, Y., Ogura, H., and Yamatsu, K., 1989, The determination of extracellular acetylcholine in the rat cerebral cortex with microdialysis method, Bull. Jap. Neurochem. Soc., 28:190.

Larkfors, L. and Ebendal, T., 1987, Highly sensitive enzyme immunoassay for β-nerve growth factor, J. Immunol. Meth., 97:41.

SM-10888, A NEW DRUG CANDIDATE FOR TREATMENT OF SENILE DEMENTIA OF THE ALZHEIMER TYPE

Tsunemasa Irie, Kazuichi Natori, Yuko Okazaki, and Junki Katsube

Research Laboratories, Sumitomo Pharmaceuticals Co., Ltd.
3-1-98, Kasugadenaka, Konohana-ku, Osaka 554, Japan

INTRODUCTION

In 1986, Summers et al.[1] reported that oral administration of 9-amino-1,2,3,4-tetrahydroacridine (THA) significantly improved the symptoms including memory deficit of senile dementia of the Alzheimer type (SDAT). However, its use was limited by adverse peripheral cholinergic side effects.

We selected SM-10888 (9-amino-8-fluoro-1,2,3,4-tetrahydro-2,4-methanoacridine citrate) from many THA derivatives in view of its selectivity for the central nervous system (CNS). We investigated the effect of SM-10888 on the selective action for the CNS and on the learning behavior in mice and compared these effects with those of THA, physostigmine,[2] amiridin[3,4] and HP-029.[5] CNS selectivity was evaluated by hypothermia (CNS cholinergic symptom) and salivation (peripheral cholinergic symptom) in mice. And, we also investgated its effects on the ACh level in the hippocampus and the jaw movement during chewing, which is considered as one of the central cholinergic behaviors in rats.

MATERIALS AND METHODS

Male ddY mice, CD-1 mice, and SD rats were used. SM-10888, amiridin, and HP-029 were synthesized in our laboratories. THA and physostigmine were purchased from Aldrich Chemical Co. and Wako Pure Chemical Co., respectively.

Inhibitory Effect on AChE activity (in vitro)

Using cortical P_2 fractions of rat brain, we determined AChE activity by the method of Ellman et al.[6]

Selective Action on CNS[7]

The selective action on the CNS was evaluated by hypothermia (CNS cholinergic symptom) and salivation (peripheral cholinergic symptom). The rectal temperature was measured before and 15, 30, 60, and 120 min after the drug (p.o.) administration. Salivation was observed and scored at 15, 30, 60, 90, and 120 min after the administration. The dose that produced 50% of the maximal effect (hypothermia or salivation) was calculated.

ACh Determination in the Mouse Cerebral Cortex

Drugs were administered p.o. to mice and the animals were killed 30 min later by microwave irradiation focused on the head (2KW for 0.6sec). The brain was removed and the cerebral cortex was separated. ACh was extracted according to the method described by Ikarashi et at.[8] ACh content was determined by high-performance liquid chromatography.[9]

Habituation of Exploratory Activity

We examined the habituation of exploratory activity according to the method described by Platel et al.[10]

ECS or CXM-induced Amnesia in Step-down Passive Avoidance Failure

In the training session, each ddY mouse was placed on a platform (10x10x0.5cm) in a transparent acrylic box (30x40cm, 30cm high) with a grid floor; and when the mouse stepped down on the foot shock (1mA) delivered grid floor or touched the grid with its forepaws three to four times, then the mouse was removed. Immediately after the training, cycloheximide(CXM, 120mg/kg, s.c.) was administered or electroconvulsive shock(ECS, 30mA, 0.2sec) was applied transcorneally. Twenty-four hours after the training session, the retention test was performed.

ACh Determination of the Rat Hippocampus by Brain Dialysis Method

A dialysis cannula was implanted into one side of the rat hippocampus according to the cordinates (A:4.2,H:2.2,L:1.5) of Pellegrino and Cushman. One day following the surgery, the area was perfused with Ringer solution containing 10 μM physostigmine under the unrestrained and unanesthetized condition. Concentration of ACh in the perfusate was determined by HPLC-ECD with an enzyme column on which ACh esterase and choline oxidase were immobilized.

Mouth movement

The chewing jaw movement was scored by the method of Salamone et al.[11]

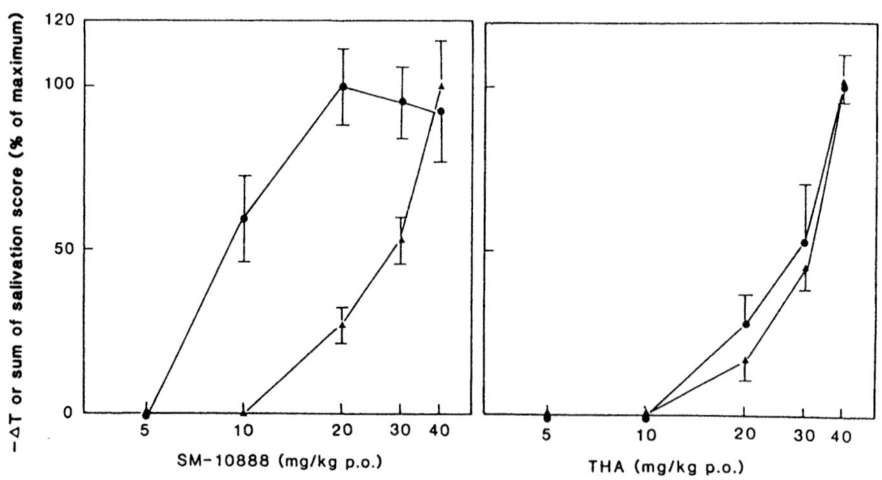

Fig. 1. Dose-response relationships for hypothermia and salivation elicited by the p.o. administration of SM-10888(left panel) and THA(right panel) to mice. ΔT is the maximum response within 2 hr at that dose level. Points are the mean ± S.E(n=4or5). Hypothermia,●;salivation,▲.

RESULTS AND DISCUSSION

First, we investigated the CNS selectivity of the drug in mice. Dose-response relationships for hypothermia and salivation induced by SM-10888 or THA are shown in Fig. 1. Hypothermia induced by SM-10888 or THA was antagonized by 15-min prior i.p. administration of atropine sulfate (a centrally-acting muscarinic antagonist), while antagonism by methylatropine bromide (a peripherally-acting muscarinic antagonist) was weak, even at a high dose. On the other hand, salivation, as distinct from hypothermia, was antagonized by either atropine sulfate or methylatropine bromide at low dose(data are not shown). Thus, hypothermia was considered to be a central cholinergic symptom; and salivation, a peripheral one. The ED_{50} value for the hypothermic effect of SM-10888 was about 3 times lower than that for its salivant effect . As for THA, those values were nearly equal. These data suggest that the action of SM-10888 is about 3 times more selective for the CNS than THA.

The results of SM-10888, THA, amiridin, HP-029, and physostigmine treatment of mice are summarized in Table 1. SM-10888(IC_{50}:2.3x10^{-7}M)

Table 1. Summary of effects of SM-10888, THA, physostigmine, amiridin and HP-029

Drugs	Inh. of AChE [1]	CNS [2]	EDmin (ACh)[7] [3]	Habit. [3][4]	CXM [3][5]	ECS [3][6]
SM-10888	2.3×10^{-7}	3.1	6	6	6	6
THA	1.9×10^{-7}	1	30	30	30	—[7] (1~15)
Physostigmine	1.6×10^{-8}	1.1	0.5	— (0.5~2)	— (0.1~1)	—[7] (0.1~0.4)
Amiridin	5.0×10^{-7}	0.9	40	40	50	—[7] (3~30)
HP-029	7.9×10^{-7}	0.8	60	50	50	— (15~60)

1) IC_{50} of inhibition of AchE (M)
2) Selective action on the central nervous system (ratio to THA)
3) Minimum effective dose (mg/kg, p.o.)
4) Habituation test
5) Cycloheximide-induced amnesia
6) Electroconvulsive shock-induced amnesia
7) Increase in brain acetylcholine level
8) Not effective, numbers in parenthesis indicate tested doses.

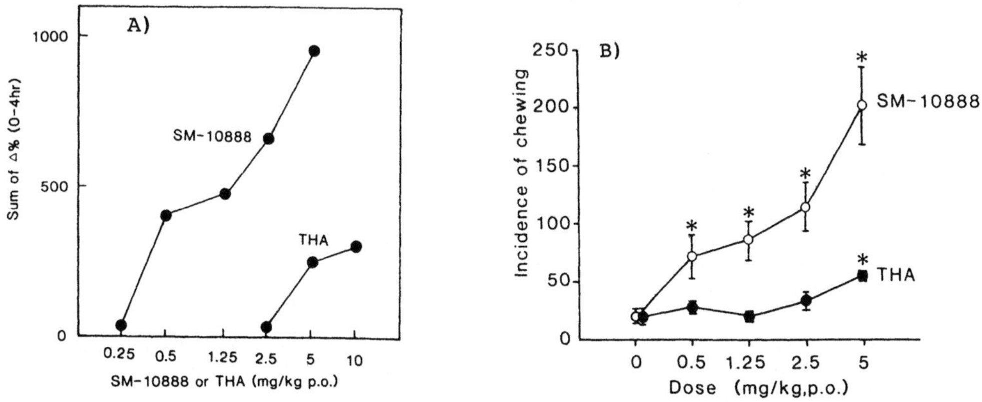

Fig. 2. Effects of SM-10888 and THA on the hippocampal ACh level(A) and the chewing jaw movements(B) in rats. (A) Data are shown as the sum of % change of hippocampal ACh from basal level between 0 and 4 hours after the administration. (B) The incidence of chewing was scored for 6 3min-piriods every 10 min for 60 min after the p.o. administration. Each point represents the mean ± S.E. of 9 rats. *; Significantly different from each 0 group (p<0.01).

inhibited AChE activity at almost the same concentration as THA. SM-10888 was about 2 and 3.5 times more potent than amiridin and HP-029. SM-10888 increased the ACh content in the mouse cerebral cortex and its minimum dose for a significant increase of ACh content [ED_{min}(ACh)] was 6 mg/kg p.o. The ED_{min}(ACh) of THA, amiridin, and HP-029 was 30, 40, and 60, respectively. The ED_{min}(ACh) of SM-10888 was 5-10 times lower than that of THA and two THA derivatives.

We examined the effect of SM-10888 on the learning behavior in mice and compared the results with those of THA and THA derivatives. Habituation of exploratory activity is considered to be a model of learning. And, ECS or CXM-induced amnesia in the step-down passive avoidance test is widely used as an animal model of amnesia. In the habituation test, the poor retention with a 6-day intersession interval was improved by SM-10888 at doses of 6 and 10 mg/kg (p.o). In the passive avoidance test, the amnestic effects of ECS and CXM were both attenuated by the treatment with SM-10888 at 6-10 mg/kg (p.o). In the habituation test and CXM-induced amnesia model, the minimum effective dose of SM-10888 was about 1/5-1/7 of that of THA, amiridin, and HP-029. There is a good correlation between the minimum effective dose of SM-10888 and THA-related compounds on the learning behavior and that increasing brain ACh level in mice. Therefore, such drugs are suggested to influence memory via activation of cholinergic mechanisms.

Finally, we investigated the effects of SM-10888 on the ACh level in the hippocampus and the chewing jaw movement in comparison with those of THA. SM-10888 dose-dependently increased the hippocampal ACh level, which was measured by the brain dialysis method under the freely moving condition, and its minimum effective dose was 0.5mg/kg,p.o.(Fig.2A). The incidence of chewing, which is reported as one of the central cholinergic behaviors, was enhanced by SM-10888 at doses of 0.5-5mg/kg,p.o.(Fig.2B). THA increased the hippocampal ACh level and the incidence of chewing at the high dose of 5mg/kg,p.o. Dose-response curves of SM-10888 and THA for the ACh level were similar to those for the incidence of chewing. Therefore, these two phenomena are supposedly closely related.

In summary, SM-10888 is about 3 times more selective for the CNS than THA, amiridin, HP-029, and physostigmine. SM-10888 has memory-enhancing and anti-amnestic action at about 5-7 times lower doses than THA and THA derivatives. Its action at lower doses is probably attributable to its high selectivity for the CNS. Thus, SM-10888 is expected to be a therapeutic drug for senile dementia of the Alzheimer type with much less peripheral cholinergic side effects than THA, amiridin, HP-029, and physostigmine.

REFERENCES

1. W. K. Summers, L. V. Majorski, G. M. Marsh, K. Tachiki, and A. Kling, Oral tetrahydroaminoacridine in long-term treatment of senile dementia, Alzheimer type, N. Eng. J. Med. 315:1241 (1986).
2. Y. Stern, M. Sano, and R. Mayeux, Long-term administration of oral physostigmine in Alzheimer's disease, Neurology 38:1837 (1988).
3. H. Kuribara, Effects of amiridin on ambulatory activity and discrete shuttle avoidance response in mice, Folia Pharmacol. Japan. 88:299 (1986).
4. T. Nabeshima, S. Yoshida, and T. Kameyama, Effects of the novel compound NIK-247 on impairment of passive avoidance in mice, Eur. J. Pharmacol. 154:263 (1988).
5. G. M. Shutske, F. A. Pierrat, M. L. Cornfeldt, M. R. Szewczak, F. P. Hunger, and G. M. Bores, (±)-9-Amino-1,2,3,4-tetrahydroacridine-1-ol. A potential Alzheimer' disease therapeutic of low toxicity, J. Med. Chem. 31:1278 (1988).
6. G. L. Ellman, D. Courtney, V. Andres, and R. M. Featherstone, A new and rapid colorimetric determination of acetylcholinesterase activity, Biochem. Pharmacol. 7:88 (1961).

7. J. M. Palacios, G. Bollinger, A. Closse, A. Enz, G. Gmelin,and J. Malonowski, The pharmacological assessment of RS-86 (2-ethyl-2,8-diazaspiro-[4,5]-decan-1,3-dion hydrobromide). A potent, specific muscarinic acetylcholine receptor agonist, Eur. J. Pharmacol. 125:45 (1986).
8. Y. Ikarashi, T. Sasahara,and Y. Maruyama, A simple method for determination of choline (Ch) and acetylcholine (ACh) in rat brain regions using high-performance liquid chromatography with electrochemical detection (HPLC-ECD), Folia Pharmacol. Japan. 84:529 (1984).
9. A. Ajima and T. Kato, Brain dialysis: detection of acetylcholine in the striatum of unrestained and unanesthetized rats, Neurosci. Lett. 81:129 (1987).
10. A. Platel and R. D. Porsolt, Habituation of exploratory activity in mice, a screening test for memory enhancing drugs. Psycopharmacology 78:346 (1982).
11. J. D. Salamone, M. D. Lalies, S. L. Channell,and S. D. Iversen, Behavioural and pharmacological characterization of the mouth movements induced by muscarinic agonists in the rat, Psycopharmacology 88:467 (1986).

INHIBITORY ACTION OF E2020, A NOVEL ACETYLCHOLINESTERASE INHIBITOR, ON CHOLINESTERASE: COMPARISON WITH OTHER INHIBITORS

Yoshiharu Yamanishi, Hiroo Ogura, Takashi Kosasa,
Shin Araki, Yoshio Sawa and Kiyomi Yamatsu

Eisai Tsukuba Research Laboratories
5-1-3 Tokodai,Tsukuba, Ibaraki 300-26, Japan

INTRODUCTION

Senile dementia of the Alzheimer type (SDAT) has been associated with specific changes in several neurotransmitter systems of the brain (Gottfries, 1985). One of the most pronounced changes has been found in the cholinergic system. Considerable evidence indicates that the loss of cholinergic function contributes to the cause of memory loss and cognitive decline in SDAT (Bartus, 1982). From this evidence, various attempts have been made to increase the activity of cholinergic systems in the brain to treat SDAT, assuming that a reduction in cholinergic function is the cause of dementia. Approaches that have been tried include the administration of precursors, acetylcholine agonists, cholinesterase inhibitors (ChE-I), and acetylcholine releasers. Physostigmine (PHY), a classical ChE-I, has been shown to improve some memory in SDAT patients. (Davis, 1982). Recently, tacrine (tetrahydroaminoacridine, THA), a centrally-acting ChE-I, produced an excellent improvement in several SDAT patients (Summers, 1986). At the present time, ChE-I therapy seems to be a promising approach for treatment of SDAT. However, the short duration of action, the low bioavailability, and peripheral side effects after oral administration of PHY have made it of limited use for the treatment of SDAT. THA was also reported to cause hepatotoxicity as a side effect. From these reasons, it was desirable to minimize the demerits of the known ChE-I's. We have found that a series of new piperidine derivatives possess a potent inhibitory action on cholinesterase. Several pharmacological aspects of E2020 (Fig. 1), one of the derivatives, have already been reported (Yamanishi, 1988). In the present study, we carried out some neurochemical studies on the pharmacological profile of E2020 for comparison of this drug with PHY and THA. We have measured the inhibition on cholinesterase *in vitro* and *ex vivo* and determined the levels of E2020 in the plasma and brain.

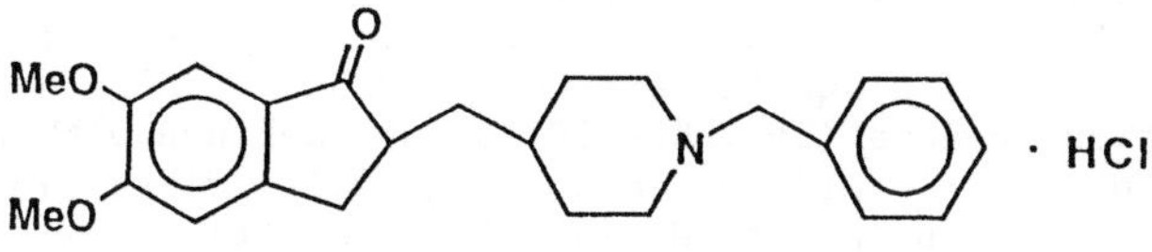

Fig. 1. Chemical Structure of E2020

Basic, Clinical, and Therapeutic Aspects of Alzheimer's and Parkinson's Diseases, Volume 2
Edited by T. Nagatsu *et al.*, Plenum Press, New York, 1990

MATERIALS AND METHODS

Chemicals and Drugs

Butyrylthiocholine iodide (ButhCh) and 5, 5'-dithio-(2-benzoic acid) (DTNB) were purchased from Wako Junyaku (Tokyo, Japan), and acetyl-thiocholine iodide (AthCh) and physostigmine sulfate (PHY), from SIGMA (St. Louis, USA). E2020 (1-benzyl-4-((5,6-dimethoxy-1-indanon)-2-yl) methyl-piperidine hydrochloride), an internal standard(1-benzyl-4-((5,6-dimethoxy-1-indanon)-2-yl) ethylpiperidine hydrochloride, and tacrine (THA, 1,2,3,4-tetrahydro-9-aminoacridine hydrochloride) were synthesized in our laboratory. All other chemicals were of analytical grade.

Animal Experiments

Drugs were orally administered to male Wistar rats(250-340 g, Japan Charles River). After one hour, blood was collected in chilled heparinized test tubes from abdominal artery under halothane anesthesia. The brains and the peripheral tissues (small intestine, heart, and pectoral muscle) were removed immediately after death and frozen in liquid nitrogen and stored at -80°C until assayed. In some experiments, the whole brains were equally divided into two parts of right and left hemisphere. Each hemisphere was prepared for assay of cholinesterase (ChE) activity and E2020 concentration. For the determination of E2020 in the plasma and brain, rats were orally given two doses of E2020(3 and 10 mg/kg) and were sacrificed as described above at various intervals.

ChE activities

ChE activity was measured according to a spectrophotometric method (Ellman, 1961). Rat brain homogenate (100 mg of the brain/2 ml of 0.1 M Na-K phosphate buffer, pH 8.0) and rat serum were used as a source of acetylcholinesterase (AChE) and butyrylcholinesterase (BuChE), respectively. AthCh and ButhCh were used as a substrate for measurement of AChE activity and BuChE activity, respectively. In brief, 300 µl of inhibitor and 1.0 ml of the enzyme homogenate or serum were mixed and preincubated for 60 min at 37°C. At the end of the preincubation period, the enzyme reaction was conducted by mixing a 130-µl aliquot of the preincubated mixture with 0.5 mM AthCh or 1.0 mM ButhCh and 0.33 mM DTNB in the phosphate buffer, pH 8.0 at 25 °C. After several concentrations of ChE-I to be tested were assayed, IC50 values were determined graphically from log concentration-inhibition curves. In ex vivo experiments, the frozen tissues were homogenized in 10 volumes of ice-cold 0.1 M Na-K phosphate buffer, pH 8.0. The ChE activity was measured with AthCh as a substrate.

Determination of E2020 level in plasma and brain

Levels of E2020 in the plasma and brain were determined by high performance liquid chromatography equipped with an ultraviolet spectrophotometer(HPLC-UV)(Sawa et al; to be published). One ml of plasma was added to 0.2 ml of 0.1 N sodium hydroxide, 1 ml of saline, and an internal standard and extracted twice with 5 ml of ethyl acetate. The two organic extracts were pooled and evaporated under a stream of nitrogen at 50 °C. The residues were reconstituted in 0.3 ml of methanol. Ten 50-µl aliquots were analyzed on a 250x4.6-mm reversed-phase analytical column (YMC.ODS-312, 5 um) combined with UV detector. The mobile phase (1 ml/min) contained 0.1% diethylamine and 60% acetonitrile. The optimum condition of UV detection was 265 nm. The detection limit was 10 ng/ml (SN ratio=5). Brain E2020 level was

Table 1. Inhibitory Effects of E2020 and Reference drugs on AChE and BuChE (in vitro)

Compound	IC50 (nM)		Ratio of IC50
	AChE Activity	BuChE Activity	(BuChE/AChE)
E2020	5.7 (5.1-6.3)	7138 (6715-7561)	1252
PHY	0.68 (0.62-0.74)	8.1 (7.1-9.1)	11.9
THA	80.6 (72.6-88.6)	73.0 (70.1-75.9)	0.9

N=4. Numbers in parentheses represent a range of 95% confidence limits.

determined by a modification of plasma procedure. Rat brain was homogenized in 10 vol. saline, and the homogenate was used for E2020 determination as described above. The peaks of plasma and brain concentration of E2020 and the area under the curve (ng·hr/ml or g) (AUC 0-24 hr) were calculated.

Statistical Analysis

Statistical significance was determined by the Student's t-test.

RESULTS

In Vitro experiments

When tested in vitro with rat brain homogenate or rat serum, the ChE-I's gave different IC50 values. As shown in Table 1, the anti-AChE activity of E2020 was 14 times more potent than that of THA, whereas the anti-AChE activity of PHY was 12 times more potent than that of E2020. The IC50 value of E2020 on BuChE activity was 7.1 μM. Thus, the anti-BuChE activity of E2020 was the least potent among ChE-I's tested. The selectivity of drugs on ChE was shown as a "ratio of IC50 " in Table 1. The selectivities of E2020, PHY, and THA toward ChE were 1300 times, 12 times and 0.9 times, respectively. PHY inhibited preferentially AChE. THA inhibited AChE and BuChE equally. Thus, E2020 was the most selective inhibitor of AChE among the three drugs (Table 1).

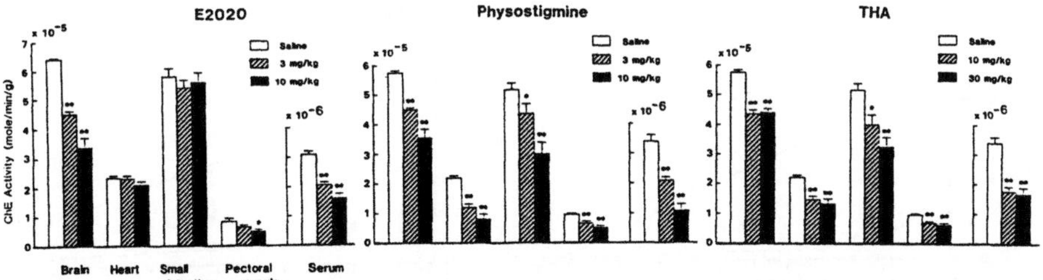

Fig. 2. Inhibitory effects of E2020 and reference drugs on ChE in the brain and peripheral tissues of rats (ex vivo). N=5-6

Ex Vivo Experiments

A comparison of the ChE values in different tissues in the rat indicates that the ChE activity was highest in brain and least in plasma. Brain contained AChE exclusively and BuChE very minutely. On the contrary, the peripheral tissues such as heart and small intestine contained BuChE predominantly, and the ChE activities were less but AChE was relatively high in pectoral muscle(data not shown). E2020 inhibited ChEs of brain and serum at the doses of 3 and 10 mg/kg significantly and dose dependently, and it also inhibited ChE's of pectoral muscle only at the higher dose. On the other hand, E2020 did not show any remarkable inhibitory effect on ChE's of heart and small intestine even at the higher dose of 10 mg/kg. THA and PHY inhibited ChE's of all the peripheral tissues tested, as well as the ChE's of brain and serum, significantly (Fig. 2).

Levels of E2020 in plasma and brain

E2020 reached its peak concentration of the plasma and brain at 0.5-2.0 hr after an oral administration. Plasma AUCs of E2020 at the doses of 3 and 10 mg/kg were also 198 and 592 ng·hr/ml, respectively. On the other hand, brain AUCs of E2020 after 3 and 10mg/kg were 1680 and 3737 ng·hr/g, respectively (Fig. 3). The AUC's of brain concentration were about 8 and 6 times higher than those of plasma at a dose of 3 and 10 mg/kg, respectively.

DISCUSSION

A series of piperidine derivatives possess a potent inhibitory action on ChE. E2020, one of the derivatives, is a novel ChE-I that is about 15 times more potent than THA, but 10 times less potent than PHY when the inhibitors are tested in vitro. Furthermore, E2020 inhibited AChE selectively with less inhibitory effect on BuChE. On the contrary, THA and PHY inhibited both AChE and BuChE. These data indicate that E2020 is the most selective AChE inhibitor among them. E2020 inhibited brain ChE without remarkable effects on ChE in peripheral tissues after an oral administration, whereas PHY and THA inhibited ChE in brain as well as in peripheral tissues. This is because brain ChE contains exclusively AChE and peripheral tissues such as heart and small intestine contain BuChE rather than AChE. In kinetics study, the brain

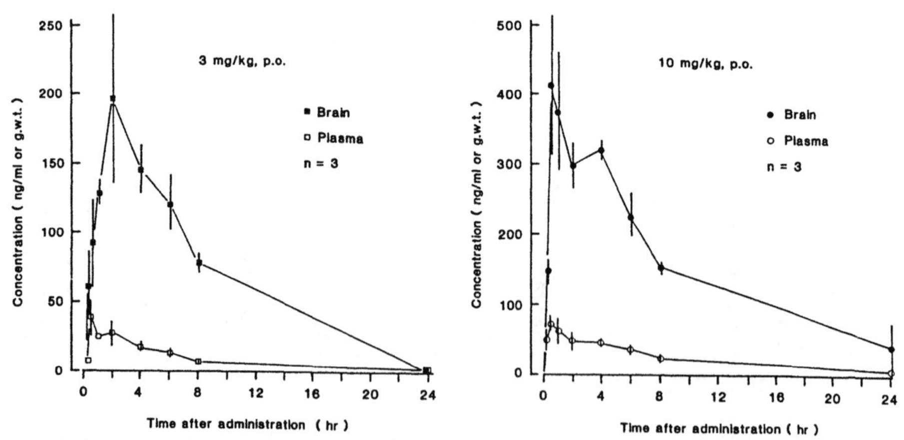

Fig. 3. Concentration of E2020 in the brain and plasma of rats.

concentration of E2020 was 5-8 times higher than that in the plasma of rats after oral doses of 3 to 10mg/kg. These findings suggest that E2020 penetrates easily into the brain. In conclusion, the present experiments show that E2020 is a novel and selective centrally-acting acetylcholinesterase inhibitor compared with PHY and THA, and suggest that E2020 may serve as a drug in the treatment of SDAT.

REFERENCES

Bartus, R. T., Dean, R. L., Beer, B. and Lippa, A., 1982, Cholinergic hypothesis of geriatric memory dysfunction, Science, 217: 408.

Davis, K. L. and Mohs, R. C., 1982. Enhancement of memory process in Alzheimer's disease with multiple-dose intravenous physostigmine, Amer. J. Psychia. 139: 1421.

Ellman, G. L., Courtney, K. D., Andre, V. Jr. and Featherstone, R. M., 1961, A new and rapid colorimetric determination of acetylcholinesterase activity, Biochem. Pharmacol., 7: 88.

Gottfries, C. G. 1985, Alzheimer's disease and senile dementia: biochemical characteristics and aspects of treatment, Psychopharmacology, 86: 245.

Yamanishi, Y., Araki, S., Kosasa, T., Ogura, H., and Yamatsu, K., 1988, Neurochemical studies of E2020, a novel centrally acting acetylcholinesterase inhibitor, 18th Annual Meeting: Soc. Neurosci. Abstr. 27:20.

INHIBITORY ACTION OF NICERGOLINE AND ITS MAJOR METABOLITES ON ACETYLCHOLINESTERASE ACTIVITY IN RAT AND MOUSE BRAIN

Yuzo Matsuoka, Tomiko Fukushima, Hiroshi Iwata,
Mamoru Matsumoto, Yumiko Shibutani, Kazutaka Kawashima,
Yukitsuka Kudo and Ryuichi Ishida

Department of Pharmacology, Safety Research Laboratory
Tanabe Seiyaku Co., Ltd., Yodogawa-ku, Osaka, Japan

INTRODUCTION

Nicergoline [(+)-10-methoxy-1,6-dimethylergoline-8β-methanol-5-bromonicotinate, NG][1,2] is known to improve memory deficits in aged rats with cerebral ischemia[3] and in mice treated with scopolamine.[4] The mechanism of its improving action has remained unclear to date. Therefore, we studied the effects of NG and its major metabolites on acetylcholinesterase(AChE) activity of rat brain crude synaptosomal fraction and on human serum cholinesterase(BuChE) activity <u>in vitro</u>. Furthermore, <u>in vivo</u> effects of NG on ChE activities in mouse brain and liver, ACh content in mouse brain, and ACh-induced neuronal excitation in rat cerebral cortex were also studied.

MATERIALS AND METHODS

Drugs and Reagents

NG and its major metabolites, 1-demethylnicergoline(1-DN), 1-methyl-10-methoxydihydrolysergol(1-MMDL), 1-hydroxymethyl-10-methoxydihydrolysergol(1-OHMMDL), 10-methoxydihydrolysergol(MDL), and 5-bromonicotinic acid(5-BNA), and reference ergot alkaloids, dihydroergotoxine(DH) and lisuride(LS), were obtained from Farmitalia Carlo Erba(Italy). Idebenone(ID) and bifemelane(BI) were synthetized in our laboratory. Physostigmine(P), tacrine(THA), and 1,5-bis(4-allyldimethylammonium-phenyl)-pentan-3-one dibromide(BW284C51) were obtained from Merck, Janssen and Sigma, respectively. Human serum BuChE, acetylthiocholine iodide(ATCh), butyrylthiocholine iodide(BuTCh), and 5,5'-dithio-bis-(2-nitrobenzoic acid)(DTNB) were purchased from Sigma.

In Vitro Experiment

The crude synaptosomal fraction(SF:2.5μg protein/ml) was obtained from the brain of male rats(Slc:Wistar, 10-week-old) by the usual method. The SF-AChE and human serum BuChE activities were assayed by Ellman's method.[5] The reaction medium contained 0.1M potassium phosphate buffer(pH7.6), 0.33mM DTNB, 0.5mM ATCh or 3mM BuTCh as substrate, 0.01ml of test drug solution, and 0.05ml of each enzyme solution in a final volume of 3ml.

In Vivo Experiment

Male mice(Slc:ddY, 7-week-old) were administered NG at 5-40 mg/kg, i.v., or at 32mg/kg, i.p., or with THA at 8mg/kg, i.p. The brain and liver were dissected 5-30 min after drug administration and homogenized in 9 volumes of ice-cold distilled water. The AChE and BuChE activities of both tissues were determined with the homogenates used as the enzyme preparation by the method mentioned above(0.1ml homogenate/3ml assay mixture). ACh contents in the cerebral cortex and hippocampus were also determined by HPLC-ECD 15 min after i.p. administration of NG or THA.

Effect of NG on ACh-induced neuronal excitation was electrophysiologically studied in the cerebral cortex of male rats(Slc:Wistar, 12-week-old). A glass microelectrode(tip diameter:5µm, filled with 2M acetate Na) attached to a four-barrelled micropipette(tip diameters:5µm, filled with 0.2M NG and 0.2 or 0.5M ACh) was inserted into the cerebral cortex(depth :500-1500µm). Extracellular action potentials of the ACh-sensitive single neuron were amplified, gated by a Window discriminator, and counted. The rate of firing was converted to analog form and displayed on a pen recorder. NG(20-50nA) and ACh(10-150nA) were applied iontophoretically to the same neuron.

RESULTS AND DISCUSSION

NG concentration-dependently inhibited the rat SF-AChE activity with an IC50 value of 0.45µM and the human serum BuChE activity with an IC50 value of 55µM (Table 1). The AChE inhibitory action of NG was equipotent to or more potent than that of P, but the BuChE inhibitory activity of NG was about 100 times less potent than that of P. Therefore, NG, unlike P and THA, is thought to be an AChE-selective ChE inhibitor. In addition, the major metabolites, except for 5-BNA, also showed an apparent inhibition of the AChE activity, though they were 15-200 times less potent than NG. DH also selectively inhibited the AChE activity with an IC50 value of 45µM. LS and ID, but not BI, showed a slight inhibition of the AChE activity at 100µM. BW284C51 showed the strongest inhibition(IC50:9nM) of the compounds tested here toward the AChE activity, but, unlike P and THA, caused no inhibitory effect on the BuChE activity, even at 10µM.

Table 1. Inhibitory Effects of Nicergoline and Its Major Metabolites on AChE and BuChE Activities in Vitro

Compound	IC_{50}(M) AChE[a]	IC_{50}(M) BuChE[b]
Nicergoline	4.5×10^{-7}	5.5×10^{-5}
Metabolites		
1-DN	7.6×10^{-6}	$>10^{-4}$ (48.1)
1-MMDL	1.0×10^{-4}	$>10^{-4}$ (46.7)
1-OHMMDL	$>10^{-4}$ (39.3)	$>10^{-4}$ (15.8)
MDL	$>10^{-4}$ (19.6)	$>10^{-4}$ (3.5)
5-BNA	$>10^{-4}$ (1.8)	$>10^{-4}$ (12.3)
References		
Dihydroergotoxine	4.5×10^{-5}	$>10^{-4}$ (31.2)
Lisuride	$>10^{-4}$ (22.0)	$>10^{-4}$ (31.1)
Idebenone	$>10^{-4}$ (29.1)	2.2×10^{-5}
Bifemelane	$>10^{-4}$ (0)	$>10^{-4}$ (32.7)
Physostigmine	1.0×10^{-6}	4.5×10^{-7}
Tacrine	9.0×10^{-8}	4.5×10^{-8}
BW284C51	9.0×10^{-9}	$>10^{-4}$ (30.9)

[a]Crude synaptosomal fraction prepared from rat brain. [b]Human serum cholinesterase. The numeral within parentheses represents % inhibition at 10^{-4}M.

Table 2. Inhibitory Effects of Nicergoline and Tacrine on ChE Activities of the Brain and Liver in Mice

Drug	Dose mg/kg	Time min	Activity(μmol/min/g wet weight) AChE Brain	AChE Liver	BuChE Brain	BuChE Liver
-	-	-	12.80±0.39 (100.0)	2.63±0.16 (100.0)	0.36±0.01 (100.0)	7.31±0.37 (100.0)
	(i.v.)					
Nicergoline	5	5	12.05±0.60 (94.1)	2.96±0.30 (112.5)	0.34±0.01 (94.4)	7.31±0.39 (111.1)
	10	5	11.73±0.24* (91.6)	3.04±0.15 (115.6)	0.35±0.01 (97.2)	8.49±0.17* (116.1)
	20	5	9.47±0.37***(74.0)	2.86±0.25 (108.7)	0.35±0.01 (97.2)	8.01±0.28 (109.6)
	40	5	7.56±0.38***(59.1)	3.22±0.17* (122.4)	0.32±0.01** (88.9)	7.93±0.28 (108.5)
-	-	-	12.74±0.08 (100.0)	2.40±0.11 (100.0)	0.33±0.01 (100.0)	7.53±0.32 (100.0)
	(i.p.)					
Nicergoline	32	15	11.21±0.37** (88.0)	2.52±0.15 (105.0)	0.34±0.02 (103.0)	7.28±0.41 (96.7)
	32	30	11.80±0.33* (92.6)	2.79±0.25 (116.2)	0.35±0.02 (106.1)	7.63±0.39 (101.3)
Tacrine	8	15	9.49±0.41***(74.5)	0.95±0.03***(39.6)	0.27±0.01** (81.8)	5.58±0.13***(76.8)
	8	30	10.29±0.46***(80.8)	1.79±0.11** (74.6)	0.29±0.01* (87.9)	6.80±0.10 (90.3)

Data are expressed as Mean±S.E. of 4-6 mice. The numeral within parentheses represents % activity of control value. *P<0.05,**P<0.01,***P<0.001: Significantly different from control(by Student's t-test).

The potent AChE inhibitory action of NG may be derived from its own chemical structure, because the metabolites and other ergot alkaloids, DH and LS, did not show so strong action as NG.

NG dose-dependently inhibited mouse brain AChE activity 5min after i.v. administration of 10-40mg/kg without substantial effects on liver AChE or brain and liver BuChE activities (Table 2). The similar selectivity for brain AChE was observed after i.p. administration of NG at 32mg/kg. It is reasonable that NG preferably inhibited brain AChE, because NG is thought to be easily and rapidly metabolized in the liver. On the other hand, THA at 8mg/kg, i.p., inhibited both ChE activities more strongly in the liver than in the brain.

The inhibition of brain AChE activity by NG was considered to be reversible, because it was lost in parallel with a decrease in the concentration of tissue added to the AChE assay system. Namely, the inhibition, observed after i.v. administration of NG at 40mg/kg, was reduced by 45% when 2% brain homogenate instead of 10% homogenate was used as an enzyme preparation, and almost completely disappeared when 0.25% homogenate was used.

When NG was intraperitoneally administered to mice at a dose of 32mg /kg, ACh contents in the hippocampus and cerebral cortex increased by 15% and 10%, respectively, 15 min after administration; while at 8 mg/kg of THA, the contents increased by 41-46% (Table 3). Furthermore, NG applied iontophoretically to ACh-sensitive single neuron of rat cerebral cortex potentiated the excitation induced by ACh in 8 out of 15 cells (Fig.1). These

Table 3. Effect of Intraperitoneal Administration of Nicergoline or Tacrine on ACh Content in the Mouse Brain

Drug	Dose mg/kg	No.of mice	ACh content (nmol/g wet weight) Cerebral cortex		Hippocampus	
-	-	6	16.1 ±1.0	(100.0)	20.0 ±0.6	(100.0)
Nicergoline	32	6	17.7 ±1.7	(109.7)	23.0 ±1.2*	(114.9)
Tacrine	8	6	23.5 ±1.1**	(145.9)	28.2 ±0.8***	(141.1)

Mice were sacrificed 15min after drug administration. Data are expressed as Mean ±S.E. The numeral within parentheses represents % of control value. *P<0.05, **P<0.01, ***P<0.001: Significantly different from control(by Student's t-test).

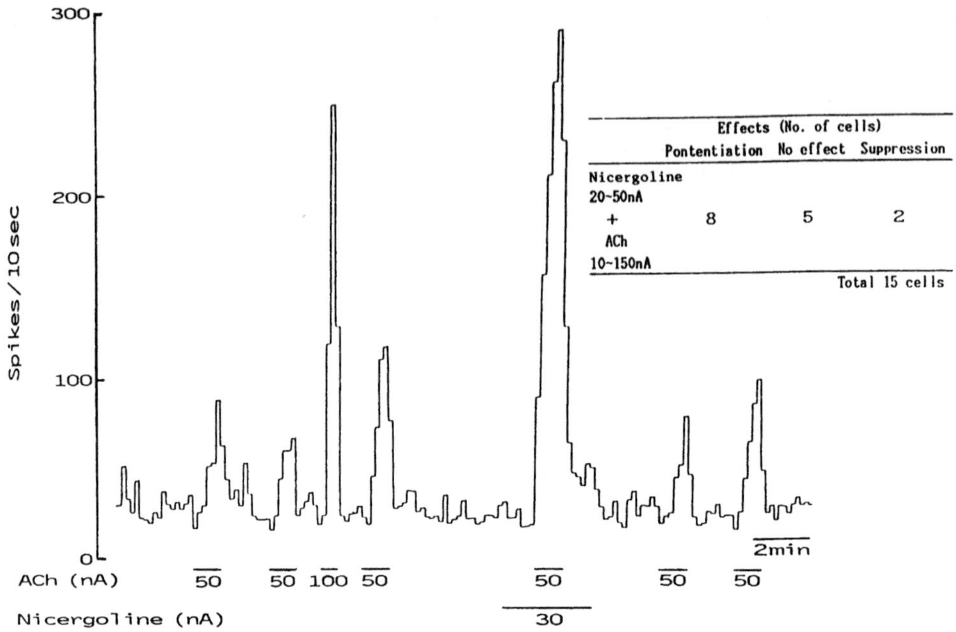

Fig. 1. Potentiation of the response of rat cortical neurons to 50nA of ACh by iontophoretical application of 30nA of nicergoline.

results suggest that NG exhibits AChE-selective ChE inhibitory action in brains of both mouse and rat.

Recently, classical non-selective ChE inhibitors such as P and THA have been emphasized for their improving effects on the memory deficits of patients with Alzheimer's disease. However, they are thought to be not ideal drugs for clinical use because of their peripheral adverse effects and very narrow safety margins. On the other hand, NG is thought to be a drug with low toxicity.[6] Therefore, NG is probably an AChE-selective and reversible ChE inhibitor with high safety.

In conclusion, the memory deficit-improving effect of NG in mice and rats may be partly due to the AChE-selective ChE inhibitory action of this drug.

REFERENCES

1. M. Lievre, M. Ollagnier, and G. Faucon, Influence of nicergoline on the cerebral blood flow and α-sympatholytical properties, Arzneim.-Forsch. 29(II): 1227 (1979).
2. K. Shintomi, Y. Ogawa, K. Yoshimoto, and H. Narita, Pharmacological study of nicergoline (III). Effects on cerebral and peripheral circulation in animals, Folia pharmacol. japon 87: 537 (1986).
3. M. Le Poncin-Lafitte, C. Grosdemouge, D. Duterte, and J. R. Rapin, Simultaneous study of haemodynamic, metabolic and behavioural sequelae in a model of cerebral ischemia in aged rats: Effect of nicergoline, Gerontology 30: 109 (1984).
4. Y. Kudo, H. Egou, and R. Ishida, Effects of nicergoline on memory retention and scopolamine-induced amnesia using one-trial passive avoidanc task in mice, Jpn. Pharmacol. Ther. 13: 6489 (1985). (in Japanese)

5. G. L. Ellman, K. D. Courtney, V. Jr. Andres, and R. M. Featherstone, A new and rapid colorimetric determination of acetylcholinesterase activity, Biochem. Pharmacol. 7: 88 (1961).
6. H. Shimazu, N, Nishimura, M. Nishikawa, Y. Ishii, K. Anabuki, S. Tamai, T. Hirakawa, H. Iwasaki, Y. Iwasawa, and T. Fujita, Subacute and chronic toxicity study of nicergoline in rats, Kiso-to-Rinsho 20: 5897 (1986).(in Japanese)

CHANGES IN DISCRIMINATION LEARNING ABILITY IN RATS WITH IBOTENIC ACID-INDUCED LESIONS IN THE NUCLEUS BASALIS MAGNOCELLULARIS

Masahiko Nomura

Department of Physiology
Saitama Medical School
Moroyama, Iruma, Saitama 350-04, Japan

INTRODUCTION

The cholinergic system is involved in much of the higher function of the central nervous system. Especially in aged humans with dementia as in Alzheimer's disease, a decrease in cholinergic function in the brain was reported. As a model of Alzheimer's disease in the present study, the nucleus basalis magnocellularis (NBM), a cholinergic nucleus, was lesioned by ibotenic acid. The changes in cholinergic function and learning ability resulting from NBM lesions were examined in 3-month-old Fischer 344 rats. Cholinergic function was assessed by the activity of cholineacetyl-transferase, and learning activity was tasked on a multiple variable interval, 15-second extinction schedule. Some of the lesioned rats were treated with the nootropic drug nicergoline. And the changes of recovery of the lesioned rats were examined.

MATERIALS AND METHODS

Animals

Thirty 3-month-old Fischer 344 rats were injected stereotaxically with ibotenic acid into the nucleus basalis magnocellularis by use an infusion pump (Gorry, 1963: Miyamoto et al., 1985). These lesioned rats were divided 3 groups: one was given no drug as a control, and two were treated with nicergoline, 1 and 4 mg/kg, intraperitoneally once a day starting one day after the acid injection. Four weeks after the lesioning, learning examination was tasked on a multiple variable-interval 15-second extinction schedule (Tsukada et al., 1977: Nomura, 1984). Another ten rats were used as a sham operated control.

Apparatus

The Skinner box (30x26x28 cm), manufactured by Ralph Gerbrands Co., was used. The front and rear walls of the box were made of aluminum sheets and the side walls of a clear acrylic resin plate. The Skinner box was provided with a window, 5 cm in diameter, in the center of its front wall at a height of 10 cm from the floor level. In the experiment, light was projected from the front side of the box to provide a brightness level of 5 x 10 foot Lamberts and a Kodak filter(ND 4.0) was used to give an attenuated brightness of 1/1,000. A feed box was located directly below the window and a lever(3x2 cm) was located at the right-hand side of the front wall 2 cm away from the right-hand corner and 2 cm above the floor level. As the rat pressed the lever, a 40-mg food pellet was dispensed into the feed box.

Procedure

The rats underwent shaping in a Skinner box according to a continuous reinforcement schedule with food pellets (40 mg) during which a brighter light (5.0 x 10 fl.) was shown on the front screen of the box. After shaping, a variable interval (VI schedule) training was interjected. The interval of pellet reinforcement was gradually increased from VI-5 sec through VI-10 to VI-15 sec extinction schedule (Nomura, 1984). The duration of the brighter brightness (S+) was 20 seconds and if the rat pressed the lever during this time period, a pellet was dispensed at a rate of variable interval-15 seconds. The duration of the less bright (1/1,000) brightness (S-) was also 20 seconds and no food pellet was dispensed even if the lever was pressed during this time period. With a 5-second interval, the two stimuli (S+) and (S-) were presented 20 times each in accordance with a Gellermann (1933) series. Animals were subjected to the session once a day for 30 consecutive days.

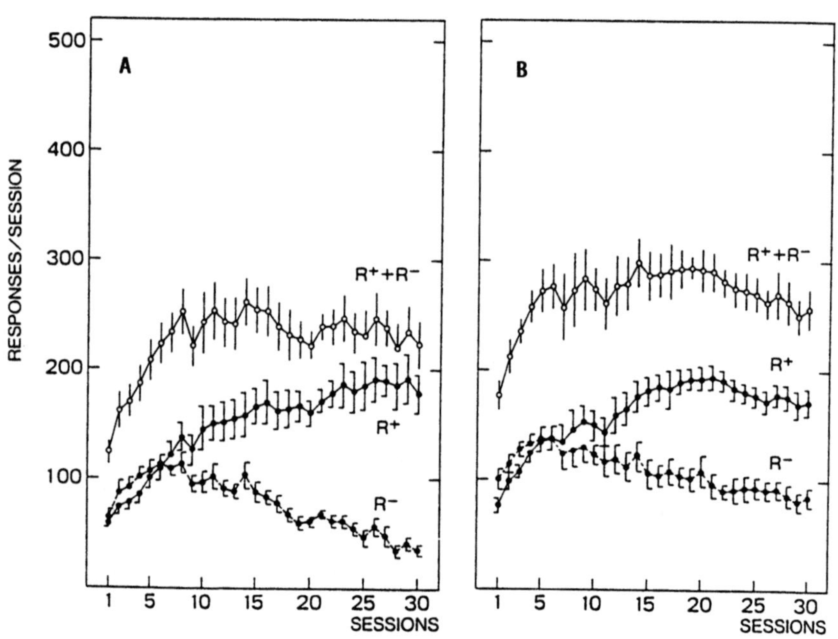

Fig. 1. Responses in brightness discrimination learning test in sham-operated (A) and ibotenic acid-lesioned (B) Fischer 344 rats.

The number of lever-pressing responses during the (S+) periods was taken as that of positive responses (R+) and the responses during the (S-) period as that of negative responses (R-). Discrimination learning and performance was then evaluated in terms of correct response ratio calculated from the formula ((R+)/(R+) + (R-))x 100. The presentation of light stimuli, dispensation of food pellets, and recording of the number of lever-pressing responses were all controlled by a microcomputer. The Skinner box, which was installed in a dark room, was monitored with a dark-field video camera placed in an adjacent room where the tester watched the monitor. In the laboratory a white noise of 60dB was produced to mask other noises. The results of learning were shown as the mean + standard error. A paired t-test using the number of responses on the first day and that on any session day in the course of learning was carried out in each of the control group and drug treatment groups (Nomura, 1985). As to drug efficacy, the number of responses in the control group and that in a test group on the same session day were compared by two-way analysis of variance (ANOVA).

RESULTS

The rats with ibotenic acid treatment (Figure 1 B) showed remarkably lower numbers of R+ and higher numbers of R- when the level of learning ability in them was compared with that of sham-operated normal rats (Figure 1 A). A low dose 1 mg/ kg of nicergoline given to the acid-lesioned rats resulted in some change in the response numbers in several sessions, but these were not significant changes. But a higher dose; 4 mg/ kg, of nicergoline yielded a remarkable increases in R+ responses and a drop in R-. Especially, the number of R- decreased after 15 sessions. So the correct response ratio was the highest in this groups. Comparison of the correct response ratio between the sham- and the ibotenic acid-treated rats is shown in Figure 2 A. And also comparison of the correct response ratios between the ibotenic acid-treated rats and nicergoline (4 mg/ kg)-treated rats is seen in Figure 2 B.

DISCUSSION

As a model of Alzheimer's disease, young (3-month-old) rats were lesioned by ibotenic acid injection into the region of the nucleus basalis magnocellularis. The changes in learning behavior in these lesioned rats were almost the same as those of aged 24-month-old rats. This means that these lesioned rats can aim as a model of the aged rat (Elias and Elias, 1976: Ingrum., 1985: Lowy et al., 1985: Nomura, 1988). On the learning aspects, especially using the operant-type discrimination learning task, the higher nervous system plays a very important role in discrimination learning performance. So, the difference between young normal and lesioned abnormal rats was remarkable. Treatment of the lesioned rats with the nootropic drug nicergoline at a high dose of 4 mg/ kg showed remarkable effects of increasing the learning ability.

Cholinergic function of the lesioned rat brain might be recovered as well as the learning ability, for the activity of cholineacetyltransferase, a marker enzyme of cholinergic activity, recovered to the same level as the normal sham-lesioned rats.

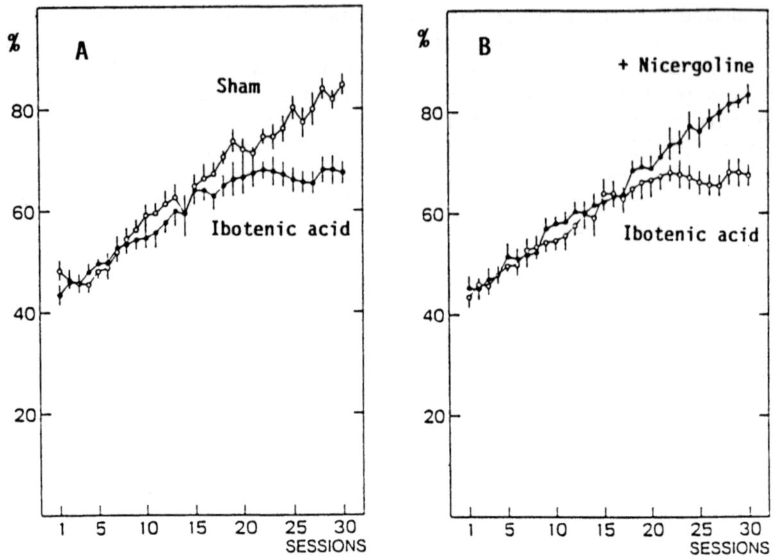

Fig. 2. Comparison of correct response ratios between sham-operated and acid-injected rats (A), and between lesioned rats with or without nicergoline (4 mg/kg) treatment (B).

REFERENCES

Elias, P. K., and Elias, M. F., 1976, Effects of age on learning ability: Contribution from the animal literature, Experim. Aging Res., 2:165.

Gellermann, L. W., 1933, Chance orders of alternating stimuli in visual discrimination experiments, J. Genetic Psychol., 42:206.

Gorry, J. D., 1963, Studies on the comparative anatomy of the ganglion basal of Meynert, Acta Anat. (Basal)., 55:51.

Ingrum. D. K., 1985, Analysis of age-related impairments in learning and memory in rodent models, Ann. New York Acad. Sci., 444:313.

Lowy, A. M., Ingrum, D. K., Olton, D. S., Waller, S. B., Reynolds, M., A., and London, E. D., 1985, Discrimination learning requiring different memory components in rats: Age and neurochemical comparisons, Behav. Neurosci., 99:638.

Miyamoto, M., Shintani, M., Nagaoka, A., and Nagawa, Y., 1985, Lesioning of the rat basal forebrain leads to memory impairments in passive and active avoidance tasks, Brain Res., 328:97.

Nomura, M., 1984, Effects of aspartame on schedule controlled behavior in rats, Res. Communi. Psychol. Psychiat. Behav., 9:373.

Nomura, M., 1985, Effect of idebenone on brightness discrimination learning in rats with central serotonergic dysfunction, Jpn. J. Psychopharmacol., 5:243.

Nomura, M., 1988, Aging effects on operant type left-right discrimination learning test in aged rats, Biomed. Geront., 12:155.

Tsukada, Y., Nomura, M., Nagai, K., Kohsaka, S., Kawahata, H., Ito, M., and Matsutani, T., 1977, Neurochemical correlates of learning ability, in "Behavioral Neurochemistry," J. M. R. Delgada and F. V. De Feudis, eds., Spectrum Pub. Inc., New York, pp.63-84.

SDZ ENA 713, AN ACETYLCHOLINESTERASE-INHIBITOR WITH PREFERENCE FOR CENTRAL ACTIVITY: ANIMAL STUDIES

A. Enz, R. Amstutz, A. Hofmann, G. Gmelin and P.H. Kelly

Preclinical Research, SANDOZ Ltd.
CH-4002 Basle/Switzerland

INTRODUCTION

The well established cholinergic deficit in the brains of patients suffering from Alzheimers's disease (AD), and the fact that experimentally-induced cholinergic hypofunction results in cognitive impairments, provides the basis for cholinergic strategies for the treatment of AD (Bartus et al. 1982, Davis et al.1982, Drachman et al. 1977). The most promising results to date have been obtained with acetylcholinesterase-inhibitors (AChE-I) such as physostigmine (Physo) and Tacrine (THA), although the usefulness of these compounds is limited by problems of peripheral side-effects, bioavailability, duration of action and toxicity. One way to overcome some of these problems is to develop brain-selective drugs. We describe here some pharmacological properties, obtained from animal studies, of SDZ ENA 713 ((+)(S)-N-Ethyl-3-[(1-dimethyl-amino) ethyl]-N-methylphenylcarbamate) an AChE-I with high bioavailability, long duration of action and which produces a greater inhibition of AChE activity in brain regions than in peripheral organs.

METHODS

Chemistry

Racemic (+/-) (S)-N-Ethyl-3-[(1-dimethyl-amino) ethyl]-N-methylphenyl-carbamate was prepared according to M.Weinstock et al. (1986). The separation of the (-) isomer was achieved by crystallisation of the racemate with (-) di-O,O'-para-toluoyl-tartaric acid from methanol/water 2:1, $[\alpha]_D$ (free base)= -32° ($c=5, CH_3OH, >99\%ee$) (Amstutz et al. in preparation).

Pharmacology

AChE assay: Activity of AChE was measured according to the method described by Ellman et al. (1961). Briefly, rat brain tissue was homogenized in cold 0.25 mM phosphate buffer, pH 7.3 containing 1 % of Triton X-100. After centrifugation, aliquots of the clear supernatant were used as enzyme source. Rats received SDZ ENA 713 p.o. or Physo s.c. before sacrifice by decapitation.

Water-Maze: A swimming pool (diameter 1.33 m, diameter of central region 0.9 m, water temp. 23°C) and image-analysing system was used. A computer calculated for each minute of the test the distance swum and the amount of the time spent inside and outside the central region (P.H. Kelly et al., in preparation).

Basic, Clinical, and Therapeutic Aspects of Alzheimer's and Parkinson's Diseases, Volume 2
Edited by T. Nagatsu *et al.*, Plenum Press, New York, 1990

Hippocampal EEG: Rhythmical slow wave activity (RSA) in hippocampal EEG was recorded in anaesthetized male Sprague Dawley rats according to the methods described by P.Bevan (1984).

Cardiovascular Parameters, Cat: The experiments were performed on male and female cats (ca 2 kg) as described by Scholtysik et al. (1975) except for a slight modification of the anaesthesia (chloralose, 43 mg/kg i.m. / urethane 430 mg/kg i.m.).

RESULTS

AChE inhibition ex vivo in different rat brain regions

The pseudoirreversible mechanism of AChE inhibition by the two carbamates (Main, 1973), SDZ ENA 713 and physostigmine allows the measurement of their inhibitory effects ex vivo. Thirty minutes after administration of different doses, the remaining AChE activity in different rat brain regions was measured ex vivo. From these dose-response curves the following ED_{50} values were calculated; SDZ ENA 713 (mg/kg p.o): striatum 1.8, cortex 0.5, hippocampus 1, pons/medulla 2 and heart 7.5; ED_{50} Physo (mg/kg s.c.): striatum 0.3, cortex 0.2, hippocampus 0.3, pons/medulla 0.3 and heart 0.2. The AChE inhibition elicited by the oral administration of SDZ ENA 713 (2.5 mg/kg) peaked 15 minutes after drug application and decreased during the next 6 hours. The effect on brain enzymes was more pronounced than that on peripheral enzymes (Fig.1).

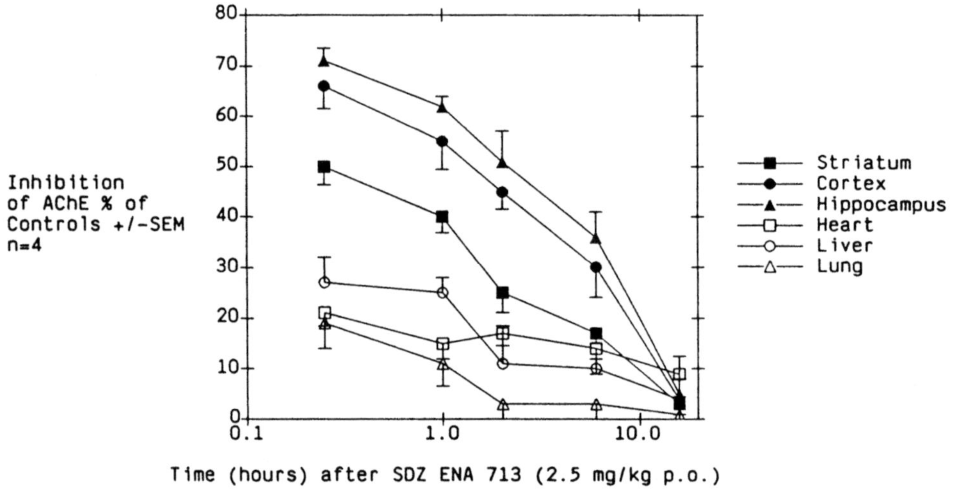

Fig. 1. AChE inhibition in rat brain and peripheral organs various times (0.25, 1, 2, 6 and 16 hours) after oral administration of 2.5 mg/kg SDZ ENA 713. The remaining enzyme activity was measured after extraction as described in Methods

Water-Maze Exploration Test

During a 5-minute exposure to a swimming pool, control rats initially swam around the perimeter of the pool, and then spent more time in the central region. Scopolamine (0.3 mg/kg, s.c., 30 min) affected this pattern of exploration, and this scopolamine effect was partly reversed by SDZ ENA 713 administered s.c. 15 minutes after scopolamine (Table 1). Qualitatively and quantitatively similar results were obtained with THA at doses three times higher (data not shown). In addition, control rats showed a memory of their first swim in that, during the first minute of a second swim three days later without any treatment, they spent a higher proportion of time in the central region than naive rats. This index of memory formation was reduced in the scopolamine-treated group, and this effect was reversed by SDZ ENA 713 with a bell-shaped dose response curve.

Tab.1 EFFECTS OF SDZ ENA 713 IN WATER MAZE EXPLORATION SWIM TEST

Treatment	Swim 1 Minute 1	2	3	4	5	Swim 2 Minute 1
Saline/saline	7 ±1	18 ±2*	31 ±3*	36 ±3*	40 ±4*	33 ±3*
Scopolamine/saline	5 ±1	3 ±1	6 ±1	8 ±2	12 ±3	15 ±2
Scop/ENA 713 (1 mg/kg)	3 ±1	5 ±1	12 ±3	18 ±3*	22 ±4*	17 ±2
Scop/ENA 713 (3 mg/kg)	5 ±1	8 ±2	12 ±2*	23 ±4*	27 ±4*	23 ±3*
Scop/ENA 713 (10 mg/kg)	3 ±1	7 ±2	17 ±3*	20 ±3*	26 ±4*	17 ±3

Percentage of time spent in the central region of a swimming pool during consecutive minutes of a 5-min swim, and during the first minute of a second swim 3 days later. Mean values ± SEM of groups of 31-33 rats.
* $p<0.05$ vs scop/saline (Mann-Whitney U-Test, 2-tailed).

Hippocampal Electroencephalogram in Rat

A characteristic effect of central muscarinic stimulation either by muscarinic agonists or by ACh is the induction of RSA in the hippocampal EEG (synchronization of theta-waves). After i.p. and p.o. administration of SDZ ENA 713, an initiation of theta waves (1.2 - 4.2 Hz) and an increase in their amplitude was found. The activation of the energy in the theta-frequency band was dose dependent (range 0.1 to 3 mg/kg p.o.) and completely blocked by pretreatment with scopolamine, but not by the peripheral muscarinic receptor blocker N-methylscopolamine (data not shown). The effect of SDZ ENA 713 on hippocampal EEG lasts longer than 6 hours after a single oral dose of 0.75 mg/kg (Fig.2).

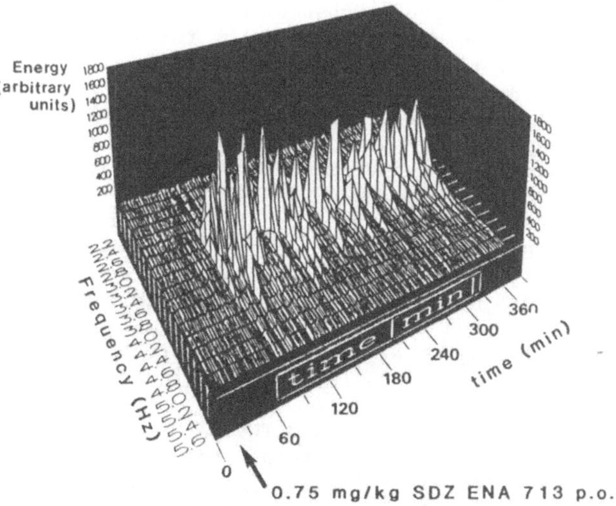

Fig.2. Energy of hippocampal EEG (theta waves) after a single dose of SDZ ENA 713 (0.75 mg/kg p.o.) as a function of frequency and time.

SDZ ENA 713 tested at a cumulative dose (0 - 1.5 mg/kg i.v.), exerted only minimal effects on circulatory parameters in the anaesthetised cat. At the dose of 0.75 mg/kg i.v. SDZ ENA 713 induced central effects manifested by strong tremor and slight cramps. Such central effects occurred after physostigmine at the dose of 0.14 mg/kg i.v., but in comparison to SDZ ENA 713 its influence on circulatory parameters was also more potent and already seen after 0.33 mg/kg i.v.. SDZ ENA 713 tested in the cat showed no influence on cardiovascular parameters at doses at which clear central effects such as strong tremor could be demonstrated.

DISCUSSION AND CONCLUSIONS

SDZ ENA 713 is a potent, long-lasting, orally active AChE-I. In ex vivo studies in the rat it shows, in contrast to Physo a preferential inhibition of AChE in brain compared to peripheral organs. Behavioral disturbances produced by scopolamine have been suggested as a putative model of those in AD. A scopolamine-induced alteration of exploration pattern in a swim-maze was partly reversed by both SDZ ENA 713 and THA, SDZ ENA 713 exhibiting a 3-fold higher potency. SDZ ENA 713 synchronized hippocampal EEG (theta waves) which reflects an increase of muscarinic activity in the hippocampus. At doses inhibiting brain enzymes by about 50-60% no clear peripheral effects such as salivation and diarrhoea were seen (data not shown). Moreover in doses up to 1.5 mg/kg i.v. SDZ ENA 713 had no effects on circulatory parameters in the anaesthetized cat, despite evidence of marked central cholinergic activation. Compounds like SDZ ENA 713, which preferentially interact with central cholinergic mechanisms should have advantages over AChE-I's which are presently available for the treatment of AD.

References

Bartus, R.T., Dean, R.L., Beer, B. and Lippa, A.S., 1982, The cholinergic hypothesis of geriatric memory dysfunction. Science, 217: 408-417

Bevan, P., 1984, Effect of muscarinic ligands on the electrical activity recorded from hippocampus:a quantitative approach. Br.J.Pharmacol., 82: 431-440.

Davis, K.L. and Mohs, R.C., 1982, Enhancement of memory processes in Alzheimer's disease with multiple-dose intra-venous physostigmine. Am.J.Psychiatry, 139: 1421-1424.

Drachman, D.A., 1977, Memory and cognitive function in man: Does the cholinergic system have a specific role? Neurology, 27: 783-790.

Ellman, G.L., Courtney, K.D., Andres, V. and Featherstone, R.M., 1961, A new and rapid colorimetric determination of acetylcholinesterase activity. Biochem. Pharmacol., 7: 88-95.

Main, A.R., 1973, Kinetics of Active-Site-Directed irreversible inhibition, in: "Essays in Toxicology", Vol.4, Wayland and Hagen, ed., pp.59-105. New York: Academic Press.

Scholtysik, G., Lauener, H., Eichenberger, E., Buerki, H., Mueller-Schweinitzer, E. and Waite, R., 1975, Pharmacological actions of the antihypertensive agent N-amidino-2-(2,6-dichlorophenyl)acetamide hydrochloride (BS 100-141). Arzneimittelforschung. Drug Res., 25: 1483-91.

Weinstock, M., Razin, M., Chorev, M. and Thasma, Z., 1986, Pharmacological activity of novel anticholinesterase agents of potential use in the treatment of Alzheimer's disease, in: "Advances in behavioral biology", Vol.29, A. Fisher, I. Hanin and C. Lachman, ed., pp.539-549. Plenum Press, New York.

DEVELOPMENT OF A COGNITIVE DRUG WITH THE ABILITY TO MODULATE GLUTAMATE (Glu) and ASPARTATE (Asp)

Choichiro Miyazaki, Kenji Matsuyama, Masataka Ichikawa,[1] Kazuhiko Suyama, Masaki Kurihara and Kazuo Mori[2]

Departments of [1]Hospital Pharmacy and [2]Neurosurgery, Nagasaki University Hospital, Sakamoto-machi 7-1, Nagasaki 852, Japan

INTRODUCTION

The acidic amino acids aspartate (Asp) and glutamate (Glu) are major excitatory transmitters in the central nervous system. When present in high concentration they are neurotoxic and could play a role as an "excitotoxin", especially in the CA_1 region of the hippocampus in ischemic stage,[1] suggesting their participation in vascular dementia.

Recently, Greenamyre proposed the hypothesis that acidic amino acids could lead to the degeneration seen in Alzheimer's disease[2] on the basis of the fact that injection of Glu into the cerebral cortex produced retrograde degenerative changes in the basal forebrain, the major sourse of cortical cholinergic innervation.[3]

It is well known that pyrrolidinone derivativea, i.e., piracetam or aniracetam, improve cognitive functions, and have a pronounced protective action against the memory-impairing effects of cycloheximide and scopolamine in experimental animals.[4] However, these drugs can not modulate Glu and Asp in the brain. In our previous studies,[5,6] valproate (VPA) was found to reduce Glu and Asp in the barin because it inhibits GABA-transaminase (GABA-T), resulting in a significant increase in brain GABA levels. With these observations in mind, we synthesized an acidic amino acid-modulating pyrrolidinone, namely, valproyl-2-pyrrolidinone (VP).

MATERIALS AND METHODS

Microdialysis Experiment

Male SHR rats (12-15 weeks old, BP: 180-200 mmHg) were used in the experiment. The vertebral artebral arteries were occluded bilaterally by electrocauterization one day before. For the experiment each rat was anesthetized with 2% halothane under controlled ventilation. Temperature on the temporal muscle was maintained at 37±0.5°C and $PaCO_2$ was kept at 35-40 mmHg. Cortical blood flow (CoBF) was measured by Laser Doppler flowmetry through a frontal bone window in all animals. The extracellular amino acids in the dosal hippocampus were collected through a microdialysis probe (2-mm membrane, BAS) and analyzed by an HPLC equipped with a fluorescence detector according to the method described previously.[5] Rats were subjected to global ischemia by temporary clamping of bilateral carotid arteries for 20 min.

Protective Effect against Cerebral Anoxia in Mice

Groups of 10 mice (ddY, male, 28-34 g) were used for each dose of drug. At 1 h after oral drug administration, all mice placed in a transparent plastic container (13x13x16 cm) equiped with an inhalant and exhaust ports. They were exposed to a gas mixture of 4% oxgen and 96% nitrogen with a flow of 5 liters/min. The survival time, measured as the time from the introduction of anoxia to the respiratory failure of the animals, was recorded.

Passive Avoidance Test

Training was carried out in an apparatus consisting of two compartments, one illuminated (25x11x12 cm) and the other darkened (25x12x30 cm). The compartments were separated by a guillotine door. The darkened chamber had an electrifiable grid floor, and a shock was delivered to the animal's feet with a shock generator scrambler. One h before acquistion of the passive avoidans response, rats (Wistar, male, 195-235 g) were trained to explore the step-through passive avoidance apparatus. Each rats was placed in the illuminated chamber and 10 sec later the door was opened. After a rat entered into the dark compartment, it was allowed to remain there for 10 sec. The acuisition test was performed similar to the training, inescapable foot shock (4.5 mA, for 1 sec) was given to the rat's feet when it stepped through to the darkened chamber. Thirty min before the acquisition test, each rat was injected with scopolamine (0.5 mg/kg, s.c.). VP emulsified in 0.5% (W/V) CMC solution was given immediately after the acquisition test ata a dose of 30, 100, or 300 mg/kg (p.o.). Twentyfour h after the acquisition test, the retention test was performed and the step-through latency was measured.

Synthesis of VP

A soln. of 8.5 g of pyrrolidinone and 20 g of valproyl chloride in 200 ml of anhydrous benzene was stirres at 60°C overnight. When the reaction was complete, the mixture was washed by 200 ml of dil. sodium bicarbonate soln., dried over anhydrous sodium sulfate, and freed of benzene *in vacuo*. The residue was distilled under reduced pressure of 3 torr. Fourteen grams (recovery: 57%) of colorless liquid (b.p. 111-113 °C/3 torr.) was obtained.

RESULTS AND DISCUSSION

Brain metabolism is highly dependent upon oxgen supply to the brain. When the blood circulation becomes insufficient, cerebral function rapidly cease; or even in the survivor, neurological injury occurs, resulting in neural necrosis. The CA_1 pyramidal cells in the hippocampus are particulary of susceptible has been well established morphologically in ischemic animal models, the pathogenesis of selective vulnerability remains unclear. We examined alterations in contents of the extracellular amino acids before, during, and after ischemia using spontaneously hypertensive rats (SHR). Figure 1 shows the effect of ischemia on the contents of extracellular acidic amino acids in the hippocampus of SHR. The contents of extracellular Glu and Asp were significantly increased during ischemia, and returned to baseline by 30 min after recirculation.

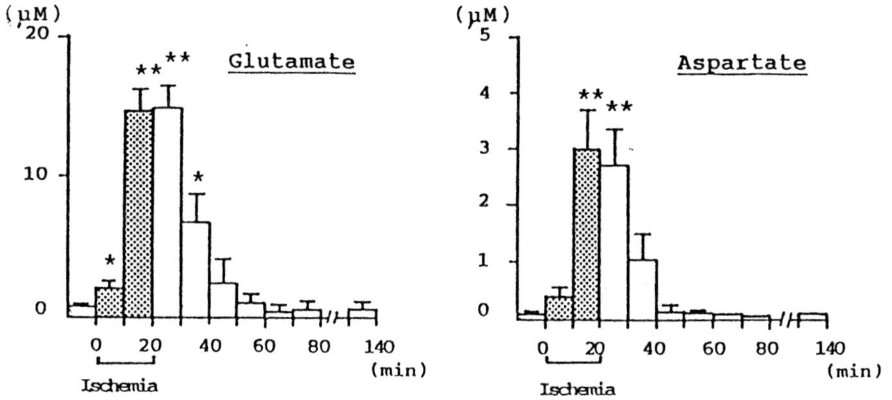

Fig. 1 Sequential changes in the contents of extracellular acidic amino acids in the hippocampus of spontaneously hypertensive rats during and after ischemic insult.
Statisitical signifcance: *$p<0.05$, **$p<0.01$ by Student's t-test.

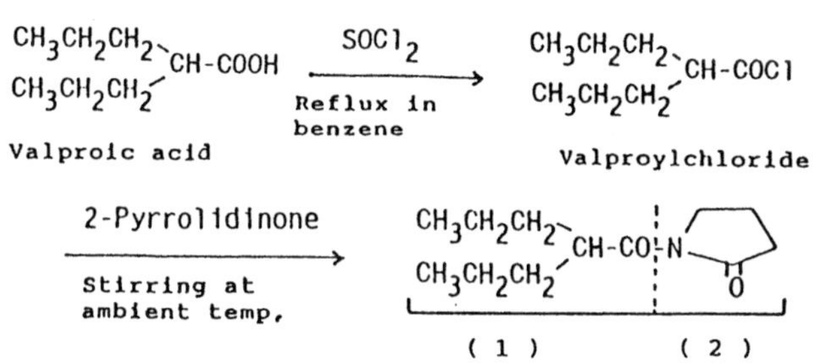

Chart 1. Procedure for synthesis of valproyl-2-pyrrolidinone (VP) and expected effect.
(1) Valproyl moiety, expected to show an anti-excitotoxin effect.
(2) 2-pyrrolidinone, expected to show a cognitive effect.

As described in the introduction, VPA was found to reduce brain Asp and Glu wherea it increased brain GABA levels. On the basis of these results, we synthesized VP from valproylchloride and 2-pyrrolidinone (Chart 1).

VP was submitted to pharmacological testing. Figure 2 shows the protective effects of calcium hopantenate (HOPA), VP, and aniracetam against cerebral anoxia. The intrperitoneal administration of HOPA at a dose of 300 mg/kg showed a significant prolongation of the survival time (the closed bar). However, in the case of oral administration, it did not prolong the time even at the dose of 1000 mg/kg. Furthermore, aniracetam did not show the protective effect against cerebral anoxia. In contrast to HOPA or aniracetam, VPA prolonged the survival time following the oral administration at the doses of 100 and 300 mg/kg. The protective effect of VP might be correlated with changes in brain acidic amino acids. Figure 3 shows the changes in the content of Asp and Glu in the hippocampus of rats after intraperitoneal administration of VP at a dose of 300 mg/kg. VP gave rise to a significant reduction in Asp and Glu at 2 h after the administration. The decrease in brain Asp and Glu may be attributed to the reduction in the cerebral metabolism through the mechanism of the malate-aspartate shuttle,[7] which might contribute to the protective effect of VP against the cerebral anoxia. Judging from the fact that VPA caused a significant changes in Asp and

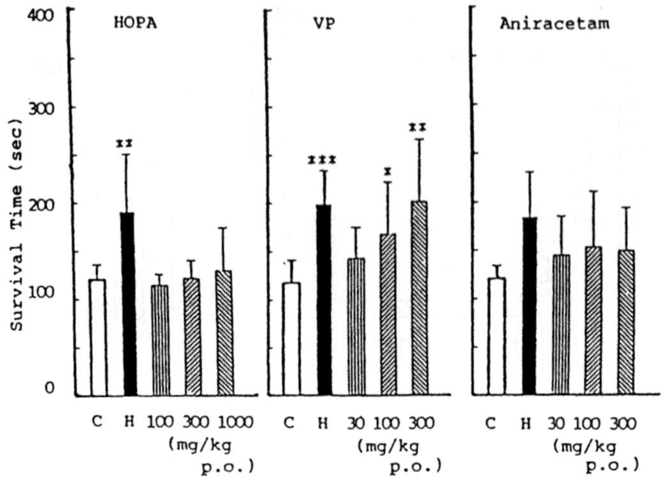

Fig. 2 Protective effects of drugs of HOPA, VP and aniracetam against cerebral anoxia in mice
The protective effect of drugs on the anoxia was investigated by mesuring the survival time in the state that mice were exposed to a gas mixture consisting of 4% O_2 and 96% N_2.
C, control. H, HOPA, 300 mg/kg, i.p..
Statistical signifcance: *p 0.05, **p 0.01, ***p 0.001 by Student's t-test.

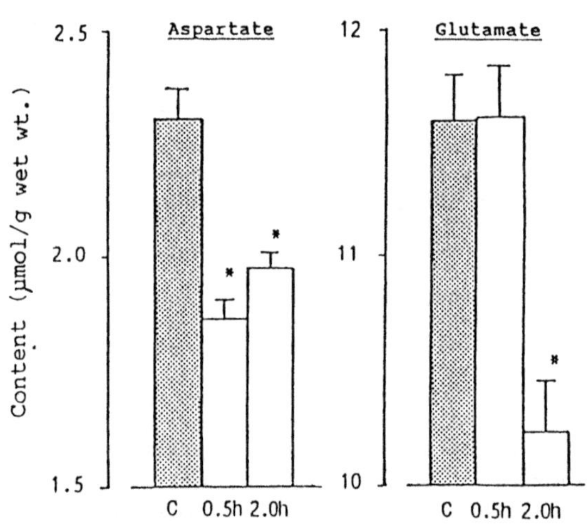

Fig. 3. Sequential changes in the content of Asp and Glu in the hippocampus of rats after the i.p. administration of VP at a dose of 300 mg/kg.
Statistical significance: *p 0.01 vs. control(C) by Student's t-test.

Glu,[6] the significant changes in Asp and Glu levels in the hippocampus are thought to be due to valproic acid derived from the hydrolysis of VP.

Figure 4 shows the comparative effect of HOPA, VP, and aniracetam on the prolongation of latency time in the passive avoidance test. VP and aniracetam showed a significant effect in the test when compared with HOPA. It is noteworthy that a bell-shaped dose-response pattern was

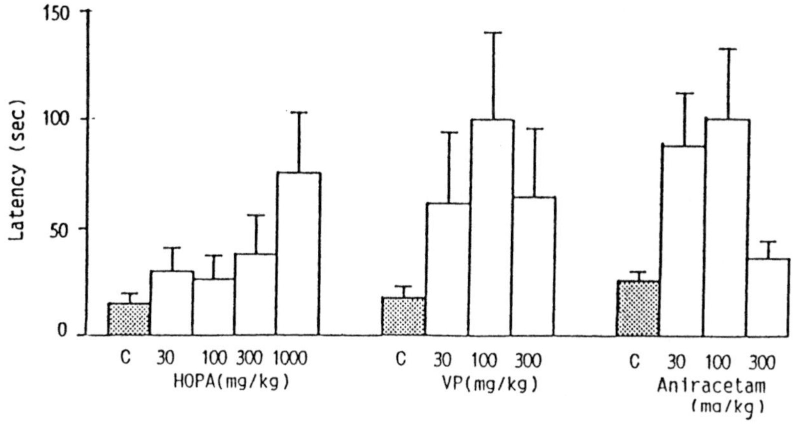

Fig. 4 Reversal of scopolamine-induced transient disruption of the memory of a passive avoidance response in rats. The anmesic agent scopolamine (0.5 mg/kg, s.c.) was injected 30 min before the passive avoidance training. HOPA, VP, or aniracetam was administered orally immediately after exposure to ECS (4.5 mA, 1 sec). The time elapsed between passive avoidance training and retention test was 24 h.

obtained for both VP and aniracetam, which is similar to pattern obtained with other drugs that are considered to improve cognitive functions.[8] From these results, VP is defined as a new pyrrolidinone drug with the ability to modulate acidic amino acids.

In adults, global ischemia sometimes leads to preferential damage of the cerebral cortex and hippocampus, areas rich in receptors for Glu, suggesting that the damage is mediated by endogenous acidic amino acids. Based in these findings, a wide variety of antagonists of Asp or Glu have ability to modulate acidic amino acids.

In adults, global ischemia sometimes leads to prefential damages of the cerebral cortex and hippocampus, areas rich in receptors for Glu, suggesting that the damage is mediated by endogenous acidic amino acids. Based in these findings, a wide variety of antagonists of Asp or Glu have been developed. However, a problem has arisen in the actual use of these agents from the point of view of safety. The value of LD_{50} of VP is found to be more than 2000 mg/kg. Along this line, further studies are now in progress.

REFERENCES

1. H. Benveniste, J. Drejer, A. Schousboe, and N. H. Diemer, Elevation of the extracellular concentration of glutamate and aspartate in rat hippocampus during transient cerebral ischemia monitored by intracerebral microdialysis, J. Neurochem. 43:1369 (1984).
2. J. T. Greenamyre, The role of glutamate in neurotrnasmission and in nurologic disease, Arch. Neurol. 43: 1058 (1986).
3. M. V. Sofroniew and R. C. A. Pearson, Degeneration of chlinergic neurons in the basal nucleus following kainic or N-metyl-D-aspartate application to the cerebral cortex in the rat, Brain Res. 339:186 (1985).
4. R. Cumin, E. F. Bandle, E. Gamuz, and E. W. Haefely, Effect of the novel compound aniracetam (Ro 13-5057) upon impaired learning and memory in rodents, Psychopharmacology 78:104 (1982).

5. C. Miyazaki, M. Ogasawara, M. Ichikawa, K. Matsuyama, and S. Goto, Determination of γ-aminobutyric acid and associated amino acids in mouse brain by liquid chromatography, J. Pharmacobio-Dyn. 11:202 (1988).
6. C. Miyazaki, K. Matsuyama, M. Ichikawa, and S. Goto, Effect of sodium valproate (VPA) on cerebral amino acid: Mechanism of γ-aminobutyric acid (GABA) elevation and possible causal relation of VPA-induced encephalopathy and glutamine levels, Chem. Pharm. Bull. 36:3589 (1988).
7. B. Hindfelt, F. Plum, and T. E. Duffy, Effect of acute ammonia intoxication on cerebral metabolism in rats with portacaval shunts, J. Clin. Invest. 59:386 (1977).
8. J. F. Flood, D. W. Landry, and M. E. Jarvik, Cholinergic receptor interactions and their effects on log-term memory processing, Brain Res. 215:177 (1981).

NERVE GROWTH FACTOR AND STAUROSPORINE FACILITATE RECOVERY FROM BASAL FOREBRAIN LESION-INDUCED AMNESIA IN RATS

Shin-ichi Ogawa, Toshitaka Nabeshima, Hidenao Nishimura
Kazuyuku Fuji, and Tsutomu Kameyama

Department of Chemical Pharmacology, Faculty of Pharmaceutical Sciences, Meijo University, Nagoya 468 Japan

INTRODUCTION

Nerve growth factor (NGF) is the most fully characterized neurotrophic factor among the various putative neuronal growth factors in the peripheral and central nervous system (CNS) (Levi-Montalcini and Angeletti, 1968; Freed, 1976; Thoenen and Barde, 1980). The influence of NGF on the central cholinergic neurons is especially significant in light of the involvement of cholinergic projection in Alzheimer's disease (AD). The degeneration of cholinergic neurons in the basal forebrain (BF) has been shown to be a major component of the neuropathology of AD. Appel (1981) and Hefti (1983) have suggested that NGF, or NGF-like trophic factors, may be essential for the maintenance of central cholinergic neurons and that the loss of such trophic factors could be the basis for cholinergic degeneration in AD. The corresponding cholinergic pathways have been suggested to be important for memory and learning. Interestingly, Morioka et al. (1985) reported that staurosporine (ST), which was isolated from the culture broth of *Streptomyces staurosporeus*, strongly induced differentiation in human neuroblastoma cell line. ST caused elongation of neurites and cell enlargement. Moreover, ST (0.02 μM) brought about these changes in one hour. The present study was undertaken to test the hypothesis that the administration of NGF and ST may reverse some of the effects of experimentally-induced lesions of the BF in the rats.

MATERIALS AND METHODS

Male Wistar rats (290-320 g) received ibotenic acid (5 μg/side) lesions of the BF (Bregma -1.5, ML$\pm$2.8, DV -7.5). Water maze task was used to investigate the effect of NGF and ST. The water maze consisted of a circular water tank with an invisible circular platform, and the tank was filled with tap water. At the start of a training session, a rat was placed at one of the five start locations and a switch was pressed to start the timing and data collection by a computer. When the rat found the platform, the rat was then allowed to remain on the platform for 30 s (a maximum recording time of 200 s). The dependent variable for analysis was latency in finding the platform. The data were analyzed by an analysis of variance (ANOVA), Mann-Whitney U-test.

Basic, Clinical, and Therapeutic Aspects of Alzheimer's and Parkinson's Diseases, Volume 2
Edited by T. Nagatsu *et al.*, Plenum Press, New York, 1990

For NGF, cannulae attached to modified mini-osmotic pumps were implanted into the lateral ventricle of the rat. In the NGF-treated rat, the pump was filled with artificial cerebrospinal fluid (CSF) containing 10-30 μg/ml of 7S-NGF. In the control and the BF-lesioned rats, the pump was filled with artificial CSF. The estimated NGF dose was 2-6 μg/2 weeks. The rat was infused with the NGF immediately after or 4 weeks after being the BF-lesioned. Behavioral test was started 10 days after implantation of NGF pump. Training was conducted on 5 consecutive days, the rat receiving 2 training sessions on each day. To assess the acute effects of ST, we injected rats with the drug (0.01, 0.03, 0.1, or 0.3 mg/kg, i.p.) 30 min prior to every training session. For sub-chronic effects, the rats were administered ST (0.05, 0.1, or 0.2 mg/kg, i.p.) for 14 days before training. Behavioral test was started 4 weeks after the BF-lesion. Training was conducted 14 consecutive days, rats receiving training once a day.

CAT activity was assayed after last training, and the data were analyzed by Student t-test.

RESULTS

Experiment 1. Effect of NGF on Basal Forebrain Lesion-induced Amnesia

In the water maze task, the latency in finding the platform for all experimental groups is presented in Fig.1. All the lesioned rats took longer to find the platform than control rats. When NGF infusion was started immediately after the BF-lesioning, the NGF (6 μg/2 weeks) treated rat found the platform significantly faster than the BF-lesioned rats ($F(1,120)=151.4$, $p<0.01$). But, when the NGF infusion was started 4 weeks after the BF-lesion, the NGF (2-6 μg/2 weeks)-treated rat found the platform more slowly. The two groups did not differ significantly. Levels of CAT activity in the parietal cortex of rats treated with NGF (6 μg/2 weeks) immediately after the BF-lesioning were measured. Although CAT activity decreased significantly in BF-lesioned animals as compared with that in the control rats (74.2 % of the control), administration of NGF protected against a significant decrease of CAT activity (86.3 % of the control).

Experiment 2. Effect of ST on Basal Forebrain Lesion-induced Amnesia

ST (0.05-0.1 mg/kg)-treatment improved the impaired performance in rats that had been the BF-lesion (0.05 mg/kg, $F(1,168)=74.9$, $p<0.01$., 0.1 mg/kg, $F(1,168)=110.7$, $p<0.01$). Administration of ST protected against the significant decrease in CAT activity in the parietal cortex induced by the BF-lesion to 88.7 % of the control ($p<0.05$).

DISCUSSION

Our results showed that subacute administration of NGF and ST, which has NGF-like activity, facilitated recovery from the BF-lesioned amnesia in rats. Moreover, NGF and ST induced attenuation of depletions of cortical CAT activity induced by the BF-lesion.

Hefti(1986) has reported that an increase in the availabity of NGF to cholinergic neurons may prevent or reduce degeneration after fimbrial transection. Montero and Hefti (1988) have suggested also that the continuous presence of NGF during more than 4 weeks is

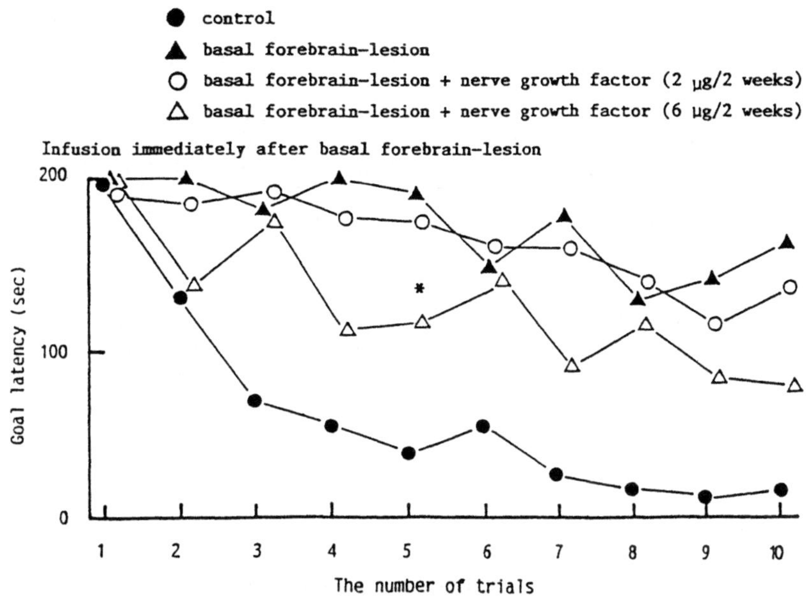

Fig.1. Effect of NGF on the basal forebrain (BF) lesion-induced amnesia in the water maze task. Rats were infused with NGF immediately after being the BF-lesioned. Behavioral test was started 10 days after implantation of the NGF pump.
* $p<0.05$ vs. the BF-lesioned rat.

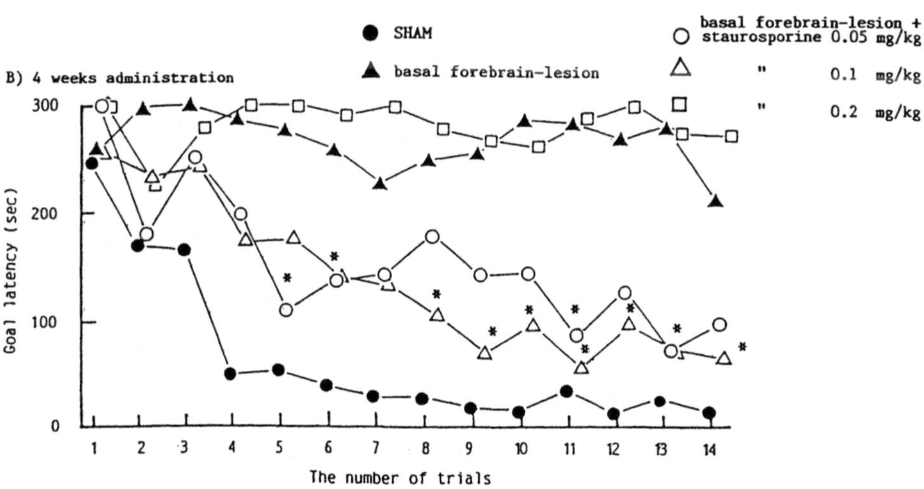

Fig.2 Effect of staurosporine (ST) on the basal forebrain (BF)-lesion-induced amnesia in water maze task. The rat was administered ST for 14 days before training and immediately after every training session. Behavioral test was started 4 weeks after the BF-lesion.
* $P<0.05$ vs. the BF-lesioned rats.

required to prevent the degeneration of cholinergic cells. In addition, infusion of NGF in our experiments led to a significant recovery of cortical CAT and acetylcholinesterase activity (Haroutunian et al., 1986). Perhaps these permit the infused NGF to be taken up by the axons of the cholinergic neurons, and witch cells then synthesizes CAT (Vahram et al., 1986). The cholinergic system in the BF has a well-defined role in the correlation that exists between neuronal choline deficits and memory dysfunction in aging and AD. Furthermore, it has been suggested that aging and AD may be related to a deficient supply or a reduced utilization of a trophic agent (Appel, 1981). In addition, lower levels of NGF are observed in the brains of aged rats than in adults (Larkfors et al., 1987)

The above results suggest that the administration of NGF and ST promotes the survival of a subpopulation of the BF neurons that have undergone ibotenic acid-induced degeneration. The evidence for NGF and NGF-like trophic factor-induced enhancement of cholinergic activity in the lesioned system may lead to novel therapeutic approaches to AD. Moreover, our results suggest that ST has a possibility to become a prototype drug for AD.

ACKNOWLEDGMENT

We are grateful to Asahi Chemical Industry Co., LTD., for the supply of staurosporine.

REFERENCES

Appel, S. H., 1981, A unifying hypothesis for the cause of amyotrophic lateral sclerosis, parkinsonism, and Alzheimer disease, Ann. Neurol., 10:499.

Freed, W., 1976, The role of nerve growth factor (NGF) in the central nervous system, Brain Res.Bull., 1:393.

Hefti, F., 1983, Alzheimer's disease caused by a lack of nerve growth factor ?, Ann.Neurol., 14:109.

Hefti, F., 1986, Nerve growth factor promotes survival of septal cholinergic neurons after fimbrial transections, J.Neurosci., 6: 2155.

Haroutunian, V., Kakof, P. D. and Davis, K, 1986, Partial reversal of lesion-induced deficits in cortical cholinergic markers by nerve growth factor, Brain.Res., 386:397.

Larkfors, L., Ebendal, T., Whittemore, S. R., Persson, H., Hoffer, B. and Olson, L., 1987, Decreased level of nerve growth factor (NGF and its messenger RNA in the aged rat brain, Mol.Brain Res., 3: 55.

Levi-Montalcini, R. and Angeletti, P. U., Nerve growth factor, Physiol. Rev., 48:534.

Montero, C. N. and Hefti, F.,1988, Rescue of lesioned septal cholinergic neurons by nerve growth factor: specificity and requirement for chronic treatment, J.Neurosci., 8:2986.

Morioka, H., Ishihara, M., Shibai, H. and Suzuki, T., 1985, Staurosporine-induced differentiation in a human neuroblastoma cell line, NB-1, Agric.Biol.Chem., 49:1959.

Thoenen, H. and Barde, Y. A., 1980, Physiology of nerve growth factor, Physion.Rev., 60:1284.

Vahram, H., Philip, D. K. and Kenneth, L. D., 1986, Partial reversal of lesion-induced deficits in cortical cholinergic markers by nerve growth factor, Brain Res., 386:397.

EFFECTS OF NOOTROPIC DRUGS ON EXPERIMENTALLY INDUCED AMNESIA OF RATS IN THE THREE-PANEL RUNWAY TASK

Tsuneyuki Yamamoto, Masuo Ohno, Iwao Kitajima, and Showa Ueki

Department of Pharmacology, Faculty of Pharmaceutical Sciences, Kyushu University 62, Fukuoka 812, Japan

INTRODUCTION

Animal models of dementia are important for an understanding of the neurobiological basis of memory impairment, and for the preclinical evaluation of nootropic drugs. Therefore, explicit criteria to evaluate the relative validity of different models can help us comprehend human memory. Recently, much research has been directed toward the development of drugs for the treatment of dementia and cognitive dysfunction in the elderly.

Recently we reported a new method using a 3-panel runway apparatus for the study of working memory in the rat (Furuya et al., 1988). Using this method, we investigated the effects of various nootropic drugs on amnesia produced by intraperitoneal scopolamine, by lesions of the dorsal hippocampus, and by cerebral ischemia.

METHODS

Subjects

Eight- to ten-week-old male Wistar strain rats were used. Initial free feeding weights were 230-250 g, and were maintained at approximately 80% of this level for the experiment.

3-Panel runway task

Working memory was assessed by use of the 3-panel runway apparatus (Furuya et al., 1988). The apparatus (175 x 36 x 25 cm, length x width x height) is composed of a start box, a goal box, and four consecutive intervening choice points. Each choice point consists of a gate with three panels (12 x 25 cm, width x height). Rats are prohibited from passing through two of the three panels in the gate (incorrect panel-gates), and also prohibited from returning to either the start box, or a previous choice point at all gates. When rats reach the goal box, they can obtain two food pellets (about 50 mg).

The rats were trained in six consecutive trials at 2-min intervals (one session) per day. The locations of the correct panel-gates were held constant within a session, but differed among the sessions. Thus, twelve patterns of correct panel-gate locations were used in this experiment. The number of times an animal pushed an incorrect panel-gate and the time required for the animal to take food pellets were recorded for each rat on every trial of a session. Rats were used for the experiment if they achieved a criterion of less than 12 errors summed across 6 trials of a session, for each of three consecutive sessions.

Amnesic models

Scopolamine (0.56 mg/kg) was injected i.p. into the rats 20 min before the runway test.

Lesions of the dorsal hippocampus were produced with a lesion generator in which the temperature of the electrode tip could be determined. Under pentobarbital anesthesia, the electrode (0.25mm in diameter, 1.0-mm exposed tip) was inserted into the dorsal hippocampus (anterior: -3.3mm from the bregma, lateral: 2.6mm, horizontal: 2.8mm from the dura) according to the rat brain atlas of Paxinos and Watson (1982). The bilateral hippocampal lesions were induced by an increase in the tissue temperature to 70°C within 15 s and then maintenance of this temperature for another 30 s. Rats were allowed 5 days to recover from surgery before being used in the experiment.

Cerebral ischemia was induced by the four-vessel occlusion method (Pulsinelli and Brierley, 1979). The carotid arteries of the rats were temporarily occluded with clips, following permanent occlusion of the vertebral arteries by electrical cauterization. Rats that lost the righting reflex during the ischemia were assigned to the ischemic group. The runway test was performed 24 h after the ischemia.

Drugs

The following drugs were used: NIK-247 [amiridin; 9-amino-2,3,5,6,7,8-hexahydro-1H-cyclopenta(b)-quinoline hydrochloride hydrate], tetrahydroaminoacridine hydrochloride hydrate (THA), physostigmine salicylate, minaprine [3-(2-morpholinoethylamino)-4-methyl-6-phenylpyridazine dihydrochloride], S-adenosyl-L-methionine sulfate tosylate (SAM; FO-1561), and (-)scopolamine hydrobromide. SAM was dissolved in a phosphate buffer solution, and the pH was then adjusted to 4-6.5 with NaOH solution. The other drugs were dissolved in distilled water. Amiridin and THA were

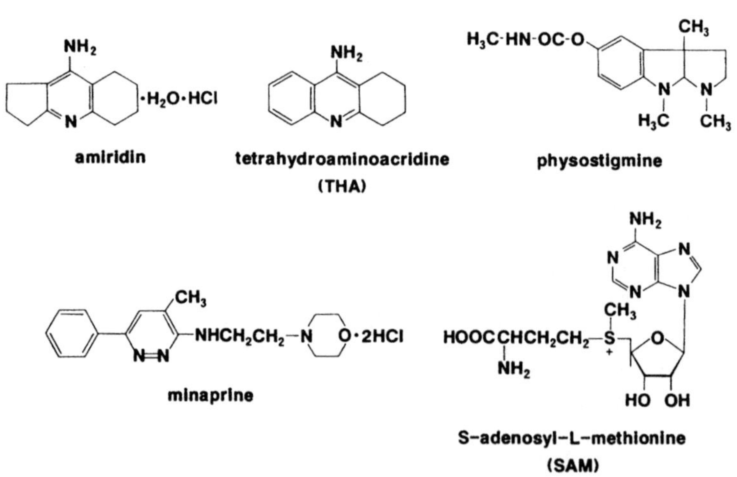

Fig. 1. Chemical structures of nootropic drugs.

administered p.o. 25 min before the runway test. Rats were treated p.o. or i.p. with physostigmine and minaprine 1 h or 30 min before the test session. SAM was injected i.p. 1 h before testing. In the ischemia experiment, the drugs were administered immediately after blood flow recirculation and then again before the test session.

RESULTS

The i.p. injection of scopolamine (0.56 mg/kg), lesions of the dorsal hippocampus, and cerebral ischemia significantly increased the number of errors and prolonged the latency in the 3-panel runway task.

Amnesia induced by scopolamine, as indicated by the increase in number of errors, was antagonized dose-dependently by cholinesterase inhibitors such as amiridin (18 mg/kg p.o.), THA (3.2-10mg/kg p.o.), and physostigmine (0.1-0.32 mg/kg i.p.). Treatment with these doses of the three cholinesterase inhibitors also blocked the increase in latency. Minaprine at doses of 10-32 mg/kg i.p. also caused a dose-dependent reduction in the increase of errors, but not in the prolongation of latency in scopolamine-treated rats. SAM, at doses of 32-180 mg/kg i.p., did not improve the scopolamine-induced amnesia. On the contrary, a significant prolongation of latency was observed in rats treated with SAM 180 mg/kg.

Similarly, amnesia induced by dorsal hippocampal lesions was also ameliorated significantly by amiridin (32 mg/kg p.o.), physostigmine (0.1 mg/kg p.o.), and minaprine (10 mg/kg p.o.), but not by SAM (180 mg/kg i.p.).

On the other hand, 100 and 180 mg/kg of SAM significantly reduced the increase in errors and the prolongation in latency induced by a 5-min period of cerebral ischemia. Minaprine (3.2-10 mg/kg i.p.), and the cho-

Table 1. Effects of Nootropic Drugs on Amnesia Induced by Scopolamine (0.56 mg/kg), Dorsal Hippocampal (DH) Lesions, and Cerebral Ischemia (5 min) in the 3-Panel Runway Task in Rats

Drug	Dose (mg/kg)	Treatment		
		Scopolamine	DH lesions	Ischemia
Non-treatment		8.3±1.1 (8)	7.5±0.8 (6)	6.0±0.9 (6)
Saline	—	24.9±2.1** (8)	18.0±0.4** (6)	21.2±1.6** (6)
Amiridin	3.2	20.4±3.3 (8)	—	12.7±2.4* (6)
	10	18.9±2.6 (8)	15.2±1.1 (6)	9.8±1.2** (6)
	18	12.4±1.8** (8)	—	—
	32	—	8.7±0.9** (6)	—
THA	1.0	22.6±1.6 (8)	—	—
	3.2	12.9±1.5** (8)	—	9.5±0.7** (6)
	10	11.3±1.9** (8)	—	—
Physostigmine	0.1	16.1±2.0** (8)	10.7±1.4** (6)	9.5±1.5** (6)
	0.32	9.1±1.4** (8)	—	—
Minaprine	3.2	22.6±2.3 (8)	—	15.0±2.0* (6)
	10	15.3±2.9* (8)	9.8±1.0** (6)	8.3±1.3** (6)
	32	12.0±2.1** (8)	—	—
SAM	32	20.3±2.4 (6)	—	—
	100	20.4±2.3 (5)	—	11.8±3.1* (8)
	180	25.6±3.1 (5)	17.0±1.4 (6)	7.6±1.5** (8)

Each value is the mean ±S.E. of errors summed across 6 trials of a session. Numerals in parentheses represent the number of animals used. The significance of differences from non-treated rats (**$P<0.01$) and from saline-administered rats with amnesic treatments (*$P<0.05$, **$P<0.01$) was determined by Student's t-test.

linesterase inhibitors such as amiridin (3.2-10 mg/kg p.o.), THA (3.2 mg/kg p.o.), and physostigmine (0.1 mg/kg i.p.) also significantly reduced the increase in both parameters in the ischemic rats.

DISCUSSION

Cholinergic function has been found to play an important role in learning and memory in both humans (Sitaram et al., 1978) and animals (Watts et al., 1981). One of the most consistent and striking features of Alzheimer's disease is the reduction in the activity of choline acetyltransferase in the cerebral cortex and hippocampus (Davies and Maloney, 1976; Price et al., 1982); this may be related to the memory deficits of this dementing disorder.

We reported earlier that our experimental procedure using the 3-panel runway apparatus would be useful for studying working memory (Furuya et al., 1988). In the runway task, scopolamine and a cholinergic neurotoxin (AF64A) produced a disruption of working memory. The present study clearly indicated that physostigmine, a well-known cholinesterase inhibitor, exerted an ameliorating effect on the impairment of working memory induced by scopolamine. Furthermore, amiridin and THA, which are centrally active cholinesterase inhibitors, as well as candidates for the treatment of senile dementia of the Alzheimer type (Summers et al., 1986), also attenuated scopolamine-induced amnesia. Minaprine has also been found to be effective in the treatment of senile dementia (Passeri et al., 1985). In the present experiment, minaprine also improved scopolamine-induced memory deficits. Recently, Muramatsu et al. (1988) reported that minaprine increases the release of acetylcholine from the hippocampus by blocking the presynaptic 5-HT_2 receptor on the cholinergic neuron. Therefore, this stimulatory action of minaprine on central cholinergic neurons may contribute to its ameliorating effect on the scopolamine-induced amnesia. In contrast, SAM did not improve the impairment of working memory induced by scopolamine. This suggests that SAM does not stimulate the cholinergic system.

The septo-hippocampal system has been found to play an important role in learning and memory. In clinical studies, the parallel between the pattern of memory deficits in Alzheimer's and hippocampus-damaged patients suggests that the function of the hippocampal formation is impaired in patients with Alzheimer's disease. In this study, lesions of the dorsal hippocampus produced a marked increase in the number of errors in the 3-panel runway task. The impairment of working memory induced by hippocampal lesions was also ameliorated by the drugs possessing a stimulatory action on the central cholinergic system, but not by SAM. These results strongly suggest that the memory deficits in the dorsal hippocampal lesioned rats are attributed to lowering of the cholinergic function. It is of great interest that there are specific changes in the cholinergic system as indicated by profound decreases in choline acetyltransferase and acetylcholinesterase activity in the hippocampal formation of Alzheimer's patients (Perry et al., 1978).

In the present experiment using the method of 4-vessel occlusion, impairment of working memory in the 3-panel runway task was successfully produced in rats by a 5-min period of cerebral ischemia. Both the cholinergic stimulant drugs and SAM improved impairment of working memory induced by cerebral ischemia. It has been reported that cerebral ischemia causes a marked reduction in brain ATP level and neuronal degeneration. SAM has beneficial effects on these ischemia-induced changes. Thus, the reversals of memory deficits produced by SAM may be attributable to its action in accelerating the recovery of ATP to normal levels, and preventing disturbances of membrane phospholipids that would occur in the ischemic brain. On the other hand, the cholinergic dysfunction, i.e., a significant decrease in acetylcholine and a marked increase in choline, has been noted in the brain of rats with ischemia (Kakihana et al., 1984). In the present experiment, the cholinesterase inhibitors and the indirect cholinergic stimulant drug minaprine reversed amnesia in an ischemic model. These findings suggest that activating actions of these drugs on the central cholinergic system play an important role in the reversals of memory deficits in rats with ischemia.

In conclusion, our findings suggest that the ameliorating effect of these nootropic drugs on amnesia produced by dorsal hippocampal lesions and scopolamine is predominantly mediated by cholinergic activation but that the effect on ischemia-elicited amnesia is not solely due to a stimulant action on the cholinergic system.

ACKNOWLEDGMENT

This study was partly supported by a Grant-in-Aid for Scientific Research (No. 01570108 and No. 63480477) from the Ministry of Education, Science, and Culture of Japan.

REFERENCES

Davies, P., and Maloney, A. J. F., 1976, Selective loss of central cholinergic neurons in Alzheimer's disease, Lancet, 2:1403.

Furuya, Y., Yamamoto, T., Yatsugi, S., and Ueki, S., 1988, A new method for studying working memory by using the three-panel runway apparatus in rats, Japan. J. Pharmacol., 46:183.

Kakihana, M., Yamazaki, N., and Nagaoka, A., 1984, Effects of idebenone (CV-2619) on the concentrations of acetylcholine and choline in various brain regions of rats with cerebral ischemia, Japan. J. Pharmacol., 36:357.

Muramatsu, M., Tamaki-Ohashi, J., Usuki, C., Araki, H., and Aihara, H., 1988, Serotonin-2 receptor-mediated regulation of release of acetylcholine by minaprine in cholinergic nerve terminal of hippocampus of rat, Neuropharmacology, 27:603.

Passeri, M., Cucinotta, D., De Mello, M., Storchi, G., Roncucci, R., and Biziere, K., 1985, Minaprine for senile dementia, Lancet, 1:824.

Paxinos, G., and Watson, C., 1982, "The Rat Brain in Stereotaxic Coordinates", Academic Press, New York.

Perry, E. K., Blessed, G., and Tomlinson, B. E., 1978, Changes in brain cholinesterases in senile dementia of Alzheimer type, Neuropath. App. Neurobiol., 4:273.

Price, D. L., Whitehouse, P. J., Struble, R. G., Clark, A. W., Coyle, J. T., Delong, M. R., and Hedreen, J. C., 1982, Basal forebrain cholinergic systems in Alzheimer's disease and related dementias, Neurosci. Comment, 1:82.

Pulsinelli, W. A., and Brierley, J. B., 1979, A new model of bilateral hemispheric ischemia in the unanesthetized rat, Stroke, 10:267.

Sitaram, N., Weingartner, H., and Gillin, J. C., 1978, Human serial learning: Enhancement with arecholine and choline and impairment with scopolamine, Science, 201:274.

Summers, W. K., Majovski, L. V., Marsh, G. M., Tachiki, K., and Kling, A., 1986, Oral tetrahydroaminoacridine in long-term treatment of senile dementia, Alzheimer type, N. Engl. J. Med., 315:1241.

Watts, J., Stevens, R., and Robinson, C., 1981, Effects of scopolamine on radial maze performance in rats, Physiol. Behav., 26:845.

EFFECTS OF KW-6055, A NOVEL BENZYLPYRIDINE DERIVATIVE, ON BRAIN MONOAMINERGIC SYSTEMS OF RATS

Akihisa Mori, Masako Kurokawa, Shizuo Shiozaki, Akio Ishii and Katsuichi Shuto

Pharmaceutical Research Laboratories
Kyowa Hakko Kogyo Co., Ltd., 1188 Shimotogari, Nagaizumi-cho
Sunto-gun, Shizuoka-ken, 411 Japan

INTRODUCTION

A novel benzylpyridine derivative KW-6055 [α -(p-butyrylamino-o-nitrobenzyl)pyridine] is found to have an ameliorative effect on learning and memory using the passive avoidance response task in various experimental amnesia models in rodents : KW-6055 (5-40 mg/kg, p.o.) prevented the scopolamine-induced amnesia in rats (1). KW-6055 significantly prevented the scopolamine induced reduction of rat brain norepinephrine (NE) levels in the frontal cortex, septum, hippocampus, and other brain regions (2).

Monoamine neurotransmitters, especially NE and serotonin (5-HT), are known to play important roles in cognition (3,4). Biochemical changes in brains of patients with Alzheimer's disease and senile dementia include reduced activity in monoaminergic systems (5). In the present study, to elucidate the mechanism of the antiamnesic effect of KW-6055, we investigated the effects of the compound on brain monoaminergic systems of rats.

MATERIALS AND METHODS

Male Wistar rats, weighing 200-300 g, were used. KW-6055 was dissolved in 0.3 % sodium carboxymethylcellulose(CMC). In this study, determination of the levels of NE, 5-HT, dopamine (DA), 5-hydroxyindole-3-acetic acid (5-HIAA), 3,4-dihydroxyphenyl-acetic acid (DOPAC), homovanillic acid (HVA), 3-methoxytyramine (3-MT), dihydroxyphenylalanine (DOPA) and 5-hydroxytryptophan (5-HTP) was performed by means of high-performance liquid chromatography with electrochemical detection as described previously with some modifications (6).

RESULTS AND DISCUSSION

Monoamine Metabolism

The contents of monoamines and their metabolites. The contents of monoamines (NE, 5-HT and DA) and their metabolites (5-HIAA, DOPAC, HVA and

3-MT) in various brain regions were determined at 1 hr after administration of KW-6055 (5-160 mg/kg, p.o.). KW-6055 increased NE and 5-HT contents in the frontal cortex, hippocampus and striatum and also significantly increased DA content in the striatum (Fig. 1). The contents of their metabolites were decreased in various brain regions. Especially in the striatum, KW-6055 induced a significant and dose-dependent decrease in DOPAC and HVA contents. However, KW-6055 induced a marked and dose-dependent increase in striatal 3-MT content (390-940 % of the control level).

<u>The extracellular levels of monoamine metabolites</u>. The extracellular levels of DOPAC, HVA and 5-HIAA in rat striatum were measured by brain microdialysis method as described previously (7). A rapid and dose-dependent decline in DOPAC and HVA in dialysates was observed after the

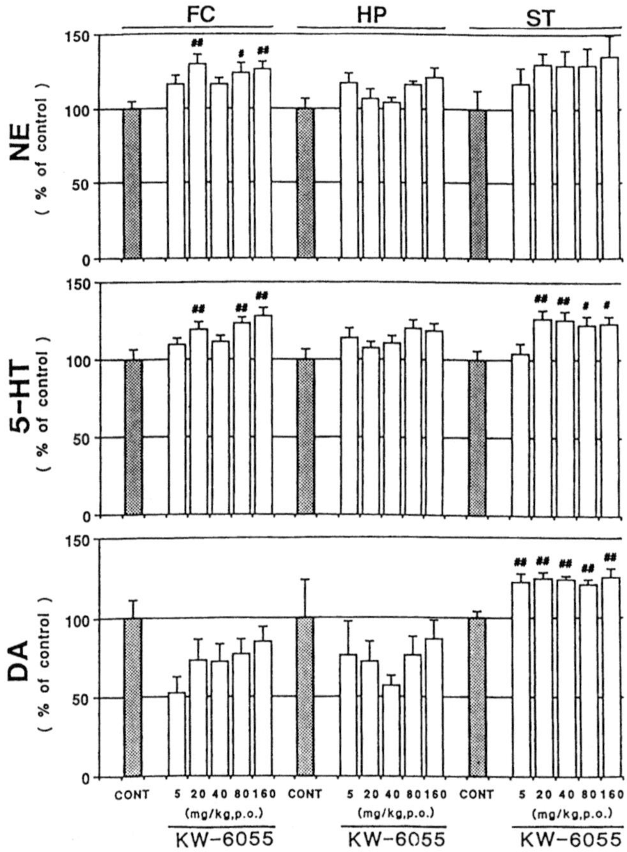

Fig. 1. The effects of KW-6055 on the contents of NE, 5-HT and DA in the frontal cortex(FC), hippocampus(HP) and striatum(ST). The results are expressed as mean±S.E.M. of % of control (CONT). # p<0.05, ## p<0.01 vs. CONT group. The NE, 5-HT and DA levels in controls (ng/g) were : FC, 174±9, 417±23, 251±29, HP, 257±16, 408±26, 94±21, ST, 102±13, 463±23, 6590±251, respectively. n=7-14.

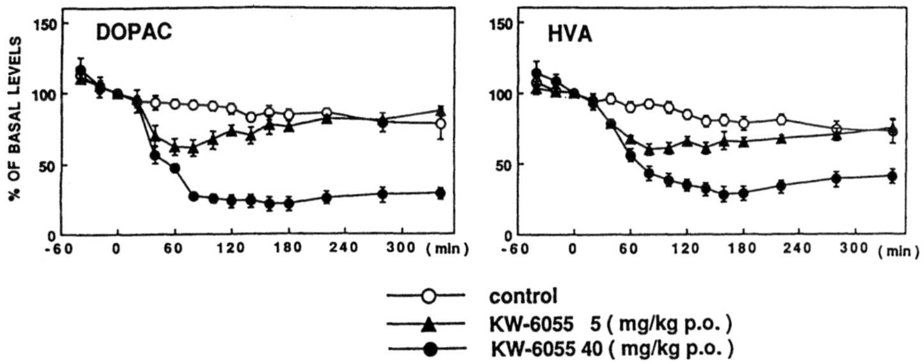

Fig. 2. Effects of KW-6055 on DOPAC and HVA in rat striatum dialysates. 0.3 % CMC or KW-6055 was injected at time zero. Dialysates were collected every 20 min. The results are expressed as % of the absolute amounts in the dialysate collected immediately before injection (mean±S.E.M.; n=4-5.). The preinjection dialysate levels (pmol/20 min fraction) were : DOPAC, 12.9±1.3, HVA, 8.5±0.8.

administration of KW-6055 (5 and 40 mg/kg, p.o.). However, in the KW-6055 5 mg/kg treated group, DOPAC and HVA were recovered to control levels within 6 hr (Fig. 2). In the KW-6055 40 mg/kg treated group, 5-HIAA in dialysates slightly decreased during 40 - 80 min post-drug period.

These results suggest that KW-6055 inhibits monoamine turnover in rat brain. 3-MT measurements have been proposed to be a biochemical indicator of DA release (8). The marked increase of 3-MT suggests that KW-6055 may partly have DA releasing effect.

Monoamine Synthesis

The accumulation of DOPA and 5-HTP 30 min after injection of NSD1015 (100 mg/kg, i.p.), the aromatic amino acid decarboxylase inhibitor, was used to evaluate the rate-limiting step in the synthesis of monoamines in rat brain (9). The levels of DOPA and 5-HTP in various brain regions at 1 hr after administration of KW-6055 (5, 20 and 80 mg/kg, p.o.) were determined. In the striatum, KW-6055 significantly decreased DOPA accumulation in a dose-dependent manner. 5-HTP accumulation was not affected in various brain regions (Table 1.). These results suggest that KW-6055 does not facilitate the biosynthetic activity of monoamines.

Table 1. Effects of KW-6055 on DOPA and 5-HTP accumulation in rat brain induced by NSD1015

Treatments (mg/kg,p.o.)	Frontal cortex		Hippocampus		Striatum	
	DOPA ng/g	5-HTP ng/g	DOPA ng/g	5-HTP ng/g	DOPA ng/g	5-HTP ng/g
N +CMC	124 ± 14	125 ± 6	80 ± 13	142 ± 8	784 ± 44	154 ± 11
N +KW-6055(5)	100 ± 14	119 ± 10	98 ± 11	141 ± 9	704 ± 38	162 ± 13
N +KW-6055(20)	90 ± 9	120 ± 8	87 ± 13	150 ± 8	614 ± 23##	157 ± 16
N +KW-6055(80)	92 ± 11	118 ± 8	75 ± 14	150 ± 10	578 ± 49##	164 ± 12

All values are the mean ± S.E.M., n=7-12. KW-6055 was administered 30 min prior to NSD1015(N). Rats were killed after another 30 min. ## P<0.01 vs. N +CMC.

Rotational Behavior in 6-Hydroxydopamine (6-OHDA) Lesioned Rats

Effects of KW-6055 on DA receptor in vivo were investigated using rotational model in rats with unilateral 6-OHDA lesions of the dopaminergic nigrostrial pathway as described previously (10). In 6-OHDA lesioned rats, apomorphine (1 mg/kg, s.c.)-induced contra-lateral rotational behavior was observed. Neither contra-lateral nor ipsi-lateral rotational behavior was produced up to 1 hr after the administration of KW-6055 (40 mg/kg, p.o.) under the same experimental condition. These results suggest that KW-6055 has no stimulating effect on DA receptor.

CONCLUSION

As demonstrated in the present results, KW-6055 increased the contents of brain monoamines, especially NE and 5-HT, at antiamnesic dose of the compound. These effects on brain monoaminergic systems of rats may partly be involved in the mechanism of the antiamnesic effect of KW-6055.

REFERENCES

1. S. Shiozaki and K. Shuto, Effects of KW-6055, a novel benzylpyridine derivative, on experimental amnesia in rodents, Japan. J. Pharmacol., 49:268P (1989).
2. K. Iwasaki, Y. Matsumoto and M. Fujiwara, Effect of KW-6055 on spatial cognitive disruption in 8-arm radial maze performance in rats, Japan. J. Pharmacol., 49:267P (1989).
3. G. M. Anlezark, T. J. Crow and A. P. Greenway, Impaired learning and decreased cortical norepinephrine after bilateral locus coeruleus lesions, Science, 181:682-684 (1973).
4. H. J. Altman, D. A. Nordy and S. O. Ögren, Role of serotonin in memory : facilitation by alaproclate and zimeldine, Psychopharmacology, 84:496-502 (1984).
5. C. G. Gottfries, Alzheimer's disease and senile dementia: biochemical characteristics and aspects of treatment, Psychopharmacology, 86:245-252 (1985).
6. M. Warnhoff, Simultaneous determination of norepinephrine, dopamine, 5-hydroxytryptamine and their main metabolites in rat brain using high-performance liquid chromatography with electrochemical detection, J. Chromatogr., 307:271-281 (1984).
7. T. Sharp, T. Zetterström and U. Ungerstedt, An in vivo study of dopamine release and metabolism in rat brain regions using intracerebral dialysis, J. Neurochem., 47:113-122 (1986).
8. P. L. Wood and A. Altar, Dopamine release in vivo from nigrostrial, mesolimbic, and mesocortical neurons: utility of 3-methoxytyramine measurements, Pharmacological Reviews, 40:163-187 (1988).
9. A. Carlsson, J. N. Davis, W. Kehr, M. Lindqvist and C. V. Atack, Simultaneous measurement of tyrosine and tryptophan hydroxylase activities in brain in vivo using an inhibitor of the aromatic amino acid decarboxylase, Naunyn-Schmiedeberg's. Arch. Pharmacol., 275:153-168 (1972).
10. M. Herrera-Marschitz and U. Ungerstedt, Evidence that apomorphine and pergolide induce rotation in rats by different actions on D1 and D2 receptor sites, Eur. J. Pharmacol., 98:165-176 (1984).

EFFECTS OF KW-6055, A NOVEL BENZYLPYRIDINE DERIVATIVE, ON VARIOUS EXPERIMENTAL AMNESIA MODELS

Shizuo Shiozaki, Akio Ishii and Katsuichi Shuto

Pharmaceutical Research Laboratories, Kyowa Hakko Kogyo Co. Ltd., 1188 Shimotogari, Nagaizumi-cho, Sunto-gun Shizuokaken, 411, Japan

INTRODUCTION

A series of benzylpyridine derivatives have been synthesized in our laboratories. Among them KW-6055 [α -(p-butyrylamino-o-nitrobenzyl) pyridine] was found to have not only anti-reserpine but also anti-anoxic properties in mice (unpublished data). Recently, KW-6055 has been reported to improve learning behavior: reversed scopolamine (SCOP)-induced spatial disruption in rats (1). The present study describes the facilitatory effects of KW-6055 on learning behavior in intact animals and experimental models compared with several drugs.

MATERIALS AND METHODS

KW-6055, bifemelane, oxiracetam and idebenone were synthesized at Kyowa Hakko Kogyo Co.. Reference antidepressant drugs were purchased from Sigma Chemical Co..

KW-6055 was suspended in 0.3% sodium carboxyl methyl cellulose solution and administered orally in male ddY mice (24 - 27 g) and male Wistar rats (250 - 280 g).

Passive avoidance task. Mouse: These tests were performed with a step-through type passive avoidance apparatus. As experimental apparatus, two compartments (bright and dark) with automatic management system were used. The automatic management system consisted of bright compartment (15 x 9 x 11 cm) with a 4W white fluorescent light and dark compartment (15 x 14 x 18 cm). The compartments were separated by a guillotine door (3 x 3 cm). In the acquisition trial, a mouse placed in the bright compartment could enter, through the door, into the dark compartment that had a grid on the floor. As soon as the mouse entered the dark compartment, a scrambled foot-shock (0.18 mA) was delivered to the floor grid for 2 sec. Rat: As experimental apparatus, bright compartment (40 x 40 x 40 cm) and dark compartment (19.5 x 15 x 20 cm) were used. The dark compartment had a grid floor of stainless rods used to give a foot shock delivered by shock generator (3 mA, 2 sec). In the test trial, given 24 hr after the acquisition trial, the animal was again placed in the bright compartment and the response latency to enter the dark compartment was measured. Maximum measurement time was for 600 sec, and the latency of animals which did not move into the dark compartment during the observation period was

Basic, Clinical, and Therapeutic Aspects of Alzheimer's and Parkinson's Diseases, Volume 2
Edited by T. Nagatsu *et al.*, Plenum Press, New York, 1990

calculated to be 600 sec. Electroconvulsive shock (ECS)-induced amnesia: Immediately after the acquisition trial, ECS was loaded on mice (25 mA, 0.2 msec, 2000 V) and rats (80 mA, 0.2 sec, 2000 V). SCOP-induced amnesia : Thirty minutes prior to the acquisition trial, SCOP (1.5 mg/kg) was intraperitoneally administered in the rats and the test trial was performed 24 hr later. Basal forebrain (BF) lesioned rats: Rats were injected bilaterally with 0.2 µg kainic acid in 1 µl solution at the following coordinates: 1 mm posterior to bregma, 2.8 mm lateral to the midsagittal suture and 6.8 mm from the skull. Sham-operated rats had similar operations without kainic acid injection.

8-arm radial maze task: By using 8-arm radial maze apparatus, training was done for the food deprived rat once a day. A rat was placed in the central platform of radial maze (length of arm 57.5 cm, width 10 cm, height 45 cm) and training was conducted until the animal could take efficiently 50 ul of milk which was put at the tip of arm. When a rat entered an arm that it had not previously entered and ate the milk, this was defined as a correct choice. When a rat re-entered an arm that it had previously selected and looked into the cup before turning back, this was defined as an erroneous choice. Running time was measured up to 10 min. The spatial cognitive disruption was induced by administering SCOP intraperitoneally 30 min prior to the test trial.

RESULTS

Passive avoidance response in intact mice: When KW-6055 was administered per os 60 min prior to the training, the response latency of acquisition trial of KW-6055 treated group was not different from the latency of normal control. When KW-6055 was administered 60 min prior to the test trial, the response latency in the test trial was prolonged remarkably.

ECS-induced amnesia in mice or rats : When KW-6055 (1.25 - 160 mg/kg, i.p.) was administered 30 min prior to the test trial, it showed a significant antagonism to ECS-induced amnesia in a broad dose range (2.5 - 40 mg/kg). In case KW-6055 (5 - 160 mg/kg, p.o.) was administered 60 min prior to the test trial, significant latency prolongation effect was observed in broad dose range. Bifemelane (5 - 80 mg/kg, i.p.) administered 30 min prior to the test trial showed significant latency prolongation effect at doses of 20 mg/kg or more. When oxiracetam (10 - 160 mg/kg, i.p.) was administered 30 min prior to the acquisition trial, significant latency prolongation effect was observed at 40 mg/kg only. Amitriptyline (10 - 80 mg/kg, p.o.), imipramine (10 - 80 mg/kg, p.o.) and desipramine (10 - 80 mg/kg, p.o.) did not show significant latency prolongation effect on ECS-induced short latency. Furthermore, in the KW-6055 (5 - 80 mg/kg, p.o.) treated rats, significant latency prolongation effect was observed by the administration immediately after loading ECS on the rat.

SCOP-induced amnesia in rats: In the rats administered with SCOP (1.5 mg/kg, i.p.), significant shortening of latency was observed compared with the latency obtained in normal control of the test trial. KW-6055 possessed significant latency prolongation effect on SCOP-induced amnesia in a dose range of 5 mg/kg or more.

BF lesioned rats: The response latency of the test trial in the sham operated group was a little shorter than the response latency in normal group, but not significantly. The response latency of BF-lesioned rats was significantly shorter compared with that in the sham operated group. KW-6055 (2.5 mg/kg or more) showed a significant latency prolongation effect on cognitive disrupted rats caused by BF-lesion. On the other hand, idebenone (10 - 40 mg/kg, p.o.) did not show latency prolongation effect. In consecutive administration experiment for 9 days, significant latency prolongation effect was observed at KW-6055 (10 mg/kg, p.o.). Remarkable latency prolongation effect was also seen at idebenone (2.5 mg/kg, i.p.) (Fig. 1).

8-arm radial maze task in rats: In 5 mg/kg of KW-6055, the number of correct choice was significantly increased and the number of erroneous choice was significantly decreased. Similar tendency was observed in other doses of KW-6055. The improvement effect of (a reference drug) physostigmine was observed (Fig. 2).

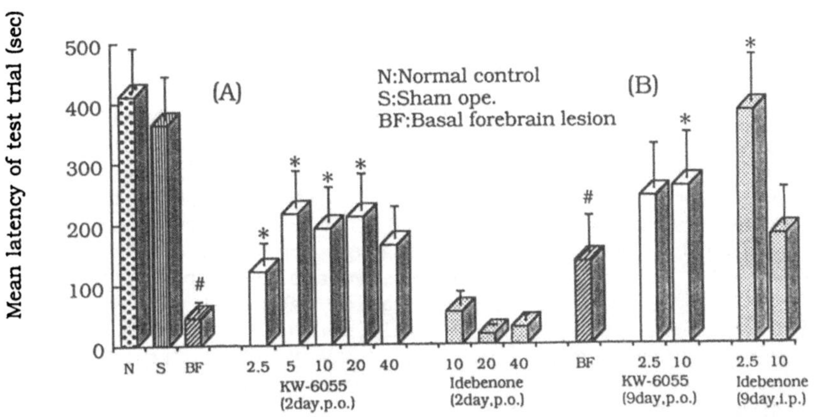

Fig. 1 Effects of KW-6055 and idebenone on impairment of passive avoidance response in the BF-lesioned rats. (A): KW-6055 or idebenone was administered per os 60 min before acquisition trial and the test trial (2nd day). (B): KW-6055 (p.o.) or idebenone (i.p.) was given repeatedly once a day from the first day to the test day (9th day). # $p<0.001$: Significant difference vs. sham ope. * $p<0.05$: Significant difference vs. BF-control. (Mann-Whitney U-test) (n=13 - 22)

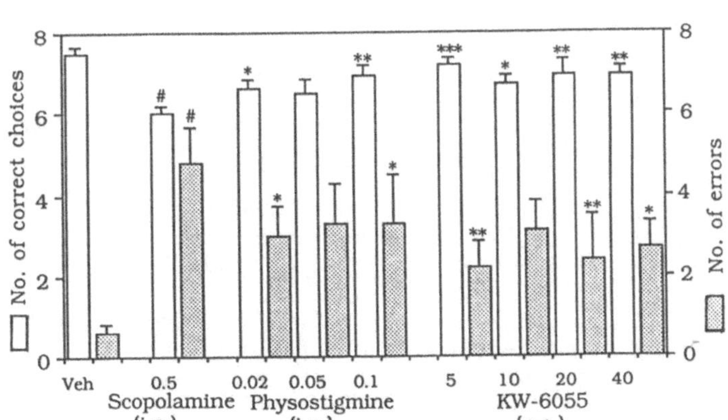

Fig. 2 Effects of KW-6055 and physostigmine on SCOP-induced spatial cognitive disruption in the radial arm maze. Left column: No. of correct choices. Right column: No. of errors. # $p<0.001$: Significant difference vs. vehicle. * $p<0.05$, ** $p<0.01$, *** $p<0.0001$: Significant difference vs. SCOP-control. (Mann-Whitney U-test) (n=10 - 15)

DISCUSSION

The anti-amnesic effects of KW-6055 on the performance of passive avoidance response task and a radial maze task were assessed. When KW-6055 was administered 60 min prior to the test trial in normal mice, the latency was facilitated. Furthermore, in the ECS-induced amnesia mice, amelioratory effects of KW-6055 were observed. In addition, KW-6055 did not affect spontaneous movement at doses which affected cognitive disturbance (5 - 40 mg/kg, p.o.). Thus, these results suggest that the effects of KW-6055 on cognitive disturbances may be due to the facilitation of cognitive function. KW-6055 possessed anti-depressant activity (anti-reserpine activity). Therefore, the effects of KW-6055 on learning were compared with those of amitriptyline, desipramine and imipramine. While imipramine showed no influence on learning, amitriptyline and desipramine blocked learning, presumably in part, due to its anti-cholinergic activity (2). Thus, these results suggest that KW-6055 has a pharmacological profile distinct from that of antidepressants in enhancing learning activity.

In patients with Alzheimer's dementia, the nucleus basalis of Meynert projecting cholinergic neurons to cortical areas degenerates (3); this suggests an important role of the cholinergic system in the learning and memory (4). Rats with lesions in the BF, which corresponds to the nucleus basalis of Meynert in humans, also show a marked impairment of memory (5). In the same models, KW-6055 (2.5 - 20 mg/kg, p.o.) improved the performance of memory task. In addition to the passive avoidance task, the effect of KW-6055 on spatial memory using 8-arm radial maze task was assessed in rats with SCOP-induced spatial cognitive disruption. KW-6055 (5 - 40 mg/kg, p.o.) caused significantly increased number of correct choices and significantly decreased number of errors in SCOP treated rats, i.e. improved spatial cognitive disruption.

In conclusion, KW-6055 ameliorated various experimentally induced amnesia models in rodents. The present results suggest that KW-6055 may be useful in the treatment of various symptoms in patients with cerebral vascular diseases, senile dementia and other organic brain syndromes.

REFERENCES

1. K. Iwasaki, Y. Matsumoto and M. Fujiwara, Effect of KW-6055 on spatial cognitive disruption in 8-arm radial maze performance in rats, Japan. J. Pharmacol. 49 (Supple):267P (1989)

2. S. Tachikawa, M. Harada and H. Maeno, Pharmacological and biochemical studies on a new compound, 2-(7-indenyloxymethyl)morpholine hydrochloride(YM-08054-1), and its derivatives with potential antidepressant properties. Arch. int. Pharmacodyn. 238:81-95 (1979)

3. P.J. Whitehouse, D.L. Price, R.G. Strube, A.W. Clerk, J.T. Coyle and M.R. Delong, Alzheimer's disease and senile dementia: Loss of neurons in the basal forebrain. Science 215:1237-1240 (1982)

4. C. Flicker, R.L. Dean, D.L. Watkins, S.K. Fisher and R.T. Bartus, Behavioral and neurochemical effects following neurotoxic lesion of a major cholinergic input to the cerebral cortex in the rat. Pharmacol. Biochem. Behav. 18:973-980 (1983)

5. M. Miyamoto, M. Shintani, A. Nagaoka and Y. Nagawa, Lesioning of the rat basal forebrain leads to memory impairments in passive and active avoidance tasks. Brain Res. 328:97-104 (1985)

VASOPRESSIN FRAGMENTS IMPROVE LEARNING PROCESS OF RATS IN THE RADIAL ARM MAZE TASK

Tohru Kojo, Kunihiko Ikeguchi, and Mitsuo Yoshida

Department of Neurology, Jichi Medical School
Tochigi-ken, Japan

INTRODUCTION

Over the past ten years expectation has grown that arginine vasopressin (AVP) may be a potent substance with a function of improving memory disturbances. Furthermore, the fragment of AVP or its derivatives could also be a candidate for a possible amnesia drug (Burbach et al., 1983). These and many other substances have been successfully applied in learning experiments on animals (Bresova 1982; Legros 1978; Sahgal 1986; Strupp 1989). We used peptide fragments shorter than AVP(4-9) , under a different way of the radial arm maze from Hiraga (1984) and Olton (1979).

MATERIALS AND METHODS

Wistar-Imamichi male rats, trained to take a bait placed on an elevated 8 arm radial maze, were divided into six groups. Fragments of AVP were administered to 5 of the groups and the 6th, the control group, received a saline solution. During the first training, baits were placed on all eight arms. The first training was a selection process to obtain enough rats for the six groups of the experiment from a field of 80 candidate rats without any treatment. In the selection process, a total of 80 rats were trained. For 2 weeks prior to the begining of training, the rats aged 8 weeks (mean weight 220 grams) were kept in a condition of slight hunger with their body weight of about 10% under normal. One trial was performed per day. The target performance point was considered to be a score of at least 7 correct choices out of the first eight over a continuous period of 5 days. After a period of 3-4 days (3-4 trials) during which the rats became accustomed to the technique of taking the baits, about 60% of the rats achieved the target, or 47 rats. This selection period took 18 + 3.9 trials (mean+SD). The successful candidates were then divided into 6 groups for the second training, which began 2 weeks later. During the second training (40 trials for all the rats selected), baits were placed on only 4 of the arms. -- the same 4 arms were used throught the experiment. Test and control substances were both administered subcutaneously 30 minutes before each trial. The concentration of them were 0.001ug/kg except for F2(0.1ug/kg).

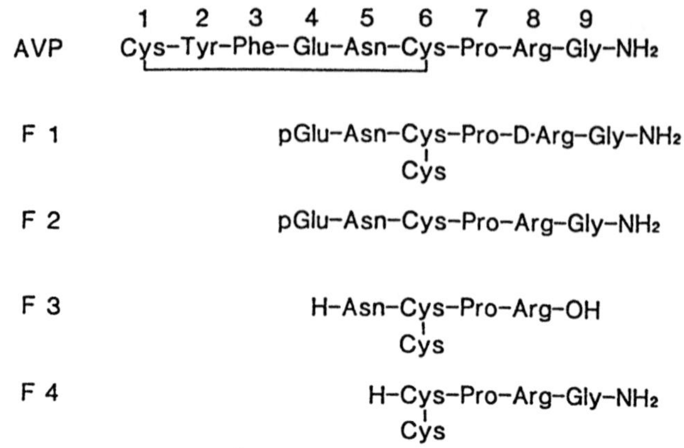

Fig.1 Arginine vasopressin and derivatives.

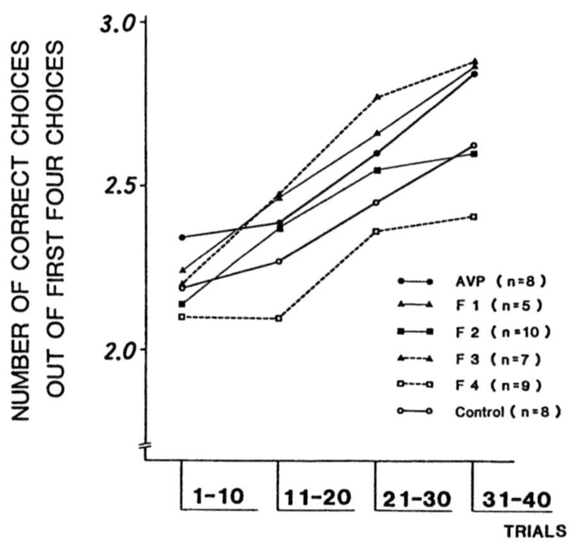

Fig.2 Each point represents the average score of all rats in the group over the ten trials numbered below, e.g. 1-10, 11-20, etc.

Next, 2.0 represents the statistically probable score based on the fact that 4 baits were placed on the 8 arms; two out of four choices, or 50%.

The slope of the graph of F3 is greater than the other slopes, and represents a greater improvement over the 40 trials.

RESULTS AND CONCLUSION

Comparing the ratio of correct choices in the first 4 choices for each group, the group which received the fragment [Cyt6]AVP(5-8) scored higher than control and the other 4 groups. Differences in the rate of correct choices would seem to reflect an enhancement of the spatial memory. The fragment shorter than [pGlu4,Cyt6]AVP(4-9) could have a function in modulating the learning process.

REFERENCES

1. Burbach,J.P., Kovacs,G.L., de Wied,D., van Nispen,J.W.,& Greven,H.M. A major metabolite of arginine vasopressin in the brain is a highly potent neuropeptide. Science, 221:1310-1312,1983.
2. Bresova,O.,& Skopkova,J. Vasopressin analogues and spatial working memory in the 24-arm radial maze. Peptides, 3:725-727,1982.
3. Hiraga,Y.& Iwasaki T. Effects of cholinergic and monoaminergic antagonists and tranquilizers upon spatial memory in rats. Pharmacol.Biochem.Behav., 20:205-207,1984.
4. Legros,J.J., Gilot,P., Seron,X., Claessens,J., Adam,A., Moegen,J.M., Audibert,A.,& Berchier,P. Influence of vasopressin on learning and memory. Lancet, 1:41-42,1978.
5. Olton,D.S. Mazes, maps and memory. Amer. Psychol., 34:583-596,1979.
6. Sahgal,A. Effects of vasopressin, desglycinamide vasopressin and amphetamine on a combined passive and active avoidance task. Psychopharmacol., 90:494-498,1986.
7. Strupp,B.J. Improvement of memory by a vasopressin fragment: Importance of individual differences in mnemonic function. Behav.Neurosci., 103:743-754,1989.

EVALUATION OF VASOPRESSIN FRAGMENTS AND THEIR ANALOGUES ON ACTIVE AVOIDANCE RESPONSES OF RATS WITH LESIONS OF THE DORSAL NOREPINEPHRINE BUNDLE

Kunihiko Ikeguchi, Ken-ichi Fujimoto, and Mitsuo Yoshida

Department of Neurology
Jichi Medical School, Tochigiken, Japan

INTRODUCTION

The nonapeptide arginine vasopressin (AVP) has been shown to act as a neurotransmitter or as a neuromodulator in the central nervous system. It is also known that the concentration of AVP in the Alzheimer brain is reduced (Mazurek et al., 1986). Systemic or intracerebroventricular administration of AVP immediately after a learning trial or shortly before the retention session results in facilitation of passive avoidance in later retention sessions (Adler and DeWied, 1972).

A hexapeptide (pGlu4,Cyt6)AVP(4-9) is a major metabolite of AVP and has recently been identified as a highly potent neuropeptide devoid of pressor activity and able to modify the behavior of rats in a passive avoidance task (Burbach et al., 1983; DeWied et al., 1984; Kovács et al., 1986). It has been suggested that in the brain AVP may serve as a precursor of fragments with more selective and potent behavioral effects. A number of other vasopressin fragments that are products of *in vitro* processing of AVP by brain membrane peptidases have been identified. The present studies were designed to evaluate the potency of these various vasopressin fragments and their analogues to influence learning and memory processes. Shuttle box avoidance behavior was used as the test procedure. The results demonstrate that some of the C-terminal vasopressin fragments and their analogues potently affect various aspects of learning and memory.

MATERIAL AND METHODS

Animals

Adult male Wistar rats weighing 250 to 280g and supplied by Saitama Laboratory Animal Center Co. were used. Under anesthesia with sodium pentobarbital, lesioning of the bilateral DNB was carried out by injections of 6-hydroxy-dopamine hydrobromide(6-OHDA-HBr) solution (4µg/2µl/side). The coordinates were 6.3mm posterior to bregma, 1.0mm lateral on both sides of the midline, and 5.0mm below the surface of the brain. Sham rats underwent similar surgical procedures except that vehicle without 6-OHDA was injected.

Basic, Clinical, and Therapeutic Aspects of Alzheimer's and Parkinson's Diseases, Volume 2
Edited by T. Nagatsu *et al.*, Plenum Press, New York, 1990

Active avoidance

Eighteen days following the injection, each rat was tested for active avoidance responses by a two-way shuttle box. A buzzer and light attached to the ceiling were used to present the conditioned stimuli. A discrete avoidance schedule (Hoffman, 1966) was used. Each trial consisted of a 25-sec intertrial interval, a 5-sec conditioned stimulus and a 3-sec unconditioned stimulus. The latter was an electric foot shock (0.5mA, scramble, maximum 3 sec). If the rat ran to the opposite compartment during application of the conditioning stimuli (for 5 sec), the rat thus avoided the electric shock, and such behavior was recorded as an avoidance response. One hundred and twenty trials were given daily for 5 days. Intertrial responses, which were unrelated to avoidance of or escape from the electric shock and indicated the spontaneous locomotor activity, were also recorded.

Shock sensitivity

The rats were also evaluated for differences in shock sensitivity using the same shock grid and source as had been used in the active avoidance task. The scrambled shock was delivered at 5 different intensity levels between 12.5 and 100V. Vocalization or a jump response was recorded for each subject at each shock level.

Drug administration

The following peptides were used: (pGlu4,Cyt6)AVP(4-9), (pGlu4,Cyt6,D-Arg8)AVP(4-9), (pGlu4,Cyt6)AVP(4-7), (Cyt6)AVP(5-8), and (Cyt6)AVP(6-9). Thirty minutes before the avoidance session, groups of animals were subcutaneously injected with graded doses of the peptides dissolved in 1ml of saline. The doses of peptides were selected on base of pilot experiments. Sham and DNB-lesioned control groups were treated with vehicle only.

Statistics

The data obtained were statistically compared by ANOVA. When P values were less than 0.05, they were defined to indicate significant differences.

RESULTS

Fig.1 shows the mean avoidance rates of the sham-operated group and lesioned group. The avoidance rate indicates the ratio as a percentage of the number of avoidance response to the number of trials. Each point represents the avoidance rate obtained from 20 sequential trials. In sham-operated rats, the rate was 8.1±8.1% in the initial 20 trials and increased up to 53.4±36.0% by the 6th 20-trial sequence on the first day. The rate of the initial 20 trials on the successive days increased gradually and reached as high as 30.6±35.7% on the fifth day. In DNB-lesioned rats, the rate was 11.6±13.0% in the first 20 trials and increased up to 48.2±38.1% by the 6th 20-trial set on the first day. But, the rate of the initial 20 trials of the successive days showed little increase, and that for the fifth day remained as low as 17.6±31.2%. Statistical significance was observed between sham-operated and DNB-lesioned groups (*:$p<0.05$, **:$p<0.01$, ANOVA).

Fig.2 shows avoidance rates in DNB-lesioned rats with administration of (pGlu4,Cyt6)AVP(4-9) or vehicle. The avoidance rate of the rats

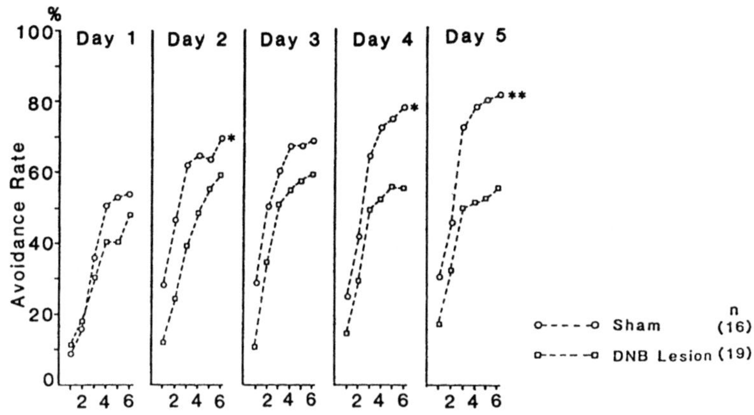

Fig. 1 Avoidance rate (number of avoidance responses / 20 trials indicated as %) in sham-operated and DNB-lesioned rats.

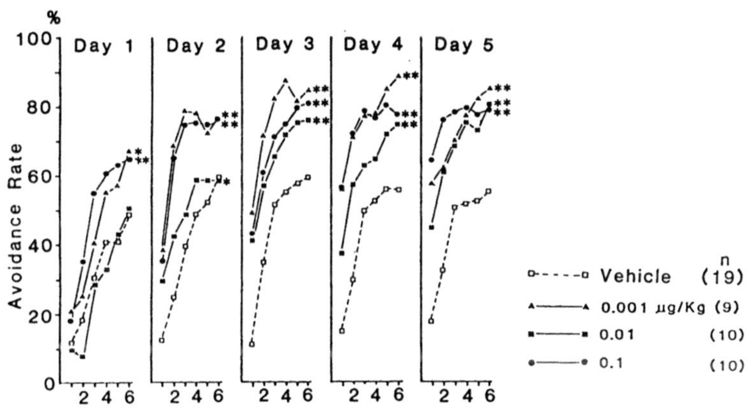

Fig. 2 Avoidance rate in DNB-lesioned rats with administration of (pGlu4,Cyt6)AVP(4-9) or vehicle.

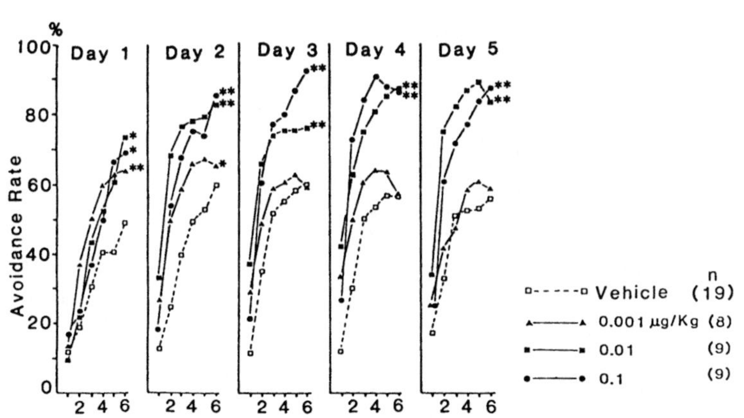

Fig. 3 Avoidance rate in DNB-lesioned rats with administration of (pGlu4,Cyt6,D-Arg8)AVP(4-9) or vehicle.

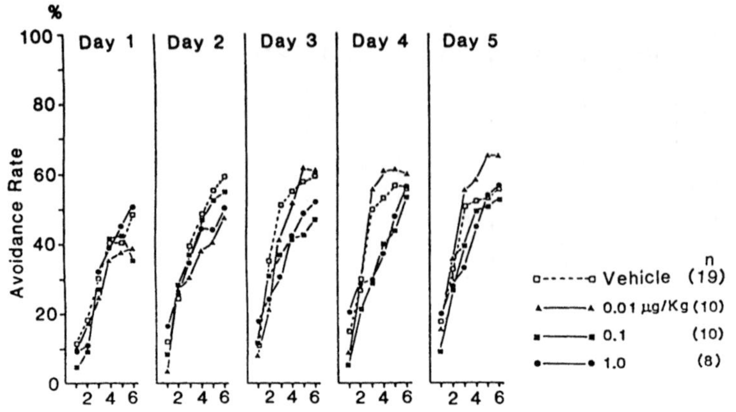

Fig. 4 Avoidance rate in DNB-lesioned rats with administration of (pGlu4,Cyt6)AVP(4-7) or vehicle.

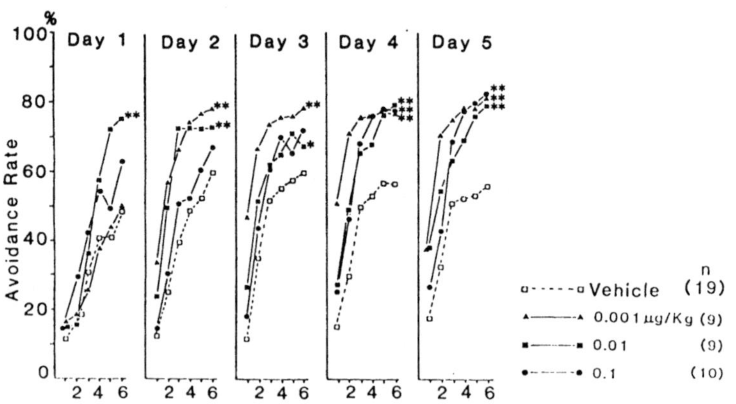

Fig. 5 Avoidance rate in DNB-lesioned rats with administration of (Cyt6)AVP(5-8) or vehicle.

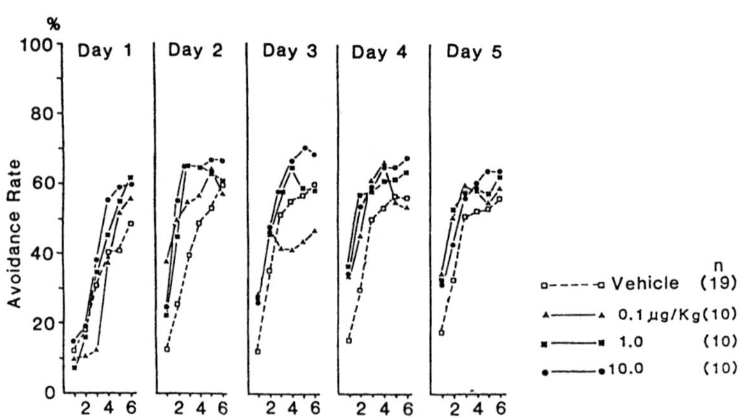

Fig. 6 Avoidance rate in DNB-lesioned rats with administration of (Cyt6)AVP(6-9) or vehicle.

Table 1. Summary of the results

	0.001	0.01	0.1	1.0	10.0 µg/kg
VPF(4-9)	+	+	+		
VPF(4-9)(D-Arg)	-	+	+		
VPF(4-7)		-	-	-	
VPF(5-8)	+	+	+		
VPF(6-9)			-	-	-

treated with 0.001, 0.01, or 0.1µg/kg of (pGlu4,Cyt6)AVP(4-9) increased significantly when compared with that of the vehicle-treated, lesioned control (*:p,0.05, **:$p<0.01$, ANOVA). When (pGlu4,Cyt6,D-Arg8)AVP(4-9) was used at 0.01 or 0.1µg/kg, the avoidance rate again increased significantly (*:$p<0.05$, **:$p<0.01$, ANOVA),as seen in Fig.3.

However the peptide (pGlu4,Cyt6)AVP(4-7) showed no significant effect (Fig.4). Fig.5 shows avoidance rates of DNB-lesioned rats with administration of (Cyt6)AVP(5-8) or vehicle. The avoidance rate increased significantly when the peptide was given at 0.001, 0.01 or 0.1µg/kg (*:$p<0.05$, **:$p<0.01$, ANOVA).

Finally, Fig.6 shows avoidance rates of DNB-lesioned rats administered (Cyt6)AVP(6-9) or vehicle. As in the case of (pGlu4, Cyt6)AVP(4-7), there were no significant differences between the peptide-treated and vehicle-treated animals. The intertrial responses of all groups showed no statistical difference.

In the shock sensitivity test, thresholds of vocalization and jump response were 50 and 75V, respectively, in all groups and no significant differences were observed between groups.

Table 1 summarizes the results.

DISCUSSION

The principal projection from the locus ceruleus to the cerebral cortex and hippocampus is the dorsal ascending noradrenergic bundle (Ungerstedt, 1971). Local injection of 6-OHDA was used to destroy this projection. By this procedure, the cortical NE level is reduced more than 90% without any significant change in the level of other transmitters such as dopamine, acetylcholine, and serotonin (Mason and Iversen, 1979).

Many methods are applied for evaluation of central acting drugs on learning and memory; however, the active avoidance procedure is one of the best methods in general use (Kuribara et al., 1985). It is generally accepted that the avoidance of hazardous stimuli such as electric shock in a shuttle box is one of the behaviors indicating learning and memory processes (Bovet et al., 1969).

In our study DNB-lesioned rats showed impairment of active avoidance responses. Since shock sensitivity and locomotor activity evaluated by the intertrial response were almost the same between the sham-operated group and DNB-lesioned group, the impairment was not a consequence of sensorio-motor disturbance.

The increase in the avoidance rate in the initial 20-trial sequence of each successive day is thought to reflect the memory function of the rats. In the DNB-lesioned rats, this increase was hardly recognized, so this memory dysfunction was probably due to the destruction of the DNB. By administration of (pGlu4,Cyt6)AVP(4-9), (pGlu4,Cyt6,D-Arg8)AVP(4-9), or (Cyt6)AVP(5-8), an increased of avoidance rate was noted. Especially daily improvement of the avoidance rate of the initial 20-trial sequence was noted in the treated groups. This improvement suggests that vasopressin fragments affect memory consolidation of the active avoidance response.

There were no significant differences in the shock sensitivity and locomotor activity between the groups treated with the peptides and the vehicle-treated control, so these improvements are not attributable to a change in sensorio-motor function.

In conclusion, vasopressin fragments with a behavioral effect on active avoidance must have the 5th through 8th amino acid residues, and the (4-9) analogue whose 8th amino acid residue is replaced with D-Arg does not lose its potency. These effects suggest that such fragments might be beneficial in the treatment of demented patients with deficiency of the NE system.

REFERENCES

Adler, R,, and DeWied, D., 1972, Effects of lysine vasopressin on passive avoidance learning, Psychon. Sci., 29:46.

Bovet, D., Bovet-Nitti, F., and Oliverio, A., 1969, Genetic aspects of learning and memory in mice, Science, 163:139.

Burbach, J. P., Kovacs, G. L., and DeWied, D., 1983, A major metabolite of argnine vasopressin in the brain is a highly potent neuropeptide, Science, 221:1310.

DeWied, D., Gaffori, O., van Ree, J. M., and deLong, W., 1984, Central target for the behavioral effects of vasopressin neuropeptides, Nature, 308:276.

Hoffman, H. S., 1966, The analysis of discriminated avoidance, in: "Operant behavior: Areas of research and application," W. K. Honing, ed., Appleton-Century-Crofts, New York.

Kovacs, G. L., Veldhuis, H. D., Versteeg, D. H. G., and de Wied, D., 1986, Facilitation of active avoidance behavior by vasopressin fragments microinjected into limbic-midbrain structures, Brain Res., 371:17.

Kuribara, H., Haraguchi, H., and Tadokoro, S., 1985, Comparisons between discrete lever-press and shuttle avoidance responses in mice: Acquisition processes and effects of psychoactive drugs, Jpn. J. Pharmacol., 38:141.

Mason, S. T., and Iversen, S. D., 1979, Theories of the dorsal bundle extinction effect, Brain Res. Rev., 1:107.

Mazurek, M. F., Beal, M. F., Bird, E. D., and Martin, J. B., 1986, Vasopressin in Alzheimer's disease: a study of postmortem brain concentrations, Ann. Neurol., 20:665.

Ungerstedt, U., 1971, Stereotaxic mapping of the monoamine pathways in the rat brain, Acta. Physiol. Scand. Suppl., 367:1.

ANTI-AMNESTIC AND ANTI-ANOXIC EFFECTS OF DROXIDOPA (L-DOPS) IN ANIMAL MODELS

Tohru Tatsuno, Michiko Okazaki, Yoshimi Maruoka, Mitsutaka Nakamura, and Junki Katsube

Research Laboratories, Sumitomo Pharmaceuticals Co., Ltd.
3-1-98, Kasugadenaka, Konohana-ku, Osaka 554, Japan

INTRODUCTION

Droxidopa(L-threo-3,4-dihydroxyphenylserine, L-DOPS) is converted directly to (-)norepinephrine(NE) by aromatic L-amino acid decarboxylase (AADC) in vitro[1] and in vivo,[2] and the clinical efficacy of L-DOPS has been reported in some neurological disorders linked to NE deficiency such as familial amyloid polyneuropathy,[3] Parkinson's disease,[4] and Shy-Drager syndrome.[5] In various studies,[6,7] L-DOPS has been shown to improve the hypofunctions of noradrenergic neurotransmission in several NE-depleted animal models. Functional disorder of noradrenergic pathways has been also considered to be possibly involved in memory and cognitive impairment in the case of Alzheimer's disease.[8,9] This suggests the possible therapeutic potential of the central NE precursor L-DOPS for several types of dementia. Here, to evaluate this possibility, we studied the effects of L-DOPS on electroconvulsive shock(ECS)-induced amnesia in the step-through passive avoidance test and on KCN- and curare-induced anoxia in mice.

MATERIALS AND METHODS

Male ddY mice(4.5-5 weeks old) were used. L-DOPS was synthesized in the laboratory of Sumitomo Pharmaceuticals Co., Ltd. L-DOPS and benserazide (Hoffman-La Roche) were suspended in 0.5 % methylcellulose solution. KCN and pancuronium bromide(Mioblock®, Sankyo Co.) were dissolved in saline.

ECS-induced Amnesia in Step-through Passive Avoidance Test

In the acquisition session(AS), each mouse was placed in the illuminated compartment(20 X 25 X 20 cm) and allowed to enter the dark compartment(20 X 25 X 20 cm) of the apparatus. As soon as the mouse entered the dark compartment, the door was closed and an unescapable foot shock(4 mA, 2 sec) was delivered through the grid floor. Immediately after the AS, the mouse was received a transcorneal ECS(20 mA, 0.2 sec). In the retention session(RS), which was performed 24 hr after the AS, each mouse was again placed in the illuminated compartment, and the response latency to enter the dark compartment(step-through latency) was measured until the cutoff time(300 sec). When the mouse avoided longer than 300 sec, the latency of the mouse was taken to be 300 sec.

KCN- and Curare-induced Anoxia

Mice were given 16 mg/kg(s.c.) of KCN or 0.6 mg/kg(s.c.) of pancuronium bromide one hour after the injection of L-DOPS. The survival time was determined until 30 min and 15 min in KCN- and pancuronium-induced anoxia, respectively.

RESULTS AND DISCUSSION

First, we studied the effect of L-DOPS on ECS-induced amnesia in the passive avoidance test. In normal mice (-ECS), the median (range) step-through latency, which was 19.5 (10.0-27.7) sec in the AS, was prolonged to 275.3 (114.3-300) sec in the RS. When ECS was given immediately after the AS, the step-through latency in the RS was significantly reduced(Fig. 1.[A]). The administration of L-DOPS(25-200 mg/kg, i.p.) 30 min before the AS significantly prevented the ECS-induced reduction in step-through latency in the RS without effects on the step-through latency in the AS. This observation indicates that L-DOPS has anti-amnestic action in this model. L-DOPS(100-200 mg/kg, i.p.) injected immediately after the ECS that followed the AS also improved the ECS-induced amnesia(Fig. 1.[B]), suggesting that the anti-amnestic effect of L-DOPS does not result from a reduction in sensitivities to foot shock in the AS and to ECS. However, L-DOPS(25-200 mg/kg, i.p.) injected 6 hr after ECS or 30 min before the RS had no effect on the ECS-induced reduction in step-through latency in the RS(data not shown). These results indicate that the ameliorative effect of L-DOPS on the ECS-induced amnesia is probably at least in part due to the improvement of the ECS-induced disorder of memory consolidation in the passive avoidance test.

On the other hand, L-DOPS has been shown to act as a NE precursor both

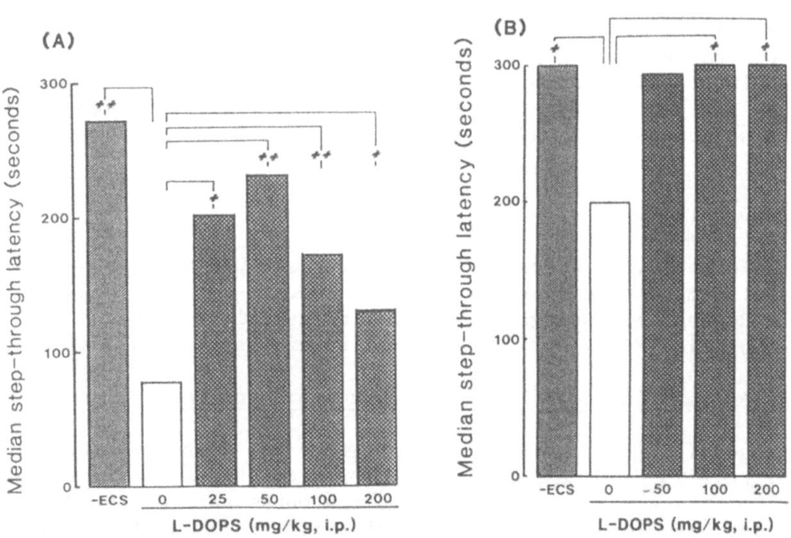

Fig. 1. Effect of L-DOPS on electroconvulsive shock(ECS)-induced impairment of passive avoidance response in mice. Each value is median step-through latency in the retention session. (A): L-DOPS was given 30 min before the acquisition session. N=10(-ECS), 9(L-DOPS 25 mg/kg), 19(L-DOPS 100 and 200 mg/kg), and 20(L-DOPS 0 and 50 mg/kg). (B): L-DOPS was given immediately after the acquisition session. N=14 (-ECS) and 16(others). #P<0.05, ##P<0.01 (Mann-Whitney's U-test).

in the peripheral system[10] and in the central system,[6,7] but the anti-amnestic effect of L-DOPS was not inhibited by the peripheral AADC inhibitor benserazide(Fig. 2). At the dose of 1 mg/kg(i.p.) benserazide has been previously shown in mice to inhibit the elevation by L-DOPS of the serum level of the NE metabolite 3-methoxy-4-hydroxyphenylethyleneglycol but not that of the level in brain.[2] Moreover, Liang et al.[11] have reported that retention of a passive avoidance response is enhanced by intra-amygdala injections of NE immediately after training. Therefore, the central action of L-DOPS as a NE precursor may contribute to the anti-amnestic effect rather than its peripheral action.

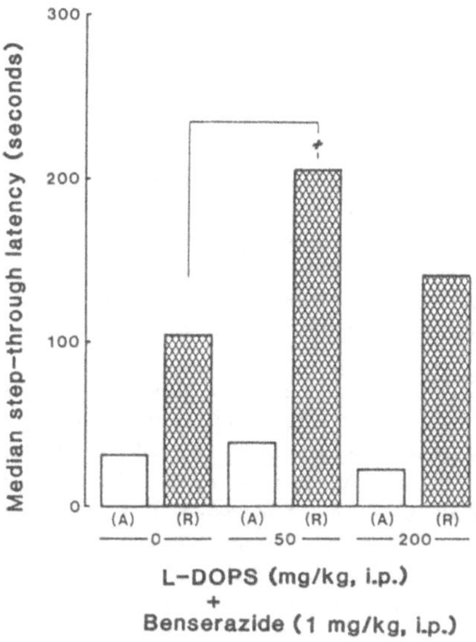

Fig. 2. Effect of the peripheral decarboxylase inhibitor benserazide on the anti-amnestic action of L-DOPS. L-DOPS and benserazide were simultaneously given 30min before the acquisition session. Step-through latencies in the acquisition session (A) and in the retention session (R) are shown. N=19 or 20. #P<0.05 (Mann-Whitney's U-test).

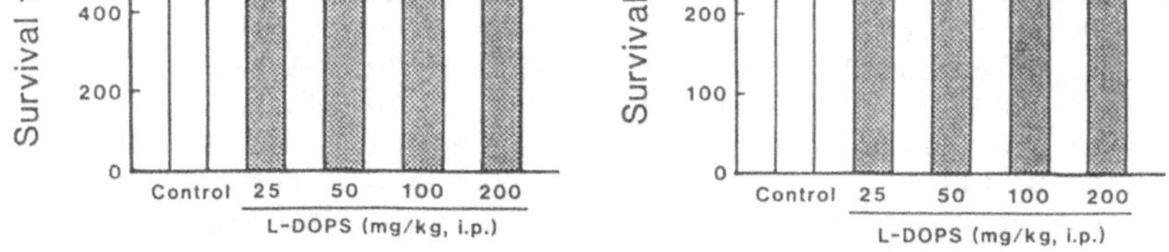

Fig. 3. Effect of L-DOPS on the survival time of mice subjected to KCN(A)- and curare(B)-induced anoxia. L-DOPS was given one hour before the exposure to anoxia. Values are means ± SE (N=20). *P<0.05, **P<0.01 vs. control (Duncan's multiple comparison test).

Secondly, we also examined the effects of L-DOPS on experimental anoxia. L-DOPS(25-200 mg/kg, i.p.) dose-dependently prolonged the survival time of mice subjected to the KCN- and curare-induced anoxia(Fig. 3). Thus, L-DOPS showed the anti-anoxic activity in these animal models. Besides, L-DOPS has been shown to increase the cerebral blood flow in rat hippocampus and striatum.[12] This effect may be at least in part attributable to the anti-anoxic effect of L-DOPS.

Clinically, it has been reported that patients with dementia of various etiologies[13] and with Korsakoff's disease[14] showed some improvement after treatments with L-DOPS and with D,L-DOPS, respectively. Taken together, the present results further support the therapeutic possibility of L-DOPS for the treatment of several types of dementia.

REFERENCES

1. C. Inagaki and C. Tanaka, Characteristics of enzymic decarboxylation of L-threo-3,4-dihydroxyphenylserine using hog renal L-aromatic amino acid decarboxylase, Biochem. Pharmacol. 27:1081 (1978).
2. N. Karai, T. Kato, M. Katsuyama, M. Nakamura, and J. Katsube, Effect of L-threo-3,4-dihydroxyphenylserine(L-threo-DOPS) on brain and serum MHPG levels in mice: evidence for NE formation in CNS, Life Sci. 40:2261 (1987).
3. T. Suzuki, I. Tsuge, S. Higa, and A. Hayashi, Effect of infused L-threo-dihydroxyphenylserine on adrenergic activity in patients with familial amyloid polyneuropathy, Eur. J. Clin. Pharmacol. 17:429 (1980).
4. H. Narabayashi, T. Kondo, A. Hayashi, T. Suzuki, and T. Nagatsu, L-threo-3,4-dihydroxyphenylserine treatment for akinesia and freezing of parkinsonism, Proc. Japan. Acad. 57:351 (1981).
5. T. Kachi, S. Iwase, T. Mano, M. Saito, M. Kunimoto, and I. Sobue, Effect of L-threo-3,4-dihydroxyphenylserine on muscle sympathetic nerve activities in Shy-Drager syndrome, Neurology 38:1091 (1988).
6. T. Kato, M. Katsuyama, N. Karai, M. Nakamura, and J. Katsube, Studies on the central action of L-threo-3,4-dihydroxyphenylserine(L-threo-DOPS) in FLA-63-treated mice, Pharmacol. Biochem. Behav. 26:407 (1987).
7. A. Hirose, M. Sasa, Y. Ohno, and S. Takaori, Inhibitory effects of L-threo-DOPS, an L-noradrenaline precursor, on locus coeruleus-originating neurons in the caudate nucleus, Japan. J. Pharmacol. 48:435 (1988).
8. A. J. Cross, T. J. Crow, E. K. Perry, R. H. Perry, G. Blessed, and B. E. Tomlinson, Reduced dopamine-beta-hydroxylase activity in Alzheimer's disease, Br. Med. J. 282:93 (1981).
9. C. M. Yates, I. M. Ritchie, J. Simpson, A. F. J. Maloney, and A. Gordon, Noradrenaline in Alzheimer-type dementia and Down syndrome, Lancet ii:39 (1981).
10. H. Araki, C. Tanaka, H. Fujiwara, M. Nakamura, and I. Ohmura, Pressor effect of L-threo-3,4-dihydroxyphenylserine in rats, J. Pharm. Pharmacol. 33:772 (1981).
11. K. C. Liang, R. G. Juler, and J. L. McGaugh, Modulating effects of posttraining epinephrine on memory: Involvement of the amygdala noradrenergic system, Brain Res. 368:125 (1986).
12. E. Sato, T. Irie, and J. Katsube, Effect of L-threo-3,4-dihydroxyphenylserine (L-DOPS), an immediate precursor of norepinephrine, on the cerebral blood flow in rats, Japan. J. Pharmacol. 43:91 (1987).
13. N. Shimizu, M. Otsuka, Y. Mizuno, and M. Yoshida, L-threo-3,4-dihydroxyphenylserine treatment for dementia of various etiologies, Proc. Japar Acad. 62,Ser.B:205 (1986).
14. P. J. Langlais, R. G. Mair, P. J. Whalen, W. McCourt, and W. J. McEntee, Memory effect of DL-threo-3,4-dihydroxyphenylserine(DOPS) in human Korsakoff's disease, Psychopharmacology 95:250 (1988).

AMELIORATION OF DELAYED NEURONAL DEATH OF THE HIPPOCAMPUS BY NERVE GROWTH FACTOR

T. Shigeno[1], T. Mima[2], K. Takakura[2], G. Katoh[3], Y. Hashimoto[4], S. Furukawa[4], and D.I. Graham[5]

[1]Saitama Medical School, [2]University of Tokyo [3]Chiba College of Health Science, [4]National Institute of Neuroscience, Japan and [5]University of Glasgow, U.K.

INTRODUCTION

The hippocampus is a unique structure in that the neural circuit is excitatory and its function contributes to memory, possibly via the mechanism of long-term potentiation. Very recently, with the aid of magnetic resonance imaging, bilateral hippocampal damage has been successfully visualized in patients with amnesia.[1] It is known that after temporal cessation of cerebral circulation in men such as occurs during cardiac arrest, the patient develops memory disturbance with the presence of hippocampal neuronal death.[2] In experimental animals such as rodents, the hippocampal neuronal death, particularly CA1 sector, develops more than two days after transient global cerebral ischemia.[3] This model might resemble the characteristics of pathology of Alzheimer's disease or vascular dementia.

The initiating event that leads to delayed neuronal death of the hippocampus is possibly neural excitation by glutamate and subsequent calcium entry into the cell.[4] However, this mechanism cannot fully explain why the phenomenon is delayed and why the occurrence is limited to the CA1 sector of the hippocampus. Here we propose a new understanding such that this is a reflection of disturbed recovery processes in the neural circuit. There is evidence that nerve growth factor (NGF) is required in the central nervous system for cell survival and repair.[5] Chronic neuronal cell loss could be caused by the lack of such trophic factors. Therefore, we determined if NGF has any influence on neuronal survival and repair in the hippocampus.

MATERIAL and METHODS

Adult Mongolian gerbils were subjected to transient forebrain ischemia by occlusion of both carotid arteries for 5 min. Neuropathological studies were undertaken in four groups of animals: (1) occlusion alone, (2) intraventricular injection of 10 µl of artificial CSF, (3) intraventricular injection of NGF (Sigma, 7S type, 10 µg in 10 µl of artificial CSF) just before ischemia, and (4) the same treatment but 15 minutes after ischemia. To determine where the pathological changes started, we utilized the Golgi technique to examine dendrites and synapses 3 hours, 1 day, 2 days, and 1 week after ischemia in both non-treated and treated animals. We also measured tissue NGF content using the sensitive two-site enzyme immunoassay developed by Furukawa et al.[6]

RESULTS

Table 1, summarizing the results of neuropathology, shows that both pre and post-ischemia treatment with NGF markedly ameliorated the occurrence of delayed neuronal death. Artificial CSF injection had also some beneficial effect in preventing the occurrence of delayed neuronal death.

Table 1. Occurrence of Delayed Neuronal Death

	Extent of Neuronal Loss in the CA1 sector		
	more than 90%	40 - 60%	less than 5%
Occlusion alone	11(100%)	0(0%)	0(0%)
Artificial CSF	7(39%)	2(11%)	9(50%)
NGF Pre-treatment	0(0%)	1(5%)	20(95%)
NGF Post-Treatment	2(12%)	0(0%)	15(88%)

Number indicates the number of animals.

Table 2 shows changes in NGF content in the hippocampus after ischemia. NGF significantly decreased two days after ischemia but recovered to the control level in one week.

Table 2. NGF Content in the Hippocampus

Control	3 Hours	2 Days	1 Week
793±194	600±199	465±197*	718±192
(7)	(9)	(5)	(10)

Values are mean±sd (pg/g).
* Significant different from control, $p<0.05$.
Number in parenthesis indicates the number of samples.

In the Golgi study, we found that changes occurred in dendrites and synapses as early as 3 hours even though the soma appeared to be entirely normal (Fig. 1). After one day there was progressive disappearance of both basal and apical dendrites. By one week, the CA1 sector had been replaced by reactive astroglia in the non-treated animals. Following NGF treatment, these dendritic changes were minimal, with only a slight astroglial reaction.

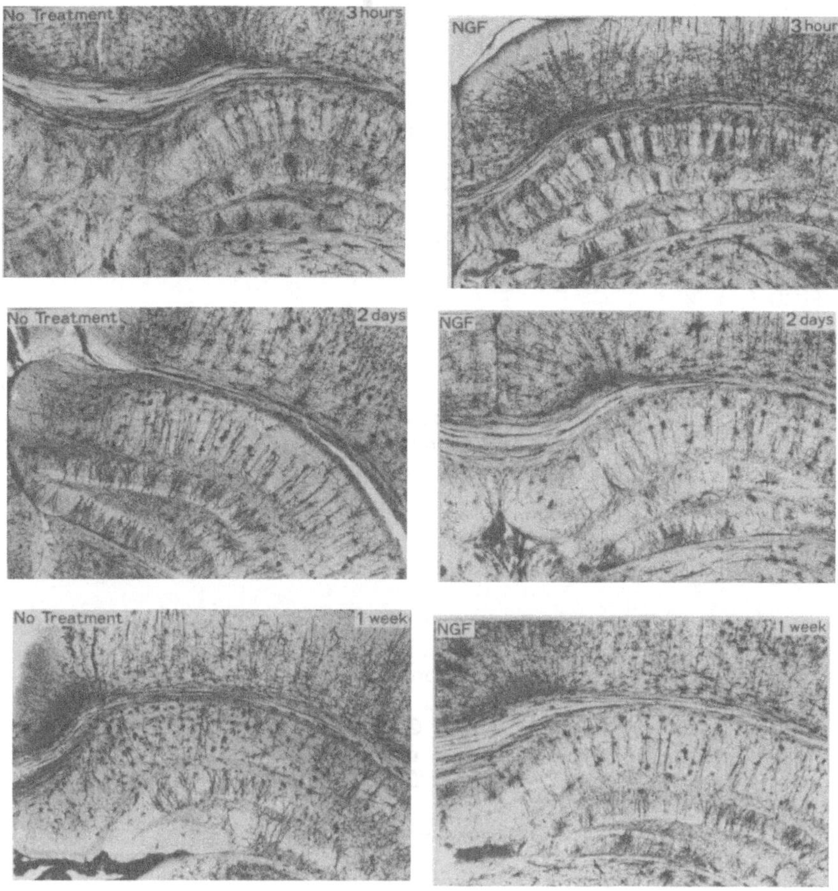

Fig. 1. Changes in dendrites of the hippocampus after ischemia. (Golgi staining technique)

DISCUSSION

The delayed neuronal death of the hippocampus by cerebral ischemia could be a model of dementia-related pathology. The delayed phenomenon suggests that this is a reflection of disturbed recovery processes in the neural circuit. We found that intraventricular injection of NGF completely prevented the development of delayed neuronal death. Using the Golgi technique, we found that destruction of dendrites began as early as 3 hours and that it could be ameliorated by NGF. This suggests that the injury is initiated at the level of synapses and dendrites. NGF is known to be produced in the hippocampus and to be retrogradely transported from the axonal terminal of the basal forebrain neurons such as in the septal nucleus.[7] We assume this mechanism in turn has protective effect on the synaptic terminal of the CA1 neuron after injury. The secondary increase in tissue NGF content might be evidence of intrinsic repair processes, particularly astroglia. The reactive astroglia observed one week after ischemia might be responsible for the production of NGF, because Furukawa et al. have demonstrated NGF production by cultured astroglia.[8]

Furthermore, the beneficial effect even with injection of artificial CSF might result from the production of NGF or other trophic factor(s) as a response to the slight tissue trauma caused by the needle puncture.

NGF is receiving widespread interest in terms of treatment of Alzheimer's disease and other dementia.[9] Although we cannot completely simulate the disease conditions seen in man by our model, the relationship between hippocampus and NGF seems to be a core issue for our understanding of neuronal cell damage and repair in memory-related pathophysiology.

ACKNOWLEDGMENTS

We thank Mr. L. Miller (University of Glasgow), Mrs. R. Matsuura (University of Tokyo), Mrs. K. Kusaka (Chiba College of Health Science), Mr. S. Imura (Yoshitomi Pharmaceuticals), and Mr. K. Nakajima (Yoshitomi Pharmaceuticals) for help in this investigation.

REFERENCES

1. G. A. Press, D. G. Amarai, L. R. Squire: Hippocampal abnormalities in amnestic patients revealed by high-resolution magnetic resonance imaging. Nature 341:54-57 (1989)
2. J. B. Brierley, D. I. Graham: Hypoxia and vascular disorders of the central nervous system, in "Greenfield's Neuropathology," Adams JH, Corsellis JAN, Duchen LW, eds., John Wiley & Sons, New York (1984) pp125-207
3. T. Kirino: Delayed neuronal death in the gerbil hippocampus following ischemia. Brain Res. 239:57-69 (1982)
4. D. M. Barnes: NMDA receptors trigger excitement. Science 239:254-256 (1988)
5. J. L. Marx: Nerve growth factor acts in brain. Science 232:1341-1342 (1986)
6. S. Furukawa, I. Kamo, Y. Furukawa, S. Akazawa, E. Satoyoshi, K. Itoh, K. Hayashi: A highly sensitive enzyme immunoassay for mouse beta nerve growth factor. J. Neurochem. 40:734-744 (1983)
7. R. Levi-Montalcini: The nerve growth factor 35 years later. Science 237:1154-1162 (1987)
8. S. Furukawa, Y. Furukawa, E. Satoyoshi, K. Hayashi: Synthesis and secretion of nerve growth factor by mouse astroglial cells in culture. Biochem. Biophys. Res. Comm. 136:57-63 (1986)
9. Ad Hoc Working Group on Nerve Growth Factor and Alzheimer's Disease (National Institute on Aging, Bethesda): Potential use of nerve growth factor to treat Alzheimer's disease. Science 243:6 (1989)

LONG-TERM ACTIONS OF NICOTINIC RECEPTOR STIMULATION IN NUCLEUS BASALIS LESIONED RATS: BLOCKADE OF TRANS-SYNAPTIC CELL LOSS

Nelida N. Sjak-Shie, Jacob N. Burks and Edwin M. Meyer

University of Florida College of Medicine
Department of Pharmacology and Therapeutics
Gainesville, FL 32610

INTRODUCTION

One of the earliest and most pronounced deficits consistently associated with Alzheimer's disease is the marked cholinergic hypofunction in the cerebral cortex, which seems to be due to the loss of neurons originating from the nucleus basalis of Meynert (1,2). Later stages of the disease are characterized by the atrophy of several brain regions including the neocortex (3). We previously investigated the possible role of early cholinergic hypofunction in causing later neocortical atrophy by lesioning the analogous nucleus basalis magnocellularis (nbm) in the rat (4). Based on the observation that these lesioned animals showed a significant cortical atrophy by 14 months post-lesioning, we hypothesized that the early cortical cholinergic hypofunction in Alzheimer's disease could induce the trans-synaptic loss of neurons in the neocortex as well (5). In the nbm-lesioned rat, this trans-synaptic cell loss appeared to begin around 5 months post-lesioning (4,11).

If neocortical neurons are destroyed subsequent to a loss of cholinergic transmission, one might predict a selective loss of those cholinergic receptors used to mediate transmission between the ascending cholinergic and neocortical neurons. A likely candidate for this receptor-subtype is the nicotinic receptor, since its levels are consistently reduced in Alzheimer's disease while the levels of muscarinic receptors can remain unchanged (6,7).

Nicotinic receptors comprise only a small fraction of the total number of cholinergic receptors in the neocortex. Nonetheless, they may be disproportionately highly involved in cholinergic neurotransmission in that brain region because, unlike their muscarinic receptor counterparts, they are cation-channels and therefore never non-functioning spare receptors requiring second messenger-coupling. Nicotinic receptors do however desensitize rapidly in the presence of some agonists, so that depolarization can be followed by a period of relative refractoriness. In addition to their role in transmission, nicotinic receptors may be important for the long term viability of some cell types, as evidenced by the atrophy of striated muscle that occurs following lesions of cholinergic nerves innervating it. Whether chronic nicotinic transmission is important for maintaining cell viability in the brain is not presently known.

We therefore examined some of the long term neurochemical and cell-density effects of chronic nicotine administration in rats receiving bilateral, excitotoxin-induced, nbm-lesions. In order to assure that the dose of nicotine used was behaviorally active, preliminary studies were performed to determine which drug-dose improved memory-related passive avoidance behavior optimally. Choline acetyltransferase activity (CAT) was measured in the frontal cortex to assess the degree of presynaptic cholinergic hypofunction. Nissl-staining cell bodies were quantified in layer 2 of the parietal cortex to assess the amount of cell-loss following nbm-lesions. Nicotinic receptor levels were measured in the occipital cortex to correlate changes in the number of these receptors with changes in neocortical cell counts.

METHODS

Nbm lesions

Male Sprague Dawley rats (300-350 g) received bilateral infusions of 5 ug ibotenic acid in to the nbm. Control animals underwent sham-surgery, but received no infusion of ibotenic acid. Starting 5.5 months post-lesioning, sham-operated and lesioned animals (n=3) received daily IP injections (1ml/kg) of 0.9% saline or 0.2 mg/kg nicotine until their sacrifice 8 months post-lesioning.

Neurochemical assays

Frontal cortex CAT activity (8) and occipital cortex nicotine receptor binding (9) were assayed as previously described. CAT activity measurements used choline and [^{3}H]acetylcoenzyme A (New England Nuclear Co., Boston, MA) as substrates. For receptor binding, [^{3}H]acetylcholine (New England Nuclear Co.) was used in the presence of atropine and physostigmine; preliminary Scatchard analyses found this ligand to bind with a single Kd of about 20 nM. Carbachol (100 μM) was used to determine nonspecific binding. The results were normalized for the amount of protein present in each sample. All values were expressed as the mean ± S.E.M. of at least three animals per group, each assayed in duplicate.

Histology

After decollation, brains were removed and fixed in 10% buffered formalin. Frozen sections (30 micron) of the entire brain were generated using a sliding microtome. The sections were placed on gelatinized slides and stained with 0.1% cresyl violet dye, a Nissl-selective stain. Neuronal densities were assessed at 500x magnification using a Nikon microscope with a drawing arm attached. Neurons were counted in a 0.09 mm^2 square region within the second layer of the parietal cortex. Only those neurons with a distinct nucleolus and with more than half of their perikarya within the boundaries of the square were counted. For each animal 5 consecutive brain sections were analyzed in this fashion. The data were expressed as the mean ± S.E.M. of all sections counted in each experimantal treatment group.

Behavior

Passive avoidance behavior was measured as described previously (10). Briefly, this involved placing the animals 5 min after drug or saline injection (1 ml/kg IP) into a well lit compartment connected to a dark chamber. Immediately after the animal entered the dark-room, the door was shut and the animal received a short (1 s) pulse of electricity (0.8 mA); this consisted of the training period. A 72 hr time interval was used for testing the acute effects of drugs in normal animals. Animals were injected again 5 min before testing. The default interval for the training period and for testing was 5 min.

RESULTS AND DISCUSSION

In order to determine an appropriate dose of nicotine for counteracting the effects of cholinergic deafferentation in the neocortex, we chose a dose that improved a memory related behavior known to be dysfunctional in nbm-lesioned animals. As shown in Figure 1, the highest dose of nicotine tested that enhanced passive avoidance behavior was 0.2 mg/kg; this dose was used in our subsequent long-term experiments.

Upon termination of the 2.5 month daily nicotine-treatment (5.5-8 months post-lesioning) all animals were assayed for frontal cortex CAT activity, occipital cortex nicotinic receptor binding, and neuronal density in layer 2 of the parietal cortex. As shown in Table 1, bilateral nbm lesions at 8 months post-lesioning caused: 1) a reduction in CAT activity; 2) no change in nicotinic receptor binding, as opposed to a significant 40% elevation at 5 months post-lesioning (G. W. Arendash and E. M. Meyer, unpublished data); and 3) a 17% decrease in neuronal density. The table also shows that chronic nicotine treatment: 1) decreased CAT activity in sham-operated and lesioned animals, but did not enhance the lesion-induced decrease in CAT activity; 2) increased nicotinic receptor binding by an additional 30% in the lesioned animals; and 3) blocked the lesion-induced decrease in neuronal density.

The present results are significant because they show for the first time that chronic nicotine treatment can exert a variety of neurochemical and cellular actions in nbm-lesioned animals, including a potentially trophic one in the parietal cortex. That the ibotenic acid infusions lesioned the nbm is demonstrated by the significant loss of CAT activity in the frontal cortex. Typically, these lesions reduce acetylcholinesterase-staining in the nbm by over 90% while neocortical CAT activity is reduced by 30-40% (4); the residual CAT activity in the neocortex apparently derives primarily from intrinsic neurons that are not affected by the lesions. Chronic nicotine administration surprisingly reduced the levels of CAT activity even in the sham-operated animals. Whether this cholinergic deficit occurs in extrinsic or intrinsic neurons in the neocortex is not clear, since both types of neurons may possess nicotinic receptors (10).

The hypothesis that early cortical cholinergic hypofunction could induce trans-synaptic cell loss is supported by our observation that bilateral nbm lesions significantly reduce neuronal density 8 months post-lesioning. This deficit in Nissl staining cell bodies is less than that observed 14 months post-lesioning (4), suggesting that this trans-synaptic cell loss is a slow but progressive process. It should be noted that we quantified Nissl staining bodies only in layer 2 of the cortex, so that changes in other brain regions or cortical layers may be different from those described here. Another limitation of this experimental procedure was that our sections were 30 microns thick, which would allow for some variability in the degree of Nissl staining due to variability in the ease of penetration of the cresyl violet dye. Nonetheless, the observation that chronic nicotine-treatment blocked the lesion-induced cell loss supports the hypothesis that this drug may prevent trans-synaptic degeneration caused by cholinergic hypofunction. This observation may have clinical ramifications given the similarity between the cholinergic hypofunction in the early stages of Alzheimer's disease and that induced by nbm lesions in the rat. Of course, it is conceivable that chronic nicotine treatment has different actions in species other than the rat.

Nbm-lesions appear to exert a complicated, time-dependent effect on cortical nicotine receptor levels. By 5 months post-lesion, receptor levels are elevated by about 40% compared to sham-operated controls (Arendash and Meyer, unpublished observation). These levels decrease to only 19% above control values (not significant at the $p < 0.05$ level) by 8 months post-lesion. Therefore, it appears that neocortical nicotinic receptor binding levels begin to decrease when cortical atrophy begins, about 5 months post-lesioning. The question remains whether nicotinic receptor levels will continue to decrease as greater cortical atrophy occurs (e.g., 14 months post-lesioning) to sub-control levels as seen in Alzheimer's disease. Nicotine treatment elevated nicotinic receptor levels in the lesioned but not sham-operated group. This observation suggests that nicotine administration may preserve those neocortical neurons that normally express these receptors, consistent with the hypothesis that nicotine is a trophic agent for these neurons. At present, however, we cannot rule out the possibility that nicotine administration acts elsewhere in the brain to block indirectly the cerebral cortical atrophy otherwise caused by the nbm lesions.

Another question presented by these results pertains to the possible role that depolarization may play in the cell-protective effect of nicotine. Depolarization has been suggested by Eugene Johnson (Washington University, St. Louis, MO) to reduce the need for trophic factors in some systems, perhaps by increasing intracellular calcium concentrations (personal communication). Nicotine also depolarizes those cells expressing nicotinic receptors, so it is possible that this drug preserves cortical neurons indirectly by reducing the need for an endogenous trophic agent normally provided by extrinsic nbm neurons. If so, nicotine may become the prototype for a new type of agent, one that readily crosses the blood brain barrier and, while perhaps having no trophic action of its own, modulates the activity of other, endogenous growth factors which are themselves not readily administered into the CNS. While this possibility is exciting, it will be very difficult to differentiate from the aforementioned hypothesis that nicotine has a trophic action of its own.

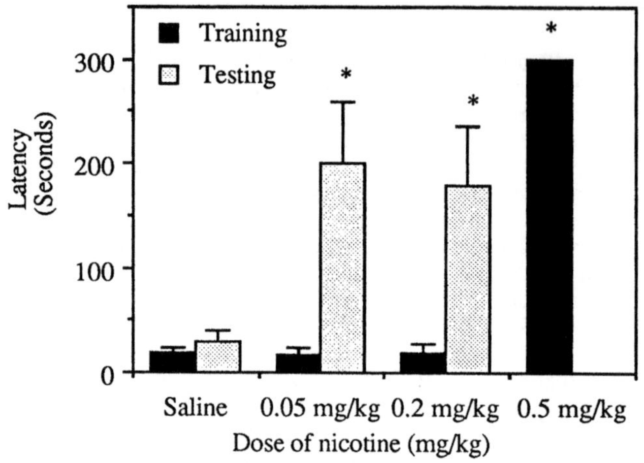

Fig. 1 A dose response curve of the effect of acutely administered nicotine on passive avoidance behavior. Normal, unoperated, male, Sprague Dawley, albino rats (n=6), received IP injections 5 min prior to trainig and testing (72 hours later). Values were expressed as the mean ± S.E.M. (*p<0.05 compared to saline).

Table 1 The effects of chronic nicotine treatment of bilaterally nbm-lesioned rats. CAT activity, nicotinic receptor binding, and neuronal density were assayed in sham-operated and lesioned male Sprague Dawley albino rats (n=3). Animals received daily IP injections of nicotine (0.2 mg/kg) or saline starting 5.5 months post-lesioning, until their sacrifice 2.5 months later (8 months post-lesioning). All values are expressed as the mean ± S.E.M.

	Treatment			
	Saline		Nicotine	
Markers	Sham	Lesioned	Sham	Lesioned
CAT activity	8.1	4.9	5.4	4.5
(S.E.M.)	0.0	0.5	0.3	0.6
(nmol ACh/mg protein)				
Nicotinic receptor binding	10.4	12.4	8.63	25.6
(S.E.M.)	2.2	3.9	2.5	4.4
(fmol ACH/mg protein)				
Neuronal density	830	690	880	880
(S.E.M.)	27	28	56	49
(cell counts/mm^2)				

ACKNOWLEDGEMENTS

This work was funded in part by the Taiho Pharmaceutical Co., Japan, and the American Council for Tobacco Research.

REFERENCES

1) Bartus, R. T. , Dean, R. L. ,Beer, B., and Lippa, A. S. (1982), Alzheimer's disease: the cholinergic hypothesis of geriatric memory dysfunction, Science 217: 408-419.
2) Coyle, J. T., Price, D. L., and DeLong, M. R. (1983), Alzheimer's disease: a disorder of cortical cholinergic innervation, Science 219: 1184-1190.
3) Terry, R. D., Peck, A., DeTeresa, R., Schechter, R., and Horoupian, D. S. (1981), Some morphometric aspects of the brain in senile dementia of the Alzheimer's type, Ann. Neurol. 10: 184.
4) Fibiger, H. C. (1982), The organization and some projections of cholinergic neurons of the mammalian forebrain, Brain Res. Rev. 4: 327-388.
5) Arendash, G. W., Millard, W. J., Dunn, A. J., and Meyer, E. M. (1987), Long-term neuropathological and neurochemical effects of nucleus basalis lesions in the rat, Science 238: 952-956.
6) Flynn, D. D., and Mash, D. R., (1986), Characterization of I-[3H]-nicotine binding in human cerebral cortex: comparison between Alzheimer's disease and the normal, J. Neurochem. 47: 1948-1954.
7) Whitehouse, P. J., Martino, A. M., Antuono, P. G., Lowenstein, P. R., Coyle, J. T., Price, D. L., and Kellar, K. J. (1986), Nicotinic-acetylcholine binding sites in Alzheimer's disease, Brain Res. 371: 146-151.
8) Meyer, E. M., Momol, A. E., and Baker, S. P. (1985), Atrial presynaptic cholinergic activity in young and senescent Rats, Mech. Aging Dev. 30: 221-225.
9) Schwartz, R. D., Lehman, J. and Kellar, K.J. (1984), Presynaptic nicotinic cholinergic receptors labelled by [3H]-acetylcholine on catecholamine and serotonin axons in the brain, J. Neurochem. 43: 1495-1498.
10) Meyer, E.M., Arendash, G. W., Judkins, J. H., Ying, L., Morse, K. and Kem, W.R. (1987), Effects of nucleus basalis lesions on the muscarinic and nicotinic modulation of [3H]-ACh release in the rat cerebral cortex. J. Neurochem. 49: 1758-1762.
11) Arendash, G.W., Strong, P., Millard, W.J. and Meyer, E.M. (1989), Long term neuropathological and behavioral effects of nucleus basalis lesions in the rat. In: Novel Approaches to the Treatment of Alzheimer's Disease, (eds: Meyer, Simpkins, and Yamamoto), Plenum Press, N.Y..

MEASURING EFFICACY OF REVERSIBLE CHOLINESTERASE INHIBITORS FOR ALZHEIMER'S

Kathleen A. Sherman, Kevin F. Harms and Elizabeth A. Doyle

Dept. of Pharmacology, Southern Illinois University School of Medicine, Springfield, IL 62794

INTRODUCTION

Shortly after the discovery that cholinergic innervation to cortex and hippocampus is selectively vulnerable in Alzheimer's disease (AD), attempts began to ameliorate the cognitive dysfunction characteristic of AD with agents intended to enhance central cholinergic transmission. Early trials emphasized precursors of acetylcholine (ACh), choline and lecithin, although animal studies provide little evidence that central cholinergic transmission is altered by elevating circulating choline[1]. Trials of i.v. or oral physostigmine (PHYSO) in which the doses were individually titrated for each AD patient resulted in modest, but significant, improvement of performance on cognitive tasks[2,3]. Although results with this acetylcholinesterase (AChE) inhibitor are encouraging, the degree of improvement varied widely among the AD patients tested, and few patients show a "clinically significant" degree of change[2,6]. The poor response to this carbamate AChE inhibitor appears to relate, at least in part, to failure to achieve adequate biological efficacy. Several studies relate the magnitude of improvement to biological measures of drug effect: Thal et al.[4] found a striking relationship to the degree of AChE inhibition in cerebrospinal fluid (CSF); Mohs et al.[2] found a significant correlation with the plasma cortisol response and Gustafson et al.[5] related change in reaction time to regional increase in cerebral blood flow after PHYSO. We have reported that even after high oral doses of PHYSO (60 μg/kg) the peak inhibition of plasma ChE averaged only 25% (12-33%) and that beyond this level of inhibition patients experience adverse side effects, principally nausea and vomiting. The inhibition of plasma ChE was transient after an acute oral administration, and red cell AChE inhibition was only 10% (4-21%). These findings suggest that failure to achieve the degree of AChE inhibition required to potentiate central cholinergic transmission might be a major factor in the limited clinical response to PHYSO at doses tolerated by most AD patients without adverse side effects.

Summers et al.[7] reported much more dramatic and consistent findings when Alzheimer dementia was treated with oral tacrine (THA) and lecithin. In subsequent trials the results have not been as striking; nonetheless, a significant proportion of AD patients appear to respond[8,9]. Again, part of the variation in response appears to relate to pharmacokinetic differences and measures of biological efficacy (CSF ACh level)[7,9]. Although THA is known as an inhibitor of AChE[10-13], much of the recent research on this

Basic, Clinical, and Therapeutic Aspects of Alzheimer's and Parkinson's Diseases, Volume 2
Edited by T. Nagatsu *et al.*, Plenum Press, New York, 1990

drug has focused on additional properties which might account for efficacy in AD relative to PHYSO. For example, THA interacts with ion channels[14-18] and nicotinic and muscarinic binding sites *in vitro*[19-22]. However, typically these actions are observed at concentrations of THA well above the IC_{50} for AChE inhibition, unlikely to occur during AD therapy. We suggest an alternate explanation of the improved results with THA in AD compared to PHYSO: In addition to improved pharmacokinetic profile and some brain regional selectivity at low doses, THA may inhibit AChE by a preferred mechanism of action. THA inhibits AChE allosterically by a reversible binding to the γ-anionic hydrophobic regions of the enzyme surface[12,13]. Carbamates like PHYSO, on the other hand, bind to the catalytic site of AChE which, when carbamylated, is irreversibly inactivated; the free enzyme is then regenerated relatively quickly after carbamylation by PHYSO[23]. THA is more potent at low substrate concentrations, and thus may be relatively selective for areas of cholinergic deficiency at low doses, while being less effective in normally innervated brain regions. Thus, reversible allosteric inhibitors such as THA may act preferentially to reverse cholinergic hypofunction in hippocampus and cortex where age-related decline in ACh release and degenerative losses due to AD coexist, without resulting in cholinergic hyperactivity and the associated side effects due to actions on intact cholinergic innervation. The reversible nature of THA-induced AChE inhibition, however, presents a special problem for estimating *in vivo* action because the drug dissociates upon dilution of tissue for *ex vivo* assay of AChE activity. We have therefore modified the radioenzymatic procedure to permit quantification of inhibition with minimal tissue dilution for assay. The technique has proven useful in estimating ChE inhibition in human plasma samples after oral administrationof THA, and provides a basis for comparing a biological measure of efficacy which can be obtained in clinical trials to the actions of the drug in brain. Lastly, we report that the technique is useful for other reversible inhibitors such as E2020, a novel piperidine derivative under development for Alzheimer dementia.

METHODS

Cholinesterase activity was measured according to a modification of the radioenzymatic method of Johnson and Russell[25] which utilizes [^{3}H]-ACh as substrate for hydrolysis. Tissue was homogenized by Polytron or by Kontes pestle (for regions weighing less than 80 mg) in 0.5% Triton X-100 containing 10 mM EDTA, pH 7.4 and stored at -70°C for assay. Fifty or eighty μl samples were quantified by positive displacement (Eppendorf Repeater). Incubation was started by addition ^{3}H-ACh (2.5 mM final concentration unless specified: 0.1 μCi of 100 mCi/mmol ^{3}H-AChI from New England Nuclear, Boston, MA, adjusted with ACh·Cl from Sigma, St. Louis, MO) in reaction media containing 120 mM NaCl and 50 mM Tris, pH 7.4 (final volume: 100 μl). The interval of incubation at 37°C (10 to 60 sec) was adjusted to utilize less than 20% of the substrate at the tissue concentration tested. At the end of incubation, the reaction is stopped and the product of hydrolysis ([^{3}H]-acetate) is protonated by 1 M chloroacetic acid and extracted into toluene-based scintillation cocktail containing 10% isoamyl alcohol. Unreacted ^{3}H-ACh is retained in the aqueous stopping solution (1 M chloroacetic acid, 0.5 M NaOH and 2 M NaCl). Blanks (buffer without tissue; 3,000-4,000 DPM) are subtracted.

Plasma cholinesterase (ChE) was measured by the same technique without Triton X/EDTA. To investigate the effect of sample dilution, plasma volume was varied from 80 μl to 0.328 μl by serial dilution to yield final dilutions at assay from 1.25-305-fold. Incubation time was increased proportionately to 120 min for the most dilute samples. The effect of dilution was compared in plasma from rats treated with 5 mg/kg THA or 2 mg/kg E2020 (s.c.) referenced to saline controls and in human plasma at varied times after oral THA.

Young adult male Sprague-Dawley rats (Harlan Industries, Indianapolis, IN) were group housed in plastic cages under standard conditions of 12 hr light:dark cycle and ad lib access to food and water for at least one week prior to use.

RESULTS

The activity of ChE present in plasma samples after systemic administration of reversible inhibitors is clearly related to the dilution of sample used for assay. As shown in Figure 1, the activity of THA samples increases as a function of increasing dilution, whereas activity in control samples increases only slightly from 1.25 to 5-fold dilution and is relatively stable thereafter. Cholinesterase activity was reduced 67% in the THA samples at 1.25-fold dilution and inhibition declined as an approximately linear function of log dilution, approaching control values after about 200-to 300-fold dilution. A probit plot of the apparent inhibition also linearizes the data from 2- to 300-fold dilution but deviates as the *y*-axis is approached (not shown). Estimates of the "true" ChE inhibition at 1-fold (*y*-intercept) are 70% by the arithmetic scale plot and 76-77% by the probit scale plot. Full reversal of THA inhibition (*X*-intercept) is predicted to occur at 180-900-fold dilution, respectively, by the two plot methods.

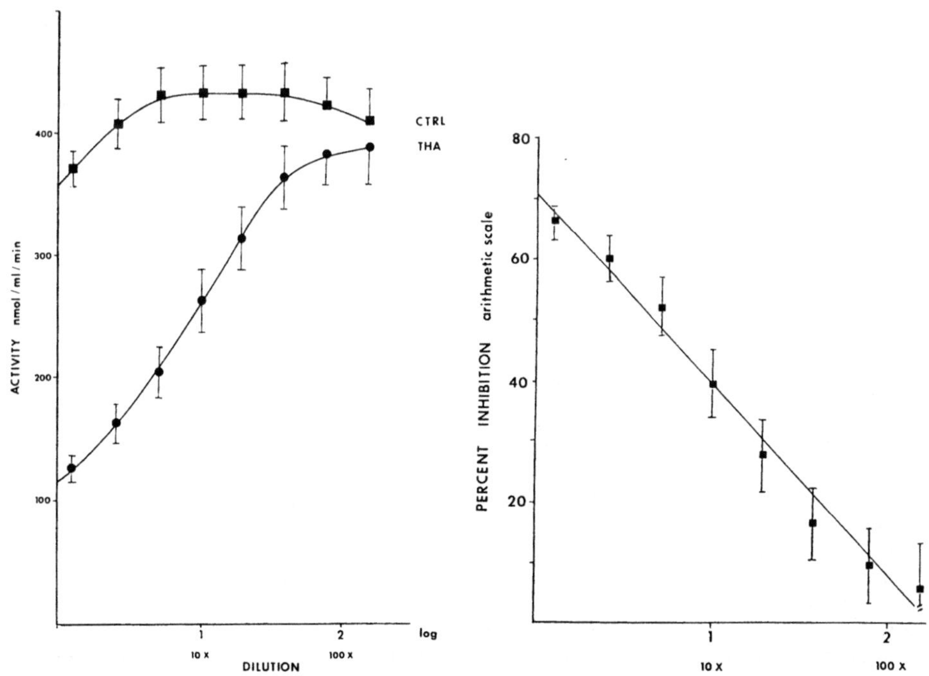

Figure 1. Inhibition of plasma ChE after systemic THA as a function of sample dilution at assay. Rats were pretreated with 5 mg/kg THA s.c. or saline 60 min before sacrifice (N=6/group). Plasma was separated from heparinized trunk blood and assayed at dilutions varied from 1.25- to 305-fold. Results are expressed as group mean ^{3}H-ACh hydrolyzed/ml plasma/min ± S.E.M. (left) or as percent activity in THA-treated plasma relative to control at each dilution (right). Abscissa is a log scale of dilution.

Virtually identical results were obtained when plasma ChE inhibition was examined in E2020-treated rats, except that the inhibition at 1.25-fold dilution was less (52% ± 2.2% of control) than after THA. By 320-fold dilution of plasma at assay, the inhibition by E2020 was reversed. We then tested the predictive power of using the 320-fold diluted activity as the animal's own control. This is conceptually analogous to the reversal of PHYSO inhibition by prolonged incubation[23] used by Thal et al.[4] to quantify the effect of PHYSO on CSF AChE of treated AD patients. In the case of reversible inhibitors, dilution of sample is used to reverse inhibition and a ratio is generated of activity at 1.25-fold dilution to activity at 320-fold dilution for each sample. Percent inhibition is calculated with reference to the same ratio of activity at the two dilutions in control plasma samples. Using this method, inhibition after 2 mg/kg E2020 was calculated as 53% at 1.25-fold dilution, which is quite close to the value cited above. "Selective" inhibitors were used to define AChE (BW284C51-inhibitable; occurring in the presence of ISO-OMPA) and butyryl-cholinesterase (ISO-OMPA inhibitable; occurring in the presence of BW284C51). Control activity is approximately 60% AChE at the ACh concentration used (2.5 mM). E2020 had a marked effect on AChE (63-76% reduction) and relatively little effect on the nonspecific enzyme (8-29% reduction, dependent on method of calculation - BW insensitive or ISO-OMPA inhibitable).

We used the reversal of inhibition by dilution method to obtain control values for a comparison of plasma ChE inhibition and drug levels in samples from humans treated with THA. The inhibition (estimated with reference to subjects with undetectable drug levels) increased as a function of log drug concentration from about 16% in samples with 25 nM drug to 40 ± 2% in subjects with 0.1 μM plasma drug level (Sherman et al., in preparation).

We have also used minimum dilution of tissue to estimate AChE inhibition in specific brain regions after systemic administration of THA as a function of dose and time. These results indicate that at low doses, THA displays some regional selectivity which diminishes at higher doses. For example, Figure 2 shows little inhibition in pons-medulla or olfactory tubercle after 0.625-1.25 mg/kg THA, whereas activity in cortex and hippocampus was reduced 20-25%. All brain regions and plasma showed dose-dependent inhibition which was maximal (60-70%) from 10-20 mg/kg THA s.c.

The time course of enzyme inhibition after systemic THA has also been characterized after 5 mg/kg THA. In plasma, ChE inhibition quickly reaches near maximal levels by 5 min and is maintained up to 3 hr before slowly returning to control between 7 and 9 hrs. Inhibition of AChE in cortex increases more slowly to a maximum by 30-60 min and also is maintained up to 3 hr, returning to control with a time course much like plasma.

For these _ex vivo_ determinations of enzyme inhibition we used 2.5 mM ACh which is the optimum of the bell-shaped substrate concentration curve under our assay conditions. However, the potency of THA is inversely related to ACh concentration and thus the apparent inhibition is likely to be an underestimate, both because of the tissue dilution and because ACh concentration may be lower _in vivo_, particularly in the case of elderly with AD. Figure 3 indicates the dependency of THA inhibition in cortex on the concentration of ACh. For example, when ACh is decreased from 2.5 mM to 0.26 mM (the K_m for AChE), the THA inhibition curve is shifted leftward about 3-fold. These results emphasize the need to develop measures with which to define the neurobiological efficacy of THA under the conditions relevant to AD brain.

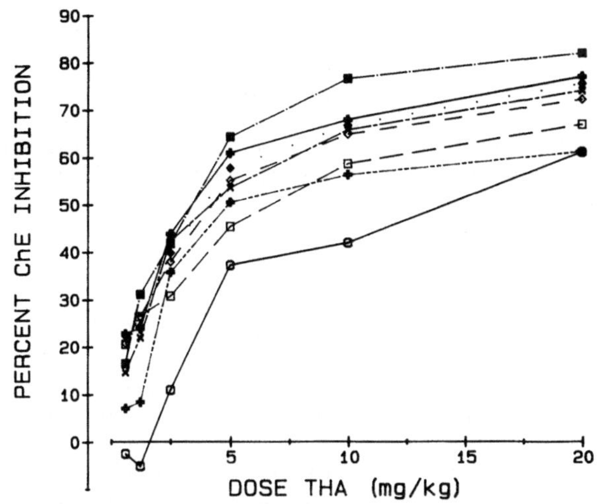

Figure 2. Dose-response curves for AChE inhibition in specific brain regions 2 hr after THA s.c. are expressed as percent activity in THA samples relative to saline-treated controls (N=5-6/group). Final dilution at assay was 2-fold for plasma (■); 6-fold for pons-medulla (O); 6.25-fold for striatum (□); 7.5-fold for hippocampus (♦) and olfactory tubercle (+-broken line); 8-fold for cortex (+-solid line) and cerebellum (x); and 9.6-fold for hypothalamus (◊).

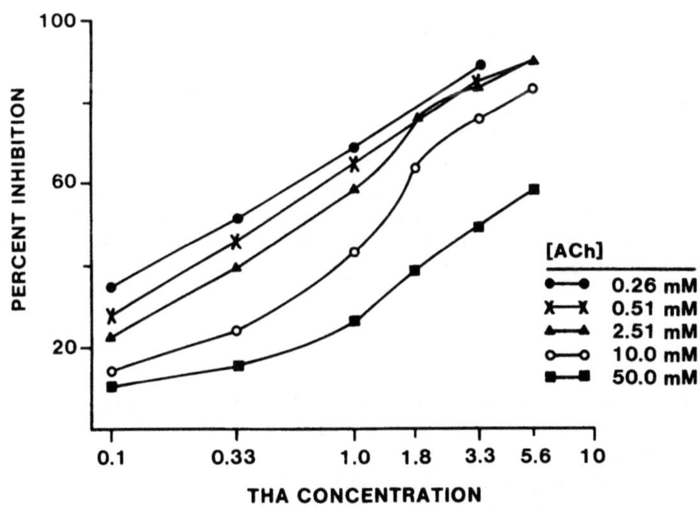

Figure 3. Effect of ACh concentration on THA potency. Percent inhibition is plotted as a percent of unpsiked tissue to show the effect of varying [^{3}H-ACh] from 0.26 to 50 mM; IC_{50} varied from 0.29 to 3.3 μM THA.

DISCUSSION

Relatively few studies tried to quantify AChE inhibition after systemic administration of reversible inhibitors such as THA. Our results show THA to be a potent and long-lasting inhibitor of brain AChE. The magnitude of effect agrees with that found in mouse whole brain after 4 mg/kg (-70%)[26], but after 5 mg/kg THA the inhibition of cortical AChE was more long lasting in our study. Consistent with Rosic and Milosevic[27], we observed inhibition which lasts 5-6 hr. Our magnitude of AChE inhibition was somewhat greater than in that report on whole rat brain, perhaps reflecting the regional differences in degree of inhibition at low doses of THA. In both plasma and red cells, we observed inhibition after THA administration in rats which far exceeds that reported by Bajgar et al. (22-45% in rat whole blood after 8-64 mg/kg vs. 49-53% decrease after 0.4-4.3 mg/kg PHYSO)[28]. On the other hand, in human plasma we observe less inhibition after oral THA than has been estimated after i.v. or i.m. administration[10,11].

Of the pharmacological approaches currently under consideration for palliative treatment of cholinergic deficiency in Alzheimer dementia, those targeted at pre-synaptic function have the advantage that they preserve "normal" timing and spatial distribution of cholinergic influence, and increase ACh availability to both muscarinic and nicotinic receptors (which may be affected in AD). The reversible inhibitors of AChE may have the further advantage at low doses of selective enhancement in brain regions involved in AD. There is a concern that with progression of cholinergic degeneration, the efficacy of drugs targeted at pre-synaptic sites will be lost. However, the results of Nordberg et al.[20] indicate that THA can augment ACh release in autopsied cortical slices to near control levels; the concentrations of THA used were, however, far above those expected <u>in vivo</u>. In rats with experimental deafferentation of the cortex, chronic THA improved performance and restored cortical ACh content to within normal limits [24,29]. Thus, while further work needs to be done to establish the doses or concentrations of THA necessary to ameliorate cholinergic deficiency in aged, denervated brain, there is now documentation supporting the rationale for use of these compounds in AD. The problem of marked individual variation in drug level and in occurrence of adverse side effects (including hepatotoxicity) remains. However, determination of plasma ChE inhibition, in conjunction with further investigation of the relationship to the central actions of drugs like THA, could serve as a guide to individualization of dosage to provide uniform biological efficacy of treatment. When this is obtained, the role of other factors such as severity of illness and involvement of additional neurotransmitter systems could be addressed more systematically.

REFERENCES

1. Sherman, K.A., Zigmond, M.J. and Hanin, I., Fluphenazine-induced decline in striatal acetylcholine is not abolished by exogenous choline, <u>Neuropharmacol.</u> 20:921-924 (1981).
2. Mohs, R.C., Davis, B.M., Greenwald, B.S., Mathe, A.A., Johns, C.A., Horvath, T.B. and Davis, K.L., Clinical studies of the cholinergic deficit in Alzheimer's disease, <u>J. Amer. Geriatr. Soc.</u> 33:749-757 (1985).
3. Thal, L.J., Masur, D.M., Blav, A.D., Fuld, P.A. and Kaluber, M.R., Chronic oral physostigmine without lecithin improves memory in Alzheimer's disease, <u>J. Amer. Geriatr. Soc.</u> 37:42-48 (1989).
4. Thal, L.J., Masur, D.M., Sharpless, N.S., Fuld, P.A. and Davies, P. , Acute and and chronic effects of physostigmine and lecithin in Alzheimer's disease, <u>Prog. Neuropsychopharmacol. Biol. Psychiat.</u> 10:627-638 (1986).

5. Gustafson, L., Edvinsson, L., Dahlgren, N., Hagberg, B., Risberg, J., Rosen, I. and Ferno, H., Intravenous physostigmine treatment of Alzheimer's disease evaluated by psychometric testing, regional cerebral blood flow (rCBF) measurement and EEG, Psychopharmacol. 93: 31-35 (1987).
6. Sherman, K.A., Kumar, V., Ashford, J.W., Murphy, J.W., Elble, R.J. and Giacobini, E., Effect of oral physostigmine in senile dementia patients: Utility of blood cholinesterase inhibition and neuroendocrine responses to define pharmacokinetics and pharmacodynamics, in: Central Nervous System Disorders of Aging: Clinical Intervention and Research (Aging, Vol. 33), R. Strong, W.G. Wood, W.J. Burke, eds., Raven Press, NY, pp. 77-90 (1987).
7. Summers, W.K., Majorski, L.V., Marsh, G.M., Tachiki, K. and King, A., Oral tetrahydroaminoacridine in long term treatment of senile dementia, Alzheimer type, N. Engl. J. Med. 315:1241-1245 (1986).
8. Gauthier, S., Masson, H., Gauthier, L., Bouchard, R., Collier, B., Bacher, Y., Bailey, R., Becker, R., Bergman, H., Charbonneau, R., Dastoor, D., Gayton, D., Kennedy, J., Kissel, C., Krieger, M., Kushnir, R., Lamontagne, A., St. Martin, M., Morin, J., Nair, N.P.V., Neirinck, L., Ratner, J., Suissa, S., Tesfaye, Y. and Vida, S., Tetrahydroaminoacridine and lecithin in Alzheimer's disease, in: Current Research in Alzheimer Therapy: Cholinesterase Inhibitors, E. Giacobini and R. Becker, eds., Taylor and Francis Intl., NY, pp. 237-245 (1988).
9. Nyback, H., Nyman, H., Ohman, G., Nordgren I. and Lindstrom, B., Preliminary experiences and results with THA for amelioration of symptoms of Alzheimer's disease, in: ibid, pp. 231-236 (1988).
10. Barrow, M.E.H. and Johnson, J.K., A study of the anticholinesterase and anticurare effects of some cholinesterase inhibitors, Brit. J. Anaesth., 38:420-431 (1966).
11. Benveniste, D., Hemmingsen, L. and Juul, P., Tacrine inhibition of serum cholinesterase and prolonged succinylcholine action, Acta Anaesth. Scand., 11:297-309 (1967).
12. Patocka, J., Bajgor, J., Bielavsky, J., and Fusek, J., Kinetics of inhibition of cholinesterase by 1,2,3,4-tetrahydro-9-aminoacridine in vitro, Coll. Chechoslov. Chem. Commun. , 41:816-824 (1976).
13. Tonkopii, V.D. and Ardabeva, T.V., Effect of reversible inhibitors on the reaction capacity of functional sites of the active center of acetylcholinesterase, Biull. Eksp. Biol. Med., 86:441-444 (1978).
14. Bradley, R.J., Edge, M.T., Moran, S.G. and Freeman, A.M., Effects of cholinergic drugs used in Alzheimer therapy at the mammalian neuromuscular junction, in: Current Research in Alzheimer Therapy: Cholinesterase Inhibitors, E. Giacobini and R. Becker, eds., Taylor and Francis Intl., NY, pp. 199-209 (1988).
15. Drukarch, B., Kits, K.S., van der Meer, E.G., Lodder, J.C. and Stoof, J.C., 9-Amino-1,2,3,4-tetrahydroacridine (THA), an alleged drug for the treatment of Alzheimer's disease, inhibits acetylcholinesterase activity and slow outward K^+ current, Eur. J. Pharmacol., 141:153-157 (1987).
16. Halliwell, J. and Grove, E., 1989, 9-Amino-1,2,3,4,-tetrahydroacridine (THA) blocks agonist-induced potassium conductance in rat hippocampal neurones, Eur. J. Pharmacol., 163:369-372 (1989).
17. Rogawski, M.A., 1987, Tetrahydroaminoacridine blocks voltage-dependent ion channels in hippocampal neurons, Eur. J. Pharmacol., 142:169-172.
18. Schauf, C.L. and Sattin, A., Tetrahydroaminoacridine blocks potassium channels and inhibits sodium inactivation in Myxicola, J. Pharmacol. Exp. Ther., 243:609-613 (1987).
19. Druckarch, B., Leysen, J.E. and Stoof, J.C., Further analysis of the neuropharmacological profile of 9-amino-1,2,3,4-tetrahydroaminoacridine (THA), an alleged drug for the treatment of Alzheimer disease, Life Sci., 42:1011-1017 (1988).

20. Nordberg, A., Nilsson, L., Adem, A., Hardy, J. and Winblad, B., Effect of THA on acetylcholine release and cholinergic receptors in Alzheimer brains, in: Current Research in Alzheimer Therapy: Cholinesterase Inhibitors, E. Giacobini and R. Becker, eds., Taylor and Francis Intl. Pub., NY, pp. 247-257 (1988).
21. Perry, E., Smith, C., Court, J., Bonham, J., Rodway, M. and Atack, J., Interaction of 9-amino-1,2,3,4-tetrahydroaminoacridine (THA) with human cortical nicotinic and muscarinic receptor binding in vitro, Neurosci. Lett., 91:211-216 (1988).
22. Potter, L., Ferrendelli, C., Hanchett,H., Hollifield, M. and Lorenzi, M., Tetrahydroaminoacrdine and other allosteric antagonists of hippocampal M_1 muscarinic receptors, Molec. Pharmacol., 35:652-660 (1988).
23. Groff, W.A., Ellin, R.I. and Skalsky, R.L., Quantitative assay of physostigmine in human whole blood., J. Pharm. Sci. 66:389-391 (1977).
24. Yamanishi, Y., Araki, S., Kosasa, T., Ogura, H. and Yamatsu, K., Neurochemical studies of E2020, a novel centrally acting acetylcholinesterase inhibitor, Soc. Neurosci. Abst., 14(I):59 (1988).
25. Johnson, C.D. and Russell, R.L., A rapid, simple radiometric assay for cholinesterase, suitable for multiple determinations, Analyt. Biochem., 64:229-238 (1975).
26. Tonkopii, V.D. and Padinker, E.P., Effect of reversible inhibitors on thermal denaturation of cholinesterases, Biull. Eksp. Biol. Med., 79:51-52 (1975).
27. Rosic, N. and Milosevic, M.P., The effect of tetrahydroaminoacridine on cholinesterase activity in brain of the rat, Yugosl. Physiol. Pharmacol. Acta, 3:43-47 (1967).
28. Bajgar, J., Fusek, J., Patocka, J. and Hrdina, V., In vivo kinetics of blood cholinesterase inhibition by 9-amino-1,2,3,4-tetrahydroacridine, its 7-methoxy derivative and physostigmine in rats, Physiol. Bohemoslov. 28:31-34 (1979).
29. Ueki, A. and Miyoshi, K., Effects of cholinergic drugs on learning impairment in ventral globus pallidus-lesioned rats, J. Neurol. Sci., 90:1-21 (1989).

QUANTITATIVE PHARMACO-ELECTROENCEPHALOGRAPHIC STUDY ON ACETYL-L-CARNITINE IN GERIATRIC PATIENTS WITH DEMENTIA

Kenji Nobuhara, Toshihiko Kinoshita, Yoshiyasu Okajima, Toshiaki Isotani, Akemi Saito, Takami Yagyu, Naomi Saito, Takahiro Nishimura, Yoshiki Ohashi, Masato Fukushima and Masani Saito

Department of Neuropsychiatry, Kansai Medical University
Fumizono-cho 1, Moriguchi, Osaka 570, Japan

INTRODUCTION

A rapid increase in the elderly population is one of the most serious problems in Japan. Treatment and prevention of psychogeriatric disorders, especially dementia, are one of the current important issues. Various substances such as amines, enzymes, hormones, and other neuropeptides have been applied to elderly patients, and their effects on dementia have been discussed for the last ten years. To our regret, however, none of them have shown any favorable effects on senile dementia of the Alzheimer type(SDAT) or on Alzheimer's disease(AD). Acetyl-L-carnitine chloride(ALC) , developed by Sigma-Tau in Italy, is a substance regarded as a probable new remedy for mental deterioration. Cholinergic actions and the promoting effects on energy metabolism and ability to learn were demonstrated in animal studies. Moreover, since ALC is a bioactive substance, side effects must be very little when it is applied in clinical treatment. Safety and tolerance of ALC in normal men were substantiated in a phase 1 study.

We administrated ALC to AD and SDAT patients in order to assess by computerized EEG techniques its therapeutic effects on the central nervous system and to study the correlation between the induced EEG changes and the clinical course.

MATERIALS AND METHODS

Four patients ranging in age from 58 to 90 years were selected for this trial; one was a hospitalized AD patient(Case 1) and three(Cases 2,3,4) were ambulatory SDAT patients. The diagnosis was based on the diagnostic standard of the NINCDS-ADRDA work group. Before starting the trial, we obtained written consents from either the patients or their family after explaining the purpose of the trial.

The patients were given ALC (1.8 grams t.i.d.) orally for at least four weeks. Eight-channel EEGs (F3, F4, C3, C4, P3, P4, O1, O2 against ipsilateral ear electrodes according to the International 10-20 system) were recorded before and every four weeks after the start of the trial

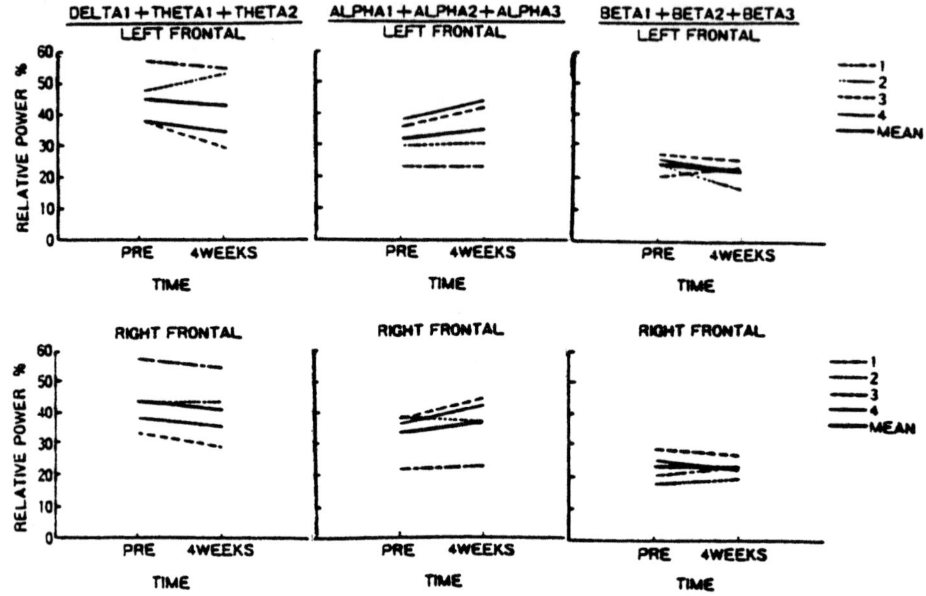

Fig. 1. EEG changes induced by ALC.

and afterwards analyzed off-line by power spectral analysis. The relative power of slow (delta and theta), alfha, and fast (beta and gamma) activities were utilized to examine the EEG changes. Additionally, clinical symptoms were assessed by Gottfries, Brane and Steen(GBS) rating scale, Hasegawa's dementia scale (HDS), and Mini Mental State Test(MMS). Dynamic brain mapping by HZI and single photon emission computed tomography(SPECT) were used for reference.

RESULTS

Of all the recorded EEG's, significant changes seen after four weeks' treatment occurred only in the frontal derivations illustrated in Figure 1. Alpha waves increased in all the SDAT patients, but no significant changes were seen in the AD patient. Fast waves decreased in the SDAT patients, and increased in the AD patient.

Figure 2 shows the absolute highest activity mapping of Case 2. Theta waves were seen in left frontal area, while the rest was covered by alpha waves before the trial. After four weeks' treatment, alpha activity increased and became diffusely dominant. The characteristic of EEG changes induced after chronic ALC administration in the SDAT patients were summarized as a decrease in slow waves associated with an increase of alpha activity and a decrease in fast waves. On the other hand, ALC induced a decrease in slow waves, no change in alpha activity, and an increase in fast waves in the AD patient. This discrepancy of the EEG changes between Case 1 and the others may be related to the predominancy of alpha activity before the treatment.

Figure 3 shows the successive changes of slow activities and fast waves in Case 1. Delta and slow theta waves gradually decreased, and fast waves gradually increased over a 16-week period. These results suggest possible effects of ALC on CNS, which they could be more prominent when a longer term trial is made.

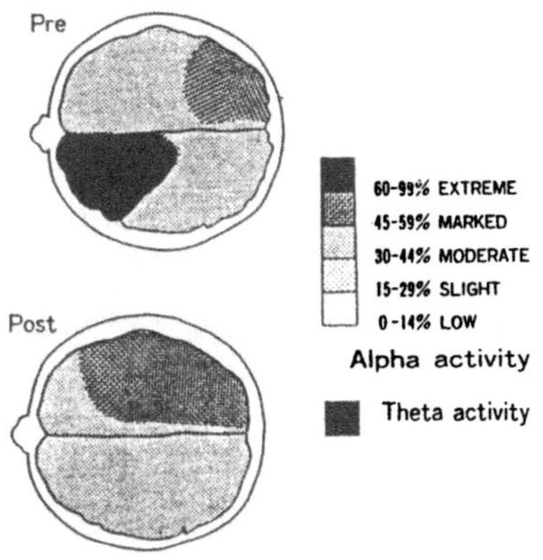

Fig. 2. Absolute highest activity

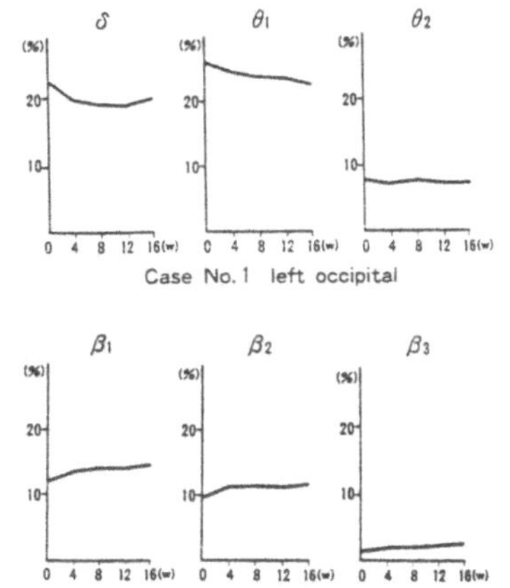

Fig. 3. Case No. 1 left occipital

Table1. Single Photon Emission Computed Tomograph

	PRE		POST (16W)	
AREA	left	right	left	right
ACA territory (frontal)	28.5	25.5	28	25
MCA territory (temporal)	29.5	24	30	27
MCA territory (parietal)	26	22.5	27	24
PCA territory (occipital)	30.5	26	31	24

CBF (ml/100mg/min)

Table2.Clinical Symptoms (Case 2-4)

PATIENT No.		MMS	HS	GBS RATING SCALE
2	PRE	12	15.5	
	POST	13	15.0	motor function : (+)
3	PRE	12	13.0	
	POST	13	10.0	
4	PRE	13	18.0	
	POST	13	26.0	spontaneous activity : (+)

MMS : MINI MENTAL STATE TEST (+) : slightly improvement
HS : HASEGAWA'S DEMENTIA SCALE
GBS : GOTTFRIES, BRANE, STEEN RATING SCALE

Table3. Clinical Symptoms (Case 1)

	MMS	HS	GBS RATING SCALE
PRE	4	2.5	
POST-4-WEEK	4	2.0	emotional function : (+)
POST-8-WEEK	3	2.0	
POST-12-WEEK	4	2.0	spontaneous ability : (+)
POST-16-WEEK	3	2.0	

MMS : MINI MENTAL STATE TEST (+) : slightly improvement
HS : HASEGAWA'S DEMENTIA SCALE
GBS : GOTTFRIES, BRANE, STEEN RATING SCALE

Table 1 shows the volume of intracerebral blood flow measured by SPECT with 133-Xenon gas inhalation in Case 1. When a comparison was made with normal age-matched controls, there was a marked decrease in blood flow over the brain before the trial, and ALC was apparently ineffective to improve it.

Table 2 shows the results of clinical evaluations after four weeks of ALC administration to the SDAT patients. HDS and MMS scores showed no significant changes. However, when studied in detail, ALC appeared to possess certain effects of improving talkativeness, spontaneous activity, and motor function. Case 2 patient, who needed an aide in buttoning before the treatment, became able to manage it by herself without help after treatment.

GBS scores of the AD patient shown in Table 3 indicate a slight improvement of emotional function after the first four weeks and spontaneous activity after twelve weeks, but there were no significant changes in HDS amd MMS scores.

DISCUSSION AND CONCLUSION

A substance that possesses certain actions on the central nervous system has an influence on brain functions and therefore produces changes in the EEG. This was the hypothesis that first led to the establishment of pharmaco-EEG. Based on the results and facts accumulated by many investigators, this hypnothesis has became widely accepted and the EEG is now known as one of the best tools to assess psychotropic effects and to monitor brain functions. As well known, even though EEG was recorded from scalp electrodes, substances that possess similar psychotropic effects provoke a similar EEG changes, and therefore psychotropic drugs can be actually classified solely by the EEG changes induced by them. According to Itil as well as Kinoshita and colleagues, drug-induced EEG profiles have been classified into four groups, namely, central depressants (major tranquilizers), anxiety relievers (minor tranquilizers), mood elevators (antidepressants), and vigilance enhancers (psychostimulants).

However, soon after the new substances grouped "nootropics" were developed and introduced, this classification of psychotropic drugs based on EEG changes became a little more complicated. Nootropics would present certain effects to cognitive functions such as memory, learning and behavior, while it might scarcely show the common traits of psychotropic agents. Itil has stated that "nootropics" should be separated as alpha enhancer from psychostimulants. Kinoshita have expanded the vigilance-enhancer categoly to incldue anti-dementia substance.

In the present trial, ALC increased alpha waves and decreased slow waves as well as fast waves in the SDAT patients, and it could be considered as an alpha enhancer. Moreover, ALC improved more or less several symptoms of dementia, such as emotional function, motor fuction, and spontaneous activity correlatively. We can conclude that ALC is a member of the nootropic substances from this trial.

ALC did not produce either any noticeable increase in alpha activity in the EEG or any particular improvement of psychogeriatric symptoms in the AD patient. This lack of effect may be related to the severity of the dementia, since only this patient was severely demented: his EEG was diffusely delta dominant at the beginning of the trial, and his HDS and MMS scores were below five points. This case suggests that the severer the dementia is, the less the drug effects will be.

We have a personal communication that ALC given at 3.6 grams was more effective in improving dementia without serious side effects in clinical trial with SDAT patients carried out in Italy. This communication also stressed that it would be necessary to continue the medication at least for six months before the appearance of clinical effects in cases of severer dementia. In the present trial, the number of subjects were limited and the daily dose was fixed at 1.8 grams. A longer follow-up study with more subjects is essential to estimate the efficacy of ALC in severely demented patients.

REFERENCES

1. T.M.Itil, CNS effects of ISF-2522, a new nootropic; a phase I safety and CNS efficacy study with quantitative pharmaco-EEG and pharmaco-psychology, Current Therapeutic Research. 5:525 (1979).
2. T.Kinoshita, A.Saito, and M.Saito, Computer-assisted EEG study of nootropic drug effects on cerebral functions, in 'Cerebral Dynamics, Laterality and Psychopathology', Elsevier Science Publishers, B.V., Amsterdam (1987).
3. C.Giuregea, The "nootropic" approach to the pharmacology of the integrative activity of the brain, Conditional Reflex 8:108 (1973).

MEGAVITAMIN THERAPY (COENZYME Q_{10} AND VITAMIN B_6) IN ALZHEIMER'S DISEASE AND SENILE DEMENTIA OF ALZHEIMER TYPE

Masaki Imagawa

Neuropsychiatric Clinic
Hyogo Prefecture Amagasaki Hospital
Hyogo, Japan

INTRODUCTION

An effective therapy for Alzheimer's disease (AD) or senile dementia of the Alzheimer type (SDAT) has not been established yet. We tried megavitamin therapy (the combination of Coenzyme Q_{10} with vitamin B_6) on patients with these diseases who did not respond to vasodilators. The symptoms and signs of these patients were relieved or eliminated by the vitamin therapy.

SUBJECTS AND METHOD

Seven males (70.4 ± 5.9 years old) and 3 females (77.3 ± 1.2 years old) were treated. AD had been diagnosed in 2 of the malis and SDAT in 5 males and 3 females. Diagnosis and the grade of the patient's dementia was based on DSM-Ⅲ-R. Therefore the middle grade was 4; and the severe grade 6. The clinical course was followed by two methods: Hasegawa's Dementia Score and the Modified Miyasaki's ADL scale (Figure 1). This scale is constructed from the patients daily behavior, that is, sleep (nighttime sleep, daytime sleep), abnormal behavior, and habitual or natural behavior monitored by an observer. These items are classified by three colors (sleep: blue, abnormal behavior: red, habitual behavior: green). The time

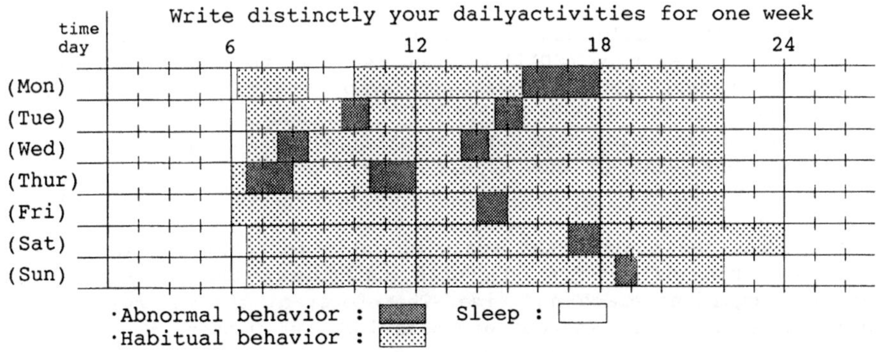

Fig.1. Sample of ADL form

Basic, Clinical, and Therapeutic Aspects of Alzheimer's and Parkinson's Diseases, Volume 2
Edited by T. Nagatsu *et al.*, Plenum Press, New York, 1990

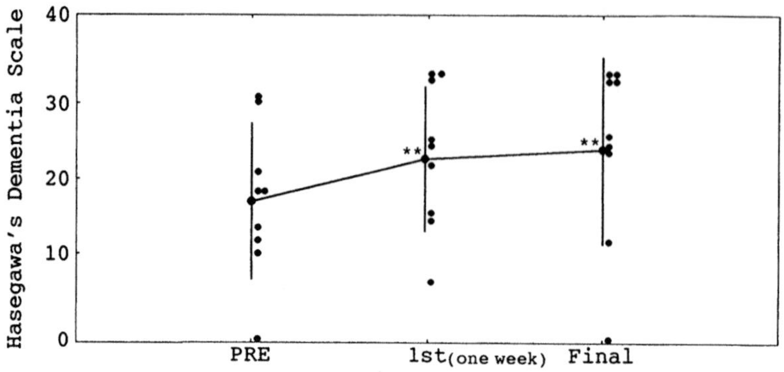

Fig.2. Changes is the Hasegawa Dementia Scale

period of the observation is one week (7 days × 24hr). These data were taken as the pre-value before drug administration (CoQ_{10}, B_6), the first value after taking drugs, and the final value. The dose of the drugs taken by the patients (Coenzyme Q_{10}, B_6) both ranged from 120 mg to 240 mg.

RESULTS

The values of the Hasegawa Dementia Score were pre., 17.1 ± 9.4; first, 22.6 ± 8; final, 23.6 ± 11.0 (M ± SD, n=9), and these latter two scores were significantly increased over the former (Figure 2). The ADL scores were measured in terms of total time spent on a siner activity for one week (24hr × 7days). By this method abnormal behavior registered 10.0 ± 12.1 (pre) vs 4.3 ± 6.5 (final); and daytime sleep, 7.9 ± 10.2 (pre) and 0.4 ± 0.9 (final) ($P<0.01$, n=8), as listed in Table 1.

DISCUSSION

In AD and SDAT, which are decreased in the brain. Vitamin B_6 is a coenzyme required for the production of these neurotransmitters, and this vitamin depends on ATP for decarboxylation. ATP utilized for this reaction in turn depends on CoQ_{10} in mitchondria. The disappearance of the symptoms and signs (for example disturbance or low power of memory and learning, wandering, delirium, loss of dialogue, loss of habitual beavior) of AD and SDAT may be data to support this theory.

CASE REPORT

A 78-year-old female had been diagnosed at age 74 as having senile dementia of Alzheimer type (SDAT) and had received therapy by a doctor. In spite of the therapy (vasodilators), the symptoms (for example, impairment in speech due to the lower function of language, wandering, and the stereotype of the behavior, the loss of habitual action) progressed.

This patient visited our clinic in the summer of 1988. From the judgment of her clinical course at that time, megavitamin (combination of vitamin B_6 with coenzyme Q_{10}) therapy was tried.

At present the patient can speak slightly for a short time, but her response was bad during the doctor's interview. She sometimes loses orientation regarding place and time. Therefore, at the interview she was anxious and repeated the stereotypic behavior, but without the depressive mood. Her visual line was not constant and so her expression of emotions was not vivid. Memory and thought (especially, sophisticated thinking) was

Table 1. ADL Scores Indicating Effective therapy whih Coenzyme (Q_{10}/B_6).

Drugs	each time (24hrs × 7days)	pre.	1st	final
VB_6+CoQ_{10}	night time sleep	81.8 ± 14.5	70.7 ± 20.7	63.2 ± 10.1
	daytime sleep	7.9 ± 10.2	**7.1 ± 9.6	**0.4 ± 0.9
	abnormal behavior	10.0 ± 12.1	**7.9 ± 7.8	**4.3 ± 6.5
	habitual and natural behavior	71.5 ± 19.5	86.1 ± 14.2	96.4 ± 7.7

M ± S.D.
** : $p<0.01$, Wilcoxon test

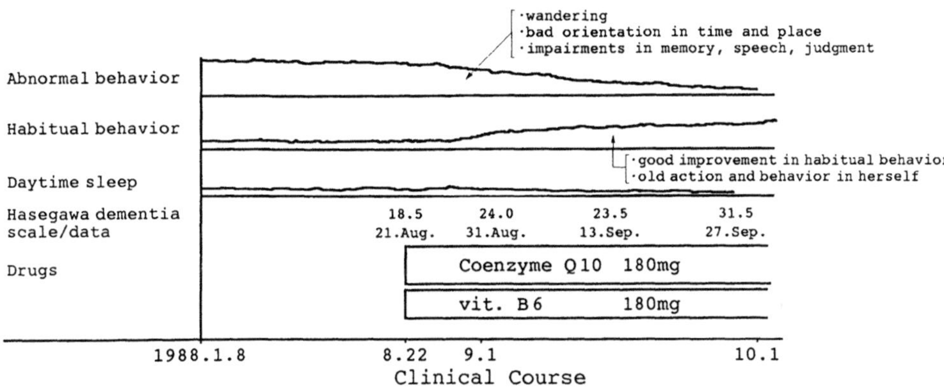

Fig.3. Clinical course of patient.

very weak in power or concentration. Other aspects of her case are as follows:

∘Physical examination: She complained only of tenderness in the epigastric area or and displayed slow movement.

∘Laboratory data: normal ranges.

∘CT-scan and EEG: The EEG showed 9~10Hz and 30~40μV waves with low-voltage, slow burst bitemporaly. CT scan revealed mild atrophy.

∘Past history no apperance.

∘Family history no apperance.

∘Clinical course: See fig.3.

REFERENCES

M.MATSUMOTO, H.SEKIMOTO, Y.GORIYA, T.NAKANO, H.IKEGAKI, M.MATSUMOTO, H.TSUCHIYA, J.MATSUBARA and T.SHIMIZU, Relationship Between Carotid Blood Flow Volume and Degree of Dementia in the Elderly Studied with a Two-Dimensional Echographically Guided Ultrasonic Doppler Flow Volumeter : A Pilot Study, J. Cardiovascular Ultrasonography 315 : 7(4). (1988)

ASSESSMENT OF NOVEL SUSTAINED RELEASE ANTIPARKINSONIAN THERAPIES IN MPTP-TREATED PRIMATES

N.M.J. Rupniak, S.J. Tye, A.E. Loper*, J.V. Bondi*, T. Naruse[+], M. Miyaji[+], and S.D. Iversen

Merck, Sharp, and Dohme Research Laboratories, Neuroscience Research Centre, Essex, U.K.; *West Point, PA, U.S.A. and [+]Banyu Pharmaceuticals Co. Ltd., Tokyo, Japan

The ability of the neurotoxin MPTP to induce parkinsonism in primates has provided great impetus for fundamental advances in the treatment of Parkinson's disease. This valuable animal model has been employed to further our understanding of the role of basal ganglia structures in cognition (Schneider et al, 1988) and mediation of drug-induced dyskinesias (Crossman et al, 1987), to develop neural transplantation technology (Redman et al, 1988) and to explore the possible neuroprotective effects of monoamine oxidase inhibitors (see Langston, 1985). Recently we have used the MPTP primate model for preclinical evaluation of novel drug delivery formulations.

L-DOPA has been the mainstay therapy for Parkinson's disease for the past two decades. Despite its efficacy in alleviating parkinsonism, long-term therapy with L-DOPA is blighted by a progressively brittle response with dramatic, unpredictable swings in neurological status ("on-off" effect) and debilitating dyskinesias. These problems appear at least partially attributable to the erratic gastrointestinal absorption and short half-life of L-DOPA since considerable benefits have been obtained using continuous intravenous infusions of L-DOPA to maintain stable plasma drug levels (Quinn et al, 1982). However, complications caused by the invasive nature of this procedure and the poor solubility of L-DOPA necessitate the development of more practical sustained release drug formulations for routine use. We have examined two such preparations in parkinsonian monkeys: NB-355, a novel prodrug for L-DOPA with improved gastrointestinal absorption characteristics in rodents, and a rate-controlled transdermal delivery system for the selective D_2 agonist (+)-PHNO (Tye et al, 1989; Rupniak et al, 1989).

A group of seven squirrel monkeys was rendered parkinsonian by treatment with the neurotoxin MPTP. Following subsequent intermittent exposure to L-DOPA, these animals developed marked dyskinesias in the limbs (particularly chorea) which coincided with the time of peak locomotor stimulation. In order to assess levels of motor activity continuously over several days, monkeys were housed in individual cages equipped with infra-red photocell units, monitored remotely by a microcomputer. Locomotor activity in these animals was reduced to around 30% of that in normal monkeys.

Dyskinesias were assessed in separate experiments by direct observation using a clinically-based rating scale. NB-355 (L-3-(3-hydroxy-4-pivaloyloxyphenyl)alanine) is a novel mono ester of L-DOPA with superior bioavailability and duration of action compared with L-DOPA in rodent studies (Ihara et al, 1989). Unlike L-DOPA, which was rapidly and almost entirely absorbed from the duodenum via the large neutral amino acid transport system, absorption of NB-355 occurred predominantly from the ileum, avoiding the initial surge in plasma drug levels seen with conventional L-DOPA which may cause peak-dose dyskinesias.

In the presence of carbidopa, oral administration of either L-DOPA (10-40mg/kg) or NB-355 (20-80mg/kg dopa equivalent) caused a dose-related increase in locomotor activity over a 4.5h period. Unlike previous studies using rats, approximately twice the dopa equivalent dose of NB-355 was required to stimulate motor activity to levels similar to those induced by L-DOPA in monkeys (Table 1). This apparent difference in bioavailability may reflect known species variations in esterase activities for specific substrates. However, at doses matched for total locomotor stimulation over the 4.5h period, there was a more gradual rise in motor stimulation by NB-355 accompanied by less severe dyskinesias than those induced by conventional L-DOPA (Table 1; at the highest dose of L-DOPA dyskinesias were inhibited by the emergence of climbing). Thus, NB-355 may be a useful therapeutic agent with reduced risk for inducing peak-dose dyskinesias. It is not yet known whether this agent would also have improved bioavailability and duration of action in man.

Table 1 Effect of oral administration of L-DOPA or NB-355 on locomotor activity and dyskinesias in primates

Treatment	Dopa Equivalent Dose mg/kg)	Activity Counts/4.5h	Dyskinesia score/4h
Vehicle	-	1500 $\pm$ 594	ND
L-DOPA	10	3243 $\pm$ 894	11.5
	20	5057 $\pm$ 1652	57
	40	7350 $\pm$ 1472	36
NB-355	20	1980 $\pm$ 959	11.4
	40	4735 $\pm$ 1503	13.5*
	80	6394 $\pm$ 1435	38

* $p < 0.05$ compared to L-DOPA 20 mg/kg, Mann Whitney U-test.

A second approach to providing stable plasma concentrations of antiparkinsonian drugs which is particularly elegent is by transdermal absorption through a rate-controlling membrane. Such drug delivery devices typically have the appearance of small adhesive plasters which can be conveniently worn behind the ear, for example. Few drugs are able to reach sufficiently high plasma concentrations to be of therapeutic value following cutaneous exposure, but this route of administration has been successfully employed for scopolamine and nitroglycerine. The selective D_2 agonist (+)-PHNO is a unique antiparkinsonian agent for which the transdermal route of administration is possible. Using the same group of MPTP-treated monkeys we examined the ability of a novel rate-controlled transdermal delivery device to stimulate locomotor activity throughout a 24h period.

The release rate of (+)-PHNO from the transdermal system _in vitro_ was 2.6μg/cm^2/h. Each patch covered a skin surface area of 2.4cm^2; a layer of face adhesive ensured good skin contact. Application of patches to the shaved chests of parkinsonian monkeys was able to fully restore locomotor activity to the same levels seen in normal animals. Following an initial lag of some 9h, activity levels remained in the normal range during the subsequent 24h period after transdermal absorption from 4.9-19cm^2 of the drug delivery system (Table 2). Our findings indicate that therapeutically meaningful levels of (+)-PHNO may be delivered continuously by rate-controlled transdermal absorption. However, this therapeutic system for Parkinson's disease should be regarded strictly as a prototype. Clinical development of (+)-PHNO has been terminated due to poor oral efficacy and unacceptable clinical safety profile.

Table 2 Effect of rate-controlled transdermal absorption of (+)-PHNO on locomotor activity in MPTP-treated squirrel monkeys

	Skin surface area available for absorption (cm^2)						
	-	2.4	4.8	7.2	9.6	14.3	19
Locomotor activity/24h (% of levels in normal untreated animals)	15	29	105	99	88	108	122

Advances in drug delivery technology over the next decade are expected to privde substantial improvements for the treatment of neurological disease. Preclinical evaluation of novel drug formulations in MPTP-treated primates could provide important information for subsequent clinical developments in Parkinson's disease.

References

Crossman A.R., Clarke C.E., Boyce S., Robertson R.G., Sambrook M.A. (1987) MPTP-induced parkinsonism in the monkey: neurochemical pathology, complications of treatment and pathophysiological mechanisms. Can. J. Neurol. Sci. 14: 428-435.

Ihara M., Tscuhiya Y., Sawasaski Y., Hisaka A., Takehana H., Tomimoto R., Yano M. (1989) A new prodrug to improve the duration of L-DOPA: L-3-(3-Hydroxy-4-pivaloyloxyphenyl0alanine. J. Pharm. Sci. in press.

Langston J.W. (1985) Mechanism of MPTP toxicity: more answers, more questions. Trends Pharm. Sci. 6: 375-378.

Quinn N., Parkes J.D., Marsden C.D. (1982) Complicated response fluctuations in Parkinson's disease: response to intravenous infusions of L-DOPA. Lancet 1: 412-415.

Redmond D.E., Sladek J.R., Roth R.H., Collier T.J., Elsworth J.D., Taylor J.R., Deutch A.Y., Haber S.N. (1988) Transplanted fetal dopamine neurons: a future treatment for Parkinson's disease? Progress in Catecholamine Research: Central Aspects p 117-123.

Rupniak N.M.J., Tye S.J., Jennings C.A., Loper A.E., Bondi J.V., Hichens M., Hand E., Iversen S.D., Stahl S.M. (1989) Antiparkinsonian efficacy of a novel transdermal delivery system for (+)-PHNO in MPTP-treated squirrel monkeys. Neurology 39: 329-395.

Schneider J.S., Unguez G., Yuwiler A., Berg S.C., Markham C.H. (1988) Deficits in operant behaviour in monkeys treated with n-methyl-4-phenyl-1,2,3,6-tetrahydropyridine (MPTP). Brain 111: 1265-1285.

Tye S.J., Rupniak N.M.J., Naruse T., Miyaji M., Iversen S.D. (1989) NB-355: a novel prodrug for L-DOPA with reduced risk for peak-dose dyskinesias in MPTP-treated squirrel monkeys. Clin. Neuropharmacol. in press.

BEHAVIORAL EFFECTS OF DOPAMINE-AGONISTS ON MPTP-INDUCED PARKINSONISM IN CYNOMOLGUS MONKEYS

Sadako Kuno,[1,2] Tetsuo Akai,[2,3] Motonori Yamaguchi,[2,3] and Eiji Mizuta [1,2]

[1]Department of Neurology and [2]Clinical Research Center
Utano National Hospital, Kyoto; [3]Research Department
Nihon Schering K.K., Osaka, Japan

INTRODUCTION

The existence of at least two subtypes of dopamine (DA) receptors, termed D-1 and D-2, is generally accepted.[1] The specific role of these two receptors and their interaction in treatment of Parkinson's disease are still uncertain.[2,3] In this study, antiparkinsonian effects of dopamine agonists were investigated in MPTP-induced parkinsonian monkeys in relation to the expression of abnormal behaviors such as irritation and excitement associated with the treatment.

METHODS

Male cynomolgus monkeys (Macaca fascicularis, 4-5 kg) were treated with 3 injections of 0.3 mg/kg (i.v.) of MPTP hydrochloride at an interval of 3-4 days, followed by several injections of the same dosage at 7-day intervals. Four monkeys with persistent parkinsonian symptoms were used one month after the last MPTP injection. Parkinsonism and abnormal behaviors (irritation, excitement, and aggression) were scored by the rating scales (0-2) with the methods of Clarke et al.[4] and Mizuta et al.[5], respectively.

RESULTS

Effects of Apomorphine and SKF 38393 on MPTP-induced Parkinsonian Monkeys

Apomorphine (D-1, D-2 agonist, 0.03-0.1 mg/kg, s.c.) improved dose-dependently the parkinsonism and induced abnormal behaviors including irritation, excitement, and aggression (Fig. 1). These abnormal behaviors disappeared when the parkinsonism was re-established within 2 hr. SKF 38393 (partial D-1 agonist, 15 mg/kg, s.c.) initially induced irritation and improved the parkinsonism 5-15 min after the injection (Fig. 2). This was followed by deterioration of the parkinsonism and by induction of eyelid closure, both of which disappeared within 3 hr. Lower doses of apomorphine (0.01 mg/kg, s.c.) and SKF 38393 (5 mg/kg, s.c.) did not show any improvement of parkinsonism or abnormal behaviors.

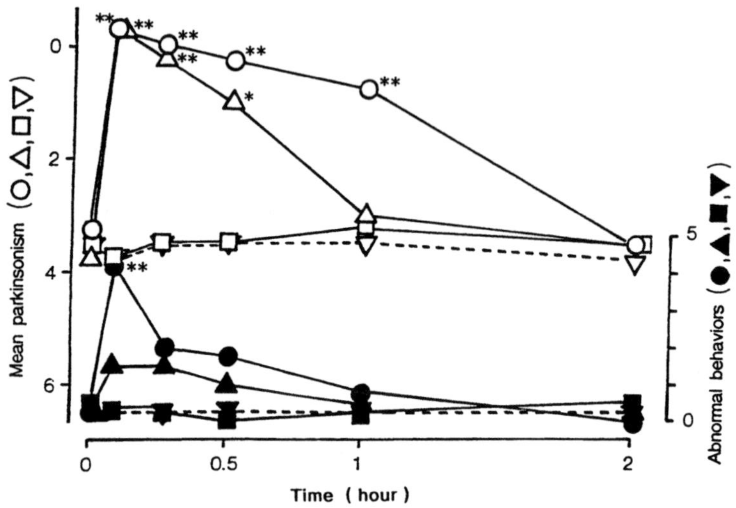

Fig. 1. Effects of apomorphine on MPTP-induced parkinsonian monkeys. Parkinsonism and abnormal behaviors (irritation, excitement and aggression) were scored, and the values given are the mean scores of 4 monkeys per condition. ○,●, 0.1; △,▲, 0.03; □,■, 0.01 mg/kg s.c. of apomorphine; ▽,▼, saline (s.c.). *p < 0.05, **p < 0.01 vs. saline.

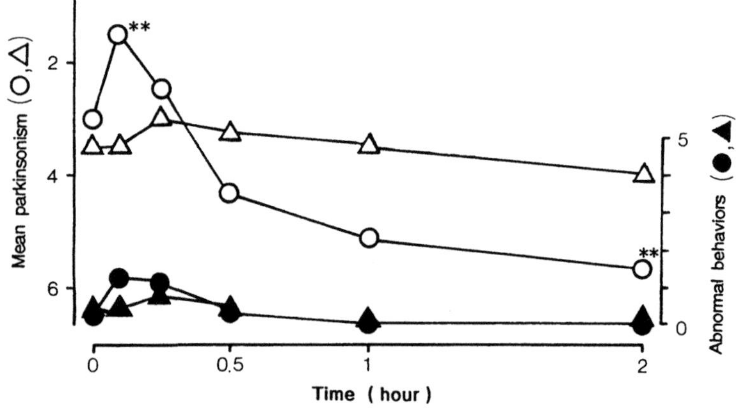

Fig. 2. Effects of SKF 38393 on MPTP-induced parkinsonian monkeys. Parkinsonism and abnormal behaviors (irritation, excitement and aggression) were scored, and the values given are the mean scores of 4 monkeys per condition. ○,●, 15; △,▲, 5 mg/kg s.c. of SKF 38393. **p < 0.01 vs. saline.

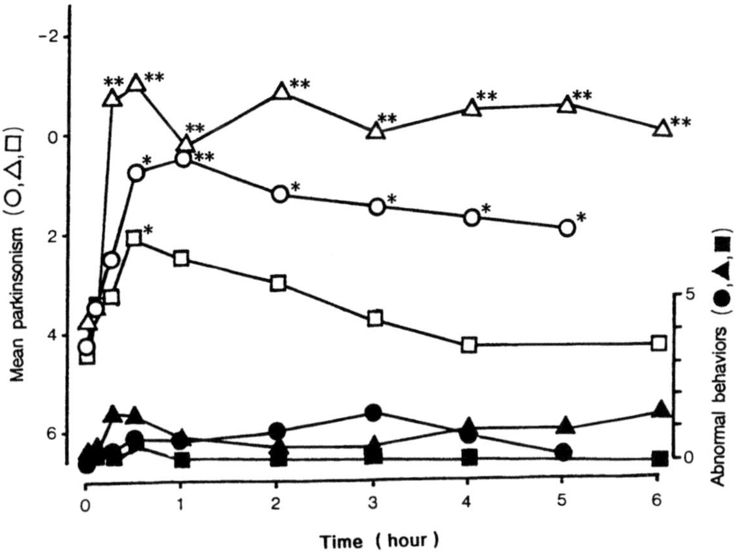

Fig. 3. Effects of lisuride, proterguride, and terguride on MPTP-induced parkinsonian monkeys. Parkinsonism and abnormal behaviors (irritation, excitement, and aggression) were scored, and the values given are the mean scores of 4 monkeys per condition. ○,●, lisuride (0.005 mg/kg, s.c.); △,▲, proterguride (0.0025 mg/kg, s.c.); □, ■, terguride (0.03 mg/kg s.c.) $*p < 0.05$, $**p < 0.01$ vs. saline.

Effects of Lisuride, Proterguride, and Terguride on MPTP-induced Parkinsonian Monkeys

Lisuride, proterguride and terguride[6] bind with high affinity to the D-2 receptors (Kd = 2.0, 1.6, and 3.5 nM, respectively), and the ratios of their affinities of D-2 receptors to those of D-1 receptors in rat brain are found to be 44, 480, and 24, respectively (data not shown). Lisuride (0.0125 mg/kg, s.c.), after a latent period of 15 min, improved markedly the parkinsonism and induced irritation, excitement, and aggression. The duration of amelioration of parkinsonism was more prolonged, and the degree of abnormal behaviors was less prominent than those with apomorphine. At a lower dosage of lisuride (0.005 mg/kg, s.c.), the parkinsonism was improved with slight abnormal behaviors for a period of more than 5 hr (Fig. 3). Proterguride (0.0025-0.01 mg/kg, s.c.) improved markedly the parkinsonism (Fig. 3). The antiparkinsonian effect continued more than 8 hr, being longer than that of lisuride. Proterguride also induced slight irritation, excitement, and aggression during the improvement of parkinsonism. Terguride (partial DA agonist, 0.01-0.1 mg/kg, s.c.) improved the parkinsonism without any sign of induction of abnormal behaviors (Fig. 3). However, the antiparkinsonian effect was weaker than that of apomorphine, lisuride, and proterguride.

DISCUSSION

L-DOPA, the direct precursor of dopamine, still is the most effective treatment of motor dysfunction in Parkinson's disease, but has been noticed to induce psychiatric side effects such as delusion, hallucination, and organic confusional states. Two subclasses of dopamine receptors (D-1 and D-2) are now recognized. Although the specific role of these receptors in the pathophysiology and pharmacotherapy of Parkinson's disease remains uncharacterized, concurrent D-1 and D-2 stimulation in the L-DOPA therapy appears to result in some side effects.

In this study, antiparkinsonian effects of various dopamine agonists, and their induction of abnormal behaviors such as irritation were investigated in MPTP-induced parkinsonian monkeys in order to clarify the role of D-1 and D-2 receptor stimulation in the induction of psychiatric side effects in Parkinson's disease. Apomorphine induced abnormal behaviors in association with amelioration of the parkinsonism. Lisuride, proterguride, and terguride improved the parkinsonism with slight induction of abnormal behaviors. These observations suggest that stimulation of D-1 receptors exaggerates side effects and that stimulation of D-2 receptors serves to ameliorate the motor dysfunction in Parkinson's disease.

The specific role of D-1 receptor stimulation in the treatment of Parkinson's disease is still controversial.[2,3] In this study, the D-1 partial agonist, SKF 38393, deteriorated slightly the parkinsonism after transient induction of irritation. This result is consistent with the previous report[2] and may be due to the partial agonistic property of this drug.

REFERENCES

1. J. W. Kebabian and D. B. Calne, Multiple receptors for dopamine, Nature 277:93 (1979).
2. M. Nomoto, P. Jenner, and C. D. Marsden, The dopamine D-2 agonist LY 141865, but not the D-1 agonist SKF 38393, reverses parkinsonism induced by 1-methyl-4-phenyl-1,2,3,6,tetrahydropyridine (MPTP) in the marmoset, Neurosci. Lett. 57:37 (1985).
3. J. A. Temlett, P. N. Chong, W. H. Oertel, P. Jenner, and C. D. Marsden, The D-1 dopamine receptor partial agonist, CY 208-243, exhibits antiparkinsonian activity in the MPTP-treated marmoset, Eur. J. Pharmacol. 156:197 (1988).
4. C. E. Clarke, M. A. Sambrook, I. J. Mitchell, and A. R. Crossman, Levodopa-induced dyskinesia and response fluctuations in primates rendered parkinsonism with MPTP, J. Neurological Sciences 78:273 (1987).
5. E. Mizuta, M. Yamaguchi, and S. Kuno, Behavioral effects of 8-hydroxy-(di-n-propylamino)tetralin (8-OH-DPAT) in monkeys, Eur. J. Pharmacol. (in press).
6. H. Wachtel, Central dopaminergic and antidopaminergic effects of ergot derivatives structurally related to lisuride, in: "Lisuride and Other Dopamine Agonists," D. B. Calne, A. Horowski, R. J. McDonald, and W. Wuttke, eds., Raven Press, New York (1983).

EFFECT OF BASIC FIBROBLAST GROWTH FACTOR ON MIDBRAIN AND OTHER CNS NEURONS IN PRIMARY CULTURE DERIVED FROM POSTNATAL AND FETAL RAT BRAIN

Susumu Matsuda, Hiroshi Saito and Nobuyoshi Nishiyama

Department of Chemical Pharmacology, Faculty of Pharmaceutical Sciences, University of Tokyo, Tokyo, Japan

INTRODUCTION

Basic fibroblast growth factor (bFGF) is a polypeptide found in adult brain tissue as a potent mitogenic protein capable of inducing cell division in a wide variety of cell type (1). Recently, bFGF has been demonstrated to increase the survival of neurons and the growth of neurites in cultures of various fetal rat brain regions (2,3,4,5).

Fetal and neonatal CNS neuron cultures have been generally adopted to investigate the properties of CNS neurons, whereas studies utilizing postnatal CNS neuron culture have been limited by technical difficulties. Postnatal CNS neuron culture can demonstrate various properties of mature neurons, and should give us important information for applications in vivo.

The present studies show that bFGF can support postnatal neuronal survival in culture.

MATERIALS AND METHODS

Fetal Neuron Culture

All neurons were dissected aseptically from embryonic day 16 Wistar rats. Tissues from 8 fetal rats were dissociated by tryptic digestion and pipetting. The dissociated cells were plated into 48-well plates coated with polylysine at a density of 2.0 - 2.5 x 10^5 cells in 500 ul of medium. After 24 hr culture in serum containing medium at 37°C, the medium was removed and changed to a serum-free medium containing of Dulbecco's modified Eagle's medium/Ham's nutrient mixture F-12 (DMEM/F-12 ; 1 : 1, vol/vol.) supplemented with transferrin (20 nM), selenite (60 nM), penicillin (50 U/ml), and streptomycin (50 U/ml) in the presence or absence bFGF. After 4 days in culture, the cells were fixed with 4 % paraformaldehyde and stained for neurofilament (Labsystem, mouse monoclonal antibody). The stained neurons were counted in about 10 % of the culture area under a phase contrast microscope.

Postnatal Neuron Culture

Cortical glial cells for a feeder layer were prepared from embryonic day 18 Wistar rats. The tissues from cortex were dissociated and plated as described above with a slight modification ; the dissected tissue after tryptic digestion was triturated vigorously through a plastic pipette with cold medium to kill and eliminate neurons, which are fragile. The plated glial cells were harvested and replated after 3 days in culture at 37°C under 5% CO_2. After a further 11 days in culture without a medium change, the glial cells were used in the studies.

All neurons were dissected aseptically from postnatal 15 day Wistar rats. The dissected tissues were immediately cut into pieces with a knife and pooled in L-15 medium cooled in ice and bubbled with 95 % O_2 and 5 % CO_2. The pooled tissues were digested by incubation at 37 C in papain (Cooper Biomed, 15 U/ml) and 0.1 % DNase solution, and triturated with a plastic pipette. After removal of the cell aggregates, single viable cells were counted under a microscope and plated into 48-well plates at a density of about 1.5×10^5 cells per 500 µl of medium (1.5×10^5 cells per 1 cm^2). The neurons were grown in DMEM/F-12 medium containing bFGF with 10 % fetal bovine serum and 1 % heat-inactivated rat serum (prepared in our laboratory). The medium with fresh bFGF was changed after 3 days. After 6 days in culture the cells were fixed with 4% paraformaldehyde and stained with antisera to neurofilament. The stained cells were counted in 5-10 % of the culture area under a phase-contrast microscope.

RESULTS AND DISCUSSION

CNS neurons from fetal rat brain (cortex, septum, striatum, midbrain, thalamus, hippocampus, colliculus, and cerebellum) were cultured to examine the effect of bFGF on neuron survival. The number of surviving neurons after 4 days in culture is shown in Fig. 1. All CNS neurons, except those of cerebellum, survived significantly better than the control after bFGF treatment. Furthermore, the effect of bFGF was dose-dependent. However, NGF did not affect the neuron survival of any CNS region tested.

CNS neurons from postnatal rat brain (cortex, septum, striatum, midbrain, hippocampus, and cerebellum) were cultured on a feeder layer of the above-mentioned glial cells. The numbers of surviving neurons after 6 days in culture are shown in Fig. 1. Survival of all CNS neurons examined, with the exception of those of cerebellum, was increased by addition of bFGF (Fig.2). The increase of postnatal neuronal survival showed clear dose dependency with bFGF treatment; 2-4 fold more than the control. At high concentrations of bFGF (more than 10 ng/ml), however, all cultured cells became fibrous, and sometimes cells peeled off together with a feeder layer (Fig.2). Especially in the case of cortex, as little as 1 ng/ml bFGF made the cells too fibrous to allow detection and counting of immunologically stained neurons. In contrast, cultured striatal and septal neurons were only slightly fibrous in the presence of 10 ng/ml of bFGF and the neurons were detectable as stained cells for neurofilament. Both striatal and septal neurons survived maximally at bFGF concentrations between 0.1-1 ng/ml. Most (80-90%) of the cultured cells from each brain region were stained immunologically with neurofilament, which indicated that the surviving cells were mostly neurons. All of the immunologically stained cells using anti-neurofilament antibody exhibited a process-bearing morphology. NGF was effective at higher doses (10 and 100 ng/ml) on striatal and septal neuron survival, but the degree of increase of surviving neurons was less than that with bFGF. We failed to culture postnatal CNS neurons without a feeder layer, with either serum-free or serum-containing medium, and no neurons remained alive even if bFGF was added (data not shown). These results indicated that a feeder layer is necessary to culture postnatal CNS neurons.

A feeder layer was cultured as a control without inoculation of postnatal cells in the presence and absence of bFGF or NGF. The feeder layer contained 1.5×10^4-1.8×10^4 neurons per cm^2 (1-2% of all cells) which would have been derived from fetal cortex. Basic FGF or NGF did not

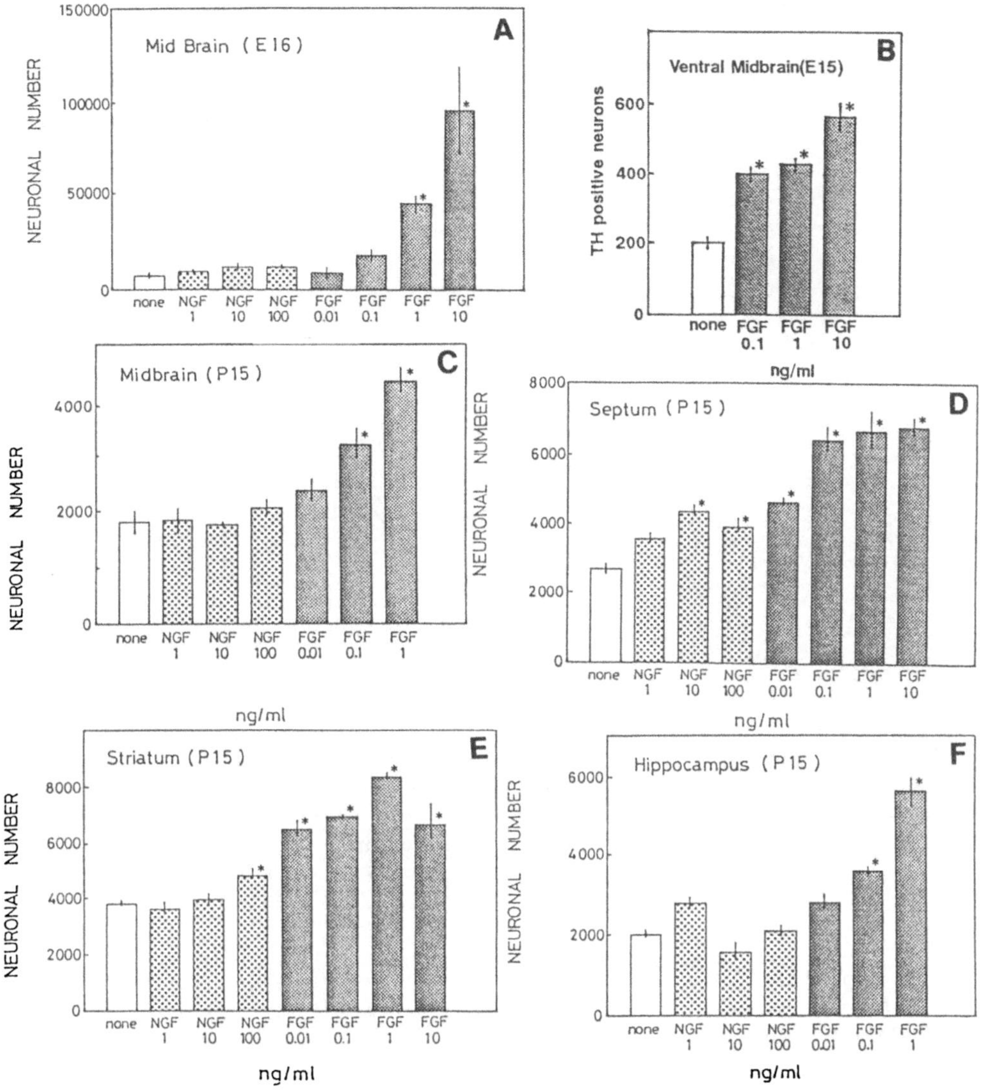

Figure 1. Typical data of neuron survival *in vitro*. Fetal (midbrain(A)) and postnatal (midbrain(C), septal(D), striatum(E) and hippocampus(F)) neurons were cultured with bFGF and NGF. Fetal ventral midbrain(B) was cultured with bFGF and stained with anti-tyrosine hydroxyrase monoclonal antibody. The data are expressed as cell per cm^2. Bars are means ± SEM for 6 cultures. *$p < 0.01$ versus control (Duncan's method)

increase the number of these contaminating neurons, or cause any morphological alterations of glial or neuronal cells. .

We cultured embryonic (E15) ventral midbrain including substantia nigral neurons with bFGF and stained with anti-tyrosine hydroxyrase monoclonal antibody. The positive stained neurons increased dose dependently with bFGF (Fig.1). These results show bFGF is also effective for increase of neuronal survival on dopaminergic neurons from embryonic substantia nigra and suggest a possibility of bFGF application for a transplantation therapy in Parkinson's disease.

These studies have indicated that bFGF is an important trophic factor for not only fetal but also postnatal neurons from midbrain and other

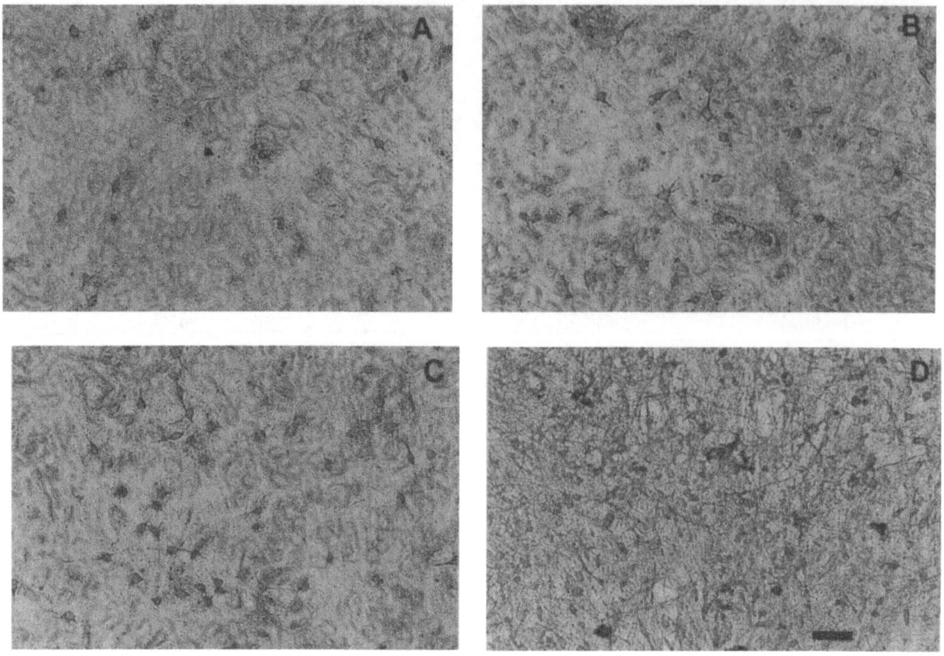

Figure 2. Phase-contrast photographs of 6-day-old neuron cultures from midbrain(A, B and C) and cortex(D). Culture in A illustrate control condition ; B, in the presence of NGF at 100 ng/ml; C and D, in the presence of bFGF at 1 ng/ml. Neurons were stained with anti-neurofilament monoclonal antibody. Fibrillar morphology of cortical culture treated with bFGF is shown in D. Bar = 10 µm.

regions of the CNS. Further, the present observations of postnatal neuron culture suggest that there is a good possibility of bFGF application in vivo.

REFERENCES

1. Baird, A., Esch, F., Mormede, P., Ueno, N. Ling, N., Bohlen, P., Ying, S-N., Whrenberg, W.B. and Guillemin, R., Molecular characterization of fibroblast growth factor: Distribution and biological activities in various tissues, Recent Prog. Horm. Res., 42(1986) 143-205.
2. Morrison, R.S., Sharma, A., Vellis, J.de, and Bradshaw, R.A., Basic fibroblast growth factor supports the survival of cerebral control neurons in primary culture, Proc. Natl. Acad. Sci. USA, 78(1986) 7205-7209.
3. Morrison, R.S., Keating, R.F. and Moskal, J.R., Basic fibroblast growth factor and epidermal growth factor exert differential trophic effects on CNS neurons, J. Neurosci. Res., 21(1988) 71-79.
4. Walicke, P.A., Cowan, W.M., Ueno, N., Baird, A. and Guillemin, R., Fibroblast growth factor promotes survival of dissociated hippocampal neurons and enhances neurite extension, Proc. Natl. Acad. Sci. USA, 83(1986) 3012-3016.
5. Walicke, P.A., Basic and acidic fibroblast growth factor have trophic effects on neurons from multiple CNS regions, J. Neurosci, 8(1988) 2618-2627.

TOWARDS A STRATEGY FOR PREVENTING PARKINSON'S DISEASE

J. William Langston

California Parkinson's Foundation
2444 Moorpark Avenue, Suite 316
San Jose, California 95128

Introduction

Will it eventually be possible to prevent Parkinson's disease? Just a few years ago this idea would have seemed beyond the realm of speculation. However, virtually all of the diagnostic and therapeutic approaches needed for such a therapeutic strategy are now being actively investigated. Interestingly, many of the ideas and observations that led to this current state of affairs resulted from the discovery of a toxin (1-methyl-4-phenyl-1,2,3,6-tetrahydropyridine, or MPTP), an event that understandably raised few if any therapeutic aspirations at the time. In this article, each of these ideas will be discussed, and an attempt will be made to show how, when considered together, they form the basis for a new therapeutic strategy for the prevention of Parkinson's disease.

Predicting Parkinson's Disease

Unless one is prepared to treat an entire population, preventing a disease by pharmacological means requires a two armed-approach. First, there must be a way to identify individuals at risk for the disease. Secondly, one must have a way to slow or halt the disease's appearance. The earliest origins of the idea that it might be possible to predict Parkinson's disease can be found in the theoretical musings of several investigators who were interested in its cause. In 1973, Finch suggested that a normal age-related decline in striatal dopamine might be a precondition for the development of Parkinson's disease (Finch, 1973). Carlsson and Winblad (1976) subsequently speculated that the combination of an early insult to the substantia nigra, combined with the age-related decline in striatal dopamine concentrations, might eventually lead to the clinical expression of Parkinson's disease. This hypothesis has since undergone many combinations and permutations (Calne and Langston, 1983; Eldridge and Ince, 1984), but all have two things in common: they depend upon the fact that an 80% depletion of striatal dopamine is needed before parkinsonism will become clinically evident (Riederer and Wuketich, 1976), and each assumes the existence of a prolonged preclinical stage of the disease.

Can individuals destined to develop Parkinson's disease be identified during this prolonged preclinical phase? Investigating this question is likely to be difficult, because there is no ready way to identify an at risk population to study. However, humans exposed to the parkinsonogenic neurotoxin MPTP may provide a unique opportunity to explore this question. In 1983, we reported a group of patients who had mistakenly used this MPTP intravenously under the impression it was a synthetic heroin (Langston et al, 1983). The neurotoxic effects of this compound faithfully replicated the features of Parkinson's disease. Patients affected by MPTP exhibited cogwheel rigidity, bradykinesia, loss of facial expression, a fixed stare, and even such subtle findings as facial seborrhea. Virtually all of them had an abnormal posture and a disturbance in voice; in fact, two of the patients were actually mute, and in three of these individuals the parkinsonism was so severe that they were

Basic, Clinical, and Therapeutic Aspects of Alzheimer's and Parkinson's Diseases, Volume 2
Edited by T. Nagatsu *et al.*, Plenum Press, New York, 1990

non-ambulatory (Ballard et al, 1985). Three exhibited a classical parkinsonian rest tremor and, in one, it was the predominant finding. Interestingly, several had a paresis of upward gaze (Ballard et al, 1985).

We were able to obtain samples of the heroin used by several of these individuals and discovered that one of these contained almost pure MPTP, suggesting that this rather simple pyridine was the likely offending agent. This was confirmed shortly thereafter by Burns et al (1983) and later by our own group (Langston et al, 1984a), both of whom showed that MPTP induces a relatively selective degeneration of neurons in the substantia nigra, thereby replicating the central neuropathologic feature of Parkinson's disease.

It is now its widely accepted that MPTP can cause severe parkinsonism in humans (and non-human primates), but it is less well known that there is a much larger group of individuals that were exposed to MPTP during 1982 who have not yet developed overt evidence of parkinsonism (Ruttenber et al, 1986). This unique cohort represents, at least to the best of this author's knowledge, the only large group of individuals with a preclinical striatal dopamine deficiency that is available for clinical investigation. Thus, they provide a unique opportunity to identify clinical markers of preclinical parkinsonism.

How does one document a preclinical dopamine deficiency in humans? Our approach was to employ positron emission tomography (PET), using 6-fluorodopa, to determine striatal dopamine content in subjects exposed to MPTP who were still asymptomatic. This work was done in collaboration with Calne and colleagues (1985) in Vancouver. The results from this study were striking, as our asymptomatic subjects demonstrated a degree of dopamine depletion intermediate between normal controls and patients with Parkinson's disease. These studies suggest humans can experience a partial depletion of striatal dopamine and remain asymptomatic, and that it is possible to identify this condition in living subjects. A particularly fascinating observation was recently made at the University of British Columbia during a study of striatal dopamine concentrations using PET scans in normal individuals of different ages. One individual was found to have a striatal dopamine concentration that was well below the rest of group, raising the possibility that a case of preclinical parkinsonism had been inadvertently identified (Donald Calne, personal communication, 1989). However, only time will tell if this speculation will prove to be correct.

While these studies demonstrate the feasibility of identifying preclinical disease, they fall far short of providing a clinical tool for screening the general population. I will return to this important issue later. However, even if PET scanning was widely available as a potential screening tool, predicting a progressive and incurable disease makes little sense unless something can be done about it. Once again, studies with MPTP have may have pointed the way out of this diagnostic and therapeutic dilemma.

Attempting to Slow the Course of Parkinson's Disease

While efforts to unravel the mechanism of action of MPTP have proceeded at a fervent pace, few would have predicted that the results of these studies would lead to a new therapeutic strategy for Parkinson's disease in such a short time. To explain this, it is necessary to briefly review the sequence of events that occur after systemic administration of MPTP. The compound is rapidly biotransformed into 1-methyl-4-phenylpyridinium (MPP+) (Langston et al, 1984b; Markey et al, 1984), which is generally thought to be the actual toxic metabolite. A critical observation, which was first made *in vitro*, was that this biotransformation was mediated by the enzyme monoamine oxidase (MAO) B (Chiba et al, 1984). This discovery quickly led a number of laboratories to test MAO inhibitors *in vivo* to determine if they would prevent the biological effects of MPTP (Heikkila et al, 1984; Markey et al, 1984; Cohen et al, 1985; Langston et al, 1984c). Without exception, each of these experiments were successful. For example, our own group found that pretreatment with the non-selective MAO inhibitor pargyline completely prevented the appearance of parkinsonism in primates after exposure to toxic doses of MPTP (Langston et al, 1984c). Postmortem studies showed that the nigral degeneration induced by MPTP was entirely prevented, as was the biotransformation of MPTP to MPP+.

While these experimental results were of interest, there was still not a strong rationale for extending these results to Parkinson's disease, as MPTP had not (and still has not) been directly linked to the idiopathic disorder. For this reason, an article that appeared in 1983 was of particular interest to us. Birkmayer and colleagues (1983) had noted that treating patients with the selective MAO B inhibitor deprenyl, which was being given as an adjunct to levodopa, had led to prolongation of L-dopa efficiency; their results also indicated that patients receiving deprenyl plus levodopa appeared to be living longer than those who had received levodopa alone. In fact, these investigators suggested that "the progress of the evolution of Parkinson's disease with long-term deprenyl treatment shows, for the first time, that the degeneration of dopaminergic nigral striatal fibers can be depressed to some extent". These investigators subsequently extended and strengthened this association in a larger series of patients in 1985 (Birkmayer et al, 1985), although they were careful to point out that their results could only be regarded as tentative, because this clinical study was open and uncontrolled. Nonetheless, these results provided an empirical rationale for testing an MAO B inhibitor in a controlled manner in Parkinson's disease, to see if the course of the illness could be favorably altered.

Finally, there is a theoretical rationale for testing an MAO inhibitor as a form of prophylactic therapy in Parkinson's disease. For every mole of dopamine oxidized in the striatum, a mole of hydrogen peroxide is generated. Of particular interest here is that fact that this reaction is mediated by MAO B. This observation has long intrigued scientists interested in mechanisms of cell death, as it has been suggested repeatedly that the generation of free radicals in the striatum might eventually overwhelm the cell's antioxidant capacity, leading to neuronal degeneration (Cohen, 1983; Graham, 1978). Further, any event that could lead to increased dopamine turnover, such as partial damage to the substantia nigra due to exposure to an environmental toxin (Calne and Langston, 1983), accentuate this process. This has been demonstrated experimentally by Cohen and Spina (1989), who provoked increased dopamine turnover in mice by administering haloperidol. The levels of oxidized glutathione subsequently tripled, indicating a marked increase in oxidative stress in this model.

These three lines of reasoning suggested that a therapeutic trial of the selective MAO B inhibitor deprenyl as a preventive agent was warranted. Therefore, in March of 1986, we began a controlled, double-blind study of deprenyl in patients with early, untreated Parkinson's disease, to see if the course of the illness could be favorably altered (Tetrud and Langston, 1987). While our study was still in the planning stages, Drs. Ira Shoulson and Stanley Fahn began organizing a much larger clinical trial that was to include not only deprenyl, but the free radical scavenger alpha-tocopherol (Lewin, 1985). This study is know by the acronym DATATOP (Deprenyl and Alpha-Tocopherol Antioxidant Therapy of Parkinsonism). Our study was completed in March of 1989; the DATATOP study was initiated in the late summer of 1987 and still continues, although the results with deprenyl were so significant as to necessitate a premature breaking of the code for that arm of the study in 1989 (Parkinson Study Group, 1989).

Both studies were highly analogous in design; in fact, our study in many ways became a pilot study for the much larger DATATOP study. We chose to use deprenyl, rather than other MAO inhibitors, for three reasons. First, as a selective MAO B inhibitor, it appears to be a safer drug than non-selective inhibitors, which can cause severe hypertension when subjects are exposed to large dietary loads of tyramine (the so-called "cheese effect") (Elsworth et al, 1978). Second, the experimental work with MPTP suggested that MAO B was the most important form of the enzyme to block. Third, deprenyl was the drug that appeared to have been effective in patients with the idiopathic disease (Birkmayer et al, 1983, 1985).

Inclusion criteria were quite similar in both trials and included the following: (1) age range from 30 to 80 years; (2) the absence of features suggesting an atypical form of parkinsonism; (3) a disease duration of less than 5 years (our rationale was that the earlier the disease was treated, the more cells there would be to protect and thus a preventive effect could more easily be detected); and (4) subjects could not be receiving any antiparkinsonian medication. There were two reasons for this last requirement. First, symptomatic treatment with antiparkinsonian drugs would immensely confound the interpretation of the results; second, there was the concern that levodopa, by accentuating dopamine turnover, might actually be harmful to the remaining nigrostriatal neurons. Finally, patients were excluded from the study if there were major medical conditions or psychiatric disease.

Two methods were used to determine the effects of deprenyl on disease progression. The first was to measure time elapsed until the patient required levodopa therapy. This was defined as the major endpoint of both studies, and was chosen because it is a finite point, and one with which all neurologists are familiar. In order to apply this end-point in a uniform manner, well-defined criteria for the initiation of levodopa therapy were used; these included threatened loss or significant change in employability, threatened inability to handle domestic responsibilities, postural stability score of 2 or greater, a gait score of 3 more on the Unified Parkinson's Disease Rating Scale (UPDRS) motor rating scale, or a rating on the Hoehn and Yahr (Hoehn and Yahr, 1967) scale of Stage III or more. In essence, this approach can be viewed as a survival analysis, designed to determine how long patients survived until they needed levodopa. Second, in order to assess the rate of clinical progression, a number of assessment scales were employed, including the UPDRS, the Schwab and England Activities of Daily Living (ADL) scale, Hoehn and Yahr staging, and the Webster step test. Similar rating scales were employed in DATATOP as well.

In our study (Tetrud and Langston, 1989), 54 patients were enrolled over an 18-month period, 27 in the deprenyl group and 27 in the placebo group. Baseline characteristics were quite similar in both groups; ages were highly comparable, as was the age at time of diagnosis and the duration of symptoms at the time of enrollment (approximately 2 years in each group). Five of the patients in the placebo group were under the age of 55 and over the age of 65 as compared to 6 and 7 respectively for the deprenyl group.

Our results showed that deprenyl significantly prolonged the period until patients needed treatment with levodopa: the time to endpoint in the placebo group was 312 days, whereas the patients in the deprenyl group went almost twice as long, 549 days, before levodopa was required ($P > 0.002$). This result is reminiscent of the survival data presented in the second paper of Birkmayer and colleagues (1985), in which they showed that the probability of survival was higher in the deprenyl-treated group as opposed to a group of patients not treated with deprenyl (both groups were on levodopa therapy). Interestingly, this effect was more prominent in younger patients in their study, suggesting that the earlier the drug is started, the more pronounced the effect may be. The clinical monitors of disease status in our study indicated that deprenyl had an equally striking effect on rate on the course of illness, which was slowed by approximately 50%.

A major potential criticism of this study is the possibility that deprenyl might have simply been exerting a symptomatic effect. If this were case, the drug would be little different from drugs such as amantadine or trihexyphenidyl, which can be used early to delay the need for levodopa. Such an interpretation would be quite different from an effect on the underlying disease process. To exclude this possibility, we carried out two specific evaluations (this was also done in DATATOP). First, patients were carefully re-evaluated one month after starting deprenyl to see if this drug was indeed producing a symptomatic effect and none was found. However, this "wash-in" study did not exclude the possibility that such an effect might have developed slowly, perhaps over a period of three to six months, in which case it would have appeared that the pace of the disease was slowing, when in fact we were actually seeing a slowly developing symptomatic effect. To assess this possibility, when patients reached endpoint, the study drug was stopped completely for one month ("wash-out") to see if there was any change in clinical status. Had deprenyl been exerting a symptomatic effect at that point, one would have expected a deterioration in clinical status. Importantly, there was no deterioration in the deprenyl (or placebo) group, even though the deprenyl-treated patients had survived almost twice as long without levodopa as the placebo group. We believe that this "wash-out" study convincingly rules out the possibility that the results of this study were simply due to a symptomatic effect of deprenyl.

It is worth noting that a 50% reduction in rate of progression, if sustained, would have a major impact on functional status over time. For example, extrapolating our data out for 10 years, a patient on deprenyl would progress to Hoehn and Yahr stage 2.5 (relative mild disability) (Hoehn and Yahr, 1967), whereas a placebo-treated patient would have reached stage 4.5 to 5 (severe disability). While this is speculative, based on the assumption of a linear progression of disease in both groups, it nonetheless demonstrates the potential impact of even a partial slowing of disease progression.

There are two potentially serious criticisms of our study. First, the number of patients was rather small. For this reason, the results of DATATOP are crucial. Twenty-eight institutions participated in this much larger clinical trial, which successfully enrolled 800 patients. In addition to testing deprenyl, DATATOP included the free radical scavenger alpha-tocopherol. The design of the study was as follows: 25% of the patients received deprenyl, 25% tocopherol, 25% deprenyl and tocopherol, and 25% placebo. As noted earlier, because the results were highly significant, the deprenyl arm of DATATOP was recently reported (Parkinson Study Group, 1989). During the first 12 months of the study, 176 patients in the placebo group reached end-point, as compared to 97 patients in the deprenyl-treated group; thus the risk of reaching end-point in patients receiving deprenyl was reduced by 57% ($p<10^{-10}$). Hence, there is virtually no doubt that early use of deprenyl significantly delays the need for levodopa treatment.

However, neither DATATOP nor our smaller study can be used to answer the most critical question: Is deprenyl truly slowing the process of nigral cell degeneration? To answer this question unequivocally, a method of determining nigral cell counts during life is needed, and of course there is no way of doing this at the present time. On the other hand, should a pharmacologic means eventually be found that stops progression entirely, these doubts would likely be put to rest. But, for the time being, debate is likely to continue on this crucial question.

A Strategy for Prevention

In this final section, an attempt will be made to assemble a coherent strategy for prevention of Parkinson's disease. We have already demonstrated humans can remain asymptomatic in spite of a partial striatal dopamine depletion, and that this dopamine depletion can be demonstrated during life (Calne et al, 1985). Thus, it should be possible to identify patients with pre-symptomatic Parkinson's disease. Now that there is evidence that chronic MAO inhibition slows the course of the illness, the obvious strategy is to identify individuals with pre-clinical disease, and start them on prophylactic therapy as soon as possible. Current data suggest that we should at least be able to retard the clinical appearance of the disease. This could be very significant if the effects of deprenyl are increasingly pronounced the earlier that it is started.

As attractive as this strategy may sound, however, there are still several major stumbling blocks to the practical initiation of this approach. Most importantly, PET scanning still is far from becoming a screening tool. It seems more likely, therefore, that it will be necessary to develop a two-tiered screening process. The goal would be to find a sensitive, but perhaps less specific, marker of preclinical disease that could be administered to at-risk individuals (i.e., those over the age of 40 or 50 on a screening basis). A positive result on this hypothetical screening test might then lead to a PET scan. What types of screen tests might be developed? It seems to me there are many possibilities. We have found a relatively specific neuropsychological profile in patients with very mild MPTP-induced parkinsonism (Stern et al, 1990), so one screening battery might be an easily administered psychometric examination. Another alternative may lie within the realm of neurophysiology. Perhaps long-latency reflexes or some other aspect of neurophysiologic function will be found to antedate the onset of clinical symptomatology, much as visual evoked responses can be used to detect a subclinical optic nerve lesion.

How does one go about the search for markers of preclinical striatal dopamine depletion? First, it is necessary to prospectively identify a population of individuals who are so affected. Since there are few if any such populations with preclinical Parkinson's disease that have been identified, the large cohort of asymptomatic MPTP-exposed individuals mentioned earlier may prove crucial for the study of this question. Indeed, this group, most of whom almost certainly have a subclinical striatal dopamine deficiency, represent a veritable "gold mine" for the search for preclinical markers.

A final, and potentially exciting, possibility relates to another recent finding associated with MPTP. The putative toxic metabolite of MPTP, MPP+, appears to be a potent mitochondrial toxin, affecting site I of mitochondrial respiration (Kindt et al, 1986). While several investigators have recently described a similar defect in mitochondria from the substantia nigra (Schapira et al, 1989) and striatum (Mizuno et al, 1989) of patients who died with Parkinson's disease, Parker et al (1989) have reported similar abnormalities in mitochondrial function in platelets of patients with Parkinson's disease. This latter observation, although limited to only 10 patients, and therefore in need of confirmation, raises the hope of a true biomarker for Parkinson's disease, something which could provide a screening test for the disorder. Obviously, this would tremendously aid the search for a way to identify patients at risk for Parkinson's disease, and would represent a major step on the road to the development of a preventive strategy for Parkinson's disease. While some, if not all, of the concepts reviewed in this article may sound futuristic, it is the author's opinion that all of the elements are there to vigorously pursue, and it seems not impossible that we will eventually be able to eradicate this disease, perhaps in our lifetime.

REFERENCES

Ballard P.A., Tetrud J.W., Langston J.W., 1985, Permanent parkinsonism in humans due to 1-methyl-4-phenyl-1,2,3,6-tetrahydropyridine (MPTP): Seven cases. Neurology, 35: 949-956.

Birkmayer W., Knoll J., Riederer P., Youdim M.B.H., 1983, (-)-Deprenyl leads to prolongation of L-dopa efficacy in Parkinson's disease. Mod Probl Pharmacopsychiat, 19: 170-176.

Birkmayer W., Knoll J., Riederer P., Youdim M.B.H., Hars V., Marton J., 1985, Increased life expectancy resulting from addition of L-Deprenyl to Madopar treatment in Parkinson's disease: A longterm study. J Neural Trans, 64: 113-127.

Burns R.S., Chiueh C.C., Markey S.P., Ebert M.H., Jacobowitz D. M., and Kopin I. J., 1983, A primate model of parkinsonism: selective destruction of dopaminergic neurons in the pars compacta of the substantia nigra by n-methyl-4-phenyl-1,2,3,6-tetrahydropyridine. Proc Natl Acad Sci USA, 80: 4546-4550.

Calne D.B., Langston J.W., 1983, Aetiology of Parkinson's disease. Lancet, 2: 1457-1459.

Calne D.B., Langston J.W., Martin W.R.W., Stoessel A.J., Ruth T.J., Adam M.J., Pate B.D., Schulzer M., 1985, Positron emission tomography after MPTP: Observations relating to the cause of Parkinson's disease. Nature, 317: 246-248.

Carlsson A., Winblad B., 1976, Influence of age and time interval between death and autopsy on dopamine and 3-methyoxytyramine levels in human basal ganglia. J Neural Transm, 38:271-276.

Chiba K., Trevor A., Castagnoli Jr. N., 1984, Metabolism of the neurotoxic tertiary amine, MPTP, by brain monoamine oxidase. Biochem Biophys Res Commun, 120: 574-578.

Cohen G., 1983, The pathobiology of Parkinson's disease: Biochemical aspects of dopamine neuron senescence. J Neural Transm., Suppl. 19: 89-103.

Cohen G., Pasik P., Cohen B., Leist A., Mytilineou C., and Yahr M.D., 1985, Pargyline and deprenyl prevent the neurotoxicity of 1-methyl-4-phenyl-1,2,3,6-tetrahydropyridine (MPTP) in monkeys. Eur J Pharmacol, 106: 209-210.

Cohen G., Spina M.B., 1989, Deprenyl suppresses the oxidant stress associated with increased dopamine turnover. Annals of Neurology, 26: 689-690.

Eldridge R., Ince S.E., 1984, The low concordance rate for Parkinson's disease in twins. Neurology, 34: 1354-1356.

Elsworth J.D., Glover V., Reynolds G.P., Sandler M., Lees A. J., Phuapradit P., Shaw K.M., Stern G.M., Kumar P., 1978, Deprenyl administration in man: A selective monoamine oxidase B inhibitor without the "cheese effect". Psychopharmacology, 57: 33-38.

Finch C.E., 1973, Catecholamine metabolism in the brains of aging male mice. Br Res, 52: 261-276.

Graham D. G., 1978, Oxidative pathways for catecholamines in the genesis of neuromelanin and cytotoxic quinones. Mol Pharmacol, 14: 633-643.

Heikkila R.E., Manzino L., Cabbat F.S., and Duvoisin R.C., 1984, Protection against the dopaminergic neurotoxicity of 1-methyl-4-phenyl-1,2,5,6-tetrahydropyridine by monoamine oxidase inhibitors. Nature, 311: 467-469.

Hoehn M.M., Yahr M.D., 1967, Parkinsonism: onset, progression and mortality. Neurology, 17: 427-442.

Kindt M.V., Nicklas W.J., Sonsalla P.S., Heikkila R.E., 1986, Mitochondria and the neurotoxicity of MPTP. Trends in Pharmacological Sciences, 7: 473-475.

Langston J.W., Ballard P., Tetrud J.W., Irwin I., 1983, Chronic parkinsonism in humans due to a product of meperidine-analog synthesis. Science 219: 979-980.

Langston J.W., Forno L.S., Rebert C.S. and Irwin I., 1984a, Selective nigral toxicity after systemic administration of 1-methyl-4-phenyl-1,2,3,6-tetrahydropyridine (MPTP) in the squirrel monkey. Brain Res, 292: 390-394.

Langston J.W., Irwin I., Langston E.B. and Forno L.S., 1984b, 1-Methyl-4-phenylpyridinium ion (MPP^+): Identification of a metabolite of MPTP, a toxin selective to the substantia nigra. Neurosci Lett, 48: 87-92.

Langston J.W., Irwin I., Langston E.B. and Forno L.S., 1984c, Pargyline prevents MPTP-induced parkinsonism in primates. Science, 225: 1480-1482.

Lewin R., 1985, Clinical trial for Parkinson's disease? Science, 230: 527-528.

Markey S.P., Johannessen J.N., Chiueh C.C., Burns, R.S., Herkenham M.A., 1984, Intraneuronal generation of a pyridinium metabolite may cause drug-induced parkinsonism. Nature, 311: 464-467.

Mizuno, Y., Ohta, S., Tanaka, M., Takamiya, S., Suzuki, K., Sato, T., Oya, H., Ozawa, T., Kagawa, Y., 1989, Deficiencies in Complex I subunits of the respiratory chain in Parkinson's disease. Biochem Biophys Res Comm, 163(3): 1450-1455.

Parker, W.D. Jr., Boyson, S.J., Parks, J.K., 1989, Abnormalities of the electron transport chain in idiopathic Parkinson's disease. Ann Neurol, 26: 719-723.

Riederer, P., and St. Wuketich, S., 1976, Time course of nigrostriatal degeneration in Parkinson's disease. J Neural Trans, 38: 277-301.

Ruttenber A.J., Garbe P.L., Kalter H.D., Castro K.G., Tetrud J.W., Porter P., Irwin I., Langston J.W., 1986, Meperidine analog exposure in California narcotics abusers: Initial epidemiologic findings. In: *MPTP: A neurotoxin producing a parkinsonian syndrome*, Markey SP, Castagnoli Jr. N, Trevor AJ, Kopin IJ (eds.). Academic Press, New York, 339-353.

Schapira, A.V.H., Cooper, J.M., Dexter, D., Jenner, P., Clark, J.B., Marsden, C.D., 1989, Mitochondrial Complex I deficiency in Parkinson's disease. Lancet, 1: 1269.

Shoulson I., Fahn S., Oakes D., Odoroff C., Lang A., Langston J.W., LeWitt P., Olanow W., Penney J.B., Tanner C. and the Parkinson's Study group, 1989, Deprenyl delays disability in early Parkinson's disease: A controlled clinical trial. N Eng J Med, 321: 1364-1371.

Stern Y., Tetrud J.W., Martin W.R.W., Kutner S.J., Langston J.W., 1990, Cognitive change following MPTP exposure. Neurology, 40: 261-264.

Tetrud J.W., Langston J.W., 1987, L-deprenyl as a possible protective agent in Parkinson's disease. J Neural Trans, Suppl 25: 1-20.

Tetrud J.W., Langston J.W., 1989, The effect of deprenyl (selegiline) on the natural history of Parkinson's disease, Science, 245: 519-522.

Langston J.W., Ballard P., Tetrud J.W., Irwin I., 1983. Chronic parkinsonism in humans due to a product of meperidine-analog synthesis. Science 219, 979-980.

Langston J.W., Forno L.S., Rebert C.S. and Irwin I., 1984a. Selective nigral toxicity after systemic administration of 1-methyl-4-phenyl-1,2,3,6-tetrahydropyridine (MPTP) in the squirrel monkey. Brain Res. 292, 390-394.

Langston J.W., Irwin I., Langston E.B. and Forno L.S., 1984b. 1-Methyl-4-phenylpyridinium ion (MPP+): identification of a metabolite of MPTP, a toxin selective to the substantia nigra. Neurosci. Lett. 48, 87-92.

TETRAHYDROBIOPTERIN THERAPY FOR JUVENILE PARKINSONISM

Tomoyoshi Kondo[1], Yoshiaki Furukawa[1], Hideto Miwa[1], Yoshikuni Mizuno[1] and Hirotaro Narabayashi[2]

1: Department of Neurology, Juntendo University School of Medicine and 2: Neurological Clinic, Tokyo, Japan

INTRODUCTION

6(R)-L-erythro-5,6,7,8-tetrahydrobiopterin (R-THBP) is one of the rate-limiting factors of tyrosine hydroxylase (TH), which is the key enzyme in the formation of dopamine (DA) in the brain. In one animal experiment, TH activity in the dopaminergic neurons was reported to be activated by R-THBP administration (Kettler et al., 1974). Another report also has suggested that enhanced release of DA from the nerve terminals results from R-THBP administration (Koshimura et al., 1989). Although TH activity is decreased in the parkinsonian brain, it is conceivable that the remaining TH activity may be enhanced by the administration of R-THBP. Several clinical trials of R-THBP for parkinsonism were examined, but those results were conflicting (Moore et al., 1987; Narabayashi et al., 1982). In this paper, we examined the pharmacokinetics and clinical efficacy of R-THBP in order to evaluate its usefulness for the treatment of patients with juvenile parkinsonism.

SUBJECTS and METHODS

Six patients with juvenile parkinsonism (Yokochi, 1979) were examined in this study. Two of them (Case 1 and 2) had never received levo-dopa or DA agonist therapy. In the other four patients, all medicines were withdrawn at least seven days prior to examination.

Total biopterin (T-BP) in the plasma was measured in five patients after they had received a single dose of 800 mg of R-THBP. Blood samples were taken before R-THBP administration and afterwards every hour for the first 4 hours, then at 6, 8, and 12 hours. Three of the six patients were examined to determine whether the substance accumulated in the plasma during administration of 800-1000 mg daily dosage lasting from three to seven days. Plasma T-BP was measured during daily administration, in the morning and 4 hours after taking the medicine, and then once again in the morning three days after cessation of administration. Plasma R-THBP was measured as T-BP by the method of Fukushima and Nixon (1980), with slight modification.

Clinical efficacy was evaluated by the scoring of clinical symptoms. Akinesia, rigidity, tremor, gait disturbance, and pulsion were graded from zero to 4 (zero: no symptoms; 1: slight symptoms; 2: moderate symptoms; 3: intermediate grade between moderate and severe; 4: severe). Gait disturbance was monitored by counting of steps and the time needed to walk twice back and forth five meters. The time needed for rolling over and standing up as well as the size of written letters were also used as indices of parkinsonism. Mechanical movement time was measured in one patient (Case 2) by use of two separated on-off switches.

One patient with slight parkinsonism (Case 1; Yahr's Stage I) had been treated with a daily 1000 mg dosage of R-THBP for eleven months.

RESULTS

Plasma T-BP levels after a single 800-mg dose of R-THBP: Basal levels of T-BP in plasma were 3.0 to 19.5 pmoles/ml. Plasma T-BP levels peaked between 2 to 4 hours after administration. Peak levels of T-BP in each case varied from 127.0 to 943.3 pmoles/ml. The high levels of T-BP, which were at least ten times higher than those prior to administration, persisted for 6 hours.

Accumulation of R-THBP in the plasma: When 800 mg of R-THBP was administered, half in the morning and half at night, T-BP levels in the morning prior to administration showed a tendency to accumulate for the first two days but, thereafter, T-BP levels in the plasma became stabilized. From the second day after stopping administration of R-THBP, T-BP levels returned to former levels.

Clinical efficacy: Changes in the score of each symptom are shown in Table 1. Administration of 800 mg of R-THBP resulted in a slight decrease in the scores of five of the six patients. Although the changes in the scores were small, improvements in the levodopa naive subjects (Case 1 and 2) were significant. Standing up from the supine position prior to taking R-THBP was impossible, but the time necessary to do so after R-THBP administration in Case 2 became significantly short. Movement time was also improved in Case 2 for 6 hours (Figure 1). Similar improvement was seen in Case 3. The main symptoms of Case 1 consisted of rigidity in his extremities, especially in the right hand; by R-THBP administration, stretch reflex recorded by EMG and repetitive finger movement evaluated by a hand-held number counter were improved (Figure 2). These improvements were sufficient enough to allow for daily work. Treatment with a 1000-mg daily dose of R-THBP for the past eleven months has resulted in continuing efficacy without any side effects. Case 6 did not show any improvement.

Discussion

An 800-mg dose of R-THBP increased T-BP levels in the plasma to at least 10 times higher that those prior to administration and these increased levels persisted for 6 hours. As shown by the improvements in the symptoms of Case 2, the clinical efficacy continued for a similar duration. Clinical improvements in 5 out of the 6 cases were remarkable but not so dramatic. In patients with severe symptoms, therefore, R-THBP therapy may not be effective enough to result in the performing of daily activities.

Of course, although levodopa has proven to be the most effective treatment for parkinsonism, recent clinical results have pointed out the adverse effects such as severe wearing-off and dyskinesia in juvenile parkinsonism related to the use of the drug. Postponing its administration during the course of therapy and/or reducing its dosage when used in combination with other agents have recently become accepted practices.

Our results indicate that R-THBP produces mild but significant effects in the early stage of juvenile parkinsonism and, when compared with starting therapy with levodopa, may lead to a more favorable prognosis.

Table 1. Changes in the Score of Each Symptom

	Case 1	Case 2	Case 3	Case 4	Case 5	Case 6
akinesia	0→0	3→2	3→2	1→1	2→2	4→4
rigidity	2→1	3→2	4→3	2→1	2→1	4→4
tremor	0→0	0→0	1→1	2→2	2→2	2→2
gait dist.	1→1	3→2	2→2	2→2	2→2	4→4
pulsion	0→0	3→3	2→2	1→1	1→1	4→4
rolling over	0→0	4→2	2→2	2→1	2→2	4→4

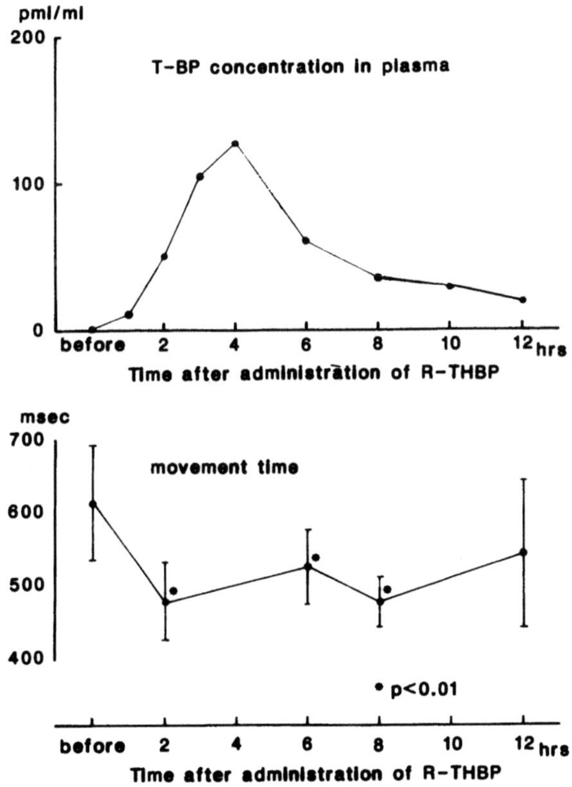

Fig. 1. T-BP concentration in plasma and mechanical movement time in Case 2. The mechanical movement time became shorter during the same period in which T-BP plasma levels were elevated.

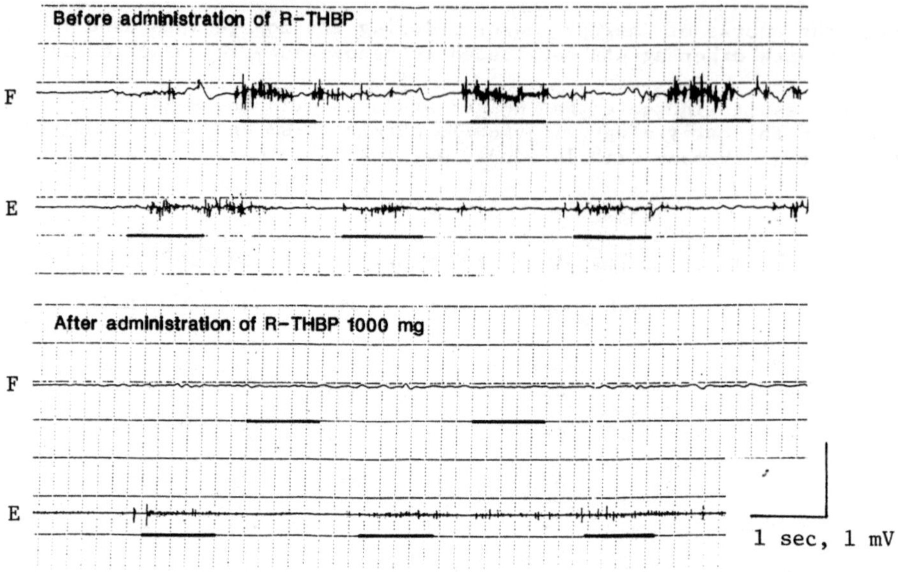

Fig. 2. EMG record for Case 1. Discharge levels of stretch reflex in wrist flexor (F) and wrist extensor (E) muscles were reduced during a 1000-mg daily dosage of R-THBP.

REFERENCES

Kettler, R., Bartholini, G., and Pletcher, A., 1974, In vivo enhancement of tyrosine hydroxylation in rat striatum by tetrahydrobiopterin, Nature 249:476.

Koshimura, K., Miwa, S., Lee, K., Fujiwara, M., and Watanabe, Y., 1989, Dopamine releasing action of natural form of tetrahydrobiopterin in vivo.--an examination by microdialysis, Abstract of the Thirtieth Congress of Societas Neurologica Japonica, 230.

Fukushima, T. and Nixon, J. C., 1980, Analysis of reduced forms of biopterin in biological tissues and fluids, Anal. Chem., 102:176.

Moore, A. P., Behan, P. O., Jacobson, W., and Armarego, W. L., 1987, Biopterin in Parkinson's disease, J. Neurol. Neurosurg. Psychiat, 50:85.

Narabayashi, H., Kondo, T., Nagatsu, T., Sugimoto, T., and Matsuura, S., 1982, Tetrahydrobiopterin administration for parkinsonian symptoms, Proc. Jpn. Acad., 58(B):283,.

Yokochi, M., 1989, Juvenile Parkinson's disease, Pt. 1. Clinical aspects, Adv. Neurol. Sci., 23:1048.

APPLICATION OF A LOW-PROTEIN DIET FOR PARKINSON'S DISEASE

Alain Destee*, Luc Defebvre*, and
Monique Rousseaux-Romon**

Departments of *Neurology and **Endocrinology
Chu Lille, France

After 10 years of L-Dopa therapy more than 80% of Parkinsonian[1-2] patients experience motor fluctuations which are responsible of an important handicap : end of dose akinesia, loss of efficacy, "on-off" effect, onset and end-of-dose dyskinesias, peak-dose dyskinesias.

For some patients some fluctuations can be improved even eliminated by the continuous intravenous administration of levodopa[3] suggesting that they may be due to fluctuations of levodopa delivery to the central nervous system.

As levodopa compete with certain neutral amino acids not only at the intestinal absorption's level but also for the transport across the blood-brain barrier[4], it has been supposed that some fluctuations mainly those observed after eating could be improved by dietary protein modification[5-6].

METHODS

Subjects

Eleven adults referred to our department with idiopathic Parkinson's disease (PD) participated in this clinical investigation after giving informed consent.

They were experiencing motor fluctuations which had not responded to readjustement in L-dopa dosage, agonists, adjuvants.

There were 6 women and 5 men, ranging from 43 to 82 years of age, with a mean age of 57 years. They had PD of 5 to 25 years' duration, with a mean of 13,5 years. The age of onset for PD ranged between 38 and 66 years with a mean age of onset of 48,9 years.

At the time of the clinical investigation all the patients received levodopa (range : 300-1150 mg/d ; mean : 490 mg) with benserazide ; ten patients received bromocriptine (range 7,5 to 40 mg/d ; mean : 23 mg) and the eleventh received trihexyphenidyl (10 mg/d).

The fluctuations were entirely predictable (end-of-dose akinesia) in three patients, predictable (end-of-dose akinesia : 5 ; end-of-dose dystonia : 2) and/or unpredictable in 8 ; there was loss of efficacy in one patient.

Four patients presented with peak-of-dose dyskinesia.

Protocol

All the subjects were first admitted as in-patients for objective evaluation. During the first 2 days they were on a regular hospital diet ; then a "low-protein diet" was applied with 15g of only vegetable protein before 4 p.m. with high-protein lunch and dinner. Diets were eucaloric with lipid and carbohydrate isocaloric intakes distributed between the 3 meals and 2 lunches (10 a.m. and 4 p.m.).

During these two periods, each day, the patients filled in a fluctuation chart, which consisted of a record of "on" or "off" and the occurence of dyskinesia and/or dystonia or tremor every hour (between 7 a.m. and 10 p.m.). All the patients were also evaluated by the same physician according to the Unified Parkinson's Disease Rating Scale[7] (part I+II+III ; max : 124) during "on" and "off" states. The patients were examined, always by the same physician, 15 days later (as out-patients) then 42 days later (as in- patients).

The clinical evaluation of PD (C.E.P.D) was quantified during regular hospital diet (D_0) during the first low-protein diet day (D_3) and during the 42th day (D_{42}) by an equation :

$$CEPD = \frac{\text{UPDRS "on"}}{124} \times \text{"on" d} + \frac{\text{UPDRS "off"}}{124} \times \text{"off" d}$$

where : "on" d = percentage of "on" state through the day
"off" d = percentage of "off" state through the day.

During the study the treatment was not modified.

Amino acid analyses

At D_0, D_3 and D_{42} plasma amino acids were measured every hour first between 8 a.m. and 4 p.m., secondly between 8 p.m. and 12 p.m.

RESULTS

Ten of the 11 patients completed the study : one female patient (case n°5) gave up the low-protein diet at the end of the first week.

CEPD appears on table.

The low-protein diet was well tolerated and only one female patient (case n°9) lost 3 kg between D_0 and D_{42} but she was simultaneously on a low-caloric diet....

For the 10 patients who completed the study there was a clinical improvement sometimes "dramatic" with a CEPD diminution between 26 and 78% (mean : 50%). The AIM duration rised only from 16,4% to 18%.

For each patient the clinical improvement has been compared to the modification of plasma amino acid concentration. At D_3 there is a clear correlation between clinical improvement and the lowering of 4 plasma amino-acid concentration : valine, isoleucine, leucine and phenylalanine.

At D_{42} it was necessary to lower levodopa dosage in 4 patients.

COMMENTS

The levodopa's intestinal absorption or transport across the blood-brain barrier is upon the dependance of the 2L-transport system ; so at these levels there is a competition between levodopa and large neutral amino-acids.

In 1975 Mena and Cotzias[5] suggested a low-protein diet to improve the levodopa efficacy.

Table 1

CEPD before and during low-protein diet

Cas N°	CEPD D_0	CEPD D_3	CEPD D_{42}	E1	E2
1	0,28	0,15	0,095	46%	66%
2	0,26	0,077	0,061	70%	76%
3	0,45	0,24	0,31	46%	31%
4	0,73	0,35	0,46	52%	36%
5	0,73	0,54		26%	
6	0,66	0,39	0,41	41%	38%
7	0,17	0,05	0,083	70%	51%
8	0,23	0,10	0,17	56%	26%
9	0,098	0,020	0,021	79%	78%
10	0,16	0,10	0,085	37%	46%
11	0,51	0,19	0,22	62%	56%

Legend : CEPD = "Clinical Evaluation of Parkinson Disease"
E1 = variation of CEPD between D_3 and D_0

E2 = variation of CEPD between D_{42} and D_0

Since it was well established that the dietary protein affects the efficacy of levodopa at a central level[8,9,10]. Our results are in good agreement with those of others[9,10]. As Riley[9] we think that low-protein diet is inexpensive, free of significant morbidity and quickly active and so has to be used for some patients.

REFERENCES

1. MARSDEN C.D., PARKES J.D.: Success and problems of long-term levodopa therapy in Parkinson's disease. Lancet. 1977; i: 345-349.
2. RINNE U.: Problems associated with long-term levodopa treatment of Parkinson's disease. Acta Neurol. Scand. 1983; 95:19-26.
3. QUINN N., MARSDEN C.D., PARKES J.D.: Complicated response fluctuations in Parkinson's disease: response to intravenous infusion of levodopa. Lancet. 1982; ii:412-415.
4. WADE L.A., KATZMAN R.: Synthetic amino acids and the nature of L-dopa transport at the blood brain barrier. J. Neurochem. 1975; 25:831-842.
5. MENA J., COTZIAS G.: Protein intake and treatment of Parkinson's disease with levodopa. N. Engl. J. Med. 1975; 292:181-184.
6. PINCUS J.H., BARRY K.: Influence of dietary protein on motor fluctuations in Parkinson's disease. Arch. Neurol. 1987; 44: 270-272.

7. FAHN S., ELTON R.L. and members of the UPDRS Development Committee. Unified Parkinson's Disease Rating Scale. in: "Recent developments in Parkinson's disease".
FAHN S., MARSDEN C.D., CALNE D., GOLDSTEIN M. Edts., 1987 ; pp 153-163. Mac Millan Healthcare Information, USA.
8. NUTT J.G., WOODWARD W.R., et al : The "on-off" phenomenon in Parkinson's disease.
N. Engl. J. Med. 1984 ; 310 : 483-488.
9. RILEY D., LANG A.E. : Practical application of a low protein diet for Parkinson's disease.
Neurology. 1988 ; 38 : 1026-1031.
10. TSUI J.K., ROSS S., et al : The effect of dietary protein on the efficacy of L-dopa : a double-blind study.
Neurology. 1989 ; 39 : 549-552.

MARPOTILINE THERAPY FOR FREEZING OF PARKINSON'S DISEASE

Shinichi Iwasaki, Yohsuke Narabayashi, Mamoru Takakusagi
Akira Iwasaki, and Katsuhiko Hamaguchi

Department of Neurology, Saitama Medical School
Moroyama-machi, Saitama-ken 350-04, Japan

INTRODUCTION

Freezing phenomenon is one of major disabling symptoms in Parkinson's disease. Recently, L-threo-DOPS, a norepinephrine precursor, has been reported by H. Narabayashi[1,2] to relieve freezing in parkinsonian patients. Thus, central nervous norepinephrine (NE) depletion is suggested to underlie the pathomechanism of freezing. Many studies,[3,4] however, failed to demonstrate an increase in concentration of NE and its metabolites in cerebrospinal fluid following administration of L-threo-DOPS. Hence, it is not clear whether the favorable effect of L-threo-DOPS on freezing is produced through formation of central NE or not. We investigated efficacy of maprotiline, a potent inhibitor of NE reuptake,[5] in freezing to know whether replenishing of central NE improves freezing in Parkinson's disease.

METHODS AND PATIENTS

Maprotiline was administered orally at a dose of 25 mg given 2 times per day. The dose was increased and maintained to 75 mg daily from the 3rd week to the 6th week. Other therapeutic regimens involving anti-parkinsonian agents was unchanged during the trial. Using self-assessment, neurological examinations, and electromyographic examinations of lower limbs of walking patients, the patients were evaluated for freezing in gait (frozen gait), speech (stuttering), writing (micrographia) before and during the 6-week trial. They were also evaluated for other symptoms such as akinesia, rigidity, tremor, postural reflex, and side effects. Freezing and other symptoms were graded according to H. Narabayashi[2] as follows: most severe, severe, moderate, slight, none. Improvement was evaluated by the change of severity as follows: marked, moderate, slight, none.

Four outpatients, 3 males and 1 female, with Parkinson's disease were studied (Table 1). Most of them were over 60 years old and duration of the disease was more than 5 years except for Case 2. They were at stage III or IV of the Hoehn-Yahr scale and were on levodopa with a peripheral decarboxylase inhibitor (DCI) or bromocriptine. Daily dose of levodopa with DCI ranged from 300 mg to 400 mg. Duration of levodopa with DCI was more than 5 years. Dyskinesia was observed in Case 4 (off-period dystonia). No wearing-off or on-off phenomenon was seen in the 4 cases, nor was dementia noted.

Table 1. Patients

Case	Age/Sex	Duration of PD (years)	Hoehn-Yahr scale	Anti-parkinsonian drugs (mg)	Freezing		
					Gait	Speech	Writing
1) M.H.	71/M	10	IV	L-dopa/carbidopa 400/40	+	+	+
2) H.O.	62/M	1	III	Bromocriptine 15	+	+	+
3) S.K.	60/M	17	III	L-dopa/benserazide 300/75	+	–	+
4) K.S.	59/F	9	III	L-dopa/benserazide 300/75	+	–	+

RESULTS

Effect of maprotiline on freezing in gait, speech, writing, and other motor symptoms in the 4 cases is shown in Table 2. Freezing in gait and speech was improved in all 4 cases. Freezing in gait, in particular, was the most improved symptom. In contrast, freezing in writing was not responsive to the drug. Akinesia was benefited by maprotiline; while rigidity, tremor, and retropulsion were not successfully treated. Akinesia, however, was improved to a lesser degree than freezing in gait.

Here, we will report one representative case out of 4 cases on maprotiline therapy. The patient, Mr. M.H., age 71, had a 10-year history of Parkinson's disease; his rigidity and tremor had been controlled by 400/40 mg levodopa plus carbidopa daily. He had developed start hesitation of walking and had come to be petrified at the sight of someone in front of him when walking 2 years ago. He was unable to walk unassisted and tended to fall down several times a day. However, he could climb a staircase. The electromyographic recordings of frozen gait were characterized by very low amplitude of discharge of the gastrocnemius. Stuttering and micrographia were marked. Postural reflex was absent in the pull-backward test. The freezing phenomenon was not successfully treated even when a maximum daily dose of levodopa plus carbidopa of 700/70 mg was reached. Maprotiline was started at a dose of 50 mg daily for 2 weeks and was then increased to 75 mg from the 3rd week, at which time the freezing in gait was dramatically improved: He could easily step out and turn in a narrow corner unassisted. Stuttering was slightly improved, while micrographia was unaltered. Akinesia in general was slightly benefited by the drug. Rigidity, tremor, and postural instability were not responsive to maprotiline. His moderately depressed mood was improved. No side effects were observed.

Table 2. Effect of Maprotiline on Freezing and Other Symptoms

Case No.	Dose of Maprotiline (mg)	Improvement of						
		Freezing			Akinesia	Rigidity	Tremor	Retropulsion
		Gait	Speech	Writing				
1)	75	marked	slight	none	slight	none	none	none
2)	75	moderate	slight	none	slight	none		none
3)	75	slight		none	none	none	none	none
4)	75	slight		none	none	none	none	none

DISCUSSION

The efficacy of maprotiline in the alleviation of freezing was observed in the present study. As far as we know, our study is the first one to demonstrate such efficacy of an agent other than L-threo-DOPS. Maprotiline is considered a drug of choice for freezing. This antidepressant blocks potently reuptake of NE into the presynaptic nerve endings but does not inhibit reuptake of other biogenic amines.[5] Hence, clinical benefits by this drug in freezing was thought to be due to an increase of central NE. Thus, our present study also substantiates the argument[1,2] that L-threo-DOPS is effective in freezing by replenishing central NE.

Since it is not known how central NE deficiency leads to the freezing phenomenon, an explanation for the effect of maprotiline on freezing may be only speculative at present. It is, however, noteworthy that in the representative case, Mr. M.H., whose freezing was dramatically improved by maprotiline, his depressed mood was normalized. It is difficult to exclude the possibility that the depressive mood was favorably modified because freezing was successfully treated. However, concentrated attention to walking might be related to the improvement of frozen gait, although restoration of central adrenergic function might beneficially influence the motor program or rhythm formation of gait. To address whether diminished alertness or dull attention to movement is involved in freezing, one should conduct a study using a grading of these psychological factors.

The response to maprotiline therapy was different between frozen gait and freezing of other kinds. This finding may reflect a different pathogenetic role of NE among freezing in gait, speech, and writing. This possibility needs further investigation.

REFERENCES

1. H.Narabayashi,T.Kondo,T.Nagatsu, et al., DL-threo-3,4-Dihydroxyphenylserine for freezing symptom in Parkinsonism, in:"Advances in Neurology," Vol.40,R.G.Hassler and J.F.Christ, eds., Raven Press, New York (1984).
2. H.Narabayashi,T.Kondo,F.Yokochi, et al., Clinical effects of L-threo-3,4-dihydroxyphenylserine in cases of Parkinsonism and pure akinesia, in:"Advances in Neurology," Vol.45, M.D.Yahr and K.J.Bergmann, eds., Raven Press, New York (1986).
3. T.Suzuki,S.Sakoda,M.Ueji, et al., Treatment of Parkinsonism with L-threo-3,4-dihydroxyphenylserine: a pharmacokinetic study, Neurology (Cleveland) 34:1446 (1984).
4. A.W.Teelken,G.A.van den Berg,F.A.J.Muskiet, et al., Catecholamine metabolism during additional administration of DL-threo-3,4-dihydroxyphenylserine to patients with Parkinson's disease, J.Neural Transm. P-D Sect. 1:177 (1989).
5. E.Richelson and M.Pfenning, Blockade by antidepressants and related compounds of biogenic amine uptake into rat brain synaptosomes: most antidepressants selectively block norepinephrine uptake, Eur.J. Pharmacol. 104:277 (1984).

PHARMACOKINETIC AND PHARMACODYNAMIC STUDIES WITH CONTROLLED-RELEASE CARBIDOPA/LEVODOPA

P.A. LeWitt, M.V. Nelson, R.C. Berchou, P. Schlick, M.P. Galloway, D. Kareti, and N. Kesaree

Sinai Hospital, Wayne State University
6767 W. Outer Dr., Detroit, Michigan 48235

INTRODUCTION

Levodopa (LD) has proven to be the most rational and efficacious therapy for symptomatic control of parkinsonism[1]. Most patients taking LD (combined with decarboxylase inhibition) achieve at least several years of continuing and readily adjustable anti-parkinsonian effect, though the quality of symptomatic control often declines after months to years. The rise and fall of benefit after each dose of LD often becomes more pronounced and shows waning duration of relief. Motoric fluctuations in several patterns can develop after continuous LD use. Patients often evolve enhanced sensitivity to small doses of LD, sometimes with a narrowed therapeutic "window" below which adverse effects are initiated.

Therapeutic strategies to improve motor fluctuations have been targeted at the peripheral pharmacokinetics of LD. A patient can experience considerable variability from dose to dose, due possibly to alterations in gastric emptying, pH, intestinal motility, other amino substances competing for absorption, and other less understood phenomena altering GI absorption[2]. Response fluctuations may be due to changes in the peripheral handling of LD. Factors of systemic metabolism, and transfer of LD to the CNS can often become critical for determining whether optimal effect can be maintained in LD-responsive patients. This has been demonstrated most convincingly by a lessening of motoric fluctuations during continuous intravenous delivery of LD[3,4]. Many investigations have shown that methods enhancing the constancy of blood LD concentrations result in improved antiparkinsonian control for parkinsonians experiencing problems with chronic LD therapy.

Since the introduction of LD and decarboxylase inhibitors, several controlled-release preparations have been studied. One of these, Sinemet CR-4 (CR), a mixture of 50 mg of CD and 200 mg of LD in a polymer matrix, has shown the most favorable release profile for providing gradual release of the constituents. Our studies to be reported here[5] include results of a "double-

Basic, Clinical, and Therapeutic Aspects of Alzheimer's and Parkinson's Diseases, Volume 2
Edited by T. Nagatsu *et al.*, Plenum Press, New York, 1990

blind" crossover trial comparing optimal regimens of CR to Sinemet 25/100 (25/100), and correlations of pharmacokinetic and pharmacodynamic profiles with each preparation.

1) "Double-Blind" Crossover Study Comparing CR to 25/100

To assess the clinical efficacy of CR, we conducted a "double-blind" crossover study of 25 patients, each requiring 4 or more daily doses of medication. Attempts to improve dose-by-dose effects by using smaller, more frequent LD doses had given no additional benefit. Patients underwent a careful assessment of optimal dose scheduling with 25/100, and adopted the most effective regimen (even if as many as 12 doses per day were needed). After optimal benefit was achieved for 4 weeks, patients switched to CR for an equivalent open-label trial at optimal dose scheduling.

Next, patients were assigned randomly to begin either 25/100 or CR, using "dummy" pills to maintain identical intake schedules. Eight weeks after use of the first preparation, patients were switched to the other. Ratings were conducted by the same clinician during the "double-blind" phase, using a variety of rating and disability scales.

Nineteen of 25 patients completed the "double-blind" trial. The mean daily number of 25/100 doses needed for optimal control was 10.2 (range: 5 - 19), reduced to 5.4 (range: 3 - 7) with CR. Intervals between doses of 25/100 averaged 2.1 hours (range: 1.0 - 3.75); for the CR phase, mean interdose spacing was extended to 3.8 hours (range: 2.7 - 6.0). The total intake of carbidopa/levodopa for 25/100 was 335 ($\pm$ 201.1)/1340 ($\pm$ 804.5) and for CR, 445.5 ($\pm$311.3)/1781.5 ($\pm$1245.0). The motor portion of the Unified Parkinson Disease Rating Scale also revealed a small but significant improvement in mean ratings on the CR regimen[5]. The total mean "on" time on the sustained-release regimen, as assessed at home by patients, was increased, as compared to results with 25/100, although at the expense of more occurrence of dyskinesia. Comparison of disability status while in an "on" state was unchanged between the 2 drug forms, although while "off", the overall mean disability score was significantly less on the CR preparation.

2) Pharmacokinetic and Pharmacodynamic Studies Comparing CR to 25/100

Nine patients participated in pharmacokinetic studies with CR and, on another occasion several weeks later, 25/100. Each trial involved 2 oral doses, given at 4-hour intervals. For each patient, LD dose had been determined from prior clinical experience as the amount giving optimal symptomatic control; in 4 instances, the same LD dose was ingested on each occasion.

Patients remained at minimal activity for the 8-hour study. A forearm intravenous catheter was placed, from which blood samples were removed and studied by HPLC[5]. At each time point and before blood sampling, patients underwent a brief clinical examination assessing of tremor, dyskinesia, rigidity, and overall parkinsonian state, as well as rate of tapping between 2 levers for 60 seconds, and time for arising and walking.

As anticipated, mean time to peak LD concentration was significantly delayed with CR compared to that of 25/100 (2.3 versus 1.1 hrs). Data from plasma LD determinations and response measurements were analyzed as both the mean per-cent deviation from mean values over the entire 8-hour study, and as mean per-cent deviation from the mean over the second dose period. LD oral clearance was estimated by dividing total dose ingested over the 8-hour period by the total area under the plasma LD-time curve (determined by the trapezoidal rule). The mean bioavailability of CR versus that of 25/100 was estimated to be 0.89 ± 0.18 (range: 0.60 - 1.20). As compared to 25/100 formulation, the range of LD concentrations from CR over the 8 hours was also significantly less, despite a substantially larger administered dose. The second 4-hour period's mean percent deviation from the mean plasma LD concentration was significantly smaller for the CR regimen.

The clinical responses showed, in most instances, a strong relationship to both magnitude and timing of plasma LD peaks. The times to 2 maximal clinical effects (tapping rate and global assessment score) were significantly longer for CR. The mean per-cent deviation from the mean tapping rate was significantly greater for 25/100 than CR over the 8-hour dosing period, but this was not found for the other response variables. When the 2 preparations were compared, two-way ANOVA revealed significant timewise differences for the mean per-cent deviation over the 8 hours. In analyzing the second 4-hour period, the mean percent deviation from the mean was consistently smaller for the CR regimen; this difference reaches statistical significance for walking time and tapping rate measures.

DISCUSSION

Though some dyskinesia and "on-off" problems may be unresponsive to alterations in LD supplementation, many studies suggest that stabilizing plasma LD can improve motoric fluctuations. Sustained-release of LD leads to a less abrupt decline in blood concentration and, consequently, less attenuation of its effect. Delayed release of LD to the absorptive surfaces of the gut will also lessen the magnitude of peak concentrations, and so may reuce occurrence of those adverse effects initiated upon reaching a particular concentration threshold (such as involuntary movements).

As expected, CR achieved more sustained plasma concentrations by delayed release. Though sustained-release preparations of LD are subject to the same factors governing distribution and degradation that regulate conventional forms of LD, the greater duration of the decay portion in the time-concentration curve reflects the continuation of absorption during the post-peak concentration period. The decreased variability in plasma LD concentration after the second dose suggests that the sustained-release form may offer improved chance for maintaining plasma LD within a closer therapeutic window. Among our patients, all of whom showed prominent dose-by-dose fluctuations, there was decreased variability in global parkinsonian scores as well as in performance at specific motor tasks closely related to parkinsonian disability. Though delayed in onset of its clinical effect, benefits from the CR regimen lasted longer. The clinical implications from these studies support the utility of CR as an adjunct or alternative to other levodopa therapy.

REFERENCES

1. P.A. LeWitt. The pharmacology of levodopa in treatment of Parkinson disease, in: "Handbook of Experimental Pharmacology, Volume 88", D.B. Calne, ed., Springer-Verlag, Berlin (1989)

2. J.G. Nutt and W.R. Woodward, Levodopa pharmacokinetics and pharmacodynamics in fluctuating parkinsonian patients. Neurology 36:739 (1986)

3. R.J. Hardie, A.J. Lees, and G.M. Stern, On-off fluctuations in Parkinson's disease. Brain 107:487 (1984)

4. N. Quinn, J.D. Parkes, and C.D. Marsden, Control of on/off phenomenon by continuous intravenous infusion of levodopa. Neurology 34:1131 (1984)

5. P.A. LeWitt, M.V. Nelson, R.C. Berchou, M.P. Galloway, N. Kesaree, D. Kareti, P. Schlick: Controlled release carbidopa/levodopa (Sinemet 50/200 CR-4). Clinical, pharmacokinetic, and pharmacodynamic studies. Neurology (in press, 1989)

RESULTS OF A MULTICENTRE STUDY COMPARING THE EFFECTS OF A CONTROLLED RELEASE PREPARATION (MADOPAR CR) WITH CONVENTIONAL LEVODOPA

D.G. MacMahon (1), D. Sachdev (2), H.G. Boddie (3), C.J.K. Ellis (4), B.R. Kendal (5), and N.A. Blackburn (6)

(1) Barncoose Hospital, Redruth, Cornwall, (2) Oldchurch Hospital, Romford, (3) North Staffordshire Royal Infirmary Stoke-on-Trent, (4) Poole General Hospital, Poole (5) Leicester Royal Infirmary, Leicester, (6) Medical Department, Roche Products Ltd., Welwyn Garden City, UK

OBJECTIVE

To investigate the effects of the introduction of a controlled release formulation of levodopa with benserazide (Madopar CR), in Parkinsonian patients experiencing symptoms of end-of-dose deterioration or on-off fluctuations on conventional levodopa with dopa decarboxylase inhibitor (L+DDCI) therapy.

SUBJECTS

87 patients with Parkinson's disease, all of whom were currently receiving conventional L+DDCI therapy and experiencing fluctuations in their condition, were recruited from 7 centres in the UK.

METHOD

Conventional Madopar Phase		Conventional Madopar + Madopar CR Phase	
Open titration of conventional Madopar to optimum	Stable assessment	Introduction of Madopar CR. Open titration to optimum dose with adjustment of conventional Madopar as necessary	Stable assessment of response to combined drug therapy regimen
2 - 6 weeks	4 weeks	2 13 weeks	4 weeks

Diaries of fluctuations and involuntary movements are kept by patients or carers throughout the trial where possible.

RESULTS

Details of the 87 patients entered are shown in Table 1, and their progress through the study is charted in Figure 1.

Table 2 gives details of the doses of Madopar and Madopar CR taken during the study.

Since many patients and their carers were unable to distinguish consistently between dyskinesia, dystonia and tremor, the diary data for dyskinesia was felt to be unreliable by investigators, and hence there is no data on this parameter. However, there was no obvious increase in dyskinesia observed by investigators, despite teh increased intake of levodopa in the Madopar CR phase of the trial (see Table 2).

The investigators felt that the introduction of Madopar CR was advantageous in 83% and disadvantageous in 11% of patients. Statistical analyses of the 37 evaluable diaries showed an increase in 'on' time (p=0.016), and a reduction in 'off' time (p=0.029) in the Madopar CR phase of the trial compared with conventional Madopar alone (see Figure 2).

Side Effects

Side effects were typical of those seen in patients taking combined levodopa + DDCI therapy. there were no significant differences in the nature, severity or frequency of side effects in the Madopar CR phase of the trial as compared with the conventional Madopar Phase.

Sleep

The overall quality of sleep appeared to be satisfactory throughout the trial and no significant differences were found between the 2 phases of the study.

TABLE 1. PATIENT DETAILS

Number entered	87 (55M : 32F)
Mean age (range)	64.5 years (44 - 87)
Mean duration of disease (range)	9.6 years (2 - 22)
Mean duration of therapy (range)	8.1 years (1 - 18)
Hoehn & Yahr rating : I	6
II	28
III	36
IV	15
unknown	2

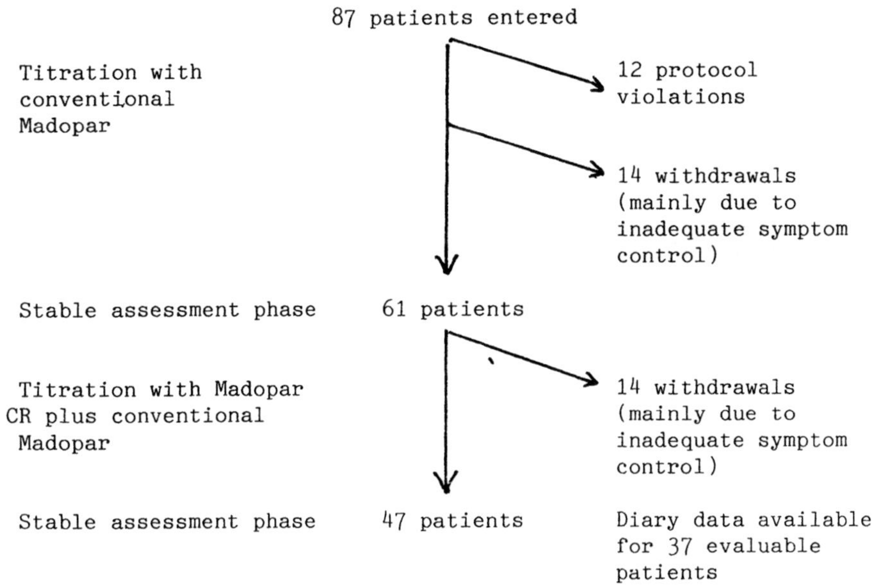

FIGURE 1. PROGRESSION OF PATIENTS THROUGH STUDY

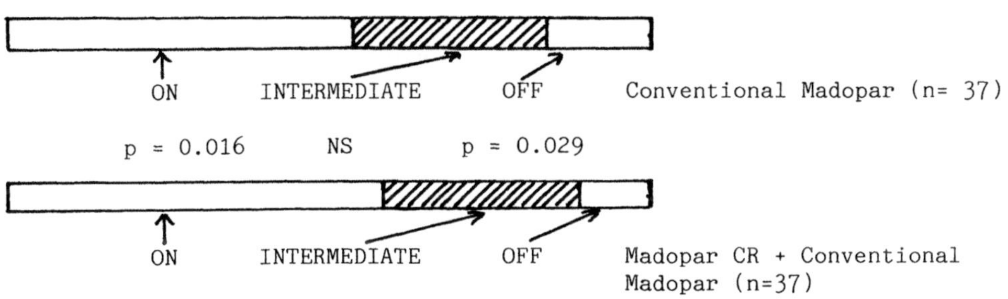

FIGURE 2. MEAN PROPORTION OF WAKING TIME 'ON', 'OFF', OR 'INTERMEDIATE'

CONCLUSIONS

1. Madopar CR is a useful adjunct to conventional Madopar in a proportion of Parkinsonian patients, increasing 'on' time and reducing 'off' time.

2. Supplemental doses of conventional Madopar need to be taken in the majority of cases.

3. An apparent increase in dose is required (33% in this trial) but is achieved without increase in dyskinesia.

4. Dose titration was not simple, and 28 patients were unable to complete the trial. It was not possible to predict outcome before entry.

TABLE 2. OPTIMUM DOSE OF LEVODOPA + BENSERAZIDE

	Conventional Madopar (n = 61)	Madopar CR + Conventional Madopar (n = 47)
Mean (range) number of doses per day	6.4 (2 - 18)	5.2 (2.6)
Mean (range) total dose per day (mg) of drugs	820 (125-2125)	1088 (250-3500)

ACKNOWLEDGEMENTS

Dr. R. Capildeo, Dr. L.J. Findley, Dr. R.J. Prowse, Dr. P. Millac, Dr. L.A. Loizou, Dr. R.L.G. Sutcliffe and Dr. T. Steiner for their assistance in monitoring patients in the study; Dr. S. Little for data handling and statistical analysis.

AN OPEN STUDY OF EVENING DOSES OF SLOW-RELEASE L-DOPA (MADOPAR HBS) IN PARKINSON PATIENTS WITH NOCTURNAL DISABILITIES

M.Wagner[1], C.Trenkwalder[2], W.H.Oertel[2], and W.Poewe[1]

1) University Hospital for Neurology, Innsbruck, Austria
2) Dept. of Neurology, Klinikum Großhadern, Munich, FRG

INTRODUCTION

Nighttime problems are common in Parkinson's disease and include sleep fragmentation, nocturnal immobility, pain and dystonic cramps[1,2]. They may continue as early morning akinesia and dystonia until the beneficial effects of the first daily dose of L-Dopa finally bring relief. In chronically treated patients nocturnal difficulties seem to be partly related to the wearing-off effect of L-Dopa during the drug-free interval prior to the morning dose and nighttime doses of L-Dopa[3] or prolongation of the effect of evening doses by selegiline[4] have been reported to ameliorate nocturnal disabilities. The present study was performed to evaluate the potential of single bedtime doses of a sustained-release formulation of L-Dopa/Benserazide (Madopar HBS) in parkinsonian patients with clinically relevant degrees of nocturnal and early morning disability.

PATIENTS AND METHODS

16 patients (6 females, 10 males; mean age 62,5 (48-70) yrs.) with idiopathic Parkinson's disease (mean duration 9.5 (4-20) yrs.) and sustained treatment with conventional L-Dopa (mean duration 6.8 (2-14); mean dose 685 (250-1250) mg/d) received single bedtime doses of sustained-release L-Dopa/Benserazide (Madopar HBS) in an open-label long-term trial. All but two had developed response fluctuations of a predictable wearing-off type (mean Hoehn & Yahr stage 1.8 (1-3) when 'on' and 3.1 (2-4) when 'off') and complex ('random') oscillations in motor performance did not occur. Nocturnal and early morning disabilities were prominent in all with 16 and 14 patients reporting troublesome nighttime and early morning immobility, respectively, and 14 suffering from nocturnal dystonic cramps and pain.

Patients received single doses of one to three capsules of Madopar HBS (100/25mg of sustained-release L-Dopa/Benserazide per capsule) immediately before retiring and subsequent dose adjustments to a maximum of three capsules were made during the first 4 weeks of the study when necessary.

For clinical assessment the following were recorded during out-patient visits at baseline, days 3,10,14 and at the end of months 1,2,3,4,6,9,12: normal nocturnal mobility problems for turning around in bed, switching on the light, getting out of bed, going to the toilet, nocturnal cramps and pain, as well as early morning akinesia, each item rated on a 3-point severity scale from mild, moderate to severe; in addition sleep duration without awakenings, number of awakenings and duration of waking period per night were also recorded. In order to provide better estimates of these clinical parameters patients were asked to fill in a self-scoring diary on every morning of the first study month and for one week prior to the next visit during subsequent months indicating hours of sleep and waking as well as periods of nocturnal immobility, cramps and dyskinesia when awake. A pre-study phase of two weeks served to train patients in the use of these diaries and also to assess their baseline disability.

RESULTS

Dosages found necessary for a sufficient clinical effect by the end of the first month of treatment comprised 200 mg (5 cases) to 300 mg (7 cases) in most, while three patients reported benefit from as little as 100 mg of L-Dopa given in form of the sustained-release preparation.

Presently results are available for the first two months of treatment. The numbers of patients experiencing mild, moderate and severe nocturnal mobility problems at baseline and follow-up visits until the end of month 2 are depicted in figure 1 for the four common activities assessed. As can be seen about 80% had moderate degrees of impairment of these functions at baseline and this figure decreased by about 30% at months 2, when 50% of those still treated reported only minor residual nocturnal immobility.

The number of patients experiencing nocturnal dystonic cramps and pain also decreased during treatment with single bed-time doses of Madopar HBS as did the severity of these symptoms.

Degree of impairment
Nocturnal immobility - common activities

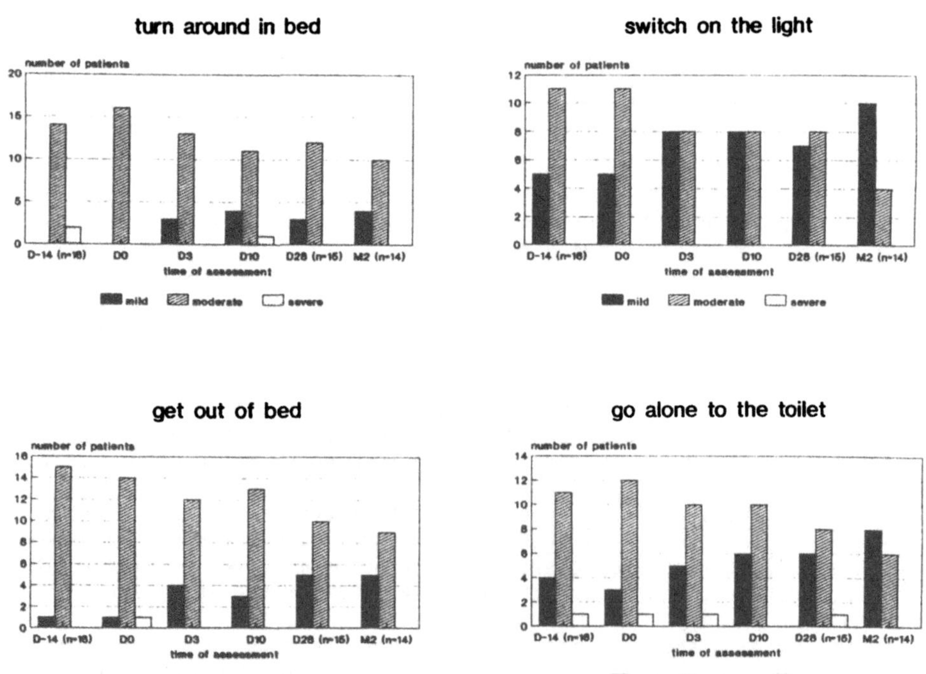

Figure 1

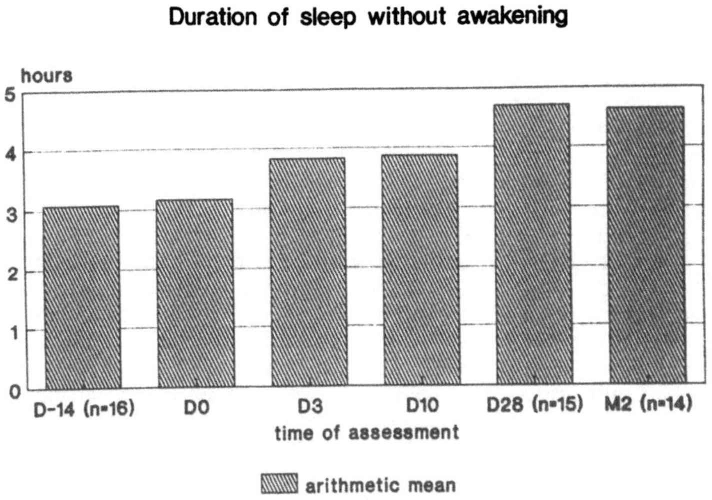

Figure 2

While half the patients had moderate to severe nocturnal cramps and pain at baseline this was only the case in one of the 14 still treated at the end of months 2, respectively.

Sleep quality showed some improvement with an increase of the mean duration of uninterrupted sleep by about 1.5 hours at months 1 to 2 compared to baseline, although this was still less than 5 hours (figure 2). The number of awakenings per night decreased from a mean of 3.3 at baseline to 2.4 after two months and the duration of each waking period became shorter.

Improvement was also noted with respect to early morning akinesia, which had been present in moderate (9 cases) to severe (3 cases) form in 12 patients at baseline. At the end of month 2 only six of the remaining 14 still fell into this category (5 moderate, 1 severe case), 4 others reported only mild early morning disability.

Two patients discontinued treatment during the first two months, one of them after two weeks because of nausea following bedtime dosing, while another complained of compromised daytime mobility after addition of Madopar HBS to his regimen. Interestingly, two further cases reported similar daytime problems but continued treatment because of the nocturnal benefit they derived. Other side effects following the introduction of bedtime doses of sustained-release L-Dopa included vivid dreaming (4 cases), nocturnal dyskinesias (2 cases) and early morning or daytime drowsiness (4 cases).

DISCUSSION

The results of the present study confirm the beneficial effects of single bedtime doses of sustained-release L-Dopa (Madopar HBS) on nocturnal disabilities of patients with Parkinson's disease as previously reported by others[5,6,7]. After two months of treatment 14 out of 16 patients showed a positive response and improvement concerned nighttime mobility as well as the incidence and severity of nocturnal cramps and pain. Prolonged therapeutic plasma concentrations of L-Dopa during the night following bedtime doses of a sustained-release preparation are a likely explanation for these effects[7]. Early morning disabilities are less responsive to the type of treatment employed in this study and this would be expected from the pharmacokinetic profile of Madopar HBS which does not lead to therapeutic plasma levels of L-Dopa for more than 5 hours.

Improved sleep quality has been a consistent finding in this and other studies of night-time Madopar HBS[5,6,7] and this is probably secondary to improved nocturnal mobility with decreased rigidity, dystonic cramps and pain. In addition direct effects of L-Dopa on parkinsonian sleep have also been proposed[8]. Positive effects of bedtime sustained-release L-Dopa on early morning mobility as observed in some patients of this study could well be secondary to improved sleep since sleep benefit is common in Parkinson's disease[5].

Potentially troublesome adverse effects of this form of treatment include nocturnal dyskinesias and vivid dreaming and special care should be taken when treating patients with a history of L-Dopa induced confusion and hallucinosis. The possible mechanisms responsible for daytime drowsiness and decreased daytime mobility following evening doses of Madopar-HBS in individual patients of this study are difficult to understand and deserve further investigation. Nevertheless the present results provide further evidence for the usefulness of single bedtime doses of sustained-release L-Dopa in the treatment of parkinsonian night-time disabilities. The stability of the observed clinical benefit with long-term treatment will have be followed.

REFERENCES

1. P.A. Nausieda, W.J. Weiner, L.R. Kaplan, S. Weber, H.L. Klawans, Sleep disruption in the course of chronic levodopa therapy: an early feature of levodopa psychosis, Clin. Neurpharmacol. 5:183 (1982)
2. A.J. Lees, N.A. Blackburn, V.L. Campbell, The nighttime problems of Parkinson's disease, Clin. Neuropharmacol. 11:512 (1988)
3. A.L. Leeman, C.J.A. O'Neill, P.W. Nicholson et al., Parkinson's disease in the elderly: response to an optimal spacing of night time dosing with levodopa, Br. J. Clin. Pharmacol. 24:637 (1987)
4. A.J. Lees, L.J. Kohout, K.M. Shaw, G.M. Stern, J.D. Elsworth, M. Sandler, M.B.H. Youdim, Deprenyl in Parkinson's disease, Lancet 2:791 (1977)
5. A.J. Lees, A sustained-release formulation of l-dopa (Madopar HBS) in the treatment of nocturnal and early morning disabilities in Parkinson's disease, Eur. Neurol. 27 (suppl.1):126 (1987)
6. E.N.H. Jansen, J.D. Meerwaldt, Madopar HBS in Parkinson patients with nocturnal akinesia, Clin. Neurol. Neurosurg. 90:35 (1988)
7. H. Baas, P.A. Fischer, H.J. Roth, Therapeutic effect of a levodopa / benserazide controlled-release preparation on nocturnal akinesia in Parkinsonism, in: D. Bartko et al. (ed.), New trends in clinical neuropharmacology, John Libbey & Co.Ltd., 311 (1988)
8. J.J.M. Askenasy, M.D. Yahr, Reversal of sleep disturbance in Parkinson's disease by antiparkinsonian therapy: a preliminary study, Neurology 35:527 (1985)

TO THE PRINCIPLES OF LEVODOPA THERAPY INCLUDING SLOW RELEASE FORM

E. Csanda*, M. Tárczy, and Annamária Takáts

Institute of Neurology, Semmelweis Medical University, Budapest, Hungary, and *1083 Budapest, Balassa u. 6, Hungary

INTRODUCTION

Since the introduction of levodopa fluctuations has become the major challenge in the treatment of Parkinson's disease. Various forms of dopamine agonists and selegiline are mainly used in the current praxis but the successful therapy is still unsolved.

At our Institute about 600 Parkinsonian patients have been treated and cared regularly as outpatients in every 3 month since 1975. The number of in-patients in a year is also about 600.
Our experiences concerning levodopa treatment is summarised in table 1.

TABLE 1

Principles of Levodopa Therapy in PD

- Start with levodopa as late as possible
- Under in-patient conditions only
- Use levodopa with DCI, never alone
- Optimal levodopa – DCI rate = 4:1
- Start with low dose, stepwise increase
- Stop at the lowest effective dose
- Early combination with selegiline or bromocriptine

In the course of long therm levodopa therapy some complications, like wearing-off, on-off phenomena, nocturnal dystonias became more and more evident.

Controlled release preparates like Madopar HBS (hydrodinamically balanced system) are suggested to be a new possibility especially in those forms of fluctuations which correspond to the alterations of plasma levodopa level (2, 3, 4, 5).

This is the first trial with Madopar HBS performed in Hungary.

PATIENTS AND METHODS

34 patients were selected to take part in the study and were followed for at least one year. The age of patients was 39-68 (mean: 56,8) years, the duration of disease varied between 5-16 (mean: 9,2) years. Duration of levodopa therapy lasted for 4-12 (mean: 7,1) years the daily dose of standard Madopar (STM) being 250-2000 (in average: 912) mgs. Standard Madopar was changed to Madopar HBS following the instructions of Hoffmann-LaRoche protocol from one day to another. Daily dosage schedule was initiated in every case with the standard form and the optimal level of Madopar HBS was carefully titrated within 1-3 weeks. Patients were followed under hospitalised and ambulatory conditions. Clinical evaluation was based on the regular use of Webster score and North Western University Disability Scale. Fluctuations were registered by the aid of motor performance scale.

Main clinical problems of the patients are listed in the table 2.

TABLE 2

nocturnal dystonia	: 28
wearing off effect	: 20
dose dependent fluctuations with AIMs	: 10
non-dose dependent fluctuations	: 4
peak-dose dyskinesia	: 8
end-of-dose dyskinesia	: 3
end-of-dose dystonia	: 2

RESULTS

1. In the course of dosage optimalization the daily dose of Madopar HBS had to be augmented to 500-3750 mgs, in average to 1642 mgs. Thus the ratio of HBS: STM proved to be 1,8 in our material, showing that rather high quantities of the HBS formule should be given to get an optimal response.

2. Madopar HBS proved to be highly effective to reduce nocturnal and early morning dystonias. These forms of dystonia correspond to low plasma levodopa levels and the probable increase of the drug level enabled the reduction or elimination of these complaints. If the only or leading problem of a patient was nocturnal dystonia the evenings standard Madopar was changed to HBS (for example: 125 STM to 250 mg HBS) with unanimously good results. In all 28 patients nocturnal dystonia disappeared.

3. 14 of 20 "wearing off" patients showed a significant improvement based on the data of motor performance scale. The use of this scale means the hourly assessment of standing up + gait (mobility), tremor and abnormal involuntary movements for 12 hours in a day. Analysis of daily diagrams showed that "on" periods became longer and the grade of hypokinesia in the "off" periods was less remarkable.

4. End-of-dose dyskinesias and dystonias were considerably reduced in a paralell way as nocturnal dystonias diminished.

5. In dose-dependent fluctuations the use of HBS resulted in decreasing the abnormal involuntary movements but usually hypokinesia was

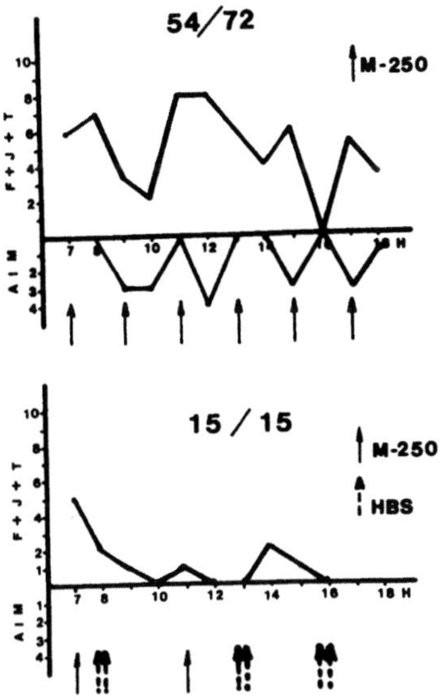

Fig. 1.

Improvement of a patient with dose-dependent regular daily fluctuation using the alternating administration of STM and HBS. Daily diagram based on the data of motor performance scale. First number = mobility desired from the values of standing up (F) and gait (J). The second number = total disability also contains the values of tremor (T) and abnormal involuntary movements (AIM).

prolonged. This was observed especially in patients who showed a regular oscillation between considerable hyperkinesia and akinesia. In 6 of 10 patients with dose-dependent fluctuations better results were achieved if alternating administration of standard and HBS Madopar was performed. This regimen needs the very careful observation of the patients and a high degree of collaboration.

6. HBS proved to be ineffective in non-dose dependent fluctuations. Previously in these patients other drug regimens (selegiline and/or dopamine agonists) were also tested without any success.

7. In some patients after the period of continously free motor performance sudden, unpredictable dyskinesias developed mainly in the late afternoon hours. This observation suggests a probable cumulative effect of HBS.

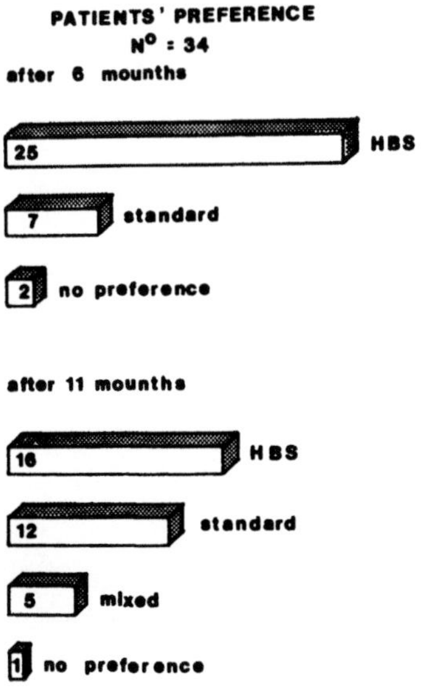

Fig. 2.

After 11 months the number of patients preferring HBS to STM decreased (from 25 to 16) while 6 patients voted for the alternating administration of STM and HBS. The main clinical problem of these patients was dose-dependent fluctuation.

8. After 6 months the beneficial effect of continous HBS treatment decreased, so either the daily dose must be increased or the alternating administration had to be introduced.

Patients' preference was asked after six months and after 11 months of treatment.

DISCUSSION

HBS was given to 34 patients with various forms and elementary symptoms of fluctuation in daily motor performance.

The best results were achieved in nocturnal and early morning dystonia as practically all patients reacted well to the administration of HBS. The "wearing off" effect was considerably reduced, while in dose-dependent fluctuations with AIMs the alternating administration of the STM and HBS formules proved to be succesful.

Practically all forms of dyskinesias decreased due to a rather stable plasma levodopa level, but it should be mentioned that in some patients unpredictable abnormal involuntary movements developed. After about six moths the efficacy of HBS tends to decrease and revision of the dosage schedule is advisable.

REFERENCES

1. Csanda E, Tárczy M: Acta neurol Scand Suppl 95:117-122 (1983)

2. Lees AJ: Sustained-Release Formulation of L-Dopa (Madopar HBS) in the Treatment of Nocturnal and Early-Morning Disabilities in Parkinson's Disease. Eur J Neurol. 27 Suppl 1:126-134 (1987)

3. Poewe WH, Lees AJ, Stern GM: Clinical and Pharmacokinetic Observations with Madopar HBS in Hospitalized Patients with Parkinson's Disease and Motor Fluctuations. ibid 93-97

4. Rinne VK: Madopar HBS in the Long-Term Treatment of Parkinsonian Patients with Fluctuations in Disability. ibid 120-125

5. Siegfried J: Therapeutic Value of Madopar HBS: Judgment after 2 Years Experience. ibid 98-104

A NEW PRODRUG OF L-DOPA TO ATTAIN PROLONGED PLASMA CONCENTRATION

Akihiro Hisaka, Masaki Ihara, Yoshimi Tsuchiya,
Yoshio Sawasaki, Masakatsu Ohtawa, and Mitsuo Yano

Central Research Laboratories
Banyu Pharmaceutical Co.,Ltd., Tokyo, Japan

INTRODUCTION

It has been reported that unpleasant oscillations of L-DOPA efficacy, such as dyskinesia, involuntary movements, and end-of-dose deterioration, can be correlated with the fluctuation of plasma concentrations of L-DOPA.[1,2] These problems occur in more than 50% of patients receiving L-DOPA therapy for up to five years. Therefore, development of prodrugs or formulations able to achieve a stable L-DOPA plasma concentration should be advantageous for the clinical treatment of parkinsonism. For this purpose, we screened L-DOPA derivatives and found that L-3-(3-hydroxy-4-pivaloyloxyphenyl)alanine (NB-355, 1) provides a preferable plasma profile in experimental animals after oral dosing.[3,4] The drug is anticipated to improve L-DOPA therapy, and is currently undergoing clinical trials.

Plasma L-DOPA concentrations were significantly prolonged with reduced Cmax and moderately increased AUC by the prodrug in rats and dogs.[3] On the other hand, the plasma concentrations of intact prodrug were very low or undetectable,[3,5] suggesting that the prodrug was transformed into L-DOPA before reaching systemic circulation, in other words, at the absorption step. Full investigation on the absorption is, therefore, needed for understanding its transformation and the mechanism of prolonged plasma L-DOPA. This article characterizes the prodrug 1, focusing on its absorption.

EXPERIMENTAL METHODS

Animals, Materials, and Assay Methods

Male SD strain rats of 7 to 12 weeks of age were used throughout the study. Rats were fasted for more than 16hr before dosing or surgery but allowed free access to water. Prodrug 1 was synthesized in our laboratory.[3,6] The concentration of 1 was determined by an HPLC method with fluorescence detection after precolumn derivatization with o-phthalaldehyde and N-acetyl-L-cysteine.[3,5] The concentration of L-DOPA was assayed by an HPLC method with electrochemical detection.[3,5] The structures of prodrug 1 and L-DOPA are shown as follows:

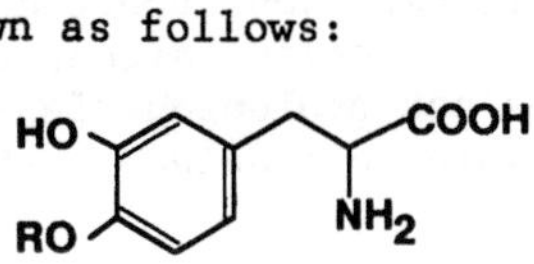

compound	R
L-DOPA	H
NB-355 (1)	$COC(CH_3)_3$

In Vitro Hydrolysis Study

Prodrug 1 (50μM) was incubated in 50mM phosphate buffer (pH2.1 to 7.4, 37°C) for various time intervals. For the enzymatic hydrolysis study, the drug was incubated in the homogenate of rat small intestine (prepared with 9 volumes of 50mM phosphate buffer containing 100μM carbidopa, an inhibitor of peripheral aromatic L-amino acid decarboxylase).

In Vitro Uptake by Rat Everted Intestinal Tissue Segments

Upper small intestine of rat was isolated, everted, and cut into ring-like slices 3 to 5 mm in length. The segments were pre-incubated for 5min at 37°C in oxygenated buffer (pH6.0) containing 0.2mM carbidopa, 0.1% ascorbic acid, and appropriate concentrations of inhibitors (L-Phe, 10mM; dinitrophenol, 0.1mM; and ouabain, 3mM). The uptake was initiated by addition of 1 or L-DOPA (500μM) to the medium. After the incubation (5 to 15min, 37°C), the segments were collected and homogenized in 19 volumes of 0.5N perchloric acid (PCA) containing 1mg/ml EDTA. Aliquots of supernatant were collected after centrifugation and stored until assay (-20°C). We added fluorescein isothiocyanate dextran (FID, 0.1mg/ml) to the pre-incubation medium occasionally. Drug concentrations in the tissue segments were corrected for extracellular space calculated from the FID concentration. FID concentrations were measured by a fluorometric method.

In Vivo Absorption Profile in Rats

Rats received 1 (10mg/kg, L-DOPA equivalents) with carbidopa (2mg/kg) orally. Under light anesthesia with ether, portal venous and abdominal arterial blood were collected and centrifuged to obtain plasma. The gastrointestinal tracts were removed immediately, and divided into 5 parts (stomach, duodenum, upper jejunum, lower jejunum, and ileum). Intestinal contents were collected by washing of the segments with ice-chilled 0.1M citrate buffer (pH4.0). Samples were deproteinized by addition of 0.5N PCA containing 1mg/ml EDTA, or by addition of 0.1N HCl:ethanol.

RESULTS AND DISCUSSION

In vitro hydrolysis of the prodrug to L-DOPA was investigated under variable pH conditions (Fig.1). Prodrug 1 was stable at acidic pH (less than pH6.5), but underwent hydrolysis spontaneously at neutral pH. Half-life time of the hydrolysis was 0.37hr at pH7.4 at 37°C. It is noteworthy that di-O-pivaloyl-L-DOPA or O-pivaloyl-L-tyrosine did not undergo hydrolysis at this pH (data not shown). The neighboring hydroxy group on compound 1 may play an essential role in the highly pH-dependent hydrolysis of the prodrug.

The *in vitro* hydrolysis rate of 1 was moderately enhanced by addition of the intestinal homogenate to the medium (Fig.1). The enhancement was canceled by further addition of diisopropylfluorophosphate(DFP). These results suggest that prodrug 1 was hydrolyzed not only chemically but also enzymatically by DFP-sensitive esterases. However, the enzymatic hydrolysis rate of 1 was far less than that of carboxy esters of L-DOPA. In our experiment, the enzymatic hydrolysis rate of 1 was only one eightieth of that of L-DOPA carboxy methylester, while its chemical hydrolysis rate was about 20 times higher than that of methylester at pH7.4. It has been reported that some carboxy esters of L-DOPA provided higher plasma concentrations of L-DOPA,[7] however, the shape of its plasma curve was not prolonged. To achieve prolonged plasma concentrations, it seems reasonable to conclude that the enzymatic hydrolysis rate must be slow.

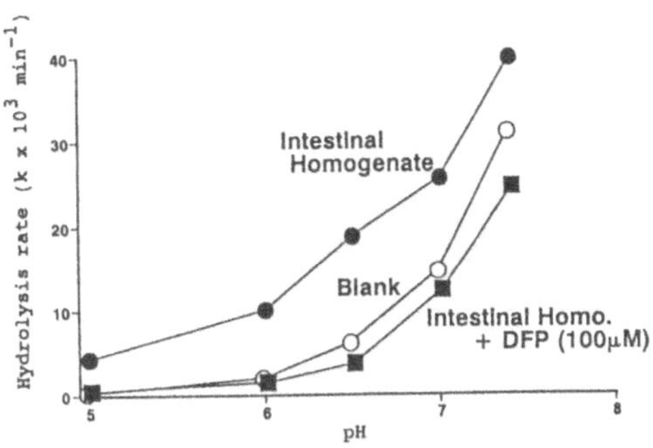

Fig. 1. **pH-dependency of hydrolysis rate of NB-355(1).**

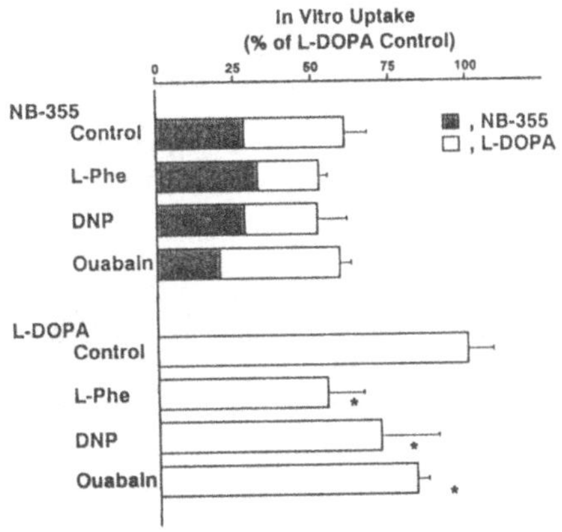

Fig. 2. **Effects of inhibitors on *in vitro* uptake by everted tissue segments of rat small intestine.** *Significant difference from control, $p<0.05$.

Prodrug 1 probably generates L-DOPA little by little in the small intestine both chemically and enzymatically, and this may be concerned with the prolonged plasma concentration to some extent. From this viewpoint, prodrug 1 is similar to controlled-release device designed to release drugs gradually in the small intestine. However, another factor should be taken into consideration in the case of prodrugs, that is, the absorbability of the intact prodrug. If the intact prodrug is absorbed more rapidly than the parent drug, sustained absorption can not be maintained.

We measured *in vitro* uptake by rat everted intestinal tissue rings to investigate the absorbability of the intact 1. The incubation was carried out at pH6.0 to prohibit the chemical hydrolysis. Prodrug 1 was taken up by the ring time-dependently, but to a lesser extent than L-DOPA (Table 1). It is well known that L-DOPA is efficiently transported across the intestinal membranes by an active transport system for L-amino acids.[1,2] However, the uptake of 1 was not affected by inhibitors of the active transport system (Fig.2), suggesting that the intact 1 is absorbed slowly by simple diffusion. Consequently, prodrug 1 is different from simple pH-dependent controlled-release devices in that the intact form is also absorbable by a different uptake mechanism. In the tissue of intestinal segments, considerable amounts of L-DOPA were generated from 1. This conversion was enzymatic under these conditions, since DFP remarkably reduced the generation.

Rats received oral 1, and levels of 1 and L-DOPA were simultaneously measured at various gastrointestinal sites and in portal vein and aortic blood. Prodrug 1 offered prolonged plasma concentrations of L-DOPA in the aortic blood as previously reported (Fig. 3).[3,4] In the lumen, almost all of the drug was in the intact form; Fig. 4 shows total of 1 and L-DOPA remaining in the lumen, although L-DOPA level was less than 3% of the dose. By

Table 1. ***In Vitro*** **Uptake by Everted Tissue Segments of Rat Small Intestine.**

(pmole/g wet tissue/min)

	NB-355 (1)#	L-DOPA
- DFP	13.4 ± 2.1*	22.0 ± 4.4
+ DFP(0.1mM)	10.9 ± 5.5*	21.9 ± 2.2

#calculated from total of 1 and L-DOPA in the segments.
*significant difference from L-DOPA ($p<0.05$).

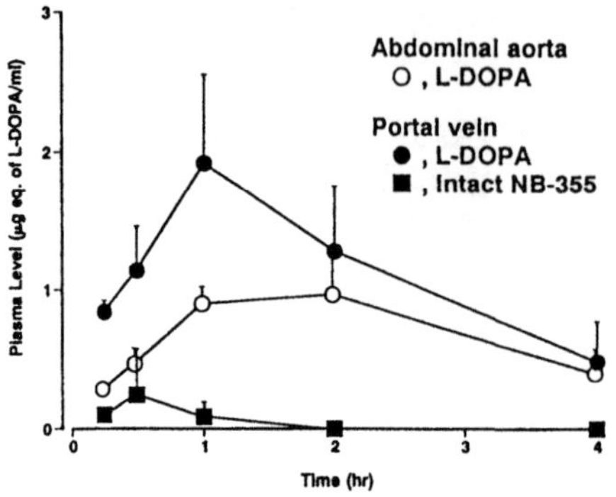

Fig. 3. Plasma concentrations of L-DOPA and intact NB-355(1) after oral dosing of NB-355 in rats.

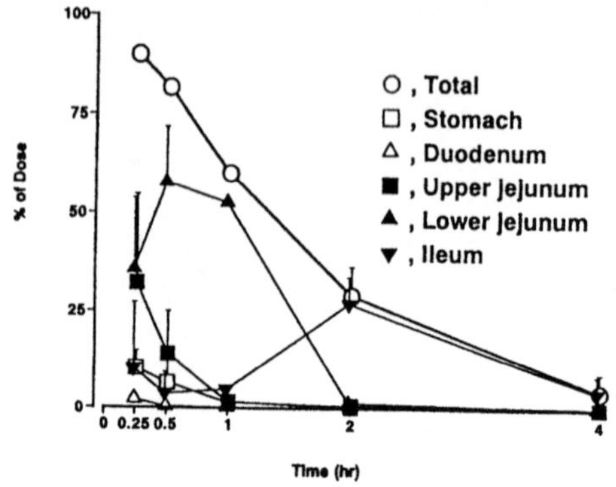

Fig. 4. Remaining amounts of drug in the gastrointestine after oral dosing of NB-355(1) in rats.

striking contrast, concentrations of intact 1 were far less than that of L-DOPA in the portal venous blood (Fig. 3). These data suggested that *in vivo* conversion of 1 to L-DOPA occurred primarily in the intestine. Fig. 4 indicates that 60% of 1 reached the lower jejunum and absorbed mainly from there. This result is contradistinctive to L-DOPA which is exclusively absorbed from the upper small intestine.[1,2] The present study demonstrates a broader absorption window of prodrug 1 than that of L-DOPA.

CONCLUSIONS

Prodrug 1 attained a prolonged L-DOPA plasma profile owing to its sustained absorption from the intestine. Two absorption pathways were proposed; first is pH-dependent hydrolysis in the intestinal lumen and uptake in the form of L-DOPA, second is uptake of intact prodrug by simple diffusion with ensuing enzymatic or chemical hydrolysis in the intestinal wall.

REFERENCES

1. J. R. Biadnchine and G. M. Shaw, *Clinical pharmacokinetics of levodopa in parkinson's disease*, Clin. Pharmacokinetics 1:468 (1976).
2. J. G. Nutt and J. H. Fellman, *Pharmacokinetics of levodopa*, Clin. Neuropharmacology 1:33 (1984).
3. M. Ihara, Y. Tsuchiya, Y. Sawasaki, A. Hisaka, H. Takehana, K. Tomimoto, and M. Yano, *A new potential prodrug to improve the duration of L-DOPA, L-3-(3-hydroxy-4-pivaloyloxyphenyl)alanine*, J.Pharm. Sci. **78**:525 (1989).
4. A. Hisaka, M. Ihara, Y. Tsuchiya, Y. Sawasaki, M. Ohtawa, and M. Yano, *A new prodrug of L-DOPA, prolonged plasma L-DOPA by sustained absorption*, Proceedings of the 16th International Symposium on Controlled Release of Bioactive Materials 195 (1989).
5. A. Hisaka, S. Kasamatsu, N. Takenaga, and M. Ohtawa, *Quantification of L-3-(3-hydroxy-4-pivaloyloxyphenyl)alanine (NB-355) by high-performance liquid chromatography using o-phthalaldehyde/N-acetyl-L-cysteine derivatization*, J.Chromatogr. **494**:183 (1989).
6. M. Ihara, S. Nakajima, A. Hisaka, Y. Tsuchiya, Y. Sakuma, K. Kitani and M. Yano, *Hydrolysis and acyl migration of a catechol monoester and L-DOPA: L-3-(3-Hydroxy-4-pivaloyloxyphenyl)alanine*, J.Pharm.Sci. *in press*.
7. N. Bodor, K. B. Sloan, T. Higuchi, K. Sasahara, *Improved delivery through biological membranes. 4. prodrugs of L-dopa*, J.Med.Chem. 20: 1435 (1977).

FRENCH OPEN CLINICAL STUDY OF MODOPAR LP (HBS) IN PARKINSONIANS

Alain DESTEE

Neurologic Department
C.H.U Lille-France

The use of l-dopa in the treatment of Parkinson's disease constituted 20 years ago a major event.

The discovery of this therapy transformed the prognosis of the disease, allowing patients to lead a quite normal life for several years, and reducing the mortality rate by 50%. Unfortunately the salutary effect of drug does not last and motor fluctuations appear after several years of treatment ; they are due to resurgence of parkinsonian symptoms or to involontary movements. These two phenomena presently represent the principal difficulty in long term treatment. To alleviate these phenomena, a pharmacokinetic approach is possible (Ref.).

Previous clinical trials have partially confirmed this idea, either by dividing doses and using low doses, or by administering l-dopa in continuous perfusion. Unfortunately this injectable form cannot be used routinely because of the high acid pH and the practical difficulties of treating out patients by intravenous route.

GALENIC PRINCIPLES OF THE HBS FORM

This form is based on the principle of the hydrodynamically balanced system as is indicated by the name Madopar[R] HBS. It consists of a gelatin capsule filled with a dry powder composed of a mixture of fillers, the active substance and hydrocolloids. Gastric fluids enter in contact with and dissolve the gelatin capsule. Then the active substance and the hydrocolloids are transformed into a gelatinous mass. The active substance is slowly released through the gelatinous structure.

At the outset only the outside part of the powder is transformed into a gel while the centre remains dry : little by little the humidity reaches the core of the capsule transforming the whole of the capsule into a gel.

Several pharmacokinetic parameters are to be explored in the assessment of the controlled release form. These are essentially the height of the plasmatic peak (C. Max) and its onset (T. Max) and the half-life of the product and its bioavailability. The half-life or HBS Madopar[R] is increased approximately 4

Basic, Clinical, and Therapeutic Aspects of Alzheimer's and Parkinson's Diseases, Volume 2
Edited by T. Nagatsu *et al.*, Plenum Press, New York, 1990

fold compared to standard Madopar[R]. The major interest of this new form is the high reduction of plasmatic peaks and maintenance of more stable levels.

PROTOCOL FOR STUDY OF MADOPAR[R] HBS

This protocol was an open, multicenter study and the aim was to evaluate the efficacy and the tolerability of a controlled-release form of l-dopa and benserazide (Madopar[R] HBS). The duration for the study was 1 year and 5 examinations were planned.

CLINICAL RESULTS (Table)

The present results constitute an intermediate analysis because the inclusions in the protocol were closed in July 1989 and a lot of patients have not yet completed a one year follow-up.

Seventy patients were included in the study. Five of them were excluded because they were not fulfilling the inclusion criteria (presence of cancer or diabetes mellitus, Hoehn and Yahr less than stage II even during off periods).

. 6 patients were lost to follow-up before six months
. 1 died from cancer
. 16 dropped out of the protocol
    * 9 for lack of efficacy
    * 5 for side-effects
    * 2 for lack of efficacy + side-effects

Forty two patients were followed-up for more than 6 months.

. 5 were lost to follow-up before 12 months
. 8 dropped out
    * 5 for lack of efficacy
    * 2 for side-effects
    * 1 for lack of efficacy + side-effects

Twenty nine patients were still in treatment after one year of follow-up. All patients were ranked in Hoehn and Yahr staging, stage I to IV in ON periods, stage III to V in OFF periods.

Student tests were used for statistical analysis.

The tolerability of treatment was good, only ten patients dropped out because of side-effects
. 6 for appearance or dramatic increase of abnormal movements
. 2 for orthostatic hypotension
. 1 for nausea
. 1 for intestinal occlusion

The results of this study confirm the present knowledge of Madopar[R] HBS.

1) Because of bioavailability of HBS form it is necessary to increase the intake of l-dopa (38% at 6 months and 45% at 12 months).

2) The switch from standard dopa therapy to Madopar[R] HBS is the major difficulty particularly in the first two months.

3) Patients who have some benefit of Madopar[R] HBS after two months continue to show improvement between 6 and 12 months, particularly, with respect to :

. Duration of ON periods

. Duration of abnormal movements

The increase of ON periods and the decrease of duration of abnormal movements between 6 and 12 months are not proportional to the increased dosage of l-dopa.

This is perhaps due to a delay of action of Madopar[R] HBS.

4) The tolerability of Madopar[R] HBS is good, only 10 patients dropped out because of side-effects.

Six of them dropped out because of an increase of abnormal movements. However, this phenomenon may be explained by the lack of efficacy or on the contrary by an overdosage.

After the difficulties of adaptation at the beginning of treatment 23 patients (35,4%) dropped out of the protocol for various reasons (lost to follow-up, cancer, etc...). Nevertheless, at 6 months 42 patients (64,6%) were still in treatment and for 36 of them the efficacy of Madopar[R] HBS was very clear. At 12 months 29 out of 42 patients at six months were still followed-up with very clear efficacy for 25 of them.

Table 1

RESULTS

	D O (N = 65)	6 MONTHS (N = 42)	12 MONTHS (N = 29)
Duration of "ON" Periods (hours)	6.7 $\pm$ 2.3	8.5 $\pm$ 3.2 + 20% - $P < 0.01$	9.4 $\pm$ 3.2 + 40% - $P < 0.01$
Total Score "ON" (0-12)	4.2 $\pm$ 2.7	3.4 $\pm$ 2.7 + 19% - N.S.	3.6 $\pm$ 3.5 - 14% - N.S.
Total Score "OFF" (0-12)	11.9 $\pm$ 2.7	9 $\pm$ 4.2 - 24% - $p < 0.001$	10.9 $\pm$ 4.8 - 8% - N.S.
Duration of AM (hours)	2.8 $\pm$ 2.7	2.4 $\pm$ 2.8 - 14% - N.S.	1.6 $\pm$ 1.6 - 43% - $p < 0.01$
Intensity of AM (0-4)	1.5 $\pm$ 1.2	1.3 $\pm$ 1.2 - 13% - N.S.	1.2 $\pm$ 1.1 - 20% - N.S.
Gait + Bradykinesia (Columbia)	5 $\pm$ 1.7	4.2 $\pm$ 1.7 - 16% $p < 0.02$	4.1 $\pm$ 1.6 - 18% - $p < 0.02$
NWUDS	36 $\pm$ 7.2	37.4 $\pm$ 8.2 4% - N.S.	36.8 $\pm$ 8.5 2% - N.S.
Dosage of Levodopa/ Day (MG)	580 $\pm$ 240	800 $\pm$ 280 + 38% $p < 0.001$	840 $\pm$ 270 - 45% - $p < 0.001$
Number of Intakes/ Day	5.7 $\pm$ 2.4	5.4 $\pm$ 1.9 - 5% - N.S.	5.7 $\pm$ 1.7 0% - N.S.

REFERENCE

MOURADIAN M.M., CHASE T.N. (1988).
Central Mechanisms and Levodopa Response Fluctuations in Parkinson's Disease.
Clin Neuropharmacol, 11 : 378-385.

FOUR YEARS OF EXPERIENCE IN THE TREATMENT OF PARKINSONIAN PATIENTS WITH A CONTROLLED-RELEASE LEVODOPA (HBS)

Carlos Chouza, Ruth Aljanati, Alejandro Scaramelli, José L. Caamaño, Ofrenda de Medina, Ricardo Buzó, Alfredo Fernández and Susana Romero

Section Extrapyramidal Diseases. Instituto de Neurología
Facultad de Medicina, Montevideo, Uruguay

INTRODUCTION

Fluctuations of mobility, dyskinesias and dystonias represent the most frequent complications and invalidating symptoms in parkinsonian patients under chronic dopatherapy. Peripheral factors, particularly variations in plasma levels of dopaminergic agents seem to be the main cause of fluctuations, although central factors are also involved. Long-term standard levodopa administration progressively tends to induce sharp peaks of plasmatic levodopa and consequently a shorter duration of the antiparkinsonian effect following each intake of the drug. Dyskinesias are also pathophysiologically related with these steep and periodically repeated variations of the plasmatic level of the drug.

A more stable plasmatic concentration of levodopa would induce smoother clinical changes and more protracted improvement periods, due to a more physiological delivery of the dopaminergic agent to the specific neurons.

New modes of antiparkinsonian drug administration with a longer half-life and more steady plasma levels were searched for several years.

A new formulation of levodopa with a controlled-release pattern, the hydrodinamically balanced system or "floating capsule" (levodopa/benserazide HBS) was developed (3).

Twenty two parkinsonian patients, most of them suffering severe and long-dating fluctuations, were included in a trial, substituting standard levodopa by the new sustained-release form.

A long-term follow-up of 4 years comprising 12 patients on this group enabled us to conclude about the results under chronic treatment with levodopa HBS.

In a pilot study we have also administered this new formulation to 5 "de novo" parkinsonians during 3 years.

Our long-term study of both groups of parkinsonian patients had the aim of defining:
1) The real potential of the controlled-release levodopa to improve complicated patients under chronic dopatherapy.

2) Its capability to prevent or delay the evolution of benign fluctuations to severe motor oscillations of the on-off type.
3) The possible prevention or delay of the appearance of dyskinesias and fluctuations in "de novo" patients.

In a third stage of the trial, selegiline was associated to the controlled-release levodopa with the object of: 1) improving the long-term results enhancing and prolonging the antiparkinsonian response. 2) reducing the dose of levodopa. 3) reducing the latency of the therapeutic effect in the morning. 4) investigating the effect of selegiline on dyskinesias and dystonias.

PATIENTS

Two groups of patients were included in the trial. Group A consisted of 22 parkinsonian out-patients with fluctuations under chronic dopatherapy, who substituted standard levodopa by levodopa HBS. Ten of these subjects could not complete 4 years of follow up due to the following reasons: 3 patients, unsatisfactory response to levodopa HBS; 1,abdominal surgery (volvulus); 1,accidental death; 2,metastatic neoplasm; 3,extraclinical reasons. We therefore analyzed the remaining 12 patients who achieved 4 years of treatment with the controlled-release form. There were 8 males and 4 females, whose mean duration of the disease had been 13 years (range: 2-24 years). The mean duration of treatment under standard levodopa had been 10.1 years (range: 1-16 years). According to Hoehn and Yahr, 2 patients were in stage II, 4 patients in stage III and 6 in stage IV. The mean Webster Score was 19.4 (range: 10-27), and the type of fluctuations was an "end of dose impairment" in 5 patients and "on-off" oscillations in 7.

Group B consisted of 5 "de novo" parkinsonian patients who were included 1 year after the beginning of the trial. There were 4 males and 1 female. The mean duration of the disease had been 6.4 years (range: 2-15 years). One patient was in stage II of Hoehn and Yahr, 2 in stage III and 2 in stage IV. The mean Webster Score was 17 (range: 10-20).

METHOD

Our method included the quantitative assessment of: 1) Degree of parkinsonism by means of Webster Score obtained every 30 minutes during 12 hours, therefore collecting 25 values, in each control. The sum of these scores yields on overall value of parkinsonism during the day ("Day Webster Score"). 2) "Early morning parkinsonism" (the Webster Score in the morning before the first intake of the drug). 3) Duration of ON periods, summing the ON periods in minutes. 4) Dyskinesias and dystonias were continuously recorded during the 12 hour-periods, according to type, intensity, onset, duration and frequency.

These controls were carried out from 8 a.m. to 8 p.m. once a month during 2 years and every 3 months thereafter.

The Stages of the long term trial for Group A included: A) a dose adaptation phase, substituting standard levodopa by the same dose of levodopa HBS. As an impairment of parkinsonism was then usually observed, an increase of the dose was required during the following days or weeks. B) a long term follow-up. Here, the size and distribution of levodopa HBS doses were adjusted until optimal clinical benefit (improvement of parkinsonism and fluctuations, and reduction of dyskinesias). C) association of selegiline during 9 months (from the 39th. month until the end of the trial). This was carried out in the 12 patients of group A who were analyzed

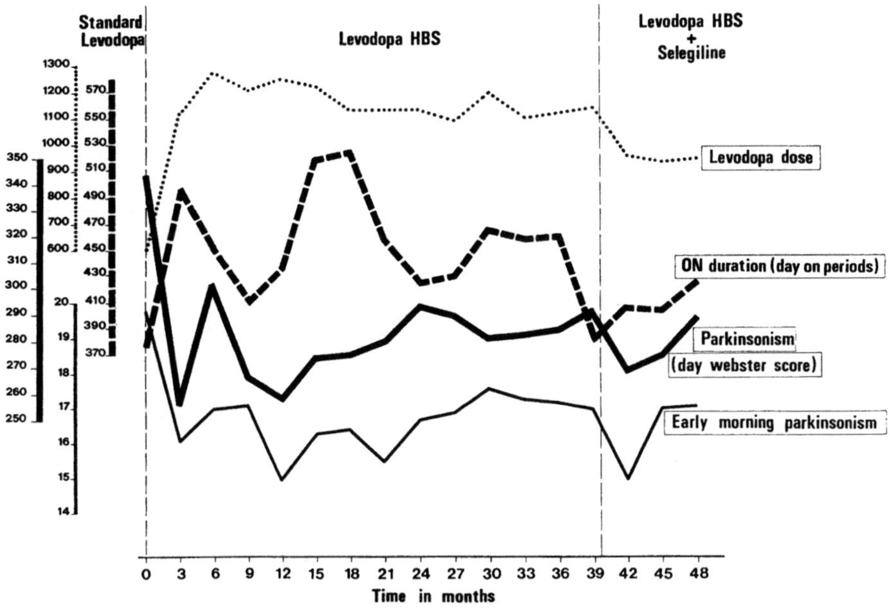

Fig. 1. Evolution of Daily Parkinsonism (DWS), Early Morning Parkinsonism, ON Duration (Day on Periods) and Levodopa HBS Dose During 4 Years.

during 4 years and in 1 "de novo" parkinsonian patient. The dose of selegiline varied between 5 and 10 mg/day, while reducing levodopa HBS.

For group B, monthly evaluations were carried out, quantifying parkinsonism and paying special attention to the appearence of dyskinesias, dystonias or fluctuations. The initial dose was 300 mg/day of levodopa HBS, followed by a gradual increase up to the optimal dose.

Statistical studies were carried out applying the t test and Wilcoxon test.

RESULTS

I) For Group A

1) The following changes after substitution were observed (Fig.1): a) Reduction of the Day-Score of Parkinsonism in 19.2% of patients. b) On periods were longer with a mean increase of 51%. c) Early morning parkinsonism practically disappeared in "end of dose impairment" cases and reached a reduction of 13.2% in ON-OFF cases, etc. d) Fluctuations showed a remarkable decrease or disappearance in benign fluctuations but achieved only a mild-moderate decrease in ON-OFF cases. e) The latency of therapeutic effect after the first drug intake in the morning was 62% longer. f) A moderate to marked decrease of dyskinesias was observed in 6 out of 8 patients. Five out of 7 parkinsonians also showed reduction or disappearence of dystonias.

2) During the long-term follow-up, the best results with the maximal peak of improvement were recorded (Fig.1): -For parkinsonism (Day Webster Score) at 3 months (25.9%) and 12 months (25%) of the substitution. -For early morning parkinsonism at 12 months (24.2%) (Fig.2). -For ON periods duration at 18 months (39%).

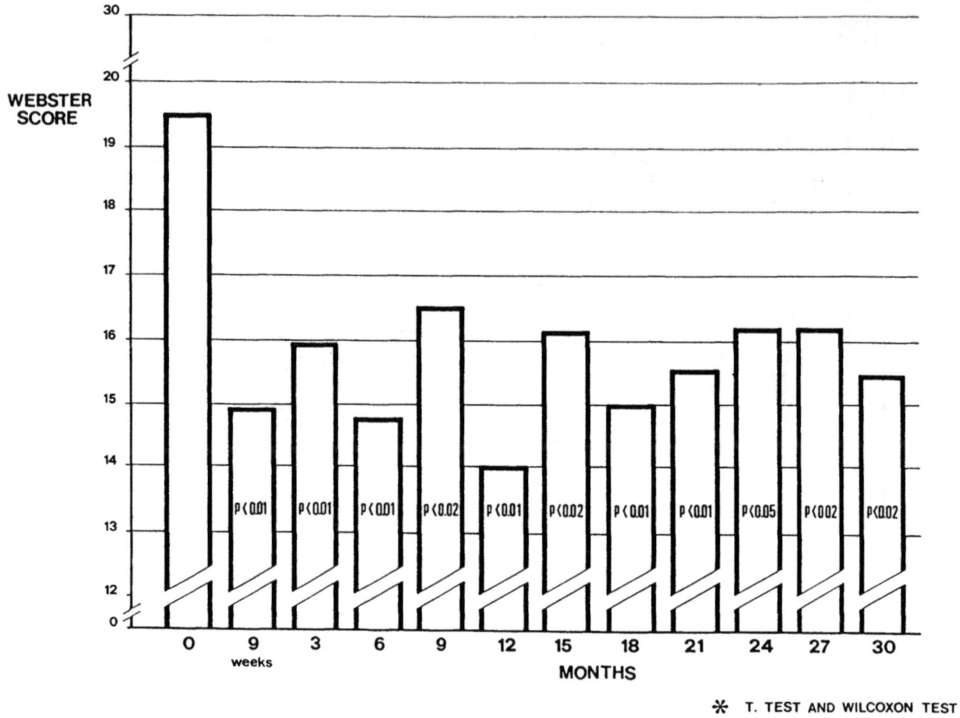

Fig. 2. Long-term Treatment with Madopar HBS: Evolution of Morning Webster Score (average values).

Table 1. Group B "de novo" patients treated with levodopa HBS during 3 years

Patient	Maximal Improvement %	Final Improvement %	Final dose (in mg of levodopa)	Fluctuations	Dyskinesias
P.P.	53	53	700	-	-
J.G.	70	*	800	-	-
M.P.	37	11	900	-	-
P.L.	50	60	800	-	-
A.A.	68	79	600	Yes	Yes

* Patient J.G. died after 11 months of treatment due to abdominal carcinomatosis

The final results at 39 months under levodopa HBS alone were: -For parkinsonism, 15.3%; -early morning parkinsonism, 11% and ON duration, 1%.

Concerning the doses of Levodopa HBS, the maximal mean dose was administered at 6 months (1280±380 mg/day), which means an increase of 113% in relation to standard levodopa. The minimal mean dose under levodopa HBS alone was recorded at 27 months: 1090±160 mg/day (81.6% increase).

II) For Group B ("de novo" patients, the results are summarized in table 1.

Tolerance was excellent. The recording of vital signs and the laboratory tests performed did not show any significant alteration.

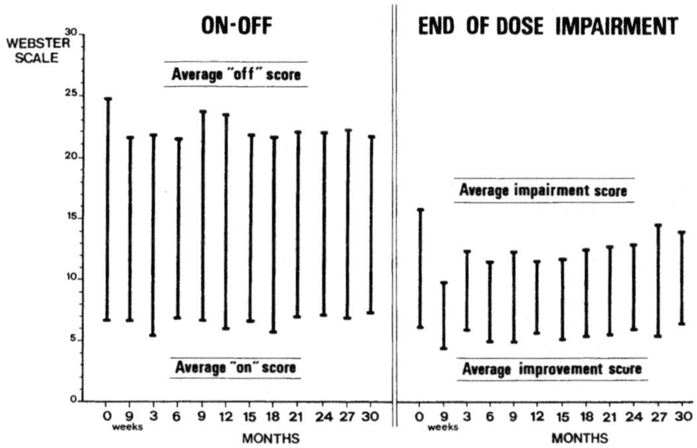

Fig. 3. Evolution of the "ON" and "OFF" Values: Comparative Study of the Motor oscillations between "on-off" and "end of dose impairment" groups.

III)The results after the association of selegiline, showed a mild overall improvement of parkinsonism in spite of reduction of levodopa doses, as well as non significant longer ON periods. The early morning parkinsonism was significatively reduced (24.2%), reaching the same maximal decrease achieved at 1 year of treatment under levodopa HBS alone. End of dose impairment cases achieved better results than "ON-OFF" cases. The best result was obtained in the "de novo" patient with juvenile parkinsonism with an improvement in Webster Scores and a reduction in fluctuations and dyskinesias.

The decrease of levodopa doses were 15.8%, 18% and 17.5% at 42, 45 and 48 months respectively.

A mild decrease of dyskinesias and a mild-moderate enhancement of dystonias ware observed.

Concerning tolerance, non-serious side-effects (somnolence, insomnia, anorexia, headache) were observed, with a drop-out because of epigastralgia.

CONCLUSIONS

Our previous results evaluating the efficacy of levodopa HBS appear elsewhere (1,2).

The outstanding beneficial effects after substitution of standard levodopa/benserazide by the sustained-release levodopa HBS were: enhancement of the antiparkinsonian effect, improvement or disappearance of motor oscillations with longer on periods, shorter and less severe off periods, as well as smoother changes from "on" to "off" states and viceversa. Likewise, a decreased or absent early morning parkinsonism and a more sustained nocturnal antiparkinsonian response, including a reduction of nocturnal pain and dystonias, with a better sleep were also observed.

"End of dose impairment" cases showed better results than "on-off" cases (Fig. 3).

"Predictible on-off"oscillations also showed greater improvements than the "unpredictible" ones.

The most positive results were achieved after 1 year of treatment with levodopa HBS. A sustained tendency to mild-moderate deterioration was recorded after this peak of improvement. However, practically all the analyzed parameters remained within statistically significant values of improvement including the last clinical control at 48 months, when compared to standard levodopa. We stress the relatively preserved benefits in most patients during this long period of time, considering that most parkinsonians were suffering from a prolonged and severe"on-off"effect. This long term follow up implies 4 extra years of aging, of the course of the disease as well as of a potentially iatrogenic treatment.

No new cases of severe fluctuations or "on-off" oscillations evolving from the group of benign fluctuations were recorded during this long period of time. This fact probably means that a sustained release levodopa could prevent or delay the development of "on-off" states. In this sense, three out of the 4 "de novo" parkinsonians who reached 3 years of treatment under levodopa HBS showed neither fluctuations nor dyskinesias or dystonias, indicating a promising evolution. However, the remaining patient, who had a juvenile parkinsonism, developed dyskinesias and fluctuations after 7 months of treatment. The sustained-release levodopa was not capable of preventing the appearance of these complications in this early and particular form of the disease.

The negative aspects of the administration of levodopa HBS are: a) a longer latency of the therapeutic response in the morning. This can be compensated by the partial substitution of levodopa HBS in the morning by standard levodopa. b) according to our experience, and that of other authors (1,2,4,5), a lower bioavailability of the sustained-release form, which demanded an increase of the dose. In spite of such increase, ranging from 80 to 100%, diskinesias did not show a worsening, which in its turn allowed the mentioned increase. The avoidance of sharp and repeated plasmatic peaks of levodopa and their lower but more sustained levels under levodopa HBS are the probable reasons to explain the reduction of both peak-dose and biphasic dyskinesias. The decrease or disappearance of dystonias, generally related to the impairment periods and linked with low plasmatic levels of levodopa, may be explained by the more sustained dopaminergic effect.

As levodopa HBS is a sustained-release preparation, we initially reduced the number of daily intakes. Nevertheless we had to increase it again in order to obtain a further improvement. In a few cases with poor responsiveness to levodopa HBS, a higher number of intakes was required to achieve the best results.

The association of selegiline during the last 9 months of follow-up, was associated with a significant decrease of early morning parkinsonism and reduction of motor disability. "Wearing-off" cases also showed a better response than the "on-off" cases. A mean reduction of 16 to 18% in the dose of levodopa could be achieved. A mild reduction of dyskinesias and mild-moderate increase of dystonias were recorded.

Levodopa HBS does not solve the problem of fluctuations but it induces the improvement of several aspects in most patients. It could be a useful mean to mitigate the unconfortable and invalidating every day suffering of these complicated patients.

The ON-OFF effect is an irreversible state that responds poorly to the sustained-release levodopa. However, the administration of this form of levodopa in cases of benign fluctuations, in those under dopatherapy without fluctuations and in "de novo" patients, could render certain degree of prevention or a delay in the development of long-term complications. Longer and larger trials are needed to confirm these assumptions.

REFERENCES

1. Chouza C., Romero S., De Medina O., Aljanati R., Scaramelli A., Caamaño J.L., González Panizza V. Substitution of Standard Madopar by Madopar HBS in Parkinsonians with Fluctuations. Eur. Neurol. 1987: 27; suppl. 1, 59-67.
2. Chouza C., Aljanati R., Caamaño J.L., De Medina O., Scaramelli A., Buzó R., Plachin V., Fernández A., Romero S. Long-Term Treatment with Madopar HBS in parkinsonians with fluctuations. "Advances in Neurology." New York. Raven Press. 1989 (In press).
3. Erm W., Held K. The Hydrodynamically Balanced System: A Novel Principle of Controlled Drug Release. Eur.Neurol.1987; 27: suppl.1, 21-27.
4. Jansen E.N.H., Meerwaldt J.D., Heersema Th., Van Manem J., Speelman J.D. Open Multicenter Trial with Madopar HBS in Parkinsonian Patients. Eur. Neurol. 1987; 27: suppl. 1, 88-92.
5. Rinne U.K. Madopar HBS in the Long-Term Treatment of Parkinsonian Patients with Fluctuations in Disability. Eur. Neurol. 1987; 27: suppl. 1, 120-125.

ANTIPARKINSONIAN EFFECTS OF LISURIDE AND BINDING OF THE DRUG TO D-2 AND $5\text{-}HT_{1A}$ RECEPTORS IN CYNOMOLGUS MONKEYS

Motonori Yamaguchi,[1] Tetsuo Akai,[1] Eiji Mizuta,[2] and Sadako Kuno [2]

[1]Research Department, Nihon Schering K.K., Osaka, and [2]Department of Neurology, Utano National Hospital, Kyoto Japan

INTRODUCTION

Lisuride, a derivative of ergot alkaloid, has been shown to have serotonin $5\text{-}HT_{1A}$, as well as dopamine D-2, agonistic activity in rats.[1] Lisuride has been used in Europe for treatment of Parkinson's disease and migraine,[2] and in Japan for treatment of emotional deficits such as depression and anxiety that often accompany cerebrovascular diseases.[3] In this study, behavioral effects of lisuride in MPTP-induced parkinsonian monkeys were investigated, as well as binding of lisuride to D-2 and $5\text{-}HT_{1A}$ receptors in the monkey brain.

METHODS

Ten male cynomolgus monkeys (Macaca fascicularis, 4-5 Kg) were used. Five of them were treated with 3 or 4 injections of 0.35 mg/kg (i.v.) of MPTP hydrochloride at an interval of 3 or 4 days. Monkeys with persistent parkinsonian symptoms were used one month after the last MPTP injection. Five other monkeys were used as a control for behavioral studies. Parkinsonism and behavioral changes were scored by the rating scales (0-2) with the methods of Clarke et al.[4] and Mizuta et al.[5], respectively.

In vitro autoradiographical analysis was made of [^{3}H]-lisuride binding to D-2 and $5\text{-}HT_{1A}$ receptors in the brains of MPTP-treated parkinsonian and non-treated monkeys. [^{3}H]-lisuride (0.125-4 nM) was incubated with brain sections at room temperature for 1 hr. Specific bindings for D-2 and $5\text{-}HT_{1A}$ receptors were defined as binding inhibited by 10 µM sulpiride and 1 µM 8-OH-DPAT, respectively. Labeled sections were apposed to Amersham Hyperfilm for 100 days at 4°C. Receptor imaging and quantitative analysis of lisuride binding were carried out with an Amersham imaging analyzer (RAS3000).

RESULTS

Binding of Lisuride to 5-HT1A and D-2 Receptors

The autoradiographical binding pattern of [^{3}H]-lisuride in the

Basic, Clinical, and Therapeutic Aspects of Alzheimer's and Parkinson's Diseases, Volume 2
Edited by T. Nagatsu *et al.*, Plenum Press, New York, 1990

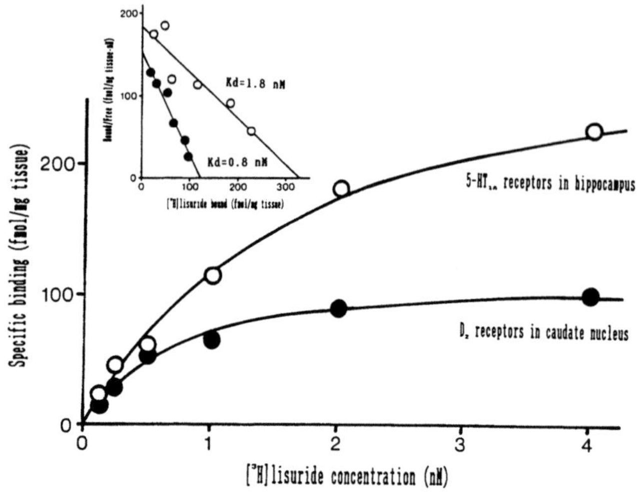

Fig. 1. Saturation curves and Scatchard plots of [^{3}H]lisuride binding to D-2 and 5-HT_{1A} receptors in the MPTP-induced parkinsonian monkey brain. Specific bindings to D-2 (●) and 5-HT_{1A} (○) receptors were indicated in the caudate nucleus and hippocampus, respectively.

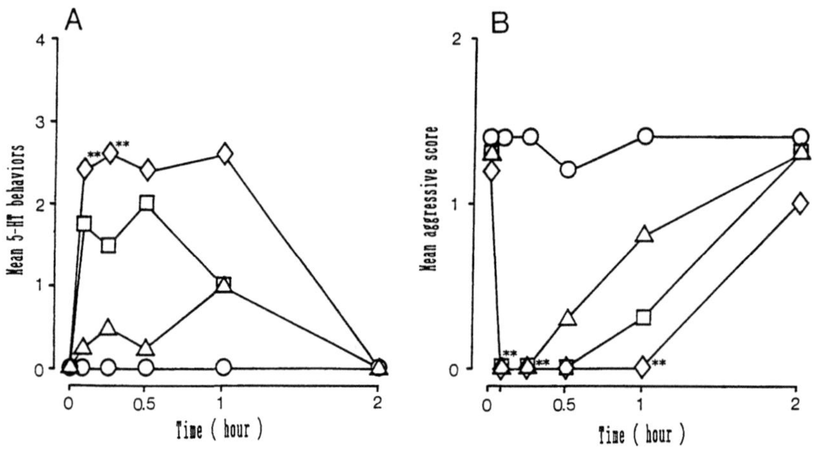

Fig. 2. Behavioral effects of lisuride on non-treated monkeys. A, 5-HT behaviors (head weaving, upper limb fluttering and lower limb extension). B, aggressiveness. The behaviors were scored by rating scales (0-2), and the values given are the mean scores of 5 monkeys per condition. $**p < 0.01$ vs. saline. ○, saline (s.c.), △, 0.05, □, 0.1, ◇, 0.2 mg/kg (s.c.) of lisuride.

parkinsonian monkey's brain showed high densities of sites displaceable by the 5-HT_{1A} agonist 8-OH-DPAT in the hippocampus (particularly CA1) and cortex (superficial and deep layers) in addition to a lesser density of sites displaceable by a D-2 antagonist, sulpiride, in the basal ganglia (particularly the caudate and putamen) . The distributions of lisuride binding to 5-HT_{1A} receptors and D-2 receptors were similar to those as determined by autoradiographs with [^{3}H]-8-OH-DPAT and [^{3}H]-spiperone, respectively, in the parkinsonian monkey brain. Scatchard analysis of

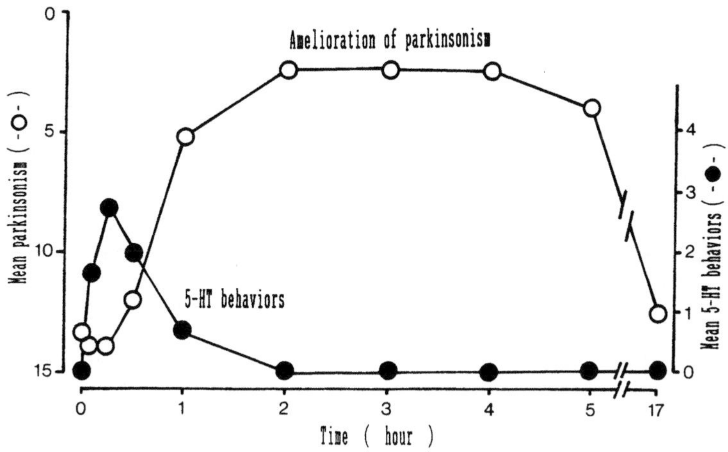

Fig. 3. Time course of behavioral effects of lisuride (0.05 mg/kg, s.c.) on MPTP-induced parkinsonian monkeys. The scorings for 5-HT behaviors (head weaving, upper limb fluttering and lower limb extension) and parkinsonism were assessed by rating scales (0-2). The values given are the mean scores of 3 monkeys per condition.

lisuride bindings in the hippocampus and caudate showed high-affinity binding of lisuride to both $5\text{-}HT_{1A}$ and D-2 receptors (Kd = 1.8 nM for $5\text{-}HT_{1A}$ receptor, Kd = 0.8 nM for D-2 receptor, Fig.1). Almost the same distribution pattern and affinity of lisuride binding in the brains from non-treated monkeys were obtained, although higher densities of binding sites were observed.

Behavioral Effects of Lisuride on Non-treated Monkeys

Lisuride (0.05-0.2 mg/kg, s.c.; Fig. 2-A) as well as 8-OH-DPAT[5] induced dose-dependently distinct behaviors characterized by head weaving, lower limb extension and upper limb fluttering, which appeared to be similar to the 5-HT behavioral syndrome previously noted in rodents. The aggressiveness in monkeys was also decreased in a dose-dependent manner (Fig. 2-B). DA agonists such as apomorphine (0.1-0.6 mg/kg, s.c.) and bromocriptine (5 mg/kg, s.c.) did not induce such behavioral effects but did induce irritation and excitement. These results suggest that the characteristic behaviors induced by lisuride are mediated by its $5\text{-}HT_{1A}$ agonistic action. Lower doses of lisuride (0.0125-0.025 mg/kg, s.c.) did not produce any sign of behavioral change.

Antiparkinsonian Effects of Lisuride in MPTP-treated Monkeys

Lisuride (0.05-0.2 mg/kg, s.c.) induced the 5-HT behaviors such as head weaving and lower limb extension for 30-60 min, and this was followed by marked amelioration of the parkinsonian symptoms for a period of more than 5-6 hr (Fig. 3). Lower doses of lisuride (0.005-0.0125 mg/kg, s.c.), apomorphine (0.03-0.3 mg/kg, s.c.), and terguride[6] (partial DA agonist, 0.01-0.1 mg/kg, s.c.) improved the parkinsonism without any sign of the above 5-HT behaviors.

DISCUSSION

Recently $5\text{-}HT_{1A}$ receptor agonists have been studied with special reference to their anxiolytic and antidepressant effects, because the $5\text{-}HT_{1A}$ agonists buspirone, gepirone, and ipsapirone have been shown to have anxiolytic effects. Such drugs are undergoing trials as antidepressants.[7] In this study lisuride was shown to have potent agonistic activity toward $5\text{-}HT_{1A}$ receptors in the monkey brain. Lisuride as well as 8-OH-DPAT was shown to have taming effect on aggressive monkeys, which is used as an animal model for screening of anxiolytics. These results are consistent with biochemical and behavioral studies of lisuride in rats.[1,8] Lisuride has been reported to have anxiolytic-like and antidepressant-like effects in water-lick conflict tests[1] and forced swimming tests[8] in rats. Thus, the present results together with previous observations suggest that lisuride has clinical anxiolytic and antidepressant activities that may be mediated by its $5\text{-}HT_{1A}$ action. Lisuride has already been shown to be useful for treatment of depression and anxiety that often accompany cerebrovascular disorders.[3]

The main symptoms in Parkinson's disease are well known to stem from extrapyramidal motor dysfunction and can be treated by supplementation therapy with dopamine. However, in addition to the main symptoms, patients with Parkinson's disease often also display psychiatric disorders such as depression and anxiety, which do not consistently respond to any of the usual forms of therapy including tricyclic antidepressants and benzodiazepine anxiolytics. Recent biochemical observations in patients with Parkinson's disease suggest that a reduction in 5-HT level in extrastriatal brain regions may be related to the disorder of mood and affect seen in Parkinson's disease.[9] In this study, lisuride was found to have not only dopaminergic effects (D-2 agonist) but also serotonergic effects ($5\text{-}HT_{1A}$ agonist) in the MPTP-induced parkinsonian monkeys. Clinical trials of other serotonergic as well as dopaminergic agonists may be warranted for treatment of depression and anxiety in Parkinson's disease.

REFERENCES

1. T. Akai, Y. Nakada, Y. Ikoma, M. Yamaguchi, H. Nakagawa, and N. Oshino, Anxiolytic effects of lisuride and its binding to $5\text{-}HT_{1A}$ receptor in rat brain, Japan.J.Pharmacol. 49: 63P (1989).
2. D. B. Calne, A. Horowski, R. J. McDonald, and W. Wuttke, "Lisuride and Other Dopamine Agonists," Raven Press, New York (1983).
3. Y. Goshima, Y. Tazaki, and E. Ootomo, Clinical effects of Eunal in treatment of cerebrovascular disorders, Geriat.Med. 16: 757 (1978).
4. C. E. Clarke, M. A. Sambrook, I. J. Mitchell, and A. R. Crossman, Levodopa-induced dyskinesia and response fluctuations in primates rendered parkinsonian with MPTP, J.Neurol. Sci. 78: 273 (1987).
5. E. Mizuta, M. Yamaguchi, and S. Kuno, Behavioral effects of 8-hydroxy-(di-n-propylamino)tetralin (8-OH-DPAT) in monkeys, Eur.J. Pharmacol. (in press).
6. H. Wachtel, Central dopaminergic and antidopaminergic effects of ergot derivatives structurally related to lisuride, in: "Lisuride and Other Dopamine Agonists, D. B. Calne, R. Horowski, R. J. McDonald, W. Wuttke, eds., Raven Press, New York, (1983).
7. A. J. Sleight and S. J. Peroutka, Drug interactions with $5\text{-}HT_{1A}$ receptors, DN & P 2 (5): 261 (1989).
8. K. Nakamura, Y. Ikoma, Y. Nakada, S. Kobayashi, M. Yamaguchi, and H. Nakagawa, Effects in animal models of depression of lisuride alone and upon co-administration with antidepressants, Folia pharmacol. Japan 94: 81 (1989).
9. O. Hornykiewicz and S. J. Kish, Biochemical pathophysiology of Parkinson's disease, in: "Advances in Neurology," Vol. 45, M. D. Yahr and K. J. Bergmann, eds., Raven Press, New York (1986).

TREATMENT OF MOTOR FLUCTUATIONS IN PARKINSON'S DISEASE WITH SUBCUTANEOUS APOMORPHINE

W. Poewe[1], M. Wagner[1], B. Kleedorfer[1], and W.H. Oertel[2]

1: Univ.-Klinik für Neurologie, A-6020 Innsbruck
Anichstrasse 35
2: Neurologische Klinik, Klinikum Großhadern, D-8000
München, FRG

INTRODUCTION

More than 50% of patients with idiopathic Parkinson's disease will develop oscillations in motor performance after sustained treatment with L-Dopa for more than 5 years [1]. The pathogenetic mechanisms responsible for these motor fluctuations are incompletely understood but peripheral and central factors of L-Dopa pharmacokinetics are likely to be involved[2]. Therapeutic efforts to control L-Dopa response oscillations have recently concentrated on modes of more constant drug delivery and subcutaneous infusions of dopamine agonists have been introduced as a means of treating patients with motor fluctuations refractory to conventional treatment. Lisuride was first to be succesfully employed as continuous subcutaneous infusions via minipumps in such patients [3], and followed more recently by the use of apomorphine [4,5].

PATIENTS AND METHODS

31 patients with idiopathic Parkinson's disease and motor fluctuations with sustained oral L-Dopa substitution have so far been treated with subcutaneous apomorphine. Their clinical data are summarised in table 1.

Table 1

PATIENT DATA
(N = 31)

	PEN	PUMP
N	17 (5 f.; 12 m.)	14 (3 f.; 11 m.)
Age (yrs.)	58,5 (45-76)	55 (44-72)
Duration (yrs.)		
- disease	9.9 (3-18)	13 (5-24)
- L-Dopa	9.0 (2-15)	10.2 (2-18)
Hoehn + Yahr stage		
- 'on'	1.4 (1-3)	1.5 (1-3)
- 'off'	3.6 (3-4)	3.8 (3-5)
L-Dopa dose (mg/d)	1013 (500 - 2250)	1269 (600 - 2400)

Basic, Clinical, and Therapeutic Aspects of Alzheimer's and Parkinson's Diseases, Volume 2
Edited by T. Nagatsu *et al.*, Plenum Press, New York, 1990

Apomorphine HCl solution (10 mg/ml) was administered either intermittently via pen injectors if there were less than 5 incapacitating off-periods per day or continuously via minipumps when more frequent daily oscillations occured. 24 hours prior to commencing apomorphine treatment with oral domperidone (20mg t.d.s.) was instituted and initial doses given as single injections were 2 mg while initial infusion rates were 2mg/h administered over 24 hours. Dose adjustments were made under in-patient conditions until a sufficient clinical response had been reached, including concomitant reductions of the daily L-Dopa dose.

Clinical efficacy was measured by means of self-scoring on-off diaries which were kept by all patients or a close relative, starting at least one week prior to apomorphine and throughout the entire in-patient dose finding phase as well as for one week prior to each out-patient follow-up visit performed at 6-weekly intervals until month 3 and 3-monthly thereafter.

RESULTS

Patients in both treatment groups (intermittent and continuous apomorphine) have now been followed up for up to 15 months and dosing parameters are summarised in table 2. Significant reductions of total daily infusions while L-Dopa requriements did hardly change with additional intermittent apomorphine. 2 patients in the pump group were able to discontinue oral treatment completely with continuous infusions of 9 and 11mg/h of apomorphine.

The number of daily hours with poorly controlled parkinsonism ('off') could be markedly reduced with either type of apomorphine administration (table 2), but complete abolition of motor oscillations was only observed in patients with continuous infusions (figure).

Table 2

TREATMENT RESULTS

	PEN (N = 17		PUMP (N = 14)	
Follow-up (mths.)	7.2	(3-15)	2.5	(0.5-13)
Apomorphine dose (mg)				
- single	3.8	(2.5-8.0)		
- hourly	-		4.6	(3.0-11.0)
- daily	12,2	(5-32)	110	(63-264)
L-Dopa dose (mg/d)				
- before Apo.	1013	(500-2250)	1269	(600-2400)
- with Apo	936	(500-2250)	617	(300-1200)
Hours 'off' per day				
- before Apo.	4.7	(2-7)	6.1	(2-10)
- with Apo.	1.7	(0,5-4)	1.3	(0-5

In addition to reducing the total daily number of hours 'off'-period dystonia with onset of effect usually occuring within 5 to 15 minutes. Duration of 'on' phases produced by single subcutaneous apomorphine doses was between 45 and 90 minutes in most cases.

Peak-dose dyskinesias increased in duration parallel to increases in daily 'on' time in those patients affected but were not enhanced in severity with the type of dose titration employed in this study.(i.e. parallel L-Dopa dose reductions with increasing apomorphine wherever possible).

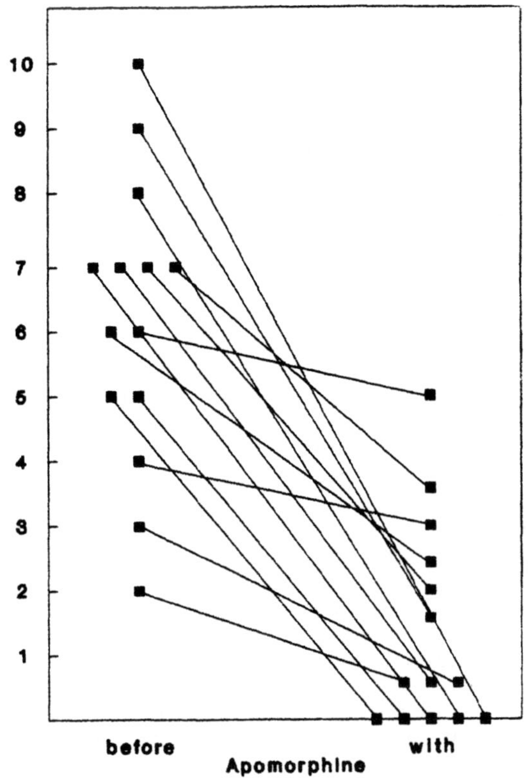

Figure 1

Daily hours 'off' before and after adding continuous subcutaneous apomorphine infusions in 14 patients with Parkinson's disease.

Other side effects included subcutaneous nodules at needle sites which were easily tolerated by all patients and mild and transient blood eosinophilia in some cases. Nausea and vomiting could be prevented by the coadministration of oral domperidone in all patients and most developed tolerance to the emetic effects of apomorphine so that domperidone could be reduced or discontinued after 2 or 3 months. So far no patient has developed apomorphine induced psychosis except one with transient visual delusions under high dose continuous infusions of 10mg/h.

Over follow-up periods of up to 15 months no signs of declining apomorphine efficacy or increasing dose requirements have yet been clearly identified.

DISCUSSION

The present results confirm previous reports about the safety and efficacy of subcutaneous apomorphine in the treatment of Parkinson patients with motor fluctuations refractory to conventional therapy [4,5,6].

Intermittent subcutaneous injections of apomorphine as first described to be effective in Parkinson's disease more than 30 years ago [7] were employed via portable penjects in those patients with a limited number of daily off-periods. Although a marked reduction in the mean number of daily hours spent in the 'off'-state could be achieved in this group, an equally important aspect of the clinical benefit experienced by these patients was the possibility to quickly and reliably interrupt intervening off-periods with or without painful dystonia by apomorphine self-administration. Contrary to recently published experimental evidence for the rapid development of tolerance with repeated dosing of subcutaneous apomorphine [8] this was not observed in this or other reported groups of patients using multiple daily doses via penjects over up to 15 months [4].

The reduction of response oscillations was even more impressive in those patients receiving continuous subcutaneous infusions of apomorphine via minipumps with complete abolition of motor swings and discontinuation of oral treatment in some. In individual cases there was even a tendency for slightly reduced dose requirements and/or decreasing peak-dose dyskinesia with long-term subcutaneous infusions, further supporting the notion that continuous dopaminergic stimulation might reverse pharmacodynamic receptor changes induced by long-term intermittent stimulation with oral L-Dopa [2]. On the other hand peak-dose dyskinesias, albeit not increased in severity when pump treatment was accompanied by L-Dopa dose reductions, nevertheless occupied greater parts of the waking period in most of those affected so that definite conclusions about the feasability of continuous apomorphine administration in patients with disabling involuntary movements will have to come from further studies.

Psychotic reactions, which have been of major concern with subcutaneous infusions of lisuride [3,9], were virtually absent in this group of apomorphine treated patients [6]. The reasons for this apparent difference between the two agonists so far used for subcutaneous treatment are still unclear although criteria of patient selection may have played a role in previous studies.

Until further progress in the field of improved dopaminergic drugs and corresponding delivery systems has been made subcutaneous apomorphine appears as a safe and efficious approach to the treatment of fluctuating Parkinson's disease.

REFERENCES

1. K. M. Shaw, A. J. Lees, G. M. Stern, Low-dose therapy in Parkinson's disease: A 6-year follow-up study. Neurology 36: 1528 (1986).
2. J. G. Nutt, On-off phenomenon: Relation to levodopa pharmacokinetics and pharmacodynamics, Ann. Neurol. 22:535 (1987).
3. J. A. Obeso, M.R. Luquin, and J. M. Martinez-Lage, Lisuride infusion pump: a device for the treatment of motor fluctuations in Parkinson's disease. Lancet 1:467 (1986).
4. C.M.H. Stibe, A. J. Lees, P. A. Kempster, and G. M. Stern, Subcutaneous Apomorphine in Parkinsonian on-off oscillations. Lancet 1:403 (1989).
5. W. H. Poewe, B. Kleedorfer, F. Gerstenbrand and W. H. Oertel, Subcutaneous Apomorphine in Parkinson's disease. Lancet 1: 943 (1988).
6. W. Poewe, B. Kleedorfer, M. Wagner, T. Benke, T. Gasser, and W. Oertel, Side-effects of subcutaneous apomorphine in Parkinson's disease Lancet 1: (1989).
7. R. S. Schwab, L. V. Amador, and J. Y. Levine, Apomorphine in Parkinson's disease. Trans. Am. Neurol. Ass. 76:251 (1951).
8. F. Grandas and J. A. Obeso, Motor response following repeated apomorphine administration is reduced in Parkinson's disease. Clin. Neuropharm. 12:14 (1989).
9. P.H.S. Critchley, F. Grandas-Perez, N. P. Quinn, J. D. Parkes and C. D. Marsden, Continuous subcutaneous lisuride infusion in Parkinson's disease. J. Neural Transm. (suppl.) 27:55 (1988).

LISURIDE AND PROTERGURIDE REVERSE MPTP-INDUCED PARKINSONISM IN THE COMMON MARMOSET

P.-A. Löschmann[2], K.J. Rettig[2], H. Wachtel[2], R. Dorow[2], P. Jenner[1] and C.D. Marsden[3]

[1]Department of Pharmacology, King's College, London W3, UK, [2]Research Laboratories of Schering AG Berlin, FRG [3]Institute of Neurology, The National Hospital, London, UK

INTRODUCTION

Administration of MPTP to man or non-human primates (1,2,3) including common marmosets (4) induces a parkinsonian syndrome due to the selective destruction of nigro-striatal neurons. Thus, the MPTP-treated monkey is the animal model of choice for Parkinson's disease , as it resembles very closely the histopathology as well as clinical features of the disease in humans. In pharmacological experiments MPTP-treated common marmosets respond to classical antiparkinsonian drugs such as L-Dopa in combination with a peripheral decarboxylase inhibitor (4), or dopamine receptor agonists such as (+)-PHNO (5), Quinpirol (6) and (-)-N-0437 (7). We used this model to compare the antiparkinsonian activity of the clinically effective 8-α-aminoergoline lisuride with the structually very similar ergoline-derivative proterguride which has a higher potency and longer duration of action in rodents.

MATERIAL AND METHODS

Eight common marmosets (Callithrix jacchus) of either sex (250 - 390 g body weight, 3-8 years old) were used. The animals were rendered parkinsonian by subcutaneous (s.c.) administration of 1-2 mg/kg 1-methyl-4-phenyl-1,2,3,6-tetrahydropyridine (MPTP) for up to six days. The cumulative doses administered ranged between 10-12 mg/kg. Four animals were treated again with 2 mg/kg MPTP s.c. on four consecutive days one year later. Following MPTP-treatment the animals were allowed to recover from the acute effects for 6 weeks before behavioural testing began. Lisuride (3-(9,10-didehydro-6-methyl-8α-ergolinyl)-1,1diethyl urea hydrogen maleate) and proterguride (1,1-diethyl-3(6-n-propyl-8α-ergolinyl)-urea) were suspended in 0.9% sterile saline containg 10% w/v Cremophor EL (polyethoxylated castor oil, BASF, FRG) and administered in a volume of 1 ml/kg intraperitoneally (i.p.) or by oral gavage (p.o.).

Edited by T. Nagatsu *et al.*, Plenum Press, New York, 1990

Domperidone was suspended in a few drops of ethanol (70%) and diluted to volume (2 ml/kg body weight) with 10% sucrose/water solution and administered p.o. 30 min prior to test compounds. Groups of four animals were treated with vehicle or one of the three doses of DA-agonists in weekly intervals. A latin square design was used for the allocation of treatments within the groups. Locomotor activity was measured simultaneously in four aluminium cages identical to the home cages but equipped with eight infrared photocells. The number of light beam interruptions due to the animal's movements were accumulated in 10 min intervals for up to 480 min using a Commodore CBM 4032 computer. In addition the animals were observed during the experiments through a one-way mirror.

RESULTS

Lisuride (3.75, 7.50 and 15.0 µg/kg i.p.) dose-dependently stimulated locomotor activity of MPTP-treated common marmosets for 2 hours (fig. 1). When administered p.o. (62.5 and 250 µg/kg) lisuride stimulated locomotor activity for 2 and 7 hours, respectively (fig. 3). Proterguride given i.p. (37.5, 75.0 and 150 µg/kg) was active for more than 4 hours in MPTP-treated monkeys (fig. 2). After peroral gavage proterguride (9.4 and 37.5 µg/kg) induced hyperlocomotion in a biphasic manner; this effect lasted for more than 8 hours (fig. 3).

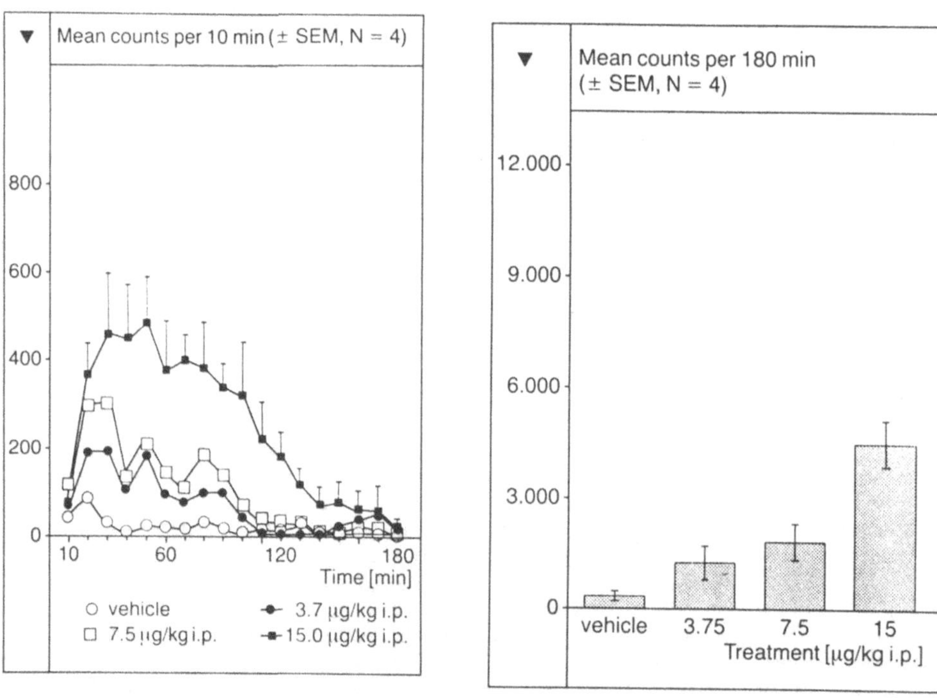

<u>Fig. 1.</u> Time course of locomotor activity of MPTP-treated common marmosets (left panel) pretreated with 2 mg/kg domperidone p.o. (t -30 min) and lisuride (3.75, 7.50 and 15.0 µg/kg i.p.) or vehicle (t 0 min). Activity counts accumulated over 3 hours are shown on the right panel.

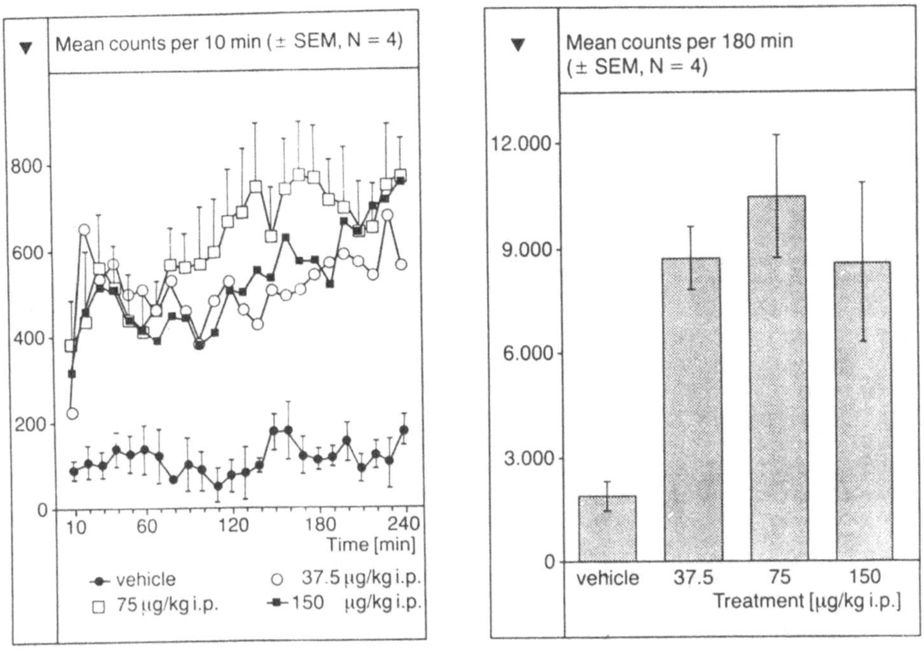

Fig. 2. Time course of locomotor activity of MPTP-treated common marmosets (left panel) pretreated with 2 mg/kg domperidone p.o. (t -30 min) and proterguride (37.5, 75 and 150 μg/kg i.p. or vehicle (t 0 min). Activity counts accumulated over 3 hours are shown on the right panel.

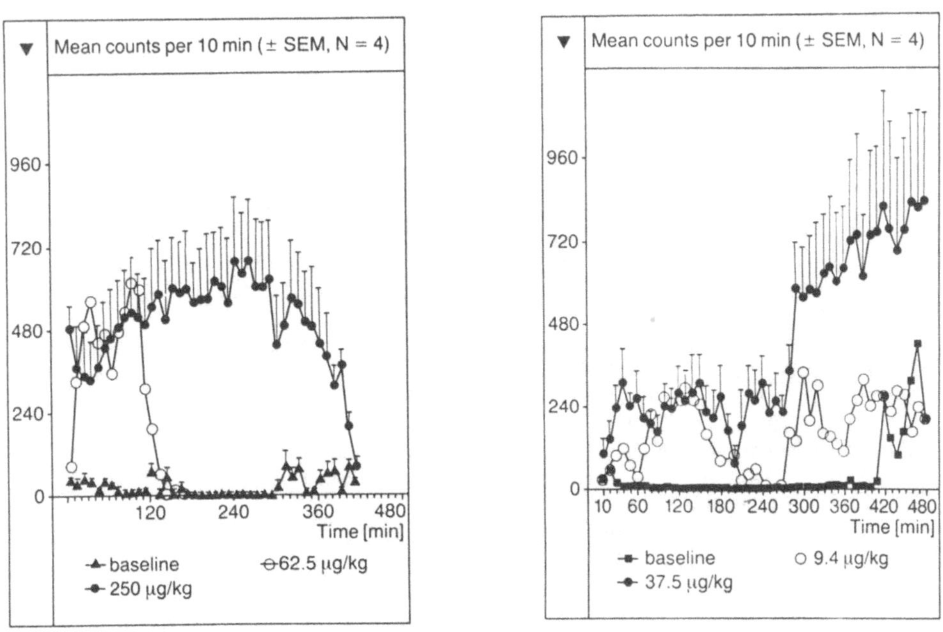

Fig. 3. Time course of locomotor activity of MPTP-treated common marmosets pretreated with 2 mg/kg domperidone p.o. (t -30 min) alone (baseline) and lisuride (62.5 and 250 μg/kg p.o., t 0 min, left panel) or proterguride (9.4 and 37.5 μg/kg p.o., t 0 min, right panel).

DISCUSSION

These data are consistent with the assumption that lisuride is an effective antiparkinsonian agent and confirmes the predictive value of the MPTP-treated common marmoset for therapeutic effects in patients. When administered p. o. lisuride (62.5 µg/kg) reversed akinesia for 2 hours. A 4-fold higher dose led to a prolonged duration of action rather then to a more pronounced stimulation of locomotor activity. In addition after 250 µg/kg p.o. symptoms like stereotypies, head-twitches, wet-dog shakes and scratching were observed, possibly indicating serotonergic effects of the compound when administered in very high doses. Proterguride also had strong antiparkinsonian activity after parenteral and enteral administration in this model. A very low dose of 9.4 µg/kg p.o. reversed symptomatology for up to eight hours in a biphasic manner. Interestingly, 37.5 µg/kg p.o. had a comparable effect during the first 4 hours followed by a remarkable higher stimulation in the second 4 hour period. Thus, locomotor activity had not returned to baseline values at the end of the experiment. This could indicate a slow absorbtion of the compound or formation of an active metabolite. The latter is rather unlikely since the compound is effective after parenteral administration in a monophasic manner. In conclusion these data suggest antiparkinsonian activity of proterguride in humans and with a longer duration of action when compared to lisuride. In healthy volunteers it has been shown that low doses of proterguride exert a long lasting prolactin lowering effect (8).

REFERENCES

(1) Davis G.C., Williams A.C., Markey S.P., Ebert M.H. Caine E.D., Reichert C.M. and Kopin I.J.; Psychiat. Res., 1, (1979), 249

(2) Langston J.W., Ballard P., Tetrud J.W. and Irwin I.; Science, 225, (1983), 1480

(3) Burns R.S., Chiueh C.C., Markey S.P., Ebert M.H., Jacobowitz D.M. and Kopin I.J.; Proc. Natl. Acad. Sci. U.S.A., 80, (1983), 4546

(4) Jenner P., Rupniak N.M.J., Rose S., Kelley E., Kilpatrick G., Lees A. and Marsden C.D.; Neurosci. Lett., 50, (1984), 85-90

(5) Nomoto M., Stahl S., Jenner P. and Marsden C.D.; Movement Disorders, 2, (1987), 37-45

(6) Nomoto M., Jenner P. and Marsden C.D.; Neurosci. Lett., 57, (1985), 37-41

(7) Löschmann P.-A., Chong P.N., Nomoto M., Tepper P.G., Horn A.S., Jenner P. and Marsden C.D.; European J. Pharmacol., 166, (1989), 373-380

(8) Dorow, R.; personal communication

TERGURIDE, A DOPAMINE PARTIAL AGONIST FOR THE TREATMENT OF PARKINSON'S DISEASE

P.-A. Löschmann[2], K.J. Rettig[2], R. Horowski[2], H. Wachtel[2], K.W. Lange[1], P. Jenner[1] and C.D. Marsden[3]

[1]Department of Pharmacology, King's College, London W3, UK, [2]Research Laboratories of Schering AG Berlin, FRG [3]Institute of Neurology, The National Hospital, London, UK

INTRODUCTION

Current treatment of striatal dopamine deficit in Parkinson's disease (PD) with indirect or direct dopamine (DA) agonists (i.e. L-Dopa, lisuride or bromocriptine) is limited by stimulation of other dopaminergic systems causing side effects. Rapid tolerance develops against the initial adverse reactions such as nausea and vomiting, but long term treatment with L-Dopa leads to fluctuations in motor performance and involuntary movements (1). Psychosis is another limiting side effect of all dopaminergic therapies, especially in elderly patients. A selective stimulation of dopaminergic receptors lacking endogenous innervation could lead to a more specific therapeutic effect and reduction of side effects. Dopamine partial agonists, with terguride being the prototype compound, are promising for such an approach. Terguride, a derivative of the dopaminergic ergoline lisuride, acts as a DA-antagonist on normosensitive receptors but does show DA-agonistic properties at supersensitive DA-receptors (2).

RESULTS

In rats treated with saline for 14 days terguride dose-dependently inhibits locomotor activity, but stimulation occurs after subchronic pretreatment with reserpine (0.5 mg/kg i.p., cf. fig. 1). Similarily, in rats treated acutely with reserpine (7.5 mg/kg i.p., 24 hours before) terguride does not induce stereotypies, whereas animals reserpinized for 14 days (0.5 mg/kg i.p.) do show stereotypies (cf. fig. 2). On the other hand the compound dose-dependently antagonizes stereotypies in rats (cf. fig. 3) induced by DA-agonists like amphetamine (1.56 mg/kg s.c.), apomorphine (0.39 mg/kg s.c.) or lisuride (0.1 mg/kg s.c.). Furthermore, and in common with classical DA-agonists, terguride suppresses hyperprolactinaemia in female rats pretreated with reserpine (2 mg/kg i.p. 16 hours before,

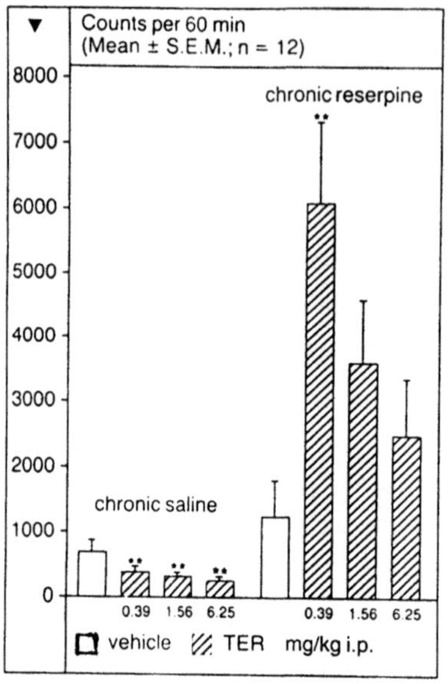

Fig. 1. Effect of various doses of terguride i.p. upon locomotor activity of rats pretreated with saline or reserpine (0.5 mg/kg i.p.) once daily for 14 days.

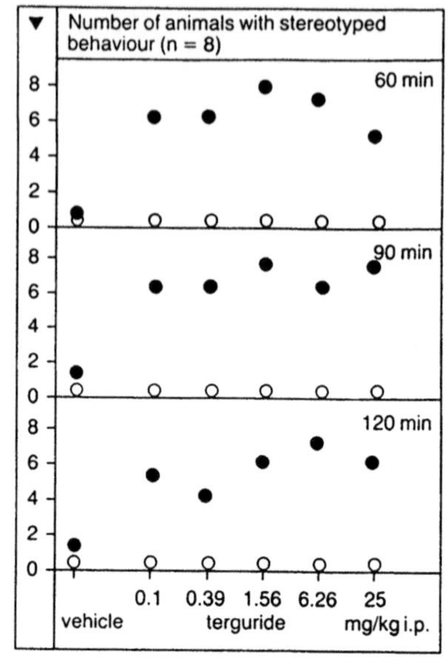

Fig. 2. Time course of stereotypies following various doses of terguride in rats pretreated with reserpine acutely (7.5 mg/kg i.p.; 24 h, O), or subchronically with 0.5 mg/kg i.p. once daily for 14 days (●).

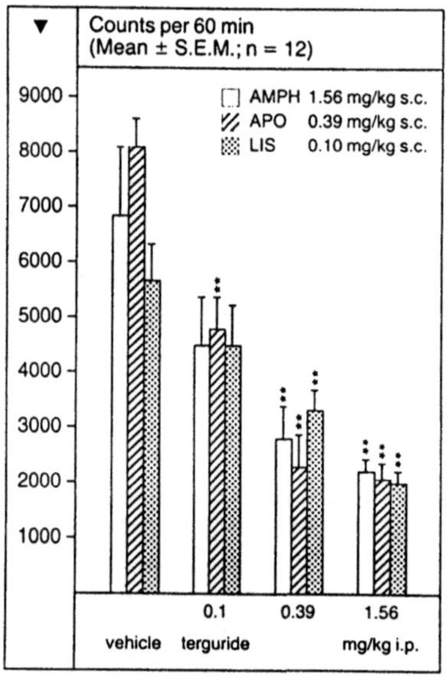

Fig. 3. Influence of various doses of terguride i.p. upon stereotypies induced by amphetamine (AMPH), apomorphine (APO) or lisuride (LIS) in rats.

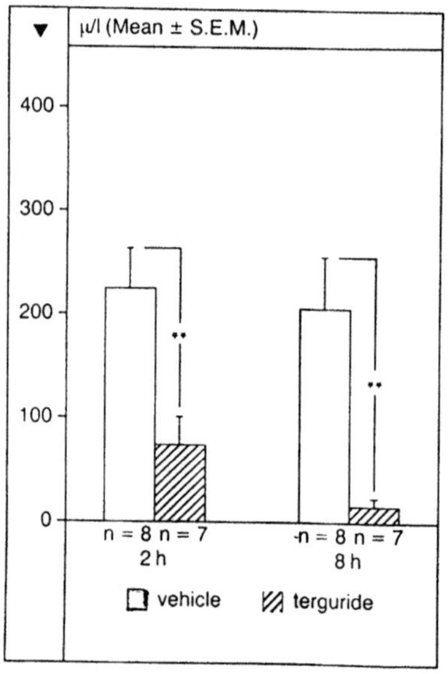

Fig. 4. Serum prolactin levels of female rats pretreated with reserpine (2 mg/kg i.p.; 16 h) at various intervals following s.c. administration of 0.05 mg/kg terguride.

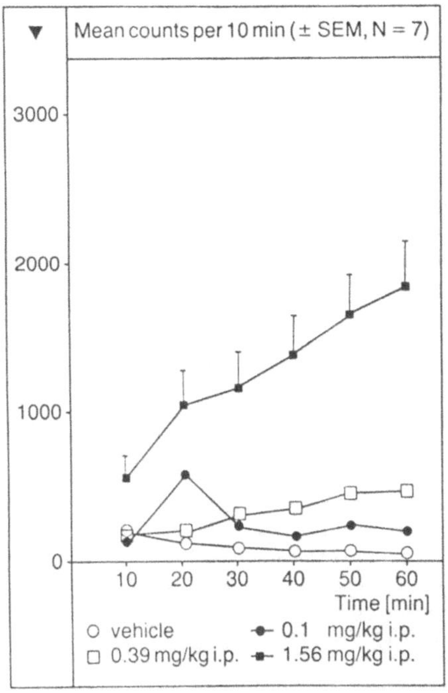

Fig. 5. Contralateral rotations in unilateral S. nigra lesioned rats (16 µg 6-OHDA/4µl) induced by indicated i.p. doses of terguride.

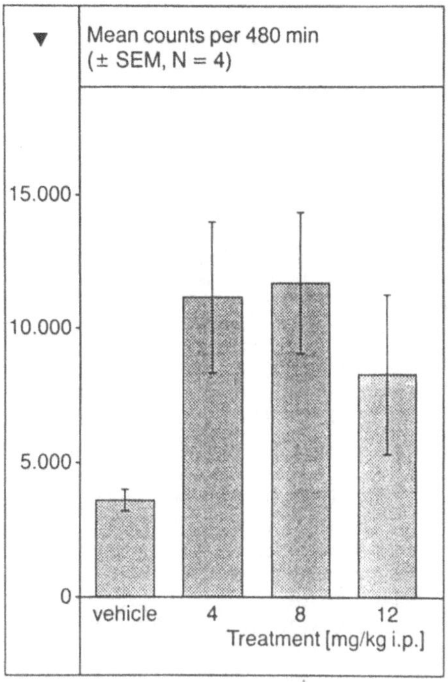

Fig. 6. Locomotor activity of MPTP-treated common marmosets (10 mg/kg s.c.) 3 month prior to i.p. administration of terguride.

cf. fig. 4) and induces contra- lateral rotations in unilateral 6-OHDA nigra-lesioned rats (cf. fig. 5). In common marmosets, rendered parkinsonian by subcutaneous administration of 10 mg/kg 1-methyl-4-phenyl-1,2,3,6-tetrahydropyridine (MPTP) terguride stimulated locomotor activity (cf. fig. 6).

DISCUSSION

In acute rodent experiments or in animals treated with saline terguride has DA-antagonistic properties resulting in a reduction of locomotor activity and in catalepsy when higher doses are administered. In line with these findings the locomotor stimulatory effects of classical DA-agonists such as apomorphine, amphetamine and lisuride are antagonized by terguride. However, in animals pretreated subchronically with reserpine terguride acts as a DA-agonist stimulating locomotor activity and inducing stereotypies and contralateral rotations in unilateral nigra-lesioned rats. In the MPTP-treated primate, the best model for Parkinson's disease available so far, terguride has antiparkinsonian effects. Thus, the compound stimulates locomotor activity of MPTP-treated common marmosets and induces contralateral rotations in monkeys rendered hemiparkinsonian by intracarotid infusion of MPTP (3).

This atypical pharmacological profile combining DA-agonistic and DA-antagonistic properties in one molecule has been described as dopamine partial agonism. Terguride therefore has an interesting range of possible indications. It lowers prolactin levels like classical DA-agonists: in humans it produces identical endocrine effects to those seen following bromocriptine but with clearly fewer side effects (4) and thus may be useful for the therapy of hyperprolactinaemia (5). In an open clinical trial, including 11 patients with chronic schizophrenia and marked negative symptoms, terguride effectively improved symptomatology in 8 patients on the Scale for Assessment of Negative Symptoms (SANS). In none of the patients were psychotic symptoms elicited and the compound was well tolerated (6). Clinical trials in patients with Parkinson's disease suggest a considerably lower frequency of adverse reactions (7), especially in elderly patients (8) and good antiparkinsonian efficacy (9). In a double-blind placebo-controlled study antidyskinetic and antiakinetic efficacy of terguride has been shown using high and low dose L-Dopa infusion in parkinsonian patients (10). In conclusion DA-partial agonists may have good antiparkinsonian activity with less side effects, especially in patients susceptible to motor fluctuations, psychosis or enhanced sensitivity towards dopaminergic drugs. Furthermore, early combination with L-Dopa could prevent complications observed with L-Dopa monotherapy. Both hypotheses are currently tested in controlled clinical trials.

REFERENCES

(1) Marsden C.D. and Parkes J.D.; Lancet 1, (1976), 292
(2) Wachtel H. and Dorow R.; Life Sciences 32, (1983), 421-432
(3) Brücke T., Bankiewicz K., Harvey-White J., and Kopin I. European Journal of Pharmacology, 148, (1988), 445-448
(4) Ciccarelli E., Touzel R., Besser G.M., Grossman A. Fertility and Sterility, 49, 4, (1988), 589-595
(5) Dallabonzana D., Liuzzi A., Oppizzi G., Cozzi R., Verde G., Chiodini P., Rainer E., Dorow R., Horowski H. Journal of Clinical Endocrinology and Metabolism 63, 4, (1986), 1002-1007
(6) Olbrich R. and Schanz, H.; Pharmacopsychiatry, 21, (1988), 389-390
(7) Riederer P., Danielczyk W., Suchy I., Brücke T. Drugs of the Future, 11, 4, (1986), 288-290
(8) Brücke T., Danielczyk W., Simányi M., Sofic E., Riederer P.,; "Advances in Neurology", Vol. 45, M.D. Yahr and K.J. Bergmann (ed.), Raven Press, New York, (1986), 573-576
(9) Corsini G.U., Bonuccelli U., Rainer E., Del Zompo M., Journal of Neural Transmission, 64, (1985), 105-111
(10) Baronti F., Mouradim M.M., Brughitta S., Conant K.E., Giaffra M., Chase T.N.; Neurology, 39, 2, (1989), 556

THE BENEFICIAL EFFECT OF LISURIDE HYDROGEN MALEATE IN THE LONG TERM TREATMENT OF PARKINSON'S DISEASE

J.M. Rabey, T. Treves, M. Streifler, and A.D. Korczyn

Department of Neurology, Tel-Aviv Medical Center, Sackler School of Medicine, Tel-Aviv University, Tel-Aviv, Israel

INTRODUCTION

In Parkinson's Disease (PD) patients treated with long-term levodopa therapy, progressive degeneration of the nigrostriatal pathway might impair the use of exogenous L-Dopa, leading to a decrease in the therapeutic efficacy and thereby a worsening of the symptoms (1). Parkinsonian patients might therefore benefit from drugs which bypass the nigrostriatal system and directly stimulate the dopamine receptors. Among the different drugs available, lisuride hydrogen maleate, N-(D-6-methyl-8-isoergolenyl-N', N'-diethylcarbamide), a semisynthetic ergot alkaloid, was shown to be beneficial in human studies (2,3). In 1985 we reported our experience in the use of lisuride compared to levodopa in a 6 months PD trial (4). We now summarize data in a long term five year follow-up.

Patients and methods

Candidates for the study were PD patients who had responded previously to levodopa, but in whom the treatment was later changed because of increased motor disability with diminishing response to levodopa, marked fluctuations of symptoms and/or dyskinesia. They never received previously direct dopaminergic agonists. Upon admission, 44 ambulatory patients (24 males) were clinically evaluated. Laboratory screening included hemogram, liver, renal and thyroid function tests. Patients were then randomly allocated into two groups: lisuride or levodopa.

1) The lisuride group consisted of 24 patients at the start. Patients were kept on a fixed amount of levodopa and decarboxylase inhibitor (Dopicar, Assia, Israel) and were given increasing doses of lisuride hydrogen maleate (Dopergin, Schering, Berlin), starting with 0.1 mg twice daily. Increments were given gradually over a period of 4-6 weeks to a maximal daily dose of 5 mg lisuride. At the beginning of the study patients of the lisuride group were taking levodopa + DCI (mean dose of 590 mg). Other adjuvants were kept on fixed dose during the trial. Soon after initiation of lisuride, three patients withdrew because there was no improvement and they suffered side effects (acroparesthesia, vertigo and confusional episodes) and one because of severe dizziness. Of the remaining 20 cases, one patient died of bronchopneumonia after 3 years and five others abandoned the treatment because of severe dizziness (1 patient) and persistent confusion (4 patients). Thus, five years after being put on lisuride, 14 patients were available for comparison (Table 1, 2).

2) The levodopa group included 20 patients taking levodopa in a mean dose of 531 mg + 53 mg carbidopa before the medication was increased. They also took other coadjuvant drugs. Patients in this group were given increased doses of levodopa/carbidopa (Dopicar, Assia, Israel). Of the 20 patients, two died (one from acute MI within the second year and one from bronchopneumonia during the third year); two were lost to follow-up and in three, a toxic psychosis urged us to considerably reduce levodopa doses. After 5 years, 13 patients were available for comparison (Table 1, 2).

Clinical Evaluation

The patients were graded according to a modification of the Columbia University scale (5). Dyskinesia was evaluated on a separate scale ranging from 0 (no dyskinesia) to 4 (severe dyskinesia). The "off" episodes were evaluated by the patients and their relatives who were requested to score their mobility during the wake hours, as follows: 0 - normal; 1 - impaired gait; 2 - no gait; a score of 2 was considered "off". The total duration of akinetic episodes (mean of five different days, expressed in minutes) before and after the 5-year trial, was scored and compared in each patient. For statistical analysis, we used the two-tailed t-test, paired or unpaired, as required.

RESULTS

Within the first 6 months patients on lisuride were taking the following doses: 11 patients were on a dose of <1 mg/day; 3 patients took 1-3 mg/day and 4 patients >3-5 mg/day. The mean final dose by this time was 1.95 mg/day. After completing 5 years of treatment, patients on lisuride were taking the following doses: 11 patients were on a final dose of < 1mg/day and three patients were taking 1 - 2 mg/day. The mean final dose of lisuride was 1.03 mg. After 5 years of treatment patients on levodopa were taking a mean daily levodopa dose of 625 mg (range 187.5-750), instead of 788.4 mg/day, by the end of the first six months.

The side effects observed with lisuride were: 9 patients showed confusional episodes and agitation, which were alleviated with diminution of the dose in four, but which obliged the interruption of the treatment in five others. Three patients reported dizziness, three reported finger paresthesia, four patients increased appetite with weight gain (about 5%). The lisuride and levodopa groups were similar concerning personal and clinical characteristics at the start of the trial.

Comparison of the two groups (Table 2) revealed that lisuride was as effective as levodopa in controlling bradykinesia, rigidity and gait but was more effective in treating "off" episodes (about 33% relief against 15% in the levodopa group) and dyskinesia (about 23% improvement). The laboratory screening (CBC and SMA 12) remained normal in each patient during the five years evaluation period.

DISCUSSION

Lisuride is the first 8-alpha-aminoergoline in clinical use. In contrast to bromocriptine, the effects of lisuride are not inhibited or impaired by depletion of endogenous dopamine levels induced by pretreatment with reserpine and alpha-methyl-p-tyrosin pretreatment but may even be enhanced in reserpinized animals (6,7). In addition, lisuride is 10-20 times more potent on a miligram basis than bromocriptine. Another advan-

Table 1. Characteristics of patients before the trial

	Lisuride (n=24)	Levodopa (n=20)
Age (years - mean)	64.7 (48-76)	66 (52-83)
Disease duration (years - mean)	7.5 (5-12)	7 (2-10)
Stage (Hoehn-Yahr median)	3	3
Levodopa dose before trial (mg)	590 (250-750)	531 (250-750)

tage of lisuride over bromocriptine is that the latter drug may act as a partial rather than a full agonist in the motor system, as clinical improvement with very high plasma bromocriptine levels is not always as great as that accompanying lower doses (8).

In our study, lisuride was as effective as levodopa in controlling bradykinesia, rigidity and gait. Tremor was substantially unchanged by either medication, but lisuride was considerably better for the management of "off" episodes and dyskinesia, compared to levodopa (Table 2). This effect could be due to the more prolonged and stable stimulant activation of striatal receptors by lisuride (9).

An interesting observation first reported in our previous study (4) and now reconfirmed, is the weight gain described by four of our patients (about 5%). It is known that lisuride is also a central serotonin agonist (10). Serotonin has been related to eating behavior. Therefore the weight gain may be explained by increased food ingestion (11). This observation is clinically important as it is well known that many parkinsonian patients lose weight during the progression of the disease (12).

The most common side effect of lisuride is the confusion and/or psychotic behavior which may disappear when the dose of the drug is reduced. Therefore by the fifth year the mean dose of lisuride was about 1 mg daily, which is about half of the mean daily dose taken by the patients at the end of the first 6 months of the study (2 mg). It is important to stress that at low dose, lisuride was still a potent dopamine agonist after a long term follow-up, suggesting the possible potentiation of its effect after chronic administration as was also found in animal studies

Table 2. Comparison of the relative effects of lisuride and levodopa on motor parameters

Motor function	Lisuride	Levodopa
Total disability score	-9.9	-7.3
Bradykinesia	-0.74	-0.7
Rigidity	-2.3	-2.7
Tremor	-0.27	-0.1
Gait	-0.8	-0.8
"OFF" (min.) **	-69.5 (33%)	-27.5* (15%)
Dyskinesia **	-2.5 (23%)	-1.0* (10%)

The effect is expressed as the difference of the mean score for each group at baseline and on completion of the trial.

* - Statistically significant at 0.005 level.

** - The values were obtained as the mean of daily evaluations over five different days before and after five years of treatment

(6). This persistent dopaminergic effect observed with lisuride contrasts with the tachyphylaxis reported in PD treatment with bromocriptine (13) and pergolide (14).

Thus, our results show that during a long term follow-up, combined treatment with low doses of lisuride and levodopa is more effective than levodopa alone, supporting the idea that the stimulation of different populations of receptors seems not only to improve the parkinsonian disability but also diminishes the development of motor fluctuations and dyskinesia.

References

1. Bartholini, G. and Tissot, P., Mecanismes pharmaco-biochimiques et therapeutique moderne des syndromes parkinsoniens. Rev. Med. Liege XXV 18:591 (1970).
2. Gopinathan, G., Teravainen, H. and Dambrosia, J.M., Lisuride in parkinsonism. Neurology 31:371 (1981).
3. Lieberman, A.N., Goldstein, M. and Neophytides, A., Lisuride in Parkinson's disease: efficacy of lisuride compared to levodopa. Neurology 31:961 (1981).
4. Rabey, J.M., Treves, T.A., Streifler, M. and Korczyn, A.D., 1986, Comparison of efficacy of lisuride hydrogen maleate with increased doses of levodopa in Parkinsonian patients. In:"Advances in Neurology", Parkinson's Disease, M.D. Yahr, K.J. Bergman, eds., Raven Press, NY.
5. Calne, D.B., Stern, G.M., Laurence, D.R., Sharkey, J. and Armitage, P. L-dopa in post encephalitic parkinsonism. Lancet i:744 (1969).
6. Horowski, R. and Watchel, H., 1979, Pharmacological effects of lisuride in rodents mediated by dopaminergic receptors: Mechanism of action and influence of chronic treatment with lisuride. In:"Dopaminergic Ergot Derivatives and Motor Function", K. Fuxe, D.B. Calne (eds) Pergamon Press, Oxford.
7. Carruba, M.O. and Mantegazza, P., 1983, Behavioral pharmacology of ergot derivatives. In:"Lisuride and other dopamine agonists: basic mechanism and endocrine and neurological effects", D.B. Calne, R. Horowski, R.J. Mc Donald, W. Wuttke (eds), Raven Press, New York.
8. Price, P., Debono, A., Parkes, J.D., Marsden, C.D. and Rosenthaler, J., Plasma bromocriptine levels, clinical and growth hormone response in parkinsonism. Br. J. Clin. Pharmacol. 6:303 (1978).
9. Burns, R.S. and Calne, D.B., 1983, Disposition of dopaminergic ergot compounds following oral administration. In:"Lisuride and other Dopamine Agonists: Basic Mechanism and Endocrine and Neurological effects", D.B. Calne, R. Horowski, R.J. Macdonald, W. Wuttke (eds), Raven Press, NY.
10. Silbergeld, E.K. and Hruska, R.E., Lisuride and LSD: Dopaminergic and serotonergic interactions in the "serotonin syndrome". Psychopharmacology (Berlin) 65:233 (1979).
11. LeWitt, P.A., Gopinathan, G. and Ward, C.D., Lisuride versus bromocriptine treatment in Parkinson diseae: A double blind study. Neurology 32:69 (1982).
12. Vardi, J., Oberman, Z., Streifler, M., Rabey, J.M., Ayalon, D. and Herzberg, M., Weight loss in elderly Parkinsonian patients on long term levodopa therapy. J. Neurol. Sci. 30;33 (1976).
13. Grimes, J.D., Bromocriptine in Parkinson's disease. Results obtained with high and low doses therapy. Can. J. Neurol. Sci. 11:225 (1984).
14. Lieberman, A.N., Goldstein, M. and Leibowitz, M., Long term treatment with pergolide: decreased efficacy with time. Neurology 34:223 (1984).

SELEGILINE IN THE TREATMENT OF DAILY FLUCTUATIONS IN DISABILITY - A DOUBLE-BLIND STUDY

E.H. Heinonen[1,2] and U.K. Rinne[2]

1. Department of Neurology, University of Turku; 2. Farmos Group Ltd, Research Center Turku, Finland

INTRODUCTION

Selegiline, an irreversible inhibitor of monoamine oxidase type B (MAO-B), has been reported to be effective in the treatment of fluctuations of disability. Many of the trials have, however, been open (Heinonen and Rinne, 1989). We have therefore carried out a double-blind study to investigate the clinical efficacy of selegiline in the treatment of fluctuations in disability of parkinsonian patients with long-term levodopa treatment.

PATIENTS AND METHODS

Inclusion criteria for the trial were idiopathic Parkinson's disease with end-of-dose type fluctuations in disability and daily frequency of administration of levodopa over six dosages. Patients with marked dementia, any disease that might influence the absorption of drugs, known allergy to the drugs under investigation, previous treatment with selegiline or dopamine agonists or age over 80 years were excluded from the trial.

19 patients with advanced Parkinson's disease and end-of-dose type fluctuations in disability entered the study. The mean age of the patients was 61.6 ± 9.6 years, the mean duration of disease 13.3 ± 3.3 years and the duration of levodopa therapy 11.0 ± 2.8 years. All the patients had end-of-dose akinesia. 12 had also early-morning akinesia, 11 nocturnal akinesia, 6 had random and 7 dose-related freezing episodes, 14 had peak dose dyskinesias, 5 had early morning and 7 had dose-related dystonia. The mean dose of levodopa at the beginning of the study was 834.2 ± 236.9 mg with the mean dosing frequency of 7.4 ± 2.0 per day. There were 7 patients in Hoehn and Yahr stage III, and 12 were on stage IV. The mean disability score (Columbia University Disability Rating Scale, CUDS) was 47.1 ± 8.7.

Basic, Clinical, and Therapeutic Aspects of Alzheimer's and Parkinson's Diseases, Volume 2
Edited by T. Nagatsu *et al.*, Plenum Press, New York, 1990

The design of the study was a randomized, double-blind, cross-over trial. Both study periods lasted for 12 weeks. The study was started with a two week run-in period, during which it was checked that the dose of levodopa was optimal. The patients were then randomized to receive either selegiline or a matching placebo tablet and switched to the other treatment after the first period. The dose of selegiline was 10 mg divided into two dosages, one given in the morning and the second at lunchtime. The daily levodopa dose and dosing frequency were kept as unchanged as possible.

The parkinsonian disability and symptoms were evaluated using CUDS. The assessments were carried out by the same investigator at the beginning of the study and after 2, 4, 6 and 12 weeks during both study periods. The patients kept a daily diary where they marked the time when they were "on" and "off" and also the severity of fluctuations. From these diaries scoring of the fluctuations (scale 0-4) and the duration of action of levodopa dose was evaluated. Blood pressure and heart rate were measured at each visit. Laboratory values including hematological values, alkaline phosphatase, serum aspartate aminotranspherase, serum gamma glutamyltranspherase and serum creatinine were measured at the beginning of the trial and at the end of both study periods. The study protocol was accepted by the local ethical committee and carried out in accordance with the principles of the Helsinki Declaration.

RESULTS

The parkinsonian disability and the main parkinsonian symptoms of the patients were significantly less during selegiline treatment (Table 1).Also the various types of fluctuations in disability were less severe during selegiline treatment and the duration of action of levodopa dose was significantly longer (Table 2). The blind investigator's opinion on the efficacy of the drug was significantly ($p < 0.03$) better during selegiline treatment. The dose of the levodopa and the frequency of dosing did not differ significantly, which was one of the purposes of the trial. The dose was, however, smaller during selegiline treatment (779.8 $\pm$ 228.7 mg and 7.0 $\pm$ 1.7 during selegiline ; 808.3 $\pm$ 245.3 mg and 7.2 $\pm$ 2.0 during placebo treatment, respectively).

Table 1. Parkinsonian disability of the patients during selegiline and placebo treatments (mean $\pm$ SD).

Variable	Selegiline	Placebo	p-value
Total CUDS	38.6 $\pm$ 8.5	43.4 $\pm$ 8.9	0.002
Tremor	2.4 $\pm$ 1.9	3.1 $\pm$ 2.2	0.010
Rigidity	11.3 $\pm$ 3.3	12.6 $\pm$ 3.0	0.027
Hypokinesia	22.0 $\pm$ 4.1	24.6 $\pm$ 4.7	0.004
Posture	0.6 $\pm$ 0.6	0.8 $\pm$ 0.7	0.043

Table 2. The severity of various types of fluctuations in disability and the mean duration (h) of action of levodopa dose during selegiline and placebo treatments.

Variable	Selegiline	Placebo	p-value
1. Freezing episodes:			
Random	0.7 ± 1.0	0.7 ± 1.0	1.000
Dose-related	0.5 ± 0.7	0.8 ± 1.0	0.219
2. End-of-dose disturbances:			
End-of-dose akinesia	1.6 ± 0.8	2.1 ± 0.6	0.013
Early-morning akinesia	0.8 ± 0.8	1.2 ± 1.1	0.039
Nocturnal akinesia	0.8 ± 0.9	1.0 ± 1.1	0.375
3. Dyskinesias:			
Dose-related dystonia	0.5 ± 0.7	0.7 ± 1.0	0.140
Early-morning dystonia	0.3 ± 0.5	0.6 ± 1.0	0.063
Peak-dose dyskinesia	1.6 ± 0.9	1.3 ± 0.9	0.016
4. Mean duration of action of levodopa (h)	2.7 ± 0.6	2.2 ± 0.5	0.002

The side-effects during the selegiline and placebo periods did not differ significantly. Dry mouth was the most common side-effect (14 during selegiline vs 9 during placebo treatment), 12 had tiredness and 11 sweating during both treatments, lightheadedness occurred in both groups (9 vs 8), as did orthostatic hypotension (9 vs 6), obstipation (6 vs 7), insomnia (6 vs 6) and nausea (4 vs 5). There were more peak-dose dyskinesias during selegiline treatment (Table 2). There were no significant differences in the blood pressure or heart rate in lying or standing position. In the laboratory values no significant differences were detected.

DISCUSSION

Levodopa still remains the cornerstone of the treatment of Parkinson's disease. Unfortunately after 5-7 years of levodopa treatment the efficacy of levodopa begins to wear off. The duration of action of an individual levodopa dose becomes shorter and the dosing frequency of the drug has to be increased. Fluctuations in clinical disability appear. Some of these fluctuations such as end-of-dose type fluctuations are associated with the serum concentrations of levodopa, while some, like random fluctuations, are not.

In many open studies selegiline has been shown to reduce the fluctuations in clinical disability in advanced Parkinson's disease. Selegiline inhibits MAO-B in the human brain, and it also inhibits the uptake of catecholamines and increases the synthesis of dopamine (Knoll 1978). Thanks to these properties it prolongs the duration of action of dopamine. In our study it was clearly shown that addition of selegiline to the prevailing levodopa treatment resulted in significant improvement of clinical disability and the main symptoms of the disease. Fluctuations of disability, especially the end-of-dose type, were significantly reduced and the mean duration of action of levodopa dose was longer as compared to the placebo. In our trial the dose of levodopa and dosing frequency were kept as unchanged as possible. This is in contrast with many other trials where the effect of selegiline is investigated, but significant reductions in levodopa dose and dosing frequency are allowed. The results of a clinical trial are easier to interpret if there are only fewer changing variables - in this case the disability of the patient.

In the advanced state of Parkinson's disease the reduction of levodopa dose is not as important as the clinical well-being of the patient. In the earlier state of the disease the possibility to reduce the levodopa dose may, however, be important, with a view to possibly preventing the long-term side-effects of levodopa therapy. Furthermore, it has been shown that when given to de novo patients selegiline prolongs the time that a patient can manage without levodopa (Tetrud and Langston,1989).

In conclusion our study showed that selegiline potentiated and prolonged the beneficial effect of levodopa in the treatment of advanced Parkinson's disease:

1. The duration of action of levodopa was significantly longer during selegiline treatment;

2. The clinical disability and the main parkinsonian symptoms were significantly improved without increasing the dose of levodopa;

3. The end-of-dose type fluctuations in disability were significantly improved.

REFERENCES

Heinonen E.H., Rinne U. K., 1989 , Selegiline in the treatment of Parkinson's disease, Acta Neurol Scand (in press).

Heinonen E.H., Rinne U.K., Tuominen J., 1989, Selegiline in the treatment of daily fluctuations in disability of parkinsonian patients with long-term levodopa treatment, Acta Neurol Scand (in press).

Knoll J.,1978, The possible mechanism of action of (-)-deprenyl in Parkinson's disease. J Neural Transm, 43:177-198.

Tetrud J.W., Langston J.W., 1989, The effect of deprenyl (selegiline) on the natural history of Parkinson's disease. Science , 245:519-522.

PHARMACOLOGICAL ACTIVITY OF A NOVEL DOPAMINE D_2 AGONIST (SK&F 101468) IN MAN

J.M. Bottomley and N. Sree Haran

Clinical Investigation Unit, Smith Kline and French Research Ltd., The Frythe, Welwyn, Hertfordshire, AL6 9AR, UK

INTRODUCTION

Dopamine agonists are useful adjuncts to L-Dopa in the treatment of Parkinson's disease. The currently available agonists (e.g. bromocriptine, pergolide) induce considerable side effects. Of these, some effects (i.e. nausea and orthostasis) are considered related to peripheral dopaminergic activity while the others (e.g. dyskinesias, confusion, hallucinations) originate in the central nervous system, occur commonly with ergoline compounds and are due to dopaminergic or serotinergic activity. It has been postulated that a non ergoline dopamine agonist with selectivity to dopaminergic (D_2) receptors would reduce the incidence of unacceptable central side effects.

SK&F 101468 (ropinirole) is a non ergoline dopamine agonist which binds selectively to central D_2 and peripheral DA_2 receptors without significant alpha or beta adrenergic and serotinergic activity. It has been demonstrated to have significant anti-Parkinson activity in the MPTP treated marmoset model.

MATERIALS AND METHODS

Healthy volunteers (n = 72) and Parkinsonian patients (n = 25) received single oral doses of SK&F 101468 ranging from 10 to 2500 mcg and 200 mcg to 6 mg respectively in several clinical studies. Evidence for "peripheral" dopaminergic activity was sought using serum prolactin as the pharmacodynamic marker and by evaluating the prevalence of nausea and orthostasis.

A preliminary study in eight Parkinsonian patients evaluated the effect of a chronic dosing regimen on peripheral dopaminergic activity.

RESULTS

Evidence for pharmacological activity

(i) Serum Prolactin. Pharmacological activity was identified in a placebo controlled dose rising study involving 14 healthy male volunteers (1). A dose proportional suppression of serum prolactin was observed within the studied range, 80 to 1000 mcg. This effect was seen in both basal (Figure 1) and food stimulated prolactin secretion (Figure 2) and was shown to persist for at least eight hours post dosing.

(ii) Other Effects Nausea and orthostatic hypotension were identified as the limiting side effects to single oral doses of SK&F 101468. The incidence of these effects occurred to a clinically significant extent only at doses of 800 mcg and above in both healthy volunteers and patients with Parkinson's disease (Table 1). On the basis of this, 600 mcg has been selected as the starting dose of an incremental chronic dosing regimen in Parkinsonian patients.

The peripheral dopaminergic nature of the above side effects was elucidated by reducing their occurrence significantly by pre-treatment with 20 mg of domperidone. In a double blind cross over study, orthostatic faintness with nausea and malaise observed following single oral doses of 0.8 mg of SK&F 101468 in five out of nine healthy subjects was prevented by domperidone pre-treatment in all but one subject. Domperidone pre-treatment in 16 Parkinsonian patients who received 66 single oral doses in the 1 to 6 mg range similarly abolished the occurrence of orthostasis without, however, abolishing the incidence of nausea (13/66), most of which was of a mild and transient nature. This effect of domperidone may indicate a differential binding to dopamine receptors (2).

Development of Tolerance to the Peripheral Dopaminergic Effects

Single doses of SK&F 101468 have been demonstrated to be hypotensive in spontaneously hypertensive rats. Repeat dosing however, resulted in a rapid development of tolerance to the hypotensive effects (3).

An open and uncontrolled study (4) in eight patients with Parkinson's disease demonstrated that an incremental dose titration regimen from 600 mcg to 2 mg b.i.d. of SK&F 101468 over four weeks, without the use of concomitant domperidone did not induce significant nausea or orthostasis, indicating that tolerance can be induced to the peripheral dopaminergic effects of SK&F 101468. Anti-Parkinson effects were observed in all eight patients. Central side effects were not observed.

CONCLUSIONS

- o Single doses of SK&F 101468 (ropinirole) demonstrate potent pharmacological activity in man.
- o Suppression of serum prolactin and the occurrence of nausea and orthostasis are indicative of peripheral dopaminergic activity.

Table 1. Incidence of Peripheral Dopaminergic Effects per Number of Single Oral Doses of SK&F 101468

	< 0.8 mg		≥ 0.8 mg	
	Volunteers	Patients	Volunteers	Patients
Nausea	1/92 (1.1%)	0/30	14/63 (22.2%)	2/11 (18.2%)
Orthostasis	2/80 (2.5%)	0/30	11/51 (21.6%)	2/11 (18.2%)
n*	50	10	53	6

* Number of subjects receiving SK&F 101468

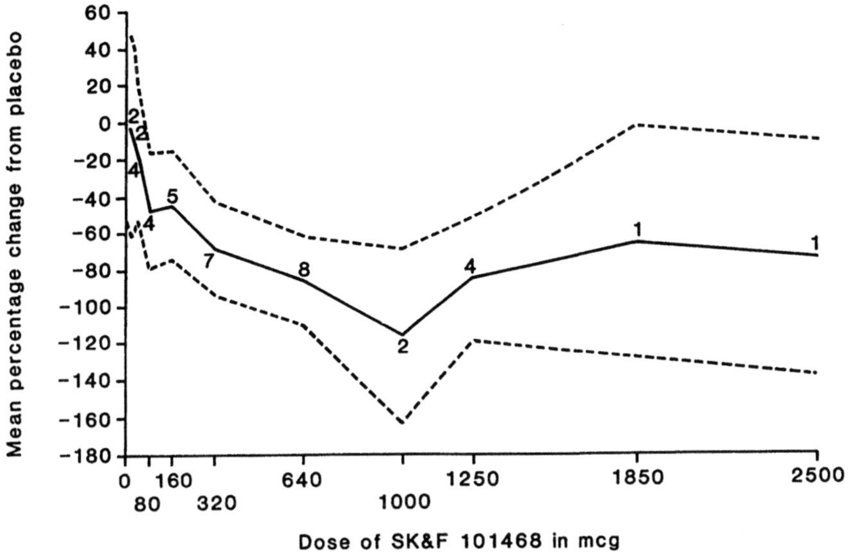

Fig.1. Effect of SK&F 101468 on mean basal serum prolactin from 0 to 4 hours post dosing. The results are displayed as mean percentage change from placebo (——) with the 95% confidence interval (---). The number of subjects receiving each dose is also indicated.

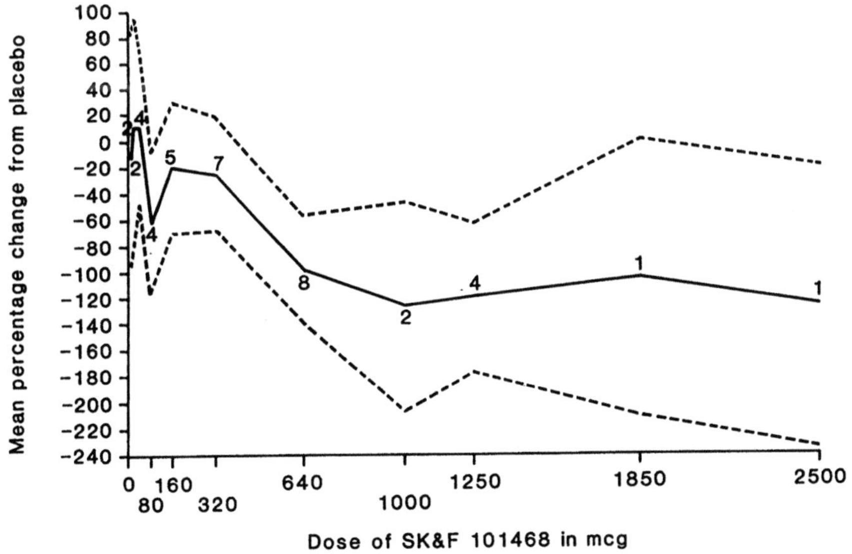

Fig.2. Effect of SK&F 101468 on mean food stimulated serum prolactin observed from 4 to 6 hours post dosing. The results are displayed as mean percentage change from placebo (——) with the 95% confidence interval (---). The number of subjects receiving each dose is also indicated.

- These dose limiting side effects could be effectively reduced by pre-treatment with 20 mg domperidone or averted by using an incremental chronic dosing schedule.

- SK&F 101468 (ropinirole) is a non ergoline, selective dopamine (D_2) agonist in clinical development as an anti-Parkinson agent.

REFERENCES

1. C. Broom, G. Acton, C. Gill and P.E. Sparrow. Effects of SK&F 101468, a novel dopamine D_2 agonist, on supine plasma prolactin and haemodynamics in normal man. Br. J. Clin. Pharm, 26:654P (1988)
2. C. Nappi, E. Colace, G.F. DiRenzo, M. Taglialatela, S. Amoroso, L. Annunziato and U. Montemagno. Domperidone antagonises bromoergocriptine-induced nausea and vomiting without affecting its inhibition of prolactin secretion in puerperal women. Eur. J. Clin Pharmacol. 32:457-460 (1987)
3. R.J. Eden, C.A. Harvey, D.A.A. Owen and S.G. Parker. Tolerance to the cardiovascular effects of 101468, a dopamine DA_2-receptor agonist, in spontaneously hypertensive rats. Br. J. Pharm, 95:791P (1988)
4. R. Kapoor, Z. Pirtosek, J.P. Frankel, G.M. Stern, A.J. Lees, J.M. Bottomley and N. Sree Haran. Treatment of Parkinson's disease with a novel dopamine D_2 agonist SK&F 101468. Lancet, i:1445-6 (1989)

SK&F 101468, A NOVEL DOPAMINE D_2 AGONIST IN PARKINSON'S DISEASE

R. Kapoor[1], Z. Pirtosek[1], J. Frankel[1], G.M. Stern[1],
A.J. Lees[1], J.M. Bottomley[2] and N. Sree Haran[2]

[1]Department of Neurology, The Middlesex and University College Hospital School of Medicine, London, UK and
[2]Clinical Investigation Unit, Smith Kline and French Research Ltd, The Frythe, Welwyn, Herts, AL6 9AR, UK

INTRODUCTION

SK&F 101468, ropinirole, is a selective central and peripheral D_2/DA_2 dopamine receptor agonist with demonstrable CNS and cardiovascular activity in several animal species and in man.

In studies with marmosets infused with 1-methyl-4-phenyl-1,2,3,6-tetrahydropyridine (MPTP) into the substantia nigra to induce motor deficits characteristic of Parkinson's disease, SK&F 101468 fully restored normal motor function. Ropinirole is currently in Phase II development as an anti-Parkinson agent.

We report a preliminary open study of its use, in combination with Levodopa, in idiopathic Parkinson's disease (PD).

STUDY AIM

To assess the safety, pharmacokinetic profile and efficacy of chronic dose administration of SK&F 101468 in Parkinson's disease.

STUDY DESIGN

Patients were studied in four groups of two. Each group received a different dosing regimen of SK&F 101468 ranging from 0.6 mg b.i.d. up to 2 mg b.i.d. over a four week incremental dosing period (see Table 1).

Parkinson's disease was assessed using the Modified Webster Rating Scale (1) and by semi-objective timed exercises (Finger Tapping, Walking Time). For full details see Figure legends.

Vital signs (supine and standing blood pressure and pulse) and ECG were measured daily. Routine haematological, biochemical and urinalysis profiles were performed weekly.

Venous blood samples were drawn for pharmacokinetic analysis. The plasma profiles of Day 1 versus Day 7 of each dose level were compared to

Basic, Clinical, and Therapeutic Aspects of Alzheimer's and Parkinson's Diseases, Volume 2
Edited by T. Nagatsu *et al.*, Plenum Press, New York, 1990

Table 1. SK&F 101468 Dose* on Study

Group**	Week 1	Week 2	Week 3	Week 4
1	0.6			
2	0.6	1.0		
3	0.6	1.0	1.4	
4	0.6	1.0	1.4	2.0

* mg b.i.d.
** Two patients per group

assess drug accumulation. Samples were analysed for SK&F 101468 and SK&F 104557 (the major metabolite of SK&F 101468 in man) by radioimmunoassay.

PATIENTS

Eight patients (5M, 3F) with PD (Hoehn and Yahr, Stages II-IV) and mean age 62.5 years (range 56 to 68), mean duration of disease 8.5 years (range 2 to 11) and mean duration of Levodopa therapy (with decarboxylase inhibitor) of 4.5 years (range 2 to 5) entered the study. Patients continued to receive their levodopa (mean daily dose 440 mg); in addition four patients received anticholinergics, and three received L-Deprenyl. (See Table 2 for patient details.)

RESULTS

Efficacy

All eight patients demonstrated clinical improvement. An improvement greater than 33% on the Modified Webster Rating Scale (Figure 1) was observed in five patients; in one of these five patients a greater than 33% effect in both the finger tapping score (Figure 2) and walking time (Figure 3) was also demonstrated, in a further three a 33% improvement in the finger tapping score was noted and in the fifth patient, walking time was improved by greater than 33%. In addition, two patients reported a reduction in painful "off" period dystonias.

Adverse Effects

SK&F 101468 was generally well tolerated without any significant peripheral dopaminergic effects. One patient with a history of postural giddiness had a transient and asymptomatic drop in blood pressure. Mild

Table 2. Patient Demographics at Study Entry

Patient	Age	Sex	Duration of PD (yrs)	Duration of of L-Dopa (yrs)	Mean daily dose L-Dopa(+DCI) (yrs)	Stage (Hoehn & Yahr)
1	56	F	6	5	437.5	III
2	68	M	5	5	375	IV
3	59	M	2	2	500	III
4	68	F	7	5	250	III
5	67	M	7	5	375	IV
6	67	M	10	5	625	IV
7	57	F	11	3	375	II
8	58	M	10	5	375	III

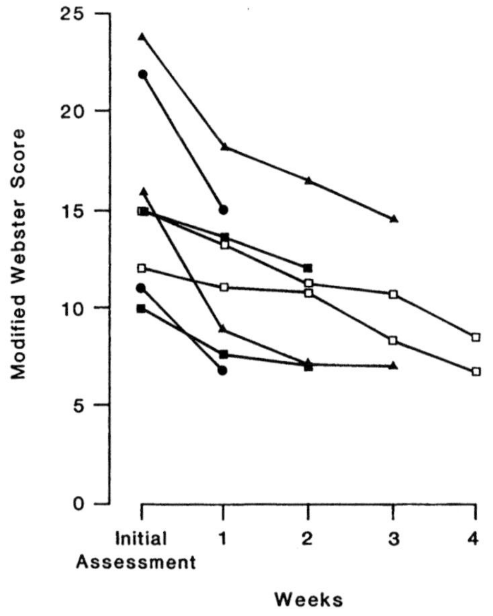

Fig. 1. Improvement in Modified Webster Scores with SK&F 101468

Disability (Total Score) was assessed daily prior to the first dose and 3 hours later.

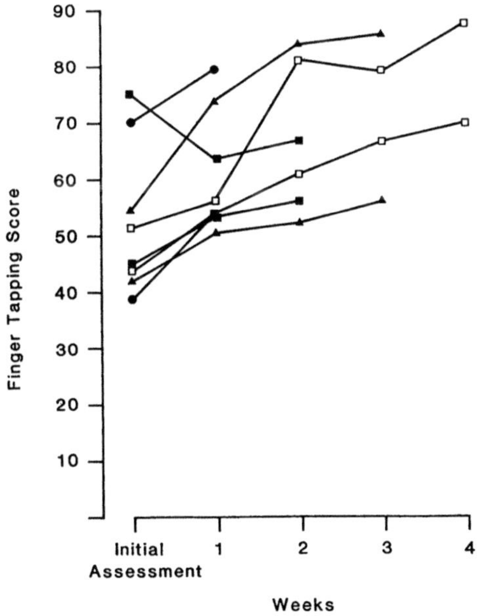

Fig. 2. Response in Finger Tapping Score with SK&F 101468

The finger tapping exercise was performed daily prior to first dose and 3 hours later. The number of finger taps made in 30 seconds between two digit counters mounted 20 cm apart was recorded.

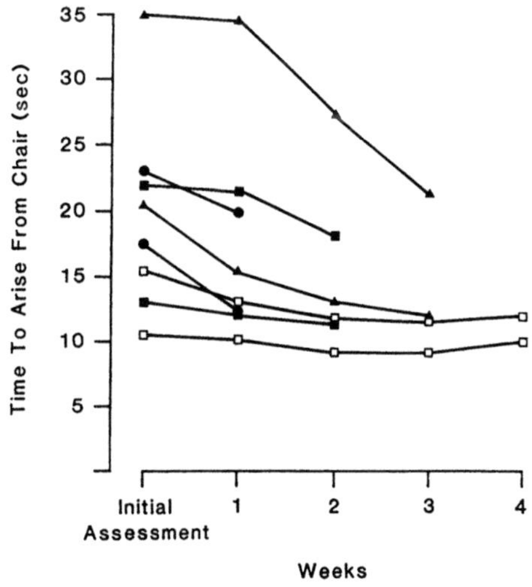

Fig. 3. Time to Arise from a Chair for 8 patients on SK&F 101468

The exercise was performed daily prior to the first dose and 3 hours later. The time taken to arise from a chair, walk 6 metres and back and resume the seated position was recorded.

nausea occurred in two additional patients. All symptoms resolved despite continued therapy. Mild nausea was also reported in tow patients the day following last administration of SK&F 101468.

Pharmacokinetics

Preliminary evaluation indicated a linear increase in systemic exposure with dose. Multiple oral dosing showed no significant accumulation of drug or metabolite (SK&F 104557) in plasma.

FIGURES

In Figures 1 to 3 the initial assessment was a single measurement made prior to first administration of SK&F 101468 on Day 1. The assessments made during the last three days of each week on study were averaged to assess treatment effect.

CONCLUSION

Findings from this open and uncontrolled pilot study suggest that SK&F 101468 has anti-Parkinson activity in the absence of significant short term adverse effects. Additional studies are ongoing to profile further the use of SK&F 101468 in Parkinson's disease.

REFERENCE

1. T.J. Christmas, P.A. Kempster, C.R. Chappel, J.P. Frankel, A.J. Lees, G.M. Stern and E.J.G. Milroy. Role of subcutaneous apomorphine in Parkinsonian Voiding Dysfunction, Lancet, ii:1451-53 (1988).

Pharmacological Evaluation of SK&F 101468-A, in MPTP Treated Marmosets.

B.Costall*, A.M.Domeney*, R.J.Eden, P.A.Gerrard*, C.A.Harvey, and D.A.A.Owen

Smith Kline & French Research Ltd, The Frythe, Welwyn
Hertfordshire, England
*Neuropharmacology Research Group, University of Bradford
Bradford, Yorkshire, England

INTRODUCTION

Previous studies in rodents have shown SK&F 101468-A, Ropinirole, to be a specific, directly acting dopamine-D_2 receptor agonist with activity in the extrapyramidal system suggesting a possible utility for treatment of Parkinson's Disease [1]. These studies were designed to investigate the effect of Ropinirole on Parkinsonian deficits induced by the neurotoxin 1-methyl-4-phenyl-1,2,3,6-tetrahydropyridine (MPTP) in a primate.

MATERIALS AND METHODS

Marmosets (Callithrix jacchus 320 - 400g) of either sex were used for these studies. Unilateral infusion of MPTP into the zona compacta of the substantia nigra was effected using subcutaneous Alzet osmotic minipumps, primed to deliver 20 µg MPTP/24 hr. in a volume of 0.48 µl/kg, via a previously implanted cannula. The cannulae were implanted 10-14 days prior to the infusion under Saffan-general anaesthesia, using stereotaxic techniques. MPTP was infused into the substantia nigra continuously for 13 days. Motor deficits were evident within 3-4 days and persisted for several weeks post-infusion. The infusion of MPTP reduced spontaneous locomotor activity, speed of movement and number of jumps between perches. Additionally, behavioural deficits such as lack of facial expression and interest in surroundings, and inability to elevate head and slowness of movement were assessed using the following scoring system; absent=0, mild=1, moderate=2 or marked/severe=3. All studies were performed blind, using a random crossover design.
Assessment of acute drug effects commenced within a 14 day period following the infusion. The effect of chronic treatment with SK&F 101468-A was assessed on days 8 - 11 of the infusion.

RESULTS

During and following the infusion of MPTP the marmosets presented with an overall bradykinesia and hypokynesia, with loss of facial expression and interest in surroundings. All these deficits were antagonised by SK&F 101468-A, 0.05-1.0 mg/kg s.c. or 0.1 mg/kg p.o. Onset of activity was within 10-20 minutes of administration, with duration greater than 2hr. Time spent in spontaneous locomotion was reduced from 57.3% to 9.3% by MPTP and restored to 49.3% by oral administration of SK&F 101468-A, 0.1 mg/kg (Fig 1 & 2). Chronic b.i.d. treatment with SK&F 101468-A, 0.5 mg/kg p.o., for 4 days showed no diminution of the reversal of the MPTP-induced deficits (Fig 3).
Similar reversals of MPTP effects were seen following L-dopa (50 mg/kg p.o.) in combination with benserazide (12.5 mg/kg s.c.), however, l-dopa treatment was always accompanied by emesis whereas antagonism of the deficits could be shown with SK&F 101468-A in the absence of this side effect (Fig 4).

Basic, Clinical, and Therapeutic Aspects of Alzheimer's and Parkinson's Diseases, Volume 2
Edited by T. Nagatsu *et al.*, Plenum Press, New York, 1990

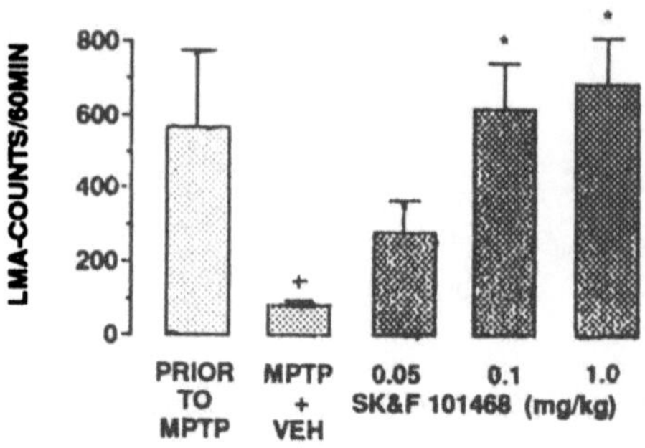

Figure 1 Reversal, by SK&F 101468-A, of MPTP-induced reduction in locomotor activity in the common marmoset.

SK&F 101468-A (SKF), 0.05 or 0.1 mg/kg, or vehicle (VEH) s.c. was administered 40 min prior to assessment. The animals were placed in photocell cages and activity measured for 60 minutes. Values represent mean ± s.e.m. (n=3 or 4).

Statistical significance was determined using ANOVA plus Dunnett's t-test .

+P<0.05 MPTP-induced reduction of activity.

*P<0.05 reversal of MPTP-induced deficit.

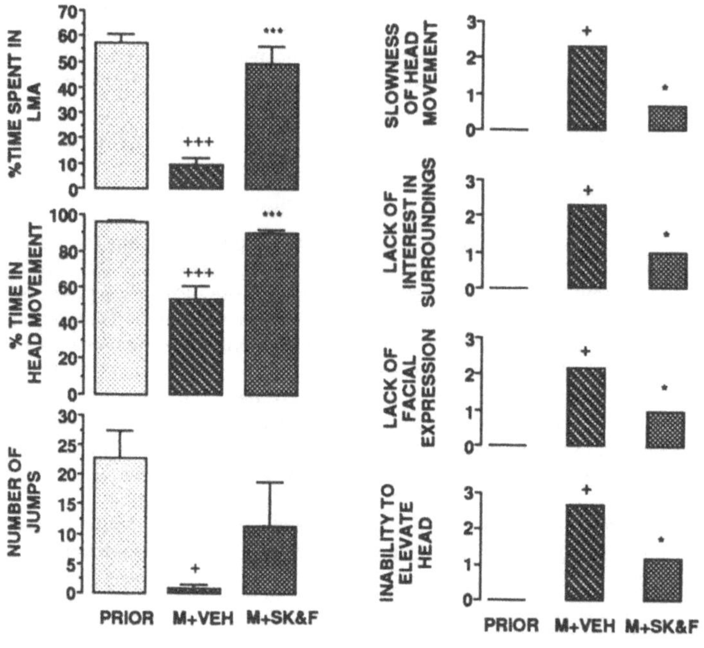

Figure 2 Reversal, by SK&F 101468-A, of MPTP-induced locomotor and behavioural deficits in the common marmoset.

The studies were carried out in the home cage. PRIOR= responses prior to MPTP treatment, M+VEH= vehicle response following treatment with MPTP and M+SKF= response following treatment with SK&F 101468-A 0.1 mg/kg p.o., 60 minute pretreatment (n=6).

Statistical significance was determined using sign test.

+P<0.05 MPTP-induced deficit.

*P<0.05 reversal of MPTP-induced deficit.

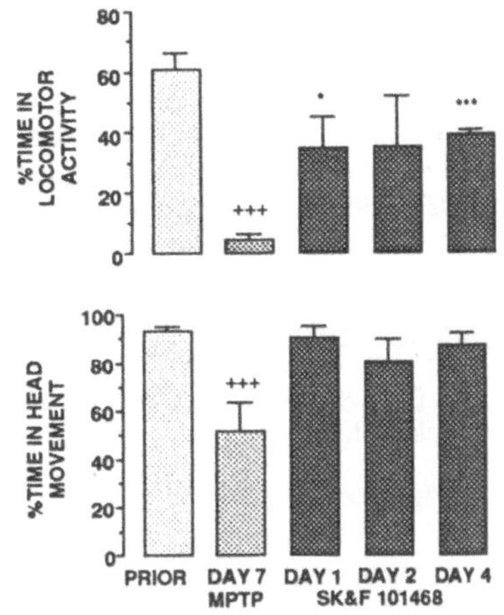

Figure 3 The effect of chronic administration of SK&F 101468-A (0.5mg/kg p.o. b.d.) on MPTP-induced deficits in the marmoset.
Treatment with SK&F 101468-A was started on day 8 of the MPTP infusion. Values represent mean ± s.e.m. of percentage time the animals spent in locomotor activity and head movement during two separate 2 min test periods (n=3 or 4)
Statistical significance was determined using paired t-test.
+++P<0.001 MPTP-induced reduction of activity.
*P<0.05 reversal of MPTP-induced deficit. ***P=0.001 reversal of MPTP-induced deficit.

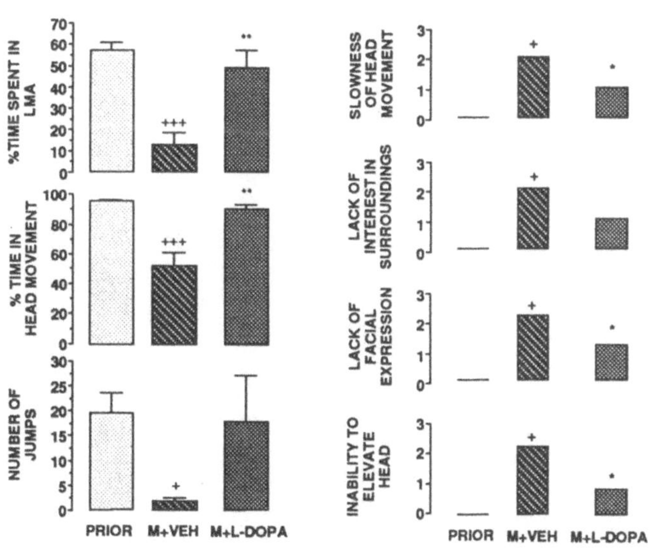

Figure 4 Reversal, by l-dopa, of MPTP-induced locomotor and behavioural deficits in the common marmoset.
The studies were carried out in the home cage. PRIOR= responses prior to MPTP treatment, M+VEH= vehicle response following treatment with MPTP and M+l-dopa = response following treatment with l-dopa 50 mg/kg p.o., 60 minute pretreatment following benserazide 12.5 mg/kg s.c. 30 minute pretreatment (n=6).
Statistical significance was determined using sign test.
+P<0.05 MPTP-induced deficit. *P<0.05 reversal of MPTP-induced deficit.

CONCLUSIONS

Human and non-human primates are exquisitely sensitive to MPTP in whom the neurotoxin induces symptoms characteristic of Parkinson's Disease. These MPTP-induced deficits, in monkey or man are reversed by current anti- Parkinsonian chemotherapy[2]. Thus, these results confirm direct dopamine-D_2 receptor agonist activity of SK&F 101468-A in the most appropriate animal model and support clinical assessment of this compound for the treatment of Parkinson's Disease.

References

1. B.Costall*, R.J.Eden, C.A.Harvey, M.E.Kelly* and D.A.A.Owen (1989) this meeting.
2. J.W.Langston and I.Irwin (1986) Clinical Neuropharmacology, 9 485-507.

SK&F 101468-A, A Possible Drug for the Treatment of Parkinson's Disease

B.Costall*, R.J.Eden, C.A.Harvey, M.E.Kelly* and D.A.A.Owen

Smith Kline & French Research Ltd, The Frythe
Welwyn, Hertfordshire, England

*Neuropharmacology Research Group, University of Bradford
Bradford, Yorkshire, England

INTRODUCTION

Dopamine D_2-receptor agonists, given as adjuncts, have been shown to ameliorate on-off fluctuations in Parkinsonian patients on long term levodopa therapy[1] and selective agonists are proving to be efficacious as monotherapy for the treatment of Parkinson' Disease[2]. Ropinirole, SK&F101468-A,(4-[2-(dipropylamino) ethyl]-1,3-dihydro-2H-indol-2-one HCl) is a novel, selective dopamine D_2-receptor agonist[3], which is currently in clinical development for the treatment of Parkinson's disease. These studies describe the pharmacological profile of SK&F 101468-A in the CNS.

MATERIALS AND METHODS

Specific binding of SK&F 101468-A to receptors from a variety of tissues was measured using radio ligand binding techniques. Six point displacement curves were used to determine K_i values or EC_{50}s at each receptor type.

Spontaneous locomotor activity was measured in mice (male, BKW strain, 25-35g) using Perspex cages fitted with photocells located 2.5 cm above the floor. Locomotor activity was recorded for 30 minutes after dosing and expressed as a cumulative count.

Post-synaptic dopamine agonist activity was demonstrated in mice (male, BKW strain, 25-30g) 7-25 days after unilateral administration of 6-OHDA into the substantia nigra using stereotaxic surgery under general anaesthesia. A simplified scoring system was used, from 0, no asymmetry to 3, intense body asymmetry pivoting in circles nose to tail. In further studies circling behaviour (rev/min) was also used as a measure of agonist activity.

Ability to induce stereotyped behaviour was assessed by observation of mice placed individually in Perspex cages. Animals were scored 0= no stereotypy, 1= periodic sniffing, head or limb movements, 2= continuous sniffing, head or limb movements, 3= discontinuous biting, gnawing, licking, 4= continuous biting, gnawing, licking. Assessments were made at 5 minute intervals for 30 minutes and effects were followed for up to 2.5 hours.

RESULTS

High affinity radioligand binding in human caudate revealed a K_i for D_2 receptors of 2.9×10^{-9} M (Table 1). Further studies in rat striatum showed SK&F 101468-A to bind selectively to mixed high and low affinity dopamine D_2- with no affinity for D_1-receptors at concentrations up to 10^{-4}M.

SK&F 101468-A was weakly active at inhibiting binding to α_2-adrenoceptors in human platelets (EC_{50} 9×10^{-6} M) and 5-HT_2 receptors in rat frontal cortex (EC_{50} 5×10^{-5} M). The compound was essentially devoid of activity at 5-HT_1, benzodiazepine and GABA receptors or α_1 and β-adrenoceptors.

SK&F 101468-A (dose-range 0.01-100 mg/kg i.p.) elicited biphasic spontaneous locomotor activity in mice suggesting presynaptic agonist activity at the lower and postsynaptic activity at the higher doses (Figure 1). Direct postsynaptic D_2 agonist activity, evidenced by contralateral turning relative to the lesion, was shown in unilateral 6-OHDA lesioned mice (Table 2). Chronic dosing with SK&F 101468-A, 1 mg/kg i.p., b.d. for 14 days, showed no diminution in this response (Table 3). Amphetamine (1.25 mg/kg i.p.) caused ipsilateral responses in the 6-OHDA lesioned mice reflecting the indirect, presynaptic agonist activity of this compound.

In contrast to amphetamine and apomorphine, SK&F 101468-A did not cause marked or dose-related stereotypies (Figure 2).

TABLE 1 Affinity of SK&F 101468-A in various receptor binding assays.

Receptor site	Tissue	Ligand	Ki (nM)
Dopamine-D_2	Human caudate*	^{3}H-Spiperone	29 ± 7 (n=6)
Dopamine-D_2	Rat striatum	^{3}H-Sulpiride	424.6, 377.7 (2 runs)
Dopamine-D_1	Rat striatum	^{3}H-SCH 23390	No affinity at 10^{-4}M
α_2-adrenoceptors	Human platelets*	^{3}H-Yohimbine	$IC_{50}=9\times10^{-6}$M
β-adrenoceptors	Human temporal* cortex	^{3}H-CGP 12177	No affinity at 10^{-4}M
5-HT_1 receptors	Rat forebrain	^{3}H-5 HT	No affinity at 10^{-4}M
5-HT_2 receptors	Rat forebrain*	^{3}H-Ketanserin	$IC_{50}=5\times10^{-5}$M
Central BZ	Bovine Cortex*	^{3}H-Flunitrazepam	No affinity at 10^{-5}M
Peripheral BZ	Human platelets*	^{3}H-PK 11195	No affinity at 10^{-4}M
GABA receptors	Bovine frontal cerebellum	^{3}H-GABA	No affinity at 10^{-5}M

TABLE 2 Asymmetry induced by acute administration of apomorphine, amphetamine or SK&F 101468-A in mice with 6-OHDA-induced unilateral lesions of the substantia nigra.

Values represent the mean ± s.e.m. of n=6 animals

Treatment	Dose (mg/kg) Route	Direction of Asymmetry*	Asymmetry Score
Apomorphine	0.5 s.c.	Contralateral	2.0 ± 0.0
Amphetamine	1.25 i.p.	Ipsilateral	1.6 ± 0.3
SK&F 101468-A	0.001 i.p.	None	None
SK&F 101468-A	0.01 i.p.	Contralateral	1.3 ±.0.4
SK&F 101468-A	0.1 i.p.	Contralateral	2.0 ± 0.0
SK&F 101468-A	1.0 i.p.	Contralateral	1.8 ± 0.2

*Designated with respect to side of the lesion.

TABLE 3 Contralateral asymmetry and circling induced by acute administration of apomorphine or chronic (b.d.) treatment with SK&F 101468-A in mice with 6-OHDA-induced unilateral lesions of the substantia nigra.

Values represent the mean ± s.e.m. of n=13 animals

Treatment (day)	Dose (mg/kg) Route	Asymmetry Score	Circling (Rev/min)
Apomorphine (-1)	0.25 s.c.	2.0 ± 0.2	6.1 ± 0.6
SK&F 101468-A (1)	1.0 i.p.	1.8 ± 0.2	6.0 ± 0.7
SK&F 101468-A (5)	1.0 i.p.	2.0 ± 0.0	8.4 ± 1.0
SK&F 101468-A (10)	1.0 i.p.	2.0 ± 0.1	7.4 ± 0.9
SK&F 101468-A (14)	1.0 i.p.	1.9 ± 0.2	7.5 ± 0.7

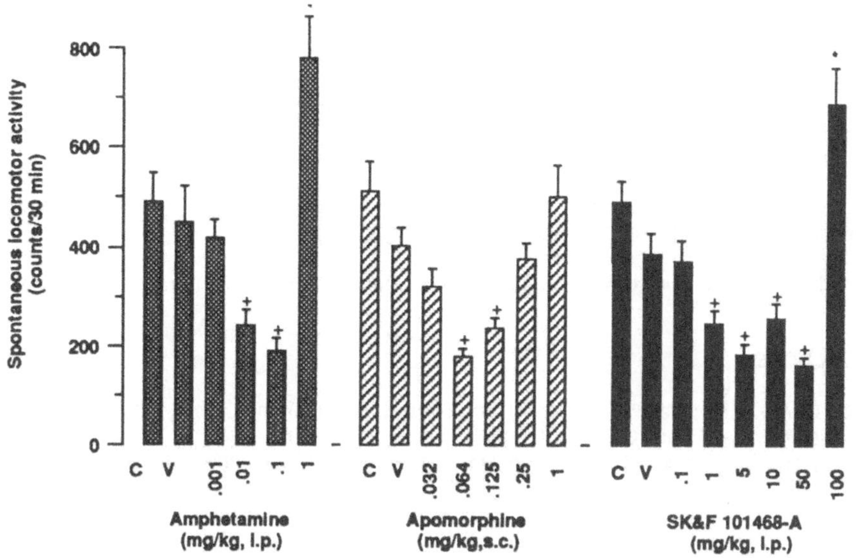

FIGURE 1 Comparison of the effects of SK&F 101468-A, amphetamine or apomorphine on spontaneous locomotor activity in mice, as measured in individual photocell cages.

Values represent mean ± s.e.m. (n=6-10)

Statistical significance was determined using one way ANOVA followed by Dunnett's t-test for multiple comparisons .
+P<0.01-0.001 significant reduction from vehicle (V) or control (C) activity.
*P<0.001 significant increase from vehicle (V) or control (C) activity..

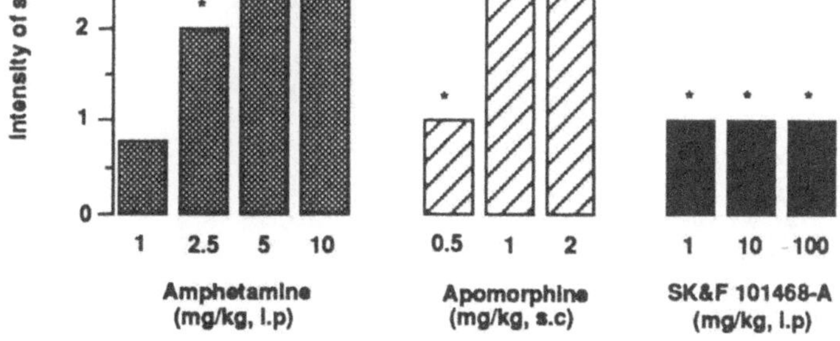

FIGURE 2 Comparison of stereotyped behaviour exhibited by mice following treatment with amphetamine, apomorphine or SK&F 101468-A.

Values represent mean scores (n=10), s.e.m.s <12%.

All values of 1 or greater are statistically significant P< 0.001 (Mann-Whitney U test).

CONCLUSIONS

These results are consistent with SK&F 101468-A being a specific, directly acting dopamine-D_2 agonist in the extrapyramidal system. SK&F 101468-A was shown to have similar post synaptic potency to apomorphine in the 6-OHDA-lesioned mice but caused significantly less stereotyped behaviour. These data suggest a possible utility for the treatment of Parkinson's Disease with a reduced liability to cause dyskinesias.

ACKNOWLEDGEMENT

We would like to thank Dr. R. Horton, of St. Georges Hospital, London. for the binding data marked*.

REFERENCES

1. Stibe C.M.H., Kempster P.A., Lees A.J. and Stern G.M. (1988) The Lancet, February.
2. Stoessl A.J., Mak E. and Calne D.B. (1985) The Lancet, December.
3. Gallagher G.J., Lavanchy P.G., Wilson J.M., Hieble J.P. and De Marinis (1985) J. Med. Chem., 28 1533-1536.

GENERAL DISCUSSION

Moderated and summarized by

T. Nagatsu[1], A. Fisher[2], I. Hanin[3] and M. Yoshida[4]

[1]Department of Biochemistry, Nagoya University School of Medicine, 65 Turuma-Cho, Showa-Ku, Nagoya, 466 JAPAN
[2]Israel Institute for Biological Research, P.O.Box 19 Ness-Ziona 70450, ISRAEL
[3]Department of Psychiatry, Western Psychiatric Institute and Clinic, University of Pittsburgh School of Medicine Pittsburgh, Pennsylvania 15213, USA
[4]Department of Neurology, Jichi Medical School Kawachi-Gun, Tochigi, 329-04 JAPAN

This Conference was capped by a lively discussion which was moderated by Drs. Nagatsu, Fisher, Hanin and Yoshida. It focused on 10 key questions, originating from discussions generated throughout the conference from its inception. The topics and the corresponding discussions are summarized in the following narative.

QUESTION 1: WHAT IS THE ROLE OF AL^{+++}, CA^{++}, FE^{++}, ETC IN THE ETIOLOGY OF ALZHEIMER'S AND PARKINSON'S DISEASES?

Dr. Fisher Dr. Khachaturian talked about the calcium hypothesis. Dr. Youdim mentioned iron accumulation in Parkinson's disease. **Dr. Mizuno** Why does iron accumulate in the substantia nigra and does not go out? **Dr. Agid** We did the same kind of experiment with Dr. Marsden. Every one agrees that iron is elevated in the substantia nigra of Parkinson's disease. Is it primary or secondary? When cells are dying, free radicals are produced and iron is accumulated. We also found that ferritin is decreased in the substantia nigra. These might explain why iron is accumulated in Parkinson's disease. Another question is which class of iron is involved, Fe^{+++} or Fe^{++}? Apparently Fe^{++} is more dangerous. Dr. Riederer found Fe^{++} to be increased. And I don't know if the opposite is dangerous or not. **Dr. Youdim** I don't know why iron accumulates in the substantia nigra, but I mentioned some possibilities for this. There may be destruction of blood brain barrier at the level of the substantia nigra. Turnover of iron metabolism may be increased. Recent work by Kastia in New Orleans studied influx and eflux of iron. Eflux is slower than influx. This may increase iron in the brain. Infection can produce increase of iron in the cytoplasma or cell. I do not think that cell degeneration per se is responsible for an increase of iron in the brain. Otherwise we would see increases of other metals as well (e.g. copper or zinc), which do not occur.

QUESTION 2: BETA-AMYLOID, CAUSE OR ETIOLOGY?

Dr. Nagatsu We have heard a lot of exciting presentations on β-amyloid protein, especially from the molecular biological point of view. But the big question is whether the deposition of β-amyloid is really the cause or the consequence of the disease. Any commments? **Dr. Yoshida** Arendash's work is, I think, very important in this regard. They showed decrease of the cortical neurons and also found some inclusion bodies in the rats 14 months after lesioning of the nucleus basalis. Some Japanese researchers did similar experiments, but they failed to confirm the changes. Can anyone comment on this? **Dr. Meyer** I can tell you a little bit of what Dr. Arendash is doing. I am collaborating with him, but I am not a member of his group. He found some pre-plaques, and the incidence of these bodies that you are referring to is something like 3 times per brain. Dr. Terry evaluated these substances and according to him these could be artefacts. I don't think it's important that they occurred 2 or 3 times per brain. So I think it is more important to look for transsynaptic cell loss or neurochemical changes occurring transsynaptically. **Dr. Hanin** For the sake of discussion, I wonder if the rat is the best model to use when looking for plaques. For instance aluminum will cause tangles in the rabit, but will not do anything in the rat. If Arendash's group had used another animal species they might have seen plaques. Is that possible? **Dr. Meyer** I agree. On the 1st day, Dr. Sakaki presented an excellent experiment in which he mentioned that in the mouse, another rodent, there is no age-related change in the expression of amyloid gene. In primates and humans we know that there is an age-related increase. So there may be a species difference in observed relationships between the age-related change and expression of β-amyloid. This brings back the question whether or not the expression of β-amyloid causes the disease. This is going to be very difficult to determine, because differential expression of β-amyloid could either result from changes of neurotransmission altering protein kinase C, which modulates the AP1 site, or from genetic changes. **Dr. Khachaturian** An important question is what the function of the protein is, because relationship between the presence of the protein and dementia is really correlational. There are many cases where you have large amounts of these lesions with no dementia. The important thing to understand is how these proteins are formed, what their nature is, and what they really do. Recently there were 2 distinctly different papers coming from Cotman and Selkoe, indicating that a fragment of the protein got a promoting effect in cell cultures and another one by indicating that it has a toxic effect. These are fine studies, but they don't completely answer the question, because they were only dealing with fragments of these proteins, not entire sequence, as it would occur in the cell.

QUESTION 3: IS THE CHOLINERGIC HYPOTHESIS OF AD ALIVE AND WELL?

Dr. Hanin From the beginning of this conference, people have been talking about the cholinergic hypothesis. Dr. Davies started out by talking about acetylcholine, but as the conference progressed you may have noticed, as I did, that every time people spoke about the cholinergic hypothesis they almost acted as if to say: "if it is true". It is clear that one can not talk anymore, as we did even 3-4 years ago, solely about the cholinergic hypothesis. We have to consider other neurotransmitters as well in the etiology of Alzheimer's disease. **Dr. Agid** One must consider in this context what is the role of the lesion of the cholinergic system in the symptomatology. The famous innominato-cortical and septo-hippocampal cholinergic systems are not the only cholinergic system of the brain. Therefore we have to look for other cholinergic systems as well. Moreover, some cholinergic pathways are also involved in Parkinson's disease or progressive supranuclear palsy. Secondly, we must consider why some cholinergic neurons are damaged while others are not. **Dr. Giacobini** We should not use the term "cholinergic hypothesis", because it is a very limited type. Rather, we are interested in the involvement of cholinergic systems in AD. If I were to list the diseases in which the Meynert nucleus is involved, at least 10 can be named in which the pathology is described. Thus the scope of cholinergic involvement is much wider. I think that there are 3 reasons why the cholinergic hypothesis is alive. One is that the cholinergic system is particularly linked with memory. Unfortunately we have not done much pharmacologically about the relationship between the cholinergic system and memory in the human being. Secondly, we have a good marker, choline acetyltransferase. Third, there has been some, but minimal success with cholinergic drugs in the treatment of AD. The other thing is NGF. As we heard this morning NGF provides the basis for a modified cholinergic hypothesis. Galanin is related to the cholinergic system and also has been implicated in AD, and also we heard about the link between the cholinergic and glutamatergic systems in the pyramidal cells. **Dr. Hunter** Before I came to this conference I thought some cases of AD had a pure cholinergic deficit and younger ones have a more mixed deficit. If you have a subgroup of patients with a "pure" cholinergic deficit, then they should be more amenable to treat with a cholinergic drug. Actually the relationship between memory and acetylcholine is important, and cholinergic treatment is aimed at treating the symptoms rather than the cause of AD. The other thing of interest in this conference was the lack of any talk about the role of serotonin in non-cognitive symptoms in AD. **Dr. Mizuno** It is true that in AD there exists marked loss of cholinergic neurons, but other neurotransmitter systems are also involved. In Parkinson's disease we know the clinical symptom are due to the deficiency of dopamine. When talking about Alzheimer's disease, we also have to know the clinical counterparts corresponding to the deficiency of various transmitters, such as acetylcholine, noradrenaline, somatostatin and so on. **Dr. Gottfries** I am involved in the study of other neurotransmitters. We know that there are multiple changes in the cortex of AD. Of course the intellectual disturbance is important, but if you look more carefully, there also are motor disturbances in AD, emotional as well as endocrinological disturbances. Also depression occurs. One cannot explain all of these symptoms by a cholinergic deficit alone. As Prof. Giacobini said the cholinergic hypothesis is still alive, but feeling not well. Cholinergic deficit is related to the memory disturbance. Moreover, together with Dr. Carlsson, we have studied vascular dementia. In these patients, not only an acetylcholine deficit but also a deficit of serotonin system was observed. Recently Dr. Kogure demonstrated dementia in animals with repeated ischemic attacks. This indicates that the vascular factor may also be important for the cause of cholinergic deficit. If the cholinergic hypothesis is reformulated for memory disturbances, one could safely state that it is still alive and valid. But too little attention is paid to the more painful symptoms of AD, such as emotional disturbances, anxiety, aggressiveness and depression. We can treat these symptoms, if you use drugs against serotonin deficiency. **Dr. Agid** In Parkinson's Disease, the cholinergic system is also involved in the so called frontal lobe like symptomatology. Although the septo-hippocampal lesion contributes to memory loss, the memory loss is not only due to the septo-hippocampal lesion but also to other histopathological stigmata. The famous glutamatergic entorhinal cortex is also largely damaged. All circuits between the hippocampus and the entorhenal cortex are in fact damaged. Therefore if I were responsible for the drug development, I don't think I would push the cholinergic idea so much for the treatment of AD. **Dr. Fisher** I would predict that this type of exclusionary approach, if applied to the treatment of PD at the time before L-DOPA was discovered may have resulted in not trying L-DOPA as a treatment drug. We may have thus lost out in the discovery of this excellent medication for PD! **Dr. Agid** The nigro-striatal pathway is quite unique and the nigral neuron is massively damaged. **Dr. Fisher** But the nigro-striatal pathway is not the only structure involved in Parkinson's disesase. **Dr. Agid** Let's say that about 25% of the patients shows purely dopaminergic deficit. **Dr. Fisher** That's exactly what I wanted to say with regard to AD. In AD about 25% of the patients show a purely cholinergic deficiency. **Dr. Agid** I doubt that. The diagnosis of Alzheimer's disease is made on the basis of both clinical features and presence of senile plaques. **Dr. Fisher** At the meeting organized recently by Dr. Wurtman in Zurich Dr. Elaine Perry showed that 20-25% of Alzheimer's patients show a pure cholinergic hypofunction. They also had tangles and plaques. This group may represent a subpopulation, and we could perhaps selectively treat these patients by cholinergic therapy. **Dr. Heldman** Depolarizations of neurons by activation of cholinergic neurons are tonic and occur at random. If we develop cholinergic drugs, we might be able to stimulate neurons in a tonic way. By this tonic activation of neurons, meaningful informations could be achieved between neurons. Thus tuning of neurons may play an important role, and one should consider the possibility that the cholinergic system exerts an important influence in the central neuron system through such a tonic mechanism.

QUESTION 4: IS THE DOPAMINERGIC HYPOTHESIS OF PD ALIVE AND WELL?

Dr. Yoshida I think the answer is clear; the hypothesis is alive and it is indeed very well. We should, therefore, use "theory" instead of "hypothesis" in this regard. I would now like to open the floor to the discussion on the reason why the dopaminergic cell is selectively affected in PD. **Dr. Mizuno** Dr. Youdim nicely showed the evidence of involvement of complex I of the mitochondrial electron transport system in PD. Complex I is inhibited by MPTP in the experimental model of parkinsonism, as well as in the brain of Parkinson's disease. In my mind, energy crisis is the most likely etiology of PD. Perhaps even not only a single substance or toxin is responsible, but number of substances present around us might be involved. So there are at least two theories in regards to pathogenesis of Parkinson's disease; the iron theory and the mitochondrial theory. I think both are valid. But I have some difficulty in determining how to unite these 2 hypotheses. **Dr. Youdim** I don't really have experimental data, but mitochondria may be damaged secondarily to the toxicity of iron. Iron induces mobilization of calcium and cytotoxicity of calcium induces the degeneration of metal enzymes or metal endopeptidases. This may, therefore, be one way by which iron accumulating within the mitochondria, induces the cytotoxicity of mitochondria. **Dr. Mizuno** Iron exists in either neuromelanin or cytoplasma. I don't think that iron goes into mitochondria very quickly. Iron accumulation is outside of mitochondria. So I still have some difficulty combining your theory and mine. **Dr. Youdim** We have set up electron microscopy to study iron-overload thalassemia. But we have not yet been successful in developing this technique in the brain. Iron must be complexed with other unknown molecules, possibly ferritin, or some other small molecules. But we also must determine where in the brain such complexes occur. **Dr. Chen** I think that dopamine is also decreased in the adreno-medulla of Parkinson's patients. Also neuromelanin is involved in this disease. Mention has also been made about free radicals. These things should be discussed more. **Dr. Agid** I agree totally with Dr. Biao Chen. With Dr. Graybiel we have counted the number of damaged cells in the dopaminergic system. Dopaminergic cells in the central gray are not damaged, these cells do not contain neuromelanin. This fact, together with the fact that lipid peroxidation is increased, and iron as well as copper is increased, indicates that enzymes involved in the catabolism of free radicals are damaged. For example, superoxide dismutase (SOD), as recently published by Dexter et al, seemed to be decreased in the soluble fraction, but not in the nuclear fraction. SOD mRNA is in the dopaminergic neurons of the substantia nigra. All these data indicate that these cells are more vulnerable than others. These cells suffer more easily from environmental factors, and easily make free radicals. Concerning iron, we have looked at iron of the substantia nigra cells of the Parkinson's patients, using microscopy. Indeed iron per cell is increased, and iron is located all over the cell. **Dr. Nagatsu** Prof. Narabayashi and Iizuka described an autopsy Parkinsonian case in which only neuromelanin had disappeared, but there was no cell death. **Dr. Chen** Neuromelanin is related to Lewy bodies. The dopamine neuron death may be secondary to the loss of neuromelanin. **Dr. Narabayashi** In the Juvenile case which Dr. Nagatsu mentioned, reduction of neurons of the substantia nigra compacta does exist, but it is not so severe and the number of reduced cells was relative. The loss was severe in the ventrolateral part of the substantia nigra and much less in the medial part. **Dr. Yoshida** Structures outside of the substantia nigra are also involved in Parkinson's disease. **Dr. Dostert** Dopamine is very oxidizable and also noradrenaline is oxidizable, and these could give rise to neuromelanin. Neuromelanin occurs not only in the dopaminergic system, but also in the noradrenergic system. **Dr. Agid** We have counted nerve cells in the dopaminergic and also noradrenergic systems. We were surprised to see that there was no correlation with cell death between the nuclei. You have 90% cell death in the substantia nigra pars compacta with only 50% loss in the locus coeruleus and vise versa; we studied this in 7 brains. If the environmental factors are important for the etiology of Parkinson's disease, the damage should be more diffuse. So other factors must also be playing a role. **Dr. Kopin** You have just said 90% loss of the locus coeruleus and only 50% cell loss in the substantia nigra. Was the patient in which you made these observations primarily with Parkinson's disease without any depression or was the patient also depressed? **Dr. Agid** I can not remember now the relationship between this case and depression. **Dr. Yoshida** Since noradrenaline story has been raised, Dr. Narabayashi would like to show a film concerning the therapy of parkinsonism with the norepinephrine precursor, L-threo-DOPS. **Dr. Narabayashi** As I described in my paper this morning, juvenile-onset or early-onset parkinsonism (JP) is interpreted as the condition involving mostly a nigro-striatal dopamine (DA) deficiency. When such cases are treated properly by medicine (levodopa with/without DA-agonist) or by microstereotaxic surgery in alleviating rigidity and/or tremor, they can function almost normally for many years without presenting akinesia or further deterioration in ADL (activity in daily life and in jobs). On the other hand, patients with classical Parkinson's disease (PD) always show progression of the difficulty and tend to deteriorate in rigidity and tremor even after similar surgical treatment. The symptoms which start to appear and worsen in the chronic stage under long-term levodopa treatment are: difficulty of postural balancing and gait, freezing in gait, and deterioration in emotional and attention level. By measuring the activity of Dopamine-β-hydroxylase (DBH), the enzyme for the biosynthesis of norepinephrine (NE), in CSF of PD patients, it was found that DBH activity was much lowered in the longstanding PD cases. We therefore used L-threo-DOPS (L-threo-3,4-dihydroxyphenylserine), an industrial precursor of NE, in an attempt to compensate for the NE deficiency, and to improve the above symptoms. The film presents two cases with severe postural difficulties and inability to walk without support, which have been treated favourably by L-thero-DOPS. These two cases indicate that NE deficiency might be the reason for difficulty of postural control and of gait. Compensation of this deficiency by L-threo-DOPS apparently improves this difficulty. Difficulties in the later stages of the disease in older patients may therefore be understood as the result of metabolic error of both DA and NE. Psychological analysis also has indicated that the lowered emotional level such as depressive state or the lowered attention level are also improved and activated by L-threo-DOPS. We suggest that when some endogenous toxic substance affects the human brain to cause parkinsonism, the resulting changes might be different in each case, depending on the grade of aging of CNS. In younger patients changes might be more limited to the DA system, while both DA and NE systems might be affected in the older patient. It seems important to note that L-threo-DOPS usually produces no change or improvement in the younger patients in which a NE deficiency is not considered to be important. From these observations it would appear that the clinical picture of Parkinson's disease is strongly influenced and manipulated by the degree of aging of the CNS.

QUESTION 5: PREVENTION VS TREATMENT

Dr. Dostert I have some difficulty in understanding the strategy used in the prevention of the progress of Parkinson's disease. Also I am not clear about the use of deprenyl as an MAO inhibitor, because dopaminergic nerve terminals do not contain MAO-B. Also MAO-B is not present in the cells of the substantia nigra. During this conference many authors suggested toxicity of oxygen free radicals due to increased dopamine. Therefore, increase of dopamine levels by an MAO-B inhibitor will not be a useful treatment strategy. Indeed, it might lead to damage of dopamine neurons. **Dr. Mizuno** For deprenyl to be effective to retard progression of Parkinson's disease, you first have to postulate some toxins which are converted into an oxidized form by MAO-B. MAO-B exists in astrocytes, not in dopamine neurons. With regard to your 2nd question, oxygen radicals may be toxic, but in my opinion they are not the primary cause for Parkinson's disease, because I favor the mitochondrial hypothesis. Once an energy crisis occurs, oxygen can not be converted completely into water. The partially released oxygen species, therefore, cannot be reduced completely, resulting in superoxide anions which are produced and may act as toxic substrates for the mitochondrial membrane and enzymes. **Dr. Dostert** I agree with you. In that sense you have to generate another hypothesis. **Dr. Tipton** We have to distinguish between what we know and what we guess. The elegant presentation by Dr. Langston indicated that use of deprenyl delays the stage of the disease to use L-DOPA. We have no evidence so far that cell death is prevented; rather it is delayed. I think we need more postmortem studies. We have to compare the brains of the patients treated with L-DOPA alone and L-DOPA + deprenyl. **Dr. Nagatsu** Together with Dr. Narabayashi we have studied biochemically many human brains of Parkinson's disease and measured dopamine β hydroxylase, a marker of noradrenaline. I think that in the early stage of the disease damage of dopamine neurons occurs and later on damage of noradrenaline neurons occurs. Therefore, I think that the alteration of dopaminergic neurons is more severe than that of noradrenaline neurons. **Dr. Agid** Dr. Chan-Palay reported in her poster that the mean decrease of number of noradrenaergic cells of the locus ceruleus was more than 90%. Although cases she looked at had very long-duration, the number of cells degenerated was greater in the locus ceruleus than in the dopaminergic cells. **Dr. Kopin** The locus ceruleus is involved in axiety and depression, but I don't know how these patients respond to stress. On another issue I would like to address a question to Dr. Tipton: Dr. Langston withdrew deprenyl. Do you think that the withdrawal of deprenyl was long enough? **Dr. Tipton** I think Dr. Langston's study was a very good controled experiment. But the half life of MAO in the rat's brain is about 10 days. Therefore they should have washed out deprenyl for a longer time period. **Dr. Kopin** Would you be satisfied if deprenyl was washed out for a month? **Dr. Tipton** Yes, but nobody knows its half life in the human brain. **Dr. Narabayashi** It is important to do postmortem studies of the cases who died at a relatively early stage of the disease. There is no report in the literature to show which region shows an earlier degeneration of nerve cells, the locus ceruleus, or the substantia nigra. **Dr. Youdim** There are many factors which produce oxygen radicals. It is dangerous to put too much emphasis on MAO for the formation of radicals. At least, as far as lipid peroxidation is concerned, inhibition of MAO in the rat cortex did not cause basal formation of free radicals. **Dr. Hyden** People keep talking about the toxicity of dopamine-produced radicals. But we think that the existence of dopamine is good.

QUESTION 6: OPTIMAL RX STRATEGIES IN AD AND PD: DESIRABLE FEATURES?

Dr. Mizuno For Parkinson's disease, we have relatively satisfactory ways of treatment. For Alzheimer's disease there are some trials of treatment, but none of them are, to my mind, satisfactory as yet. In Alzheimer's disease many transmitters and neuromodulators are decreased. Therefore we have to combine the ways to increase these many transmitters. Has anyone in the audience ever tried a combined treatment employing more than 2 transmitters? **Dr. Fisher** I agree that the polypharmaceutical approach would be neccessary eventually. The problem is that as the starting point there are very few cholinergic drugs. For example, physostigmine is so short acting, and has side effects. THA has hepatotoxicity even if you decrease the dose. Concerning pilocarpine, people forget that it can cause significant dystonia at least in animals. Bethanecol is a very weak M_2 agonist, and it is therefore irrational to use bethanecol. Arecoline is also short acting. Choline and lecithin treatment have no rationale behind them; perhaps the earlier use of these materials was based on an analogy to L-DOPA treatment in the Parkinson's disease. What I want to say is that there still does not exist a good (long acting, minimal side effects) clinically evaluated cholinergic compound. Of course it is very important to combine other neurotransmitter drugs with cholinergic drugs in the treatment of AD, such as noradrenergic, 5-HT analogues and perhaps a nicotinic agonist as well. **Dr. Gottfries** I don't understand why you should start from the cholinergic system, Dr. Fisher. The 5-HT system is also damaged. If you have a deficit in the serotonergic system, you can have aggressiveness, depressiveness, axiety, mood disturbances, and endocrinological disturbances. That is why we use Citalpram, which is a perfect drug which activates 5-HT system. Also 60% of the patients have disturbances of the dexamethasone suppression test. We could influence these disturbances by such a drug. **Dr. Agid** We have serotonergic and noradrenergic agents, and uptake inhibitors of these transmitters. But we are still waiting for agonists for noradrenergic, serotoninergic, glutamatergic and also GABAergic systems. This is the same for blocking agents. We have neuroleptic drugs, but other antagonists are necessary. **Dr. Ogawa** I don't understand why people think that the replacement therapy is important. L-DOPA therapy for Parkinson's patient is exceptional, since the striatum is a very specific structure compared to the other structures. In Parkinson's disease, you don't have to think about the neuronal network; an increase in dopamine is enough. But in Alzheimer's disease, if you look at the frontal cortex, the laminar structures are completely destroyed. Therefore even if you would be successful in increasing acetylcholine, already the structures which receive the acetylcholine are destroyed, including second messenger systems. So we have to try to develop drugs to protect the degeneration of the cells. **Dr. Potter** I don't think that nicotine, 5-HT, clonidine and so on are tested correctly. As Dr. Ogawa said, even in the advanced cases of Parkinson's disease receptors are still there, and they respond well to dopamine. In Alzheimer's disease, M_1 receptors also are preserved, although they may be rearranged. So still there is some expectation for a cholinergic agonist. **Dr. Ogawa** Yes, what I said was with regard to the advanced case of Alzheimer's disease. In early stage of AD replacement therapy may indeed be successful. **Dr. Fisher** Citalpram is good for emotional disturbances, but is it good for the memory disturbances? No. So still we are facing a major problem. That is why I still think that cholinergic agonists are important for the treatment of memory disturbances. **Dr.** One crucial question is the side effect. How many, and how serious side effects should be allowed for drugs which are used for the treatment of dementia? This question has to be discussed with some consideration. **Dr. Giacobini** I think that patients in the final stages of AD can be compared to patients with cancer. When you look at cancer, you don't talk about side effects. If we can get some success, the patient will accept a certain amount of side effects. **Dr.** I think that the FDA has a completely different point of view on that matter. **Dr. Agid** There are 3 kinds of symptoms in this disease. For psychiatric symptoms we can do something. Secondly, there are memory disorders and attentional defects. These are symptoms for which at present we cannot do anything. Thirdly, aphasia, apraxia and agnosia exist, and these are directly related to the lesion of the cerebral cortex of the specific areas. For these we cannot do anything either. **Dr. Gottfries** In AD, the memory deficit occurs first, but the patient does not go to doctors. When depression, aggressiveness and restlessness appear, patients go to doctors. Memory disturbance is of course important, but the emotional disturbances are more important for the patients and their relatives. **Dr. Fisher** Do you think that you can treat all these symptoms with one drug? Moreover, one crucial problem is that drug companies do not want to make "cocktails". **Dr. Nagatsu** Dr. Gottfries, do you think that depression is dependent on a deficit of serotonin? **Dr. Gottfries** There are many debates concerning the relationship between serotonin and depression. We diagnose depression by the presence of emotional disturbances, anxiety, aggressiveness and restlessness. Some people call it "atypical" depression. What we need is an emotional stabilizer for these patients. **Dr. Giacobini** You can treat emotional disturbances symptomatically without developing new drugs. Memory and cognitive disturbance, however are progressive. This is the case in AD. **Dr. Gottfries** I don't agree with what you said. Tricyclic antidepressants cannot be used in AD, because they cause confusion in these patients. There is a new type of drug, a 5-HT reuptake blocker, Citalpram. This is much better from a toxicological view point. We have treated 98 patients with this compound. It has significant effects on the kinds of symptoms which one sees in AD. **Dr. Kopin** I have a question from a pharmacological point of view. Dopamine re-uptake blockers are not used for the Parkinsonian patient, since the dopaminergic neurons do not release dopamine any more. With regard to the lack of serotonin, how does a serotonin re-uptake blocker act? **Dr. Gottfries** There is serotonin left. Using imipramine binding about 50% decrease of serotonin has been observed in brain tissue of Alzheimer's patients. **Dr. Kopin** You have to have 80% depletion of dopamine, if you take an analogy of Parkinson's disease, for the patient to develop the symptoms. Then I don't think that the re-uptake blocker helps that much. **Dr. Gottfries** Depression might come earlier than following 80% depletion of serotonin. **Dr. Nagatsu** We have found a 50% decrease of tryptophan hydroxylase in the AD brain. **Dr. Yoshida** I think during this session two levels of discussion have been mixed up. Dr. Gottfries mentioned really important current ways of treating Alzheimer's patients. I am also a clinical neurologist and what he mentioned was very useful for our daily practice. On the other hand, Dr. Fisher was trying to emphasize the importance of developing a new cholinergic drug. Since L-DOPA appeared about 20 years ago, the history of treatment of Parkinson's disease was completely changed. Therefore, we really need at least one very good cholinergic drug.

QUESTION 7: DRUGS IN SEARCH OF A MECHANISM OF ACTION

Dr. Fisher We have had here many interesting posters and presentations concerning so called nootropic drugs, which apparently have no side effects. A major question is what is the action of these drugs. A related question is whether these drugs can be approved at all, because as far as I know the FDA does not approve drugs for which the mechanisms of action are not known. **Dr. Giacobini** In the 1st Alzheimer meeting in Eilat, there were many participants from pharmaceutical companies. They all sat in the back of the room and were very silent. At that time, there were no drugs for AD, but there was an excellent market for such drugs. I recently have read a pharmaceutical bulletin stating that next year the Japanese market for drugs for treatment of senile dementia will be 1 billion dollors. So the market is successful. Please tell me now the mechanism of action of these miracle drugs. Presumably some of these drugs are memory consolidators; others are activators of cerebral metabolism; anti-amnestic, anti-anoxic, etc. Please tell me the secret of these drugs. **Dr. Kopin** If these drugs are non-toxic, people think that they perhaps work. I think that this reflects the poverty of our pharmacopeia. Sometimes people laugh or sometimes cry with the placebo effect.

QUESTION 8: ANIMAL MODELS IN AD AND PD: DESIRABLE FEATURES?

Dr. Hanin This question should probably come up before all the previous questions, because we are so unclear about treatment strategies. What should be the minimum requirement for an animal model in AD or in PD? **Dr. Nagatsu** In the Parkinson's disease, we have a very nice model in the MPTP-monkey. On the other hand, we don't have any good model for ALzheimer's disease. The best model we have so far is the AF64A model. **Dr. Giacobini** The model of amyloid is very important. I believe that transgenic mice will be a very good model. Soon I think we will have such a model in which we have production and accummulation of amyloid. **Dr. Nagatsu** I think that Prof. Sakaki's group is trying to produce such transgenic mice to express β-protein. **Dr. Agid** Alzheimer's disease has several lesions, such as dopaminergic, cholinergic, noradrenergic and serotoninergic lesions. New models should be made as having several lesions. It is too simple idea to have a single lesion. **Dr.** You are trying to have a perfect model. But if you look at the Parkinson's model, Dr. Carlsson started 30 years ago to make such a model. This was a reserpine model, and people thought that this was a imperfect model. But as time goes on, a better model such as the MPTP-treated animal appeared. Therefore, I think that at the beginning the model should be a simple one. **Dr. Fisher** That is exactly what I wanted someone to say. **Dr. Potter** I think the simpler model is necessary for testing memory, since we don't know so far how the brain works. To test memory, you have to test memory. We have tried nicotine, we have tried propranolol which did not have much effect, we have done lot of things and don't have to feel so badly. I think that it is of major importance to continue to test new drugs; not only cholinergic, but also other drugs are necessary for testing memory function.

QUESTION 9: BIOLOGICAL MARKERS IN AD AND PD: DESIRABLE FEATURE?

Dr. Nagatsu We can measure directly dopamine, its metabolites or others in CSF. Use of plasma and even urine is much better. Of course, PET is promising for Parkinson's disease, but it is very expensive. **Dr. Gottfries** So far as CSF is concerned, we can not get consistent data from CSF. Monoamine metabolites are not good markers for AD. Monoamine oxidase in platelets may be another marker. But so far I think that vitamin B_{12} deficiency would be a most promising marker in these patients. Now we are investigating gangliosides in CSF as a marker of the gray matter. **Dr. Nagatsu** Is vitamin B_{12} a specific marker for Alzheimer's disease? **Dr. Gottfries** No, but there is a subgroup of patients with senile dementia of Alzheimer type, in whom vitamin B_{12} was reduced, yet they don't have anemia. Accurately, they should be diagnosed "secondary dementia". But today they are still categorized as SDAT, according to DSM III. **Dr.** I like to mention about the paper by Dr. which was presented in the last Neuroscience Meeting in Arizona. He studied 28 Alzheimer's brains anatomically, staining cells in the frontal cortex. He grouped these patients into 3 categories. In some cases, cells were dropped in one layer and in another case the cell drop was in another layer, and in the 3rd group there were cell losses in all layers. There were no good correlations between the cell loss and the deposition of plaques and tangles. **Dr. Mizuno** I think that this question should be broken down into 3 levels. One is the clinical diagnostic level, while the patient is still alive. The 2nd is the post-mortem level. The 3rd will be a future problem. At the clinical level we don't have any satisfactory markers, and therefore clinical judgement by our neurologists is most important. At the 2nd level, we have satisfactory factors, such as Lewy bodies and decreased dopamine for Parkinson's disease, and tangles and senile plaques for Alzheimer's disease. For the 3rd level, we should like to have definite markers, perhaps related to genes or DNA. **Dr. Nagatsu** I am sure that molecular diagnosis will be promising for the future in developing promising biological markers in AD.

QUESTION 10: DEMENTIA IN PD: ETIOLOGY? TREATMENT?

Dr. Yoshida The dementia associated with PD has been so often discussed. I think there are at least 4 pathologically related dementias. One is the dementia of PD associated with the Alzheimer type change. The 2nd one is due to excessive deposits of Lewy bodies in the PD brain. The 3rd is that medially located nigral cells are related to the dementia which has recently been reported by Dr. Rinne. The 4th is some combination of VD (vascular dementia) with PD. There may be some other categories to explain the dementia in PD. **Dr. Mizuno** Dr. Yamamura at the Metropolitan Geriatic Hospital has conducted a lot of pathological studies. They found a PD group with dementia and the pathological change was found only in the substantia nigra and the locus ceruleus. **Dr. Narabayashi** It has been stated that about 90% of PD have dementia. I believe that they are not having dementia, but rather it is a kind of hypofunction of mental activity. I would call this behavior bradyphrenia instead of dementia. **Dr. Mizuno** I am hesitant to contradict Prof. Narabayashi, but I must say that only 10% of my patients with PD have dementia. If I include bradyphrenia, the percentage will increase up to approximately 30%. Also, it is very difficult to distinguish between dementia and bradyphrenia in PD. Thus, I believe that dementia is an integral part of the symptoms of PD. **Dr. Rinne** In a recent paper, we reported that a decline of mental activity or dementia in PD is partly related to a decrease in the medial neurons of the pars compacta of the substantia nigra. This is based on the topographical oraganization of the neurons of the substantia nigra, since we know that the lateral neurons project to the putamen, which is related to the motor and the medial ones to the caudate nucleus, which is associated with the frontal cortex. Lack of motivation and attention in the patient would be related to the locus ceruleus, which projects to the frontal cortex. So the cognitive impairment or dementia in PD is different from that of AD. **Dr. Kuno** We have unpublished data to indicate a decreased second messenger system in Parkinson's disease. IP_3 was markedly decreased in the brains of PD patients with dementia, compared with those without dementia. Therefore, we believe that the dementia seen in a group of PD patients is due to the decreased 2nd messenger system. **Dr. Agid** We have to be cautious when talking about dementia in PD. Some people imply transmitters, while others blame this on nerve cell loss. But this does not mean that these are the causes of dementia. Since all of these patients die after a course of PD for 15 or 20 years, how can one be sure that the dementia is not due to the duration of the disease instead? I think that dementia is extremely complicating and polyfactorial. As far as dopamine is concerned, the MPTP -intoxicated patients have almost no dementia. Also, early onset PD patients with a marked response to L-DOPA have no dementia, although there may be a marked defiency of dopamine in their whole brain. **Dr. Rinne** Although there may be multiple factors which cause dementia, one cannot ignore the involvement of dopamine for cognitive function. **Dr. Giacobini** If major dementia is not discriminated from that of Alzheimer's dementia, where is the truth? If you say that multifactors are involved in PD, are there 2 diseases or even 3 different subcategories of diseases in PD? **Dr. Yoshida** I think this problem is really a difficult one.

INDEX

AD = Alzheimer's Disease
PD = Parkinson's Disease

GPSR Compliance

The European Union's (EU) General Product Safety Regulation (GPSR) is a set of rules that requires consumer products to be safe and our obligations to ensure this.

If you have any concerns about our products, you can contact us on ProductSafety@springernature.com

In case Publisher is established outside the EU, the EU authorized representative is:

Springer Nature Customer Service Center GmbH
Europaplatz 3
69115 Heidelberg, Germany

Batch number: 10684222

Printed by Printforce, the Netherlands